UNITEXT for Physics

UNITEXT for Physics series publishes textbooks in physics and astronomy, characterized by a didactic style and comprehensiveness. The books are addressed to upper-undergraduate and graduate students, but also to scientists and researchers as important resources for their education, knowledge, and teaching.

Sergio Cecotti

Quantum Mechanics

A Concise Textbook with Emphasis on Exact Methods

 Springer

Sergio Cecotti
Quantum Fields and String Theory
BIMSA (Beijing Institute for Mathematical
Sciences and Applications)
Beijing, China

ISSN 2198-7882 ISSN 2198-7890 (electronic)
UNITEXT for Physics
ISBN 978-3-031-98823-3 ISBN 978-3-031-98824-0 (eBook)
https://doi.org/10.1007/978-3-031-98824-0

This Springer imprint is published by the registered company Springer Nature Switzerland AG
The registered company address is: Gewerbestrasse 11, 6330 Cham, Switzerland

If disposing of this product, please recycle the paper.

jo i ghi darài chistu libri,
parsè ch'al podrà capì
la so novitàt: obediensa
e disobediensa, insièmit.

Pier Paolo Pasolini

Preface

This is the third textbook arising from the author's lectures of Physics at the Qiuzhen College, the *elite* Chinese Institution for the most talented math students in the country. *Elite* students require *elite* textbooks: the ambition is to write textbooks that are introductory and self-contained but, at the same time, present a deeper and more modern perspective on the subject.

Being addressed to talented math students, this book aims to present Quantum Mechanics in a mathematically precise and elegant way, while providing physical insight on the meaning and implications of the formal statements. The Qiuzhen College wants its students to become accomplished mathematicians but also to develop a good "physical intuition."

Contrary to typical Quantum Mechanics textbooks with a mathematical leaning, in this book Functional Analysis is kept at its minimum, while stressing the Algebraic, Geometric, and Representation Theoretic aspects, including some deep stories never told in Quantum Mechanics textbook such as: the application of the Picard-Vessiot theory of differential Galois groups to the Schrödinger equation, the Prüfer formulation of Sturm-Liouville problems, the relation of the Riemann-Roch theorem with Quantum Mechanics in presence of magnetic fields, the Riemann-Hilbert correspondence vs. the Bohm-Aharonov effect, the theory of harmonic polynomials, the use of Hilbert spaces of holomorphic functions, the geometric origin of the Berry phase, and so on.

The math tools we introduce reflect the emphasis of the book: contrary to most of the physically oriented introductions to Quantum Mechanics, we focus on the *exact methods* rather than on approximation schemes (which are nevertheless described in Chap. 8 for the sake of completeness). Many of the techniques and examples we discuss have never been part of an introductory treatment of the subject. To name a few such examples: the analogue of the Landau levels for a charged particle moving in the hyperbolic plane, or in the field of a monopole, or the free motion in a spherical triangle. The harmonic oscillator is used as an illustration of the several exact methods and their mutual interplay: in the textbook there are more than a dozen solutions of the harmonic oscillator using diverse Hilbert space formulations as well as a number of alternative path integral representations. The particle moving in the central potential $-\alpha/r$ is solved in four different ways (including exact path integral) in an *arbitrary number* of spatial dimensions, not just in the standard $d = 3$. This is done to illustrate the deep elegant structures underlying this fundamental quantum system, structures that cannot be fully appreciated looking to just the special case of three dimensions.

Most of the exact methods are not described in any Quantum Mechanics textbook and some seem to be totally novel, as, for instance, the exact computation of non-Gaussian path integrals by inductive methods. The presentation of the general Schrödinger representation in the language of Differential Geometry allows us to state and prove a more general and explicit version of the Noether theorem for quantum systems with a Lie group symmetry. In the context of symmetry in Quantum Physics, we introduce *supersymmetry* and its Representation Theory (including the Witten index and its relation to Morse theory) in the Hilbert space formulation as well as in the path integral one, including its connections with stochastic differential calculus. Supersymmetry is also exploited as a math trick to solve *non*-supersymmetric quantum systems.

Path integrals are introduced and studied in great detail from several viewpoints and with diverse techniques. Chapter 6 contains a long list of exact methods to compute them. Various approximate schemes to compute path integrals are described in Chap. 8, including a rigorous derivation of the *dilute instanton gas* whose textbook treatments had been a major source of criticism in the literature.

A primer in open quantum systems, entanglement, von Neumann entropy, Quantum Information, and all that, is given in Chap. 7. While the treatment is limited to the basic aspects, there is an effort to precision, and, as in the rest of the book, an emphasis on the algebraic and geometric aspects of the theory, including the beautiful *Bures geometry* of the space of quantum states in an open quantum system.

Beijing, China
May 2025

Sergio Cecotti

Declarations

Competing Interests The author has no competing interests to declare that are relevant to the content of this manuscript.

Contents

Chapter 1
Quantum Mechanics: *A New Paradigm*

Classical mechanics has been developed continuously from the time of Newton and applied to an ever-widerring range of dynamical systems, including the electromagnetic field in interaction with matter. The underlying ideas and the laws governing their application form a simple and elegant scheme, which one would be inclined to think could not be seriously modified without having all its attractive features spoilt. Nevertheless it has been found possible to set up a new scheme, called Quantum Mechanics, which is more suitable for the description of phenomena on the atomic scale and which is in some respects **more elegant and satisfying than the classical scheme**.

P.A.M. Dirac, incipit of THE PRINCIPLES OF QUANTUM MECHANICS [1]

1.1 The Quest for a New Paradigm

As a mathematical construct Classical Physics is as consistent and elegant as any mathematical theory can possibly be. Mathematically speaking, Classical Mechanics is symplectic geometry, Maxwell theory harmonic analysis, and classical Statistical Physics probability theory: three math topics with fully rigorous foundations. Yet as a *physical* theory Classical Physics is *inconsistent*—not merely inconsistent with the facts (experiments), but also *logically* inconsistent. This reflects the fundamental fact that—contrary to the popular prejudice—the internal logic consistency of a physical theory is a much stronger requirement than the mere consistency of its math apparatus. As stressed by Eddington [2], a physical theory—independently of its agreement or not with the experiments—must be consistent with the Second Law of Thermodynamics. Classical Physics does not. As Boltzmann showed in 1884 [3], the Second Law predicts that a black body at temperature T irradiates an energy σT^4 per unit time and area (σ is an universal quantity called the *Stefan constant* [4]); on the other hand, as Lorentz suggested in 1903, and Einstein proved in 1905, Classical Physics predicts a *linear* law ρT for the black body radiation, not a *fourth* power of T: what is worse, it also predicts that

© The Author(s), under exclusive license to Springer Nature Switzerland AG 2025
S. Cecotti, *Quantum Mechanics*, UNITEXT for Physics,
https://doi.org/10.1007/978-3-031-98824-0_1

the proportionality constant ρ is divergent! Thus, according to the classical theory, to heat up anything to a positive temperature $T = \epsilon > 0$ requires, at equilibrium, an infinite amount of energy, which means that no positive temperature may exist.[1] Life (intelligent or otherwise) cannot exist in a world ruled by Classical Physics.

There are many other reasons why Classical Physics is *untenable.* The most obvious one is the existence and stability of matter (including ourselves). We know as an empirical fact that matter is made of atoms which are formed by a positively charged heavy nucleus and negatively charged light electrons rotating around the nucleus. The hydrogen atom is just a negatively charged electron orbitating in the Coulomb potential $V = -\alpha/r$ generated by a (much heavier) positively charged proton. The electron is subjected to the Coulomb attractive force, hence is accelerating, and, according to the rules of Classical Physics, it should radiate electromagnetic fields, thus loosing energy. As its energy goes down, the average distance of the electron from the proton decreases. Eventually the classical electron would collapse in the nucleus: in Problem 1.1 the reader is invited to show that a classical hydrogen atom cannot survive for more than 10^{-10} s. Classical Physics rules out stable matter!

To describe the real world we need physical principles radically different from the classical ones. The resulting new scientific paradigm is Quantum Physics. Yet Classical Physics is not "trivially wrong": it is an elegant and rigorous mathematical apparatus which explains many tricky physical phenomena, such as the motion of the planets in the sky. As implied by the opening quotation from Dirac, all beautiful and deep aspects of Classical Physics should *and will* be incorporated in Quantum Physics which indeed, is a much more elegant and satisfying scheme which *explains* the nice aspects of the classical theory: Classical Physics is a "limit" of Quantum Physics and inherits all its beauty and usefulness from its quantum parent theory. Quantum Physics also explains the puzzling elements of the classical theory which looked "out of place" from a purely classical perspective. For instance, in chapter 11 of [5] it is shown that the dynamical equations of a classical mechanical system may be written as equations describing the propagation of *waves.* What is the meaning of these "mechanical" waves? Classically they look a mere math artifact without physical significance. Quantum Mechanics gives the deep answer.

1.2   *History-Fiction:* Deformation Quantization

All scientific revolutions advance along a tortuous path of tries and errors, which is full of false starts and misconceptions. The transition from classical to quantum physics during the first three decades of the twentieth century is *no exception.* Readers wanting to understand *historically* how that happened, may have a look to the book [6]. In this section we instead present a purely fictional version of the

---

[1] For a discussion see, for instance, the last paragraph of §. 2.8 in [4].

transition between the old and the new paradigm which describes as history *should* have gone in a world of (non-existing) ideal physicists fully aware of what they are doing.

Mathematically speaking, Classical Mechanics is the geometry of a smooth manifold $\mathcal{W}$ (the *phase space*) equipped with a symplectic form ω [5]. The classical physical quantities (also called *classical observables*) are just the functions[2] $f \in \Omega^0(\mathcal{W})$ which form an algebra (over $\mathbb{R}$ but let's us extend the ground field to $\mathbb{C}$). The algebra of observables is a commutative, associative, $*$-algebra[3] with unit 1. $*: f \to f^*$ is just complex conjugation. The elements of the form ff^* are non-negative[4] for all f. In addition the $*$-algebra of observables is endowed with a Lie algebra structure given by the Poisson bracket $[\cdot, \cdot]_{\mathrm{PB}}$; the two algebra structures on $\Omega^0(\mathcal{W})$ are related by a compatibility condition (the Leibniz formula) see §. 6.4 of [5].

Although history did *not* go this way, at the end of the nineteenth century smart people could/should have posed the natural

*Question 1.1 How **unique** is this algebraic description of the observables?*

It is an algebraic theorem[5] that the pair $(\mathcal{W}, \omega)$ defines an *universal* family of *pair-wise non-isomorphic* associative $*$-algebras $\Omega^0(\mathcal{W})_\hbar$ with unit 1—endowed with a compatible Lie algebra structure—which are parametrized by a real variable $\hbar \geq 0$. When $\hbar = 0$ one recovers the algebra of classical physical observables. *Universal* means that *any* continuous $*$-algebra deformation of the classical algebra, which is still associative with unit, is isomorphic to precisely one of the $\Omega^0(\mathcal{W})_\hbar$'s. The underlying vector space of $\Omega^0(\mathcal{W})_\hbar$ is still the space of smooth complex functions on $\mathcal{W}$, but now equipped with a different associative bilinear product

$$(f, g) \mapsto f \star g, \tag{1.1}$$

---

[2] Here and throughout the book we freely use the math terminology, notations, and conventions introduced and defined in chapter 2 of [5] to which we refer the reader. For instance, $\Omega^k(\mathcal{M})$ stands for the vector space of smooth differential k-forms on the manifold $\mathcal{M}$ cf. [5] chap. 2.

[3] Recall that a $\mathbb{C}$-algebra with unity A is a $*$-*algebra* if it has an anti-linear involution $*: A \to A$, acting as complex conjugation on the scalars, which is an antiautomorphism, i.e. such that for all $x, y \in A$

$$(x + y)^* = x^* + y^*, \quad (xy)^* = y^*x^*, \quad 1^* = 1, \quad (x^*)^* = x.$$

[4] When this is the case we also say that $*$ is a *positive* anti-linear involution.

[5] See e.g. ref. [7].

uniquely defined by ω and $\hbar$ (up to algebra isomorphism) called the *Moyal product* or the *star product*.[6] One has

$$f \star g = fg + \frac{i\hbar}{2}[f, g]_{\mathrm{PB}} + O(\hbar^2) \tag{1.2}$$

where $[\cdot, \cdot]_{\mathrm{PB}}$ is the Poisson bracket defined by the symplectic structure ω see [5] chaps. 6 and 7. In Problem 1.2 the diligent reader is invited to write down the closed form of the Moyal product $\star$ to all orders in the deformation parameter $\hbar$. The $\star$ product, while associative, is not commutative when $\hbar \neq 0$

$$f \star g - g \star f = i\hbar[f, g]_{\mathrm{PB}} + O(\hbar^2). \tag{1.3}$$

Two observables commute (to the first order in $\hbar$) iff they are in involution. Note that $(f \star g)^* = g^* \star f^*$, where $*$ is the ordinary complex conjugation in the underlying space of complex functions on $\mathcal{W}$, so the deformed algebra is a genuine $*$-algebra. In the late nineteenth century a smart guy could have said:

> *The fundamental geometric structures of Mechanics define a family of theories of the physical observables which is parametrized by a single* universal *constant* $\hbar$. *What is the value of* $\hbar$ *in the real world? Let us measure it!*

If they had measured it, they would have discovered that $\hbar$ is not zero but

$$\hbar = 6.62607015 \times 10^{-34} \,\mathrm{m}^2 \,\mathrm{kg/s}, \tag{1.4}$$

pretty small in everyday units, but not zero.

Unfortunately history did not go that way. Quantum Physics was not discovered *algebraically*. One reason is that setting $\hbar \neq 0$ is not an innocent endeavor: it requires a rather radical departure from the classical thinking. To see how radical, consider a particle moving on the line $\mathbb{R}$ with position variable $q \in \mathbb{R}$ and conjugate momentum p [5]. We have

$$q \star q = q^2, \quad p \star p = p^2, \quad q \star p = qp + \frac{i\hbar}{2}, \quad p \star q = qp - \frac{i\hbar}{2}. \tag{1.5}$$

When $\hbar = 0$ the algebra is commutative, and it makes sense to assign numerical values to its elements: at time $t = 0$ the particle has a given position $q_0 \in \mathbb{R}$ and a definite momentum $p_0 \in \mathbb{R}$. A sharp initial datum (q_0, p_0), on the contrary, is inconsistent when $\hbar \neq 0$, since simultaneous precise values for q and p contradict the algebraic relation $q \star p - p \star q = i\hbar \neq 0$.

Let us make this observation a bit more precise (we shall return to this issue when we have more tools in Sect. 2.13 and then make the story fully rigorous). We claim that an element of $\Omega^0(\mathcal{W})_\hbar$ of the form $f \star f^*$ is non-negative. We stress

---

[6] For more details see chap. 4 of [8] or ref.[9].

that $f \star f^*$ is "non-negative" as an element of the deformed algebra: the underlying function may not *look* non-negative.[7] Then we have[8]

$$0 \leq \left(q - q_0 + it(p - p_0)\right) \star \left(q - q_0 - it(p - p_0)\right)$$
$$= (q - q_0)^2 + t^2(p - p_0)^2 + t\hbar \tag{1.6}$$

for all $t \in \mathbb{R}$, which means that the discriminant of the quadratic polynomial in t in the RHS is non-positive, that is,

$$(q - q_0)^2 (p - p_0)^2 \geq \frac{\hbar^2}{4}. \tag{1.7}$$

We may quantify the precision of a measure of the position q (resp. momentum p) in terms of its standard deviation from the mean q_0

$$\Delta q \equiv \sqrt{(q - q_0)^2} \qquad (\text{resp. } \Delta p \equiv \sqrt{(p - p_0)^2}), \tag{1.8}$$

so that in the $\hbar$-deformed algebra

$$\Delta q \, \Delta p \geq \frac{\hbar}{2}, \tag{1.9}$$

which means that there is a limit on the precision of simultaneous measures of q and p. We can measure q with arbitrary high precision, making a very small error $\Delta q = \epsilon$, but we pay for this luxury with a very poor measure of the momentum, making a huge error $O(\hbar/\epsilon)$, and viceversa we may measure p quite accurately if we accept of knowing the position with a big uncertainty. The argument shows that only quantities that are in Poisson involution can be measured precisely simultaneously.[9] We conclude:

> **Indetermination Principle**
> *Two physical quantities which are not in involution can be measured simultaneously and exactly if and only if $\hbar = 0$, i.e. in the* undeformed *classical theory*

---

[7] The relevant notion of "non-negativity" will be clear from the Hilbert space formulation of Quantum Mechanics in Chap. 2.

[8] Here q_0 and p_0 are real numbers, that is, elements of $\Omega^0(W)_\hbar$ of the form $q_0 \cdot 1$, $p_0 \cdot 1$ with 1 the identity in $\Omega^0(W)_\hbar$ and $q_0, p_0 \in \mathbb{R}$.

[9] The statement holds to first order in $\hbar$.

Therefore, if we want to preserve the elegant geometric (and algebraic) structures of mechanics and yet go beyond the classical theory—which is logically untenable—we must accept that we cannot make simultaneous measures of position and velocity with arbitrary high accuracy, but only within the finite precision given by the bound (1.9). This statement is the *Heisenberg indetermination principle* and is a cornerstone of Quantum Physics. We shall revisit it in a more systematic way in Sect. 2.13 and elsewhere through the book.

A New Notion of Physical State
In Classical Mechanics the *state* of the system at time t_0 was specified by giving the (precise) values of all positions q^i and momenta p_j at time $t = t_0$. Then the Hamilton equations of motion allowed us to compute (in principle) the state $(q^i(t), p_j(t))$ at any other time t. When $\hbar \neq 0$ this definition of initial state is untenable because it requires the specification of *a contradictory set of data*. To solve the conundrum we have to introduce a brand new *quantum* notion of *state* which does not require the specification of both q's and p's. Yet the *state* at time $t = t_0$ should contain all the information needed to predict (using the dynamical equations of motion) the state at any other time t. This may look impossible to the classical mind, but a diligent student who reads the Classical Mechanics book [5] carefully, may already guess how this is possible. The new notion of *state* will be described in detail below.

A first consequence of this situation is that when $\hbar \neq 0$ the notion of "trajectory" of a particle is meaningless: by definition the trajectory is the time-evolution $(q^i(t), p_j(t))$ of the state in the classical sense. Since the classical notion of state becomes contradictory as soon as $\hbar \neq 0$, *a fortiori* the "trajectory" of a particle makes no sense.

The approach to Quantum Mechanics (QM) through the deformation of algebras and the Moyal product is called *Deformation Quantization*. It is not the most convenient or deep viewpoint on QM, and we shall not pursue it any further in this textbook. We shall focus on the traditional Dirac formulation of Quantum Mechanics and its close variants and generalizations such as path integrals.

1.3 Polarization and Interference

Quantum Physics is based on very precise mathematical *Principles* ("axioms"). To make these axioms "intuitive", physicists do not state them out of the blue, but motivate them through the analysis of some basic experiments. These experiments were invented *a posteriori* for the didactical purpose of explaining the *Principles* to students, and played no role in the historical development of the subject. Most of them were technologically impossible until recent years, and the description of their results in textbooks rested only on faith (which was eventually vindicated by actual experiments when the relevant technology became available).

This section is motivational, hence a-technical (a mere *cartoon*).

Consider *light*. We know that it can be described in terms of electromagnetic waves, governed by Maxwell theory, and also by "quantum particles", the *photons*. The two descriptions cannot contradict each other.[10] It is an experimental fact that the light presents the typical phenomena associated with a transverse wave: *polarization* and *interference*. Yet it is made of "particles".

Polarization

Light may be linearly polarized along any line in the plane transverse to its direction of propagation, as well as circularly or, more generally, elliptically polarized. Let the light propagate in the z direction; the two non-zero components of the electric field are[11]

$$E_x = C_x \, \mathrm{e}^{\mathrm{i}(\omega t - kz)} + \text{c.c.}, \qquad E_y = C_y \, \mathrm{e}^{\mathrm{i}(\omega t - kz)} + \text{c.c.} \tag{1.10}$$

where C_x and C_y are *complex* constants. If $C_y = 0$ (resp. $C_x = 0$) the light is linearly polarized along the x-axis (resp. y-axis). If $C_y/C_x = m \in \mathbb{R}$, the light is polarized along a line which makes an angle $\phi = \arctan m$ with the x-axis. In the general case where $C_y/C_x \notin \mathbb{R}$ the polarization is elliptic (circular if $|C_x| = |C_y|$). Note that all polarization states can be obtained by taking linear combinations of the two basic linear polarizations along the x- and y-axes with complex coefficients C_x, C_y.

This pattern applies also to a very weak light beam, which consists of just a single photon. Thus polarization should be a property of the *single* photon, not a collective property of the light beam. We conclude that a photon has infinitely many polarization states parametrized by a pair (C_x, C_y) of complex numbers with the proviso that $C_x = C_y = 0$ is no light at all, so it does *not* represent any polarization state of the photon. Moreover, multiplying the two constants by any non-zero complex number λ,

$$(C_x, C_y) \rightsquigarrow (\lambda C_x, \lambda C_y), \qquad \lambda \in \mathbb{C}^\times \equiv \mathbb{C} \setminus \{0\}, \tag{1.11}$$

the polarization state does not change. We conclude that the different polarizations of a photon are parametrized by the complex manifold

$$(\mathbb{C}^2 \setminus \{0\}) \Big/ \big\{ (z_1, z_2) \sim \lambda(z_1, z_2) \big\} \equiv \mathbb{P}^1(\mathbb{C}) \tag{1.12}$$

that is, by the *complex projective line*, also called the *Riemann sphere* $\mathbb{P}^1(\mathbb{C}) \simeq S^2$ [10–12], while all polarization states can be written as linear combinations of the two linear polarizations $(1:0)$ and $(0:1)$ with complex coefficients.

---

[10] In chapter 3 of [4] it is explained in detail why and how this magical "duality" works.

[11] Here and through the book the shorthand " + c.c." stands for "plus complex conjugate".

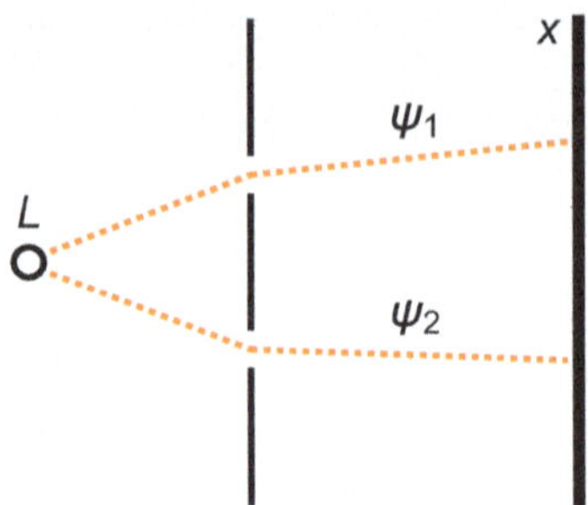

Fig. 1.1 The interference experiment described in the main text. The thick vertical line is the screen on which we observe the interference pattern

Interference: Photons

We consider a wall with two slits parallel to a screen as in Fig. 1.1. Before the wall we put a light source L in a position symmetric with respect to the two slits. What will we observe on the screen?

As for polarization, we have two physical pictures of what is going on. In the wave picture we have two waves, ψ_1 and ψ_2, which go through the first and the second slit, respectively. To the right of the wall each wave has the (rough) form

$$\psi_a(t, \mathbf{x}) = A \cos\big(k(t - d_a(\mathbf{x}))\big) \quad (a = 1, 2) \tag{1.13}$$

where $d_a(\mathbf{x})$ is the distance of the point $\mathbf{x}$ from the a-th slit (whose size we take to be negligible). Since Maxwell equations are linear, the electromagnetic waves satisfy a *superposition principle,* i.e. the linear combination of two solutions is again a solution of the equations. The full Maxwell wave after the wall is then

$$\psi_1(t, \mathbf{x}) + \psi_2(t, \mathbf{x}). \tag{1.14}$$

The image we see is given by the amplitude of the wave (1.14) when the point $\mathbf{x}$ varies on the screen. If x is the coordinate along the screen,

$$d_1(x) - d_2(x) \approx xd/D, \tag{1.15}$$

where D is the distance of the wall from the screen, and d is the distance between the two slits (we assume $x, d \ll D$). In the wave picture the wave amplitude at the position x on the screen is then

$$\psi(x) = \psi_1(x) + \psi_2(x) \approx 2A \cos\left(\frac{kxd}{D}\right) \cos\big(k(t - D)\big), \tag{1.16}$$

while the light density we see on the screen is the time-average[12] of the norm-square of the wave

$$\overline{|\psi|^2} \approx 2A^2 \cos^2\left(\frac{kxd}{D}\right) = 2\cos^2\left(\frac{kxd}{D}\right)\left(\overline{|\psi_1|^2} + \overline{|\psi_2|^2}\right). \qquad (1.17)$$

From this equation we conclude that we see bright lines in the screen separated by dark ones: their positions are

$$\begin{aligned} \text{bright lines} \quad & x = n\pi D/kd \\[2mm] \text{dark lines} \quad & x = (n + \tfrac{1}{2})\pi D/kd, \quad \text{with } n \in \mathbb{Z}. \end{aligned} \qquad (1.18)$$

This is the *interference effect:* the waves passing through the two slits interfere with each other either constructively, enhancing the light intensity by a factor 2, or destructively by making the intensity zero. From the wave side this phenomenon has a simple explanation, but how we interpret interference *from the photon perspective?*

In the classical theory a particle would follow a *definite* trajectory, so it goes through either the first or the second slit. One would be led to think that the interference pattern arises because a photon which goes through the first slit interferes with one coming from the second hole in the wall, with the consequence of producing either four or zero photons, depending on the position x on the screen where they meet. However this naive idea cannot be reconciled with the conservation of energy. The only sensible conclusion is that *each photon interferes with itself.*

This would be impossible if the trajectory of a photon—and its ending position on the screen—was well defined in general. We already gave up that assumption. We replace it with the notion that, in general, the photon has not passed through a definite slit, nor has been absorbed in a specific point in the screen. We may think of a photon which goes through the first (second) slit as being in a "state" which is described by the "wave" $\psi_1(t, \mathbf{x})$ (resp. $\psi_2(t, \mathbf{x})$). These waves satisfy the Maxwell equations. Since a linear combination of electromagnetic waves is still a solution to the equations, the general state of photon is a *linear superposition* (with complex coefficients) of the two "basic" states described by $\psi_1(t, \mathbf{x})$ and $\psi_2(t, \mathbf{x})$. This proposal cannot work in the classical theory where $\hbar = 0$, because in that set-up the notion of trajectory of the photon is well defined, and it makes no sense to say that a particle moves in a "superposition" of two distinct trajectories. However, as argued in Sect. 1.2, there is *no* contradiction in this idea when $\hbar \neq 0$ because the deformed algebra of observables is *non commutative.*

---

[12] The operation of taking the average over time of a physical quantity $Q(t)$ is denoted by a bar over the symbol representing that quantity, that is, $\overline{Q} := \lim_{T \to \infty} T^{-1} \int_0^T dt\, Q(t)$.

Interference: Electrons
Above we discussed the interference of photons. Interference is a general quantum phenomenon which does not depend on the nature of the particle or physical system. Indeed in 1989 Tonomura et al. [13, 14] carried out the two slits interference experiments with *electrons*, and they did find the expected interference pattern, thus showing that the electrons do not pass through a definite slit, but are in a general superpositions of the two "basic" states which describe electrons going through one of the two openings in the wall.

Polarization and interference illustrate the most fundamental principle of Quantum Physics, the *superposition principle*, that we now introduce.

1.4 The Superposition Principle

Quantum States
A *state* in Quantum Mechanics is defined as in all physical theories: it is the specification of the data at a given time t_0 which are needed and suffice to predict (in principle) the state (that is, the same collection of data) at all other times t using the dynamical equations of motion. The set of data specifying a state should be consistent, i.e. without internal contradictions. For instance, we have argued above that the simultaneous specification of the position and velocity of a particle is inconsistent when $\hbar \neq 0$, and therefore exact positions and velocities cannot both enter simultaneously in the specification of a quantum state.

Superposition of Quantum States: Ket Vectors
The discussion of experiments in the previous section (actual experiments as well as *gedanken* ones) shows that quantum states have a special property: given two states of a quantum system we can form new states by *superposing them*. For instance, in the interference experiment the general state of the photon (resp. electron) is a superposition of the "basic" states where the particle went through a definite opening in the wall. In the polarization experiment we saw that all polarization states of a photon are described as the superposition of two basic polarization states. The possible polarization states are given by complex linear combinations of two vectors with the proviso that two vectors v and λv ($\lambda \in \mathbb{C}^\times$) represent the same polarization state, while the zero vector does not represent any state.

These experiments, and other considerations, lead us to the notion that quantum states may be superposed to get new states, and that the superposition of two states satisfies algebraic properties akin to taking linear combinations of two vectors in a complex vector space. This statement is the most fundamental principle of Quantum Physics:[13] the *superposition principle*.

---

[13] Quantum Physics is much more general than Quantum Mechanics. Quantum Mechanics dwells with non-relativistic quantum systems with finitely many degrees of freedom, while there are

The mathematical formulation of the quantum *superposition principle* states that quantum states are represented by vectors in a suitable space (*state vectors*). Following Dirac [1], we represent the state vectors by the notation $|A\rangle$—called a *ket (vector)*—where the symbol A stands for the label of the particular state which may consist of one or several symbols as the need may be. Two kets $|A\rangle$ and $|B\rangle$ may be *superposed* to get a third ket vector. This corresponds to taking a linear combination of them

$$|C\rangle = c_1|A\rangle + c_2|B\rangle \tag{1.19}$$

where c_1, c_2 are constants. The constants c_1, c_2 may take arbitrary complex values: this is required to be able to reproduce all linear and elliptic polarization states out of the two basic linear polarizations of the photon. More generally, if $|x\rangle$ is a family of kets depending on a continuous variable x, we may consider the superposition[14]

$$\int dx\, c(x)\, |x\rangle \tag{1.20}$$

where $c(x)$ is a complex function.

What happens when we superpose a state $|A\rangle$ with itself

$$c_1|A\rangle + c_2|A\rangle = (c_1 + c_2)|A\rangle\ ? \tag{1.21}$$

Clearly the superposition of a state with itself should give back the same state. We saw this to be the case with the polarization states, cf. Eq. (1.11). The vector $c|A\rangle$ represents the same state as $|A\rangle$ for all $c \in \mathbb{C}^\times \equiv \mathbb{C} \setminus \{0\}$. This also entails that the zero vector *does not* represent any state, just as in the case of polarization. We formalize our conclusions in the *fundamental principle* (axiom) of Quantum Physics:

Superposition Principle 1 *For each quantum system there is a "ket" vector space* $\mathcal{H}$, *defined over* $\mathbb{C}$, *such that its quantum states are identified with points in the* ***projectivization*** $\mathbb{P}(\mathcal{H})$ *of* $\mathcal{H}$

$$(manifold\ of\ quantum\ states) = \mathbb{P}(\mathcal{H}) \overset{\mathrm{def}}{=} (\mathcal{H} \setminus \{0\})/\sim, \tag{1.22}$$

where $\sim$ *is the equivalence relation*

$$|A\rangle \sim c|A\rangle \quad for\ all\ c \in \mathbb{C}^\times\ and\ |A\rangle \in \mathcal{H} \setminus \{0\}. \tag{1.23}$$

---

quantum systems which are relativistic and/or with several degrees of freedom. Quantum Physics includes, besides Quantum Mechanics, Quantum Field Theory, Quantum Statistics, the Theory of Condensed States, Quantum Information theory, and Quantum Gravity. The superposition principle lays at the foundation of all these physical theories.

[14] The precise definition of the integral (1.20) is rather tricky and will be spelled out in Chap. 2.

The ket vector space

$$\mathcal{H} \xrightarrow{\pi} \mathbb{P}(\mathcal{H}) \tag{1.24}$$

is then the total space of the tautological complex line bundle over the manifold $\mathbb{P}(\mathcal{H})$ *of quantum states.*

We say that a non-zero vector $|A\rangle \in \mathcal{H}$ represents *the quantum state A if* $|A\rangle$ is in the $\sim$ equivalence class of $A \in \mathbb{P}(\mathcal{H})$. In "practical" Quantum Physics one always works with ket vectors representing the states, and, by abuse of language, refers to any representative ket $|A\rangle$ as the "state" and to the vector space $\mathcal{H}$ as the "space of states". Following tradition, in this book we shall abuse language quite a lot. However one should always keep in mind that the identification of vectors and states is only modulo an essential ambiguity given by multiplication by an arbitrary non-zero complex number. This ambiguity is not harmless, since the bundle (1.24) is topologically non-trivial (see Sect. 8.11 for details). Luckily it has a nice geometry:

Fact 1.1 $\mathcal{H} \xrightarrow{\pi} \mathbb{P}(\mathcal{H})$ *is a homogeneous holomorphic line bundle with a canonical Chern connection.*

This **Fact** will be important in the formal developments at the end of Chap. 8; we refer the reader to that discussion.

Two quantum states A, B are *linearly independent* iff the kets $|A\rangle$ and $|B\rangle$ representing them are linearly independent in $\mathcal{H}$. The notion is independent of the chosen representative vectors.

All features of Quantum Physics should be consistent with the superposition principle. In particular the time evolution of the superposition of two quantum states must be the superposition of their respective evolutions

$$|C, t\rangle = c_1|A, t\rangle + c_2|B, t\rangle \quad \text{at all times } t \tag{1.25}$$

with constant coefficients c_1, c_2. This entails that the time evolution from time t_0 to time t lifts to a *linear map* $\mathcal{H} \to \mathcal{H}$ on the total space of the line bundle (1.24). More generally,

Superposition Principle 2 *Any equation/condition/property satisfied by the quantum states $A \in \mathbb{P}(\mathcal{H})$ may be written as a linear equation/condition on the ket vectors $|A\rangle \in \mathcal{H}$ representing them.*

Thus Quantum Physics is a *linear theory* (albeit a tricky one).

Corollary 1.1 *The set of vectors which represent the states of our system which enjoy a particular physical property $\mathcal{P}$ is a vector subspace $\mathcal{H}_\mathcal{P} \subset \mathcal{H}$.*

Definition 1.1 The *number of states with the property $\mathcal{P}$ is*

$$\#(\text{states with property } \mathcal{P}) \overset{\text{def}}{=} \dim_\mathbb{C} \mathcal{H}_\mathcal{P}. \tag{1.26}$$

In other words, we count only linear independent states. This definition of "number" agrees with the physical intuition about the proper way to count quantum states, and is the one entering in all physical statements.

Bra Vectors

Having a space of states is not enough to construct a physical theory. The results of experiments are *numbers,* not points in an abstract manifold of states. To construct a predictive quantum theory, we must be able to attach numbers to our quantum states. The assignment of numbers to states should be compatible with the linear structure given by the superposition principle, that is, they should be induced by linear maps

$$\mathcal{H} \to \mathbb{C}. \tag{1.27}$$

There is a universal canonical construction to accomplish this. Once we have the space $\mathcal{H}$ of ket vectors, we may introduce its dual vector space $\mathcal{H}^\vee$ whose elements are called *bra (vectors)* and denoted by the symbol $\langle B|$. The canonical pairing (inner product) of a ket $|A\rangle$ and a bra $\langle B|$ is a complex number, written as

$$\langle B|A\rangle \tag{1.28}$$

The pairing forms a *complete bra-c-ket* out of the bra and the ket. The semantics of the Dirac notation is that a complete bracket is a complex number, while an incomplete bracket is an element of a vector space, either a bra or a ket [1]. The pairing is linear in the ket by design

$$\langle B|(c_1|A_1\rangle + c_2|A_2\rangle) = c_1 \langle B|A_1\rangle + c_2 \langle B|A_2\rangle. \tag{1.29}$$

The Dirac semantics interprets the symbol $|A\rangle\langle B|$, or more generally a sum of the form $\sum_k |A_k\rangle\langle B_k|$, as a linear map $\mathcal{H} \to \mathcal{H}$

$$|C\rangle \mapsto \sum_k |A_k\rangle\langle B_k|C\rangle \equiv \sum_k (\langle B_k|C\rangle)|A_k\rangle \in \mathcal{H}, \tag{1.30}$$

or, equivalently, as an element of the vector space $\mathcal{H} \otimes \mathcal{H}^\vee$.

Now we are confronted with the embarrassment of riches: roughly speaking, we got *two copies* of the space of quantum states, $\mathcal{H}$ and $\mathcal{H}^\vee$. To solve the conundrum we just identify the two spaces. More precisely, we assume the existence of *anti-linear isomorphisms*[15]

$$\eta : \mathcal{H} \to \mathcal{H}^\vee, \quad \eta^\vee : \mathcal{H}^\vee \to \mathcal{H} \quad \text{with} \quad \eta\eta^\vee = 1, \quad \eta^\vee\eta = 1, \tag{1.31}$$

---

[15] Through this textbook we write simply 1 for the identity map in any vector space.

which set a correspondence between kets and bras. The bra $\eta(|A\rangle)$ corresponding to $|A\rangle$ is written $\langle A|$. It represents the same quantum state A. Going through the discussion of the polarization of photons as seen from the Maxwell perspective[16] (Sect. 1.3)—which we used to motivate the superposition principle—one easily sees that the correspondence η must be *anti-linear*

$$\eta(c_1|A_1\rangle + c_2|A_2\rangle) = c_1^*\langle A_1| + c_2^*\langle A_2|, \tag{1.32}$$

where the $*$ stands for complex conjugation of complex numbers. Given two kets, $|A\rangle$ and $|B\rangle$, we may form a number

$$\langle B|A\rangle \equiv \eta(|B\rangle)|A\rangle, \tag{1.33}$$

which depends linearly on $|A\rangle$ and anti-linearly on $|B\rangle$. Generalizing the properties valid for the polarization states of the photon, we require

$$\langle B|A\rangle = \langle A|B\rangle^*, \tag{1.34}$$

and also

$$\langle A|A\rangle > 0 \tag{1.35}$$

except when $|A\rangle = 0$, a vector which does not represent any state. In other words:

Superposition Principle 3 *The $\mathbb{C}$-vector space $\mathcal{H}$ is endowed with a positive-definite Hermitian product written as a* complete bracket

$$|A\rangle, \ |B\rangle \mapsto \langle B|A\rangle. \tag{1.36}$$

The product is linear in $|A\rangle$ and anti-linear in $|B\rangle$.

The pairing $\langle B|A\rangle$ is called the *quantum amplitude from state $|A\rangle$ to state $|B\rangle$* or else the *overlap* of the two states $|A\rangle, |B\rangle$. The physical rationale of this terminology will be clear in the following chapters.

A bra $\langle A|$ and a ket $|B\rangle$ are called *orthogonal* iff their product (bracket) is zero $\langle A|B\rangle=0$, while two kets $|A\rangle$ and $|B\rangle$ are *orthogonal* iff

$$\langle A|B\rangle^* \equiv \langle B|A\rangle \equiv \eta(|B\rangle)|A\rangle = 0. \tag{1.37}$$

Two quantum states represented by orthogonal vectors $|A\rangle$ and $|B\rangle$ are called *orthogonal states:* this condition is independent of the choice of representative vectors, hence intrinsic. Two states are orthogonal iff they have zero overlap, i.e.

---

[16] The statement is well known in *classical* electromagnetism. See e.g. the first footnote in §. 48 of the celebrated book [15].

the quantum amplitude between them vanishes. Since the ket is determined by the state only up to a factor, it is usually convenient to choose it to have *unit norm*, that is,

$$\| |A\rangle \|^2 \stackrel{\text{def}}{=} \langle A|A\rangle = 1. \tag{1.38}$$

A ket satisfying this condition is said to be *normalized*. Normalized representatives of states are unique only up to multiplication by *phase factors* $e^{i\alpha}$ ($\alpha \in \mathbb{R}$). In geometric language: normalized states form a principal $U(1)$-bundle

$$U(\mathcal{H}) \to \mathbb{P}(\mathcal{H}) \tag{1.39}$$

over the manifold $\mathbb{P}(\mathcal{H})$ of quantum states. $U(\mathcal{H})$ is the circle bundle associated to the tautological complex line bundle (1.24).

1.5   Closed vs. Open Quantum Systems: This Textbook

Quantum Mechanics was historically developed, and is traditionally taught, with reference to a special class of quantum systems, the *closed* ones, that is, isolated systems which we look at as external observers. In the polarization and interference experiments we used to motivate the description of quantum states as complex lines in a vector space $\mathcal{H}$, the photon was seen as a closed system observed from the outside. It is easy to see that the description of states as points in $\mathbb{P}(\mathcal{H})$ applies *only* to the states of closed quantum systems, as it does the full machinery of the Hilbert space formulation of Quantum Mechanics that is the topic of the next chapter.

The notion of a *closed* quantum system is, of course, a highly idealized one. The real physical systems interact with a plethora of other quantum systems (the measuring apparatus, the laboratory, etc.), known as the *environment*, and we, the observers, are part of the environment and subject to the same quantum effects as everything else. In other words, the actual physical systems are *open*, and their behavior is governed by principles ("axioms") which look rather different as math statements from the "axioms" valid for the ideal closed systems. For instance, the dynamical equations of a closed quantum system are invariant under time-reversal, so that all their processes are *reversible*, while the open system processes are almost never reversible (as the everyday life suggests[17]). However, at a deeper level, both behaviors arise from the same quantum physics.

---

[17] Think of reversing the breaking of an egg.

While the theory of open quantum systems has been a marginal topic for most of the history of Quantum Mechanics, nowadays it has become a central subject of Quantum Physics for two main reasons:

1. the role of the theory of open quantum systems (in particular quantum entanglement, von Neumann entropy, and all that) for Quantum Information, Quantum Computing, and all those exciting developments which are expected to lead in the near future to a major technological revolution;
2. the importance of the theory for the most important scientific problem ever: understanding Quantum Gravity.

This textbook is dedicated to talented *undergraduate* students with no previous knowledge of Quantum Physics. Therefore, in order to be introductory and self-contained—while covering the standard curricular syllabus—we follow the historical path, and formulate the Principles of Quantum Mechanics for *closed* systems, first in the Hilbert space formulation (Chaps. 2–5) and then in the "modern" path integral approach (Chap. 6). Didactically speaking, starting from the theory of the closed systems is unavoidable: the Quantum Mechanics of closed systems is a simple, mathematically elegant, formal structure which teaches us a lot about the basic Laws of Nature. The theory of the open systems is then built on the one of closed systems. The point is that we can always "complete" an open quantum system into a closed one by considering a closed system composed of two parts: our pet system and its "environment". In this way we may *deduce* the basic Laws valid for the open systems from the well-known "axioms" valid for the (ideal) closed systems.

We touch upon the theories of open systems, quantum entanglement, and Quantum Information in Chap. 7. These are advanced topics well beyond the scope of an introductory textbook on Quantum Mechanics, and our treatment of these fundamental topics is rather sketchy and limited. Our goal is to give the flavor of these modern developments, pinpoint the important issues, and outline the central notions such as *entanglement* and *entropy.* In a sense Chap. 7 should be taken as an invitation to further reading.

The emphasis in this book is on *exact* techniques. However, for the sake of completeness, we also added a chapter, the last one, on the *approximation schemes* one may use for practical computations in Quantum Mechanics. Historically this has been a major topic, hence leaving it out completely appeared to be a too drastic choice.

Problems

1.1 Estimate the time required to a *classical* electron orbitating around a classical proton on a circular trajectory of radius $\approx 10^{-8}$ cm to irradiate all its energy and fall in the center.

1.2 Write down the complete expression of the Moyal $\star$ product to all orders in $\hbar$.

References

1. P.A.M. Dirac, *Principles of Quantum Mechanics*, 4th edn. (Clarendon, Oxford, 1958)
2. A.S. Eddington, *The Nature of the Physical World* (Cambridge University Press, Cambridge, 1928)
3. L. Boltzmann, *Ableitung des Stefan'schen Gesetzes, betreffend die Abhängigkeit der Wärmestrahlung von der Temperatur aus der electromagnetischen Lichttheorie* [Derivation of Stefan's law, concerning the dependency of heat radiation on temperature from the electromagnetic theory of light]. Ann. Phys. Chem. **258**, 291–294 (1884)
4. S. Cecotti, *Statistical Mechanics. A Concise Adavanced Textbook* (Springer, Berlin, 2024)
5. S. Cecotti, *Analytic Mechanics. A Concise Textbook* (Springer, Berlin, 2024)
6. H. Kragh, *Quantum Generations. A History of Physics in the Twentieth Century* (Princeton University Press, Princeton, 1999)
7. M. Kontsevich, Deformation quantization of Poisson manifolds. I. Lett. Math. Phys. **66**, 157–216 (2003). arXiv:q-alg/9709040
8. P. Sharan, *Some unusual topics in Quantum Mechanics*. Lecture Notes in Physics, vol. 1020, 2nd edn. (Springer, Berlin, 2023)
9. N. Moshayedi, *Kontsevich's Deformation Quantization and Quantum Field Theory*. Lecture Notes in Mathematics, vol. 2311 (Springer, Berlin, 2022)
10. T. Carroll, *Geometric Function Theory. A Second Course in Complex Analysis* (Springer, Berlin, 2024)
11. H.M. Farkas, I. Kra, *Riemann Surfaces*. Graduate Texts in Mathematics, vol.71, 2nd edn. (Springer, Berlin, 1992)
12. P. Griffiths, J. Harris, *Principles of Algebraic Geometry* (Wiley, Hoboken, 1978)
13. A. Tonomura, J. Endo, T. Matsuda, T. Kawasaki, H. Ezawa, Demonstration of single-electron buildup of an interference pattern. Am. J. Phys. **57**, 117 (1989)
14. A. Tonomura, *Electron Holography*, 2nd edn. (Springer, Berlin, 1999)
15. L.D. Landau, E.M. Lifshitz, *The Classical Theory of Fields*. Landau and Lifshitz Course in Theoretical Physics, vol. 2, 4th edn. (Pergamon, Oxford, 1984)

Chapter 2
Hilbert Space Formulation of Quantum Physics

In fact, a great deal of natural philosophy is simply a process of linguistic simplification—an effort to invent languages in which half a page of equations can express an idea which could not be stated in less than a thousand pages of so-called "simple" language

W.P. Miller, Canticle for Leibowitz

In this chapter we lay the foundations of the Hilbert space formulation of Quantum Physics in the spirit of Dirac. The first ten sections contain fundamental principles and general formalisms which apply to all Quantum Physics, while the last seven sections are specific to *mechanical* quantum systems, that is, non-relativistic systems with finitely-many degrees of freedom. We shall outline an alternative formulation of Quantum Mechanics in Chap. 6.

The *superposition principle* stipulates that the states of any quantum system are points in $\mathbb{P}(\mathcal{H})$ where $\mathcal{H}$ is a vector $\mathbb{C}$-space endowed with a positive-definite Hermitian form $\langle \cdot | \cdot \rangle$. In this chapter we describe in detail the math formulation of Quantum Physics starting from the superposition principle and culminating with the Schrödinger, Heisenberg, and Dirac versions of the dynamical equations of motion.

2.1 The Hilbert Space of Quantum States

In Chap. 1 we formulated the *superposition principle* in terms of an abstract vector $\mathbb{C}$-space $\mathcal{H}$. When dealing with a specific quantum system, the abstract $\mathcal{H}$ gets replaced by a concrete $\mathbb{C}$-space which is "sufficiently big" to contain all quantum states of the system. While there are quantum systems whose space of states $\mathcal{H}$ is finite-dimensional, i.e. $\mathcal{H} \simeq \mathbb{C}^n$ for some n, most quantum systems need an *infinite dimensional* $\mathcal{H}$. We introduce some definitions which encompass both the finite and infinite dimensional situations.

© The Author(s), under exclusive license to Springer Nature Switzerland AG 2025
S. Cecotti, *Quantum Mechanics*, UNITEXT for Physics,
https://doi.org/10.1007/978-3-031-98824-0_2

Since the Hermitian product $\langle \cdot | \cdot \rangle$ is *positive-definite,* it induces a *norm*

$$\| \cdot \| : \mathcal{H} \to \mathbb{R}_{\geq 0} \tag{2.1}$$

by the formula

$$\big\| |A\rangle \big\|^2 \stackrel{\text{def}}{=} \langle A | A \rangle. \tag{2.2}$$

One easily checks that $\| \cdot \|$ satisfies the axioms of a norm in a topological vector space. A sequence $|A_n\rangle$ ($n \in \mathbb{N}$) is a *Cauchy sequence* iff, for all $\epsilon > 0$, there is a $N \in \mathbb{N}$ such that

$$\big\| |A_m\rangle - |A_n\rangle \big\| < \epsilon \quad \text{for } m, n > N. \tag{2.3}$$

A normed vector space $\mathcal{H}$ is *complete* iff all Cauchy successions converge in $\mathcal{H}$. A vector space which is complete for a norm induced by a Hermitian product $\langle \cdot | \cdot \rangle$, as in Eq. (2.2), is called a *Hilbert space* [1]. Clearly we may always assume $\mathcal{H}$ to be complete by replacing it (if necessary) with its completion. Hence, with no loss of generality, we declare

Superposition Principle 4 *The vector space of quantum states $\mathcal{H}$ is a Hilbert space defined over $\mathbb{C}$.*

We recall that the *Cauchy inequality*

$$|\langle B | A \rangle|^2 \leq \langle A | A \rangle \, \langle B | B \rangle \equiv \big\| |A\rangle \big\|^2 \, \big\| |B\rangle \big\|^2. \tag{2.4}$$

holds in all Hilbert spaces $\mathcal{H}$.

However not all Hilbert spaces will do. A Hilbert space whose dimension is *non-countable* is a wild beast where physically undesirable phenomena happen. Hence physicists' good sense forces us to make a further restriction:

Superposition Principle 5 *The dimension of the Hilbert space $\mathcal{H}$ of quantum states is at most countable.*

A Hilbert space $\mathcal{H}$ of finite dimension, $\dim \mathcal{H} = n$, is isomorphic to $\mathbb{C}^n$ with its standard Hermitian form

$$\langle (z_i) | (w_j) \rangle = \sum_{i=1}^{n} \bar{z}_i w_i. \tag{2.5}$$

A Hilbert space whose dimension is *infinite but countable* is said to be *separable.* Up to isomorphism, there is a *unique* separable Hilbert space. There are countless different concrete realizations of it, with distinct math and physical interpretations,[1] yet abstractly they are all equivalent to the space $\ell^2(\mathbb{N})$ of complex sequences with bounded quadratic norm

$$\ell^2(\mathbb{N}) = \left\{ a_n \in \mathbb{C}, \ n \in \mathbb{N}: \ \sum_{n=1}^{\infty} a_n^* a_n < +\infty \right\}, \tag{2.6}$$

endowed with its natural Hermitian form

$$\big\langle (b_n) \big| (a_n) \big\rangle = \sum_{n=1}^{\infty} b_n^* a_n. \tag{2.7}$$

We leave to the reader to show that the space $\ell^2(\mathbb{N})$ is complete for the topology defined by the quadratic norm induced by the Hermitian form (2.7).

2.2 Survey of Hilbert Space Constructions

We pause a while to give examples of different realizations of the (complex) separable Hilbert space. All these realizations play a significant role in Physics. In the context of Quantum Mechanics they are called *representations*. Although they are all abstractly equivalent, a specific computation may be easy in one representation and impossibly hard in another. Getting familiar with several representations is a smart move for a student. Yet the reader may prefer to skip this section in a first reading, and return to it when the particular construction is needed for an application.

Construction 1 Let S be a countable set. The separable Hilbert space $\ell^2(S)$ is the $\mathbb{C}$-space of maps $\phi: S \to \mathbb{C}$ with bounded ℓ^2-norm

$$\|\phi\|^2 \stackrel{\text{def}}{=} \sum_{s \in S} \phi(s)^* \phi(s) < +\infty, \tag{2.8}$$

equipped with the natural Hermitian form induced by the quadratic norm (2.8) via polarization

$$\langle \phi | \psi \rangle = \sum_{s \in S} \phi(s)^* \psi(s). \tag{2.9}$$

---

[1] See next section for a survey.

When $S = \mathbb{N}$ we recover $\ell^2(\mathbb{N})$. Other important examples are $\ell^2(\mathbb{N}^n)$ and $\ell^2(\mathbb{Z}^n)$. Any bijection of sets $S \to \mathbb{N}$ induces an isomorphism of Hilbert spaces

$$\ell^2(S) \to \ell^2(\mathbb{N}). \tag{2.10}$$

Construction 2 The Hilbert space $L^2([0, a])$ of Lebesgue integrable (complex) functions $\psi(x)$ in the interval $[0, a]$ with the boundary conditions[2] $\psi(0) = \psi(a) = 0$ and finite L^2-norm[3]

$$\|\psi\|^2 \stackrel{\text{def}}{=} \int_0^a \psi(x)^* \psi(x)\, dx < +\infty. \tag{2.11}$$

with Hermitian product

$$\langle \psi_1 | \psi_2 \rangle = \int_0^a \psi_1^*(x)\, \psi_2(x)\, dx. \tag{2.12}$$

The isomorphism with $\ell^2(\mathbb{N})$ is given by the Fourier series

$$(a_n) \mapsto \psi(x) \equiv \sqrt{\frac{2}{a}} \sum_{n=1}^{\infty} a_n \sin(\pi n x / a), \tag{2.13}$$

which is an isomorphism of Hilbert spaces by the *Parseval formula* [2]

$$\int_0^a |\psi(x)|^2\, dx = \sum_{n=1}^{\infty} |a_n|^2. \tag{2.14}$$

This example extends to the Hilbert space of L^2-functions on any n-dimensional finite box $B \equiv [0, a_1] \times [0, a_2] \times \cdots \times [0, a_n]$ with

$$\int_B |\psi|^2\, d^n x < +\infty \tag{2.15}$$

and vanishing on the boundary ∂B, which is isomorphic to $\ell^2(\mathbb{N}^d)$ via the multiple Fourier expansion.

Warning 1 In the definition of $L^2([0, a])$ we have been a bit sloppy. Properly speaking Eq. (2.11) defines a mere *semi-norm* on the vector space of integrable functions in the interval. A *semi-norm* $\| \cdot \|$ satisfies all the axioms of a norm, except

---

[2] Here we are using the *non-rigorous* language of physicists. The math description of this **Construction** is more involved. The actual Hilbert space is the closure in the norm topology of the space of finite Fourier sums of the form in the RHS in Eq. (2.13).

[3] Here and below $*$ stands for complex conjugation.

that it is not true that $\|v\| = 0$ implies $v = 0$. For instance, a function f which is non-zero only at one point (or more generally on a zero-measure subset of $[0, a]$) has $\|f\| = 0$ but $f \neq 0$. In any vector space V equipped with a semi-norm $\|\cdot\|$ which arises from a *non-negative* Hermitian product $\langle \cdot | \cdot \rangle$, the set of vectors $v \in V$ such that $\|v\| = 0$ is a linear subspace[4] $R \subset V$. Then the semi-norm $\|\cdot\|$ descends to a norm in the quotient vector space $\mathcal{H} \equiv V/R$. The norm of an element $h \in \mathcal{H}$ is the semi-norm $\|h'\|$ where $h' \in V$ is any representative of h. Thus

> *The Hilbert space of quantum states is the quotient space: its elements are the equivalence classes of finite-norm functions modulo zero-norm functions. In particular the function ψ is defined by the physical state only modulo functions vanishing almost everywhere (and the overall scale)*

This observation applies to all Hilbert spaces of the form $L^2(\mathcal{M})$ to be discussed below. We shall not repeat this **Warning** which applies implicitly to all Hilbert spaces of measurable functions.

Remark 2.1 The *smooth functions* (and even the *real analytic* ones) are *dense* in the Hilbert space $L^2([0, a])$ for the norm topology. Indeed already the finite Fourier sums are dense. This observation holds also for the other examples of Hilbert spaces of the form $L^2(\mathcal{M})$.

Definition 2.1 A *complete orthonormal system* $\{|n\rangle\}$ $(n = 1, 2, \ldots)$ in a separable Hilbert space $\mathcal{H}$, i.e. a sequence of vectors $\{|n\rangle\}_n$ such that

$$\langle m|n \rangle = \delta_{m,n} \quad \text{and} \quad \langle n|v \rangle = 0 \text{ for all } \langle n| \implies |v\rangle = 0, \tag{2.16}$$

is called a *Hilbert basis*. In other words, a Hilbert basis in the separable space $\mathcal{H}$ is an explicit isomorphism $\ell^2(\mathbb{N}) \to \mathcal{H}$

$$\ell^2(\mathbb{N}) \ni (a_n) \mapsto \sum_{n \in \mathbb{N}} a_n |n\rangle \in \mathcal{H}. \tag{2.17}$$

---

[4] **Proof**   First note that $\|v\| = 0$ if and only if $\langle w|v \rangle = 0$ for all $w \in V$. Indeed:

$$0 \leq \langle w + tv|w + tv \rangle = \langle w|w \rangle + 2 \operatorname{Re}(t \langle w|v \rangle) \quad \text{for all } t \in \mathbb{C}, \text{ and all } w \in V.$$

Now, if $v_1, v_2 \in R$, $\langle w|c_1 v_1 + c_2 v_2 \rangle = c_1 \langle w|v_1 \rangle + c_2 \langle w|v_2 \rangle = 0$ for all $w \in V$, thus $c_1 v_1 + c_2 v_2 \in R$.

Let $\{|n\rangle\}_n$ be a Hilbert basis in $\mathcal{H}$. For any $|\phi\rangle \in \mathcal{H}$ consider the vector

$$|\phi\rangle - \sum_{n=1}^{\infty} |n\rangle \langle n|\phi\rangle. \tag{2.18}$$

The vector (2.18) is orthogonal to all $|n\rangle$ by construction, and hence is zero. We conclude

$$1 = \sum_n |n\rangle \langle n| \qquad \text{(resolution of the identity).} \tag{2.19}$$

Conversely, an orthonormal sequence $\{|\psi_n\rangle\}_n$ in a separable Hilbert space $\mathcal{H}$ is *complete* ($\equiv$ a Hilbert basis) iff it resolves the identity, i.e. iff

$$1 = \sum_n |\psi_n\rangle \langle \psi_n|. \tag{2.20}$$

Example 2.1 For instance

$$|n\rangle = \sqrt{2/a}\, \sin(\pi n x/a), \quad n \in \mathbb{N} \tag{2.21}$$

is a Hilbert basis for the Hilbert space of "functions in the interval $[0, a]$ vanishing at the endpoints" (see Footnote 2).

Later we shall encounter more general resolutions of the identity.

Construction 3 Likewise we may consider the Hilbert space $L^2(S^1)$ of complex functions on the unit circle (i.e. functions on $\mathbb{R}$ periodic with period 2π) which have finite L^2-norm

$$\|\psi\|^2 \overset{\text{def}}{=} \oint |\psi(\theta)|^2 \frac{\mathrm{d}\theta}{2\pi} < +\infty. \tag{2.22}$$

In view of the Parseval identity, the Hilbert space $L^2(S^1)$ is isomorphic to $\ell^2(\mathbb{Z})$ via the exponential Fourier series

$$\psi(\theta) = \sum_{n=-\infty}^{+\infty} a_n\, \mathrm{e}^{in\theta}. \tag{2.23}$$

Multiple Fourier series extend this isomorphism to the separable Hilbert spaces

$$L^2(T^n) \simeq \ell^2(\mathbb{Z}^n) \tag{2.24}$$

of integrable complex functions on the n-dimensional torus $T^n \equiv (S^1)^n$ which have finite quadratic norm

$$\|\psi\|^2 \stackrel{\text{def}}{=} \int_{T^n} |\psi(\theta_1, \ldots, \theta_n)|^2 \, \frac{d^n\theta}{(2\pi)^n} < +\infty. \tag{2.25}$$

Construction 4   We may generalize Construction 3 to any finite-dimensional manifold M endowed with a volume measure $d\mu$. We consider the vector space of $d\mu$-measurable complex functions $\psi : M \to \mathbb{C}$ endowed with the quadratic norm

$$\|\psi\|^2 = \int_M |\psi|^2 \, d\mu. \tag{2.26}$$

The Hilbert space $L^2(M, d\mu)$ is the vector space of functions with finite norm (modulo functions of zero norm) equipped with the obvious Hermitian product induced by polarization of the quadratic norm (2.26). If $\partial M \neq \varnothing$ we must supplement the definition of $\mathcal{H}$ with suitable "boundary conditions"[5] as we did in Construction 2 when $M = [0, a]$.

Construction 5   Specializing Construction 4 to $M = \mathbb{R}$, we get the fundamental Hilbert space $L^2(\mathbb{R})$ of Lebesgue-mesurable functions $\psi : \mathbb{R} \to \mathbb{C}$ with quadratic norm

$$\|\psi\|^2 = \int_{-\infty}^{+\infty} \psi(x)^* \psi(x) \, dx < +\infty. \tag{2.27}$$

More generally, we may fix a positive function $W : \mathbb{R} \to \mathbb{R}_{>0}$ and consider the space $L^2(\mathbb{R}, W)$ of functions $\psi : \mathbb{R} \to \mathbb{C}$ equipped with norm[6]

$$\|\psi\|^2 = \int dx \, W(x) \, |\psi(x)|^2 < +\infty. \tag{2.28}$$

The function $W(x)$ is called the *weight* of the Hilbert space $L^2(\mathbb{R}, W)$ of functions on the line. The map

$$\psi(x) \rightsquigarrow \sqrt{W(x)} \, \psi(x) \tag{2.29}$$

sets an isomorphism $L^2(\mathbb{R}, W) \simeq L^2(\mathbb{R})$.

Separability of $L^2(\mathbb{R})$ It may be not obvious that the Hilbert space $L^2(\mathbb{R})$ is separable. To settle the issue, we present an explicit isomorphism $L^2(\mathbb{R}) \simeq \ell^2(\mathbb{N})$.

---

[5] Again, we are speaking in the loose physicists' language, cf. Footnote 2.

[6] When there is no danger of confusion, we write simply $\int$ for $\int_{-\infty}^{+\infty}$.

Consider the sequence $\{\psi_n(x)\}_{n\geq 0}$ of functions on the real line defined by the following generating function

$$\sum_{n=0}^{\infty} \sqrt{\frac{2^n}{n!}} \, \psi_n(x) \, z^n = \frac{1}{\pi^{1/4}} \exp\left(-\tfrac{1}{2}x^2 + 2xz - z^2\right). \tag{2.30}$$

We have the equality

$$\sum_{m,n=0}^{\infty} \sqrt{\frac{2^{m+n}}{m!\,n!}} \, w^m z^n \int_{-\infty}^{+\infty} \mathrm{d}x \, \psi_m(x)^* \, \psi_n(x) =$$

$$= \frac{1}{\sqrt{\pi}} \int_{-\infty}^{+\infty} \mathrm{d}x \, \mathrm{e}^{-(x-w-z)^2 + 2wz} = \mathrm{e}^{2wz} = \sum_{n=0}^{\infty} \frac{(2wz)^n}{n!}, \tag{2.31}$$

that is, our sequence of functions $\{\psi_n(x)\}_n$ is *orthonormal* with respect to the $L^2(\mathbb{R})$ inner product

$$\langle \psi_m | \psi_n \rangle \equiv \int_{-\infty}^{+\infty} \mathrm{d}x \, \psi_m(x)^* \, \psi_n(x) = \delta_{mn}. \tag{2.32}$$

We claim that the sequence $\{\psi_n(x)\}_{n\geq 0}$ is also *complete* in the sense that a vector $|\phi\rangle$ orthogonal to all $|\psi_n\rangle$ is the zero vector (in the sense of Warning 1, i.e. it is zero almost everywhere). Indeed for all $z \in \mathbb{C}$ we have

$$0 = \sum_{n=0}^{\infty} \sqrt{\frac{2^n}{n!}} \, z^n \, \langle \psi_n | \phi \rangle = \mathrm{e}^{-z^2} \int_{-\infty}^{+\infty} \mathrm{d}x \, \mathrm{e}^{2xz} \left(\mathrm{e}^{-x^2/2}\phi(x)\right); \tag{2.33}$$

specializing z to be $\mathrm{i}y/2$ (with $y \in \mathbb{R}$) the integral in the RHS becomes the Fourier transform of the function $\mathrm{e}^{-x^2/2}\phi(x)$. Since an integrable function which has zero Fourier transform is zero almost everywhere, we conclude that the sequence $\{\psi_n\}_n$ is a complete orthonormal sequence, that is a *Hilbert basis*, in $L^2(\mathbb{R})$. Then the isomorphism $\ell^2(\mathbb{N}) \simeq L^2(\mathbb{R})$ we are after is

$$a_n \mapsto \psi(x) \equiv \sum_{n=0}^{\infty} a_n \, \psi_n(x), \qquad \sum_{n=0}^{\infty} |a_n|^2 = \int_{-\infty}^{+\infty} \mathrm{d}x \, |\psi(x)|^2, \tag{2.34}$$

where the second equation (the *Bessel-Parseval identity*) follows from (2.32) and the completeness of the ψ_n's which, as in Remark 2.1, implies the resolution of the identity

$$1 = \sum_{n=0}^{\infty} |\psi_n\rangle\langle\psi_n|, \tag{2.35}$$

which can be rewritten in terms of the *reproducing integral kernel* $\delta(x, y)$

$$f(x) = \int dy\, \delta(x, y)\, f(y) \qquad \forall\, f(x) \in L^2(\mathbb{R})$$

$$\text{where } \delta(x, y) \equiv \sum_{n=0}^{\infty} \psi_n(x)\, \psi_n(y)^*.$$

$$(2.36)$$

An example of sub-space of smooth functions which is dense in $L^2(\mathbb{R})$ is the space of finite linear combinations of the ψ_n's. We shall describe more general dense subspaces of smooth functions below.

Construction 6 Given a separable Hilbert space $\mathcal{H}$, we can form the separable Hilbert space $\mathbb{C}^n \otimes \mathcal{H}$ whose vectors are n-tuples $(|\psi_1\rangle, \ldots, |\psi_n\rangle)$ of elements of $\mathcal{H}$ with the natural square-norm

$$\sum_{s=1}^{n} \big\||\psi_s\rangle\big\|^2.$$

$$(2.37)$$

For instance, the elements of $\mathbb{C}^n \otimes L^2(\mathbb{R})$ are functions $\psi : \mathbb{R} \to \mathbb{C}^n$ with

$$\|\psi\|^2 = \sum_{s=1}^{n} \int dx\, \psi_s(x)^* \psi_s(x).$$

$$(2.38)$$

Construction 7 If $\mathcal{H}_1$ and $\mathcal{H}_2$ are two separable Hilbert spaces, we may construct the separable Hilbert space

$$\mathcal{H}_1 \,\widehat{\otimes}\, \mathcal{H}_2,$$

$$(2.39)$$

where $\widehat{\otimes}$ stands for the norm completion of the algebraic tensor product of the two vector spaces. If $\{|\psi_n^a\rangle\}_n$ $(a = 1, 2)$ are Hilbert bases for the two factor Hilbert spaces, a Hilbert basis of $\mathcal{H}_1 \,\widehat{\otimes}\, \mathcal{H}_2$ is

$$\big\{\, |\psi_m^1\rangle \otimes |\psi_n^2\rangle \quad m, n \in \mathbb{N} \big\}.$$

$$(2.40)$$

In particular, when $\mathcal{M}_a$ are manifolds with positive measures,

$$L^2(\mathcal{M}_1) \,\widehat{\otimes}\, L^2(\mathcal{M}_2) \equiv L^2(\mathcal{M}_1 \times \mathcal{M}_2).$$

$$(2.41)$$

Construction 8 In the same fashion we may form a separable Hilbert space $\mathcal{H}_1 \,\widehat{\oplus}\, \mathcal{H}_2$. For instance we have "mixed" representations of the form

$$\ell^2(S) \,\widehat{\oplus}\, L^2(\mathcal{M})$$

$$(2.42)$$

where S is a discrete set which is at most numerable, and M a manifold of finite dimension with a positive measure.

Construction 9 The space in Construction 3, and more generally all Hilbert spaces $L^2(M)$ with M *non-simply-connected*, may be generalized by introducing a non-trivial *monodromy representation* of the fundamental group

$$m : \pi_1(M) \to U(1) \tag{2.43}$$

whose deep physical meaning will be elucidated in Sect. 3.11.

We focus on the example $M = S^1$. We have seen in Chap. 1 that the vector representing a quantum state is determined up to a non-observable phase factor $e^{i\alpha}$. This means that, when going around the circle S^1, we may return to the starting point and find that the state is now represented by a vector which differs from the original one by a non-trivial phase factor. This phase factor, or rather the group homomorphism

$$\pi_1(S^1) \simeq \mathbb{Z} \to U(1) \qquad n \mapsto e^{in\alpha}, \tag{2.44}$$

is called the *monodromy* (representation). We may think of the monodromy as a generalized periodicity condition for the functions $\psi(\theta)$ on the circle, where

$$\psi(\theta + 2\pi) = e^{i\alpha}\psi(\theta). \tag{2.45}$$

We may then consider the Hilbert space $L^2(S^1)_\alpha$ with prescribed (fixed) monodromy representation. Clearly inequivalent monodromies yield inequivalent representations of the separable Hilbert space.

Combining with Construction 6, when $\mathcal{H} = \mathbb{C}^n \otimes L^2(M)$ we may, more generally, consider Hilbert spaces defined by a non-Abelian monodromy representation of the form

$$\sigma : \pi_1(M) \to U(n) \tag{2.46}$$

where the monodromy acts of the factor $\mathbb{C}^n$ of the Hilbert space.

Construction 10 We may unify and generalize Constructions 6 and 9 by considering a complex vector bundle $V \to M$ of rank n over the manifold M (with volume form $d\mu$) endowed with a fiber Hermitian product h. We consider the vector space of locally integrable sections $\psi : M \to V$ with finite quadratic norm

$$\int_M d\mu \, \psi^* h \psi < \infty \tag{2.47}$$

(modulo zero norm sections).

Construction 11 An important special case of Construction 10 is when $\mathcal{M}$ is an oriented Riemannian manifold of dimension m and

$$\mathcal{V} = \bigoplus_{k=0}^{m} \wedge^k (T^*\mathcal{M} \otimes \mathbb{C}), \tag{2.48}$$

whose smooth sections are the complex differential forms $\alpha \in \Omega^\bullet(\mathcal{M})\otimes\mathbb{C}$, equipped with the Hodge Hermitian inner product

$$\langle \alpha_1 | \alpha_2 \rangle \equiv \int_{\mathcal{M}} \alpha_1^* \wedge *\alpha_2 < \infty, \tag{2.49}$$

where $*$ is the *Hodge dual* induced by the Riemannian metric on $\mathcal{M}$ [3, 4].

Construction 12 Let $\Omega \subset \mathbb{C}^n$ be a *bounded domain*. We may consider the space $K^2(\Omega)$ of the holomorphic functions on Ω with the inner product

$$\langle f | g \rangle = \int_{\Omega} f(z)^* g(z) \, \mathrm{d}^n z \, \mathrm{d}^n \bar{z} \tag{2.50}$$

It may not be obvious that this normed vector space is *complete*, that is, that the limit (in the norm sense) of a Cauchy sequence of holomorphic functions is holomorphic. For brevity we limit to show this fact when Ω is the unit disk[7] $\mathfrak{D}$ in $\mathbb{C}$,

$$\mathfrak{D} = \{z \in \mathbb{C} \colon |z| < 1\}. \tag{2.51}$$

For a holomorphic function $f(z) = \sum_{n \geq 0} a_n z^n$ the norm (2.50) is

$$\|f\|^2 = \sum_{n \geq 0} \frac{\pi}{n+1} |a_n|^2, \tag{2.52}$$

which is convergent when the series

$$\sum_{n \geq 0} a_n z^n \tag{2.53}$$

defines a holomorphic function in $\mathfrak{D}$, i.e. when the radius of convergence of the power series is ≥ 1. We consider a sequence of complex numbers $(a_n)_n$ such that the RHS of (2.52) is finite: to check completeness, we have to show that the corresponding power series (2.53) converges for all $|z| < 1$ then defining a

---

[7] By the Riemann mapping theorem this is the general case when $\Omega \neq \mathbb{C}$ has complex dimension 1 and is simply connected.

holomorphic function in the unit disk $\mathfrak{D}$. Indeed, using (2.4),

$$\pi \left| \sum_{n \geq 0} a_n z^n \right|^2 = \pi \left| \sum_{n \geq 0} \frac{a_n}{\sqrt{n+1}} \cdot \sqrt{n+1}\, z^n \right|^2 \leq$$
$$\leq \|f\|^2 \sum_{n \geq 0} (n+1)|z|^{2n} = \frac{\|f\|^2}{(1-|z|^2)^2}. \tag{2.54}$$

This shows that when $\|f\|^2 < \infty$ the series $\sum_{n \geq 0} a_n z^n$ converges for all $|z| < 1$, and hence our vector space $\mathcal{H}$ of *holomorphic* functions is complete for the Hilbert space norm. Equation (2.54) says that, for all points $z_0 \in \mathfrak{D}$, the *evaluation map*

$$e_{z_0} : \mathcal{H} \to \mathbb{C}, \qquad f(z) \mapsto f(z_0), \tag{2.55}$$

is *continuous* in the Hilbert space topology with norm

$$\|e_{z_0}\| = \frac{1}{\pi} (1 - |z_0|^2)^{-2} < \infty. \tag{2.56}$$

This implies that in this holomorphic set-up the *reproducing kernel* $\delta(z, \bar{w})$ defined by the property

$$f(z) = \int_D dw\, d\bar{w}\; \delta(z, \bar{w})\, f(w) \quad \text{for all } f \in \mathcal{H} \tag{2.57}$$

is a *smooth* function in $\mathfrak{D} \times \overline{\mathfrak{D}}$

$$\delta(z, \bar{w}) = \frac{1}{\pi} \frac{1}{(1 - z\bar{w})^2} \tag{2.58}$$

called the *Bergman kernel* [5–7]. The Bergman kernel has remarkable function-theoretic properties. Similar results hold, with the obvious modifications, for *all* bounded domains $\Omega \subset \mathbb{C}^n$.

Construction 13 Let $B^2(\mathbb{C})$ be the space of *entire holomorphic functions* $\phi(z)$ on $\mathbb{C}$ with finite quadratic norm

$$\|\phi(z)\|^2 \overset{\text{def}}{=} \int_{\mathbb{C}} |\phi(z)|^2 e^{-z\bar{z}}\, dz\, d\bar{z} < +\infty \tag{2.59}$$

Again, the space of entire functions is *closed* with respect to this quadratic norm: indeed the square-norm of $\phi(z) = \sum_{n \geq 0} a_n z^n$ is

$$\|\phi(z)\|^2 = \pi \sum_{n \geq 0} n! \, |a_n|^2 \tag{2.60}$$

while

$$\left| \sum_{n \geq 0} a_n z^n \right|^2 \leq \sum_{n \geq 0} \left| (\sqrt{n!}\, a_n) \left(\frac{z^n}{\sqrt{n!}} \right) \right|^2 \leq$$

$$\leq \frac{1}{\pi} \|\phi(z)\|^2 \sum_{n \geq 0} \frac{|z|^{2n}}{n!} \equiv \frac{e^{|z|^2}}{\pi} \|\phi(z)\|^2 \tag{2.61}$$

so all formal series $\sum_n a_n z^n$ with $\sum_n n! \, |a_n|^2 < \infty$ is convergent for all $z \in \mathbb{C}$ and defines an entire function. This shows that the space of entire functions with finite norm in the sense of Eq. (2.59) is *complete* for the norm topology, hence $B^2(\mathbb{C})$ is a separable Hilbert space. As in Construction 12 the evaluation map at a point $z_0 \in \mathbb{C}$ is continuous, which means that the reproducing kernel $\delta(z, \bar{w})$ is an *analytic function* of its arguments

$$\delta(z, \bar{w}) = \frac{1}{\pi} \exp(z\bar{w}) \tag{2.62}$$

$$f(z) = \int \frac{dw \, d\bar{w}}{\pi} \, e^{z\bar{w} - w\bar{w}} \, f(w). \tag{2.63}$$

This representation of the separable Hilbert space is called the *Bargmann space* (the name Bargmann-Fock is also used to emphasize its main physical application). As we shall see, it may be used to describe the states of simple quantum systems with one degree of freedom. Its advantage with respect to the more familiar space $L^2(\mathbb{R})$ is that all expressions in the Bargmann representation are highly regular, in facts *analytic*, while we may exploit the powerful methods of complex analysis.

Construction 14 The *Hardy space* of the unit disk $H^2(\mathfrak{D})$ is the vector space of the holomorphic functions in the disk, as in Example 12, but with a different Hermitian product and norm

$$\langle f | g \rangle = \sup_{0 < R < 1} \oint \frac{d\theta}{2\pi} \, f(Re^{i\theta})^* \, g(Re^{i\theta}). \tag{2.64}$$

The monomial z^n ($n = 0, 1, 2, \dots$) form an orthonormal Hilbert basis and the reproducing kernel is the Cauchy residue kernel.

2.3 Linear Operators and Physical Observables

The description of a physical system consists of a space of *states* together with a set of *physical quantities,* that is, properties of the states that we can measure. Typical physical quantities are energy, momentum, angular momentum, etc. The superposition principle says that the states are represented by (non-zero) vectors in a complex Hilbert space $\mathcal{H}$ (either finite dimensional or separable). As already remarked, all physical properties of a quantum state A should be expressed by linear operations on the representing vector $|A\rangle$. Roughly speaking, all properties of the states should be given by linear maps from $\mathcal{H}$ to itself. We shall see below that there is a subclass of (densely defined) linear operators which have the appropriate properties to be identified with the physical observables. For our purposes it is convenient to adopt the following definition of operator [8, 10]:

Definition 2.2 An *operator in the Hilbert space $\mathcal{H}$* is a linear map

$$A : D(A) \to \mathcal{H} \tag{2.65}$$

where $D(A) \subset \mathcal{H}$ is a vector subspace which is *dense* in the Hilbert space topology i.e. $\overline{D(A)} \equiv \mathcal{H}$. The subspace $D(A)$ is called the *domain* of the operator A. *Linear map* means that

$$A(a|V\rangle + b|W\rangle) = a\,A|V\rangle + b\,A|W\rangle$$
$$\text{for all } |V\rangle,\ |W\rangle \in D(A) \text{ and } a, b \in \mathbb{C}. \tag{2.66}$$

The (operator) *norm* of A is

$$\|A\| \overset{\text{def}}{=} \sup_{0 \neq |V\rangle \in D(A)} \frac{\||A|V\rangle\|^2}{\||V\rangle\|^2}. \tag{2.67}$$

When the Hilbert space is finite-dimensional all linear maps (operators) are continuous. When $\mathcal{H}$ is *separable* we need to distinguish two kinds of operators:

Bounded: They have finite norm $\|A\| < +\infty$ and hence can be extended by continuity to genuine continuous linear maps

$$A : \mathcal{H} \to \mathcal{H} \tag{2.68}$$

defined in all $\mathcal{H}$. The space $\mathcal{B}(\mathcal{H})$ of bounded operators $\mathcal{H} \to \mathcal{H}$ is a Banach algebra, a C^*-algebra, and a von Neumann algebra;

Unbounded: The norm of A is *infinite,* and A *cannot* be extended to the full $\mathcal{H}$. The operator A makes sense only when acting on the vectors belonging to some *dense* domain $D(A) \subsetneq \mathcal{H}$.

Sometimes an operator A, which was initially defined in some domain $D \subset \mathcal{H}$, may be extended to a larger domain $D' \supset D$. The operator A may be extended to act on the full Hilbert space $\mathcal{H}$ iff it is bounded. In physics it is convenient to work with (densely-defined) unbounded operators although in principle the theory may be formulated in terms of $\mathcal{B}(\mathcal{H})$ alone.

Example 2.2 Consider $\mathcal{H} = \ell^2(\mathbb{N})$ with its standard Hilbert basis $\{|n\rangle\}_n$. The linear operator which acts on the basis vectors as $|n\rangle \to n|n\rangle$ is unbounded. Its domain contains the finite linear combinations of basis vectors, which is obviously a dense subspace of $\mathcal{H}$.

Example 2.3 Let $\mathcal{H}$ be the Hilbert space $L^2(\mathbb{R})$. $\mathcal{H}$ has a Lie group $\mathbb{R}$ of automorphisms (translations) acting as $\psi(x) \mapsto \psi(x - L)$ for $L \in \mathbb{R}$. Now

$$\frac{\|x\,\psi(x-L)\|}{\|\psi(x-L)\|} \equiv \frac{\|(L+x)\psi(x)\|}{\|\psi(x)\|} = L + O(1) \qquad \text{as } L \to \infty \tag{2.69}$$

so the linear operator "multiplication by x", $\psi(x) \mapsto x\,\psi(x)$, is *unbounded* in $L^2(\mathbb{R})$.

Example 2.4 Let $\mathcal{H} = L^2(\mathbb{R})$ and consider the linear operator

$$\psi \mapsto -\mathrm{i}\frac{\partial \psi}{\partial x} \tag{2.70}$$

initially defined, say, on the dense subspace of finite linear combinations of the Hilbert basis $\{\psi_n(x)\}_n$ of Eq. (2.30). The Lie group $\mathbb{R}$ acts on $L^2(\mathbb{R})$ by the automorphisms

$$\psi(x) \mapsto \mathrm{e}^{\mathrm{i}kx}\psi(x), \quad k \in \mathbb{R} \tag{2.71}$$

$$-\mathrm{i}\frac{\partial \psi}{\partial x} \mapsto -\mathrm{i}\frac{\partial \psi}{\partial x} + k\psi, \tag{2.72}$$

and

$$\frac{\|-\mathrm{i}\frac{\partial}{\partial x}(\mathrm{e}^{\mathrm{i}kx}\psi(x))\|}{\|\mathrm{e}^{\mathrm{i}kx}\psi(x)\|} = k + O(1) \quad \text{as } k \to \infty \tag{2.73}$$

so the operator (2.70) is *unbounded* and cannot be extended to the full $L^2(\mathbb{R})$.

In physics we usually don't care too much of the distinction between bounded and unbounded operators. We content ourselves to work in suitable dense domains. Indeed, by definition, we can find an element of the domain D which approximates the given state with an error as small as we wish and, in particular, much smaller than the experimental error implicit in the best available technology. On the other hand, only linear operators defined in *dense* domains may describe physical quantities (observables): indeed, we may always envision a *gedanken* experiment to measure the observable in any state with arbitrary small (but finite) precision.

Remark 2.2 The algebra $\mathcal{B}(\mathcal{H})$ has a unity, namely the identity map $\mathcal{H} \to \mathcal{H}$ that we denote as 1. The unit yields an embedding $\mathbb{C} \hookrightarrow \mathcal{B}(\mathcal{H})$ given by $c \mapsto c \cdot 1$ whose image we write simply c. An element $c \in \mathbb{C} \subset \mathcal{B}(\mathcal{H})$ is called a *constant*, a *number*, or, when emphasis is needed, a *c-number*.

Adjoint Operator

When T is an operator and $|a\rangle \in D(T)$, we sometimes write $|Ta\rangle$ for $T|a\rangle$ iff it makes the expression less clumsy or more transparent.

If $T : \mathcal{H} \to \mathcal{H}$ is a bounded operator, there exists an unique *adjoint operator* $T^\dagger$ such that

$$\langle T^\dagger b|a\rangle = \langle b|Ta\rangle \quad \text{i.e.} \quad \langle a|T^\dagger|b\rangle^* = \langle b|T|a\rangle \quad \forall |a\rangle, |b\rangle \in \mathcal{H}, \tag{2.74}$$

as a consequence of the Riesz representation theorem [1]. The adjoint operation $\dagger$ has the following properties

$$(T^\dagger)^\dagger = T$$

$$(c_1 T_1 + c_2 T_2)^\dagger = c_1^* T_1^\dagger + c_2^* T_2^\dagger, \qquad \forall\, T_1, T_2 \in \mathcal{B}(\mathcal{H}),\ \forall\, c_1, c_2 \in \mathbb{C} \tag{2.75}$$

$$(T_1 T_2)^\dagger = T_2^\dagger T_1^\dagger \qquad\qquad \forall\, T_1, T_2 \in \mathcal{B}(\mathcal{H}),$$

that is, $T \mapsto T^\dagger$ is an anti-linear anti-automorphism of the $*$-algebra $\mathcal{B}(\mathcal{H})$ and an involution. Moreover T is invertible iff $T^\dagger$ is invertible, and

$$(T^{-1})^\dagger = (T^\dagger)^{-1}. \tag{2.76}$$

These are the same formal properties enjoyed by the adjoint conjugate of a linear operator acting on a finite-dimensional vector space, and their proof follows the same pattern as in Linear Algebra.

We wish to extend the $\dagger$ operation to operators T which are defined in a *dense* domain $D(T)$ of a separable Hilbert space $\mathcal{H}$. We set

$$D(T^\dagger) \overset{\text{def}}{=} \left\{ |b\rangle \in \mathcal{H} \ \middle|\ \sup_{\substack{|a\rangle \in \mathcal{H} \\ \langle a|a\rangle = 1}} \left|\langle b|T|a\rangle\right| < +\infty \right\} \tag{2.77}$$

It is easy to see (and left as an exercise to the reader) that there is a *unique* linear operator

$$T^\dagger : D(T^\dagger) \to \mathcal{H}, \tag{2.78}$$

called the *adjoint of T*, such that

$$\langle T^\dagger b\,|\,a\rangle = \langle b\,|\,Ta\rangle \qquad \forall |a\rangle \in D(T),\ |b\rangle \in D(T^\dagger). \tag{2.79}$$

The properties (2.75) still hold in the domains where the expressions are defined.

Remark 2.3 Uniqueness of $T^\dagger$ follows from the fact that $D(T)$ is dense.

Special Classes of Operators
An operator T is called:

(1) *Normal* if $TT^\dagger = T^\dagger T$ when acting on the domain where one side is defined;
(2) *Symmetric* (or *Hermitian*) iff

$$\langle Tb|a\rangle = \langle b|Ta\rangle \quad \forall |a\rangle, |b\rangle \in D(T) \tag{2.80}$$

in this case $D(T) \subseteq D(T^\dagger)$;
(3) *Self-adjoint* if it is symmetric and in addition $D(T) = D(T^\dagger)$ i.e.

$$T = T^\dagger \tag{2.81}$$

(4) *Essentially self-adjoint* if it is symmetric and has a unique self-adjoint extension. In this case $T^\dagger$ is self-adjoint;
(5) An *isometry* if $D(T) = \mathcal{H}$ (i.e. $T \in \mathcal{B}(\mathcal{H})$) and $T^\dagger T = 1$;
(6) *Unitary* if it is an automorphism of $\mathcal{H}$, i.e. an invertible isometry, or equivalently if T is bounded and

$$T^\dagger T = TT^\dagger = 1 \tag{2.82}$$

(7) *Non-negative* if $\langle v|T|v\rangle \geq 0$ for all $v \in D(T)$.

Remark 2.4 In the physical usage one usually is cavalier with the subtle distinctions between "symmetric", "self-adjoint" and "essentially self-adjoint" and uses the generic term "Hermitian operator" for all them, except when, for the particular system at hand, the distinction carries a physical significance.

If T is any operator, $T^\dagger T$ and $TT^\dagger$ are Hermitian and non-negative:

$$(T^\dagger T)^\dagger = T^\dagger (T^\dagger)^\dagger = T^\dagger T, \tag{2.83}$$

$$\langle v|T^\dagger T|v\rangle \equiv \langle Tv|Tv\rangle = \|T|v\rangle\|^2 \geq 0 \tag{2.84}$$

whenever the expressions are defined.

Example 2.5 We present an example of an isometry which is *not* unitary. This is not possible in finite dimension: in $\mathbb{C}^n$ all isometries are automatically unitary. Consider $\ell^2(\mathbb{N})$ and let $\{|n\rangle\}_{n\geq 0}$ be its standard (orthonormal) Hilbert basis. The *shift operator*

$$S: |n\rangle \mapsto |n+1\rangle \tag{2.85}$$

is clearly an isometry, but it is not invertible since $|0\rangle$ is not in its range. $S^\dagger$ sends $|n\rangle$ to $|n-1\rangle$ if $n \geq 1$ and to zero otherwise. So

$$S^\dagger S = 1, \qquad SS^\dagger = 1 - |0\rangle\langle 0|. \tag{2.86}$$

Example 2.6 Identifying $\ell^2(\mathbb{N})$ with $H^2(D)$ via $|n\rangle \leftrightarrow z^n$, the shift S is multiplication by z whose image is the space of holomorphic functions in the disk which vanish at the origin. More generally we have a non-invertible isometry of $H^2(D)$ whose image is the space of holomorphic functions which vanish at the point $z_0 \in D$ which is given by multiplication by

$$\frac{z - z_0}{1 - z\bar{z}_0} = z\, \frac{1 - z^{-1}z_0}{1 - z\bar{z}_0} \tag{2.87}$$

which is an isometry since this factor reduces to a phase when $|z| = 1$.

Projectors: Spectral Families

An *(orthogonal) projector* $P \in \mathcal{B}(\mathcal{H})$ is a bounded Hermitian operator which is *idempotent* i.e.

$$P^\dagger = P, \qquad P^2 = P. \tag{2.88}$$

The image $\operatorname{im} P \subset \mathcal{H}$ of a projector is a closed linear subspace of $\mathcal{H}$. Note that if P is a projector, $1 - P$ is also a projector whose image is the orthogonal complement $(\operatorname{im} P)^\perp$ to $\operatorname{im} P$.

Fact 2.1 *If S is an isometry, $SS^\dagger$ is a projection.*

Indeed $(SS^\dagger)^2 = S(S^\dagger S)S^\dagger = SS^\dagger$.

Definition 2.3 A *spectral family* is a family of projectors $P(\lambda) \in \mathcal{B}(\mathcal{H})$, parametrized by $\lambda \in \mathbb{R}$, such that for all $|v\rangle \in \mathcal{H}$

$$\lim_{\lambda \to -\infty} P(\lambda)|v\rangle = 0 \tag{2.89}$$

$$\lim_{\lambda \to +\infty} P(\lambda)|v\rangle = |v\rangle \tag{2.90}$$

$$\langle v|P(\mu)|v\rangle \leq \langle v|P(\lambda)|v\rangle \text{ when } \mu \leq \lambda \tag{2.91}$$

$$\lim_{\epsilon \to 0^+} P(\lambda + \epsilon)|v\rangle = P(\lambda)|v\rangle. \tag{2.92}$$

Example 2.7 Let $\{|\psi_n\rangle\}$ ($n \in \mathbb{N}$) be any (orthonormal) Hilbert basis in $\mathcal{H}$. We have the spectral family

$$P(\lambda) = \sum_{n \leq \lambda} |\psi_n\rangle\langle\psi_n| \tag{2.93}$$

where $P(\lambda)$ projects on the finite dimensional sub-space spanned by the first $[\lambda]$ basis vectors.

Example 2.8 Let $\chi_\lambda(x)$ be the characteristic function of $(-\infty, \lambda] \subset \mathbb{R}$. The family of bounded linear maps $P(\lambda) \colon L^2(\mathbb{R}) \to L^2(\mathbb{R})$

$$P(\lambda)\,\psi(x) = \chi_\lambda(x)\,\psi(x) \tag{2.94}$$

is a spectral family. The image of $P(\lambda)$ are the L^2-functions with support in $(-\infty, \lambda]$.

The image of $P(\mu)$ is contained in the image of $P(\lambda)$ when $\lambda \geq \mu$ so that

$$P(\lambda)\,P(\mu) = P(\mu)\,P(\lambda) = P(\mu) \quad \mu \leq \lambda. \tag{2.95}$$

A spectral family $P(\lambda)$ defines a measure $\mathrm{d}P(\lambda)$ on $\mathbb{R}$ (in the sense of Stieltjes) valued in the projectors $\mathcal{H} \to \mathcal{H}$, by the rule

$$\int_{\lambda_1}^{\lambda_2} \mathrm{d}P(\lambda) = P(\lambda_2) - P(\lambda_1), \qquad \lambda_2 > \lambda_1. \tag{2.96}$$

Indeed, the RHS is a projector

$$\begin{aligned}
(P(\lambda_2) - P(\lambda_1))^2 &= P(\lambda_2)^2 + P(\lambda_1)^2 - P(\lambda_2)P(\lambda_1) - P(\lambda_1)P(\lambda_2) \\
&= P(\lambda_2) - P(\lambda_1),
\end{aligned} \tag{2.97}$$

where we used (2.95). For instance in Example 2.7 $P(\lambda_1) - P(\lambda_2)$ is the projector on the subspace spanned by the basis vectors $\{|\psi_n\rangle\}$ with $\lambda_1 < n \leq \lambda_2$.

Equivalently, for all $|v\rangle,\ |w\rangle \in \mathcal{H}$ we have

$$\int_{\lambda_1}^{\lambda_2} \langle v|\mathrm{d}P(\lambda)|w\rangle \equiv \int_{\lambda_1}^{\lambda_2} \mathrm{d}\langle v|P(\lambda)|w\rangle = \langle v|P(\lambda_2)|w\rangle - \langle v|P(\lambda_1)|w\rangle \tag{2.98}$$

where now the LHS is an ordinary (i.e. $\mathbb{C}$-valued) Stieltjes integral.

Physical Observables

All operators L can be written as $A + iB$ with A and B Hermitian: just take

$$A = (L + L^\dagger)/2 \quad \text{and} \quad B = (L - L^\dagger)/2\mathrm{i}. \tag{2.99}$$

Thus, informally, for an operator being Hermitian is akin to being real for a complex number. Since the superposition principle requires the physical observables to be given by linear operators, while they are real quantities, it is natural to postulate the following

Principle 1 *The physical observables of a quantum system are the (essentially) self-adjoint operators T with a dense domain $D(T) \subset \mathcal{H}$*

The meaning of this **Principle** will be clarified by the spectral theorem.

Commutators: The Jacobi Identity

The *commutator* of two operators A, B

$$[A, B] \stackrel{\text{def}}{=} AB - BA, \tag{2.100}$$

is antisymmetric in its two arguments and bilinear in both of them

$$[A, B] = -[B, A] \tag{2.101}$$

$$[c_1 A_1 + c_2 A_2, B] = c_1[A_1, B] + c_2[A_2, B] \tag{2.102}$$

$$[A, c_1 B_1 + c_2 B_2] = c_1[A, B_1] + c_2[A, B_2], \tag{2.103}$$

where $c_1, c_2 \in \mathbb{C}$. The operation $[A, \cdot]$ is a *derivation* on the $\mathbb{C}$-algebra of operators which kills the constants

$$[A, c] = 0 \qquad c \in \mathbb{C}, \tag{2.104}$$

and satisfies the non-commutative version of the Leibniz rule

$$[A, BC] = [A, B]C + B[A, C], \tag{2.105}$$

indeed

$$[A, BC] = ABC - BCA = ABC - BAC + BAC - BCA =$$
$$= [A, B]C + B[A, C]. \tag{2.106}$$

The order of factors in (2.105) is important since the operator algebra is non-commutative. For general background on derivations of associative algebras see §.2.6 of [4]. As all derivations (cf. **Lemma 2.2** in [4]) the commutator satisfies the *Jacobi identity*

$$[A, [B, C]] + [B, [C, A]] + [C, [A, B]] =$$
$$= A(BC - CB) - (BC - CB)A + B(CA - AC) -$$
$$- (CA - AC)B + C(AB - BA) - (AB - BA)C = 0, \tag{2.107}$$

where in the last equality we used that the operator algebra is associative. Comparing with §. 6.4 in [4] we conclude that *the commutator of operators has the same formal properties as the Poisson bracket in Analytic Mechanics* except for two aspects:

(i) now the underlying algebra is non-commutative;

(ii) they satisfy opposite reality conditions. The commutator bracket of two Hermitian operators $A = A^\dagger$ and $B = B^\dagger$ is *anti*-Hermitian:

$$[A, B]^\dagger = (AB)^\dagger - (BA)^\dagger = B^\dagger A^\dagger - A^\dagger B^\dagger = BA - AB = -[A, B]. \tag{2.108}$$

The antisymmetric bracket which is a *real* derivation of the $*$-algebra of operators is then $- \mathrm{i}[\cdot, \cdot]$ which has the desired property

$$A, B \text{ Hermitian} \quad \Rightarrow \quad -\mathrm{i}[A, B] \text{ is Hermitian.} \tag{2.109}$$

We shall return to these issues in Sect. 2.11.

2.4 Spectrum and the Spectral Theorem

The space of bra or ket vectors when the vectors are restricted to be of finite length and to have finite scalar products is called by mathematicians a Hilbert space. *The bra and ket vectors that we now use form a* **more general space** *than a Hilbert space.*

P.A.M. Dirac, §. 10 of THE PRINCIPLES OF QUANTUM MECHANICS [9]

We started this section with a quote from Dirac to emphasize that, traditionally, physicists and mathematicians use different approaches to the spectral theory of self-adjoint operators. The two approaches are equivalent, and both fully rigorous: indeed soon after Dirac introduced his 'more general space' L. Schwartz formalized it in the mathematical theory of *distributions* [2]. The two approaches are both useful: the physicists' one is more convenient for actual computations and provides a more "intuitive" picture, while the math one is more suited to prove abstract theorems.

Mathematicians prefer to work in the actual Hilbert space of states $\mathcal{H}$. Physicists work in a larger topological vector space $\mathcal{P}^\vee \supset \mathcal{H}$ where one can give sense to *generalized* eigenvectors and eigenvalues. Of course, the generalization is needed only when $\mathcal{H}$ is infinitely-dimensional, i.e. separable: the spectral theory in a finite-dimensional Hilbert space $\simeq \mathbb{C}^n$ just says that for all $n \times n$ Hermitian matrix A we can find an orthonormal basis in $\mathbb{C}^n$ where A becomes diagonal $A_{ij} = a_i \delta_{ij}$ with real eigenvalues $a_i \in \mathbb{R}$.

Passing to the dual spaces, we have a subspace $\mathcal{P} \subset \mathcal{H}^\vee \equiv \mathcal{H}$ (since the Hilbert spaces are self-dual). Physicists work with a *spectral triple* of spaces

$$\mathcal{P} \subset \mathcal{H} \subset \mathcal{P}^\vee \tag{2.110}$$

where the subspace $\mathcal{P}$ is *dense* in the Hilbert space $\mathcal{H}$ (in particular it contains Hilbert bases of $\mathcal{H}$). One chooses $\mathcal{P}$ such that all the un-bounded self-adjoint operators A_i one is interested in are defined in $\mathcal{P}$ and map $\mathcal{P}$ into $\mathcal{P}$. Then one extends these operators to the big dual space $\mathcal{P}^\vee$ by the rule

$$\langle \psi | A \phi \rangle \overset{\text{def}}{=} \langle A \psi | \phi \rangle \quad \forall \, |\psi\rangle \in \mathcal{P}, \ \ |\phi\rangle \in \mathcal{P}^\vee. \tag{2.111}$$

It then makes sense to define generalized eigenvalues and eigenvector living in $\mathcal{P}^\vee$.

For definiteness we take the representation of the separable Hilbert space $\mathcal{H}$ to be of the form $L^2(\mathbb{R}^n)$ for some n. In this set-up $\mathcal{P}^\vee$ is a certain space of generalized complex "functions" $\psi(x)$ in $\mathbb{R}^n$ such that the integral

$$\int \mathrm{d}^n x \, |\psi(x)|^2 = \begin{cases} < +\infty & \psi(x) \in \mathcal{H} \\ +\infty & \psi(x) \in \mathcal{P}^\vee \setminus \mathcal{H}. \end{cases} \tag{2.112}$$

Physicists abuse language and speak of *normalizable states* (or functions) and *non-normalizable states* (functions) to refer, respectively, to elements of $\mathcal{H}$ and $\mathcal{P}^\vee \setminus \mathcal{H}$.

Warning 2 We stress that only *normalizable* vectors represent physically realizable states. Non-normalizable states are "idealizations" which are convenient additions to the formalism, but they are not actual states of the system

The precise choice of $\mathcal{P}$ is a matter of convenience. A natural choice is as follows. We take for $\mathcal{P}$ the space $\mathcal{S}(\mathbb{R}^n)$ of *Schwartz test functions,* i.e. smooth functions in $\mathbb{R}^n$ whose derivatives of arbitrary order vanish at infinity more rapidly than any power of the coordinates, that is,

$$\mathcal{S}(\mathbb{R}^n) = \left\{ f : \mathbb{R}^n \to \mathbb{C} \text{ s.t. } \sup_{x \in \mathbb{R}^n} \left| x_1^{j_1} \cdots x_n^{j_n} \frac{\partial^{a_1 + \cdots + a_n} f}{\partial x_1^{a_1} \cdots \partial x_n^{a_n}} \right| < \infty, \ \forall \, j_k, a_\ell \right\} \tag{2.113}$$

$\mathcal{S}(\mathbb{R}^n)$ is endowed with a topology which makes it into a Frechét space (a complete, locally convex, metrizable space). The space $\mathcal{S}(\mathbb{R}^n)$ is *dense* in $L^2(\mathbb{R}^n)$: indeed the elements of the orthonormal Hilbert basis $\{\psi_n(x)\}$ in Construction 5 are Schwartz test functions: we invite the reader to check this claim. Then the dual space of generalized states $\mathcal{P}^\vee$ gets identified with the space $\mathcal{S}(\mathbb{R}^n)^\vee$ of *Schwartz tempered distributions* [2].

We begin with a quick survey of the mathematicians' viewpoint, and then proceed to the physicists' one.

2.4.1 Math: Spectral Theory

Let A be a self-adjoint operator.

Definition 2.4 The *resolvent set* $\rho(A) \subset \mathbb{C}$ is defined as

$$z \in \rho(A) \;\Leftrightarrow\; (A - z) \text{ is invertible with bounded inverse } (A - z)^{-1} \qquad (2.114)$$

Definition 2.5 The *spectrum* is the complement of the resolvent

$$\sigma(A) \equiv \mathbb{C} \setminus \rho(A). \qquad (2.115)$$

From the definitions we see that at $z \in \sigma(A)$ two different things may happen:[8]

(a) $(A - z)$ is not invertible in the sense that $\ker(A - z) \neq 0$ in $\mathcal{H}$;
(b) $\ker(A - z) = 0$, but $(A - z)^{-1}$ is unbounded.

The set $\sigma_p(A) \subset \mathbb{C}$ where $(A - z)$ has a non-trivial kernel is called the *point spectrum*, or *discrete spectrum* in the physical language, and the set $\sigma_e(A) \subset \mathbb{C}$ where $(A - z)^{-1}$ not bounded the *essential spectrum* or the *continuous spectrum*.[9] When $\sigma_e(A) = \varnothing$ (resp. $\sigma_p(A) = \varnothing$) we say that A has a *purely discrete* (resp. *purely continuous*) spectrum.

Example 2.9 $A = |a\rangle\langle a|$ with $\langle a|a\rangle = 1$. The operator $(A - z)$ is invertible for $z \neq 0, 1$. Indeed its inverse is

$$(A - z)^{-1} = \frac{1}{z(1 - z)} |a\rangle\langle a| - \frac{1}{z}. \qquad (2.116)$$

Thus $|a\rangle\langle a|$ (which is the projector to the complex line generated by the vector $|a\rangle$) has a purely discrete spectrum consisting of two points: 0 and 1. This is the spectrum for all non-trivial (orthogonal) projectors.

---

[8] In facts *three*: there is also the possibility that z belongs to the *residual spectrum* so that $(A - z)^{-1}$ is bounded but $\overline{\text{im}(A - z)^{-1}} \neq \mathcal{H}$ see [10]. This possibility does not arise for A self-dual and we shall disregard it.

[9] To be pedantic the essential spectrum is $\sigma(A) \setminus \sigma_p(A)$ which is not exactly the same as the continuous spectrum since for some z we can have $\ker(A - z) \neq 0$ while z belongs to the continuous spectrum.

An important class of operators are the ones with a purely discrete spectrum $\sigma(A) \equiv \sigma_p(A) \subset \mathbb{R}$ such that each subspace $\ker(A - z) \subset \mathcal{H}$ has *finite* dimension.

Example 2.10 Let $|n\rangle$ $(n \in \mathbb{N})$ be a Hilbert basis and A the operator $|n\rangle \rightarrow n|n\rangle$ i.e. $A = \sum_n n|n\rangle\langle n|$. In this case $\sigma(A) \equiv \sigma_p(A) = \mathbb{N}$ and $\dim \ker(A - n) = 1$ for all n.

Example 2.11 We consider $\mathcal{H} = L^2(\mathbb{R})$ and let A be the multiplication operator $\psi(x) \mapsto x\,\psi(x)$. Clearly $\ker(A - z) = 0$ for all $z \in \mathbb{C}$. However $(A - z)^{-1}$ is unbounded for all $z \in \mathbb{R}$, so that A has a purely continuous spectrum $\sigma_e(A) = \mathbb{R}$.

Discrete Spectrum

A complex number $z \in \sigma_p(A)$ is called an *eigenvalue* of A. The subspace

$$\mathcal{H}_z \equiv \ker(A - z) \subset \mathcal{H} \tag{2.117}$$

is the *eigenspace* associated to the eigenvalue $z \in \sigma_p(A)$. A non-zero element $0 \neq |\psi, z\rangle \in \ker(A - z)$ is an *eigenvector* (or *eigenstate*[10]) of A. In other words, an eigenvector $|\psi, z\rangle$ with eigenvalue $z \in \sigma_p(A)$ is an element of $\mathcal{H}$, that is,

$$\langle \psi, z | \psi, z \rangle < \infty \tag{2.118}$$

which satisfies the *eigenvector equation*

$$A|\psi, z\rangle = z|\psi, z\rangle. \tag{2.119}$$

An eigenvalue $z \in \sigma_p(A)$ is said to be *non-degenerate* iff it has multiplicity one i.e. iff $\dim \ker(A - z) = 1$. When $\dim \ker(A - z) > 1$ we say that the eigenvalue z is *degenerate*. The integer $\dim \ker(A - z)$ is the *degeneration* number of z. If z is an eigenvalue which does not belong to the continuous spectrum, $(A - z)^{-1}$ is bounded in the subspace $\mathcal{H}_z^{\perp} \subset \mathcal{H}$ orthogonal to $\ker(A - z)$.

Lemma 2.1 *The eigenvalues of a self-adjoint operator, $A = A^{\dagger}$, are real*

$$\sigma_p(A) \subset \mathbb{R}. \tag{2.120}$$

Proof Let $|\psi, z\rangle$ be a normalized eigenvector associated to the eigenvalue $z \in \sigma_p(A)$

$$z = z\langle \psi, z | \psi, z \rangle = \langle \psi, z | A | z\psi, z \rangle = \langle \psi, z | A^{\dagger} | \psi, z \rangle^* =$$
$$= \langle \psi, z | A | z\psi, z \rangle^* = (z\langle \psi, z | \psi, z \rangle)^* = z^* \tag{2.121}$$

$\square$

---

[10] When $\mathcal{H}$ is represented as a Hilbert space of functions, the term *eigenfunction* is also used.

Lemma 2.2 *Let $z_1, z_2 \in \sigma_p(A)$, $z_1 \neq z_2$, be two* distinct *eigenvalues of a self-adjoint operator $A = A^\dagger$. The eigenvectors associated to z_1 are orthogonal to the eigenvectors associated to z_2.*

Proof Since the eigenvalues are real, the adjoint of the eigenvalue Eq. (2.447) reads

$$z\langle\psi, z| = z^*\langle\psi, z| = \langle\psi, z|A^\dagger = \langle\psi, z|A \tag{2.122}$$

i.e. the *eigenbras* $\langle\psi, z|$ of A associated to an eigenvalue z are the Hermitian adjoint of its *eigenkets* $|\psi, z\rangle$ with the same eigenvalue. Hence

$$z_1\langle\psi_1, z_1|\psi_2, z_2\rangle = \langle\psi_1, z_1|A|\psi_2, z_2\rangle = z_2\langle\psi_1, z_1|\psi_2, z_2\rangle \tag{2.123}$$

that is,

$$(z_1 - z_2)\,\langle\psi_1, z_1|\psi_2, z_2\rangle = 0. \tag{2.124}$$

$\square$

Lemma 2.3 *Let $A = A^\dagger$ be self-adjoint and $\sigma_e(A) = \varnothing$. The eigenvectors of A form a Hilbert basis of $\mathcal{H}$. If $(\lambda_n)_{n\in\mathbb{N}}$ is the increasing sequence of eigenvalues of A and $|\lambda_n, k\rangle$, $k = 1, \ldots, \dim\ker(A - \lambda_n)$, is an orthonormal basis of $\ker(A - \lambda_n) \subset \mathcal{H}$, the completeness equation (a.k.a. called Bessel-Parseval identity, a.k.a. resolution of the identity) holds*

$$1 = \sum_n \sum_k |\lambda_n, k\rangle\langle\lambda_n, k|. \tag{2.125}$$

Proof We have to show that if $|v\rangle$ is orthogonal to all $|\lambda_n, k\rangle$ it is zero. Assume $|v\rangle \notin \cup_n \ker(A - \lambda_n)$; then the expression

$$\langle v|(A - z)^{-1}|v\rangle \tag{2.126}$$

is a bounded holomorphic function in $\mathbb{C}$, hence a constant. Then

$$0 = \frac{\partial}{\partial z}\langle v|(A - z)^{-1}|v\rangle\Big|_{z=0} = \|A^{-1}|v\rangle\|^2 \tag{2.127}$$

which implies $|v\rangle = 0$. $\square$

Note that each term in the sum (2.125)

$$\sum_k |\lambda_n, k\rangle\langle\lambda_n, k| \equiv P_{\lambda_n} \tag{2.128}$$

is the orthogonal projector to the eigenspace of A of eigenvalue λ_n. Hence

$$P_{\lambda_n} A P_{\lambda_n} = A P_{\lambda_n} = P_{\lambda_n} A = \lambda_n P_{\lambda_n}, \quad P_{\lambda_n}^\dagger = P_{\lambda_n}, \quad P_{\lambda_n}^2 = P_{\lambda_n}, \tag{2.129}$$

and (2.125) is the decomposition of 1 in a spectral family of orthogonal projectors

$$1 = \sum_{\lambda_n \in \sigma_p(A)} P_{\lambda_n}, \qquad P_{\lambda_m} P_{\lambda_n} = \delta_{mn} P_{\lambda_n}, \tag{2.130}$$

while

$$A = \sum_{\lambda_n \in \sigma_p(A)} \lambda_n \sum_k |\lambda_n, k\rangle \langle \lambda_n, k|. \tag{2.131}$$

General Spectrum

We consider the general case where $\sigma_e(A) \neq \varnothing$. The situation is analogue to the one with pure discrete spectrum except that the resolution of the identity (2.125) now contains (in general) both a sum over $\sigma_p(A)$ and an integral over $\sigma_e(A)$. In the math approach this is expressed using spectral families of projectors in the sense of Definition 2.3. The precise statement is:

Theorem 2.2 (Spectral Theorem) *Let A be a self-adjoint operator with domain $D(A) \subset \mathcal{H}$ (recall that $\overline{D(A)} = \mathcal{H}$). There exists a* unique *spectral family $P(\lambda)$ such that for all $|v\rangle \in D(A)$ and $|w\rangle \in \mathcal{H}$:*

$$D(A) = \left\{ |v\rangle \in \mathcal{H} \,\middle|\, \int \lambda^2 \, \mathrm{d}\langle v|P(\lambda)|v\rangle < \infty \right\} \tag{2.132}$$

$$\langle w|Av\rangle = \int \lambda \, \mathrm{d}\langle w|P(\lambda)|v\rangle \tag{2.133}$$

$$\||A|v\rangle\|^2 = \int \lambda^2 \, \mathrm{d}\langle v|P(\lambda)|v\rangle. \tag{2.134}$$

The support of the integral measure is $\sigma(A) \subset \mathbb{R}$.

For a proof of the theorem see e.g. [10, 11].

Definition 2.6 A self-adjoint operator A is said to be *non-negative*, written $A \geq 0$, iff its spectrum is contained in $\mathbb{R}_{\geq 0}$, i.e. iff the support of its spectral measure $\mathrm{d}P(\lambda)$ belongs to the positive real axis.

Operators Satisfying an Algebraic Equation

To construct explicitly the spectral projectors of a self-adjoint operator A is in general a difficult task. A simple situation is when A satisfies an algebraic equation, that is, there is a polynomial $P(z)$ such that $P(A) = 0$. We may assume without loss that $P(z)$ is the unique monic polynomial of *minimal* degree such that $P(A) = 0$ (the *minimal polynomial of A*). We claim that the minimal polynomial of a self-adjoint operator is *square-free*, i.e. all its roots are real[11] of multiplicity 1. Indeed, write the minimal polynomial in the form

$$P(z) = (z - z_1)^n \, Q(z) \quad \text{with } Q(z_1) \neq 0 \text{ and } z_1 \in \mathbb{R}. \tag{2.135}$$

Suppose $n \geq 2$. Let $|v\rangle \in \mathcal{H}$ and set $|u\rangle = (A - z_1)^{n-2} Q(A)|v\rangle$.

$$\|(A - z_1)|u\rangle\|^2 = \langle u|(A - z_1)^2|u\rangle = \langle u|(A - z_1)^n Q(A)|v\rangle = 0, \tag{2.136}$$

so for all $|v\rangle \in \mathcal{H}$

$$(A - z_1)^{n-1} Q(A)|v\rangle \equiv (A - z_1)|u\rangle = 0 \tag{2.137}$$

contrary to the assumption that $P(z)$ is minimal. Hence $P(z) = \prod_i (z - z_i)$ with all $z_i \in \mathbb{R}$ distinct. Now set

$$P_i = \prod_{j \neq i} \frac{A - z_j}{z_i - z_j}, \qquad P_i^\dagger = P_i \tag{2.138}$$

One has

$$A \, P_i = P_i \, A = z_i \, P_i \tag{2.139}$$

and

$$P_i^2 = \prod_{j \neq i} \frac{A - z_j}{z_i - z_j} P_i = \prod_{j \neq i} \frac{z_i - z_j}{z_i - z_j} P_i = P_i, \qquad P_i P_j = 0 \quad i \neq j, \tag{2.140}$$

hence $\{P_i\}$ is a system of commuting mutually orthogonal projectors. Consider now the expression

$$\sum_i \frac{P(z)}{(z - z_i) \, P'(z_i)} - 1. \tag{2.141}$$

It is a polynomial of degree $\leq (n - 1)$ which vanishes at the n distinct values z_i; hence it is identically zero. Replacing $z \rightsquigarrow A$ in Eq. (2.141), and comparing with (2.138), we get

$$\sum_i P_i = 1, \tag{2.142}$$

---

[11] The roots z_i of the minimal polynomial belong to the spectrum, indeed $P(A) = 0$ implies that all linear factors $(A - z_i)$ are non-invertible. Since the spectrum of a self-adjoint operator is real (Lemma 2.1), the roots z_i are real.

that is, $\{P_i\}$ is a *spectral family* for A and

$$A = \sum_i z_i \, P_i. \tag{2.143}$$

Corollary 2.1 *If the self-adjoint operator A satisfies a (minimal) algebraic equation $P(A) = 0$, its eigenvalues are the roots z_i of its minimal polynomial and the z_i-eigenspace is $P_i \mathcal{H} \subset \mathcal{H}$. Conversely all self-adjoint operators whose spectrum is a finite subset $\subset \mathbb{R}$ satisfy an algebraic equation.*

2.4.2   *Phys: Non-Normalizable Eigenstates. Dirac δ-Function*

As mentioned before, physicists work with three different spaces

$$\mathcal{S} \subset \mathcal{H} \subset \mathcal{S}^{\vee} \tag{2.144}$$

the prototypical case being

$$\mathcal{S}(\mathbb{R}^n) \subset L^2(\mathbb{R}^n) \subset \mathcal{S}(\mathbb{R}^n)^{\vee} \tag{2.145}$$

where $\mathcal{S}(\mathbb{R}^n)^{\vee}$ is the space of Schwartz tempered distributions. In this construction $\mathbb{R}^n$ can be replaced by any reasonable manifold $\mathcal{M}$. For simplicity we focus on $\mathcal{M} = \mathbb{R}$, the extension to general manifolds being straightforward.

The multiplication operator $q : \psi(x) \mapsto x\,\psi(x)$ is well defined as a linear operator

$$q : \mathcal{S}(\mathbb{R}) \to \mathcal{S}(\mathbb{R}). \tag{2.146}$$

We see the distributions in the dual space $\mathcal{S}(\mathbb{R})^{\vee}$ as generalized functions of the real variable x, and write the canonical pairing

$$\langle \cdot | \cdot \rangle : \mathcal{S}(\mathbb{R}) \times \mathcal{S}(\mathbb{R})^{\vee} \to \mathbb{C} \tag{2.147}$$

as an integral

$$\langle \psi_1 | \psi_2 \rangle = \int_{\mathbb{R}} \psi_1(x)^* \, \psi_2(x) \, dx. \tag{2.148}$$

The basic idea of the Dirac-Schwartz formalism is that, while q has no eigenvectors in $\mathcal{H}$ (cf. Example 2.11), it has a complete system of eigenvectors

$\{|x\rangle\}_{x\in\mathbb{R}}$ in $\mathcal{S}(\mathbb{R})^{\vee}$ with the expected properties

$$q|x\rangle = x|x\rangle \tag{2.149}$$

$$\langle x|q = x\langle x| \quad x \in \mathbb{R} \equiv \sigma_e(q) \tag{2.150}$$

$$1 = \int_{\mathbb{R}} dx\, |x\rangle\langle x| \tag{2.151}$$

where the ket $|x\rangle \in \mathcal{S}(\mathbb{R}^n)^{\vee}$ is now normalized in a suitable generalized sense: since $|x\rangle \notin \mathcal{H}$, i.e. $\langle x|x\rangle = \infty$, the normalization condition of (generalized) eigenvectors, associated to eigenvalues in the continuous spectrum, **cannot be** $\langle x|x\rangle = 1$. In the next paragraph we shall construct the generalized eigenvectors $|x\rangle \in \mathcal{S}(\mathbb{R})^{\vee}$ and then work out the *continuous spectrum normalization condition* using the example of the operator q acting on $L^2(\mathbb{R})$ as our guide.

Wave-Functions

In the Dirac formalism the isomorphism $\mathcal{H} \simeq L^2(\mathbb{R})$ is written as

$$|\psi\rangle \mapsto \langle x|\psi\rangle \overset{\text{def}}{=} \psi(x) \quad x \in \mathbb{R}, \tag{2.152}$$

an identification consistent with the Hermitian products via Eq. (2.151)

$$\langle \psi_1|\psi_2\rangle = \int_{\mathbb{R}} dx\, \langle \psi_1|x\rangle\langle x|\psi_2\rangle = \int_{\mathbb{R}} dx\, \psi_1(x)^*\psi_2(x). \tag{2.153}$$

We call $\psi(x) \equiv \langle x|\psi\rangle \in L^2(\mathbb{R})$ the *wave function* of the state $|\psi\rangle \in \mathcal{H}$. More generally the wave function of a state in $L^2(\mathbb{R}^n)$ is[12]

$$\psi(x^1,\dots,x^n) = \langle x^1,\dots,x^n|\psi\rangle \equiv \langle \boldsymbol{x}|\psi\rangle. \tag{2.154}$$

Dirac δ-Function

We have to determine the wave-function of an eigenvector $|x\rangle$ of q

$$\langle x'|x\rangle \in \mathcal{S}(\mathbb{R})^{\vee}. \tag{2.155}$$

By translational invariance (cf. Example 2.3) $\langle x'|x\rangle$ is a tempered distribution depending only on the difference $x' - x$ that we write as

$$\delta(x' - x) \tag{2.156}$$

---

[12] Through this book we use the boldface symbol $\boldsymbol{x}$ for the vector $(x_1,\dots,x_n) \in \mathbb{R}^n$.

and call the *Dirac δ-function*. The notation is meant to emphasize that the δ-*function* is the 'obvious' generalization to the continuous spectrum of the Kronecker δ-*symbol* for the discrete spectrum: if A is a self-adjoint operator with purely discrete spectrum (with non-degenerate eigenvalues) and $\{|\psi_n\rangle\}_{n\in\mathbb{N}}$ is an orthonormal Hilbert basis made of eigenvectors

$$\langle\psi_m|\psi_n\rangle = \delta_{n-m,0} \quad \xrightarrow{\substack{\text{continuous}\\\text{spectrum}}} \quad \langle x'|x\rangle = \delta(x-x'). \tag{2.157}$$

The last equation may be seen either as the (normalized) wave-function of the eigenstate $|x\rangle$ or as the normalization condition for the generalized eigenstates $|x\rangle$ belonging to the continuum spectrum which generalizes the LHS valid for the discrete case. This continuum normalization condition applies to all continuum eigenvectors of arbitrary self-adjoint operators.

Let $\phi(x) \in \mathcal{S}(\mathbb{R})$; then, using Eq. (2.151),

$$\begin{aligned}
\phi(x)^* &\equiv (\langle x|\phi\rangle)^* = \langle\phi|x\rangle = \\
&= \int_{\mathbb{R}} dx'\, \langle\phi|x'\rangle\langle x'|x\rangle = \int_{\mathbb{R}} dx'\, \phi(x')^*\, \delta(x'-x).
\end{aligned} \tag{2.158}$$

Here $\phi(x)^*$ is just an arbitrary Schwartz tempered test function. Hence

Corollary 2.2 *The Dirac δ-function $\delta(x)$ is the tempered distribution ($\equiv$ continuous linear functional on $\mathcal{S}(\mathbb{R})$) which associates to a smooth function $f(x)$ its value at the origin $f(0)$*

$$f(0) = \int_{\mathbb{R}} dx\, \delta(x)\, f(x). \tag{2.159}$$

The distribution $\delta(x)$ has support in $\{0\}$. Equivalently $\delta(x-x')$ is the reproducing kernel in $\mathcal{S}(\mathbb{R})$, *that is,*

$$f(x) = \int_{\mathbb{R}} dx'\, \delta(x-x')\, f(x') \quad \forall\, f(x) \in \mathcal{S}(\mathbb{R}). \tag{2.160}$$

From the definition we get the basic properties of the δ-function:

$$\delta(-x) = \delta(x) \qquad\qquad \delta(ax) = \frac{1}{|a|}\delta(x) \tag{2.161}$$

$$x\,\delta(x) = 0 \qquad\qquad f(x)\,\delta(x-y) = f(y)\,\delta(x-y), \tag{2.162}$$

and more generally

$$\delta(g(x)) = \sum_r \frac{1}{|g'(x_r)|} \delta(x - x_r) \qquad \begin{array}{l} \text{where } x_r \text{ are the} \\ \text{solutions to } g(x_r) = 0. \end{array} \tag{2.163}$$

The δ-function is the derivative—in the sense of distributions—of the Heaviside step function[13] $\Theta(x)$:

$$\Theta(x) = \begin{cases} 1 & x > 1 \\ 0 & x \le 0 \end{cases} \qquad \delta(x) = \frac{d}{dx}\Theta(x). \tag{2.164}$$

Concretely this means that, for all test function $f \in \mathcal{S}(\mathbb{R})$,

$$\int_{\mathbb{R}} dx\, f(x) \frac{d}{dx}\Theta(x) = -\int_{\mathbb{R}} dx\, \frac{df}{dx}\,\Theta(x) = -\int_0^\infty dx\, \frac{df}{dx} = \tag{2.165}$$
$$= -f(\infty) + f(0) \equiv f(0) = \int_{\mathbb{R}} dx\, f(x)\,\delta(x).$$

We may consider the higher derivatives (in the sense of distributions) $\delta^{(n)}(x)$ of $\delta(x)$ of arbitrary order n. By definition we have

$$\int_{\mathbb{R}} dx\, f(x) \frac{d^n}{dx^n}\delta(x) = (-1)^n \left.\frac{d^n f}{dx^n}\right|_{x=0} \qquad \forall\, f \in \mathcal{S}(\mathbb{R}). \tag{2.166}$$

One can show that a tempered distribution with support in $0 \in \mathbb{R}$ is a *finite sum* $\sum_n c_n\, \delta^{(n)}(x)$.

Intuitively $\delta(x)$ is a "function", vanishing away from the origin, which "diverges" at $x = 0$ in such a way that its integral on the line is 1. Then $\delta(x)$ may be written as the limit of ordinary functions in many ways: some useful formulae are

$$\delta(x) = \lim_{\epsilon \to 0} \frac{1}{\sqrt{\pi}\epsilon} e^{-x^2/\epsilon^2} \qquad \delta(x) = \lim_{L \to \infty} \frac{\sin(Lx)}{\pi x} \tag{2.167}$$

$$\delta(x) = \lim_{L \to \infty} \frac{\sin^2(Lx)}{\pi L x^2} \qquad \delta(x) = \lim_{\epsilon \to 0} \frac{1}{\pi} \frac{\epsilon}{x^2 + \epsilon^2}. \tag{2.168}$$

Relation with the Fourier Transform

Since the Fourier transform interchanges x and $-i\partial_x$, we see from (2.113) that the Fourier transform

$$F\psi(p) \overset{\text{def}}{=} \int_{\mathbb{R}} \frac{dx}{\sqrt{2\pi}}\, e^{ipx}\, \psi(x) \tag{2.169}$$

---

[13] For later convenience we define the step function so that $\Theta(-x)$ is semi-continuous from the right.

of a Schwartz test function $\psi(x) \in S(\mathbb{R})$ is again a Schwartz test function

$$\psi(x) \in S(\mathbb{R}) \quad \Rightarrow \quad F\psi(p) \in S(\mathbb{R}). \tag{2.170}$$

This implies dually that *all tempered distributions* $\Phi(x) \in S(\mathbb{R})^{\vee}$ *have a Fourier transform,* which is also a tempered distribution, $F\Phi(p) \in S(\mathbb{R})^{\vee}$, defined by the rule

$$\langle \Psi | F\Phi \rangle \overset{\text{def}}{=} \langle F^{-1}\Psi | \Phi \rangle \qquad \Phi \in S(\mathbb{R})^{\vee}, \ \Psi \in S(\mathbb{R}). \tag{2.171}$$

The Parseval formula

$$\int_{\mathbb{R}} dp \, |F\psi(p)|^2 = \int_{\mathbb{R}} dx \, |\psi(x)|^2 \tag{2.172}$$

shows that the Fourier transform F, when restricted to $L^2(\mathbb{R}) \subset S(\mathbb{R})^{\vee}$, is a linear map

$$F: L^2(\mathbb{R}) \to L^2(\mathbb{R}) \tag{2.173}$$

which is a Hilbert space isometry, and being invertible, also a *unitary transformation.* Reiterating the Fourier transform one gets

$$F^2\psi(x) = \int \frac{dp \, dy}{2\pi} e^{ixp + ipy} \psi(y) = \psi(-x) \tag{2.174}$$

which implies that the unitary operator F satisfies the algebraic equation

$$F^4 = 1, \tag{2.175}$$

so its eigenvalues are ± 1 and $\pm i$ (all eigenspaces are infinite-dimensional). An example of complete Hilbert basis in $L^2(\mathbb{R})$ made of eigenvectors of F is given by the functions $\{\psi_n(x)\}_n$ defined in Eq. (2.30) (Construction 5). Indeed

$$\sum_{n=0}^{\infty} \sqrt{\frac{2^n}{n!}} F\psi_n(p) z^n = \frac{1}{\pi^{1/4}} \int \frac{dx}{\sqrt{2\pi}} e^{ipx} \exp\left(-\tfrac{1}{2}x^2 + 2xz - z^2\right) =$$

$$= \frac{1}{\pi^{1/4}} \exp\left(-\tfrac{1}{2}p^2 + 2ipz + z^2\right) = \sum_{n=0}^{\infty} \sqrt{\frac{2^n}{n!}} \psi_n(p) \, (iz)^n, \tag{2.176}$$

that is,

$$F\psi_n(p) = i^n \psi_n(p) \qquad n = 0, 1, 2, \ldots \tag{2.177}$$

Equation (2.174) may be written in the form

$$\psi(x) = \int \frac{dx' \, dp}{2\pi} \, e^{ip(x'-x)} \psi(x'), \tag{2.178}$$

comparing with Eq. (2.160) we get the *integral representation of the δ-function*

$$\delta(x - y) = \int_{-\infty}^{+\infty} \frac{dp}{2\pi} \, e^{ip(x-y)}. \tag{2.179}$$

We recall that the *(Cauchy) principal value of* $1/x$, written $P(1/x)$, is the tempered distribution defined by (for $f \in S(\mathbb{R})$)

$$\int_{\mathbb{R}} dx \, f(x) \, P\frac{1}{x} \overset{\text{def}}{=} \lim_{\sigma \to 0} \left(\int_{-\infty}^{-\sigma} dx \, \frac{f(x)}{x} + \int_{+\sigma}^{+\infty} dx \, \frac{f(x)}{x} \right). \tag{2.180}$$

Let $f(x) \in S(\mathbb{R})$:

$$\begin{aligned}
\lim_{\epsilon \to 0} \int_{-\infty}^{+\infty} \frac{f(x) \, dx}{x - i\epsilon} &= \\
&= \int_{-\infty}^{-\sigma} \frac{f(x) \, dx}{x} + \lim_{\epsilon \to 0} \int_{-\sigma}^{+\sigma} \frac{f(x) \, dx}{x - i\epsilon} + \int_{+\sigma}^{+\infty} \frac{f(x) \, dx}{x},
\end{aligned} \tag{2.181}$$

the RHS is independent of σ; taking $\sigma \to 0$ we get

$$\lim_{\epsilon \to 0} \int_{-\infty}^{+\infty} \frac{f(x) \, dx}{x - i\epsilon} = \int dx \, f(x) \, P\frac{1}{x} + i\pi f(0) \tag{2.182}$$

which yields the fundamental identity of distributions (crucial in dispersion analysis, cf. [12] §. 6.1)

$$\frac{1}{x - i\epsilon} = P\frac{1}{x} + i\pi \, \delta(x), \tag{2.183}$$

where we used the physical convention[14] that when an ϵ appears in an identity the equality is meant to be valid in the limit $\epsilon \to 0^+$.

Remark 2.5 (Weyl Criterion) $\mathcal{H}$ is dense in $S(\mathbb{R})^{\vee}$. Hence if $\lambda \in \sigma(A)$ there is a sequence $\psi_N \in \mathcal{H}$ which converges in $S(\mathbb{R})^{\vee}$ to a generalized eigenvector of A associated to the eigenvalue λ. This gives the Weyl criterion:

---

[14] Mathematicians write $i0$ instead of $i\epsilon$ with the same meaning.

Corollary 2.3 *A self-adjoint. $\lambda \in \sigma(A)$ iff there exists a sequence $\{\psi_N \in \mathcal{H}\}$ such that*

$$\lim_{N \to \infty} \|A\psi_N - \lambda \psi_N\| = 0 \qquad \|\psi_N\| = 1. \tag{2.184}$$

2.4.3 Summary: Spectrum of Self-Adjoint Operators

A general self-adjoint operator A has both a discrete spectrum $\sigma_p(A) = \{\lambda_n\}$ and a continuous spectrum $\sigma_e(A)$ (not necessarily disjoint) with eigenvectors in the generalized Dirac sense

$$A|n\rangle = \lambda_n |n\rangle, \quad A|\lambda\rangle = \lambda|\lambda\rangle, \quad \lambda_n \in \sigma_p(A), \quad \lambda \in \sigma_e(A) \tag{2.185}$$

which are normalized in either the discrete or the continuous sense, i.e.

$$\langle m|n\rangle = \delta_{mn}, \tag{2.186}$$

$$\langle \lambda'|\lambda\rangle = \delta(\lambda - \lambda'), \tag{2.187}$$

$$\langle n|\lambda\rangle = 0. \tag{2.188}$$

Assuming for simplicity that all eigenspaces are one-dimensional we have the identity

$$1 = \sum_n |n\rangle\langle n| + \int_{\sigma_e(A)} d\lambda \, |\lambda\rangle\langle\lambda| \tag{2.189}$$

$$A = \sum_n \lambda_n |n\rangle\langle n| + \int_{\sigma_e(A)} d\lambda \, \lambda|\lambda\rangle\langle\lambda|. \tag{2.190}$$

Comparing with Theorem 2.2, we see that the physicists' approach is equivalent to the math one via the identification

$$P(\lambda) = \sum_{\lambda_n \leq \lambda} |n\rangle\langle n| + \int_{-\infty}^{\lambda} d\lambda' |\lambda'\rangle\langle\lambda'| \tag{2.191}$$

In particular Theorem 2.2 guarantees that we always have enough generalized eigenvectors to express all states as *generalized* linear combination of them (a sum over the discrete spectrum plus an integral over the continuous one).

Equation (2.189) is the *completeness* condition for the generalized eigenvalues of A, also called the *resolution of the identity*. A system of generalized eigenvectors $\{|n\rangle, |\lambda\rangle\}$ satisfying (2.189) is called a *complete orthogonal system*. A complete orthogonal system gives a representation of $\mathcal{H}$ in the form

$$\mathcal{H} = \ell^2(\sigma_p(A)) \,\widehat{\oplus}\, L^2(\sigma_e(A)) \tag{2.192}$$

where, for simplicity of notation, we assumed that all eigenvalues of A are non-degenerate. In the general case, where the eigenvalues may be degenerate, Eqs. (2.189) and (2.190) become (using the math notation)

$$1 = \int_{\mathbb{R}} dP(\lambda) \tag{2.193}$$

$$A = \int_{\mathbb{R}} \lambda \, dP(\lambda). \tag{2.194}$$

The physical way of writing these equations when some generalized eigenvalues of A are degenerate will be given in Sect. 2.7.

We express Eq. (2.190), and it more general expression (2.194), by saying that *the self-dual operator A is diagonalized in this complete orthogonal system of eigenvectors,* or else that (2.192) is the *A-representation of $\mathcal{H}$.*

2.5 Commuting Operators

Let A, B be two commuting self-adjoint operators

$$A = A^{\dagger}, \quad B = B^{\dagger}, \qquad [A, B] = 0. \tag{2.195}$$

Let $V_a \subset \mathcal{S}^{\vee}$ be the subspace of generalized eigenvectors of A with eigenvalue λ_a. V_a is invariant under the action of B: indeed

$$A(B|v\rangle) = BA|v\rangle = \lambda_a(B|v\rangle) \quad \forall \, |v\rangle \in V_a \quad \Rightarrow \quad BV_a \subseteq V_a. \tag{2.196}$$

Then we have a self-adjoint linear operator $B|_{V_a} : V_a \to V_a$, to which we may apply the spectral theorem, getting (generalized) eigenvectors of B which are simultaneously eigenvectors of A

$$A|\psi, \lambda_a, \mu_b\rangle = \lambda_a|\psi, \lambda_a, \mu_b\rangle, \qquad B|\psi, \lambda_a, \mu_b\rangle = \mu_b|\psi, \lambda_a, \mu_b\rangle \tag{2.197}$$

In other words: *A and B can be simultaneously diagonalized by a common complete system of eigenvectors.* The converse statement is also true:

Corollary 2.4 *If A and B have a complete orthonormal system of (generalized) eigenvectors in common, they commute.*

In particular the two spectral families $P_A(\lambda)$, $P_B(\mu)$ commute

$$P_A(\lambda) P_B(\mu) = P_B(\mu) P_A(\lambda) \qquad \text{for all } \lambda, \mu \in \mathbb{R}. \tag{2.198}$$

These results extend to an arbitrary family $\{A_i\}$ $(i = 1, 2, \ldots, k)$ of pair-wise commuting self-adjoint operators

$$[A_i, A_j] = 0 \qquad i, j = 1, \ldots, k, \tag{2.199}$$

which can all be diagonalized simultaneously.

Normal Operators

The operator (with dense domain) N is *normal* iff $[N, N^\dagger] = 0$. Then

$$A \equiv \frac{1}{2}(N + N^\dagger), \qquad B = \frac{1}{2i}(N - N^\dagger) \tag{2.200}$$

are two self-adjoint operators which commute $[A, B] = 0$. Hence we can find a complete orthonormal system which diagonalizes them both

$$A|\lambda_A, \lambda_B\rangle = \lambda_A|\lambda_A, \lambda_B\rangle, \qquad B|\lambda_A, \lambda_B\rangle = \lambda_B|\lambda_A, \lambda_B\rangle. \tag{2.201}$$

This orthonormal system also diagonalizes the original operator N with eigenvalues

$$N|\lambda_A, \lambda_B\rangle = (\lambda_A + i\lambda_B)|\lambda_A, \lambda_B\rangle. \tag{2.202}$$

The generalized eigenvectors of N with distinct eigenvalues are orthogonal.

2.6   Functional Calculus

Let A be a self-adjoint operator. Given a function $f(z)$ we wish to make sense of $f(A)$ as an operator (unbounded in general). We assume the function $f(z)$ to be defined and piece-wise continuous on the spectrum $\sigma(A) \subset \mathbb{R}$. When all eigenvalues of A are non-degenerate, we set

$$f(A) \stackrel{\text{def}}{=} \sum_n f(\lambda_n)|n\rangle\langle n| + \int_{\sigma_e(A)} d\lambda \, f(\lambda) \, |\lambda\rangle\langle\lambda|, \tag{2.203}$$

and more generally (using the math notation)

$$f(A) \stackrel{\text{def}}{=} \int f(\lambda)\, dP(\lambda). \tag{2.204}$$

where $P(\lambda)$ is the spectral family of projectors of the operator A in the sense of Definition 2.3.

When $f(z) = \sum_k c_k z^k$ is an analytic function, and the spectrum $\sigma(A)$ is fully contained in the convergence disk of its power series, we may also write

$$f(A) = \sum_k c_k A^k, \tag{2.205}$$

and this expression coincides with (2.204). When $A \subset \mathcal{B}(\mathcal{H})$ with $\|A\|$ smaller than the convergence radius of $\sum_k c_k z^k$, the series in the RHS converges in the norm topology.

Suppose that the function $f(z)$ is *bounded* in $\sigma(A)$ (A densely defined). Then $f(A) \colon \mathcal{H} \to \mathcal{H}$ is a bounded operator defined everywhere in $\mathcal{H}$. In particular, let L be a self-adjoint operator (whose domain is dense by definition), and consider the operator

$$\exp(itL) \equiv U(t) \qquad t \in \mathbb{R}. \tag{2.206}$$

Since the function e^{itz} is bounded in $\sigma(L) \subset \mathbb{R}$ for all $t \in \mathbb{R}$, the expression (2.206) defines a one-parameter family of continuous operators in $\mathcal{H}$. They form an Abelian group

$$U(t)\, U(s) = U(t+s) = U(s)\, U(t), \qquad t, s \in \mathbb{R} \tag{2.207}$$

isomorphic to a quotient of $\mathbb{R}$. In particular

$$U(t)^{-1} = \exp(-itL) \equiv U(t)^{\dagger}, \tag{2.208}$$

so (2.206) is a group of *unitary operators* ($\equiv$ a group of automorphisms of $\mathcal{H}$). The converse statement is the Stone theorem:

Theorem 2.3 (Stone Theorem) *If $U(t)$ ($t \in \mathbb{R}$) is a strongly continuous one-parameter unitary group on $\mathcal{H}$, there exists a unique self-adjoint operator*

$$L \colon D(L) \to \mathcal{H}, \tag{2.209}$$

with $D(L)$ dense in $\mathcal{H}$, such that

$$U(t) = \exp(itL) \quad \forall\, t \in \mathbb{R}. \tag{2.210}$$

For a proof see e.g. [10, 11]. If $f(z)$ has a Fourier transform, so that

$$f(z) = \int dt\, e^{-izt}\, \hat{f}(t), \tag{2.211}$$

we can write $f(A)$ as a superposition of unitary operators

$$f(A) = \int dt\, \hat{f}(t)\, e^{-itA}. \tag{2.212}$$

In particular the spectral projectors $P(\lambda)$ are bounded operators of this form

$$P(\lambda) = \Theta(\lambda - A) \equiv \int_{-\infty}^{+\infty} \frac{dt}{2\pi i} \frac{e^{i\lambda t}}{t - i\epsilon}\, e^{-iAt} \tag{2.213}$$

where $\Theta(z)$ is the Heaviside step function (2.164).

Corollary 2.5 *A unitary operator U may be diagonalized in a generalized orthonormal system and its generalized eigenvalues are phases, i.e. there exists a spectral family $P(\lambda)$ (with $P(\lambda)P(\mu) = P(\mu)$ for $\mu \leq \lambda$)) such that*

$$U = \int e^{i\lambda}\, dP(\lambda). \tag{2.214}$$

Lemma 2.4 *If A, B are two commuting self-adjoint operators, then*

$$[f(A), g(B)] = 0 \tag{2.215}$$

for all functions $f(z)$ and $g(z)$.

The following statement is obvious when A and B have purely discrete spectrum but it hold in general.

Theorem 2.4 (von Neumann) *Let A, B be commuting self-adjoint operators. There is a self-adjoint operator C and functions $f(z)$, $g(z)$ such that $A = f(C)$ and $B = g(C)$.*

The Baker-Campbell-Hausdorff (BCH) Formula
Let A, B be two self-dual operators and consider the product

$$e^{iA}\, e^{iB} \tag{2.216}$$

which is a unitary operator, so there exists a self-adjoint operator $Z(A, B)$ such as

$$e^{iA}\, e^{iB} = e^{iZ(A,B)}. \tag{2.217}$$

$Z(A, B)$ is given explicitly by a series of multiple commutators of A, B

$$iZ(A, B) = iA + iB - \frac{1}{2}[A, B] - \frac{i}{12} \left([A, [A, B]] + [B, [B, A]]\right) -$$
$$- \frac{1}{24}[B, [A, [A, B]]] + \cdots$$
(2.218)

Equation (2.217) is called the *Baker-Campbell-Hausdorff formula* (BCH). The term homogenous of degree n in A, B in the RHS is an *universal* polynomial in the reiterated commutators called the n-th *Dynkin polynomial.*

The Trotter Formula
The Trotter formula is a very useful identity that we shall need in Chap. 6. We state it without proof (see e.g. [13] **Theorem 3.2.2**):

Theorem 2.5 *Assume that one of the following is true:*

(a) *A, B are bounded operators acting on a Hilbert space $\mathcal{H}$;*
(b) *the operators A and B are bounded below and essentially self-adjoint. Moreover $A + B$ is also essentially self-adjoint.*

Then

$$e^{A+B} = \lim_{n \to \infty} \left(e^{A/n} e^{B/n}\right)^n.$$
(2.219)

where the limit converges in the norm topology for case (a) *and in the strong topology for* (b).

2.7 Observables: Measures and Probabilities

We return to the discussion of *observables* ($\equiv$ physical quantities). We stress that the present discussion holds for *closed* quantum systems which we observe from outside. This is a highly idealized situation. Measures in *open* quantum systems will be addressed in Chap. 7.

We know that the physical observables are identified with self-adjoint operators A acting on the Hilbert space $\mathcal{H}$. One basic principle of Quantum Physics states

Principle 2

(1) *The possible outcomes of a measure of the physical quantity A are its generalized eigenvalues a_i (discrete or continuous).*
(2) *When the quantum system is in a physical state which is an eigenstate $|a_i\rangle$ of A associated to the (real) eigenvalue a_i, the measure of A gives the result a_i with 100% certainty.*

(3) *If the state is a superposition $\sum_k c_k |a_k\rangle$ of eigenstates, the result of the measure of A is one of the a_k's with non-zero coefficient $c_k \neq 0$, and the possible outcomes of the measure are distributed according to a probability law determined by the particular state.*

(4) *The expectation value A_ψ of the measure of A, computed in the probability distribution associated to the state $|\psi\rangle$, is*

$$A_\psi = \frac{\langle \psi | A | \psi \rangle}{\langle \psi | \psi \rangle}, \tag{2.220}$$

or, when the state vector $|\psi\rangle$ is normalized, simply

$$A_\psi = \langle \psi | A | \psi \rangle. \tag{2.221}$$

The consistency of the identification of physical observables with self-adjoint operators requires that all self-adjoint operators enjoy two properties:

Ob1 since the results of the experiments are real numbers, the spectrum of all self-adjoint operators should be a subset of $\mathbb{R}$;

Ob2 the quantity A may be measured in *any* state getting (with some probability) one of its possible values. Therefore *any* state should have an expansion in (generalized) eigenvectors of A, that is, the eigenvectors of A should form a complete system—equivalently the spectral family $\{P_A(\lambda)\}$ of all self-adjoint operator A should satisfy the equation

$$\int_{\mathbb{R}} dP_A(\lambda) = 1. \tag{2.222}$$

Happily both conditions are guaranteed by the spectral theorem, and the identification in Principle 2 makes perfect physical sense.

Suppose, for simplicity, that A has a purely discrete spectrum, λ is a non-degenerate eigenvalue, and $P_\lambda = |\lambda\rangle\langle\lambda|$ is the projector on the eigenspace $\mathbb{C}|\lambda\rangle$. P_λ is self-adjoint, hence an *observable*. The expectation value of P_λ in a (normalized) state $|\psi\rangle \in \mathcal{H}$ is

$$\langle \psi | P_\lambda | \psi \rangle = \langle \psi | \lambda \rangle \langle \lambda | \psi \rangle = |\langle \lambda | \psi \rangle|^2 \geq 0. \tag{2.223}$$

The meaning of this quantity is obvious: by construction, it is the expectation value of a variable which is 1 when the result of the measure is λ and zero otherwise, hence it is the *probability* that a measure of A returns the possible result λ. Therefore

Principle 3 *Let A have a purely discrete spectrum, $\sigma_e(A) = \varnothing$. The probability that a measure of A in the quantum state $|\psi\rangle \in \mathcal{H}$ returns the value $\lambda \in \sigma(A)$ is*

$$\sum_k |\langle \lambda, k | \psi \rangle|^2 \tag{2.224}$$

where the sum is over an orthonormal basis $|\lambda, k\rangle$ of the λ-eigenspace of A.

This conclusion is consistent since the sum of the probabilities of all possible results is 1: when A has a purely discrete spectrum with non-degenerate eigenvalues

$$\sum_{\lambda \in \sigma(A)} |\langle \lambda | \psi \rangle|^2 = \sum_{\lambda \in \sigma(A)} \langle \psi | \lambda \rangle \langle \lambda | \psi \rangle = \langle \psi | \psi \rangle = 1, \tag{2.225}$$

where we used the resolution of identity $1 = \sum_\lambda |\lambda\rangle\langle\lambda|$. The Hermitian product $\langle \lambda | \psi \rangle$ is also called the *transition amplitude* between the state $|\psi\rangle$ and the state $|\lambda\rangle$ or the *overlap* of the two states. The probability of the result λ is the modulus square of the overlap $\langle \lambda | \psi \rangle$.

We now consider the case where the continuous spectrum $\sigma_e(A)$ is non-empty. Again the possible outcomes are the values $\lambda \in \sigma(A)$ but, when the eigenvalue belongs to the continuous spectrum, $\lambda \in \sigma_e(A)$, the probability of getting *exactly* the value λ is, of course, zero. In this situation we have a *continuous probability distribution* with support on $\sigma_e(A)$. Putting together the discrete and continuous spectra we get the probability measure on $\sigma(A)$

$$d\Pi_\psi = d\langle \psi | P_A(\lambda) | \psi \rangle \tag{2.226}$$

where $P_A(\lambda)$ is the spectral family of projectors of A. The probability that a measure of A in the state $|\psi\rangle$ returns a result in the interval $[a_1, a_2]$ is then

$$\mathrm{Prob}(a_1 \leq A \leq a_2) = \int_{a_1}^{a_2} d\Pi_\psi. \tag{2.227}$$

Again, the total probability is 1

$$\int_{-\infty}^{+\infty} d\langle \psi | P_A(\lambda) | \psi \rangle = 1 \quad \text{for all } |\psi\rangle \in \mathcal{H}, \tag{2.228}$$

cf. Eq. (2.90).

Example 2.12 Let $|n\rangle$ ($n \in \mathbb{N}$) be an orthonormal Hilbert basis and N the observable, diagonal in this Hilbert basis, such that $N|n\rangle = n|n\rangle$. A general normalized state has the form $\sum_n c_n |n\rangle$ with $\sum_n |c_n|^2 = 1$. The probability that a measure of N in the state $\sum_n c_n |n\rangle$ returns the value n is

$$\mathrm{Prob}(N = n) = |c_n|^2. \tag{2.229}$$

Example 2.13 We consider the operator q acting on the wave-functions of $L^2(\mathbb{R})$ as multiplication by x. The probability that a measure of x returns a value in the interval $a \leq x \leq b$ is

$$\int_a^b d\langle \psi | P_q(x) | \psi \rangle \equiv \int_a^b dx \, \langle \psi | x \rangle \langle x | \psi \rangle = \int_a^b dx \, |\psi(x)|^2, \tag{2.230}$$

that is, the *probability density* of a measure of the position q between x and $x + dx$ is

$$|\psi(x)|^2 \, dx. \tag{2.231}$$

Compatible Measures

We ask when two observables A and B may be measured *simultaneously with certainty* (i.e. with an error that we can make as small as we wish). Clearly this is possible when our state $|\psi\rangle$ is a simultaneous eigenstate of both A and B. The measurements of A and B are said to be *compatible* iff *all* states can be written as superpositions of states where we can measure A and B simultaneously with certainty. In view of Corollary 2.4 this means that A and B *commute*:

Definition 2.7 The measures of a family $A_1, \ldots, A_n$ of observables are *compatible* iff the A_i's commute pairwise, i.e. $[A_i, A_j] = 0$.

Complete System of Observables

The state of the system at time t is defined as a *maximal* set of informations about the system at the given time which are *mutually consistent*. Clearly the (eigen)values $\lambda_1, \ldots, \lambda_n$ of a compatible family of observables $A_1, \ldots, A_n$ is a consistent datum about the system at a given time. In order for this datum to specify a state *uniquely* we only need that this family is *complete,* i.e. maximal:

Definition 2.8 A family of commuting observables $A_1, \ldots, A_n$ is *complete* iff the dimension of their simultaneous generalized eigenspaces

$$V_{(\lambda_i)} = \left\{ |v\rangle \in \mathcal{S}^\vee : A_i |v\rangle = \lambda_i |v\rangle, \;\; i = 1, \ldots, n \right\} \tag{2.232}$$

is 1, i.e. if the n-tuple of eigenvalues $(\lambda_1, \ldots, \lambda_n)$ uniquely determines the generalized eigenvector $|\lambda_1, \ldots, \lambda_n\rangle$ (up to overall normalization[15]). We also say that the family of observables $A_1, \ldots, A_n$ *separates the states.*

If $A_1, \ldots, A_n$ is a complete family of observables, their generalized eigenstates are conveniently written as $|\lambda_1, \ldots, \lambda_n\rangle$ and[16]

$$\mathcal{H} = \widehat{\bigoplus_{\lambda_a \in \sigma(A_a)}} \mathbb{C} |\lambda_1, \ldots, \lambda_n\rangle \tag{2.233}$$

is the representation of $\mathcal{H}$ associated to the complete family $A_1, \ldots, A_n$.

In view of von Neumann's theorem, there exist an observable L such that $A_i = f_i(L)$. The eigenspaces of L are unidimensional; hence the representation associated to the complete family $A_1, \ldots, A_n$ is more conveniently seen as the L-

---

[15] Recall that two linearly dependent vectors represent the *same* physical state.

[16] The direct sum in (2.233) is a direct integral when the spectrum is continuous.

representation with the resolution of the identity

$$1 = \sum_{\lambda_n \in \sigma_p(L)} |\lambda_n\rangle\langle\lambda_n| + \int_{\sigma_e(L)} d\lambda \, |\lambda\rangle\langle\lambda| \equiv \int dP_L(\lambda) \tag{2.234}$$

where $P_L(\lambda)$ is the spectral family of L. Then

$$A_i = \int f_i(\lambda) \, dP_L(\lambda). \tag{2.235}$$

Corollary 2.6 *Let $A_1, \ldots, A_n$ be a complete family of observables. If B commutes with all A_i, then*

$$B = F(A_1, \ldots, A_n) \tag{2.236}$$

for some function $F(z_1, \ldots, z_n)$ (equivalently $B = g(L)$).

2.8 The Schrödinger Equation

In the previous sections we presented a schematic but essentially complete mathematical description of the states of a (closed) quantum system at a given time t_0. To get a *dynamical* theory we need, in addition, equations of motion which allow us to determine the state $|\psi, t\rangle$ at time t when given the state $|\psi, t_0\rangle$ at some initial time t_0. Recall that we *defined* the state $|\psi, t_0\rangle$ to be the information which is needed and suffices for the purpose of reproducing the same amount of information about the system at any other instant t. Now it is time to check that the math definition of quantum state we gave on the basis of the superposition principle has the required property of reproducing itself in time via the dynamical equations.

By the very definition of *state,* the time evolution of the initial state $|\psi, t_0\rangle$ from time t_0 to time t should produce a state $|\psi, t\rangle$ which is of the same nature, i.e. represented by a vector in the *same* Hilbert space $\mathcal{H}$, since the physically allowed states of the system are the same now and tomorrow, or at any other time t in the future (and also in the past).

In facts, all we said in the previous sections about the Hilbert-space description of today's states, applies word-for-word to the tomorrow's states, since the basic Laws of physics are not going to change during the night. This entails that the map between the state now, $|\psi, t_0\rangle$, and the state $|\psi, t\rangle$ at any future time t should be a *Hilbert-space automorphism,* that is, an *unitary linear map.* We enshrine our conclusion in the

Principle 4 *The evolution of a closed quantum system from time t_0 to time t is given by a two-parameter family of* unitary *operators $U(t, t_0)$, $t_0, t \in \mathbb{R}$,*

$$U(t, t_0)^\dagger \, U(t, t_0) = U(t, t_0) \, U(t, t_0)^\dagger = 1, \tag{2.237}$$

which satisfy the group(oid) law

$$U(t + s, t_0) = U(t + s, t) \, U(t, t_0) \qquad t, s \in \mathbb{R} \tag{2.238}$$

$$U(t_0, t_0) = 1, \qquad U(t_0, t) = U(t, t_0)^{-1}, \tag{2.239}$$

such that the state of the quantum system at time t is

$$|\psi, t\rangle = U(t, t_0)|\psi, t_0\rangle \tag{2.240}$$

where $|\psi, t_0\rangle$ is its state at time t_0.

Remark 2.6 Principle 4 is the quantum version of the classical fact that the time-evolution in Hamiltonian mechanics is given by a family of automorphisms of the phase space (i.e. a family of symplectomorphisms [4]).

The Hamiltonian Operator H

It is more common to describe the evolution of a physical system in terms of differential equations than in terms of a group(oid) of automorphisms of the space of states. One has

$$\frac{\mathrm{d}}{\mathrm{d}t}|\psi, t\rangle = \frac{\mathrm{d}}{\mathrm{d}t}\Big(U(t, t_0)|\psi, t_0\rangle\Big) = \frac{\mathrm{d}U(t, t_0)}{\mathrm{d}t}|\psi, t_0\rangle =$$
$$= \frac{\mathrm{d}U(t, t_0)}{\mathrm{d}t}U(t, t_0)^{-1}|\psi, t\rangle \tag{2.241}$$

Consider the operator (we set $U \equiv U(t, t_0)$ and $\dot{U} \equiv \mathrm{d}U/\mathrm{d}t$)

$$\mathrm{i}\,\dot{U}U^{-1} = \mathrm{i}\,\dot{U}U^\dagger, \tag{2.242}$$

it satisfies the reality condition

$$\left(\mathrm{i}\,\dot{U}U^\dagger\right)^\dagger = -\mathrm{i}\,U\dot{U}^\dagger = -\mathrm{i}\frac{\mathrm{d}}{\mathrm{d}t}(UU^\dagger) + \mathrm{i}\,\dot{U}U^\dagger = \mathrm{i}\,\dot{U}U^\dagger \tag{2.243}$$

i.e. it is *Hermitian*. Since the operator $\mathrm{i}\dot{U}U^{-1}$ is Hermitian, it is a physical observable. To see which physical observable it is, we look at the synoptic Table 2.1 where the formal structures of classical and quantum mechanics are compared. A look to the last row leads to the obvious guess that the physical quantity $\mathrm{i}\dot{U}U^\dagger$ is the operator associated to the Hamiltonian, i.e. to the physical observable "energy". However $\mathrm{i}\dot{U}U^\dagger$ has the dimension of $(\text{time})^{-1}$ (in the usual unit system it is

Table 2.1 Comparison of the formal structure of classical and quantum mechanics

	Classical	Quantum
Physical states	Points in a symplectic manifold	Vectors in a Hilbert space
Time evolution	Automorphism of symplectic manifold	Automorphism of Hilbert space
	$\equiv$ Symplectomorphism	$\equiv$ Unitary transformation
	$\equiv$ Canonical transformation	
Generator of time evolution	Generating function $S(t, t_0)$	Unitary operator $U(t, t_0)$
	$\equiv$ Hamilton's principal function	
Time derivative of generator	$\frac{\partial S(t,t_0)}{\partial t} = -H(t)$	Physical observable $i\dot{U}U^{\dagger}$
	the Hamilton-Jacobi equation	
	$H(t) \equiv$ Hamiltonian ("energy")	

measured in s^{-1}) while the energy is usually measured in Joule (J), so we need a universal unit-conversion constant $\hbar$ with dimension $J \cdot s$ to match $i\dot{U}U^{\dagger}$ and energy. Then

$$i\dot{U}U^{-1} = H/\hbar \tag{2.244}$$

where H is the quantum operator which corresponds to the observable *energy*, that we call the *Hamiltonian operator H*. We take Eq. (2.244) as the general *definition* of the Hamiltonian operator H for a general (closed) quantum system.

The Hamiltonian *operator H* is the quantum counterpart to the Hamiltonian *function* $H(q, p)$ of Analytic Mechanics ([4] chap. 6). This interpretation is consistent with the general rule that the classical *mechanical* observables are functions $F(q, p)$ in phase space which at the quantum level get promoted to self-adjoint operators. Then we write the dynamical equations in the form

$$i\hbar \frac{d}{dt}|\psi, t\rangle = H(t)|\psi, t\rangle \tag{2.245}$$

where $H(t)$ is a Hermitian operator that we interpret as the Hamiltonian at time t. Equation (2.245) is the celebrated *Schrödinger equation* in its abstract general form. In the following chapters we shall write more explicit forms of this equation for various particular classes of quantum mechanical systems. However we stress that the abstract discussion applies to *all* closed quantum systems not just to the Quantum Mechanical ones. In a sense Eq. (2.245) is the Theory of Everything.

Time Independent Systems
When the system is invariant by translations in time, the operator $H(t)$ does not depend explicitly on t, and the equation becomes

$$i\hbar \frac{d}{dt}|\psi, t\rangle = H|\psi, t\rangle, \tag{2.246}$$

whose solution with initial condition $|\psi, t_0\rangle$ at time t_0 is simply

$$|\psi, t\rangle = \mathrm{e}^{-\mathrm{i}H(t-t_0)/\hbar}|\psi, t_0\rangle \qquad (2.247)$$

which may be rewritten as an equation for the unitary transformation U:

$$\mathrm{i}\hbar\,\dot{U} = HU \quad \Rightarrow \quad U(t, t_0) = \exp\left(-\frac{\mathrm{i}}{\hbar}(t - t_0)H\right). \qquad (2.248)$$

In general the Hamiltonian H is an *unbounded* operator defined only in some dense domain $D(H) \subset \mathcal{H}$. However this is not a problem since, in view of the Stone Theorem 2.3, its exponential $\exp(-\mathrm{i}(t - t_0)H/\hbar)$ is a *bona fide* unitarity operator defined on the *whole* of $\mathcal{H}$. This is a consistency requirement, since all physical states in $\mathcal{H}$ should have a meaningful time evolution.

The eigenvalues E_n in the discrete spectrum of H are called *energy levels*

$$H|E_n\rangle = E_n|E_n\rangle \qquad E_n \in \sigma_p(H). \qquad (2.249)$$

The continuous eigenvalues E of H are the energies in the *continuum* energy spectrum, whose generalized eigenvectors are not normalizable. The physical meaning of the distinction between discrete and continuous values of energy will be clarified in Chap. 3.

If $|E\rangle$ is an eigenstate of H with energy E, its time-evolution is

$$|E, t\rangle = \mathrm{e}^{-\mathrm{i}Et/\hbar}|E, 0\rangle. \qquad (2.250)$$

We know that the multiplication of a state-vector by a phase does not change the physical state. Thus a state which is an eigenstate of H does not change with time: these states are called *stationary*.

2.9 Quantum Pictures: Schrödinger, Heisenberg, Dirac

In §.9.1 of [4] we noted that the debate in the years 1830's about the proper formulation of classical mechanics was followed ninety years later by an analogue discussion between Schrödinger and Heisenberg on the appropriate formulation of Quantum Mechanics, until Dirac proved that the several formulations—called *pictures*—are all equivalent. In this section we outline the various formulation and show their equivalence.

2.9.1   *Schrödinger Picture*

In the *Schrödinger picture* the states evolve according to the Schrödinger equation (2.245), while the quantum observables A are *time independent*. If

$$|\psi, t\rangle \equiv U(t)|\psi\rangle \quad \text{where} \quad U(t) \equiv \mathrm{e}^{-\mathrm{i}Ht/\hbar}, \tag{2.251}$$

is the quantum state at time t, the expectation value $A_\psi(t)$ of the physical quantity A evolves with time according to the formula

$$A_\psi(t) = \langle \psi, t|A|\psi, t\rangle \equiv \langle \psi|U(t)^\dagger A U(t)|\psi\rangle, \tag{2.252}$$

and similarly for the general matrix elements of A between any two physical states

$$\langle \psi_1|U(t)^\dagger A U(t)|\psi_2\rangle. \tag{2.253}$$

The same formula holds for the probability $\mathrm{Pr}_\psi(a; t)$ that a measure of A at time t returns a definite result a, which is the special case of (2.252) with A replaced by the projector P_a on the a-eigenspace of A.

2.9.2   *Heisenberg Picture*

In the *Heisenberg picture* the operators which describe the observables evolve in time, while the states are constant in time. If we define the new *time-dependent* Hermitian operator

$$A(t) \stackrel{\text{def}}{=} U(t)^\dagger A U(t) \equiv U(t)^{-1} A U(t), \qquad A(t)^\dagger = A(t), \tag{2.254}$$

called the *Heisenberg operator* of the observable A, the expectation value (i.e. the outcome of the experiments) of the observable $A(t)$ in the *fixed* state $|\psi\rangle$

$$A_\psi(t) = \langle \psi|A(t)|\psi\rangle \equiv \langle \psi|U(t)^\dagger A U(t)|\psi\rangle \tag{2.255}$$

is identically to the one computed in the Schrödinger picture, Eq. (2.252). Hence the two pictures make the *same prediction* for all observables, that is, they are physically equivalent.

In the Heisenberg picture the equations of motion describe the evolution with time of the Heisenberg operators $A(t)$ which represent the various physical observables. One has

$$\frac{\mathrm{d}}{\mathrm{d}t} A(t) = \frac{\mathrm{d}U(t)^{-1}}{\mathrm{d}t} A U(t) + U(t)^{-1} A \frac{\mathrm{d}U(t)}{\mathrm{d}t}$$

$$= -U(t)^{-1} \frac{\mathrm{d}U(t)}{\mathrm{d}t} U(t)^{-1} A U(t) + U(t)^{-1} A \frac{\mathrm{d}U(t)}{\mathrm{d}t} U(t)^{-1} U(t)$$

$$= -\frac{1}{\mathrm{i}\hbar} U(t)^{-1} [H, A] U(t) = -\frac{1}{\mathrm{i}\hbar} [U(t)^{-1} H U(t), U(t)^{-1} A U(t)]$$

$$\equiv \frac{1}{\mathrm{i}\hbar} [A(t), H(t)], \tag{2.256}$$

where we used the Schrödinger equation in the form $\dot{U} = -\mathrm{i}HU/\hbar$.

The equations

$$\mathrm{i}\hbar \frac{\mathrm{d}A(t)}{\mathrm{d}t} = [A(t), H(t)] \equiv A(t)\, H(t) - H(t)\, A(t) \tag{2.257}$$

are called the *Heisenberg equations of motion*. In writing them we assumed that the original Schrödinger picture observable A has no *explicit* dependence on time, so that the full dependence on t of the Heisenberg operator $A(t)$ arises from the quantum evolution of the degrees of freedom of our quantum system. More generally we may have an explicit dependence of A on t: the generalized Heisenberg equations then take the form

$$\frac{\mathrm{d}A}{\mathrm{d}t} = \frac{1}{\mathrm{i}\hbar}[A, H] + \frac{\partial A}{\partial t} \tag{2.258}$$

where the symbol $\partial A/\partial t$ stands for the derivative with respect to the explicit dependence on time

$$\frac{\partial A(t)}{\partial t} \overset{\mathrm{def}}{=} U(t)^\dagger \frac{\partial A}{\partial t} U(t). \tag{2.259}$$

Under the identification of the commutator of two quantum observables with $\mathrm{i}\hbar$ times the Poisson bracket of the corresponding classical quantities

$$[A(t), B(t)] \equiv A(t)\, B(t) - B(t)\, A(t) \rightsquigarrow \mathrm{i}\hbar [A_{\mathrm{cl}}, B_{\mathrm{cl}}]_{\mathrm{PB}}, \tag{2.260}$$

the Heisenberg equations (2.258) become identical to the Hamilton equations of motion (chap. 6 of [4]). The identification (2.260) is consistent with the algebraic properties of both sides of the correspondence, as discussed in the last paragraph of Sect. 2.3. We shall elaborate more on this identification in Sect. 2.11.

Remark 2.7 Since the Schrödinger and the Heisenberg pictures are physically equivalent, using one or the other is a matter of convenience. While there are problems which are more efficiently analyzed using the Heisenberg picture, for most quantum *mechanical* problems the Schrödinger equation is more handy. This is easy to understand: consider a system whose Hilbert space is finite-dimensional $\mathcal{H} \simeq \mathbb{C}^n$. A state is described by a vector i.e. by n complex numbers ($2n$ real numbers); an operator is a $n \times n$ Hermitian matrix specified by n^2 real numbers. The Schrödinger equation is an ODE with n unknown functions, while the Heisenberg ODE contain n^2 of them. For $n \ggg 1$ it is clear that the Schrödinger description is *generically* much more economic.

Conserved Quantities ($\equiv$ Integrals of Motion)
From the Heisenberg equations we get

Corollary 2.7

(1) *A (Heisenberg) observable* $A = A(t)$ *is a* conserved quantity *iff*

$$\frac{1}{i\hbar}[A, H] + \frac{\partial A}{\partial t} = 0. \tag{2.261}$$

(2) *A time-independent observable* A *is a conserved quantity if and only if it commutes with the Hamiltonian* H

$$[A, H] = 0. \tag{2.262}$$

(3) *Since* $[H, H] \equiv 0$, *the Hamiltonian* H—*that is, the energy—is conserved if and only if* H *is time-independent, that is, iff the quantum system is invariant under time-translations.*

These three statements are identical to the corresponding classical ones (cf. §. 6.6 of [4]) under the correspondence (2.260) between the quantum commutators and the classical Poisson brackets. As in the classical case

Corollary 2.8 *(1) If* A *is a conserved observable, so is* $f(A)$ *for all functions* $f(z)$. *(2) If the two observables* A *and* B *are conserved, the (Hermitian) observable* $-i[A, B]$ *is also conserved.*

Proof **(2)** Using the Jacobi identity (2.107), we get

$$i\hbar \frac{d}{dt}[A, B] = \big[[A, B], H\big] + i\hbar\Big[\frac{\partial A}{\partial t}, B\Big] + i\hbar\Big[A, \frac{\partial B}{\partial t}\Big] =$$

$$= -\big[[B, H], A\big] - \big[[H, A], B\big] + i\hbar\Big[\frac{\partial A}{\partial t}, B\Big] + i\hbar\Big[A, \frac{\partial B}{\partial t}\Big]$$

$$= \Big[i\hbar\frac{\partial A}{\partial t} + [A, H], B\Big] - \Big[i\hbar\frac{\partial B}{\partial t} + [B, H], A\Big] = 0.$$

$$\tag{2.263}$$

$\square$

We conclude that

Corollary 2.9 *The conserved observables of any quantum system form a Lie algebra with respect to the Lie bracket* $-\frac{i}{\hbar}[\cdot, \cdot]$.

2.9.3 Dirac Pictures

The Schrödinger and Heisenberg pictures are two special instances of a family of pictures that we call generically *Dirac pictures*. We first present them in an elementary fashion and then discuss their meaning. In Sect. 2.9.4 we give an important application of them to be further exploited in Chap. 8.

We may unify the Schrödinger and Heisenberg pictures by writing the equations of motion for both states and observables in the form

$$i\hbar\frac{d}{dt}|\Psi\rangle = L|\Psi\rangle, \qquad i\hbar\frac{dA}{dt} = [A, K] \qquad (2.264)$$

where the two Hermitian operators L, K are

$$
\begin{array}{llll}
\text{Schrödinger picture:} & L = H & K = 0 & \\
\text{Heisenberg picture:} & L = 0 & K = H, &
\end{array}
\qquad (2.265)
$$

with H the Hamiltonian, i.e. the operator whose eigenvalues are the allowed energy levels. The two pictures are related by the time-dependent unitary transformation acting on operators as in Eq. (2.254) and on states as

$$|\psi, t\rangle_{\text{Heis.}} = U(t)^{-1}|\psi, t\rangle_{\text{Schr.}} \qquad (2.266)$$

where $U(t) \equiv \exp(-iHt/\hbar)$ is the time evolution operator. More generally, we may consider an arbitrary time-dependent unitary transformation $W(t)$ and define

$$|\psi, t\rangle_{\text{Dirac}} = W(t)^{-1}|\psi, t\rangle_{\text{Schr.}} \qquad (2.267)$$

$$A(t)_{\text{Dirac}} = W(t)^{-1}A(t)_{\text{Schr.}} W(t) \qquad (2.268)$$

$$W(t)W(t)^{\dagger} = W(t)^{\dagger}W(t) = 1. \qquad (2.269)$$

Clearly this unitary redefinition would not change the physical expectation values

$$A_{\psi}(t) = {}_{\text{Dirac}}\langle\psi, t|A(t)_{\text{Dirac}}|\psi, t\rangle_{\text{Dirac}} = {}_{\text{Schr.}}\langle\psi, t|A(t)_{\text{Schr.}}|\psi, t\rangle_{\text{Schr.}} \qquad (2.270)$$

$|\psi, t\rangle_{\text{Dirac}}$ and $A(t)_{\text{Dirac}}$ are, respectively, the states and the observables in the *Dirac picture* defined by the one-parameter family of unitary operators $\{W(t)\}_{t \in \mathbb{R}}$. When $W(t) = U(t)$ we get back the Heisenberg picture.

The equations of motion in the Dirac picture are

$$i\hbar \frac{d}{dt}|\psi, t\rangle_{\text{Dirac}} = -i\hbar W^{-1}\dot{W}W^{-1}|\psi, t\rangle_{\text{Schr.}} + W^{-1}H|\psi, t\rangle_{\text{Schr.}} =$$

$$= (W^{-1}HW - i\hbar\, W^{-1}\dot{W})|\psi, t\rangle_{\text{Dirac}} \tag{2.271}$$

$$\frac{d}{dt}A_{\text{Dirac}} = -W^{-1}\dot{W}W^{-1}A_{\text{Schr}}W + W^{-1}A_{\text{Schr}}\dot{W} =$$

$$= -(W^{-1}\dot{W})(W^{-1}A_{\text{Schr}}W) + (W^{-1}S_{\text{Schr}}W)(W^{-1}\dot{w}) = \tag{2.272}$$

$$= [A_{\text{Dirac}}, W^{-1}\dot{W}]$$

that is, comparing with Eq. (2.264),

$$L_{\text{Dirac}} = W^{-1}HW - i\hbar\, W^{-1}\dot{W}, \qquad K_{\text{Dirac}} = i\hbar\, W^{-1}\dot{W}. \tag{2.273}$$

These expressions should be compared with the formulae in Analytic Mechanics which describe how the Hamiltonian changes under a *time-dependent* canonical transformation (cf. [4] chaps. 8 and 9). If we identify

$$W \rightsquigarrow \exp(iS/\hbar) \tag{2.274}$$

and ignore the issue with the order of operators, the first Eq. (2.273) becomes

$$H_{\text{new}} = H_{\text{old}} + \frac{\partial S}{\partial t} \tag{2.275}$$

which is the central equation in the classical Hamilton-Jacobi theory (cf. eq.(8.72) in [4]). We conclude that the equations of motion in a general Dirac picture are the quantum counterpart to the classical canonical equations of motion as written in a general system of canonical variables which are related to the original ones by a canonical transformation whose generating function is $S \rightsquigarrow -i\hbar \log W$.

We present an alternative interpretation of the Dirac pictures. From the relativistic perspective the energy E is just the time component of the momentum, and a general gauge transformation acts on energy and momentum as

$$E \to E - \frac{\partial f}{\partial t}, \qquad p_i \to p_i + \frac{\partial f}{\partial x^i}, \tag{2.276}$$

where $f(x^i, t)$ is an arbitrary function in space-time. The first Eq. (2.273) is the quantum version of the first of these equations, so—except for the issue with non-commutativity—it is just a gauge transformation with $f \rightsquigarrow i\hbar \log W$. We shall discuss the second Eq. (2.276) in Sect. 2.11.

2.9.4 *Interaction Picture*

Solving explicitly the Schrödinger equation is typically very hard, except for a few *integrable* models which are the quantum counterpart to the classically integrable systems discussed in chap. 10 of [4]. Basic examples of quantum integrable systems will be presented in Chaps. 3, 4, and 5.

A large class of quantum systems are *quasi-integrable* in the Poincaré sense ([4] chap. 11), i.e. their Hamiltonian *operator* is the sum of two self-adjoint operators

$$H = H_0 + H_I \tag{2.277}$$

where H_0 is such that the Schrödinger equation

$$i\hbar \frac{\mathrm{d}}{\mathrm{d}t}|\psi\rangle = H_0|\psi\rangle \tag{2.278}$$

is explicitly *integrable,* while the operator H_I is "small" in some sense (a nice case is when $H_I \in \mathcal{B}(\mathcal{H})$, and an even better one when H_I is a *compact* operator). In view of the typical application, H_0 is called the *free Hamiltonian* and H_I the *interaction (Hamiltonian)*.

In this quasi-integrable situation we can construct explicitly the family of unitary operators $W(t)$ which solves the "unperturbed" integrable Schrödinger equation i.e. the ODE with initial condition

$$i\hbar \frac{\mathrm{d}}{\mathrm{d}t}W(t) = H_0 W(t), \qquad W(0) = 1, \tag{2.279}$$

whose solution we may write as

$$W(t) = \exp(-iH_0 t/\hbar) \quad \text{(the "free" evolution).} \tag{2.280}$$

The Dirac picture defined by the unitary transformations $W(t)$ is called the *interaction picture.* From Eq. (2.273)

$$L = \mathrm{e}^{iH_0 t/\hbar}(H_0 + H_I)\mathrm{e}^{-iH_0 t/\hbar} - H_0 \equiv H_I(t) \tag{2.281}$$

where

$$H_I(t) \stackrel{\mathrm{def}}{=} \mathrm{e}^{iH_0 t/\hbar} H_I \, \mathrm{e}^{-iH_0 t/\hbar} \tag{2.282}$$

is the interaction picture version of the interaction operator H_I (cf. Eq. (2.268)). The time-evolution of the states in the interaction picture is

$$i\hbar \frac{\mathrm{d}}{\mathrm{d}t}|\psi, t\rangle_I = H(t)_I|\psi, t\rangle_I. \tag{2.283}$$

When the operator $H(t)_I$ is "small" (as it is in the typical applications), the interaction picture states $|\psi, t\rangle_I$ are "almost stationary", and we can solve (approximately) the dynamical Eq. (2.283) "order by order" in the small "perturbation" $H(t)_I$. However the interaction picture Hamiltonian $H(t)_I$ is now explicitly time-dependent and this makes the Schrödinger equation (2.283) harder to solve. We turn to this issue.

H Time-Dependent: Time-Ordered Exponentials

The solution to the time-evolution equation

$$i\hbar \frac{d}{dt} U(t) = HU(t), \qquad U(0) = 1 \tag{2.284}$$

is simple when the Hamiltonian operator H is time-independent, i.e. the system is invariant under time-translation:

$$U(t) = e^{-iHt/\hbar} \equiv \int e^{-iEt/\hbar} \, dP(E) \tag{2.285}$$

where $\{P(E)\}$ is the spectral family of the fixed operator H. Thus, when the system is time-independent, solving the equations of motion is effectively reduced to solving the *time-independent Schrödinger equation*

$$H|E\rangle = E|E\rangle, \qquad |E\rangle \in \mathcal{S}^\vee, \qquad E \in \sigma(H) \tag{2.286}$$

i.e. to solving the generalized eigenvector equation for H in order to determine explicitly the spectrum and spectral family $\{P(E)\}$ of the Hamiltonian operator H.

When the Hamiltonian depends explicitly on time, the formal solution is slightly more subtle. There are two cases: when the Hamiltonians at different times commute

$$[H(t), H(t')] = 0 \quad \text{for all } t, t' \in \mathbb{R}, \tag{2.287}$$

they can be diagonalized simultaneously in the same generalized complete orthonormal system, equivalently, we can find a constant operator L such that $H(t) = h(L, t)$, and the solution is

$$U(t) = \exp\left(-\frac{i}{\hbar} \int_0^t H(s)\, ds\right) = \int \exp\left(-\frac{i}{\hbar} \int_0^t h(\lambda, s)\, ds\right) dP_L(\lambda) \tag{2.288}$$

where $P_L(\lambda)$ is the spectral family of L.

In the general case when $[H(t), H(t')] \neq 0$ the above expression makes no sense: one has to keep tract of the order of the various operators. There are many equivalent ways to write the solution. Again, the one used by physicists is more efficient and illuminating. We review it.

T-*Ordered Product* Let $A_1(t_1)$, $A_2(t_2)$ be two Heisenberg operators at different times $t_1 \neq t_2$. The expression

$$\mathbf{T}\big(A_1(t_1)\,A_2(t_2)\big) \tag{2.289}$$

stands for the product of the two operators *ordered* in such a way that the operator at later time is in the leftmost position, that is,

$$\mathbf{T}\big(A_1(t_1)\,A_2(t_2)\big) = \begin{cases} A_1(t_1)A_2(t_2) & \text{if } t_1 > t_2 \\ A_2(t_2)A_1(t_1) & \text{if } t_2 > t_1. \end{cases} \tag{2.290}$$

More generally, for all n-tuple of Heisenberg operators $A_r(t_r)$ $(r = 1, \ldots, n)$ the symbol $\mathbf{T}$ stands for the prescription of reordering the factors in the product $\prod_r A_r(t_r)$ in such a way that the times in the product decrease from left to right. $\mathbf{T}$ is called the *time-order operation* or simply the $\mathbf{T}$-*order.* Explicitly

$$\mathbf{T}\Big(A_1(t_1)A_2(t_2)\cdots A_n(t_n)\Big) \overset{\text{def}}{=}$$

$$\overset{\text{def}}{=} \sum_{\pi \in \mathfrak{S}_n} \Theta(t_{\pi(1)} - t_{\pi(2)})\Theta(t_{\pi(2)} - t_{\pi(3)})\cdots\Theta(t_{\pi(n-1)} - t_{\pi(n)}) \times \tag{2.291}$$

$$\times A_{\pi(1)}(t_{\pi(1)})\,A_{\pi(2)}(t_{\pi(2)})\cdots A_{\pi(n)}(t_{\pi(n)}),$$

where $\Theta(x)$ is the Heaviside step function, cf. Eq. (2.164). The sum over the permutations in the RHS selects the unique product where the operators are ordered according to their time arguments t_r: the operator at latest time being the leftmost (first) one, the second operator being at next to latest time, and so on, until the operator at the earliest time which is the rightmost one.

Remark 2.8 The $\mathbf{T}$-order is not uniquely defined since (2.290) says nothing about the case $t_1 = t_2$. Different prescriptions for the coinciding-time limit yield different definitions of the $\mathbf{T}$-order. However the difference between two $\mathbf{T}$-order prescriptions has support (as a distribution) on $t_2 - t_1 = 0$, so it is a finite sum of derivatives of $\delta(t_2 - t_1)$ called *contact terms*. This ambiguity will have no consequence in our arguments. However contact terms are very important in Quantum Field Theory.

T-*Ordered Exponentials* Let $A(t)$ be a one-parameter family of operators where we see the parameter t as time. We define the *time-ordered exponential* as

$$\mathbf{T}\exp\left(\int_0^t A(s)\,\mathrm{d}s\right) \overset{\text{def}}{=} \sum_{n=0}^{\infty} \frac{1}{n!}\mathbf{T}\left(\int_0^t A(s)\,\mathrm{d}s\right)^n \equiv$$

$$\equiv 1 + \int_0^t A(s)\,\mathrm{d}s + \int_0^t A(s_1)\,\mathrm{d}s_1 \int_0^{s_1} A(s_2)\,\mathrm{d}s_2 + \tag{2.292}$$

$$+ \int_0^t A(s_1)\,\mathrm{d}s_1 \int_0^{s_1} A(s_2)\,\mathrm{d}s_2 \int_0^{s_2} A(s_3)\,\mathrm{d}s_3 + \cdots$$

T-ordered exponentials form a group(oid) in the following sense

$$\mathbf{T}\exp\left(\int_0^{t_1+t_2} A(s)\,\mathrm{d}s\right) =$$

$$= \mathbf{T}\exp\left(\int_{t_1}^{t_1+t_2} A(s)\,\mathrm{d}s\right)\mathbf{T}\exp\left(\int_0^{t_1} A(s)\,\mathrm{d}s\right). \tag{2.293}$$

Now

Lemma 2.5 *Let $H(t)$ a family of Hermitian Hamiltonians parametrized by the time t. The solution to the OPE with initial condition*

$$i\hbar\frac{\mathrm{d}}{\mathrm{d}t}U(t) = H(t)\,U(t) \qquad U(0) = 1 \tag{2.294}$$

*is the family of unitary operators given by the **T**-ordered exponential*

$$U(t) = \mathbf{T}\exp\left(-\frac{i}{\hbar}\int_0^t H(s)\,\mathrm{d}s\right). \tag{2.295}$$

Proof Consider the expansion in reiterated integrals in the second line of (2.292) with $A(s)$ replaced by $-iH(s)/\hbar$. Taking the derivative with respect to t, we get

$$-\frac{i}{\hbar}H(t)\cdot 1 - \frac{i}{\hbar}H(t)\int_0^t\left(-\frac{i}{\hbar}H(s)\right)\mathrm{d}s + \cdots = -\frac{i}{\hbar}H(t)\,U(t), \tag{2.296}$$

so the **T**-exponential is a formal solution to (2.294). The series (2.292) is convergent for all reasonable $H(t)$ for which the solution to (2.294) exists. $\square$

2.10   Density Matrix: The Quantum Liouville Equation

The quantum states $|\psi\rangle \in \mathcal{H}$, represented by vectors in a Hilbert space $\mathcal{H}$, are called *pure states*. They may be identified with the corresponding (orthogonal) projectors of rank-1[17]

$$P_\psi = P_\psi^\dagger, \quad P_\psi \geq 0, \quad P_\psi^2 = P_\psi, \quad \mathrm{Tr}\, P_\psi = 1, \tag{2.297}$$

given by

$$P_\psi \overset{\text{def}}{=} |\psi\rangle\langle\psi| \qquad |\psi\rangle \text{ normalized as } \langle\psi|\psi\rangle = 1, \tag{2.298}$$

while the expectation value of an observable A and the transition probabilities can be written as

$$\langle\psi|A|\psi\rangle = \mathrm{Tr}(A P_\psi), \qquad |\langle\phi|\psi\rangle|^2 = \mathrm{Tr}(P_\phi P_\psi) \tag{2.299}$$

Conversely, any operator P satisfying the conditions (2.297) may be written in the form (2.298) for some $|\psi\rangle \in \mathcal{H}$.

It is natural to consider more general operators ϱ which generalize the rank-1 projectors and satisfy the conditions

$$\varrho^\dagger = \varrho, \quad \varrho \geq 0, \quad \mathrm{Tr}\,\varrho = 1. \tag{2.300}$$

They are called *density matrices* or *mixed states*. They were introduced by von Neumann as a quantum analogue of the classical Liouville density ρ in phase space ([4] §. 6.6). The theory of the mixed states will be discussed in Chap. 7 in the context of open quantum systems. Here we limit ourselves to establishing the quantum Liouville equation.

The Density Matrix ϱ

We consider a quantum dynamical system which at a given instant t_0 is in one between several possible states according to a probability law (an *ensemble* of states). For simplicity we assume that the possible states form a discrete orthogonal[18] set $\{|m\rangle\}$ labeled by a positive integer m. $|m\rangle$ stands for a normalized vector representing the m-th state in the set. The generalization to a continuous

---

[17] The identification of quantum states with rank-1 projectors is more canonical than the identifications with vectors, since it is not affected by phase ambiguities and normalization issues. Indeed the space of rank-1 orthogonal projectors $\mathcal{H} \to \mathcal{H}$ is canonically isomorphic to the projective space $\mathbb{P}(\mathcal{H})$.

[18] The general ensemble of possibly non-orthogonal states is described by the *Schrödinger mixing theorem*, see Chap. 7.

spectrum (or to a spectrum with both continuous and discrete components) is totally straightforward and left to the reader.

Let p_m be the probability that our system is in the state $|m\rangle$. We define the *density operator* (a.k.a. *density matrix*) ϱ as

$$\varrho = \sum_m |m\rangle \, p_m \, \langle m|, \qquad 0 \le p_m \le 1, \qquad \sum_m p_m = 1. \tag{2.301}$$

ϱ is a Hermitian operator acting on the Hilbert space[19] $\mathcal{H}$ whose eigenvalues are non-negative. Since the $|m\rangle$ are orthonormal (i.e. $\langle m|m'\rangle = \delta_{m,m'}$)

$$\mathrm{Tr}\,\varrho = \sum_m p_m = 1, \tag{2.302}$$

where $\mathrm{Tr}(\cdot)$ is the trace over the Hilbert space $\mathcal{H}$. Note that

$$\mathrm{Tr}\,\varrho^2 = \sum_m p_m^2 \le 1, \tag{2.303}$$

so that ϱ is a Hilbert-Schmidt operator[20] acting on the separable Hilbert space, hence a *compact* operator even when the sum is infinite or an integral over a continuum of generalized vectors. A pure state P_ψ is the special case of a density matrix with rank-1.

If $\mathcal{O}$ is a Hermitian operator ($\equiv$ observable) of the quantum system, its expectation value on the quantum distribution ϱ is

$$\langle \mathcal{O} \rangle = \mathrm{Tr}[\mathcal{O}\varrho] \equiv \sum_m p_m \, \langle m|\mathcal{O}|m\rangle. \tag{2.304}$$

The density matrix ϱ satisfies the quantum analogue of the classical Liouville equation. We start by a brief review of the classical story.

The Classical Liouville Equation

Let Ω be the symplectic form of the phase space $\mathcal{W}$ of a classic mechanical system. The *Liouville volume* is the positive $2m$-form

$$\mu = \frac{\Omega^m}{m!} \tag{2.305}$$

---

[19] Recall that all our Hilbert spaces have at most *numerable* dimension.

[20] An operator $\mathcal{O}$ is Hilbert-Schmidt iff $\mathrm{Tr}(\mathcal{O}^\dagger \mathcal{O}) < +\infty$, i.e. iff $\mathcal{O}$ is normalizable as a vector in $\mathcal{H}\hat{\otimes}\mathcal{H}^\vee$. A Hilbert-Schmidt operator is automatically compact, i.e. it maps the unit ball in $\mathcal{H}$ into a relatively compact subset of $\mathcal{H}$.

(here $2m \equiv \dim \mathcal{W}$). Suppose we have a distribution of states in the classical phase space $\mathcal{W}$ given by the non-negative $2m$-form

$$\rho(q, p, t)\, \mu \tag{2.306}$$

where $\rho(q, p, t)$ is a (possibly time-dependent) *non-negative* function on the phase space $\mathcal{W}$. Often it is convenient to normalize the distribution so that its total mass is 1; then the distribution can be interpreted as a *probability measure*

$$p(q, p, t) = \frac{\rho(q, p, t)}{Z} \quad \text{where} \ \ Z \stackrel{\text{def}}{=} \int_{\mathcal{W}} \rho\, \mu \tag{2.307}$$

Z is known as the *partition function*. The continuity equation together with the Liouville theorem yields the *classical Liouville equation*[21]

$$\frac{\partial \rho}{\partial t} + [\rho, H]_{\text{PB}} = 0, \tag{2.308}$$

see [4] chap. 6 for a proof.

The Quantum Liouville Equation
We study the time-evolution of the operator ϱ. In Eq. (2.301) the states $|m\rangle$ evolve according to the Schrödinger equation

$$i\hbar \frac{\mathrm{d}}{\mathrm{d}t}|m\rangle = H|m\rangle, \tag{2.309}$$

while the p_m's remain constant since a state cannot jump from a state solving the Schrödinger equation to a different state without violating the unitarity of time evolution. Then

$$i\hbar \frac{\mathrm{d}\varrho}{\mathrm{d}t} = i\hbar \sum_m \left(\frac{\mathrm{d}|m\rangle}{\mathrm{d}t}\, p_m\, \langle m| + |m\rangle\, p_m\, \frac{\mathrm{d}\langle m|}{\mathrm{d}t} \right) =$$
$$= \sum_m \left(H|m\rangle\, p_m\, \langle m| - |m\rangle\, p_m\, \langle m|H \right) = H\varrho - \varrho\, H. \tag{2.310}$$

The *quantum Liouville equation*

$$i\hbar \frac{\mathrm{d}\varrho}{\mathrm{d}t} + [\varrho, H] = 0 \tag{2.311}$$

reduces to the classical one (2.308) as $\hbar \to 0$ provided the commutator of quantum operators becomes in this limit the Poisson bracket of the corresponding classical

---

[21] As before $[A, B]_{\text{PB}}$ is the Poisson bracket of the two functions $A, B \in \Omega^0(\mathcal{W})$.

quantities

$$\left[A, B\right] \rightsquigarrow i\hbar \left[A_{\text{class.}}, B_{\text{class.}}\right]_{\text{PB}} \qquad \text{as } \hbar \to 0. \tag{2.312}$$

A density matrix ϱ *is in equilibrium* iff it does not evolve with time, that is, iff

$$[\varrho, H] = 0 \tag{2.313}$$

where the bracket is the quantum commutator (resp. the Poisson bracket) in the quantum (resp. classical) case.

Partial Traces

One important source of non-trivial density matrices (mixed states) are the "complex" systems composed of several subsystems.

Suppose that our quantum system is made of two sub-systems with Hilbert spaces $\mathcal{H}_1$ and $\mathcal{H}_2$, respectively. The Hilbert space of the compound system is

$$\mathcal{H} = \mathcal{H}_1 \widehat{\otimes} \mathcal{H}_2. \tag{2.314}$$

Theorem 2.6 (Schmidt Decomposition) *Given a pure state $|\Psi\rangle \in \mathcal{H}$ we can find two orthonormal sets of vectors $|j\rangle_1 \in \mathcal{H}_1$ and $|j\rangle_2 \in \mathcal{H}_2$ such that*

$$|\Psi\rangle = \sum_j c_j |j\rangle_1 \otimes |j\rangle_2 \in \mathcal{H} \tag{2.315}$$

where c_j are non-negative real numbers with $\sum_j c_j^2 = 1$.

This theorem will be discussed and proved in Chap. 7. Consider the observables A for the first subsystem. We identity them with the observables $A \otimes 1$ in the combined system. Then we have

$$\langle\Psi|(A \otimes 1)|\Psi\rangle = \sum_{i,j} (c_i \,_1\langle i| \otimes \,_2\langle i|)(A \otimes 1)(c_j|j\rangle_1 \otimes |j\rangle_2) =$$
$$= \sum_j c_j^2 \,_1\langle j|A|j\rangle_1 \tag{2.316}$$

As long as we are only interested in measuring physical quantities associated with the first subsystem—or we have experimental access only to this subsystem—all expectation values of observable of the first subsector in the combined state $|\Psi\rangle$ can be computed using the density matrix

$$\varrho_\Psi = \sum_j c_j^2 |j\rangle_1 \otimes \,_1\langle j|, \tag{2.317}$$

by the formula

$$\langle\Psi|A|\Psi\rangle = \mathrm{Tr}(A\,\varrho_\Psi) \qquad \forall\ \text{1st subsystem observable} A \tag{2.318}$$

In this situation, the fact that we describe the physics in terms of a *reduced density matrix* ϱ_Ψ instead of a pure state $|\Psi\rangle$ reflects our *ignorance* about the second subsystems and its observables we can't measure. This ignorance manifests itself in the fact that, from the viewpoint of the first subsystem *per se,* the subsystem is in a probability distribution of states $|j\rangle_1$ with probabilities c_j^2. Since the mixed states are 'partial informations' about the system, it is natural to expect that they play a pivotal role in Quantum Information theory to be discussed in Chap. 7.

One may pass from the pure-state projector $|\Psi\rangle\langle\Psi|$ to the corresponding reduced density matrix for the first sub-system, $\varrho_1 \equiv \varrho_\Psi$, by taking a *partial trace* only over the degrees of freedom of the second sub-system

$$\varrho_1 = \mathrm{Tr}_{\mathcal{H}_2}|\Psi\rangle\langle\Psi|. \tag{2.319}$$

Partial traces is a typical source of density matrices. The partial trace procedure has an "inverse" called *purification:*

Lemma 2.6 *Given a density matrix*

$$\varrho = \sum p_m\,|m\rangle\langle m| \in \mathcal{B}(\mathcal{H}) \tag{2.320}$$

we can find a pure state $|\Psi\rangle \in \mathcal{H}\widehat{\otimes}\mathcal{H}'$, *where* $\mathcal{H}'$ *is any Hilbert space with* $\dim\mathcal{H}' \geq$ rank ϱ, *such that*

$$\mathrm{Tr}(A\,\varrho) = \langle\Psi|A|\Psi\rangle \tag{2.321}$$

for all operators A acting on $\mathcal{H}$. The state $|\Psi\rangle$ is called a purification *of ϱ.*

Clearly the purification is far from being unique. A canonical choice for $\mathcal{H}'$ is a second copy of $\mathcal{H}$ with purification

$$|\Psi\rangle = \sum_m \sqrt{p_m}\,|m\rangle\langle m|. \tag{2.322}$$

Definition 2.9 Two quantum systems are *entangled* if the compounded system is in a state $|\Psi\rangle \in \mathcal{H}_1\widehat{\otimes}\mathcal{H}_2$ which cannot be written in the factorized form

$$|\psi_1\rangle \otimes |\psi_2\rangle, \qquad |\psi_a\rangle \in \mathcal{H}_a. \tag{2.323}$$

Equivalently they are entangled iff

$$\mathrm{rank}(\mathrm{Tr}_{\mathcal{H}_1}|\Psi\rangle\langle\Psi|) \equiv \mathrm{rank}(\mathrm{Tr}_{\mathcal{H}_2}|\Psi\rangle\langle\Psi|) > 1. \tag{2.324}$$

Entanglement is a fundamental property of Quantum Physics—in facts *it is its defining feature*—and the existence of entangled states is the most dramatic departure of Quantum Physics from the classical paradigm, with deep and far-reaching implications.

In Chap. 7 we shall elaborate on the partial trace, Schmidt decomposition, purification, entanglement, Quantum Information, and all that, from the viewpoint of the quantum theory of open systems. We refer the reader to that chapter and the literature quoted therein.

2.11 Quantization Conditions: Canonical Quantization

Up to now we discussed the general principles and formalism of Quantum Physics which apply to *any* quantum theory. It is time to focus on the quantum *mechanical* systems, that is, the non-relativistic quantum models with finitely many degrees of freedom. From a broader perspective these systems may be identified as *one-dimensional* Quantum Field Theories.

There is an important special sub-class of quantum mechanical systems: those which have a *classical analogue*. One also says that these quantum systems are obtained by *quantizing* the corresponding classical mechanic system. The procedure to construct a quantum model starting from its classical counterpart (when possible) is called *quantization*. We already alluded at this procedure in the fiction-history tale of Sect. 1.2.

The algebra of observables of a classical mechanical system with n degrees of freedom is generated by the $2n$ Darboux canonical coordinates

$$(q^i, p_j) \equiv (w^a) \qquad (i, j = 1, \ldots, n, \quad a = 1, \ldots, 2n) \tag{2.325}$$

of the phase space $\mathcal{W}$, which are characterized by their *canonical Poisson brackets* ([4] §. 6.4)

$$[q^i, q^j]_{\mathrm{PB}} = [p_i, p_j]_{\mathrm{PB}} = 0, \quad [q^i, p_j]_{\mathrm{PB}} = \delta^i{}_j. \tag{2.326}$$

Since the classical observables commute, we may identify the classical observable algebra $\mathfrak{A}_{\mathrm{class}}$ with the quotient of the free associative algebra generated by the w^a's by the bilateral ideal

$$R \equiv (w^a w^b - w^b w^a) \tag{2.327}$$

generated by all commutators $w^a w^b - w^b w^a$, that is,

$$\mathfrak{A}_{\mathrm{class}} = \mathbb{C}\langle w^a \rangle \big/ (w^a w^b - w^b w^a) \tag{2.328}$$

The algebra of observables of the corresponding quantum system is generated by the densely-defined Hermitian operators

$$q^i, \quad p_j \quad (i, j = 1, \ldots, n) \tag{2.329}$$

that we write collectively as w^a ($a = 1, \ldots, 2n$). The quantum algebra cannot be commutative: otherwise we get back the classical theory as we observed in Sect. 1.2. The relation ideal R should be modified in such a way that in the classical limit $\hbar \to 0$ it reduces back to the commutator ideal. In particular, in the quantum algebra $\mathfrak{A}_{\text{quan}}$ there must be no extra relations besides the ones arising from a deformation of the classical relations $w^a w^b - w^b w^a = 0$. On general grounds the deformed relations should have the form

$$w^a w^b - w^b w^a = i\hbar\, C^{ab} + O(\hbar^2) \tag{2.330}$$

for some Hermitian operators $C^{ab} = -C^{ba}$ ($a, b = 1, \ldots, 2n$). In view of the Poincaré-Birkhoff-Witt theorem, the absence of extra relations is required in order for a basis of the algebra $\mathfrak{A}_{\text{quan}}$ to be given by the *ordered* products

$$(w^1)^{k_1} (w^2)^{k_2} \cdots (w^{2n})^{k_{2n}} \qquad k_i = 0, 1, 2, \cdots \tag{2.331}$$

so that we have one independent quantum observable for each classical one. The quantum relations (2.330) are called the *quantum conditions* or *commutation relations*.

In the last paragraph of Sect. 2.3 we noticed that the commutator of quantum observables has exactly the same algebraic properties as the classical Poisson bracket: $[\cdot, \cdot]$ is antisymmetric, bilinear, kills the constants, satisfies the Leibniz rule, and obeys the Jacobi identity. It is then natural to identify commutators and Poisson brackets up to the non-trivial factor $i\hbar$ required by reality and dimensional analysis. In view of (2.331) we may always rescale the quantum generators w^a so that

$$\left[w^a, w^b\right] \equiv w^a w^b - w^b w^a = i\hbar[w^a, w^b]_{\text{PB}}. \tag{2.332}$$

This leads to the same correspondence

$$\text{quantum commutators} \longrightarrow \text{classical Poisson brackets} \tag{2.333}$$

$$[A, B] \xrightarrow{\ \hbar \approx 0\ } i\hbar[A_{\text{clas}}, B_{\text{clas}}]_{\text{PB}},$$

which relates:

(i) the quantum equations of motion in the Heisenberg picture to the classical Hamilton equations, cf. Sect. 2.9.2;

(ii) the quantum and classical Lie algebras of conserved quantities, cf. Sect. 2.9.2;

(iii) the quantum Liouville equation to the classical one, cf. Sect. 2.10.

All these considerations lead us to the *canonical commutation relations*

$$[q^i, q^j] = [p_i, p_j] = 0, \qquad [q^i, p_j] = i\hbar \, \delta^i{}_j. \tag{2.334}$$

The procedure of quantizing a classical mechanical system by replacing the commutative condition (2.327) with the canonical commutation relations (2.334) is called *canonical quantization*. Most quantum systems to be studied in this book are canonically quantized systems with a classical analogue (or limit).

Remark 2.9 Equation (2.334) can be also inferred from a dimensional argument. The operator C^{ab} in (2.330) should have dimension

$$[C^{ab}] = \begin{cases} \mathrm{kg}^{-1}\,\mathrm{s} & a, b \leq n \\ \mathrm{adimensional} & a \leq n < b \\ \mathrm{kg}\,\mathrm{s}^{-1} & a, b > n \end{cases} \tag{2.335}$$

There is no element (2.331) of the algebra $\mathfrak{A}_{\mathrm{quan}}$ of dimension $\mathrm{kg}^{-1}\,\mathrm{s}$ or $\mathrm{kg}\,\mathrm{s}^{-1}$ so the commutator of two q' or two p's should vanish. The only adimensional operator is the identity, so the commutator of a q with a p must be proportional to the identity, hence a constant.

Warning 3 Canonical quantization is less innocent than it may seem and should be used with a pinch of salt. Classically there are many systems of Darboux coordinates related by canonical transformations (chap. 8 of [4]). A general non-linear classical canonical transformation will be affected by ambiguities about the proper order of the operators, and typically will preserve the canonical commutators only up to $O(\hbar^2)$ terms. See however Sect. 2.17 for a more systematic approach to these issues.

2.11.1 Unitary Transformations

A unitary operator U

$$U^\dagger U = U U^\dagger = 1, \tag{2.336}$$

yields an automorphism of the Hilbert space $\mathcal{H}$ hence it maps self-adjoint (resp. projector, unitary, normal, etc.) operators into operators of the same kind. Since

$$(A - \lambda)|\lambda\rangle = 0 \quad \Rightarrow \quad U^{-1}(A - \lambda)U U^{-1}|\lambda\rangle = 0$$

$$\Rightarrow \quad U^{-1} A U (U^{-1}|\lambda\rangle) = \lambda (U^{-1}|\lambda\rangle), \tag{2.337}$$

a unitary transformation U

$$A \mapsto U^{-1}AU \tag{2.338}$$

preserves the spectrum $\sigma(A)$ of the operator while the spectral family $P_{U^{-1}AU}(\lambda)$ of the transformed operator $U^{-1}AU$ is

$$P_{U^{-1}AU}(\lambda) = U^{-1}P_A(\lambda)U. \tag{2.339}$$

This entails that U preserves all algebraic and functional relations between operators; for instance

$$f(U^{-1}AU) = \int_{\sigma(A)} f(\lambda)\,dP_{U^{-1}AU}(\lambda) =$$
$$= U^{-1}\int_{\sigma(A)} f(\lambda)\,dP(\lambda)\,U = U^{-1}f(A)\,U. \tag{2.340}$$

In particular, if $\boldsymbol{q}^i$, $\boldsymbol{p}_j$ is a system of canonical operators satisfying the canonical commutation relations (2.334) their unitary transforms

$$\tilde{\boldsymbol{q}}^i \equiv U^{-1}\boldsymbol{q}^i U, \qquad \tilde{\boldsymbol{p}}_j \equiv U^{-1}\boldsymbol{p}_j U \tag{2.341}$$

also satisfy the canonical commutation relations; for instance

$$[\tilde{\boldsymbol{q}}^i, \tilde{\boldsymbol{p}}_j] = [U^{-1}\boldsymbol{q}^i U, U^{-1}\boldsymbol{p}_j U] =$$
$$= U^{-1}[\boldsymbol{q}^i, \boldsymbol{p}_j]U = U^{-1}(i\hbar\,\delta^i{}_j)U = i\hbar\,\delta^i{}_j \tag{2.342}$$

The $\tilde{\boldsymbol{q}}^i$, $\tilde{\boldsymbol{p}}_j$ are new quantum canonical operators on the same footing as the original ones $\boldsymbol{q}^i$, $\boldsymbol{p}_j$. Conversely,

Lemma 2.7 *Let $\tilde{\boldsymbol{q}}^i$, $\tilde{\boldsymbol{p}}_j$ $(i, j = 1, \ldots, n)$ be self-adjoint operators acting on $L^2(\mathbb{R}^n)$ which satisfy the canonical commutator relations (2.334) while the $\tilde{\boldsymbol{q}}^i$'s form a complete system of commuting operators. Then there exists a unitary operator*

$$U \in \mathcal{B}(L^2(\mathbb{R}^n)) \tag{2.343}$$

such that

$$\tilde{\boldsymbol{q}}^i = U^{-1}\boldsymbol{q}^i U, \qquad \tilde{\boldsymbol{p}}_j = U^{-1}\boldsymbol{p}_j U. \tag{2.344}$$

We defer the proof to Sect. 2.12.1. This result is often stated as

Theorem 2.7 *All irreducible unitary representations of the canonical commutation relations (2.334), in a separable Hilbert space $\mathcal{H}$, are unitary equivalent (i.e. equivalent up to an automorphism of $\mathcal{H}$).*

In the classical theory two sets of canonical variables ($\equiv$ Darboux coordinates) are related by a canonical transformation ([4] Chap. 8). The classical canonical transformations are typically highly non-linear, and hence hard to define in the quantum set-up due to issues with the ordering of the non-commuting operators. See however, Sect. 2.17 for a precise order prescription. We see from Lemma 2.7 that the classical theory of canonical transformations gets replaced in Quantum Mechanics by the theory of unitary transformations, which are much better behaved. There is one situation in which the two theories agree on the nose, namely the canonical transformations which act *linearly* on the canonical variables.

Linear Unitary Transformations

We consider a unitary transformation U which acts linearly on the canonical operators

$$
\begin{aligned}
\tilde{q}^i &\equiv U^{-1} q^i U = A^i{}_k q^k + B^{il} p_l \\
\tilde{p}_j &\equiv U^{-1} p_j U = C_{jk} q^k + D_j{}^l p_l
\end{aligned}
\tag{2.345}
$$

where A, B, C, D are $n \times n$ matrices of real numbers. Imposing the canonical commutation relations

$$
\left[\tilde{q}^i, \tilde{q}^j\right] = \left[\tilde{p}_i, \tilde{p}_j\right] = 0, \quad \left[\tilde{q}^i, \tilde{p}_j\right] = \delta^i{}_j,
\tag{2.346}
$$

we get the following conditions on the numerical matrices A, B, C, D

$$
AB^t - BA^t = 0, \quad CD^t - DC^t = 0, \quad AD^t - BC^t = 1
\tag{2.347}
$$

which just state that the $2n \times 2n$ matrix (whose entries are $n \times n$ matrix blocks)

$$
M = \begin{pmatrix} A & B \\ C & D \end{pmatrix} \in Sp(2n, \mathbb{R})
\tag{2.348}
$$

is *symplectic*, that is, satisfies the equation

$$
M \Omega M^t = \Omega, \qquad \text{where} \quad \Omega = \begin{pmatrix} 0 & 1 \\ -1 & 0 \end{pmatrix} \in Sp(2n, \mathbb{R}).
\tag{2.349}
$$

2.12 Schrödinger and Momentum Representations

In this section we focus on systems which are the quantum analogue of classical systems with configuration space $\mathbb{R}^n$. They have the Hilbert space $L^2(\mathbb{R}^n)$. The prototypical example is a spinless[22] non-relativistic particle moving in n-dimensional Euclidean space in presence of suitable potentials. In Sect. 2.15 we shall consider more general configuration spaces $\mathcal{M}$, while in Sect. 2.14 we consider configurations spaces with boundaries, $\partial\mathcal{M} \neq \varnothing$. We mainly adopt the more efficient physicists' viewpoint. The equivalent math approach is briefly outlined in Sect. 2.12.4.

2.12.1 Schrödinger Representation

We write q^i ($i = 1, \ldots, n$) for the quantum observables which describe the position of the particle in $\mathbb{R}^n$: their eigenvalues x^i are the Cartesian coordinates of the particle. We write $x \equiv (x^1, \ldots, x^n) \in \mathbb{R}^n$ for the vector of eigenvalues of the q^i's.

The spectrum of q^i is obviously continuous,

$$-\infty < x^i < +\infty \qquad i = 1, 2, \ldots, n, \tag{2.350}$$

and the q^i's form a *complete system of commuting observables* in $L^2(\mathbb{R}^n)$ which separates the states, hence they define a *representation* of the Hilbert space in terms of Dirac's generalized "basis vectors" $|x\rangle \equiv |x^1, \ldots, x^n\rangle$ with the properties

$$q^i|x\rangle = x^i|x\rangle \qquad i = 1, 2, \ldots, n, \tag{2.351}$$

$$\langle \tilde{x}|x\rangle = \delta(\tilde{x} - x) \overset{\text{def}}{=} \prod_{j=1}^{n} \delta(\tilde{x}^j - x^j) \in \mathcal{S}(\mathbb{R}^n)^{\vee} \tag{2.352}$$

and, for $|\psi\rangle \in \mathcal{H}$,

$$\psi(x) \equiv \langle x|\psi\rangle, \tag{2.353}$$

$$\langle \psi|\psi\rangle = \int_{\mathbb{R}^n} \mathrm{d}^n x \, |\psi(x)|^2 < +\infty \tag{2.354}$$

---

[22] For the moment *spinless* means simply that $\mathcal{H} = L^2(\mathbb{R}^n)$ instead of a Hilbert space of the form $\mathcal{H}_m = L^2(\mathbb{R}^n) \otimes \mathbb{C}^m$ with $m \geq 2$. In Chap. 4 the notion of *spin* (and *spinless*) will be introduced physically and clarified mathematically.

In particular the map $|\psi\rangle \mapsto \langle x|\psi\rangle$ yields an isomorphism $\mathcal{H} \simeq L^2(\mathbb{R}^n)$. The representation (2.351)–(2.354) is called the *Schrödinger representation,* and the function $\psi(x) \in L^2(\mathbb{R}^n)$ is the *(Schrödinger) wave function* of the state $|\psi\rangle$. As stressed in Warning 1, the wave function $\psi(x)$ is defined up to functions which vanish almost everywhere. The operator q^j acts on the wave function $\psi(x)$ as multiplication by x^j

$$\langle x|q^j|\psi\rangle = x^j\langle x|\psi\rangle \equiv x^j\,\psi(x). \tag{2.355}$$

In view of our discussion of physical measurements in Sect. 2.7, the probability of finding a particle in the (normalized) state $\psi(x) \in L^2(\mathbb{R}^n)$ in the domain $K \subset \mathbb{R}^n$ is

$$\mathsf{Prob}(x \in K) = \int_K \mathrm{d}^n x\,|\psi(x)|^2. \tag{2.356}$$

Fact 2.8 *The modulo square* $|\psi(x)|^2$ *of the Schrödinger wave function* $\psi(x)$ *has the physical interpretation of the* probability density of finding the particle at $x \in \mathbb{R}^n$.

Consider the (densely defined) linear operator, to be written $\partial/\partial q^j$, which acts on the wave-function $\psi(x)$ as the derivative with respect to x^j

$$\left\langle x\left|\frac{\partial}{\partial q^j}\right|\psi\right\rangle \overset{\text{def}}{=} \frac{\partial}{\partial x^j}\langle x|\psi\rangle \equiv \frac{\partial}{\partial x^j}\psi(x). \tag{2.357}$$

If ψ_1, ψ_2 are elements of $\mathcal{S}(\mathbb{R}^n)$

$$
\begin{aligned}
\left\langle \psi_1\left|\frac{\partial}{\partial q^j}\psi_2\right.\right\rangle &= \int_{\mathbb{R}^n} \mathrm{d}^n x\;\psi_1(x)^*\frac{\partial}{\partial x^j}\psi_2(x) = \\
&= -\int_{\mathbb{R}^n} \mathrm{d}^n x\;\left(\frac{\partial}{\partial x^i}\psi_1(x)_1\right)^* \psi_2(x) = \\
&= -\left\langle \frac{\partial}{\partial q^j}\psi_1\left|\psi_2\right.\right\rangle,
\end{aligned}
\tag{2.358}
$$

where in the second equality we integrated by parts and used that the functions in $\mathcal{S}(\mathbb{R}^n)$, and their derivatives, vanish at infinity more rapidly than any power. Equation (2.358) says that the operator $\partial/\partial q^j$ is *anti*-Hermitian. We then consider the *self-adjoint* operator

$$-\mathrm{i}\hbar\,\frac{\partial}{\partial q^j}. \tag{2.359}$$

When acting on a function $f \in \mathcal{S}(\mathbb{R}^n)$,

$$\left[q^i, -i\hbar \frac{\partial}{\partial q^j} \right] f = i\hbar \left(\frac{\partial}{\partial q^j} q^i - q^i \frac{\partial}{\partial q^j} \right) f = i\hbar \, \delta^i{}_j \, f, \tag{2.360}$$

hence the operator

$$p_j + i\hbar \frac{\partial}{\partial q^j} \tag{2.361}$$

commutes with all q^i's. Since the $\{q^i\}$'s form a *complete system* of observables, by Corollary 2.6, the operator (2.361) must be some function $A_j(q)$ of the q^i's. The quantum momentum operators p_j are then

$$p_j = -i\hbar \frac{\partial}{\partial q^j} + A_j(q). \tag{2.362}$$

Now

$$0 = [p_j, p_k] = -i\hbar \left(\frac{\partial}{\partial q^j} A_k(q) - \frac{\partial}{\partial q^k} A_j(q) \right) \overset{\text{def}}{=} -i\hbar \, F_{jk}(q), \tag{2.363}$$

which means that the one-form $A_j(q)\, dq^j$ must be *closed,* hence *exact* since $H^1(\mathbb{R}^n) = 0$. Therefore there is a function $\alpha : \mathbb{R}^n \to \mathbb{R}$ such that

$$p_j = -i\hbar \frac{\partial}{\partial q^j} + \frac{\partial \alpha(q)}{\partial q^j}. \tag{2.364}$$

We **claim** that physics is independent of the function $\alpha(q)$, which then may be chosen according convenience. We present two (related) arguments.

First Argument   In Classical Mechanics the momenta change under an electromagnetic gauge transformation

$$A \rightsquigarrow A - d\alpha, \tag{2.365}$$

as (cf. [4] *Remark 3.3*)

$$p_j \rightsquigarrow p_j + \frac{\partial \alpha}{\partial x^j}, \tag{2.366}$$

so that the ambiguity of the quantum operator p_j by the gradient of a function $\alpha(q)$ of the coordinates is just the quantum version of the inherent ambiguity due to our freedom to choose any convenient gauge. Changing the gauge does not affect any physical observable, and then the function $\alpha(q)$ may be chosen as we please.

Second Argument The normalized vector representing a state $|\psi\rangle$ is unique only up to a phase factor. Therefore, in the representation of the states in terms of wave functions $\psi(x)$, we have the luxury of redefining the representative vector by a x-dependent phase factor

$$\psi(x) \rightsquigarrow \psi(x)' \equiv e^{i\alpha(x)/\hbar}\psi(x), \tag{2.367}$$

a transformation which leaves the observable probability density unchanged

$$|\psi(x)'|^2 = |\psi(x)|^2. \tag{2.368}$$

Under the redefinition (2.367) of the representative wave-function

$$-i\hbar\frac{\partial}{\partial q^j} \rightsquigarrow -i\hbar\frac{\partial}{\partial q^j} + \frac{\partial\alpha(q)}{\partial q^j}, \tag{2.369}$$

so that at the quantum level the gauge transformation (2.366) gets re-interpreted as the redefinition (2.367) of the Schrödinger representation. In other words: the second term in the RHS of (2.364) may always be set to zero by multiplying the wave-function by the appropriate x-dependent phase. We stress that (2.367) is a *unitary transformation*, hence an automorphism of the Hilbert space $L^2(\mathbb{R}^n)$.

Geometric Viewpoint The geometric description of the "ambiguity" $\partial\alpha(q)/\partial q^j$ in (2.364) is as follows. In the Schrödinger representation, the Hilbert space of states is actually the space of square-summable *sections* of a Hermitian complex line bundle[23] over $\mathbb{R}^n$, with structure group $U(1)$, equipped with a connection form $A = A_i\,dx^i$ which is unique only up to gauge transformations, i.e. up to choices of trivialization of the line bundle. In absence of background electromagnetic fields, the connection is *pure gauge* and may be set to zero by a suitable gauge transformation.

Following standard usage, we make the simplest choice, and define the Schrödinger representation of momentum operators with $\alpha(x) \equiv 0$:

$$p_j = -i\hbar\frac{\partial}{\partial q^j}. \tag{2.370}$$

---

[23] Since $\mathbb{R}^n$ is contractible, all such bundles are topologically trivial. Then the connection A is non-trivial precisely if the curvature $F \equiv dA = 0$.

SUMMARY: Schrödinger Representation

The canonical operators q^i, p_j are represented on the wave functions $\psi(x) \equiv \langle x | \psi \rangle$ in (a dense domain of) the Hilbert space $L^2(\mathbb{R}^n)$ as:

SR1   *the coordinate operator q^i by the multiplication operator*

$$\psi(x) \mapsto x^i \psi(x) \tag{2.371}$$

SR2   *the conjugate momentum operator p_j by the first order differential operator*

$$\psi(x) \mapsto -i\hbar \frac{\partial \psi(x)}{\partial x^j}. \tag{2.372}$$

The operators q^i, p_j satisfy the canonical commutation relations (2.334), and more generally,

$$[f(q), p_j] = i\hbar \frac{\partial f}{\partial q^j}, \qquad [g(p), q^j] = -i\hbar \frac{\partial g}{\partial p_j}, \tag{2.373}$$

for all smooth functions $f(z)$, $g(z)$. All other (irreducible, unitary) representation of the canonical commutation relations is unitary equivalent to **SR1**, **SR2**.

The $L^2(\mathbb{R}^n)$ Hermitian product is

$$\langle \psi_1 | \psi_2 \rangle = \int_{\mathbb{R}^n} \mathrm{d}^n x \; \psi_1(x)^* \psi_2(x), \tag{2.374}$$

$|\psi(x)|^2$ has the physical interpretation of the density of probability of finding the "particle" in the state $\psi(x)$ at the particular point $x \in \mathbb{R}^n$.

Deferred Proof of Lemma 2.7 We return to the proof that there is only one representation of the canonical commutator algebra modulo unitary equivalence. We assume to have a set of (densely defined) operators $\tilde{q}^i$, $\tilde{p}_j$ ($i, j = 1, \ldots, n$) acting on $L^2(\mathbb{R}^n)$ which satisfy the canonical commutation relations

$$[\tilde{q}^i, \tilde{q}^j] = [\tilde{p}_i, \tilde{p}_j] = 0, \qquad [\tilde{q}^i, \tilde{p}_j] = i\hbar \, \delta^i{}_j \tag{2.375}$$

while an operator which commutes with all $\tilde{q}^i$ is a function of the $\tilde{q}^i$'s. The claim of the **Lemma** is that there is a *unitary operator U* such that

$$\tilde{q}^i = U^{-1} q^i U, \qquad \tilde{p}_j = U^{-1} p_j U. \tag{2.376}$$

Proof The spectrum of each $\tilde{q}^j$ is $\sigma(\tilde{q}^j) = \mathbb{R}$ because[24] the dual operator $\tilde{p}_j$ generates a group $\mathbb{R}$ of translations $\tilde{q}^j \rightsquigarrow \tilde{q}^j - a^j$. By assumption the $\tilde{q}^j$'s form a complete system, hence we can find a representation whose generalized basis vectors $|\tilde{x}\rangle \equiv |\tilde{x}^1, \ldots, \tilde{x}^n\rangle$ are simultaneous (generalized) eigenvectors of the $\tilde{q}^j$'s

$$\tilde{q}^j|\tilde{x}\rangle = \tilde{x}^j|\tilde{x}\rangle, \qquad \langle \tilde{x}'|\tilde{x}\rangle = \delta(\tilde{x} - \tilde{x}'). \tag{2.377}$$

Now consider the function in $\mathbb{R}^n \times \mathbb{R}^n$

$$\langle \tilde{x}|x\rangle. \tag{2.378}$$

We claim that $\langle \tilde{x}|x\rangle$ is the *integral kernel* of a unitary transformation U of $L^2(\mathbb{R}^n)$ defined as

$$(U\psi)(\tilde{x}) = \int_{\mathbb{R}^n} \mathrm{d}^n x \, \langle \tilde{x}|x\rangle \, \psi(x). \tag{2.379}$$

U is unitary since

$$U^\dagger U = \int_{\mathbb{R}^n} \mathrm{d}^n x' \, \mathrm{d}^n \tilde{x} \, \mathrm{d}^n x \, |x'\rangle\langle x'|\tilde{x}\rangle\langle \tilde{x}|x\rangle\langle x| =$$
$$= \int_{\mathbb{R}^n} \mathrm{d}^n x' \, \mathrm{d}^n x \, |x'\rangle \, \delta(x' - x) \, \langle x| = 1, \tag{2.380}$$

and analogously $UU^\dagger = 1$. We have

$$\langle \tilde{x}|U^{-1}|x\rangle = \int_{\mathbb{R}^n} \mathrm{d}^n y \, \langle \tilde{x}|y\rangle\langle y|U^{-1}|x\rangle = \delta(\tilde{x} - x) \tag{2.381}$$

and dually

$$\langle x|U|\tilde{x}\rangle = \delta(x - \tilde{x}), \tag{2.382}$$

so that

$$\langle x|q^i U|\tilde{x}\rangle = x^j\langle x|U|\tilde{x}\rangle = x^j\delta(x - \tilde{x}) =$$
$$= \delta(x - \tilde{x})\tilde{x}^j = \langle x|U|\tilde{x}\rangle\tilde{x}^j = \langle x|U\tilde{q}^j|\tilde{x}\rangle. \tag{2.383}$$

Since $|x\rangle$ and $|\tilde{x}\rangle$ are both complete systems, this implies

$$U\tilde{q}^j = q^i U \quad \Rightarrow \quad \tilde{q}^j = U^{-1}q^j U. \tag{2.384}$$

---

[24] The $\mathbb{R}$-group action is as in Example 2.3. Alternatively refer to Lemma 2.8.

On the other hand

$$\langle x|p_j U|\tilde{x}\rangle = -i\hbar \frac{\partial}{\partial x^j}\langle x|U|\tilde{x}\rangle = -i\hbar \frac{\partial}{\partial x^j}\delta(x-\tilde{x}) =$$

$$= i\hbar \frac{\partial}{\partial \tilde{x}^j}\delta(x-\tilde{x}) = i\hbar \frac{\partial}{\partial \tilde{x}^j}\langle x|U|\tilde{x}\rangle = \langle x|U\tilde{p}_j|\tilde{x}\rangle, \tag{2.385}$$

hence

$$\tilde{q}^j = U^{-1}q^j U, \qquad \tilde{p}_j = U^{-1}p_j U, \tag{2.386}$$

and U is the unitary transformations which implements the transformation

$$(q^j, p_j) \rightsquigarrow (\tilde{q}^j, \tilde{p}_j) \tag{2.387}$$

between the two complete sets of operators satisfying the canonical commutation relations. $\square$

2.12.2 Momentum as the Generator of Translations

We saw in Example 2.3 that the Hilbert space $L^2(\mathbb{R}^n)$ has an Abelian Lie group $\mathbb{R}^n$ of automorphisms given by translations of the Cartesian coordinates

$$a: x^i \mapsto x^i + a^i, \qquad a \in \mathbb{R}^n, \tag{2.388}$$

which—as all Hilbert space automorphisms—are implemented by a group $U(a)$ of unitary operators. We may see them as acting either on states or on operators (these two viewpoints are akin for the *space* translations to, respectively, the Schrödinger and Heinsenberg views of *time* translations):

$$\langle x|U(a)|\psi\rangle = \langle x+a|\psi\rangle \qquad \text{on wave-functions} \tag{2.389}$$

$$\left[\begin{array}{l} U(a)^{-1}q^j U(a) = q^j - a^j \\[2mm] U(a)^{-1}p_k U(a) = p_k \end{array} \right. \qquad \text{on operators} \tag{2.390}$$

$$U(a+b) = U(a)U(b) = U(b)U(a) \qquad \text{group law.} \tag{2.391}$$

In view of the Stone Theorem 2.3, there exist Hermitian operators A_i such that[25]

$$U(a) = \exp(\mathrm{i}a^i A_i) \equiv \exp\left(\mathrm{i}\sum_i a^i A_i\right).\tag{2.392}$$

In facts

Lemma 2.8 *The group of unitary operators $U(a)$ $(a \in \mathbb{R}^n)$ which generate spatial translations in $L^2(\mathbb{R}^n)$ is given by*

$$U(a) = \exp\left(\frac{\mathrm{i}}{\hbar}a^j p_j\right).\tag{2.393}$$

Proof $U(a)$ commutes with p_k so the second Eq. (2.390) holds. Then

$$U(a)^{-1}q^k U(a) - q^k = U(a)^{-1}[q^k, U(a)] =$$
$$= \mathrm{i}\hbar\, U(a)\frac{\partial}{\partial p_k}U(a) = -a^k.\tag{2.394}$$

$\square$

Momentum is the generator of spatial translations in Quantum Mechanics exactly as it is in the classical case ([4] chaps. 3, 6).

Momentum Eigenfunctions

To simplify the notation we focus on the $n = 1$ case, giving the straightforward generalization to arbitrary n at the end of the discussion. We know from Example 2.4 that $\sigma(p) = \mathbb{R}$. In view of the spectral theorem, the momentum observable p has a complete system of generalized eigenvectors $|p\rangle \in \mathcal{S}^\vee$ with the properties

$$p\,|p\rangle = p\,|p\rangle, \qquad \langle p'|p\rangle = \delta(p - p'), \qquad 1 = \int_\mathbb{R} \mathrm{d}p\,|p\rangle\langle p|.\tag{2.395}$$

The Schrödinger representation of the momentum eigenstates is given by the wave functions $\langle x \mid p\rangle$ which solve the eigenvector differential equation

$$p\,\langle x \mid p\rangle \equiv \langle x \mid p \mid p\rangle = -\mathrm{i}\hbar\,\frac{\partial}{\partial x}\langle x \mid p\rangle,\tag{2.396}$$

whose solution is

$$\langle x \mid p\rangle = C\exp\left(\frac{\mathrm{i}\,x\,p}{\hbar}\right) \in \mathcal{S}(\mathbb{R})^\vee.\tag{2.397}$$

---

[25] We recall that throughout this book we use Einstein's convention that whenever an index is repeated in a formula it is meant to be summed over, unless explicitly stated otherwise.

The integration constant C is fixed by the continuous normalization condition

$$\delta(p - p') = \langle p'|p \rangle = \int_{\mathbb{R}} dx \, \langle p'|x \rangle \langle x|p \rangle =$$
$$= |C|^2 \int_{\mathbb{R}} dx \, e^{ix(p-p')/\hbar} = 2\pi |C|^2 \hbar \, \delta(p - p'), \tag{2.398}$$

where we used the integral representation of the δ-function, Eq. (2.179). Hence

$$\langle x \mid p \rangle = \frac{1}{\sqrt{2\pi\hbar}} \exp\left(\frac{i\,x\,p}{\hbar}\right) \in \mathcal{S}(\mathbb{R})^{\vee} \tag{2.399}$$

and, more generally, with n degrees of freedom (valued in $\mathbb{R}^n$)

$$\langle x^j \mid p_l \rangle = \frac{1}{(2\pi\hbar)^{n/2}} \exp\left(\frac{i\,x^k p_k}{\hbar}\right) \in \mathcal{S}(\mathbb{R})^{\vee}. \tag{2.400}$$

The wave function (2.400) is known as a *plane wave* moving in the direction of the vector $p_k \in \mathbb{R}^n$.

Remark 2.10 The probability density $|\langle x^j \mid p_l \rangle|^2$ of a plane wave is constant in the Euclidean space (hence the wave-function of a momentum eigenstate is *not* normalizable). This corresponds to the fact that momentum is invariant under space translations, Eq. (2.388), and hence the probability of finding a particle in a momentum eigenstate is the same everywhere.

2.12.3 The Momentum Representation

In the momentum representation the momentum operators p_j are *diagonal*. To avoid writing silly coefficients, it is convenient (and standard) to change the normalization of the momentum eigenvectors to

$$\langle p'_j|p_k \rangle = (2\pi\hbar)^n \prod_k \delta(p_k - p'_k), \tag{2.401}$$

so that we identify the Hilbert space $\mathcal{H}$ with the space $L^2(\mathbb{R}^n)$ endowed with the L^2-measure $d^n p/(2\pi\hbar)^n$, that is, we write the Hermitian product in the form

$$\langle \psi_1|\psi_2 \rangle = \int_{\mathbb{R}^n} \frac{d^n p}{(2\pi\hbar)^n} \, \Phi_1(p)^* \, \Phi_2(p), \tag{2.402}$$

where we wrote

$$\Phi_a(p) = \langle p | \psi_a \rangle \quad a = 1, 2, \tag{2.403}$$

for the momentum-representation wave-functions of the states $|\psi_a\rangle \in \mathcal{H}$. With this normalization of the measure, the Fourier transform

$$\Phi(p) \stackrel{\text{def}}{=} \int_{\mathbb{R}^n} \mathrm{d}^n x \, \mathrm{e}^{-ipx/\hbar} \, \psi(x) \equiv \int_{\mathbb{R}} \mathrm{d}^n x \, \langle p | x \rangle \langle x | \psi \rangle \tag{2.404}$$

is an isomorphism of Hilbert spaces

$$L^2(\mathbb{R}^n, \mathrm{d}^n x) \to L^2(\mathbb{R}^n, \mathrm{d}^n p / (2\pi \hbar)^n). \tag{2.405}$$

The properties of the Fourier transform as a unitary automorphism of $L^2(\mathbb{R}^n)$ were discussed at the end of Sect. 2.4.2. In the new normalization the momentum-representation resolution of the identity reads

$$1 = \int_{\mathbb{R}^n} \frac{\mathrm{d}^n p}{(2\pi \hbar)^n} \, |p\rangle\langle p|. \tag{2.406}$$

The momentum eigenstates $|p\rangle$ are *not* normalizable (i.e. $|p\rangle \notin \mathcal{H}$) and hence are not physically realizable states. The physical states are superpositions of the form

$$|\Phi\rangle = \int \frac{\mathrm{d}^n p}{(2\pi \hbar)^n} \Phi(p) |p\rangle \tag{2.407}$$

where $\Phi(p) \in L^2(\mathbb{R}^n)$ are the *momentum-representation wave functions*. States of the form (2.407) where $\Phi(p)$ has support in a small neighborhood of a particular value of p are called *wave packets* with momentum $\approx p$. They are states with definite momentum up to some non-zero "error bar" Δp.

In the momentum representation the operators p_j act as multiplication operators and, dually, the q^i act as first order differential operators

$$\langle p | p_j | \psi \rangle = p_j \Phi(p), \qquad \langle p | q^k | \psi \rangle = i\hbar \, \frac{\partial \Phi(p)}{\partial p_k}. \tag{2.408}$$

In particular the group of unitary operators

$$\exp\left(-\frac{i}{\hbar} b_k \, q^k\right) \tag{2.409}$$

implement the translations $p_j \to p_j - b_j$ in momentum space, cf. Example 2.4.

The Schrödinger and the momentum representations are *unitary equivalent*, and give two descriptions of the same physics which are related by the (unitary) Fourier transform. One uses one or the other representation according to convenience for the particular application at hand.

2.12.4 The Weyl Algebra

Mathematicians prefer to work with bounded operators which are everywhere defined and continuous in $\mathcal{H}$. They replace the $2n$ densely-defined unbounded canonical operators q^j, p_k by the family of unitary operators

$$V(a) \stackrel{\text{def}}{=} \exp(ia^i q_i /\hbar), \qquad W(b) \stackrel{\text{def}}{=} \exp(ib^j p_j /\hbar) \tag{2.410}$$

where $a, b \in \mathbb{R}^n$. The operators $\{V(a), W(b)\}$ generate the *Heisenberg Lie group* $\mathsf{H}(n)$ which is the extension of the translation group in space generated by the $W(b)$'s by the translation group in momentum space generated by the $V(a)$'s. We can read the commutation relations in the group $\mathsf{H}(n)$ from the Baker-Campbell-Hausdorff formula (2.218); we get

$$V(a)\, W(b) = \exp(-ia \cdot b)\, W(b)\, V(a) \tag{2.411}$$

$$V(a)\, V(a') = V(a')\, V(a) \tag{2.412}$$

$$W(b')\, W(b) = W(b)\, W(b') \tag{2.413}$$

The completion in $\mathcal{B}(\mathcal{H})$ of the group algebra $\mathbb{C}[\mathsf{H}(n)]$ is called the *Weyl algebra*. Clearly the commutation relations (2.411)–(2.413) are equivalent to the canonical commutators (2.334) via the Stone theorem.

Theorem 2.9 (von Neumann) *The Weyl algebra $\overline{\mathbb{C}[\mathsf{H}(n)]}$ is semi-simple and all irreducible representations of the Weyl algebra are unitary equivalent to its Schrödinger representation.*

We essentially proved this **Theorem** in Sect. 2.11.1 using physicists' language.

2.13 Heisenberg Indetermination Principle

We already stated that we cannot determine simultaneously the exact position and velocity (momentum) of a quantum system moving in $\mathbb{R}^n$. We wish to quantify the *minimal possible error* in a simultaneous measure of position and momentum, and also to characterize the states where this error is minimal. To simplify the notation we first work out the case $n = 1$ and then state the obvious generalization to any n.

For any (normalized) state $|\psi\rangle$, let

$$p_0 = \langle \psi | p | \psi \rangle \in \mathbb{R}, \qquad q_0 = \langle \psi | q | \psi \rangle \in \mathbb{R} \tag{2.414}$$

be, respectively, the expectation value of momentum and position along the line. For all $t \in \mathbb{R}$ we have

$$
\begin{aligned}
0 \leq &\|(\boldsymbol{p} - p_0 + it(\boldsymbol{q} - q_0))|\psi\rangle\|^2 \equiv \\
\equiv &\langle\psi|(\boldsymbol{p} - p_0)^2|\psi\rangle + it\langle\psi|(\boldsymbol{p} - p_0)(\boldsymbol{q} - q_0)|\psi\rangle - \\
&- it\langle\psi|(\boldsymbol{q} - q_0)(\boldsymbol{p} - p_0)|\psi\rangle + t^2\langle\psi|(\boldsymbol{q} - q_0)^2|\psi\rangle = \\
= &\langle\psi|(\boldsymbol{p} - p_0)^2|\psi\rangle + t\,\hbar + t^2\langle\psi|(\boldsymbol{q} - q_0)^2|\psi\rangle.
\end{aligned}
\tag{2.415}
$$

Then the quadratic polynomial in the RHS should have a non-positive discriminant

$$
\hbar^2 - 4\langle\psi|(\boldsymbol{q} - q_0)^2|\psi\rangle\,\langle\psi|(\boldsymbol{p} - p_0)^2|\psi\rangle \leq 0
\tag{2.416}
$$

$\langle\psi|(\boldsymbol{q} - q_0)^2|\psi\rangle$ (resp. $\langle\psi|(\boldsymbol{p} - p_0)^2|\psi\rangle$) is the *variance* of the probability distribution of the position (resp. momentum) in the quantum state $|\psi\rangle$. Writing Δq (resp. Δp) for its square-root—the *standard deviation* of the position (resp. momentum) distribution

$$
\Delta q \overset{\text{def}}{=} \sqrt{\langle\psi|(\boldsymbol{q} - q_0)^2|\psi\rangle}, \qquad \Delta p \overset{\text{def}}{=} \sqrt{\langle\psi|(\boldsymbol{p} - p_0)^2|\psi\rangle},
\tag{2.417}
$$

Eq. (2.416) yields the exact form of the *Heisenberg indetermination principle*

$$
\Delta q \cdot \Delta p \geq \frac{\hbar}{2}
\tag{2.418}
$$

which is one of the fundamental features of Quantum Physics. Mathematically Eq. (2.418) is a well-known property of the Fourier transform: if a function is sharply picked, its Fourier transform has a broad support and viceversa. E.g. the Fourier transform of the δ-function $\delta(x)$, which has support in just one point (the origin), is the constant 1.

Remark 2.11 The argument may be generalized to any pair of observables A_s ($s = 1, 2$). The standard deviations of their measures in a (normalized) state $|\psi\rangle$ is

$$
\Delta A_s \equiv \sqrt{\langle\psi|A_s^2|\psi\rangle - (\langle\psi|A_s|\psi\rangle)^2} \quad s = 1, 2,
\tag{2.419}
$$

and we have the lower bound on the precision of the simultaneous measurements of A_1, A_2 in the state $|\psi\rangle$

$$
\Delta A_1 \cdot \Delta A_2 \geq \frac{1}{2}\left|\langle\psi|[A_1, A_2]|\psi\rangle\right|.
\tag{2.420}
$$

In particular

$$\Delta q^i \, \Delta p_j \geq \frac{\hbar}{2} \delta^i{}_j \qquad i, j = 1, \ldots, n. \tag{2.421}$$

Coherent States

We are interested in the states which saturate the inequality (2.418), i.e. the ones where positions and momenta can be measured simultaneously with the greatest possible precision. Such special states are called *coherent*.[26] The inequality is saturated when the quadratic polynomial (2.415) has a double root at some $t_0 \in \mathbb{R}$. Applying Descartes' rule of signs [14] to the last line of (2.415) we see that $t_0 \equiv -\lambda < 0$. So

$$\big(\boldsymbol{p} - p_0 - \mathrm{i}\lambda(\boldsymbol{q} - q_0)\big)|\psi\rangle = 0, \quad \lambda > 0. \tag{2.422}$$

In the Schrödinger representation this equation reads

$$\left(-\mathrm{i}\hbar\frac{\partial}{\partial x} - p_0 - \mathrm{i}\lambda x + \mathrm{i}\lambda q_0\right)\psi(x) = 0, \qquad \lambda > 0, \tag{2.423}$$

whose solution is

$$\psi(x)_{\mathrm{coh.}} = C\,\mathrm{e}^{\mathrm{i}p_0 x/\hbar}\,\exp\!\left(-\frac{\lambda}{2\hbar}(x - q_0)^2\right) =$$

$$= C'\exp\!\left(-\frac{\alpha^2}{2}x^2 + \beta x\right), \tag{2.424}$$

$$\text{where}\quad \alpha^2 = \frac{\lambda}{\hbar} > 0, \quad \beta = \frac{\lambda q_0 + \mathrm{i}p_0}{\hbar}$$

for some constants C, C'. Thus $\psi(x)_{\mathrm{coh.}}$ is a Gaussian centered at q_0 and "boosted" by the momentum p_0. Since the Fourier transform of a Gaussian is still a Gaussian, the momentum-space wave function of a coherent state has the same form

$$\Phi(p)_{\mathrm{coh.}} \equiv \int_{\mathbb{R}} \mathrm{d}x\,\mathrm{e}^{-\mathrm{i}px/\hbar}\psi(x)_{\mathrm{coh.}} = C''\,\mathrm{e}^{-\mathrm{i}q_0 p/\hbar}\exp\!\left(-\frac{\lambda^{-1}}{2\hbar}(p - p_0)^2\right) \tag{2.425}$$

$\Phi(p)_{\mathrm{coh.}}$ is a wave packet centered at momentum p_0.

---

[26] The reason of the name "coherent" is explained in Sect. 3.5.6.

Comparing with Eq. (2.30), we may identify the coherent state $\psi(x)_{\mathrm{coh.}}$ with a superposition of elements of the Hilbert basis $\{\psi_n\}_n$ introduced in Construction 5

$$\psi(x)_{\mathrm{coh.}} = e^{-|\beta/\alpha|^2/4} \sum_{n=0}^{\infty} \sqrt{\frac{1}{2^n n!}}\, \psi_n(\alpha x)\,(\beta/\alpha)^n, \tag{2.426}$$

a formula whose deep physical meaning will be explained in Sect. 3.5.6.

Phase-Space Volume vs. Number of States

Roughly speaking, the Heisenberg indetermination principle says that there is no physically doable experiment which may distinguish points inside a tiny region $R \subset \mathcal{W}$ of the phase space with Liouville volume

$$\mathrm{Vol}(R) = \prod_{i=1}^{n} \Delta x^i \, \Delta p_i = O(\hbar^n). \tag{2.427}$$

Since two physical states which cannot be distinguished—even *in principle*—by *any* experiment should be considered the *same* state, heuristically we conclude that *there is one quantum state per phase-space cell of volume* $\approx \hbar^n$. We want to make this qualitative statement precise, including the exact proportionality constant implicit in the RHS of Eq. (2.427).

To do this we introduce a function counting the number of quantum states with energy $\leq E$ (cf. Definition 1.1). As explained in chapter 2 of [12], this function has the physical interpretation of the *microcanonical partition function* $\Omega(E)$. The function $\Omega(E)$ is well behaved when the configuration space $\mathcal{M}$ has finite volume:[27] in this case, typically, the spectrum of the quantum Hamiltonian H is bounded below, discrete, with no accumulation points, and all its eigenspaces are finite-dimensional. Then $\Omega(E)$ is just the trace on the Hilbert space $\mathcal{H}$ of the spectral family

$$P_H(E) \equiv \Theta(E - H) \tag{2.428}$$

of the Hamiltonian H:

$$\Omega(E) = \mathrm{Tr}_{\mathcal{H}}\,\Theta(E - H) \equiv \left[\begin{array}{c} \textbf{number of quantum} \\ \textbf{states with energy } \leq E \end{array} \right] \in \mathbb{N} \tag{2.429}$$

Now we rewrite this function as an integral over the phase-space $\mathcal{W}$. Preliminarily we explain how one may rewrite the Hilbert space trace, $\mathrm{Tr}_{\mathcal{H}}(A)$, of any observable A as an integral over the phase space $\mathcal{W}$.

---

[27] Here we use the loose terminology common in the physical literature. Properly speaking, "finite volume" should be replaced by "finite diameter".

Symbols of Quantum Operators We associate to each operator A a function $\mathsf{A}(\mathbf{x}, \mathbf{p}) \equiv \mathsf{A}(x^i, p_j)$ on $\mathcal{W}$ by the rule

$$\mathsf{A}(\mathbf{x}, \mathbf{p}) \overset{\text{def}}{=} \frac{\langle \mathbf{p}|A|\mathbf{x}\rangle}{\langle \mathbf{p}|\mathbf{x}\rangle} \equiv \langle \mathbf{x}|\mathbf{p}\rangle\langle \mathbf{p}|A|\mathbf{x}\rangle, \tag{2.430}$$

where $\mathbf{x} = (x^1, \ldots, x^n)$, $\mathbf{p} = (p_1, \ldots, p_n)$ and $|\mathbf{x}\rangle \in \mathcal{S}^\vee$ (resp. $|\mathbf{p}\rangle \in \mathcal{S}^\vee$) are the position (resp. momentum) generalized eigenvectors normalized as in Eqs. (2.352) and (2.401). The function $\mathsf{A}(\mathbf{x}, \mathbf{p}) \in \Omega^0(\mathcal{W})$ is sometimes called the *symbol* of the quantum operator A. Note that when A is a canonical operator $\boldsymbol{q}^i$, $\boldsymbol{p}_j$ the symbol is just the classical Darboux coordinate

$$\mathsf{q}^i = x^i, \qquad \mathsf{p}_j = p_j. \tag{2.431}$$

Now we may express the Hilbert-space trace of an operator as the integral over $\mathcal{W}$ of its symbol

$$\begin{aligned}
\mathrm{Tr}_{\mathcal{H}}(A) &= \int_M \mathrm{d}^n x \, \langle \mathbf{x}|A|\mathbf{x}\rangle = \int_{\mathcal{W}} \frac{\mathrm{d}^n p \, \mathrm{d}^n x}{(2\pi\hbar)^n} \, \langle \mathbf{x}|\mathbf{p}\rangle\langle \mathbf{p}|A|\mathbf{x}\rangle = \\
&= \int_{\mathcal{W}} \frac{\mathrm{d}^n p \, \mathrm{d}^n x}{(2\pi\hbar)^n} \, \mathsf{A}(\mathsf{q}, \mathsf{p}) = \int_{\mathcal{W}} \frac{\mathrm{d}\mu}{(2\pi\hbar)^n} \, \mathsf{A}(\mathsf{q}, \mathsf{p}).
\end{aligned} \tag{2.432}$$

where $\mathrm{d}\mu$ is the *canonical Liouville volume* in phase space ([4] §. 6.8).

The Quantum Unit Cell We return to the counting of quantum states. Replacing in (2.432) A with the projector $\Theta(E - H)$, we get

$$\#\big(\text{states with energy } \leq E\big) = \int_{\mathcal{W}} \frac{\mathrm{d}^n p \, \mathrm{d}^n x}{(2\pi\hbar)^n} \, \frac{\langle \mathbf{p}|\Theta(E - H)|\mathbf{q}\rangle}{\langle \mathbf{p}|\mathbf{q}\rangle}. \tag{2.433}$$

We conclude

Fact 2.10 *The (asymptotic) number of quantum states with energy $\leq E$ is the Liouville volume of the region $R(E)$ where the energy is $\leq E$ divided by the volume of the "quantum unit cell"*

$$\mathsf{Vol}(\textit{unit cell}) = (2\pi\hbar)^n, \tag{2.434}$$

where n is the number of degrees of freedom. The region $R(E)$ is defined only in an approximate sense (which becomes exact as $\hbar \to 0$ or $E \to \infty$) as the locus where the integrand of (2.433) is "essentially" non-zero. However the formula (2.433) is always exact. We call

$$\frac{\mathrm{d}^n p \, \mathrm{d}^n q}{(2\pi\hbar)^n} \tag{2.435}$$

the normalized phase-space volume form.

From the above computation we learn that the classical limit as $\hbar \to 0$ of the Hilbert-space trace of a quantum operator $\mathcal{O}$ with a smooth classical limit $\mathcal{O}_{\text{class}}$ is

$$\text{Tr}_{\mathcal{H}} \, \mathcal{O} \rightsquigarrow \int_{\mathcal{W}} \mathcal{O}_{\text{class}} \, \frac{\mathrm{d}\mu}{(2\pi\hbar)^n}. \tag{2.436}$$

Using this formula we see that the quantum density matrix ϱ (cf. Sect. 2.10) has as classical limit a Liouville density ρ in phase-space, and the quantum Liouville equation reduces to the classical one. See also [12] chap. 2.

2.14 Boundary Conditions and All that

We now consider the modifications of the Schrödinger representation when the physical configuration space M has a boundary, $\partial M \neq \varnothing$.

Torus a.k.a. Periodic Boundary Condition
We consider first the situation where the configuration space $M = T^n \simeq (S^1)^n$ is a flat n-torus with circles of lengths $2\pi L_i$ ($i = 1, \ldots, n$). Equivalently, we may say that the Schrödinger wave functions $\psi(x^i)$ are *periodic* in the x^i's of periods $2\pi L_i$

$$\psi(x^1, x^2, \ldots, x^i + 2\pi L_i, x^{i+1}, \ldots, x^n) =$$
$$= \psi(x^1, x^2, \ldots, x^i, x^{i+1}, \ldots, x^n) \quad \forall \, i = 1, \ldots, n, \tag{2.437}$$

i.e. they satisfy the so-called *periodic boundary conditions*. More generally, we may consider the Hilbert space $L^2(T^n; \boldsymbol{\alpha})$ of wave functions which are *periodic up to prescribed phases*

$$\psi(x^1, x^2, \ldots, x^i + 2\pi L_i, x^{i+1}, \ldots, x^n) =$$
$$= e^{2\pi i \alpha_i} \, \psi(x^1, x^2, \ldots, x^i, x^{i+1}, \ldots, x^n) \quad \forall \, i = 1, \ldots, n \quad \alpha_i \in \mathbb{R}, \tag{2.438}$$

where the real numbers α_i are unique mod 1.[28] Mathematically: when $\alpha_i \notin \mathbb{Z}$ the wave-functions are *sections* of a complex line bundle $\mathcal{L} \to T^n$ which is flat but *non-trivial* since it carries a non-trivial monodromy representation $\boldsymbol{\alpha} \colon \pi_1(T^n) \to U(1)$.

---

[28] In the text we assume the α_i's to be *constants*. We may consider the more general situation where α_i is a function of the x^j's with $j \neq i$. In this set-up the wave functions are sections of a *non-flat* bundle. The discussion of this more general case is deferred to Chap. 5.

The unitary Weyl operators $V_i = \exp(\mathrm{i}q^i/L_i)$ act multiplicatively in the Schrödinger representation $L^2(T^n; \boldsymbol{\alpha})$ as

$$\psi(\mathbf{x}) \mapsto \exp(\mathrm{i}x^i/L_i)\psi(\mathbf{x}) \tag{2.439}$$

and form a *complete set* of commuting operators since they separate points. The Schrödinger representation of the momentum component $\boldsymbol{p}_i$ along the i-th circle is still given by the differential operator $-\mathrm{i}\hbar\frac{\partial}{\partial x^i}$

$$\langle \boldsymbol{x}|\boldsymbol{p}_j|\psi\rangle = -\mathrm{i}\hbar\frac{\partial}{\partial x^j}\langle \boldsymbol{x}|\psi\rangle = -\mathrm{i}\hbar\frac{\partial}{\partial x^j}\psi(\boldsymbol{x}). \tag{2.440}$$

We have a Hilbert basis given by

$$\psi_{\boldsymbol{n}}(\boldsymbol{x}) = \frac{1}{V^{1/2}}\exp\left(\mathrm{i}\sum_i \frac{(\alpha_i + n_i)x^i}{L_i}\right), \quad \boldsymbol{n} = (n_i) \in \mathbb{Z}^n \tag{2.441}$$

where $V = \mathsf{vol}(T^n) = \prod_i (2\pi L_i)$. Note that we can always set $\alpha_i = 0$ by the wave-function redefinition

$$|\psi\rangle \rightsquigarrow |\psi\rangle_{\mathrm{new}} \equiv \mathrm{e}^{-\mathrm{i}\sum_i \alpha_i q^i/L_i}|\psi\rangle, \tag{2.442}$$

$$A \rightsquigarrow A_{\mathrm{new}} \equiv \mathrm{e}^{-\mathrm{i}\sum_i \alpha_i q^i/L_i} A \, \mathrm{e}^{\mathrm{i}\sum_i \alpha_i q^i/L_i}, \tag{2.443}$$

at the price of making the observables to look slightly more complicated. In particular

$$p_i^{\mathrm{new}} = -\mathrm{i}\frac{\partial}{\partial x^i} + \hbar\frac{\alpha_i}{L_i} \qquad \substack{\textbf{when acting on the}\\ \textbf{periodic functions } \psi(x)_{\mathrm{new}}} \tag{2.444}$$

The momentum operator $\boldsymbol{p}_j$ generate the unitary group of translations along the j-circle of the n-torus $T^n \simeq (S^1)^n$.

Warning 4 In Sect. 2.12 a position-dependent change of phase represented a gauge transformation, i.e. a change of trivialization of the vector bundle where the wave functions take value. This is **not true** for the change of phase in Eq. (2.442). Indeed the wave-function redefinition (2.442) is **not** a unitary automorphism of the Hilbert space $\mathcal{H}$: only *periodic* **x**-dependent phase factors preserve the boundary conditions and hence are automorphisms of $\mathcal{H}$. Rather Eq. (2.442) is an isomorphism between two **distinct** Hilbert spaces with, respectively, non-trivial and trivial periodic boundary conditions. The genuine gauge transformations on a configuration space $\mathcal{M}$ have the form

$$\psi \rightsquigarrow \mathrm{e}^{\mathrm{i}f}\psi, \qquad A \rightsquigarrow A + \hbar\,\mathrm{d}f \tag{2.445}$$

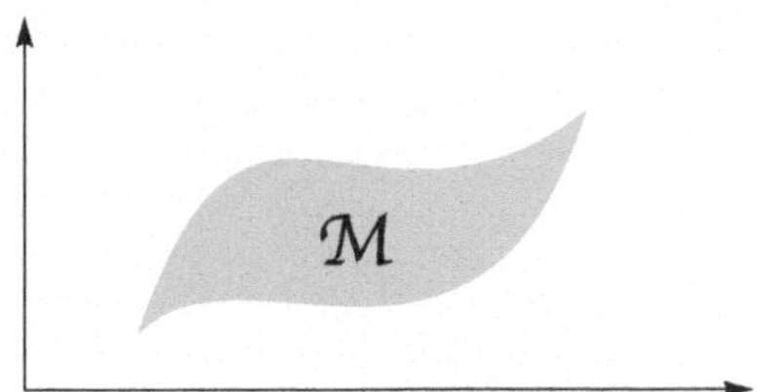

Fig. 2.1 The configuration space $\mathcal{M}$ is a domain in $\mathbb{R}^2$ (the grey region in the figure). The probability of finding the 'particle' outside $\mathcal{M}$ is zero

where $\exp(\mathrm{i}f)$ is a well-defined *global* function on $\mathcal{M}$. For $\mathcal{M} = T^n$ this requires f to be *periodic* up to 2π shifts. On the contrary the phase $\exp(-\mathrm{i}\sum_i \alpha_i x^i/L_i)$ is *multivalued* in T^n unless $\boldsymbol{\alpha} \in \mathbb{Z}^n$.

The expansion of the periodic wave-function $\psi(x)_{\text{new}}$ in the Hilbert basis $\psi_n(x; 0)$ is simply the Fourier series expansion

$$\psi(x)_{\text{new}} = \sum_{n\in\mathbb{Z}^n} a_n\, \psi_n(x;0) \equiv \frac{1}{V^{1/2}} \sum_{n\in\mathbb{Z}^n} a_n \exp\!\left(\mathrm{i}\sum_i n_i x^i/L_i\right), \qquad (2.446)$$

which, by the Parseval formula, is an isomorphism of Hilbert spaces $L^2(T^n) \simeq \ell^2(\mathbb{Z}^n)$, see Eq. (2.24).

Now the spectrum of each momentum operator[29] $\boldsymbol{p}_j$ is *purely discrete*

$$\sigma(\boldsymbol{p}_j) = \left\{ p_j \equiv \frac{n_j + \alpha_j}{L_j},\ n_j \in \mathbb{Z} \right\} \qquad (2.447)$$

as is typically the case when the configuration manifold $\mathcal{M}$ is *compact*. Let us deform continuously the exponent α_i from α_i to $\alpha_i + 1$: the final boundary conditions (2.438) coincide with the original ones, and the spectrum (2.447) returns to itself *as a set*, **but** the individual eigenvalues and the momentum eigenfunctions shift by one unit. This phenomenon is called *spectral flow* and has interesting consequences.

Boundaries

We now consider the case when $\mathcal{M} \subset \mathbb{R}^n$ is a proper sub-domain in Euclidean n-space with $\partial\mathcal{M} \neq \varnothing$. In this situation mathematicians make a lot of fuss, whereas physicists use a pragmatic approach which is both rigorous and intuitive.

We may see the system as a 'particle' moving in $\mathbb{R}^n$ which is prevented by a physical constraint to go in the complementary region $\mathbb{R}^n \setminus \mathcal{M}$, see Fig. 2.1. This means that the probability of finding the particle in $\mathbb{R}^n \setminus \mathcal{M}$ is *zero*, that is, *the wave-function is constrained to have support in* $\mathcal{M}$. By "physical continuity" the wave-functions of all nice, *finite-energy*, states should also vanish on the boundary

---

[29] Properly speaking $\boldsymbol{p}_j$ is an *angular* momentum since the dual coordinates x^i/L_i are periodic *angles*.

∂M, that is, the *nice state* wave-functions satisfy a *Dirichlet boundary condition*

$$\psi(x)\big|_{x \in \partial M} = 0. \tag{2.448}$$

In particular, the energy eigenfunctions $\langle x|E \rangle$ vanish on the boundary ∂M. We then study the action of the Hamiltonian H in a dense subspace $S \subset \mathcal{H}$ of smooth functions satisfying the Dirichlet boundary conditions.

The prototypical instance of this situation is a particle moving in an interval $[0, L] \subset \mathbb{R}$. As discussed in Construction 2, we use the Hilbert basis in $\mathcal{H}$

$$\psi_n(x) \equiv \frac{1}{\sqrt{2L}} \sin\left(\frac{\pi n x}{L}\right) \quad n = 1, 2, \dots \tag{2.449}$$

and expand the wave-functions in a Fourier series of sines

$$\psi(x) = \frac{1}{\sqrt{2L}} \sum_{n \in \mathbb{Z}} a_n \sin\left(\frac{\pi n x}{L}\right). \tag{2.450}$$

We use the dense Schwartz subspace S of Fourier series with

$$\left\{ a_n \in \ell^2(\mathbb{Z}), \ \sup_n |n^k a_n| < \infty, \ \forall \, k \in \mathbb{N} \right\} \tag{2.451}$$

as the first space in our spectral triple (2.110).

Remark 2.12 Note that in the present situation we cannot define eigenvectors of the momentum operator p, not even in the generalized sense. The point is that the momentum operators generate *space translations,* and hence their eigenstates are invariant under translation (up to phase). In the interval $[0, L]$ the translation symmetry is *explicitly broken* by the Dirichlet boundary conditions at the endpoints, and no translation-invariant state may exist. However there are eigenstates of p^2, in facts a complete system of them: they are precisely the elements of our Hilbert basis (2.449).

2.15 Schrödinger Representations on General Manifolds

We briefly comment on the case where the configuration space is a general connected Riemannian manifold $(\mathcal{M}, g_{ij})$. In the Schrödinger representation the Hilbert space is the space of L^2-functions $\psi : \mathcal{M} \to \mathbb{C}$ with

$$\|\psi\|^2 = \int_{\mathcal{M}} \sqrt{g} \, \mathrm{d}^n x \, |\psi(x)|^2 < \infty \tag{2.452}$$

or, more generally, the Hilbert space $L^2(\mathcal{M}, \mathcal{V})$ of *sections* of a complex vector bundle $\mathcal{V} \to \mathcal{M}$, endowed with a positive-definite Hermitian metric h (i.e. $\mathcal{V}$ is a *Hermitian vector bundle*). $L^2(\mathcal{M}, \mathcal{V})$ is equipped with the natural Hilbert Hermitian product

$$\langle \psi_1 | \psi_2 \rangle = \int_{\mathcal{M}} \sqrt{g}\, \mathrm{d}^n x\, \psi_1(x)^\dagger h\, \psi_2(x), \tag{2.453}$$

where, as always, $g = \det g_{ij}$. As before, the elements of $L^2(\mathcal{M})$ are equivalence classes of square-integrable functions modulo zero-norm ones. The global functions $f \in \Omega^0(\mathcal{M})$ may be interpreted as quantum operators acting by multiplication in the Schrödinger representation

$$f : \psi(x) \mapsto f(x)\, \psi(x). \tag{2.454}$$

A continuous function f corresponds to a bounded operator iff it is bounded as a function in $\mathcal{M}$.

The Free Hamiltonian

In Classical Mechanics the Darboux coordinates (q^i, p_j) are (in general) only defined *locally* in the phase-space $\mathcal{W}$; on the other hand, consistency of the time evolution requires the classical Hamiltonian H_{cl} to be a globally defined function on $\mathcal{W}$.

This observation extends to the quantum set-up. When the configuration space $\mathcal{M}$ is a general Riemannian manifold, there may not exist globally defined canonical operators corresponding to the local coordinates (q^i, p_j), but the *quantum Hamiltonian H must be a globally defined operator for all configuration spaces $\mathcal{M}$*. This follows from the fact that the time-evolution is a group of unitary transformations of $\mathcal{H}$ and the Stone theorem.

The classical free[30] Hamiltonian for a 'particle' of mass m moving in the Riemannian manifold $(\mathcal{M}, g_{ij})$ is

$$H_{\mathrm{cl}} = \frac{1}{2m} g^{ij} p_i\, p_j \tag{2.455}$$

where, as always, g^{ij} is the inverse of the metric g_{ij}. We look for the quantum operator H, acting densely on $L^2(\mathcal{M})$, which corresponds to H_{cl}. Our guiding principles are:

---

[30] In the present context *free* means *in absence of potentials*: The only forces acting on the particle are the constraint forces implied by the non-trivial geometry of the configuration space $\mathcal{M}$.

H1 H should be a densely-defined Hermitian operator acting on $L^2(\mathcal{M})$;

H2 when expressed in local coordinates x^i, H should be a second-order differential operator acting on functions whose *principal symbol* is the classical Hamiltonian (2.455);

H3 H should be a scalar operator invariant under all diffeomorphisms of $\mathcal{M}$;

H4 under a rescaling of the metric $g_{ij} \rightsquigarrow \lambda\, g_{ij}$, $\lambda \in \mathbb{R}_{>0}$

$$H \rightsquigarrow \lambda^{-1} H \tag{2.456}$$

H5 H should be proportional to $\hbar^2$.

There is a one-parameter family of operators satisfying these requirements [15]:

$$H = \frac{\hbar^2}{2m}\left(\Delta + \xi R\right) \tag{2.457}$$

where R is the scalar curvature of the metric g_{ij}, $\xi \in \mathbb{R}$ is a parameter to be determined, and Δ is the *Laplacian* operator of $(\mathcal{M}, g_{ij})$ (acting on functions)

$$\Delta\psi \stackrel{\text{def}}{=} -\frac{1}{\sqrt{g}}\frac{\partial}{\partial x^i}\left(\sqrt{g}\, g^{ij}\frac{\partial\psi}{\partial x^j}\right). \tag{2.458}$$

We see from Eq. (2.457) that to a classical system there correspond several quantum models with Hamiltonians differing by terms which vanish as $\hbar \to 0$. In the free case the family of quantum systems with the same classical limit is parametrized by the scalar curvature coupling ξ.

Adding a potential bounded below, $V\colon \mathcal{M} \to \mathbb{R}_{\geq \Lambda}$, we get the natural Hamiltonian on $\mathcal{M}$ (in absence of magnetic fields and spins):

$$H = \frac{\hbar^2}{2m}\left(\Delta + \xi R\right) + V. \tag{2.459}$$

The General Hamiltonian

In the general case—when we also have magnetic field and/or spins—the Hilbert space $\mathcal{H}$ is the space $L^2(\mathcal{M}, \mathcal{V})$ of *square-summable sections* of a *Hermitian vector bundle* $\mathcal{V} \to \mathcal{M}$ equipped with a unitary connection A [16–18]. See Construction 10. When $A = 0$ we replace Δ with the natural Laplacian acting on the sections of $\mathcal{V}$

$$\Delta_h\psi = -\frac{1}{\sqrt{g}}h^{-1}\frac{\partial}{\partial x^i}\left(\sqrt{g}\, g^{ij}\frac{\partial(h\psi)}{\partial x^j}\right), \tag{2.460}$$

and, more generally, by the natural *covariant* Laplacian Δ_A defined by the connection A. Now we may have additional quantization "ambiguities" in H, i.e.

terms vanishing in the classical limit which parametrize the family of inequivalent quantum systems arising from quantization of the same classical model:

$$H = \frac{\hbar^2}{2m}\Big(\Delta_A + \xi R + \text{terms linear in the curvature of } A\Big) + V. \qquad (2.461)$$

See Remark 5.5 for a fundamental example of these extra curvature couplings. These "extra" terms are particularly crucial in presence of spins (e.g. for the Dirac equation in an electromagnetic field). In Eq. (2.461) V is a global function on $\mathcal{M}$ (the *potential*).

Momentum Operators

The classical Hamiltonian H_{cl} and Noether charges M^a_{cl} are classically *global functions* on the phase space $\mathcal{W}$, and hence are expected to correspond to well-defined operators in the analogue quantum system with configuration space $\mathcal{M}$—in agreement with the general "abstract" theory of the Schrödinger equation in Sect. 2.8, which predicts the existence of a global generator H for the time evolution as well as for all one-parameter group of symmetries.[31] On the contrary, the classical momenta p^{cl}_i are, in general, mere local coordinates on $\mathcal{W}$ with non-trivial transformations in the overlaps between coordinates patches. At the quantum level these changes of local coordinates in $\mathcal{W}$ are troublesome since they require a prescription on the order of the non-commuting operators q^i, p_j. Although this order problem may be resolved in principle (see Sect. 2.17), in the general case p_j remains a non-intrinsically defined operator. When $\mathcal{M} = \mathbb{R}^n$ the momentum p_j is the generator of translations, and hence globally well-defined, but in absence of a translational symmetry it cannot be defined globally. Indeed, were the p_j's well defined, by the Stone theorem they would produce a group of unitary transformations which—while not necessarily symmetries of the dynamics—would be geometric automorphisms of $L^2(\mathcal{M})$, and this can be possible only if $\mathcal{M}$ satisfies very restrictive geometric conditions. For an explicit counterexample to the existence of p_j, see Remark 2.12.

In order to define meaningful global momentum operators p_j, we therefore must assume that the Riemannian manifold $\mathcal{M}$—while not necessarily *flat*—is *globally diffeomorphic* to $\mathbb{R}^n$. In particular we cover it by a single coordinate chart. The metric

$$\mathrm{d}s^2 = g_{ij}(x)\,\mathrm{d}x^i\mathrm{d}x^j \qquad (2.462)$$

may be, however, arbitrary. We have

$$\int \mathrm{d}^n x \, \sqrt{g}\, \psi_1^* \frac{\partial}{\partial x^i} \psi_2(x) = -\int \mathrm{d}^n x \, \frac{\partial}{\partial x_i}\big(\sqrt{g}\, \psi^*(x)\big)\psi_2(x) =$$

$$= -\int \mathrm{d}^n x \, \sqrt{g}\left(\frac{1}{\sqrt{g}}\frac{\partial}{\partial x_i}(\sqrt{g}\, \psi^*(x))\right)\psi_2(x), \qquad (2.463)$$

---

[31] For more precise statements see the first sections of Chap. 4.

that is,

$$\left(\frac{\partial}{\partial x^i}\right)^\dagger = -\frac{1}{\sqrt{g}}\frac{\partial}{\partial x^i}\sqrt{g} \tag{2.464}$$

This formula implies the

Lemma 2.9 *Suppose* $M \simeq \mathbb{R}^n$. *Let* $L^2(M)$ *be the Hilbert space of measurable functions* $\psi : M \to \mathbb{C}$ *satisfying Eq.* (2.452) *(modulo functions vanishing almost everywhere). The densely-defined operator*

$$p_i = -i\hbar\, g^{-1/4}\frac{\partial}{\partial x^i}g^{1/4} \tag{2.465}$$

is (essentially) self-adjoint.

Proof

$$p_i^\dagger = -i\hbar\, g^{-1/4}\left(\frac{\partial}{\partial x^i}\right)^\dagger g^{-1/4} = -i\hbar\, g^{-1/4}\frac{\partial}{\partial x^i}g^{1/4} = p_i. \tag{2.466}$$

$\square$

Up to gauge transformations, Eq. (2.465) is the most reasonable definition of *momentum operator* for a Riemannian manifold diffeomorphic to $\mathbb{R}^m$. It reduces to the usual definition when the metric is flat.

2.16   The Quantum Noether Theorem

In Classical Mechanics [4] to a one-parameter group of symmetries of the classical Hamiltonian H_{cl} there corresponds a conserved quantity $M_{cl} \equiv M(q, p)_{cl}$, that is, a function on the phase space $\mathcal{W}$ which Poisson-commutes with H_{cl}:

$$\left[M_{cl}, H_{cl}\right]_{PB} = 0. \tag{2.467}$$

When $M(q, p)_{cl}$ is a polynomial in the momenta p_i of degree one, we say that the conserved quantity M_{cl} is a *Noether charge*. A Noether charge generates a symmetry of the Lagrangian modulo total time-derivatives ([4] chap. 3), and has the general form

$$M_{cl} = m^i(q)\, p_i + a(q), \tag{2.468}$$

where $m^i(q)\partial_{q^i}$ is a Killing vector for the Jacobi metric [4]. The zero-order term $a(q)$ is often absent, e.g. when the group of symmetries is semisimple.

We wish to prove the corresponding theorem in Quantum Mechanics. For simplicity we assume the Hamiltonian to have the form (2.459), but from the proof it will be clear that the result holds in wider generality.

For any given ξ the free quantum Hamiltonian (2.457) is invariant under all isometries of the Riemannian manifold $(\mathcal{M}, g_{ij})$. A one-parameter group of isometries of $\mathcal{M}$ is generated by a global *Killing vector field* (see e.g. [4] chap. 2):

$$m \equiv m(x)^i \frac{\partial}{\partial x^i} \in \mathrm{iso}(\mathcal{M}). \tag{2.469}$$

The one-parameter group of isometries leaves invariant the interacting Hamiltonian (2.459) iff, in addition, the Killing vector m kills the potential

$$\mathscr{L}_m V = 0 \tag{2.470}$$

where $\mathscr{L}_m$ is the *Lie derivative* along the vector field m ([4] chap. 2).

Quantum Noether Theorem *Consider a quantum system with configuration space the Riemannian manifold $(\mathcal{M}, g_{ij})$ whose Hilbert space of states has the Schrödinger representation $L^2(\mathcal{M})$. Assume the Hamiltonian to have the physically natural form*

$$H = \frac{\hbar^2}{2m}\Delta + \xi \hbar^2 R + V(x). \tag{2.471}$$

(1) *To all Killing vector m which preserves the potential, Eq. (2.470), there corresponds a densely-defined, self-adjoint, conserved, Noether charge*

$$M = -i\hbar\, m^i(x)\frac{\partial}{\partial x^i} \tag{2.472}$$

which commutes with the Hamiltonian, $[M, H] = 0$, and such that the corresponding one-parameter group of symmetries of the quantum system is implemented on the Hilbert space $L^2(\mathcal{M})$ by the group of unitary operators

$$U_M(s) \equiv \exp(i s\, M/\hbar) \quad s \in \mathbb{R} \tag{2.473}$$

in accordance with the Stone theorem.

(2) *Let G be the Lie group[32] of isometries of $\mathcal{M}$ which leaves invariant the potential V, $\mathfrak{g}$ its Lie algebra, and $\{m_a\}$ a basis of $\mathfrak{g}$ with Lie bracket*

$$[m_a, m_b] = f_{ab}{}^c m_c \tag{2.474}$$

---

[32] For any Riemannian n-manifold $\mathcal{M}$, the group of isometries $\mathsf{Iso}(\mathcal{M})$ is a Lie group of dimension $\leq n(n+1)/2$, see [19]. The condition of preserving the potential selects a *closed* subgroup $G \subset \mathsf{Iso}(\mathcal{M})$ which then is also a Lie group.

for certain (real) structure constants $f_{ab}{}^c = -f_{ba}{}^c$. Let $M_a \equiv -i\hbar m_a$ be the corresponding conserved observables. They satisfy the quantum commutation relations

$$[M_a, M_b] = -i\hbar f_{ab}{}^c M_c. \tag{2.475}$$

(3) *All Noether charges which generate a **semisimple** group of symmetries of the quantum system arise in this way.*[33]

Formally the quantum conserved quantity M is obtained from its classical counterpart M_{cl} in Eq. (2.468) (with $a(q) = 0$) by the replacement

$$p_i \rightsquigarrow -i\hbar \frac{\partial}{\partial x^i}. \tag{2.476}$$

However notice that p_i is not a globally defined operator in general, while M is always globally well defined.

Example 2.14 When $M = \mathbb{R}^n$ and $m = \partial_{x^i}$ is the infinitesimal translation in the i-th direction, M is the canonical momentum operator p_i in that direction which does generate the translations $x^i \rightsquigarrow x^i + a$.

To prove the **Theorem** one has to show that M is a (densely defined) self-adjoint operator in $L^2(M)$ and that it commutes with H. The second property follows from the geometric equalities

$$\mathcal{L}_m \Delta = \Delta \mathcal{L}_m, \qquad \mathcal{L}_m R = \mathcal{L}_m V = 0, \tag{2.477}$$

valid for all Killing vector m which kills V. Now we prove self-adjointness. Let ψ_1, ψ_2 be smooth functions with compact support in the interior of M:

$$
\begin{aligned}
\int_M \sqrt{g}\, \mathrm{d}^n x\, \psi_1^* M \psi_2 &= \\
&= -i\hbar \int_M \sqrt{g}\, \mathrm{d}^n x\, \psi_1^* m^i \frac{\partial}{\partial x^i} \psi_2 = i\hbar \int_M \mathrm{d}^n x\, \frac{\partial}{\partial x^i}\left(\sqrt{g}\, m^i \psi_1^*\right) \psi_2 = \\
&= \int_M \sqrt{g}\, \mathrm{d}^n x\left(-i\hbar m^i \frac{\partial}{\partial x^i}\psi_1\right)^* \psi_2 + i\hbar \int_M \sqrt{g}\, \mathrm{d}^n x (\nabla_i m^i)\psi_1^* \psi_2 = \\
&= \int_M \sqrt{g}\, \mathrm{d}^n x\, (M\psi_1)^* \psi_2 + i\hbar \int_M \sqrt{g}\, \mathrm{d}^n x\, (\nabla_i m^i)\psi_1^* \psi_2,
\end{aligned}
\tag{2.478}
$$

where in the second equality we integrated by parts and used the fact that there is no surface term. From (2.478) we see that M is self-adjoint iff $\nabla_i m^i = 0$ (that is,

---

[33] A typical Noether charge which is not of the form (2.472) is the generator of a Galilei boost, see Sect. 4.11. The Galilei group is *not* semisimple.

iff the vector field $m^i \partial_i$ preserves the volume [4]). This holds automatically: m^i is a Killing vector which satisfies the equation ([4] chap. 2)

$$\nabla_i m_j + \nabla_j m_i = 0 \quad \Rightarrow \quad \nabla_i m^i = 0. \tag{2.479}$$

We may state the result in a different form:

Corollary 2.10 *Suppose that a classical system with time-independent Hamiltonian*

$$H_{\mathrm{cl}} = \frac{1}{2m} g^{ij} p_i p_j + V(q) \tag{2.480}$$

has a set of Noether charges M_a^{cl} $(a = 1, \ldots, s)$, i.e. conserved quantities which are linear in the momenta p_i,

$$M_a^{\mathrm{cl}} \equiv m_a^i(q) p_i. \tag{2.481}$$

Then: the corresponding quantum system, whose Hamiltonian operator has the Schrödinger representation

$$H = \frac{\hbar^2}{2m} \left(\Delta + \xi R \right) + V(q), \tag{2.482}$$

contains the quantum Noether charges

$$M_a = -i\hbar \, m_a^i(q) \frac{\partial}{\partial q^i}, \tag{2.483}$$

which are Hermitian and conserved *$[H, M_a] = 0$. The Lie algebra of the quantum Noether charges*

$$[M_a, M_b] = -i\hbar \, f_{ab}{}^c M_c \tag{2.484}$$

is isomorphic to the classical Poisson-Lie algebra i.e. the structure constants $f_{ab}{}^c$ are the same ones in the Poisson brackets of the classical Noether charges

$$[M_a^{\mathrm{cl}}, M_b^{\mathrm{cl}}]_{\mathrm{PB}} = -f_{ab}{}^c M_c^{\mathrm{cl}}. \tag{2.485}$$

Note that Eq. (2.484) is consistent with the general statement that, to the first non-trivial order in $\hbar$, the quantum commutator is the classical Poisson bracket times $i\hbar$: here we have the stronger assertion that the commutator of Noether charges has no higher order correction in $\hbar$.

Remark 2.13 The theorem holds also in presence of magnetic fields (and spins) with the Laplacian Δ replaced by the appropriate covariant second-order elliptic differential operator Δ_A acting on sections of the vector bundle $\mathcal{V}$ where the Schrödinger wave-functions take value. The crux of the matter is that if the geometric operator Δ_A is preserved by a first order differential operator m (typically a covariantized version of a Killing vector) then $-\,\mathrm{i}\hbar m$ is a Noether charge.

Remark 2.14 The situation with classical conserved quantities which are of higher order in momenta (associated to Killing *tensors* [4] §§. 6.6, 6.7) is subtler because of operator-order issues. However usually one can find a conserved quantum operator which extends the classical one. Explicit examples will be provided in Chap. 5.

2.17 The Quantum Action Principle

In classical mechanics the evolution of the states from time t_1 to time t_2 is given by an automorphism of the classical space of states i.e. by a symplectomorphism (a.k.a. canonical transformation) generated by a function $S(q_2^i, t_2; q_1^j, t_1)$ of the initial and final coordinates, q_1^i and q_2^i, which satisfies the *two* Hamilton-Jacobi equations

$$\frac{\partial S}{\partial t_2} + H\left(q_2^i, \frac{\partial S}{\partial q_2^i}\right) = 0, \qquad \frac{\partial S}{\partial t_1} - H\left(q_1^i, -\frac{\partial S}{\partial q_1^i}\right) = 0, \tag{2.486}$$

whose local solution is the *Hamilton principal function* given by

$$S(q_2^i, t_2; q_1^j, t_1) = \int_{t_1}^{t_2} L(q^i, \dot{q}^j, t)dt \tag{2.487}$$

where the integral is evaluated on the solution of the equations of motion satisfying the boundary conditions $q^i(t_1) = q_1^i$ and $q^i(t_2) = q_2^i$ see [4].

In quantum mechanics the space of states is (the projectivation of) a Hilbert space $\mathcal{H}$ and the evolution of the system from time t_1 to time t_2 is given by an automorphism of the *quantum* space of states, namely a *unitary* transformation $U(t_2, t_1)$. For a quantum system with a classical limit, we wish to understand the relation of $U(t_2, t_1)$ with the classical generating function S and Lagrangian L. When the Hamiltonian is time-independent

$$U(t_2, t_1) = \exp(-\mathrm{i}H(t_2 - t_1)/\hbar) \tag{2.488}$$

but we wish to be more general. We introduce the *integral kernel* of the time evolution

$$U(q_2^i, t_2; q_1^j, t_1) = \langle q_2^i | U(t_2, t_1) | q_1^j \rangle \tag{2.489}$$

where $\{|q^i\rangle\}$ is a complete system of generalized eigenvectors of the quantum operators q^i. The time evolution of the wave function is then

$$\psi(q_2^i; t_2) = \int d^n q_1 \, U(q_2^i, t_2; q_1^j, t_1) \, \psi(q_1^j). \tag{2.490}$$

Going to the Heisenberg picture, we may see the above kernel as the overlap ($\equiv$ inner product)

$$\langle q_2^i; t_2 | q_1^j; t_1 \rangle \tag{2.491}$$

where $|q_a^i; t_a\rangle$ is the (generalized) eigenvector of the time-dependent Heisenberg operator $q^i(t_a)$ at time t_a with eigenvalue q_a^i, that is,

$$q^i(t_a)|q_a^i; t_a\rangle = q_a^i|q_a^i; t_a\rangle, \quad a = 1, 2. \tag{2.492}$$

We introduce the *time-ordered functions* $\mathbf{T}f(q^i(t_1), q^j(t_2))$ of the Heisenberg coordinate operators $q^i(t)$ at the initial and final times. As before, *time-ordered* means that the later time operators $q^j(t_2)$ are written at the left of the early time operators $q^j(t_2)$. Explicitly, for any continuous function $f(q_1^i, q_2^j)$ of the $2n$ real variables (q_1^i, q_2^j) $(i, j = 1, \ldots, n)$, we have

$$\mathbf{T}f(q^i(t_1), q^j(t_2)) \stackrel{\text{def}}{=} \int d^n q_1 \, d^n q_2 \, f(q_1^i, q_2^j) \, P_2(q_2) \, P_1(q_1) \tag{2.493}$$

where $P_a(q_a)$ $(a = 1, 2)$ is the spectral family of the complete system of commuting (Heisenberg) observables $q_a^i \equiv q^i(t_a)$ $(i = 1, \ldots, n, a = 1, 2)$. By construction we have the identity

$$\langle q_2^i; t_2|\mathbf{T}f(q^k(t_1), q^l(t_2))|q_1^j; t_1\rangle = f(q_1^k, q_2^l)\langle q_2^i; t_2|q_1^j; t_1\rangle. \tag{2.494}$$

We define the (complex) function $\mathsf{S}(q_2^i, t_2; q_1^j, t_1)$ by the equation

$$\exp\big(i\mathsf{S}(q_2^i, t_2; q_1^j, t_1)/\hbar\big) = U(q_2^i, t_2; q_1^j, t_1) \equiv \langle q_2^i; t_2|q_1^j; t_1\rangle. \tag{2.495}$$

We claim that S is the quantum analogue of the classical action in the sense that we have the operator equations

$$p_i(t_2) = \mathbf{T}\frac{\partial \mathsf{S}(q_2, q_1)}{\partial q_2^i} \qquad\qquad H(t_2) = -\mathbf{T}\frac{\partial \mathsf{S}(q_2, q_1)}{\partial t_2} \tag{2.496}$$

$$p_j(t_1) = -\mathbf{T}\frac{\partial \mathsf{S}(q_2, q_1)}{\partial q_1^i} \qquad\qquad H(t_1) = \mathbf{T}\frac{\partial \mathsf{S}(q_2, q_1)}{\partial t_1}, \tag{2.497}$$

that is, the Heisenberg time-evolution of the operators, seen as a unitary transformation, is given by the same formulae as the classical canonical transformation with generating function of the first kind S ([4] chap. 8) *except* for the time-ordering prescription for the Heisenberg operators at different times (the $\mathbf{T}$ in front of the RHS of the above equations). The time-order prescription resolves the inherent ambiguity with the order of non-commuting operators when we look for a quantum analogue of a classical formula. We conclude that S is the quantum analogue of Hamilton's principal function S and the two coincide in the classical limit $\hbar \to 0$. These elegant observations by Dirac [9] were historically the starting point of the path integral approach to Quantum Mechanics to be developed in Chap. 6.

We prove the first equation in (2.496) the proofs of the other ones being similar. We have

$$\langle q_2^i; t_2 | \boldsymbol{p}_k(t_2) | q_1^j; t_1 \rangle = -i\hbar \frac{\partial}{\partial q_2^k} \langle q_2^i; t_2 | q_1^j; t_1 \rangle = \frac{\partial \mathsf{S}}{\partial q_2^k} \langle q_2^i; t_2 | q_1^j; t_1 \rangle \qquad (2.498)$$

Using the identity (2.494), the RHS may be written as

$$\langle q_2^i; t_2 | T \frac{\partial \mathsf{S}}{\partial \boldsymbol{q}_2^k} | q_1^j; t_1 \rangle. \qquad (2.499)$$

Since $|q_1^j; t_1\rangle$ and $\langle q_2^i; t_2|$ are complete systems of (generalized) eigenvectors, this implies the first (2.496).

Problems

2.1 Write the Schrödinger representation kernel $\langle x^i | U | y^j \rangle$ of a general unitary transformation $U \in Sp(2n, \mathbb{R})$ acting linearly on the canonical operators.

References

1. G. Botelho, D. Pellegrino, E. Teixeira, *Introduction to Functional Analysis*. Universitext (Springer, Berlin, 2025)
2. L. Grafakos, *Fundamentals of Fourier Analysis*. Graduate Texts in Mathematics, vol. 302 (Springer, Berlin, 2024)
3. J. Jost, *Riemannian Geometry and Geometric Analysis*, 7th edn. Universitext (Springer, Berlin, 2016)
4. S. Cecotti, *Analytic Mechanics. A Concise Textbook* (Springer, Berlin, 2024)
5. S. Bergman, *The Kernel Function and Conformal Mapping*, 2nd edn. (AMS, Providence, 1970)
6. S. Krantz, *Geometric Analysis of the Bergman Kernel*. Graduate Texts in Mathematics, vol. 268 (Springer, Berlin, 2013)

7. R. Beals, R.S.C. Wong, *More Explorations in Complex Functions*. Graduate Texts in Mathematics, vol. 298 (Springer, Berlin, 2023)

8. A. Teta, *A Mathematical Primer on Quantum Mechanics* (Springer, Berlin, 2018)

9. P.A.M. Dirac, *Principles of Quantum Mechanics*, 4th edn. (Clarendon, Oxford, 1958)

10. V. Moretti, *Spectral Theory and Quantum Mechanics. Mathematical Foundations of Quantum Theories, Symmetries and Introduction to the Algebraic Formulation*, 2nd edn. (Springer, Berlin, 2017)

11. M. Reed, B. Simon, *Methods of Modern Mathematical Physics*, vol. I–IV (Academic Press, Cambridge, 1970)

12. S. Cecotti, *Statistical Mechanics. A Concise Advanced Textbook* (Springer, Berlin, 2024)

13. J. Glimm, A. Jaffe, *Quantum Physics. A Functional Integral Point of View* (Springer, Berlin, 1981)

14. M. Bensimhoun, Historical account and ultra-simple proofs of Descartes's rule of signs, De Gua, Fourier, and Budan's rule. arXiv:1309.6664

15. B.S. De Witt, Dynamical theory in curved spaces. I. A review of the Classical and Quantum action principles. Rev. Mod. Phys. **29**, 377 (1957)

16. S. Kobayashi, *Differential Geometry of Complex Vector Bundles* (Princeton University Press, Princeton, 1987)

17. P. Griffiths, J. Harris, *Principles of Algebraic Geometry* (Wiley, Hoboken, 1978)

18. D. Huybrechts, *Complex Geometry. An Introduction*. Universitext (Springer, Berlin, 2005)

19. S. Kobayashi, *Transformations Groups in Differential Geometry*. Classics in Mathematics (Springer, Berlin, 1995)

Chapter 3
Schrödinger Equation I

In this chapter we study more in detail the Schrödinger equation for non-relativistic quantum mechanical systems. The topic will be further discussed in Chap. 5 with the help of the quantum theory of symmetry.

3.1 General Properties of the Schrödinger Equation

We adopt the Schrödinger picture and study the Schrödinger equation of motion

$$i\hbar \frac{\mathrm{d}}{\mathrm{d}t}|\Psi\rangle = H|\Psi\rangle. \tag{3.1}$$

We focus on quantum mechanical systems with Hilbert space $L^2(\mathcal{M})$ which in the limit $\hbar \to 0$ reduce to a classical system with a Hamiltonian quadratic in momenta

$$H_{\mathrm{cl}} = H(p_i, q^j, t) = \frac{1}{2m} g^{ij}(p_i - eA_i)(p_j - eA_j) + V, \tag{3.2}$$

where the q^i's are coordinates on the Riemannian manifold $(\mathcal{M}, g_{ij})$. A_i and V are, respectively, a one-form and a function on $\mathcal{M}$ which may have an explicit dependence on the time t. We loosely refer to such a quantum system as a *spinless* particle moving in $\mathcal{M}$ in presence of the magnetic field A_i and potential V. For the moment "spinless" is just a code-word for the statement that there are no other quantum degrees of freedom besides the ones already visible in the classical limit (namely the q^i's), so that the Hilbert space $\mathcal{H} \equiv L^2(M)$ carries an *irreducible* representation of the algebra of differential operators acting on $\Omega^0(\mathcal{M})$ in the sense that all time-independent quantum observable A which commutes with the (local) coordinate operators q^i is a function of the coordinates $A(q^j)$. We shall describe more general quantum systems *with spin* in Chap. 4.

© The Author(s), under exclusive license to Springer Nature Switzerland AG 2025   115
S. Cecotti, *Quantum Mechanics*, UNITEXT for Physics,
https://doi.org/10.1007/978-3-031-98824-0_3

In this chapter the magnetic potential is set to zero, $A_i = 0$, except in the Sects. 3.10.1 and 3.11 where we discuss the Bohm-Aharonov effect. We shall switch on the magnetic field in the final sections of Chap. 5.

We work mostly (but not exclusively) in the Schrödinger representation. We write x^i for the eigenvalues of q^i. We rewrite Eq. (3.1) as a partial differential equation (PDE) for the wave-function $\psi(x^i) \equiv \langle x^i | \psi \rangle \in L^2(\mathcal{M})$

$$i\hbar \frac{\partial}{\partial t}\psi = H\psi = \frac{\hbar^2}{2m}\Delta\psi + \hbar^2 \xi R\psi + V(x,t)\psi, \tag{3.3}$$

where Δ is the natural Laplacian operator[1] on $\mathcal{M}$

$$\Delta f \stackrel{\mathrm{def}}{=} -\frac{1}{\sqrt{g}}\frac{\partial}{\partial x^i}\left(\sqrt{g}\, g^{ij}\frac{\partial f}{\partial x^j}\right), \tag{3.4}$$

and R the scalar curvature (see Sect. 2.15). If $\partial\mathcal{M} \neq \varnothing$ we need to supplement the Eq. (3.3) with suitable boundary conditions. The admissible boundary conditions are severely restricted by the condition that H must be essentially self-adjoint.

Continuity of the Probability Density

The probability of finding the particle in a region $R \subset \mathcal{M}$ at time t is

$$\mathsf{Prob}(R,t) = \int_R \sqrt{g}\,|\psi(x,t)|^2\, \mathrm{d}^n x. \tag{3.5}$$

Using Eq. (3.3) we get

$$\frac{\partial|\psi|^2}{\partial t} = \left(\frac{\partial\psi^*}{\partial t}\psi + \psi^*\frac{\partial\psi}{\partial t}\right) = \frac{i\hbar}{2m}\left((\Delta\psi^*)\psi - \psi^*\Delta\psi\right) =$$
$$= -\frac{i\hbar}{2m}\frac{1}{\sqrt{g}}\frac{\partial}{\partial x^i}\left[\sqrt{g}\, g^{ij}\left(\frac{\partial\psi^*}{\partial x^j}\psi - \psi^*\frac{\partial\psi}{\partial x^j}\right)\right]. \tag{3.6}$$

Defining the *probability current*

$$\eta^i \stackrel{\mathrm{def}}{=} \frac{i\hbar}{2m}g^{ij}\left(\frac{\partial\psi^*}{\partial x^j}\psi - \psi^*\frac{\partial\psi}{\partial x^j}\right), \tag{3.7}$$

---

[1] Notice the overall *minus* sign in the definition of Δ. This is the math convention. Physics textbooks may have the opposite sign. Remember this sign when comparing with formulae in other books.

Eq. (3.6) takes the form of the *continuity equation*

$$\frac{\partial |\psi|^2}{\partial t} + \nabla_i \eta^i = 0 \tag{3.8}$$

In view of the Stokes theorem,[2]

$$\int_R \sqrt{g}\, \nabla_i \eta^i\, d^n x = \int_{\partial R} *\eta, \quad \text{where } \eta \equiv g_{ij}\eta^i dx^j, \tag{3.9}$$

Eq. (3.8) says that the variation in time of the probability $\mathsf{Prob}(R, t)$ of finding the particle in a region $R \subset \mathcal{M}$ is equal to the outgoing probability flux through its boundary ∂R

$$\frac{\partial}{\partial t}\mathsf{Prob}(R, t) = -\int_{\partial R} *\eta. \tag{3.10}$$

Equation (3.8) is the local form of the physical principle that the *total probability*

$$\int_{\mathcal{M}} \sqrt{g}\, |\psi(x)|\, d^n x \equiv \big\| |\psi\rangle \big\|^2 = 1 \tag{3.11}$$

is conserved in time, indeed equal 1 at all times, as a consequence of the fact that the time evolution is a unitary flow[3] in the space of states which preserves the norms.

Semiclassical Limit
We look for a solution to Eq. (3.3) of the form

$$\psi(x, t) = A(x, t; \hbar) \exp\big(iS(x, t; \hbar)/\hbar\big) \tag{3.12}$$

where $S(x, t; \hbar)$ and $A(x, t; \hbar)$ are *real* functions of their arguments which are *regular* as $\hbar \to 0$. We plug this ansatz in Eq. (3.3) and equate the two sides order by order in powers of $\hbar$. The vanishing of the $\hbar^0$ terms yields

$$\frac{\partial S}{\partial t} + \frac{1}{2m}g^{ij}\frac{\partial S}{\partial x^i}\frac{\partial S}{\partial x^j} + V(x, t) = 0, \tag{3.13}$$

while the order $\hbar^1$ terms yield

$$\frac{\partial A^2}{\partial t} + \nabla_i(A^2 v^i) = 0 \quad \text{where} \quad v^i \equiv \frac{1}{m}g^{ij}\frac{\partial S}{\partial x^j}. \tag{3.14}$$

---

[2] See [1] §. 2.9.

[3] In the physicists' usage conservation of total probability is also called *unitarity*.

The meaning of these two equations is transparent: Eq. (3.13) is the classical Hamilton-Jacobi equation [1] of the particle moving in $\mathcal{M}$ with velocity

$$v^i = \frac{g^{ij} p_j}{m} \equiv \frac{1}{m} g^{ij} \frac{\partial S}{\partial x^j}. \tag{3.15}$$

In other words: *the Schrödinger equation is the $\hbar$-deformation of the classical Hamilton-Jacobi equation.* Moreover

$$A^2 \equiv |\psi|^2 \tag{3.16}$$

is the probability density of finding the particle in some point of $\mathcal{M}$, $\eta^i \equiv A^2 v^i$ is the probability current, and Eq. (3.14) is the *continuity equation* for the probability density (at the linear level in $\hbar$).

Time-Independent Systems

When the system is *time-independent*, in classical mechanics we may separate the time variable t: we look for solutions of the Hamilton-Jacobi equation of the form

$$S(x, t) = -Et + W(x), \tag{3.17}$$

where E is a constant (the *energy* of the solution) and $W(x)$ is the *Hamilton characteristic function* which solves the stationary Hamilton-Jacobi equation

$$H\left(\frac{\partial W}{\partial x^i}, x^j\right) = E, \tag{3.18}$$

see [1] §§. 9.6, 9.7 for details. In the same fashion, when the Hamiltonian operator does not depend explicitly on time, the variable t separates in the quantum Schrödinger equation (which is the $\hbar$-deformation of the Hamilton-Jacobi one). Comparing Eqs. (3.12) and (3.17) we see that the time-separated stationary solutions to Eq. (3.3) must be of the form

$$\psi(x, t) = e^{-iEt/\hbar} \psi_E(x), \tag{3.19}$$

where $\psi_E(x)$ satisfies the *stationary Schrödinger equation*

$$H\psi_E \equiv \left(\frac{\hbar^2}{2m}\Delta + \hbar^2 \xi R + V(x)\right)\psi_E(x) = E\,\psi_E(x). \tag{3.20}$$

Equation (3.20) is the *eigenvector equation* for the Hamiltonian operator H. One has to look for solutions $\psi_E(x)$ of this PDE which belong to the appropriate space $\mathcal{S}^\vee \supset L^2(\mathcal{M})$ to be *generalized eigenfunctions* of H. The values of E corresponding to such admissible solutions form the *spectrum* of the Hamiltonian H: they are the physically allowed values of energy E for the quantum system.

We assume that $V(x)$ is such that H is essentially self-adjoint and bounded below i.e. $\sigma(H) \subset [E_{min}, \infty)$. We also assume $V(x)$ to be piece-wise continuous. An eigenstate of the Hamiltonian with the minimal possible energy E_{min} is called a *ground state*.

First Properties of Solutions

We list some general properties of the solutions to the Schrödinger PDE.

Regularity In each region where $V(x)$ is continuous the solution ψ_E is required to be of class C^2. If $V(x)$ if finite everywhere, the function ψ_E must be globally of class C^1: i.e. both the function and its first derivatives should be continuous at loci where $V(x)$ is discontinuous. The wave-functions vanish in the region R_∞ where $V = +\infty$. At the boundary of R_∞ the energy eigenfunctions are continuous but their first derivatives are not: this amounts to imposing Dirichlet boundary conditions at ∂R_∞, cf. Sect. 2.14.

Reality In absence of magnetic fields, $A_i = 0$, the Schrödinger representation of the Hamiltonian of a spinless particle is a *real* differential operator. When, in addition, H is time-independent, and $\psi_E(x)$ is an energy eigenfunction with eigenvalue E, its real and imaginary parts, $\mathrm{Re}\,\psi_E(x)$ and $\mathrm{Im}\,\psi_E(x)$,—if non-zero—are also eigenfunctions of H with the same energy. Hence each energy eigenspace has a basis whose elements are *real* functions.

Time-Reversal As in the classical set-up, when H is time independent and there are no external magnetic fields, $A_i = 0$, nor spins, the system is invariant under *time-reversal*, i.e. the evolution backwards in time is identical to the forward evolution. If $\psi(q, t)$ is a solution of the forward Schrödinger equation (3.3), its complex conjugate $\psi(q, t)^*$ is a solution of the backwards one. Hence the *time-reversal symmetry* T acts on the Schrödinger wave-function as

$$T : \psi \rightsquigarrow \psi^*. \tag{3.21}$$

We stress that T acts *anti-linearly*. In a general quantum system T—when present— is given by an *anti-automorphism* of the Hilbert space i.e. the composition of a unitary transformation and complex conjugation.

Variational Characterization The normalizable solutions to Eq. (3.20)—i.e. the strict-sense eigenvectors of the Hamiltonian in $L^2(\mathcal{M})$—have a convenient variational characterization as the extrema of the functional

$$A[\psi, \psi^*] = \int_{\mathcal{M}} \sqrt{g}\, \mathrm{d}^n x \left(\frac{\hbar^2}{2m} \partial_i \psi^* \partial_j \psi + \hbar^2 \xi R \psi^* \psi + V \psi^* \psi \right) \tag{3.22}$$

subjected to the normalization constraint

$$\langle \psi | \psi \rangle \equiv \int_{\mathcal{M}} \sqrt{g}\, \mathrm{d}^n x \, \psi^* \psi = 1. \tag{3.23}$$

Indeed, applying the method of Lagrange multipliers ([1] §. 4.4), we need to extremize the functional

$$\hat{A}[\psi, \psi^*, \lambda] = A[\psi, \psi^*] + \lambda \int_M \sqrt{g}\, d^n x\, \psi^* \psi - \lambda \qquad (3.24)$$

whose Euler-Lagrange equation is (3.20) with $E \equiv -\lambda$. The second variation operator (cf. [1] §. 4.1) is then $H - E$, and the number of its negative eigenvalues is the number of discrete energy states with energy $< E$. The absolute minimum of the functional A corresponds to the *ground state*, while an extremum of (3.24) with Morse index n gives the n-th discrete state in the non-decreasing energy order.

3.2 Bound and Scattering States

There are two kinds of eigenfunctions of H: the strict-sense eigenvectors $\psi_E \in L^2(M)$, whose eigenvalues E belong to the discrete spectrum,[4] and the generalized ones $\psi_E(x) \notin L^2(M)$. We write $\mathcal{V} \subset L^2(M)$ for the closed span of the strict-sense eigenvectors so that

$$L^2(M) = \mathcal{V} \oplus \mathcal{V}^\perp \qquad (3.25)$$

where, by the spectral theorem, the elements of the orthogonal complement $\mathcal{V}^\perp$ are superpositions of generalized eigenstates of the continuous spectrum of H. The decomposition (3.25) is preserved by the time evolution. When M is *compact*, and we have admissible boundary conditions on ∂M (if non empty),

$$\mathcal{V}^\perp = 0. \qquad (3.26)$$

Indeed in a compact M all continuous function $\psi_E(x)$ is automatically in $L^2(M)$.

The physical behavior of the states in the two summands of (3.25) is quite different. For definiteness we specialize our analysis to $M = \mathbb{R}^n$ with the flat metric, but the result is of general validity, *mutatis mutandis*.

Theorem 3.1 (Ruelle)

(1) *When the normalized wave-function $\psi(x, 0)$ belongs to the discrete-spectrum subspace $\mathcal{V} \subset L^2(\mathbb{R}^n)$, for all $\epsilon > 0$ we can find a compact $K \in \mathbb{R}^n$ such*

$$\inf_t \int_K d^n x\, |\psi(x, t)|^2 > 1 - \epsilon. \qquad (3.27)$$

---

[4] **Beware!** Recall that the discrete and continuous spectra may overlap!

(2) *When the wave-function* $\psi(x, 0) \in \mathcal{V}^{\perp} \subset L^2(\mathbb{R}^n)$, *for all compact* $K \Subset \mathbb{R}^n$

$$\lim_{T \to +\infty} \frac{1}{T} \int_0^T dt \int_K d^n x \, |\psi(x, t)|^2 = 0. \tag{3.28}$$

Statement **(2)** is a particular case of the RANGE theorem.[5] Its proof shows more: for *any* wave-function $\psi(x, t) \equiv \langle x | \psi, t \rangle \in L^2(\mathbb{R}^n)$ we have

$$\lim_{T \to \infty} \frac{1}{T} \int_0^T dt \int_K d^n x \, |\psi(x, t)|^2 = \sum_n |\langle n | \psi \rangle|^2 \int_K d^n x \, |\psi_n(x)|^2 \tag{3.29}$$

where the sum in the RHS is over all energy eigenstates $|n\rangle$ of the *discrete* spectrum and $\psi_n(x) \equiv \langle x | n \rangle$.

Sketch of Proof Setting $\hbar = 1$ we have:

$$\lim_{T \to \infty} \frac{1}{T} \int_0^T dt \int_K d^n x \, |\psi(x, t)|^2 =$$

$$= \lim_{T \to \infty} \frac{1}{T} \int_0^T dt \int dE_1 \, dE_2 \, e^{i(E_1 - E_2)t} \int_K d^n x \, \langle \psi | E_1 \rangle \langle E_1 | x \rangle \langle x | E_2 \rangle \langle E_2 | \psi \rangle \tag{3.30}$$

$$= \lim_{T \to \infty} \int dE_1 \, dE_2 \, \frac{e^{i(E_1 - E_2)T} - 1}{i(E_1 - E_2)T} \int_K d^n x \, \langle \psi | E_1 \rangle \langle E_1 | x \rangle \langle x | E_2 \rangle \langle E_2 | \psi \rangle$$

where $dE \, |E\rangle\langle E| \equiv dP(E)$ is the projector-valued spectral measure of the Hamiltonian H which is the sum of a continuous measure with support on the continuous spectrum $\sigma_e(H)$ and valued in projectors with image in $\mathcal{V}^{\perp}$, and a discrete measure with support in the point spectrum $\sigma_p(H)$ which is valued in projectors with image in $\mathcal{V}$. One can show that, for K compact, the limit of the integral in the last line of (3.30) is the integral of the limit. Now

$$\lim_{T \to \infty} \frac{e^{i(E_1 - E_2)T} - 1}{i(E_1 - E_2)T} = \begin{cases} 1 & \text{for } E_1 = E_2 \\ 0 & \text{otherwise,} \end{cases} \tag{3.31}$$

so the integrand has a support in a subset of the E_1-E_2 plane of *zero Lebesgue measure,* and the integral of any continuous measure vanishes. We remain with the contribution from the singular spectral measure of the discrete spectrum which yields Eq. (3.29). □

Ruelle's **Theorem** has a clear physical meaning: if the particle is in a state $\psi \in \mathcal{V}$ at $t = 0$, it will remain *almost certainly* forever in some (sufficiently large) compact subset $K \Subset \mathbb{R}^n$, while if $\psi \in \mathcal{V}^{\perp}$ the particle eventually will escape to spatial infinity as $t \to \pm\infty$ (with probability 1). This dichotomy leads to the central physical notion of *bound state.*

Definition 3.1 A *bound state* is a state $\psi \in \mathcal{V} \subset L^2(\mathbb{R}^n)$. The states orthogonal to all bound states $\psi \in \mathcal{V}^{\perp}$ are called *scattering states.*

---

[5] See **Theorem XI.115** of [2] volume 3.

Scattering states are superpositions of states of the continuous energy spectrum. For a prototypical situation see Example 3.1 below.

Remark 3.1 Bound states are the quantum counterpart to classical bounded motions, i.e. motions which remain in a bounded domain B of the phase space $\mathcal{W}$ for all t.

3.3 Separation of Variables

We have seen above that the Schrödinger equation is the $\hbar$-deformation of the classical Hamilton-Jacobi equation. It follows that when we are able to solve explicitly the Schrödinger equation (for all $\hbar$), we may construct a solution of the classical Hamilton-Jacobi equation by taking the limit $\hbar \to 0$. Then integrability of the two equations must be related. The most general way to solve explicitly the Hamilton-Jacobi equation is by *separation of variables* ([1] chaps. 9, 10). Correspondingly, the most general analytic[6] technique to integrate the Schrödinger equation is by separation of variables. For a classical (time-independent) system, with n degrees of freedom, this means finding a *complete* solution to the Hamilton-Jacobi equation of the form

$$S(q^i, t; \alpha^1, \ldots, \alpha^n) = -E(\alpha^1, \ldots, \alpha^n)\, t + \sum_{i=1}^{n} F_i(q^i; \alpha^1, \ldots, \alpha^n) \qquad (3.32)$$

where each function F_i depends only on a single variable q^i and the n constant parameters $\alpha^1, \ldots, \alpha^n$ (we may choose the energy E to be one of them).

The PDE (3.18) decomposes into n ordinary differential equations (ODE) one for each one of the n functions F_i. If such a complete solution exists, we say that the Hamilton-Jacobi equation is *separable*. We stress that when the Hamilton-Jacobi equation separates, it separates in a *specific class* of coordinates $\{q^i\}$; in the rare case when the equation separates in several (inequivalent) coordinates we say that the model is *superseparable* ([1] §. 9.7). The time variable t separates iff the classical system is time-independent: then we may look for solutions of the Hamilton-Jacobi equation of the form $S = -Et + W$ where W is Hamilton's characteristic function which solves the stationary Hamilton-Jacobi equation (3.18). See [1] §§. 9.6, 9.7 for more details.

Since in the limit $\hbar \to 0$ the wave-function is the exponential of the action (cf. (3.12)), the sum (3.32) becomes a product of wave-functions. This leads to the following definition:

---

[6] A general Schrödinger equation may also be studied numerically.

Definition 3.2 We say that the Schrödinger PDE (3.3) *is solved by separation of variables* (or simply that is *(fully) separable*) iff we can find a system of coordinates $\{q^i\}$ in $\mathcal{M}$ and a *complete* system of solutions in[7] $S(\mathcal{M})^\vee$ of the form

$$\psi(q^i, t; \boldsymbol{\alpha}) = \mathrm{e}^{-\mathrm{i}E(\boldsymbol{\alpha})t/\hbar} \prod_{i=1}^{n} \chi_i(q^i, \boldsymbol{\alpha}) \tag{3.33}$$

where each function χ_i depends on a single coordinate q^i and $\boldsymbol{\alpha} \equiv \{\alpha^1, \ldots, \alpha^n\}$ is a set of labels (continuous or discrete) which parametrize the family of solutions. *Complete* means that all wave-functions $\psi(q^i) \in L^2(\mathcal{M})$ may be written as superpositions of the solutions (3.33) in the form

$$\psi(q^i) = \int \mathrm{d}\mu(\boldsymbol{\alpha})\, \hat{\psi}(\boldsymbol{\alpha}) \prod_i \chi_i(q^i, \boldsymbol{\alpha}) \tag{3.34}$$

for some complex function $\hat{\psi}(\boldsymbol{\alpha})$ and measure $\mathrm{d}\mu(\boldsymbol{\alpha})$ on the label space. In other words: the family $\{\psi(q^i, t; \boldsymbol{\alpha})\}$ is *complete* iff a vector $\phi(q^i) \in L^2(\mathcal{M})$ which is orthogonal to all $\psi(q^i, 0; \boldsymbol{\alpha})$ is zero. A quantum system is *superseparable* iff its Schrödinger equation is separable in multiple inequivalent systems of coordinates.

The time-coordinate t is separable iff the quantum Hamiltonian is time independent, i.e. iff time-translations are symmetries. In this case $\psi(t) = \mathrm{e}^{-\mathrm{i}Et/\hbar}\Phi$ where Φ solves the *stationary Schrödinger PDE*

$$H\Phi = E\Phi \tag{3.35}$$

which is the quantum counterpart to (3.18). The solutions of (3.35) are the eigenvalues and eigenfuctions of the Hamilton operator H. Thus, in the time-independent case, solving the Schrödinger equation is equivalent to solving the spectral problem for the densely-defined self-adjoint operator H, cf. Sect. 2.4.

When the linear PDE (3.35) is fully separable, it decomposes into a set of linear ODEs for the various factor functions $\chi_i(q^i; \boldsymbol{\alpha})$ (i fixed) of the schematic form

$$K_i\left(-\mathrm{i}\hbar\frac{d}{dq^i}, q^i; \boldsymbol{\alpha}\right)\chi_i(q^i; \boldsymbol{\alpha}) = 0. \tag{3.36}$$

When the Schrödinger equation is separable, the $\hbar \to 0$ limit of (3.33) produces a separation of variables of the corresponding classical Hamilton-Jacobi. Thus if the Schrödinger equation is separable (resp. superseparable) the Hamilton-Jacobi is automatically separable (resp. superseparable). Then, by the Liouville theorem [1],

---

[7] By $S(\mathcal{M})^\vee$ we mean a suitable generalization of the space $S(\mathbb{R}^n)^\vee$ of tempered distributions.

the classical theory has n functionally-independent conserved quantities A_i^{cl} which are in *involution* in the Poisson sense

$$\left[A_i^{\text{cl}}, A_j^{\text{cl}}\right]_{\text{PB}} = 0 \qquad i, j = 1, \ldots, n. \tag{3.37}$$

This observation yields a necessary condition for quantum separability: the corresponding classical system must be separable. See [1] chap. 9, 10 for some classical criteria of separability.

When the A_i^{cl} are polynomials in the *Noether charges*, it follows from Corollary 2.10 that the quantum system has n conserved Hermitian operators A_i which are functionally independent and commute

$$[H, A_i] = [A_i, A_j] = 0. \tag{3.38}$$

As $\hbar \to 0$ the quantum conserved quantities A_i reduce back to their classical siblings A_i^{cl}. In the quantum set-up we can diagonalize simultaneously the Hamiltonian H and the conserved quantities A_i, that is, we may look for eigenvalues and eigenstates of H in the simultaneous eigenspaces of all the A_i's

$$V(\boldsymbol{\alpha}) = \left\{\Psi \in \mathcal{S}^\vee \;:\; A_1 \Psi = \lambda_1(\boldsymbol{\alpha})\Psi, \cdots, A_n \Psi = \lambda_n(\boldsymbol{\alpha})\Psi\right\} \subset \mathcal{S}^\vee. \tag{3.39}$$

The system of n eigenvalue equations

$$A_i \Psi = \lambda_i(\boldsymbol{\alpha})\Psi \tag{3.40}$$

is equivalent to the system of n ODE's (3.36).

The Noether charges generate a *symmetry* of the quantum system, and to describe the separation of variables in its full glory, we have preliminarily to understand *symmetry* in quantum physics. This will be done in detail in Chap. 4. Then in Chap. 5 we shall solve several physically interesting Schrödinger equations by separation of variables exploiting their symmetries.

Roughly speaking, separation of variables reduces solving our n degrees of freedom system to the solution of a sequence of n one degree of freedom systems. In this chapter we discuss one-dimensional quantum systems both for their intrinsic interest and as a step towards solving more realistic physical systems with several degrees of freedom. Before going to the general discussion of one-dimensional quantum systems we study a few basic examples which are both elementary and fundamental.

3.4   Example: The Free Particle in $\mathbb{R}^n$

A free (non-relativistic) particle moving in $\mathbb{R}^n$ has Hamiltonian[8]

$$H = \frac{p_i \, p_i}{2m} \xrightarrow{\text{Schrödinger repres.}} -\frac{\hbar^2}{2m} \frac{\partial^2}{\partial x^i \, \partial x^i} \tag{3.41}$$

This system is invariant under all isometries of $\mathbb{R}^n$, in particular under translations. Hence the n components of the momentum p_j are conserved and commuting

$$[H, p_j] = 0 \qquad [p_j, p_k] = 0. \tag{3.42}$$

The momentum operators p_j and the Hamiltonian H may be diagonalized simultaneously. The momentum is diagonal in the *momentum representation*, see Sect. 2.12. The basis vectors of this representation are the generalized ($\equiv$ non-normalizable) momentum eigenstates $|k\rangle$ (here $k = (k_1, \ldots, k_n)$)

$$p_j |k\rangle = k_j |k\rangle \tag{3.43}$$

$$\langle k'|k\rangle = (2\pi\hbar)^n \delta(k - k'). \tag{3.44}$$

In momentum representation the free Hamiltonian acts as a multiplication operator

$$H|k\rangle = \frac{|k|^2}{2m}|k\rangle. \tag{3.45}$$

Equivalently, the Schrödinger equation is fully separable in Cartesian coordinates: in other words, we have a complete system of (non-normalizable) energy eigenfunctions in the form of *plane waves*

$$\psi(x, t; k) = \exp\left(-i\frac{|k|^2 t}{2m\,\hbar}\right) \prod_{i=1}^{n} e^{ik_i x^i /\hbar} \equiv \exp\left(-i\frac{|k|^2 t}{2m\,\hbar}\right) e^{ik \cdot x/\hbar}. \tag{3.46}$$

The general solution of the Schrödinger equation in the x-representation with initial condition $\psi(x, 0) \in L^2(\mathbb{R}^n)$ is conveniently written in the form

$$\psi(x, t) = \int d^n y \, \langle x|e^{-iHt/\hbar}|y\rangle \, \psi(y, 0) \in L^2(\mathbb{R}^n) \tag{3.47}$$

---

[8] The indices $i, j, k, \ldots$ take the values $1, 2, \ldots, n$ and the sum over repeated indices is always implied unless explicitly stated otherwise.

in terms of the *evolution kernel* $\langle x|e^{-iHt/\hbar}|y\rangle$ which has the explicit expressions

$$
\begin{aligned}
\langle x|e^{-iHt/\hbar}|y\rangle &= \int \frac{d^n k}{(2\pi\hbar)^n} \frac{d^n h}{(2\pi\hbar)^n} \langle x|k\rangle\langle k|e^{-iHt/\hbar}|h\rangle\langle h|y\rangle \\
&= \int \frac{d^n k}{(2\pi\hbar)^n} \frac{d^n h}{(2\pi\hbar)^n} e^{ik\cdot x/\hbar} e^{-i|k|^2 t/(2m\hbar)} (2\pi\hbar)^n \delta(k-h) e^{-ih\cdot y/\hbar} \qquad (3.48) \\
&= \int \frac{d^n k}{(2\pi\hbar)^n} e^{ik\cdot(x-y)/\hbar - i|k|^2 t/(2m\hbar)}
\end{aligned}
$$

The last integral is best computed by analytic continuation: writing $it = s$, the integral

$$
\int \frac{d^n k}{(2\pi\hbar)^n} e^{ik\cdot(x-y)/\hbar - s|k|^2/(2m\hbar)} \qquad (3.49)
$$

is Gaussian and convergent for $\mathrm{Re}\, s > 0$. Hence the free evolution kernel in the imaginary time s is

$$
\langle x|e^{-sH/\hbar}|y\rangle = \frac{m^{n/2}}{(2\pi\hbar s)^{n/2}} \exp\left(-\frac{m}{2\hbar s}|x-y|^2\right). \qquad (3.50)
$$

As a check, note that the RHS goes to $\delta(x-y)$ as $s \to 0$ (cf. first Eq. (2.167)). Returning to real time t, the free evolution kernel is

$$
\langle x|e^{-itH/\hbar}|y\rangle = \frac{m^{n/2}}{(2\pi i\hbar t)^{n/2}} \exp\left(\frac{im}{2\hbar t}|x-y|^2\right) \qquad (3.51)
$$

By the quantum action principle (Sect. 2.17) the argument of the exponential must be

$$
\frac{iS}{\hbar} + O(1) \quad \text{as } \hbar \to 0, \qquad (3.52)
$$

with S the Hamilton principal function for the free motion

$$
S \equiv \frac{m}{2} v^2 t = \frac{m}{2t}|x-y|^2. \qquad (3.53)
$$

We see that the equality (3.52) is satisfied *exactly* without extra subleading quantum corrections of order $O(1)$. We leave to the reader to check that the kernel (3.51) does satisfy the Schrödinger PDE.

Example 3.1 (Scattering States) Suppose we have a quantum particle moving in $\mathbb{R}^n$ with Hamiltonian

$$H = \frac{\hbar^2}{2m}\Delta + V \tag{3.54}$$

where the potential function V is non-positive, bounded below, smooth, and $V \to 0$ as $|x| \to \infty$. Since the kinetic operator $\hbar^2\Delta/2m$ is *non-negative*, the spectrum of H is contained in (min $V, +\infty$).

We study the wave-functions of the *scattering states:* we know from Sect. 3.2 that a state is scattering iff the particle escapes to infinity with probability essentially 1. At infinity we may neglect the potential term in H, hence the scattering wave-functions satisfy asymptotically for large x the free Schrödinger equation, and therefore may be taken to have the asymptotic form (3.46). In particular the energy of the scattering states is $E \geq 0$. All positive energies belong to the continuous spectrum.[9] On the other hand states of negative energy—if present—should be *bound states* and hence must belong to the discrete spectrum. In this situation there is no positive energy bound state, so the bound states are precisely the ones with $E < 0$.

3.5 The Quantum Harmonic Oscillator

In this section we study in detail the quantum harmonic oscillator of frequency ω whose Hamiltonian operator is

$$H = \frac{1}{2m}P^2 + \frac{m\omega^2}{2}Q^2, \qquad [Q, P] = \mathrm{i}\hbar. \tag{3.55}$$

To make the Lie algebraic aspects more transparent, it is better to consider a more general system: we take n non-interacting copies of the system (3.55) and consider the *isotropic harmonic oscillator* in $\mathbb{R}^n$

$$H = \sum_{i=1}^{n}\left(\frac{1}{2m}P_i P_i + \frac{m\omega^2}{2}Q^i Q^i\right) \tag{3.56}$$

where (Q^i, P_j) $(i, j = 1, \ldots, n)$ are self-adjoint operators satisfying the canonical commutation relations (2.334) and $m, \omega > 0$ are constant parameters. The linear

---

[9] The presence of a negative V does not spoil this statement since adding a negative operator can only lower the eigenvalues.

unitary redefinition

$$Q^i \rightsquigarrow q^i = \sqrt{m\omega}\, Q^i, \qquad P_j \rightsquigarrow p_j = \frac{1}{\sqrt{m\omega}}\, P_j \tag{3.57}$$

sets the Hamiltonian in its Birkhoff normal form [1]

$$H = \frac{\omega}{2} \sum_{i=1}^{n} \left(p_i p_i + q^i q^i \right) \tag{3.58}$$

in terms of the new canonical operators (q^i, p_j) with commutators

$$\left[q^i, q^j \right] = \left[p_i, p_j \right] = 0, \quad \left[q^i, p_j \right] = i\hbar\, \delta^i{}_j. \tag{3.59}$$

The energy eigenvalues of the n-copy system (3.58) are the sum of the n eigenvalues of the single-copy Hamiltonians

$$H_i = \frac{\omega}{2} \left(p_i p_i + q^i q^i \right) \quad \text{not summed over } i \tag{3.60}$$

and the eigenvectors of H are the tensor products of the corresponding single-copy eigenvectors.

Given the fundamental importance of this quantum system, in this section we solve it from five different viewpoints:

- Heisenberg equations
- Lie theoretic/algebraic viewpoint
- Bargmann representation
- Schrödinger PDEs
- Meyer kernel

A sixth viewpoint (solution by separation of variables in polar coordinates) will be presented in Chap. 5 and other four solutions will be given in Chap. 6 using path integrals. In addition in the box on page 145 the harmonic oscillator is analyzed from the point of view of the *differential Galois group* [3], while in Example 4.4 we show how it can be solved exploiting its semi-group of *non-invertible* symmetries. Finally in Example 8.6 we treat it in the Bohr-Sommerfeld quantization.

3.5.1   Heisenberg Picture

The Heisenberg equations of motion read

$$\frac{d\boldsymbol{q}^i}{dt} = [\boldsymbol{q}^i, H] = \omega\, \boldsymbol{p}_i \tag{3.61}$$

$$\frac{d\boldsymbol{p}^i}{dt} = [\boldsymbol{p}^i, H] = -\omega\, \boldsymbol{q}_i. \tag{3.62}$$

These equations are linear in the canonical operators and identical in form to the classical ones. Thus their solutions have the same expressions as the classical solutions

$$\boldsymbol{q}^i(t) = \boldsymbol{q}^i(0)\cos(\omega t) + \boldsymbol{p}_i(0)\sin(\omega t),$$
$$\boldsymbol{p}_i(t) = \boldsymbol{p}_i(0)\cos(\omega t) - \boldsymbol{q}^i(0)\sin(\omega t), \tag{3.63}$$

except that now both sides of these equalities are Hermitian *operators* densely defined in $\mathcal{H}$. From these equations we see that

(1) the time evolution $U(t)|_{t=2\pi n/\omega}$ for an integral number of periods, $\omega t \in 2\pi\mathbb{Z}$, acts as the *identity* on the canonical operators $(\boldsymbol{q}^i, \boldsymbol{p}_j)$;
(2) the evolution for half a period $U(t)|_{t=\pi/\omega}$ acts as the *parity symmetry*

$$(\boldsymbol{q}^i, \boldsymbol{p}_j) \rightsquigarrow (-\boldsymbol{q}^i, -\boldsymbol{p}_j), \tag{3.64}$$

(3) the evolution for a quarter of a period $U(t)|_{t=\pi/2\omega}$ acts as the *Fourier transform*

$$(\boldsymbol{q}^i, \boldsymbol{p}_j) \rightsquigarrow (\boldsymbol{p}_i, -\boldsymbol{q}^j). \tag{3.65}$$

The time evolution $U(t)$ is a unitary transformation which acts *linearly* on the canonical operators: we know from Sect. 2.11.1 that the linear unitary transformations have the same $Sp(2n, \mathbb{R})$ form as the *linear canonical transformations* which yield the evolution of the classical harmonic oscillator. This explains why in this model the quantum and classical evolutions take the same form. It suggests to study the system in terms of the Lie theoretical properties of $Sp(2n, \mathbb{R})$ as we are going to do.

3.5.2   Lie Algebraic Considerations

Consider the group $Sp(2n, \mathbb{R})$ of linear unitary redefinitions of the canonical operators (q^i, p_j), see Sect. 2.11.1. In the block-matrix notation we write

$$M = \begin{pmatrix} A & B \\ C & D \end{pmatrix} \qquad \Omega = \begin{pmatrix} 0 & 1 \\ -1 & 0 \end{pmatrix} \tag{3.66}$$

where each entry is a real $n \times n$ matrix. $M \in Sp(2n, \mathbb{R})$ satisfies the equation

$$M \Omega M^t = \Omega. \tag{3.67}$$

The Lie Algebra $\mathfrak{u}(n)$
A maximal compact subgroup of $Sp(2n, \mathbb{R})$ is given by the elements $U \in Sp(2n, \mathbb{R})$ which commute with Ω, so that Eq. (3.67) reduces to

$$U U^t = 1 \qquad \text{(assuming } U \Omega = \Omega U \text{).} \tag{3.68}$$

Such an U has the block form

$$U = \begin{pmatrix} A & B \\ -B & A \end{pmatrix} \quad \text{where } A, B \atop \text{satisfy} \quad (A + iB)^\dagger (A + iB) = 1 \tag{3.69}$$

From the last equality we see that a maximal compact subgroup of $Sp(2n, \mathbb{R})$, $Sp(2n, \mathbb{R}) \cap O(2n)$, is isomorphic to $U(n)$. By Eq. (3.68) a unitary transformation $U \in U(n)$ acts on the $2n$-vector (q^i, p_j) by an $O(2n)$ rotation, thus leaving invariant the quadratic expression

$$\sum_i (q^i q^i + p_i p_i) \tag{3.70}$$

that is,

Fact 3.2 *The quantum Hamiltonian (3.58) is invariant under the group $U(n)$ of compact linear unitary transformations. By the Stone theorem we have n^2 self-adjoint operators J_a which generate the action of this Lie group: since the transformation of the (q^i, p_j) is* linear, *the J_a are quadratic[10] in the canonical operators (q^i, p_j).*

---

[10] This follows immediately from the quantum Noether theorem, Corollary 2.10.

From Eq. (3.69) we see that the $2n$ canonical operators (q^i, p_j) span the reducible complex representation $n \oplus \bar{n}$ of $U(n)$. It is convenient to use a complex basis of operators which splits the $U(n)$ representation in its irreducible components. We define the operators

$$a_j = \frac{1}{\sqrt{2\hbar}}(q^j + \mathrm{i}p_j), \qquad a_j^\dagger = \frac{1}{\sqrt{2\hbar}}(q^j - \mathrm{i}p_j) \tag{3.71}$$

which satisfy the complex version of the *canonical algebra* $(i, j = 1, \ldots, n)$

$$[a_i, a_j] = [a_i^\dagger, a_j^\dagger] = 0, \qquad [a_i, a_j^\dagger] = \delta^i{}_j. \tag{3.72}$$

The $a_j^\dagger$'s (resp. a_j's) transform in the fundamental (resp. anti-fundamental) representation of $U(n)$, that is,

$$U a_j U^{-1} = (A - \mathrm{i}B)_j{}^i a_i \tag{3.73}$$

$$U a_j^\dagger U^{-1} = (A + \mathrm{i}B)_j{}^i a_i^\dagger. \tag{3.74}$$

More generally, we have the action on the canonical operators $(a_i, a_j^\dagger)$ of the group $Sp(2n, \mathbb{R})$ of linear unitary transformations. They preserve the canonical algebra (3.72), see the off-text **BOX** on page 133.

The n^2 Hermitian generators J_a of $U(n)$ have the following form. Let

$$\lambda_a^{ij} = (\lambda_a^{ji})^* \qquad i, j = 1, \ldots, n \quad a = 1, \ldots, n^2, \tag{3.75}$$

be a basis of the $\mathbb{R}$-space of $n \times n$ Hermitian matrices which yield a basis of the Lie algebra $\mathfrak{u}(n)$ acting in the fundamental n representation. We define the operators

$$J_a \equiv J_a^\dagger \overset{\text{def}}{=} a_i^\dagger \lambda_a^{ij} a_j. \tag{3.76}$$

The J_a's induce the (anti-)fundamental representation of $\mathfrak{u}(n)$ on $a_k^\dagger$ (a_k):

$$[J_a, a_k^\dagger] = a_j^\dagger \lambda_a^{jk}, \qquad [J_a, a_k] = -\lambda_a^{kj} a_j. \tag{3.77}$$

Their commutator algebra is

$$\begin{aligned}
[J_a, J_b] &= [J_a, a_i^\dagger \lambda_b^{ij} a_j] = [J_a, a^\dagger]\lambda_b^{ij} a_j + a_i^\dagger \lambda_b^{ij}[J_a, a_j] = \\
&= a_i^\dagger \lambda_a^{ik} \lambda_b^{kj} a_j - a_i^\dagger \lambda_b^{ik} \lambda_a^{kj} a_j = a_i^\dagger [\lambda_a, \lambda_b]^{ij} a_j \equiv \mathrm{i} f_{ab}{}^c J_c
\end{aligned} \tag{3.78}$$

where the real coefficients $f_{ab}{}^c$ are the $\mathfrak{u}(n)$ structure constants in the basis given by the $n \times n$ Hermitian matrices λ_a, that is,

$$[\lambda_a, \lambda_b] = i\, f_{ab}{}^c \lambda_c. \tag{3.79}$$

We conclude that the map

$$\lambda_a \mapsto J_a \equiv a_i^\dagger \lambda_a^{ij} a_j \tag{3.80}$$

is an isomorphism of Lie algebras between $\mathfrak{u}(n)$ *and the symmetry algebra of the isotropic harmonic oscillator in* $\mathbb{R}^n$. This isomorphism sends the matrix 1 to

$$
\begin{aligned}
1 \mapsto \sum_i a_i^\dagger a_i &= \frac{1}{2\hbar} \sum_i (q^i - i\, p_i)(q^i + i\, p_i) = \\
&= \frac{1}{2\hbar} \sum_i \left(q^i q^i + p_i p_i + i[q^i, p_i] \right) = \\
&= \frac{1}{2\hbar} \sum_i (p_i p_i + q^i q^i) - \frac{n}{2}.
\end{aligned}
\tag{3.81}
$$

Comparing with (3.58), we write the Hamiltonian operator in the form

$$H = \hbar\omega \sum_{i=1}^n \left(a_i^\dagger a_i + \frac{1}{2} \right). \tag{3.82}$$

Up to a trivial shift and an overall constant factor, the Hamiltonian H is just the image of the identity matrix under the Lie group isomorphism (3.80). Since the matrix 1 commutes with all λ_a's, the full set of self-adjoint operators J_a is conserved, $[H, J_a] = 0$ by Eq. (3.78). We write

$$H = \sum_{i=1}^n H_i, \qquad [H_i, H_j] = 0, \quad \text{where } H_i \overset{\text{def}}{=} \hbar\omega\left(a_i^\dagger a_i + \frac{1}{2} \right) \tag{3.83}$$

H_i is the Hamiltonian of the i-th copy of the one-dimensional harmonic oscillator. The individual Hamiltonians H_i are conserved and commuting, so the system is trivially separable in the canonical coordinates q^i. We will see in Chap. 5 that this system is in facts *superseparable*.

In the rest of this section we focus with no loss on the one-dimensional harmonic oscillator ($n = 1$).

BOX: Bogoliubov Transformations

The action of the group of linear unitary transformations $Sp(2n, \mathbb{R})$ on the canonical operators a_i, $a_i^{\dagger}$ is called the group of *Bogoliubov transformations*. It is given by the defining $Sp(2n, \mathbb{R})$ representation rewritten in the Cayley rotated complex basis ([4] §. 1.5.1)

$$M = \begin{pmatrix} V & W \\ W^* & V^* \end{pmatrix} \in Sp(2n, \mathbb{R})$$

where now the complex $n \times n$ matrices V, W satisfy

$$V^t W^* - W^{\dagger} V = 0, \qquad V^t V^* - W^{\dagger} W = 1,$$

and the subgroup $U(n)$ is given by the elements with $W = 0$.
The general Bogoliubov transformation is

$$a_i^{\dagger} = V_{ij} a_j^{\dagger} + W_{ij} a_j, \qquad a_i = W_{ij}^* a_j^{\dagger} + V_{ij}^* a_j.$$

See also Problem 3.2.

3.5.3 Energy Spectrum: Algebraic Approach

The one-dimensional quantum harmonic oscillator has Hamiltonian

$$H = \hbar \omega \left(a^{\dagger} a + \frac{1}{2} \right) \tag{3.84}$$

where the operators a, $a^{\dagger}$ satisfy the canonical algebra

$$[a, a^{\dagger}] = 1, \tag{3.85}$$

while the Hilbert space $\mathcal{H}$ contains a dense subspace which is an *irreducible* representation of the (densely defined) canonical operator algebra

$$\mathfrak{A}_{\text{can}} \equiv \mathbb{C}\langle a, a^{\dagger} \rangle / ([a, a^{\dagger}] - 1), \tag{3.86}$$

that is, a quantum operator with commutes with both a and $a^{\dagger}$ is a c-number. This irreducibility condition is the algebraic version of the physical idea that the system has no other degrees of freedom besides q, i.e. that we are not missing anything by describing the system in terms of a and $a^{\dagger}$. More general situations will be discussed in Chap. 4.

We note that the Hermitian operator $a^\dagger a$ is non-negative, so *the Hamiltonian is bounded below* and the energy spectrum

$$\sigma(H) \subset \left[\tfrac{1}{2}\hbar\omega, +\infty\right). \tag{3.87}$$

In the algebraic approach we study the *unitary*[11] representations of the Lie algebra $\mathfrak{L}$ generated by the four operators $H, a^\dagger, a, 1$. Besides Eq. (3.85), the other non-trivial commutation relations in $\mathfrak{L}$ are

$$[H, a] = \hbar\omega[a^\dagger, a]a = -\hbar\omega\, a, \qquad [H, a^\dagger] = \hbar\omega\, a^\dagger. \tag{3.88}$$

Note that the universal enveloping algebra of $\mathfrak{L}$ is the canonical algebra $\mathfrak{A}_{\text{can}}$.

Lemma 3.1 *Let $|E\rangle$ be a non-zero eigenvector of the Hamiltonian (3.84) of energy E, i.e. $H|E\rangle = E|E\rangle$. The vector $a^\dagger|E\rangle$ is a non-zero eigenvector of H of eigenvalue $E + \hbar\omega$, while $a|E\rangle$,* if non-zero, *is an eigenvector with eigenvalue $E - \hbar\omega$.*

Proof The state $a^\dagger|E\rangle$ satisfies the eigenvalue equation

$$\begin{aligned} H(a^\dagger|E\rangle) &= [H, a^\dagger]|E\rangle + a^\dagger H|E\rangle = \\ &= \hbar\omega a^\dagger|E\rangle + Ea^\dagger|E\rangle \equiv (E + \hbar\omega)a^\dagger|E\rangle, \end{aligned} \tag{3.89}$$

while it is non-zero

$$\begin{aligned} \|a^\dagger|E\rangle\|^2 &= \langle E|aa^\dagger|E\rangle = \langle E|(1 + a^\dagger a)|E\rangle = \\ &= \||E\rangle\|^2 + \|a|E\rangle\|^2 \neq 0. \end{aligned} \tag{3.90}$$

Conversely,

$$\begin{aligned} H(a|E\rangle) &= [H, a]|E\rangle + aH|E\rangle = \\ &= -\hbar\omega a|E\rangle + Ea|E\rangle \equiv (E - \hbar\omega)a|E\rangle \end{aligned} \tag{3.91}$$

so, *if non-zero, $a|E\rangle$ is an eigenstate with eigenvalue $(E - \hbar\omega)$.*   □

In view of these properties, $a^\dagger$ (resp. a) is called the *raising* (resp. *lowering*) operator.[12]

Repeating recursively the argument of the **Lemma**, we conclude that the vector $a^k|E\rangle$, *if non-zero,* is an energy eigenstate of energy $E - k\hbar\omega$. For all energy eigenstate $|E\rangle$ there is a *minimal $k \in \mathbb{N}$* such that $a^k|E\rangle$ is the zero vector: otherwise we could find energy eigenstates of arbitrary low energy, while the spectrum of the

---

[11] *Unitary* means compatible with the positive-definite Hermitian product in the Hilbert space $\mathcal{H}$.

[12] In view of their role in second quantization (see [5] chap. 3), $a^\dagger$ (resp. a) is also called the *creation* (resp. *annihilation*) operator.

Hamiltonian is bounded below, Eq. (3.87). Let $|0\rangle = a^{k-1}|E\rangle \neq 0$. The vector $|0\rangle$ satisfies the equations

$$a|0\rangle = 0, \qquad H|0\rangle = \hbar\omega\big(a^\dagger a + \tfrac{1}{2}\big)|0\rangle = \tfrac{1}{2}\hbar\omega|0\rangle, \tag{3.92}$$

so that $|0\rangle$ is a state of the lowest possible energy $\tfrac{1}{2}\hbar\omega$. The states of lowest possible energy in a quantum system are called *ground states*: $|0\rangle$ is a ground state of the one-dimensional harmonic oscillator and its energy $E_0 \equiv \tfrac{1}{2}\hbar\omega$ is called the *zero-point energy* (often one sets it to zero by subtracting the constant $\tfrac{1}{2}\hbar\omega$ from the Hamiltonian). We shall see momentarily that the ground state is *unique*, i.e. the energy eigenvalue $E = \tfrac{1}{2}\hbar\omega$ is *non-degenerate*. This is in fact a general property which holds for all spinless one-dimensional quantum systems, see Sect. 3.6. For the moment we fix a normalized ground state $|0\rangle$ with $\langle 0|0\rangle = 1$ and proceed algebraically.

Consider the vectors

$$|n\rangle \overset{\text{def}}{=} (a^\dagger)^n|0\rangle \quad \text{with} \quad n = 0, 1, 2, \ldots. \tag{3.93}$$

By the **Lemma** the $|n\rangle$'s are non-zero eigenvectors of the Hamiltonian H

$$H|n\rangle = \hbar\omega\left(n + \frac{1}{2}\right)|n\rangle \quad n \in \mathbb{N}, \tag{3.94}$$

or, equivalently,

$$a^\dagger a|n\rangle = n|n\rangle. \tag{3.95}$$

The operator $N \equiv a^\dagger a$ is also called the *number operator*: its spectrum are the non-negative integers. Now

$$a|n\rangle = aa^\dagger|n-1\rangle = [a, a^\dagger]|n-1\rangle + a^\dagger a|n-1\rangle =$$
$$= |n-1\rangle + (n-1)|n-1\rangle \equiv n|n-1\rangle. \tag{3.96}$$

Hence the Hilbert space completion $\widehat{\oplus}_n \mathbb{C}|n\rangle$ of the vector space $\oplus_n \mathbb{C}|n\rangle$ is an *irreducible* cyclic representation of the canonical algebra (3.86) which is contained in $\mathcal{H}$. In the language of Representation Theory [6, 7], $\oplus_n \mathbb{C}|n\rangle$ is the *Verma $\mathfrak{L}$-module* of the highest weight vector $|0\rangle$. A unitary Verma module is *irreducible* iff it does not contain any non-zero zero-norm vector, as it is our case by Eq. (3.90). Since (by assumption) $\mathcal{H}$ is irreducible as a representation of the enveloping algebra $\mathfrak{A}_{\text{can}}$, the space $\widehat{\oplus}_n \mathbb{C}|n\rangle$ is the full Hilbert space, and $\{|n\rangle\}_{n \in \mathbb{N}}$ is a complete system of eigenvectors of the Hamiltonian (3.84). In particular $|0\rangle$ is the *only* ground state, and all energy eigenvalues are *non-degenerate*. The spectrum is *purely discrete* and all energy eigenstates are *normalizable*. In facts we have

$$\| |n+1\rangle \|^2 = \langle n+1|n+1\rangle = \langle n|aa^\dagger|n\rangle =$$
$$= \langle n|(a^\dagger a + 1)|n\rangle = (n+1)\langle n|n\rangle \equiv (n+1)\| |n\rangle \|^2 \tag{3.97}$$

a recursion relation which, for our initial condition $\langle 0|0\rangle = 1$ yields

$$\langle n|n\rangle = n! \tag{3.98}$$

Corollary 3.1 *In $\mathcal{H}$ we have the resolution of the identity*

$$1 = \sum_{n=0}^{\infty} |n\rangle \frac{1}{n!} \langle n|. \tag{3.99}$$

SUMMARY: Spectrum of the Harmonic Oscillator
The one-dimensional oscillator (3.84) has a purely discrete energy spectrum

$$E_n = \hbar\omega\left(n + \tfrac{1}{2}\right) \qquad n = 0, 1, 2, \cdots \in \mathbb{N} \tag{3.100}$$

with normalized energy eigenstates

$$|E_n\rangle \equiv \frac{1}{\sqrt{n!}}|n\rangle \equiv \frac{1}{\sqrt{n!}}(a^\dagger)^n|0\rangle \tag{3.101}$$

Remark 3.2 (Relation to Classical Adiabatic Invariants) In §. 10.3 of [1] it was shown that the energy/frequency ratio E/ω is an *adiabatic invariant* for a classical harmonic oscillator. The value of this ratio in the quantized oscillator is $\hbar(n + \tfrac{1}{2})$ in agreement with the arguments of the 1911 Solvay conference, see the quoted reference and Example 8.6.

3.5.4 Bargmann Quantization

Recall that the Bargmann Hilbert space $B^2(\mathbb{C})$ is the space of *entire functions* in $\mathbb{C}$ with the quadratic norm[13]

$$\|f(z)\|^2 = \int_{\mathbb{C}} e^{-z\bar{z}} \frac{dz\,d\bar{z}}{2\pi} |f(z)|^2, \tag{3.102}$$

---

[13] **Beware!** To avoid all misunderstand, in Eq. (3.103) we specify the measure in polar coordinates since there are different conventions about the normalization of $dz\,d\bar{z}$. Here $dz\,d\bar{z}$ stands for *twice* the usual area form in the plane $dx\,dy$.

$$\text{with measure } \mathrm{e}^{-z\bar{z}}\,\frac{\mathrm{d}z\,\mathrm{d}\bar{z}}{2\pi} \equiv \mathrm{e}^{-r^2}\mathrm{d}r^2\,\frac{\mathrm{d}\theta}{2\pi}, \tag{3.103}$$

where (r, θ) are the usual polar coordinates in the plane $\mathbb{R}^2 \simeq \mathbb{C}$ and the overbar stands for complex conjugation. In Problem 3.3 the reader checks that the linear map

$$\mathcal{H} \to B^2(\mathbb{C}), \qquad |n\rangle \mapsto z^n \tag{3.104}$$

is an isomorphism of Hilbert spaces. That is, in the Bargmann representation of $\mathcal{H}$ we have (in Dirac's notation):

$$\langle z|n\rangle = z^n, \qquad \langle n|w\rangle = \bar{w}^n. \tag{3.105}$$

The state $|w\rangle$ is the eigenstate $|w\rangle$ of a with eigenvalue $\bar{w}$. Indeed

$$\langle w|a^\dagger|n\rangle = \langle w|n+1\rangle = w^{n+1} = w\,\langle w|n\rangle \tag{3.106}$$

$$\Rightarrow \quad \langle n|a|w\rangle = \bar{w}\langle n|w\rangle \quad \Rightarrow \quad a|w\rangle = \bar{w}|w\rangle, \tag{3.107}$$

since $\{\langle n|\}$ is a dual Hilbert basis of $\mathcal{H}$.

The reproducing kernel in $B^2(\mathbb{C})$ is then (cf. Eq. (3.99))

$$\langle z|w\rangle = \sum_{n\geq 0}\langle z|n\rangle\frac{1}{n!}\langle n|w\rangle = \sum_{n\geq 0}\frac{(z\bar{w})^n}{n!} = \mathrm{e}^{z\bar{w}} \tag{3.108}$$

in agreement with (2.62). Up to overall normalization, the holomorphic function $\langle z|w\rangle \equiv \exp(z\bar{w})$ is also the Bargmann wave-function of the eigenstate $|w\rangle$ of a.

The Bargmann representation of the operators, acting on the entire functions $f(z)$, are then

$$a^\dagger = z, \qquad a = \frac{\partial}{\partial z} \tag{3.109}$$

$$H = \omega\hbar\left(z\frac{\partial}{\partial z} + \frac{1}{2}\right) \tag{3.110}$$

The Bargmann wave-functions are both simpler and—being holomorphic—more regular than the Schrödinger wave-functions to be discussed momentarily. The time-evolution kernel is particularly simple in this representation

$$\langle z|\mathrm{e}^{-\mathrm{i}Ht/\hbar}|w\rangle = \mathrm{e}^{-\mathrm{i}\omega t/2}\sum_n \langle z|n\rangle\frac{\mathrm{e}^{-\mathrm{i}n\omega t}}{n!}\langle n|w\rangle =$$

$$= \mathrm{e}^{-\mathrm{i}\omega t/2}\sum_n\frac{(\mathrm{e}^{-\mathrm{i}\omega t}z\bar{w})^n}{n!} = \exp\left[\mathrm{e}^{-\mathrm{i}\omega t}z\bar{w} - \frac{\mathrm{i}}{2}\omega t\right] \tag{3.111}$$

Note that when $\omega t \in 2\pi\mathbb{Z}$ this evolution kernel is $\pm$ the reproducing kernel ($\equiv$ identity operator), when $\omega t = \frac{3}{2}\pi + 2\pi\mathbb{Z}$ it is the Fourier transform kernel (up to an overall phase), while for $\omega t = \pi + 2\pi\mathbb{Z}$ it is the parity kernel (up to phase) in agreement with the Heisenberg picture analysis in Sect. 3.5.1.

The wave function of the time evolution of state $|w\rangle$ ($\bar{w} \in \mathbb{C}$) is

$$\langle z|w, t\rangle \equiv \langle z|e^{-iHt/\hbar}|w\rangle = e^{-i\omega t/2}\exp\!\left(e^{-i\omega t}z\bar{w}\right) = e^{-i\omega t/2}\langle z|e^{i\omega t}w\rangle,$$

$$(3.112)$$

thus, the time-evolved state $U(t)|w\rangle$ is simply

$$\exp(-iHt/\hbar)|w\rangle = e^{-i\omega t/2}|w(t)\rangle \equiv e^{-i\omega t/2}|e^{i\omega\hbar t}w\rangle, \tag{3.113}$$

that is, under time evolution the a-eigenvectors $|w\rangle$ remain a-eigenvectors, while the trajectories $\bar{w}(t)$ of the a-eigenvalues in the Bargmann plane are circles centered at the origin just as the classical trajectories in phase space of the harmonic oscillator.

Remark 3.3 The linear map

$$B^2(\mathbb{C}) \hookrightarrow L^2(\mathbb{C}) \equiv L^2(\mathbb{R}^2), \quad f(z) \mapsto \frac{1}{\sqrt{\pi}}\, f(x+iy)\, e^{-(x^2+y^2)/2}, \tag{3.114}$$

is an isometric closed immersion: the states of the harmonic oscillator may then be identified with a *special class* of states of a quantum system with two degrees of freedom living on the plane $\mathbb{R}^2$. The deep meaning of this identification will be clear when we describe the quantum mechanics of charged particles in presence of a magnetic field. At an even deeper level: the 2-degrees of freedom system is secretly supersymmetric and the special "holomorphic" states (3.114) are its BPS states.

3.5.5   The Schrödinger PDE

The Schrödinger representation in $L^2(\mathbb{R})$ of the various operators is

$$\frac{H}{\omega} = -\frac{\hbar^2}{2}\frac{d^2}{dx^2} + \frac{x^2}{2} \tag{3.115}$$

$$\sqrt{2\hbar}\,a = x + \hbar\frac{d}{dx} \tag{3.116}$$

$$\sqrt{2\hbar}\,a^\dagger = x - \hbar\frac{d}{dx}. \tag{3.117}$$

In principle, to solve the stationary Schrödinger equation, we have to solve the second order linear ODE

$$-\frac{\hbar^2}{2}\frac{\mathrm{d}^2\psi_E(x)}{\mathrm{d}x^2} + \frac{x^2}{2}\psi_E(x) = \frac{E}{\omega}\psi_E(x), \tag{3.118}$$

(the bi-confluent hypergeometric ODE) whose general solutions are known as *parabolic cylinder functions* [8, 9] see the off-text **BOX** on page 140. The quantum energy eigenfunctions of the harmonic oscillator are the *very special* solutions of (3.118) which belong to $S(\mathbb{R})^{\vee}$. Such solutions exist only for very special values of the parameter E namely for the *energy eigenvalues* ($\equiv$ the physically allowed quantized values of energy) in which case the special solution $\psi_E(x)$ is normalizable i.e. it belongs to $L^2(\mathbb{R})$.

Let us summarize the story from the viewpoint of the ODE (3.118). The second-order linear ODE (3.118) has two linearly-independent solutions. We can choose the basis of solutions so that the first solution has the asymptotic behavior as $x \to +\infty$

$$\psi_E(x) \approx e^{-x^2/2\hbar}\left(\sqrt{\frac{2}{\hbar}}\,x\right)^{\frac{E}{\hbar\omega}-\frac{1}{2}}\left(1 + \cdots\right), \qquad x \to +\infty, \tag{3.119}$$

while all other linear independent solutions behave as[14]

$$\psi(x)_{\mathrm{other}} \approx C\,e^{x^2/2\hbar}\left(\sqrt{\frac{2}{\hbar}}\,x\right)^{-\frac{E}{\hbar\omega}-\frac{1}{2}}\left(1 + \cdots\right), \qquad C \neq 0. \tag{3.120}$$

The preferred solution with asymptotic behavior (3.119) will be constructed in Sect. 3.9 (see Eq. (3.261)): it has the integral representation ($E/\omega \geq 0$)

$$\psi_E(x) = C\,e^{x^2/2}\int_0^\infty \mathrm{d}t\; t^{E/\omega-1/2}e^{-t^2/2}\,\cos\left[\sqrt{2}\,xt - \frac{\pi}{2}\left(\frac{E}{\omega} - \frac{1}{2}\right)\right]. \tag{3.121}$$

Except for this particular solution $\psi_E(x)$, all other parabolic cylinder functions, $\psi(x)_{\mathrm{other}}$ grow too rapidly as $x \to +\infty$ to be in $S(\mathbb{R})^{\vee}$. As $x \to -\infty$ the "good" solution $\psi_E(x)$ behaves asymptotically as

$$\psi_E(x) \approx \frac{\sqrt{2\pi}}{\Gamma\left(\frac{1}{2} - \frac{E}{\hbar\omega}\right)}\,e^{x^2/2\hbar}\left(\sqrt{\frac{2}{\hbar}}\,|x|\right)^{-\frac{E}{\hbar\omega}-\frac{1}{2}}\left(1 + \cdots\right)+$$

$$+ \sin\left(\frac{\pi E}{\hbar\omega}\right)e^{-x^2/2\hbar}\left(\sqrt{\frac{2}{\hbar}}\,|x|\right)^{\frac{E}{\hbar\omega}-\frac{1}{2}}\left(1 + \cdots\right). \tag{3.122}$$

---

[14] This follow from Eq. (3.167) which is valid for all 2nd order linear ODE and proven in that general context.

$\psi_E(x)$ belongs to $\mathcal{S}(\mathbb{R})^\vee$ iff the first term in the RHS vanishes. The entire function $1/\Gamma(z)$ has simple zeros along the negative axis at

$$z = 0, -1, -2, -3, \ldots, \tag{3.123}$$

and hence $\psi_E \in \mathcal{S}(\mathbb{R})^\vee$ if and only if

$$E = \hbar\omega(n + 1/2) \quad n \in \mathbb{N} \tag{3.124}$$

in agreement with the previous algebraic analysis.

BOX: Parabolic Cylinder Functions

The parabolic cylinder functions owe their name to the fact that they enter in the separation of variables for the Laplacian in $\mathbb{R}^3$—that is, in the separation of variables of the *superseparable* free quantum motion in $\mathbb{R}^3$—in the parabolic cylinder coordinates (ξ, η, z)

$$x_1 = \frac{1}{2}(\xi^2 - \eta^2), \quad x_2 = \xi\eta, \quad x_3 = z$$

instead of the more usual Cartesian (or polar) coordinates (cf. [1] §. 9.6.1). One has

$$H = \frac{p^2}{2m} = \frac{\hbar^2}{2m}\Delta = -\frac{\hbar^2}{2m}\left(\frac{\partial^2}{\partial z^2} + \frac{1}{\xi^2 + \eta^2}\left(\frac{\partial^2}{\partial \xi^2} + \frac{\partial^2}{\partial \eta^2}\right)\right)$$

Looking for energy eigenfunctions of the free system in the form

$$e^{ip_z z/\hbar}\chi(\xi)\phi(\eta)$$

we obtain the two decoupled ODEs

$$-\frac{\hbar^2}{2m}\frac{d^2\chi(\xi)}{d\xi^2} + \left(\frac{p_z^2}{2m} - E\right)\xi^2\,\chi(\xi) = \lambda\chi(\xi)$$

$$-\frac{\hbar^2}{2m}\frac{d^2\phi(\eta)}{d\eta^2} + \left(\frac{p_z^2}{2m} - E\right)\eta^2\,\phi(\eta) = -\lambda\phi(\eta)$$

both of which can be recast in the form (3.118) by a linear redefinition of the variables. The solutions $\chi(\xi)$, $\phi(\eta)$ of these equations are called *parabolic cylinder functions* [8, 9].

Close Form Solution

Solving Eq. (3.118) is greatly simplified by exploiting the conclusions of the algebraic approach. The ground state wave-function $\psi_0(x) = \langle x|0\rangle$ satisfies the 1st order differential equation

$$0 = \sqrt{2\hbar}\,\langle x|a|0\rangle = \left(x + \hbar\frac{d}{dx}\right)\psi_0(x), \tag{3.125}$$

so the normalized Schrödinger wave-function of the fundamental state is

$$\psi_0(x) = \frac{e^{-x^2/2\hbar}}{(\pi\hbar)^{1/4}}. \tag{3.126}$$

Remark 3.4 The Hamiltonian (3.84) is invariant under the interchange $q \leftrightarrow p$ and hence the Schrödinger equation written in the momentum representation is *identical* to the one written in the q-representation. Hence (3.126) is also the wave-function in the momentum representation (up to the different conventions in normalization)

$$\langle p|0\rangle = \sqrt{2}(\pi\hbar)^{1/4}\,e^{-p^2/2\hbar}. \tag{3.127}$$

In particular (3.126) is its own Fourier transform.

The other (normalized) energy eigenstates are

$$\psi_n(x) = \frac{1}{\sqrt{n!}}\langle x|(a^\dagger)^n|0\rangle = \frac{1}{\sqrt{(2\hbar)^{n+1}\,n!}}\left(x - \hbar\frac{d}{dx}\right)^n \exp\left(-\frac{x^2}{2\hbar}\right) \equiv$$

$$\equiv (-1)^n \frac{1}{(\pi\hbar)^{1/4}}\sqrt{\frac{\hbar^n}{2^n\,n!}}\,e^{x^2/2\hbar}\frac{d^n}{dx^n}e^{-x^2/\hbar}. \tag{3.128}$$

We see that the solving the 2nd order linear ODE (3.118) when $2E/\hbar\omega$ is an odd integer reduces to solving a 1st order linear ODEs: on general grounds this happens whenever the connected component of the *differential Galois group* of the equation is *solvable* (as a linear algebraic group [10]), see e.g. **Theorem 1.43** in [11] or **Theorem 5.21** in [12]. The differential Galois group for the one-dimensional Schrödinger equation with a non-constant *polynomial* potential $V(x)$ is briefly illustrated in the **BOX** on page 145.

Hermite Polynomials and All that

In the rest of this section we set $\hbar = 1$, as a choice of units, as it is customary nowadays. It is clear from the Leibniz rule that for all $n \in \mathbb{N}$ the real functions

$$H_n(x) \stackrel{\text{def}}{=} (-1)^n\,e^{x^2}\frac{d^n}{dx^n}e^{-x^2} \tag{3.129}$$

Table 3.1 Hermite polynomials for small n

n	$H_n(x)$	n	$H_n(x)$
0	1	1	$2x$
2	$4x^2 - 2$	3	$8x^3 - 12x$
4	$16x^4 - 48x^2 + 12$	5	$32x^5 - 160x^3 + 120x$
6	$64x^6 - 480x^4 + 720x^2 - 120$	7	$128x^7 - 1344x^5 + 3360x^3 - 1680x$

are polynomials $H_n(x) \in \mathbb{Z}[x]$ of degree n with the symmetry

$$H_n(-x) = (-1)^n H_n(x) \tag{3.130}$$

i.e. the function $H_n(x)$ is *even* (*odd*) for n *even* (resp. *odd*). $H_n(x)$ is called the n-th *Hermite polynomial*. The first few Hermitian polynomials are listed in Table 3.1.

> **SUMMARY: Harmonic Oscillator Wave-Functions**
> The Schrödinger wave-function of the energy eigenstate of energy $E_n = \omega(n + \frac{1}{2})$ of the one-dimensional harmonic oscillator of frequency ω (setting $\hbar = m = 1$) is
>
> $$\psi_n(x) = \frac{1}{2^{n/2}\pi^{1/4}\sqrt{n!}} H_n(x)\,e^{-x^2/2} \tag{3.131}$$
>
> $$\langle x|n \rangle = \frac{1}{2^{n/2}\pi^{1/4}} H_n(x)\,e^{-x^2/2} \tag{3.132}$$
>
> where $H_n(x)$ is the n-th Hermite polynomial.

From Eq. (3.131) we get the explicit integral kernel of the Hilbert-space isomorphism $B^2(\mathbb{C}) \to L^2(\mathbb{R})$:

$$\langle x|z \rangle = \sum_n \langle x|n \rangle \frac{1}{n!} \langle n|z \rangle = \frac{1}{\pi^{1/4}} \sum_n \frac{1}{n!} \left(\frac{\bar{z}}{\sqrt{2}} \right)^n H_n(x)e^{-x^2/2} =$$

$$= \frac{1}{\pi^{1/4}} \sum_n \frac{1}{n!} \left(-\frac{\bar{z}}{\sqrt{2}} \right)^n \frac{d^n}{dx^n} e^{-x^2/2} = \frac{1}{\pi^{1/4}} \exp\left[-\frac{1}{2}\left(x - \frac{\bar{z}}{\sqrt{2}} \right)^2 \right] \tag{3.133}$$

where in the last equality we used the Taylor expansion of the entire function $f(\bar{z}) = e^{-(x-\bar{z}/\sqrt{2})^2}$ around $\bar{z} = 0$.

The Hermite polynomials have remarkable properties: they are our first example of a sequence of *orthogonal polynomials*. Orthogonal polynomials have a deep theory [13, 14].

Proposition 3.1 *The Hermite polynomials $H_n(x)$ have the properties:*

1. Orthogonality

$$\int_{-\infty}^{+\infty} H_m(x)\, H_n(x)\, e^{-x^2}\, dx = \sqrt{\pi}\, 2^n n!\, \delta_{mn}. \tag{3.134}$$

2. Differential Equation

$$H_n'' - 2x\, H_n' + 2n\, H_n = 0 \tag{3.135}$$

3. Three-Terms Recursion Relation

$$x\, H_n(x) = \frac{1}{2} H_{n+1}(x) + n\, H_{n-1}(x) \tag{3.136}$$

4. Derivative

$$\frac{d}{dx} H_n(x) = 2n\, H_{n-1}(x) \tag{3.137}$$

5. Leading Coefficient

$$H_n(x) = 2^n x^n + O(x^{n-2}). \tag{3.138}$$

6. Generating Function

$$\sum_{n=0}^{\infty} \frac{H_n(x)}{n!} z^n = e^{2xz - z^2} \tag{3.139}$$

7. Integral Representation.

$$H_n(x) = \frac{2^{n+1}}{\sqrt{\pi}} e^{x^2} \int_0^{\infty} e^{-t^2} t^n \cos(2xt - n\pi/2)\, dt. \tag{3.140}$$

Proof Equation (3.139) is equivalent to Eq. (3.133). The other properties are either obvious or consequences of this equation and the arguments discussed in Construction 5 of Sect. 2.2.   □

Remark 3.5 The orthonormal system of wave function $\psi_n(x)$ in Eq. (3.131) is the Hilbert basis of $L^2(\mathbb{R})$ we introduced in Construction 5 of Sect. 2.2. In particular the energy eigenfunction $\psi_n(x)$ of the harmonic oscillator are eigenvectors of the Fourier transform F

$$F\psi_n(p) = i^n \psi_n(p) \tag{3.141}$$

cf. Eq. (2.177). This is a consequence of the already mentioned symmetry $q \leftrightarrow p$ of the Hamiltonian (3.84). Since the energy eigenvalues are not degenerate, the symmetry must leave invariant all energy eigenvectors up to a phase.

Proposition 3.2 (Zeros) **(1)** $H_n(x)$ *has n simple real zeros symmetric under the parity transformation $x \rightarrow -x$. In particular $x = 0$ is a zero iff n is odd.* **(2)** *Between two zeros of $H_{n+1}(x)$ there is precisely one real zero of $H_n(x)$ and these are all the zeros of $H_n(x)$.* **(3)** *Between two zeros of $H_n(x)$ there is one zero of $H_{n+1}(x)$.*

Proof **(2)** follows from **(1)** in view of Eq. (3.137), the Rolle theorem, and the fact that $H_n(x)$ has at most n real zeros. Now we show that all zeros of $H_n(x)$ are simple. Suppose that x_0 is a zero of $H_{n+1}(x)$ of multiplicity $m \geq 2$. Equation (3.137) shows that x_0 is also a zero of $H_n(x)$. Then the 3-term recursion (3.136) implies that $H_n(x_0) = 0$ for all n. By this is impossible since $H_0(x) = 1$. (Alternatively: x_0 must be a zero of the RHS of (3.139) for all z, but this function has no zero). Now consider the functions

$$f_n(x) \equiv \frac{\mathrm{d}^n}{\mathrm{d}x^n} \mathrm{e}^{-x^2} . \tag{3.142}$$

The zeros on the real line of $f_n(x)$ are the real zeros of $H_n(x)$ plus two asymptotic zeros at $x = \pm\infty$. By the Rolle theorem, there is at least one zero of $f_n(x)$ between two zeros of $f_{n+1}(x)$ (including the asymptotic ones). Now proceed by induction in n. If the statement is true for n, $f_n(x)$ has n real zeros for finite x plus the two zeros at infinity. Hence $f_{n+1}(x)$ has at least $(n + 1)$ zeros at finite x, one in each interval between two zeros of $f_n(x)$ (finite or infinite). Hence $H_{n+1}(x)$ has at least $(n + 1)$ distinct real zeros. Since the polynomial $H_{n+1}(x)$ has at most $(n + 1)$ real zeros counted with multiplicities, we conclude that it has exactly $(n + 1)$ distinct real zeros. $\qquad\qquad\square$

We express properties **(2), (3)** by saying that the (real) zeros of $H_n(x)$ and $H_{n+1}(x)$ *interlace*.

Corollary 3.2 *The wave-function of the n-th energy eigenfunction $\psi_n(x)$ has n real zeros and the zeros of $\psi_n(x)$ interlace with the ones of $\psi_{n+1}(x)$. In particular the wave-function of the ground state vanishes nowhere.*

As we shall see in Sect. 3.7 this statement holds for all one-dimensional system with a purely discrete energy spectrum. In the general context this fact is called the *oscillation theorem* (cf. Theorem 3.7). Here we checked this general result in the special case of the harmonic oscillator by elementary means.

BOX: Differential Galois Group for the Schrödinger Equation

We consider the stationary Schrödinger equation for a one-dimensional system

$$\frac{d^2}{dx^2}\psi - r(x)\psi = 0 \qquad r(x) \equiv \frac{2m}{\hbar^2}\left(V(x) - E\right) \qquad (\spadesuit)$$

where $V(x)$ is a non-constant polynomial in x (for the more general case where $V(x)$ is a rational function see [3, 11]). One has

Theorem (Liouville 1841—see [11] page 29)

(1) The differential Galois group of the linear differential equation ($\spadesuit$) with $V(x)$ a non-constant polynomial, is either $SL(2, \mathbb{C})$ or its (solvable) Borel subgroup

$$B = \left\{ \begin{pmatrix} a & b \\ 0 & a^{-1} \end{pmatrix} \ \middle|\ a \in \mathbb{C}^{\times}, \ b \in \mathbb{C} \right\}$$

(2) Solving the 2nd ODE ($\spadesuit$) reduces to solving a chain of 1st linear ODEs if and only if the differential Galois group of the equation is B.

(3) When the potential is a non-constant polynomial, the differential Galois group is B if and only if $\deg V(x) = 2n$ and there are two polynomials v and F, with $\deg v = n$ such that the function

$$u \equiv v + \frac{F'}{F} \qquad (\clubsuit)$$

satisfies the Riccati equation

$$u' + u^2 = r. \qquad (\diamondsuit)$$

Corollary (See [3] *Example 8.2*) Suppose the potential $V(x) = m\omega^2(x - x_0)^2$ in ($\spadesuit$) is a quadratic polynomial in x. Then the ODE has differential Galois group B iff

$$r = (ax + b)^2 + a(2n + 1) \quad n = 0, 1, 2, \cdots .$$

When $a \equiv -m\omega/\hbar < 0$ we get precisely the harmonic oscillator Schrödinger equation with $E = \hbar\omega(n + 1/2)$ a physically allowed energy. Indeed, eq.($\clubsuit$) is satisfied with $v = ax + b$ while

$$F = H_n\left(\sqrt{|a|}\,x - \frac{b}{\sqrt{|a|}}\right)$$

with $H_n(z)$ the n-th Hermite polynomial. Indeed in this case the Riccati equation ($\diamondsuit$) is

$$F'' + 2(-|a|x + b)F' + 2n|a|\,F = 0$$

which is (up to a linear change of variable) the ODE for the n-th Hermite polynomial, cf. Eq. (3.135).

3.5.6 Coherent States Revisited

The *coherent states* in $L^2(\mathbb{R})$ were defined in Sect. 2.13 as the wave functions $\psi(x)$ on the real line with *minimal Heisenberg indeterminacy*. Comparing our present findings with Sect. 2.13, we see that the *coherent states* in $L^2(\mathbb{R})$ are precisely the eigenvectors of the lowering operator a

$$a|\phi\rangle = \frac{\lambda}{\sqrt{2\hbar}}|\phi\rangle \quad \Leftrightarrow \quad \left(x + \hbar\frac{\mathrm{d}}{\mathrm{d}x}\right)\phi = \lambda\phi, \tag{3.143}$$

with eigenvalue $\sqrt{2\hbar}\,\lambda \in \mathbb{C}$, whose solution is

$$\phi(x) = \frac{\mathrm{e}^{-(\mathrm{Re}\,\lambda)^2/2\hbar}}{(\pi\hbar)^{1/4}}\,\mathrm{e}^{-x^2/2\hbar+\lambda x/\hbar}. \tag{3.144}$$

Comparing with Eq. (3.108) we see that the Bargmann eigenstates $|w\rangle$ of a, i.e. of $\partial/\partial z$, are precisely the coherent states. In units where $\hbar = 1$, their Schrödinger representation wave-functions are

$$\langle x|z\rangle = \frac{1}{\pi^{1/4}}\exp\left(-x^2/2 + \sqrt{2}\bar{z}x - (\mathrm{Re}\,z)^2\right) =$$

$$= \frac{1}{\pi^{1/4}}\mathrm{e}^{-x^2/2+\bar{z}^2/2-(\mathrm{Re}\,z)^2}\sum_{n=0}^{\infty}\frac{H_n(x)}{2^{n/2}\,n!}\bar{z}^n = \tag{3.145}$$

$$= \mathrm{e}^{\bar{z}^2/2-(\mathrm{Re}\,z)^2}\sum_{n=0}^{\infty}\frac{\bar{z}^n}{\sqrt{n!}}\psi_n(x).$$

Morally speaking, the coherent states—while not localized in the real line $\mathbb{R}$—are "localized at $\bar{w} \in \mathbb{C}$" in the Bargmann complex plane. This is the best we can do in terms of localization of a physical state in the Bargmann plane, so "obviously" these states should minimize the Heisenberg "error". Their physical relevance, besides being "error minimizing", stems from the fact that, in the harmonic oscillator, *coherent states stay coherent under time evolution*, that is, they do not dissipate into states with larger Heisenberg indeterminacy.

Fact 3.3 *The time evolution of the harmonic oscillator maps coherent states into coherent states.*

Proof Equation (3.113) yields

$$\exp(-\mathrm{i}Ht/\hbar)|w\rangle = \mathrm{e}^{-\mathrm{i}\omega t/2}|\mathrm{e}^{\mathrm{i}\omega\hbar t}w\rangle, \tag{3.146}$$

so, up to an irrelevant overall phase, under time evolution the coherent state $|w\rangle$ goes to the coherent state $|\mathrm{e}^{\mathrm{i}\omega\hbar t}w\rangle$. $\qquad\square$

The trajectories of the coherent states in the Bargmann plane $\mathbb{C}$ are circles centered at the origin as the classical trajectories in phase space. Informally: the coherent states are the quantum states which behave "more closely" to classical states. The outcomes of measures on a coherent state also follows a "classical probability" pattern:

Fact 3.4 *When measuring the number observable* $N \equiv a^\dagger a$ *in a coherent state* $|z\rangle$, *the probability of the various outcomes* $n \in \mathbb{N}$ *follows the* Poisson distribution *with mean* $|z|^2$ *(in units* $\hbar = 1$*), that is,*

$$|\langle n|z\rangle|^2 = \frac{|z|^{2n}}{n!}\, e^{-|z|^2}. \tag{3.147}$$

3.5.7   Mehler Formula

We choose units so that $\hbar = 1$. The Schrödinger representation time evolution kernel (a.k.a. *propagator*) of the harmonic oscillator

$$\langle x|e^{-iHt}|y\rangle \quad \text{with} \quad H \equiv \frac{\omega}{2}\left(p^2 + q^2\right) \tag{3.148}$$

was computed by F.G. Mehler in 1866 (60 years before Schrödinger!) in view of its applications in harmonic analysis. The easiest way to get the kernel is to use the isomorphism with the Bargmann description. We start with an identity also called the Mehler formula

$$\sum_{n=0}^{\infty} \frac{H_n(x)\, H_n(y)}{2^n\, n!} z^n = (1 - z^2)^{-1/2} \exp\left(\frac{2xyz - (x^2 + y^2)z^2}{1 - z^2}\right) \tag{3.149}$$

Proof Write $z = 2w^2$. The LHS may be written

$$\int e^{-t\bar{t}} \frac{dt\, d\bar{t}}{2\pi} \sum_m \frac{H_m(x)}{m!} (w\bar{t})^m \sum_n \frac{H_n(y)}{n!} (wt)^n =$$
$$= \int e^{-t\bar{t}} \frac{dt\, d\bar{t}}{2\pi} \exp\left(2xw\bar{t} - w^2\bar{t}^2 + 2ywt - w^2t^2\right), \tag{3.150}$$

where the measure is normalized as in (3.103) and in the second line we used (3.139). Computing the Gaussian integral in the RHS, we get

$$\text{RHS of (3.150)} = (1 - 4w^4)^{-1/2} \exp\left(\frac{4xyw^2 - (x^2 + y^2)(2w^2)^2}{1 - 4w^4}\right). \tag{3.151}$$

$\square$

Replacing in this equation $2w^2 \equiv z \rightsquigarrow e^{-i\omega t}$, and using Eq. (3.131), we get

$$
\langle x|e^{-iHt}|y\rangle \equiv \sum_{n=0}^{\infty} \psi_n(x)e^{-i(n+\frac{1}{2})t}\psi_n(y) =
$$

$$
= \frac{1}{\sqrt{2\pi i \sin(\omega t)}}\exp\left[i(x^2+y^2)\frac{\cos(\omega t)}{2\sin(\omega t)} - i\frac{xy}{\sin(\omega t)}\right].
$$

(3.152)

The formula is often written for the imaginary time s in the form

$$
\langle x|e^{-Hs}|y\rangle = \frac{1}{\sqrt{2\pi \sinh(\omega s)}}\exp\left[-(x^2+y^2)\frac{\cosh(\omega s)}{2\sinh(\omega s)} + \frac{xy}{\sinh(\omega s)}\right].
$$

(3.153)

3.6   One-Dimensional Systems I: Generalities

Time-independent quantum systems of the class (3.3) with a *one-dimensional* configuration manifold $\mathcal{M}$ enjoy special properties and are particularly simple. In facts, in this situation the stationary Schrödinger equation is an *ordinary* differential equation instead of a *partial* one. The special properties of one-dimensional quantum systems parallel the well-known fact that all one-dimensional time-independent classical systems are *integrable* in the Liouville sense [1]. Quantum systems which are fully separable decompose in a sequence of one-dimensional problems, just as in classical Hamilton-Jacobi theory ([1] chaps. 9, 10). In this section we review the general properties of the class of time-independent, spinless, one-dimensional quantum systems.

We may assume $\mathcal{M}$ connected. Then we have four possible geometries:

(i) $\mathcal{M} = \mathbb{R}$, the real line (ii) $\mathcal{M} = \mathbb{R}_{\geq 0}$, the half-line

(iii) $\mathcal{M} = [0, L]$, the interval (iv) $\mathcal{M} = S^1$, the circle.

When $\partial\mathcal{M} \neq \varnothing$ we impose the Dirichlet boundary conditions on the Schrödinger wave-function

$$
\psi\big|_{\partial\mathcal{M}} = 0.
$$

(3.154)

Mathematically there exist other admissible boundary conditions such that H is essentially self-adjoint, see e.g. [15–17]. However these more general conditions look rather unnatural physically, and we shall disregard them.

2nd Order Linear ODEs
The stationary Schrödinger equation is a second-order linear ordinary differential equation. We briefly review the relevant results of the theory of this class of equations.

A general second-order linear ODE has the form

$$\left(p_0(x)\frac{\mathrm{d}^2}{\mathrm{d}x^2} + p_1(x)\frac{\mathrm{d}}{\mathrm{d}x} + p_2(x)\right)\psi = 0 \tag{3.155}$$

where the $p_i(x)$'s are given real functions of x and $p_0(x) \not\equiv 0$. The identity

$$\begin{aligned}
0 &= \left(p_0(x)\frac{\mathrm{d}^2}{\mathrm{d}x^2} + p_1(x)\frac{\mathrm{d}}{\mathrm{d}x} + p_2(x)\right)\psi \equiv \\
&\equiv p_0 \left(\frac{\mathrm{d}}{\mathrm{d}x} + \frac{p_1}{p_0}\right)\frac{\mathrm{d}\psi}{\mathrm{d}x} + p_2\psi \equiv \\
&\equiv p_0\,\mathrm{e}^{-\int^x p_1/p_0\,\mathrm{d}x'}\left[\frac{\mathrm{d}}{\mathrm{d}x}\left(\exp\left[\int^x \frac{p_1}{p_0}\,\mathrm{d}x'\right]\frac{\mathrm{d}\psi}{\mathrm{d}x}\right) + \exp\left[\int^x \frac{p_1}{p_0}\,\mathrm{d}x'\right]\frac{p_2}{p_0}\psi\right]
\end{aligned}$$
$$\tag{3.156}$$

allows to rewrite the Eq. (3.155) in the standard *Sturm-Liouville form*

$$\frac{\mathrm{d}}{\mathrm{d}x}\left(A(x)\frac{\mathrm{d}\psi}{\mathrm{d}x}\right) + B(x)\psi = 0, \tag{3.157}$$

where

$$A(x) \equiv \exp\left[\int^x \frac{p_1}{p_0}\,\mathrm{d}x'\right], \qquad B(x) \equiv \exp\left[\int^x \frac{p_1}{p_0}\,\mathrm{d}x'\right]\frac{p_2(x)}{p_0(x)}. \tag{3.158}$$

The advantage is that now the equation takes the form $L\psi = 0$ where the operator

$$L = -\frac{\mathrm{d}}{\mathrm{d}x}A(x)\frac{\mathrm{d}}{\mathrm{d}x} - B(x), \tag{3.159}$$

with our boundary condition (3.154), is essentially self-adjoint in $L^2(\mathcal{M})$ with respect to the Hermitian product

$$\langle \psi_1|\psi_2\rangle = \int_{\mathcal{M}} \mathrm{d}x\; \psi_1(x)^*\,\psi_2(x). \tag{3.160}$$

Fix a point $x_0 \in \mathcal{M}$. The Eq. (3.157) has a *unique* solution satisfying the initial conditions[15]

$$\psi(x_0) = a \qquad \psi'(x_0) = b. \tag{3.161}$$

---

[15] ψ' (resp. ψ'') is a short hand for $\mathrm{d}\psi/\mathrm{d}x$ (resp. $\mathrm{d}^2\psi/\mathrm{d}x^2$).

This implies that the space of solutions to Eq. (3.157) is a vector space of *dimension* 2 (over $\mathbb{R}$ or $\mathbb{C}$). In particular, all zeros of a non-zero solution are *simple*. Indeed, if x_0 is a zero of multiplicity > 1,

$$\psi(x_0) = \psi'(x_0) = 0, \tag{3.162}$$

and the solution vanishes identically. A (necessarily simple) zero of a wave-function is called a *node*. We may choose a basis of solutions to (3.157) consisting of *real* functions since L in Eq. (3.159) is a real differential operator. Given two real solutions $\psi_1(x)$ and $\psi_2(x)$ of (3.157) we form their *Wronskian*

$$W[\psi_1, \psi_2](x) \stackrel{\text{def}}{=} A(x)\Big(\psi_1(x)\psi_2'(x) - \psi_1'(x)\psi_2(x)\Big) \tag{3.163}$$

which satisfies the equation

$$\begin{aligned}
\frac{\mathrm{d}}{\mathrm{d}x} W[\psi_1, \psi_2] &= A\psi_1'\psi_2' + \psi_1\frac{\mathrm{d}}{\mathrm{d}x}\left(A\psi_2'\right) - \frac{\mathrm{d}}{\mathrm{d}x}\left(A\psi_1'\right)\psi_2 - A\psi_1'\psi_2' \\
&= \psi_1\left(-B\psi_2\right) - \left(-B\psi_1\right)\psi_2 = 0,
\end{aligned} \tag{3.164}$$

which is nothing else than the continuity Eq. (3.8) for the stationary complex solution $\psi_1 + i\psi_2$. Thus the Wroskian $W[\psi_1, \psi_2]$ of any two solutions is a *constant*. Two solutions ψ_1 and ψ_2 are *linearly dependent* if and only if their Wroskian vanishes at one point (hence everywhere). Indeed

$$\begin{aligned}
W[\psi_1, \psi_2] = 0 \quad &\Rightarrow \quad \frac{\psi_1'}{\psi_1} = \frac{\psi_2'}{\psi_2} \quad \Rightarrow \\
&\Rightarrow \quad \frac{\mathrm{d}}{\mathrm{d}x}\log(\psi_1/\psi_2) = 0 \quad \Rightarrow \quad \psi_1/\psi_2 = \text{const.}
\end{aligned} \tag{3.165}$$

Suppose we know one solution $\psi(x)$ of (3.157). To find a second, linearly independent, solution $\tilde{\psi}(x)$, and complete the basis of the space of solutions, it suffices to solve the first order ODE

$$1 = W[\psi, \tilde{\psi}] = A(\psi\tilde{\psi}' - \psi'\tilde{\psi}) \tag{3.166}$$

which is easily done by a quadrature

$$\frac{\mathrm{d}}{\mathrm{d}x}\frac{\tilde{\psi}}{\psi} = \frac{1}{A\psi^2} \quad \Rightarrow \quad \tilde{\psi}(x) = \psi(x)\int^x \frac{\mathrm{d}y}{A(y)\,\psi(y)^2}. \tag{3.167}$$

Sturm Comparison Theorem: Prüfer Equation

A central result of the Sturm-Liouville theory, with fundamental implications for Quantum Mechanics, is

Theorem 3.5 (Sturm Comparison Theorem[16]**)** *Suppose we have two 2nd order linear ODEs set in the standard form*

$$\frac{\mathrm{d}}{\mathrm{d}x}\left(A_1(x)\frac{\mathrm{d}\psi}{\mathrm{d}x} \right) + B_1(x)\psi = 0$$

$$\frac{\mathrm{d}}{\mathrm{d}x}\left(A_2(x)\frac{\mathrm{d}\phi}{\mathrm{d}x} \right) + B_2(x)\phi = 0,$$

(3.168)

where the functions $A_a(x) > 0$ and $B_a(x)$ $(a = 1, 2)$ are piece-wise continuous in some connected domain $R \subset \mathbb{R}$. We assume

$$\frac{1}{A_1(x)} \geq \frac{1}{A_2(x)} > 0, \qquad B_1(x) \geq B_2(x). \tag{3.169}$$

(1) *The zeros of the (non-zero) solutions $\psi(x)$ and $\phi(x)$ are simple, and they do not accumulate, i.e. in any finite interval (a, b) the solutions have at most finitely-many zeros.*

(2) *Let x_1, x_2 be two distinct zeros of a non-trivial solution $\phi(x)$ of the second equation. Then any solution $\psi(x)$ of the first equation has at least one zero x_0 in the interval $x_1 < x_0 < x_2$ unless we are in the trivial case: $A_1 = A_2$, $B_1 = B_2$ almost everywhere and $\psi(x) = \lambda\,\phi(x)$.*

Sketch of Proof We change the Sturm-Liouville equation (3.157) into its *Prüfer form* (a variant of the better known *Riccati form*[17]). Writing

$$\psi(x) = \rho(x) \sin \vartheta(x),$$

$$A(x)\,\psi'(x) = \rho(x) \cos \vartheta(x),$$

(3.170)

we reduce the 2nd order linear ODE (3.157) to a 1st order non-linear ODE for the function $\vartheta(x)$

$$\vartheta' = \frac{1}{A} \cos^2\vartheta + B \sin^2\vartheta, \tag{3.171}$$

---

[16] Cf. e.g. **Theorem 6.1.1** in [15]. In physics textbooks (say §.7.2 of [18]) one finds more elementary proofs. The argument in the main text yields additional valuable information to be used later.

[17] To be studied in Chap. 8 as a convenient approach to the semiclassical approximation.

which is equivalent to the original 2nd order ODE (3.157) in the sense that if we know a solution to (3.171) we may find an one-parameter family of solutions to (3.157) by mere quadrature

$$\rho(x) = \rho(x_0) \exp\left(\int_{x_0}^{x} \left(\frac{1}{A(y)} - B(y)\right) \sin\vartheta(y)\cos\vartheta(y)\,dy\right) \tag{3.172}$$

a formula which shows that $\rho(x) > 0$ everywhere in any non-trivial solution. We conclude that the zeros of the solution $\psi(x)$ are precisely the points x_i where

$$\vartheta(x_i) = k_i\pi \quad \text{with} \quad k_i \in \mathbb{Z}. \tag{3.173}$$

Remark 3.6 Setting $Y = \cot\vartheta$, Eq. (3.171) becomes the well-known *Riccati equation*

$$Y' + A^{-1}Y^2 + B = 0. \tag{3.174}$$

Lemma 3.2 *Let $x_1 < x_2$ be two consecutive zeros of a solution $\psi(x)$ to Eq. (3.157) where $A(x) > 0$. We have*

$$\vartheta(x_2) = \vartheta(x_1) + \pi. \tag{3.175}$$

More generally, the number $Z(a, b)$ of zeros of $\psi(x)$ in the interval $[a, b]$ satisfies

$$0 \le Z(a, b) - \frac{1}{\pi}\big(\vartheta(b) - \vartheta(a)\big) \le 1. \tag{3.176}$$

Proof of Lemma We see from Eq. (3.171) that $\vartheta(x)$ is a continuous function of x such that $\vartheta(x) \notin \pi\mathbb{Z}$ for $x_1 < x < x_2$. Therefore: for $x_1 < x < x_2$ the phase $\exp[i\vartheta(x)]$ either remains in the upper half-plane $\operatorname{Im} z > 0$ or in the lower half-plane $\operatorname{Im} z < 0$, and it exits the half-plane precisely at the zeros x_i of the solution $\psi(x)$. The **Lemma** claims that, whenever x crosses a zero of $\psi(x)$ in the positive direction, the phase $\exp[i\vartheta(x)]$ enters the opposite half-plane from the positive direction as indicated by the red arrows in the figure:

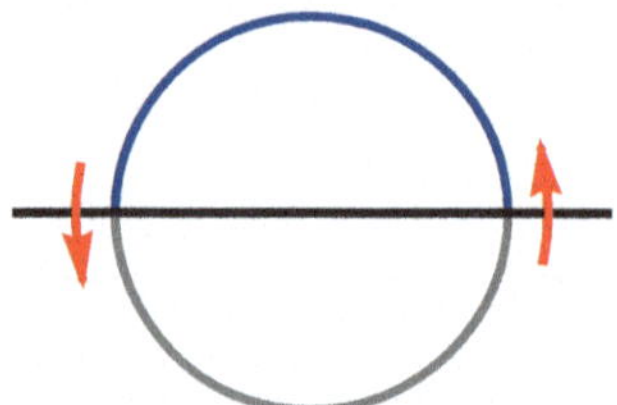

The **Lemma**: the phases in the upper (resp. lower) complex half-plane correspond to the blue (resp. gray) arc. When x crosses a zero of $\psi(x)$ in the positive direction, the phase $\exp[i\vartheta(x)]$ enters the opposite half-plane following one of the two red arrows

$$\tag{3.177}$$

Indeed, from the Prüfer equation (3.171) we see that when $\vartheta(x) \approx k\pi$ ($k \in \mathbb{Z}$) $d\vartheta(x)/dx > 0$ since $1/A > 0$. We conclude that the angle $\vartheta(x)$ *increases* when we cross a zero of ψ. Then $\vartheta(x_2) = \vartheta(x_1) - \pi$ is excluded, and we remain with only one possibility: $\vartheta(x_2) = \vartheta(x_1) + \pi$. $\square$

We go back to the proof of the **Sturm Theorem**. For part **(1)**, let a, b two arbitrary points in R

$$\frac{\vartheta(b) - \vartheta(a)}{\pi} = \frac{1}{\pi} \int_a^b \vartheta' \, dx \le$$

$$\le \frac{1}{\pi} \int_a^b \max\left\{\frac{1}{A(x)}, \max\{B(x), 0\}\right\} dx < \infty, \tag{3.178}$$

and hence the number of zeros $Z(a, b)$ is finite. Next we prove part **(2)**. We write the two equations (3.168) in the Prüfer form

$$\vartheta_1' = \frac{1}{A_1} \cos^2 \vartheta_1 + B_1 \sin^2 \vartheta_1 \equiv F_1(x, \vartheta_1),$$

$$\vartheta_2' = \frac{1}{A_2} \cos^2 \vartheta_2 + B_2 \sin^2 \vartheta_2 \equiv F_2(x, \vartheta_2) \tag{3.179}$$

Translating the angles ϑ_a by multiples of π, we may assume without loss of generality

$$0 = \vartheta_2(x_1) \le \vartheta_1(x_1) < \pi. \tag{3.180}$$

We set

$$L(x) = \frac{F_1(x, \vartheta_1(x)) - F_1(x, \vartheta_2(x))}{\vartheta_1(x) - \vartheta_2(x)}$$

$$H(x) = F_1(x, \vartheta_2) - F_2(x, \vartheta_2) \equiv \left(\frac{1}{A_1} - \frac{1}{A_2}\right) \cos^2 \vartheta_2 + (B_1 - B_2) \sin^2 \vartheta_2 \ge 0, \tag{3.181}$$

and write the differential relation

$$\vartheta_1' - \vartheta_2' = L(\vartheta_1 - \vartheta_2) + H \tag{3.182}$$

whose solution with the initial condition (3.180) is (for $x > x_1$)

$$\vartheta_1(x) - \vartheta_2(x) = \vartheta_1(x_1) \exp\left(\int_{x_1}^x L(y)dy\right) + \int_{x_1}^x dy\, H(y) \exp\left(\int_{x_1}^x L(t)dt\right) \ge 0 \tag{3.183}$$

where the inequality is saturated only in the trivial case. In the non-trivial case

$$\vartheta_1(x_2) > \vartheta_2(x_2) = \pi. \tag{3.184}$$

Since the angle $\vartheta_1(x)$ is a continuous function, Eqs. (3.180) and (3.183) imply that there is a point $x_1 < x_0 < x_2$ with $\vartheta_1(x_0) = \pi$, that is, there is at least one zero of $\psi(x)$ between two successive zeros of the second equation.   □

Remark 3.7 The bound (3.178) on the number of zeros in an interval is rather poor. For a Schrödinger equation in an interval (with Dirichlet boundary conditions) an optimal bound on the number of zeros may be obtained using a different version of the Prüfer trick, see Problem 3.8.

Rational Coefficients: Fuchsian Equations
A special case, crucial for the applications to Quantum Mechanics, is when p_a ($a = 1, 2, 3$) in Eq. (3.155) (resp. A, B in Eq. (3.157)) are polynomials (or rational functions). In this case we can write the equation in the form

$$\psi'' + q_1 \, \psi' + q_2 \, \psi = 0, \tag{3.185}$$

where the coefficients q_1 and q_2 are *rational functions*. Let $S \subset \mathbb{P}^1$ be the finite set of points in the Riemann sphere $\mathbb{P}^1$ where the functions q_1, q_2 have poles. The ODE may be extended to the complex domain $\mathbb{P}^1 \setminus S$

$$\frac{\mathrm{d}^2\psi(z)}{\mathrm{d}z^2} + q_1(z)\frac{\mathrm{d}\psi(z)}{\mathrm{d}z} + q_2(z)\,\psi(z) = 0, \qquad z \in \mathbb{P}^1 \setminus S. \tag{3.186}$$

As first realized by Riemann in 1851, it is easier and more enlightening to study the ODE (3.186) in this complex set-up, where the solutions $\psi(z)$ are multivalued holomorphic functions [12] and we may exploit the powerful tools of complex analysis. The name of the game is the *Riemann-Hilbert correspondence* which, under favorable conditions, allows us to solve the ODE by quadratures, i.e. produces an explicit integral representation of the solutions.

The points $z \in S$ are called *singularities* of the ODE. A point $z_0 \in S$, $z_0 \neq \infty$ is a *regular* singularity iff the two limits

$$\lim_{z \to z_0} (z - z_0)^a q_a(z) \qquad a = 1, 2 \tag{3.187}$$

are finite. When $z_0 = \infty$ we have to re-write the equation in terms of the good variable at infinity i.e. $w = 1/z$. Then the point ∞ is a regular singularity iff the two limits

$$\lim_{z \to \infty} z^a q_a(z) \qquad a = 1, 2 \tag{3.188}$$

are finite. An ODE with rational coefficients with only regular singularities in $\mathbb{P}^1$ (including the point at infinity) is called *Fuchsian*. There is a deep theory on *Fuchsian equations* [12, 19, 20].[18] In particular, a Fuchsian ODE with rigid monodromy can be solved by quadratures [12, 22, 23]: the method was pioneered by Riemann in 1851 who solved the hypergeometric ODE of order 2 (the higher order hypergeometric ODEs were solved soon later by Thomae using Riemann's breakthrough).

Order 2 Fuchsian equations are classified by:

Fu1   the number s and positions $\{z_1, z_2, \ldots, z_s\} \equiv S \subset \mathbb{P}^1$ of the regular singularities (modulo the automorphism group $SL(2, \mathbb{C})$ of $\mathbb{P}^1$). A singularity z_i is irrelevant iff for all pairs of solutions ψ_1, ψ_2 the limit $\lim_{z \to z_i} \psi_1(z)/\psi_2(z)$ is a non-zero constant. In this case the irrelevant singularity may be eliminated by a redefinition

$$\psi(z) \to f(z)\psi(z) \tag{3.189}$$

with $f(z)$ a suitable univalued function. We assume all irrelevant singularities have been eliminated;

Fu2   the *exponents* (α_i, β_i), $i = 1, \ldots, s$, at each singularity which describe the leading behavior as $z \to z_i$ of two independent solutions at a relevant regular singularity

$$\begin{aligned}
\alpha_i \neq \beta_i \quad & \psi_1(z) \sim (z - z_i)^{\alpha_i}, \quad \psi_1(z) \sim (z - z_i)^{\beta_i} \\
\alpha_i = \beta_i \quad & \psi_1(z) \sim (z - z_i)^{\alpha_i}, \quad \psi_1(z) \sim (z - z_i)^{\alpha_i} \log(z - z_i)
\end{aligned} \tag{3.190}$$

Fu3   other $2(s - 3)$ parameters called the *accessory parameters*.

When $s = 1$ we may assume $z_1 = \infty$ so q_1, q_2 are polynomials. The only polynomial with finite limits (3.188) is the zero polynomial. The solutions are elementary functions: $\psi = a + bz$. If $s = 2$ we may assume $z_1 = 0$ and $z_2 = \infty$. Regularity at 0 yields $q_a(z) = p_a(z)/z^a$ with p_a polynomials. Regularity at ∞ says that the p_a are constants c_a. Then the ODE is

$$0 = z^2 \frac{\mathrm{d}^2}{\mathrm{d}z^2}\psi + c_1 z \frac{\mathrm{d}}{\mathrm{d}z}\psi + c_2\psi \equiv \left(z\frac{\mathrm{d}}{\mathrm{d}z}\right)^2\psi + (c_1 - 1)z\frac{\mathrm{d}}{\mathrm{d}z}\psi + c_2\psi, \tag{3.191}$$

whose solutions are again elementary functions[19]

$$\psi_1 = z^{\alpha_1}, \qquad \psi_2 = z^{\alpha_2} \tag{3.192}$$

---

[18] For a nice survey, which puts the theory in the historic perspective and highlights the connections with several other branches of mathematics, see [21].

[19] This reflects the fact that the fundamental group of $\mathbb{P}^1 \setminus \{0, 1\}$ is Abelian.

where α_a $(a = 1, 2)$ are the two roots[20] of the quadratic equation

$$x^2 + (c_1 - 1)x + c_2 = 0. \tag{3.193}$$

When $s = 3$ we can fix the three (relevant) singularities at $\{0, 1, \infty\}$, so there is no essential freedom in the choice of S. In this case there are no accessory parameters, and the 2nd-order ODE is fully determined (up to redefinitions of the functions) by the exponents at the 3 regular singularities. The solutions for generic exponents are not elementary functions anymore (because the fundamental group of the 3-punctured sphere is non-Abelian[21]), yet the ODE can be solved by quadratures (since there are no accessory parameters). The Fuchsian equation with 3 regular singularities is the *hypergeometric equation* of Gauss [14, 21], and its general solution is due to Riemann. The standard way of writing the hypergeometric ODE is [14]

$$z(1 - z)\psi'' + [c - (a + b + 1)z]\psi' - ab\,\psi = 0, \tag{3.194}$$

where a, b, c are parameters related to the exponents [14].

More generally, we may have ODEs with rational coefficients and some *irregular singularities*. We may construct an irregular singularity at $z_0 \in \mathbb{P}^1$ starting with ℓ regular singularities at nearby points z_i and taking the limit $z_i \to z_0$ for $i = 1, 2, \ldots, \ell$. This is called a *confluent* limit: there are several such limits, depending on how many regular singularities we merge together at various points on the Riemann sphere.

For instance, by coalescing two singularities of the hypergeometric ODE (3.194) we get the *confluent hypergeometric ODE* [14] which has one regular and one irregular singularity of order 2. To get the confluent hypergeometric ODE in its standard form, called the *Kummer equation*, we replace in Eq. (3.194) z with z/b, so that now the singularities are at $z = 0, b, \infty$. Taking the limit $b \to \infty$ makes two singularities coalesce at infinity. We remain with the *Kummer equation*

$$z\,\psi'' + (c - z)\psi' - a\,\psi = 0. \tag{3.195}$$

The confluent hypergeometric equation plays a central role in Quantum Mechanics: it will be used again and again in the following sections and in Chap. 5. Other equivalent forms of this ODE will be introduced in those physical contexts.

Finally, making all 3 regular singularity to coalesce in a single singularity at ∞, we get $q_1 = (2az + b)$ and $q_2 = c$ (with a, b, c constants) and the ODE

$$\psi'' + (2az + b)\psi' + c\,\psi = 0 \qquad a \neq 0. \tag{3.196}$$

---

[20] When the two roots coincide the solutions are z^α and $z^\alpha \log z$.

[21] $\pi_1(\mathbb{P}^1 \setminus \{0, 1, \infty\})$ is isomorphic to the free group in two generators.

Shifting $z \rightsquigarrow z + b/2a$ we may set $b = 0$. Then setting $\psi(z) = \mathrm{e}^{-az^2/2}\chi(z)$, we get the *parabolic cylinder ODE*

$$- \chi'' + a^2 z^2 \chi = (c - a)\chi, \tag{3.197}$$

which for appropriate values of the parameters a, c becomes the Schrödinger equation for the harmonic oscillator we studied in the previous section.

3.7   One-Dimensional Systems II: $\mathcal{M} = \mathbb{R}$

We focus on one-dimensional Schrödinger equations of the form

$$- \frac{\hbar^2}{2m} \frac{\mathrm{d}^2}{\mathrm{d}x^2} \psi(x) + V(x)\psi(x) = E\psi(x) \tag{3.198}$$

where $x \in \mathbb{R}$ and we are interested in solutions which either belong to $L^2(\mathbb{R})$ (actual eigenfunctions associated to discrete energy eigenvalues E), or to a natural larger space $\mathcal{S}^\vee$ for the generalized eigenfunctions in the continuous energy spectrum. We call such solutions of (3.198) *physically admissible*.

Since the space of solutions to (3.198) is two-dimensional, an energy eigenvalue E is at most doubly degenerate. For the discrete eigenvalues we have the stronger

Lemma 3.3 *Let $E \in \sigma_p(H)$ be a discrete energy level of a one-dimensional system with $\mathcal{M} = \mathbb{R}$. E is non-degenerate.*

Proof Suppose that we have two linearly independent solutions ψ_1, ψ_2 of Eq. (3.198) both in $L^2(\mathbb{R})$. Their Wronskian $W[\psi_1, \psi_2](x)$ is a non-zero constant. Since ψ_1, ψ_2 go to zero at infinity, $\lim_{x \to \pm\infty} W[\psi_1, \psi_2](x) = 0$ and we get a contradiction. $\qquad\square$

Symmetries
We consider the "geometric" symmetries[22] which act on the coordinate as

$$x \rightsquigarrow x' = f(x). \tag{3.199}$$

These symmetries are the isometries of $\mathbb{R}$ which leave invariant the potential $V(x)$. Recall that $\mathsf{Iso}(\mathbb{R}) = \mathbb{R} \rtimes \mathbb{Z}_2$ where $\mathbb{R}$ acts by *translation* $T_a \colon x \mapsto x + a$ and $\mathbb{Z}_2$ as *parity* $\mathcal{P}\colon x \mapsto -x$. A non-trivial potential with $V'(x) \neq 0$ may preserve at most a discrete subgroup of $\mathrm{Iso}(\mathbb{R})$:

---

[22] In addition to the "geometric" symmetries we may have "dynamical" symmetries which mix positions and momenta. The typical example of dynamical symmetries are the $U(n)$ symmetries of the n-dimensional isotropic harmonic oscillator discussed in Sect. 3.5.2.

Pa when $V(x)$ is an *even* function, $V(x) = V(-x)$, the time-evolution conserves *parity* $\mathscr{P}$.

$$
\begin{aligned}
\mathscr{P}^2 &= 1, & \mathscr{P}\mathscr{P}^\dagger &= \mathscr{P}^\dagger\mathscr{P} = 1, \\
\mathscr{P}\boldsymbol{q}\mathscr{P} &= -\boldsymbol{q}, & \mathscr{P}\boldsymbol{p}\mathscr{P} &= -\boldsymbol{p}, \\
\langle x|\mathscr{P}|\psi\rangle &= \langle -x|\psi\rangle, & \langle p|\mathscr{P}|\psi\rangle &= \langle -p|\psi\rangle.
\end{aligned}
\tag{3.200}
$$

In this case the solutions of the Schrödinger equation (3.198) may be chosen to have definite parity

$$
\psi_\pm(-x) = \pm\psi_\pm(x);
\tag{3.201}
$$

Tr if $V(x)$ is *periodic* of period a,

$$
V(x + a) = V(x),
\tag{3.202}
$$

the *translation* T_a by a is a symmetry, and its unitary generator $\exp(ia\,\boldsymbol{p}/\hbar)$ may be diagonalized simultaneously with the Hamiltonian H.

Energy Spectrum and Eigenfunctions

We shall discuss periodic potentials in Sect. 3.12. For the moment we focus on a simpler class of potentials which satisfy the three conditions:

V1 $V(x)$ is a piece-wise continuous, locally bounded function on $\mathbb{R}$;
V2 $V(x)$ is bounded below[23] $V \geq V_{\min}$;
V3 the two limits

$$
\lim_{x\to-\infty} V(x) \equiv V_{-\infty}, \qquad \lim_{x\to+\infty} V(x) \equiv V_{+\infty},
\tag{3.203}
$$

exist finite or infinite.

An example of a potential satisfying these assumptions is shown in Fig. 3.1. When **V1, V2, V3** are satisfied we have the trivial bound on the energy spectrum $E \geq V_{\min}$ for all $E \in \sigma(H)$.

Fix the energy E. Classically the particle cannot enter in the region

$$
\mathsf{CF} \equiv \left\{x \in \mathbb{R} \mid E < V(x)\right\} \subset \mathbb{R}
\tag{3.204}
$$

which is called the *classically forbidden region* (cf. [1] §. 5.1). This is no longer true in the quantum case: *there is a small but non-zero probability of finding the particle in the region* CF. Indeed, if a solution $\psi(x)$ to Eq. (3.198) vanishes in an open

---

[23] This condition may be slightly relaxed.

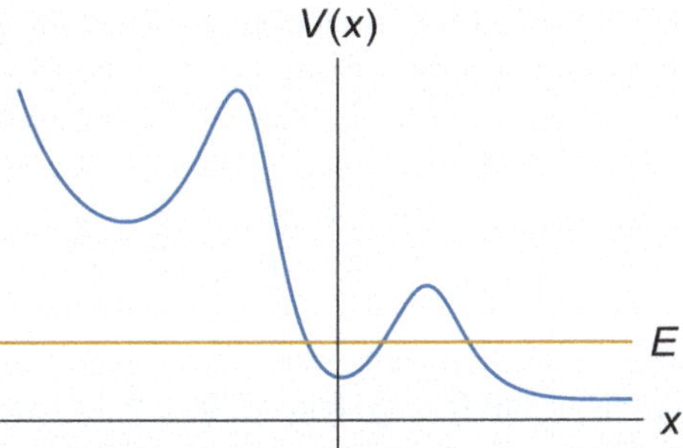

Fig. 3.1 An example of the graph of the potential $V(x)$. In this particular example we have $\lim_{x \to +\infty} V(x) = V_{+\infty} < \infty$ while $\lim_{x \to -\infty} V(x) = +\infty$. The regions where the function $V(x)$ is below (resp. above) the line of constant energy E (the *orange horizontal line*) are the *classically allowed* (resp. *forbidden*) regions at energy E

interval $I \subset \mathsf{CF} \subset \mathbb{R}$, for $x_0 \in I$ we have $\psi(x_0) = \psi'(x_0) = 0$ and the solution is identically zero, while the zero function does not represent any state (cf. Chap. 1). The fact that a quantum particle may penetrate the classically non accessible region is called the *tunneling effect* (more on this below).

Asymptotics of Energy Eigenfunctions The asymptotic form of the Schrödinger equation (3.198) as $x \to \pm\infty$ reads

$$-\frac{\hbar^2}{2m}\frac{\mathrm{d}^2}{\mathrm{d}x^2}\psi(x)_{\pm\infty} + V_{\pm\infty}\,\psi(x)_{\pm\infty} = E\,\psi(x)_{\pm\infty}, \tag{3.205}$$

whose solutions are

$$\psi(x)_{\pm\infty} = A_{\pm\infty}\exp\!\left(\sqrt{2m(V_{\pm\infty} - E)}\,x/\hbar\right) + $$
$$+ B_{\pm\infty}\exp\!\left(-\sqrt{2m(V_{\pm\infty} - E)}\,x/\hbar\right) \tag{3.206}$$

with $A_{\pm\infty}$, $B_{\pm\infty}$ constants. Equation (3.206) yields the asymptotic form of the solutions to (3.198) as $x \to \pm\infty$ under the assumption that the limits $V_{\pm\infty}$ exist *and are finite*.

When the region of large positive x belongs to the classically *allowed* region, $V_{+\infty} - E < 0$, the asymptotic solutions (3.206) are *bounded* and *oscillatory*

$$\psi(x) \approx A_{+\infty}\exp\!\left(\mathrm{i}\,\sqrt{2m|V_{+\infty} - E|}/\hbar\,x\right) + $$
$$+ B_{+\infty}\exp\!\left(-\mathrm{i}\,\sqrt{2m|V_{+\infty} - E|}/\hbar\,x\right) \quad \text{as } x \to +\infty. \tag{3.207}$$

This formula has a clear physical interpretation: at $+\infty$ the potential is constant and the asymptotic Hamiltonian differs from the free one (3.41) by an inessential additive constant. Hence the asymptotic solutions for large x are given by the plane waves

$$\psi(x) \approx A\, e^{ipx/\hbar} + B\, e^{-ipx/\hbar} \quad \text{where} \quad p \equiv \sqrt{2m(E - V_{+\infty})}. \tag{3.208}$$

Thus states with energy $E > V_{+\infty}$ are *scattering* states whose asymptotic form is given by plane waves as in the free system of Sect. 3.4. The first term in (3.208) describes a state of positive momentum, i.e. a particle outgoing to $+\infty$, whereas the second term represents a particle ingoing from infinity. Both solutions are physically acceptable: they are not normalizable, so—*if* an admissible solution exists with the asymptotics (3.208) for large positive x—it should belong to the continuous spectrum of energy, in agreement with the Ruelle theorem characterization of scattering states.

On the contrary, when the region of large positive x belongs to the classically *forbidden* region, $E < V_{+\infty}$, the solutions (3.207) are real exponentials. The first term grows exponentially and this behavior is not physically admissible.[24] Hence $A_{+\infty} = 0$, and we remain with the exponentially small asymptotic solution

$$\psi(x) \approx B_{+\infty} \exp\left(-\sqrt{2m(V_{+\infty} - E)}\, x/\hbar\right) \quad \text{as } x \to +\infty, \tag{3.209}$$

which expresses the physical fact that the wave-function decays exponentially as the particle enters deeper and deeper in the classically *forbidden* region. So, while the probability of finding the particle in classical inaccessible regions is *non zero*, it is *exponentially small*, and it vanishes quite rapidly in the classical limit $\hbar \to 0$.

Equation (3.209) holds when the limit V_∞ is finite. When $V_\infty = +\infty$ the wave function $\psi(x)$ vanishes as $x \to +\infty$ *more rapidly* than all exponentials

$$\log \psi(x) \approx -\frac{1}{\hbar} \int^x \sqrt{2m\, V(x)}\, dx \quad \text{as } x \to +\infty. \tag{3.210}$$

Corollary 3.3 *A generalized energy eigenvalue $E < \max\{V_{-\infty}, V_{+\infty}\}$ is non-degenerate.*

Indeed the Wronskian of two admissible solutions goes to zero in at least one of the two limits $x \to \pm\infty$ and hence is zero everywhere.

First Examples
We illustrate the main features in a few simple examples. To simplify the expressions we set $2m = \hbar = 1$.

---

[24] Indeed an exponentially growing function is not a tempered distribution.

Example 3.2 (Step Potential) Suppose the potential is $V(x) = V_0\,\Theta(x)$ where $\Theta(x)$ is the step function (2.164), see the following figure

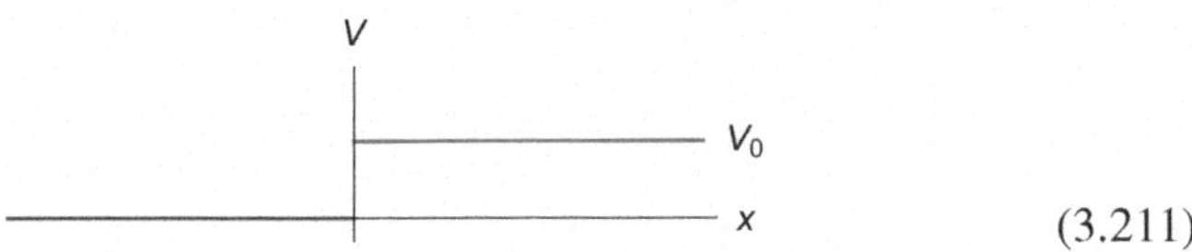

$$(3.211)$$

A wave-function with energy in the range $0 < E < V_0$ has the form

$$\psi_E(x) = \begin{cases} A\exp(-\sqrt{V_0 - E}\,x) & x > 0 \\ B\exp(\mathrm{i}\sqrt{E}\,x) + C\exp(-\mathrm{i}\sqrt{E}\,x) & x < 0. \end{cases} \tag{3.212}$$

The wave function should be of class C^1. Equating the function and its first derivative on the two sides of the origin, we get

$$C^* = B = \frac{A}{2}\left(1 + \mathrm{i}\sqrt{\frac{V_0}{E} - 1}\right). \tag{3.213}$$

Hence the properly normalized energy eigenfunction with $0 < E < V_0$ is

$$\psi_E(x) = \begin{cases} e^{\mathrm{i}(\sqrt{E}x - \phi_E)} + e^{-\mathrm{i}(\sqrt{E}x - \phi_E)} & x < 0 \\ 2\sqrt{\dfrac{E}{V_0}}\,e^{-\sqrt{V_0 - E}\,x} & x > 0 \end{cases} \tag{3.214}$$

where

$$\tan\phi_E = -\sqrt{\frac{V_0}{E} - 1}. \tag{3.215}$$

The term $e^{\mathrm{i}(\sqrt{E}x - \phi_E)}$ in the wave-function $\psi(x)$ for $x < 0$ represents the *direct* wave with positive momentum $p = \sqrt{E}$ incoming from $x = -\infty$, while the other term $e^{-\mathrm{i}(\sqrt{E}x - \phi_E)}$ is the *reflected back* wave with $p = -\sqrt{E}$ outgoing to $-\infty$.

We stress that *the direct and reflected waves have the **same** amplitude*: in view of the continuity Eq. (3.8), this means that there is no net transmission to the region of large positive x. Indeed the wave-function decays exponentially in the classically forbidden region, see the figure:

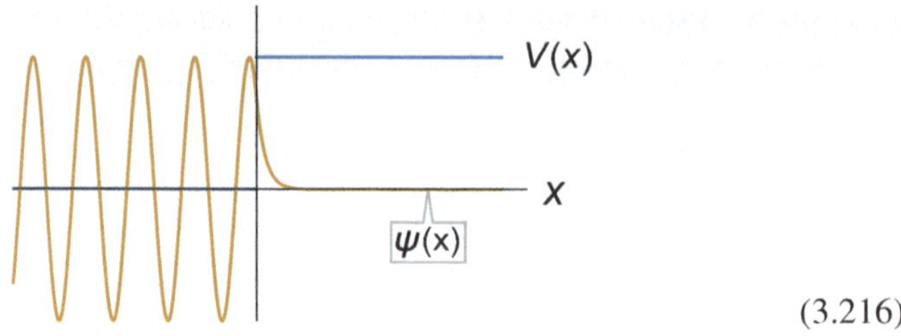

$$(3.216)$$

The reflected wave has a relative phase $\exp(2i\phi_E)$ with respect to the incoming one: this *phase shift* is a fundamental quantum effect: it says that the wave penetrates a little bit in the barrier before getting reflected back: as $V_0 \to \infty$, $\psi(x) \to 0$ for $x \geq 0$ and $\exp(2\phi_E) \to -1$ as for the classical reflection of a wave. The phase shift $\exp(2i\phi_E)$ as a function of E is a crucial physical property of a reflecting potential: they are the eigenvalues of the *scattering matrix* [24].

For the energy eigenstates with energy $E > V_0$ see Problem 3.9.

Example 3.3 Slightly generalizing Example 3.2, consider a general potential

$$V(x) = \begin{cases} 0 & \text{for } x < 0 \\ F(x) & \text{for } x > 0 \end{cases}$$

$$(3.217)$$

$$\text{such that:} \quad \lim_{x \to +\infty} F(x) = V_0 > 0.$$

We look at the states with energy $0 < E < V_0$ that we expect to describe waves incoming from $x = -\infty$ and then reflected back with the same amplitude (by unitarity) but with a relative *phase shift* $\exp(2i\phi_E)$ which depends on the energy E and the shape of the function $F(x)$

$$\psi_E(x) = e^{i(px - \phi_E)} + e^{-i(px - \phi_E)} \quad \text{for } x < 0 \text{ where } p = \sqrt{E} \qquad (3.218)$$

(in units $2m = \hbar = 1$). Indeed for $x \geq 0$ and $E < V_0$ the Schrödinger equation has a unique solution $f_E(x)$ which does not blow up exponentially as $x \to +\infty$ and $f_E(0) = 1$. Since the wave function $\psi_E(x)$ is continuous at the origin,

$$\psi_E(x) = 2\cos\phi_E \, f_E(x) \quad \text{for } E \geq 0. \qquad (3.219)$$

Requiring that the derivative is also continuous at $x = 0$, we get

$$\tan\phi_E = \frac{f'_E(0)}{p \, f_E(0)}. \qquad (3.220)$$

Example 3.4 (Liouville Potential) Consider the Schrödinger equation with the Liouville potential ($2m = \hbar = 1$)

$$-\frac{\mathrm{d}^2\psi}{\mathrm{d}x^2} + \mathrm{e}^{2x}\psi = E\psi. \tag{3.221}$$

For $x > 0$ the potential is exponentially large, while for $x < 0$ it is exponentially small. Hence, for any energy $E > 0$, the wave-function should be *extremely suppressed* at large x, while for $x < 0$ it behaves essentially as a free particle with momentum $\pm\sqrt{E}$. Only one linear combination

$$\mathrm{e}^{-\mathrm{i}\phi_E}\mathrm{e}^{\mathrm{i}\sqrt{E}x} + \mathrm{e}^{\mathrm{i}\phi_E}\mathrm{e}^{-\mathrm{i}\sqrt{E}x} \tag{3.222}$$

of the ingoing and outgoing waves will appear in the scattering asymptotic region $x \to -\infty$ because the eigenstates are non-degenerate (cf. Corollary 3.3). As in the previous examples, the two waves have the same amplitude (i.e. the coefficient $\mathrm{e}^{\pm\mathrm{i}\phi_E}$ are phases) because there cannot be any transmission to $x \to +\infty$. The only effect of the potential on the asymptotic waves is the phase shift $\exp(2\mathrm{i}\phi_E)$ between the incoming and reflected waves.

A simple integration by parts shows the identity (cf. Exercise 3.5)

$$\left(\frac{\mathrm{d}^2}{\mathrm{d}x^2} - \mathrm{e}^{2x} + E\right)\int_0^\infty \exp\left[-\mathrm{e}^x\cosh t\right]\cos(\sqrt{E}t)\,\mathrm{d}t = 0 \tag{3.223}$$

where $E \geq 0$. Then the physically admissible solution of the Liouville Schrödinger equation (3.221) is

$$\psi_E(x) \equiv \int_0^\infty \exp\left[-\mathrm{e}^x\cosh t\right]\cos(\sqrt{E}t)\,\mathrm{d}t. \tag{3.224}$$

Comparing integral representations [9], we conclude

$$\psi_E(x) \equiv K_{\mathrm{i}\sqrt{E}}(\mathrm{e}^x), \tag{3.225}$$

where $K_\nu(z)$ is the modified Bessel function[25] of second kind with index ν. Here the index is imaginary: $\nu \equiv \mathrm{i}\sqrt{E}$. As $x \to +\infty$ the function (3.225) behaves as

$$K_{\mathrm{i}\sqrt{E}}(\mathrm{e}^x) \approx \sqrt{\frac{\pi}{2}}\,\mathrm{e}^{-x/2}\exp\left[-\mathrm{e}^x\right]\left(1 + O(\mathrm{e}^{-x})\right), \tag{3.226}$$

---

[25] See e.g. refs.[9, 14, 25, 26].

Fig. 3.2 The energy
eigenfunction of the Liouville
Schrödinger equation (3.221)
with $E = 4$

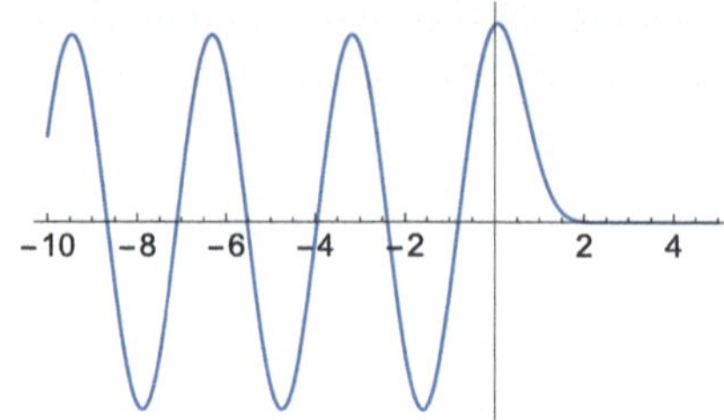

so it decays even faster than the exponential of the exponential, while for $x \to -\infty$

$$K_{i\sqrt{E}}(e^x) \approx -\sqrt{\pi}\,\frac{\sin\{\sqrt{E}(x - \log 2) - \arg \Gamma(1 + i\sqrt{E})\}}{\sqrt{\sqrt{E}\sinh(\pi\sqrt{E})}}\left(1 + O(e^{-|x|})\right)$$

$$(3.227)$$

is the sum of an ingoing and an outgoing plane wave with a relative phase

$$e^{2i\phi_E} = -\frac{\Gamma(1 + i\sqrt{E})}{\Gamma(1 - i\sqrt{E})}\,e^{2i\sqrt{E}\log 2}$$

$$(3.228)$$

See Fig. 3.2 for the graph of the eigenfunction with $E = 4$.

Bound States

The wave functions of bound states have special properties.

Fact 3.6 *A real eigenfunction which belongs to the* interior[26] *of the continuous spectrum has infinitely-many zeros. On the contrary the eigenfunction of a bound state below the threshold*

$$E < \min\{V_{-\infty}, V_{+\infty}\}$$

$$(3.229)$$

has only finitely-many zeros.

Indeed zeros do not accumulate in $\mathbb{R}$ (otherwise the solution vanishes) and when the condition (3.229) holds the wave-function is exponentially decreasing (hence non zero) outside a compact set. In the next paragraph we shall get a much stronger result (Theorem 3.7).

The expectation value of the momentum operator $\boldsymbol{p}$ vanishes in all bound states. Indeed their wave function is real and hence

$$\langle \psi | \boldsymbol{p} | \psi \rangle = \int_{\mathbb{R}} dx\, \psi(x)^* \left(-i\hbar\frac{d\psi(x)}{dx}\right)$$

$$(3.230)$$

---

[26] Eigenfunctions with energy on the boundary of $\sigma_e(H)$ are subtle and may not obey Fact 3.6. See, for instance, the solution with $E = 0$ of the Liouville model in Example 3.4.

is manifestly purely imaginary. Since the expectation value of a Hermitian operator is real, it must be zero.

Existence of Bound States A fundamental problem is to prove the (non)-existence of bound states for a given potential $V(x)$ and to understand the nature and properties of these states (when they are present). From the physical side it is crucial to know whether a given quantum system has or not *stable* bounded states that do not dissolve into several pieces.

For instance, showing that atoms and molecules exist, amounts to proving that a quantum system formed by nuclei and electrons has bound states where the several particles do not escape from each other, but stay together forming stable complex structures (atoms and molecules). It is of paramount importance to know which atoms and molecules may actually form and which ones are impossible, and to determine the physical properties of the bound systems that do exist. The structures that arise may be incredibly rich and subtle: think of the DNA molecules. All chemistry and most of biology—and indeed life itself—dwell on the theory of quantum bound states. See [27] chap. 10 for a survey of results on this fundamental problem (for several degrees of freedom). Here we limit to a very simple but useful observation:

Proposition 3.3 *Consider a one-dimensional system with $\mathcal{M} = \mathbb{R}$ and Hamiltonian $H = p^2/2m + V(q)$ where the potential $V(x)$ is a piece-wise continuous, locally bounded function such that*

$$\lim_{x \to -\infty} V(x) = \lim_{x \to +\infty} V(x) = V_0 < \infty, \tag{3.231}$$

while

$$\forall x : \ V(x) \leq V_0 \quad and \quad \exists\, x_0 \in \mathbb{R} : \ V(x_0) < V_0. \tag{3.232}$$

The system has at least one bound state of energy $E < V_0$.

The proof is based on the following obvious Lemma (cf. Problem 3.6):

Lemma 3.4 *Let $V_1(x)$, $V_2(x)$ be two piece-wise continuous, locally bounded potentials such that*

$$\lim_{x \to -\infty} V_1(x) = \lim_{x \to +\infty} V_1(x) =$$
$$= \lim_{x \to -\infty} V_2(x) = \lim_{x \to +\infty} V_2(x) = V_0 < \infty, \tag{3.233}$$

while $V_2(x) \leq V_1(x)$. If $V_1(x)$ has a bound state of energy $E_1 < V_0$, $V_2(x)$ has a bound state of energy $E_2 \leq E_1 < V_0$.

Proof of Proposition By piece-wise continuity there is a non-trivial interval $I \subset \mathbb{R}$ such that for $x \in I$, $V(x) \leq V_0 - \delta$ with $\delta > 0$. In the previous **Lemma** we set $V_2(x) = V(x)$ and

$$V_1(x) = \begin{cases} V_0 - \delta & x \in I \\ V_0 & x \notin I. \end{cases} \tag{3.234}$$

It remains to check that $V_1(x)$ has at least one bound state for all non-trivial interval I and $\delta > 0$. We refer to Sect. 3.13.1 for the explicit check. $\qquad\square$

Example 3.5 Consider the Schrödinger equation $(2m = \hbar = 1)$

$$-\frac{d^2\psi_E(x)}{dx^2} - \frac{2\,\psi_E(x)}{\cosh^2 x} = E\,\psi_E(x) \tag{3.235}$$

whose potential profile looks like

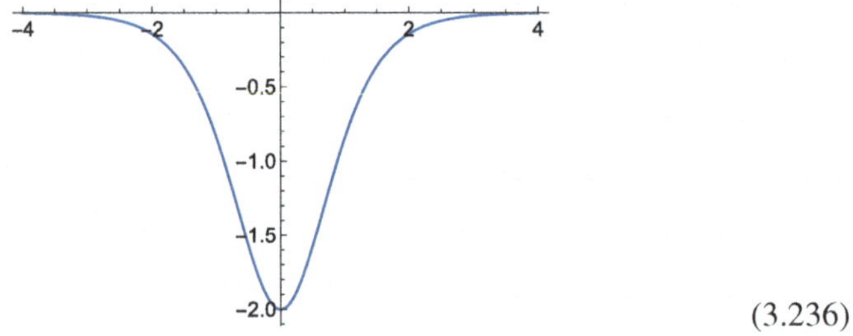

$$\tag{3.236}$$

$V(x)$ satisfies the conditions of the **Proposition**. In the Appendix to this chapter it is proven that this system has a single bound state with wave-function

$$\psi(x)_{\text{ground}} = \frac{1}{\sqrt{2}\,\cosh x}, \qquad E_{\text{ground}} = -1, \tag{3.237}$$

all the other energy eigenstates are scattering states which belong to the continuum spectrum

$$\psi_p(x) = \frac{1}{\sqrt{p^2+1}}\left(\tanh x - i p\right)e^{ipx}, \quad -\infty < p < \infty, \quad E = p^2. \tag{3.238}$$

See the Appendix for further details.

Oscillation Theorem

We suppose that our one-dimensional system with $\mathcal{M} = \mathbb{R}$ has bound states, that is,

$$\sigma(H) \cap \left(-\infty, \min\{V_{-\infty}, V_{+\infty}\}\right) \neq \varnothing, \tag{3.239}$$

while $V(x)$ is bounded below. Then the following holds

Theorem 3.7 *Let $E_0 < E_1 < E_2 < \cdots$ be the discrete energy eigenvalues listed in increasing order. The k-th eigenfuction $\psi_k(x)$ of energy E_k has exactly k nodes and between two successive nodes of $\psi_k(x)$ there is exactly one node of $\psi_{k+1}(x)$ i.e. the zeros of $\psi_k(x)$* interlace *those of $\psi_{k+1}(x)$. In particular the ground state wavefunction $\psi_0(x)$ may be taken to be everywhere positive.*

Sketch of Proof The Prüfer form of the Schrödinger equation for the k-th eigenfunction ψ_k is

$$\vartheta' = \cos^2\vartheta + (E - V(x))\sin^2\vartheta \quad \text{with } E = E_k. \tag{3.240}$$

Since $\psi_k \in L^2(\mathbb{R})$, the wave-function vanishes at $-\infty$, and we need to solve Eq. (3.240) with the boundary condition

$$\lim_{x \to -\infty} \vartheta(x) = 0. \tag{3.241}$$

The wave-function vanishes also at $+\infty$, hence

$$\lim_{x \to +\infty} \vartheta(x) = \ell(k)\pi \qquad 1 \le \ell(k) \in \mathbb{N} \tag{3.242}$$

where we used the argument in Fig. (3.177) to give a lower bound to the integer $\ell(k)$. That argument yields the more precise formula

$$\ell(k) = \#(\text{nodes of } \psi_k) - 1. \tag{3.243}$$

In particular $\ell(k)$ is finite by Fact 3.6. Now we increase continuously the parameter E until we get to the next eigenvalue E_{k+1}. From the Sturm comparison theorem

$$\frac{\partial}{\partial E}\vartheta(+\infty, E) > 0. \tag{3.244}$$

We get a solution in $L^2(\mathcal{M})$, i.e. an eigenvalue, whenever $\vartheta(+\infty, E) \in \pi\mathbb{Z}$. By continuity, the first such value we encounter when we increase E from E_k is

$$\ell(k + 1) = \ell(k) + 1. \tag{3.245}$$

We may use the argument in reverse: decreasing E we get a new eigenfunction with $\ell(k - 1) = \ell(k) - 1$. We keep going down until we reach $\ell(k_0) = 1$, which corresponds to a function with no zeros. Clearly if we decrease the energy further we do not find any other eigenfunction, since the number of zeros cannot be negative. Hence $\ell(k) = k + 1$ and the theorem follows. $\qquad\square$

Example 3.6 We saw in Sect. 3.5 that the wave-function of the k-th energy eigenstate of the harmonic oscillator is a polynomial of degree k with k distinct real roots, and that the zeros of two successive eigenfunctions interlace. Now we learn that this is a *general property* of the eigenfunctions in the discrete spectrum.

Parity

Suppose that the potential $V(x)$ is *even*

$$V(-x) = V(x), \tag{3.246}$$

so that the Hamiltonian preserves parity $[H, \mathscr{P}] = 0$. For simplicity, we assume that $V(x)$ is *confining* (meaning that $V_{\pm\infty} = +\infty$ and all stationary states are bound states) as was the case for the harmonic potential

$$V(x) = \frac{1}{2} m \, \omega^2 x^2. \tag{3.247}$$

If $\psi(x) \in L^2(\mathbb{R})$ is a solution to the stationary Schrödinger equation, $\psi(-x)$ is also a solution in $L^2(\mathbb{R})$. Since the discrete energy levels are non-degenerate, $\psi(-x)$ should be a multiple $\lambda\psi(x)$ of $\psi(x)$. Now

$$\psi(x) = \psi(-(-x)) = \lambda\psi(-x) = \lambda^2\psi(x) \quad \Rightarrow \quad \lambda = \pm 1, \tag{3.248}$$

i.e. the discrete energy eigenfunctions are either *even* or *odd*. The parity transformation $\mathscr{P}$: $\psi(x) \mapsto \psi(-x)$ leaves the zeros invariant, so that the zeros are distributed symmetrically around the origin. By the oscillation theorem the k-th eigenfunction has k zeros. If k is odd, the only symmetric arrangement is that one zero is at the origin and then we have $(k-1)/2$ positive and $(k-1)/2$ negative zeros. Since all zeros are simple, and the zeros of the even wave-functions interlace with the odd wave-function zeros, we have

Fact 3.8 *If the potential preserves parity $\mathscr{P}$, the $(2\ell+1)$-th bound wave-function is odd while the 2ℓ-th bound wave-function is even*

$$\psi_k(-x) = (-1)^k \, \psi_k(x). \tag{3.249}$$

In particular the ground state wave-function, being positive, must be even. We already observed these properties in the example of the harmonic oscillator in Sect. 3.5.

Summary

Under the present assumptions on the potential $V(x)$, an admissible solution with energy $E \neq V_{-\infty}, V_{+\infty}$, either vanishes exponentially or oscillates as $x \to +\infty$ and, respectively, $x \to -\infty$. Putting everything together, we conclude:

Spectrum of One-Dimensional Spinless System with $\mathcal{M} = \mathbb{R}$
Assume the potential $V(x)$ satisfies **V1, V2, V3**.

(1) The energy eigenfunctions with

$$E_{\min} \leq E < \min\{V_{-\infty}, V_{+\infty}\} \qquad (3.250)$$

cannot escape to $x = \pm\infty$ and hence represent *bound states* belonging
to the *discrete spectrum* of the Hamiltonian H. These energy levels are
non-degenerate and have at most *finitely many nodes* (zeros): the wave-
function of the k-th bound state has k zeros;

(2) Energy eigenfunctions with

$$\min\{V_{-\infty}, V_{+\infty}\} < E < \max\{V_{-\infty}, V_{+\infty}\} \qquad (3.251)$$

escape to infinity in one direction only (either $x = +\infty$ or $x = -\infty$).
These *scattering states* belong to the continuous spectrum, have *infinitely
many zeros*, but are *non-degenerate*. In the region at infinity where the
particle can escape the wave-function has the asymptotic form $A\,e^{ipx} +
B\,e^{-ipx}$ with $|A| = |B|$ since the reflected probability current is equal
to the direct one. The difference in phase, $-\,i \log B/A$ is the *phase shift*
between the two waves;

(3) Energy eigenfunctions with

$$E > \max\{V_{-\infty}, V_{+\infty}\} \qquad (3.252)$$

escape to infinity in both directions. These scattering states belong to
the continuous spectrum, have infinitely many zeros, and are *doubly
degenerated.*

(4) The continuous spectrum of the Hamiltonian H is separated from the
discrete one in the sense that $\sup \sigma_p(H) \leq \inf \sigma_e(H)$.

In particular when $V_{-\infty} = V_{+\infty} = +\infty$ the spectrum is *purely discrete* and *all*
states are bounded. This is the case, say, for the harmonic oscillator with potential
$V(x) = m\omega^2 x^2/2$, as we saw in Sect. 3.5.

Remark 3.8 (Energy Eigenstates at Threshold) The cases where one of the
bounds (3.251) is saturated are trickier and the answer depends on how fast
$V(x)$ reaches its asymptotic value. These energy eigenvalues, if present in the
spectrum, should be analyzed model by model. A bound state with energy $E =
\min\{V_{-\infty}, V_{+\infty}\}$ is called a *bound state at threshold*. Bound states at threshold exist,
but they are quite rare since their stability is not guarded by any energy barrier,

so they must be stabilized by a fancier physical mechanism which requires the system to have special properties. Bound states at threshold, when they exist, are very subtle. A more common phenomenon is that the energy $E \equiv \min\{V_{-\infty}, V_{+\infty}\}$ is an *accumulation point* of discrete energy levels, that is, we have infinitely many discrete energy levels $E_n < \min\{V_{-\infty}, V_{+\infty}\}$ and

$$\lim_{n \to \infty} E_n = \min\{V_{-\infty}, V_{+\infty}\}, \tag{3.253}$$

while above the threshold energy $\min\{V_{-\infty}, V_{+\infty}\}$ we have a continuous spectrum. There is a simple criterion: suppose that $V(x)$ is *even*, $V(-x) = V(x)$, monotonically increasing for $x \geq 0$, and $\lim_{x \to +\infty} V(x) = V_\infty < \infty$. We claim that the threshold energy $E \equiv V_\infty$ is an accumulation point of discrete energy levels $E_n < V_\infty$ if

$$\lim_{x \to +\infty} \int_a^x \sqrt{V(x) - V(y)}\, \mathrm{d}y = \infty \qquad a > 0. \tag{3.254}$$

This happens, for instance, when $V(x)$ is asymptotic for large x to $- C/x^\alpha$ with $0 < \alpha < 2$ (C is a positive constant). See Example 3.7 and Problem 3.7.

Example 3.7 We check the prediction that a system with a potential $V(x) = -1/x^\alpha$ with $0 < \alpha < 2$ has infinitely many bound states. For large x the zero-energy Schrödinger equation takes the asymptotic form

$$- \psi'' - \psi/x^\alpha \approx 0 \tag{3.255}$$

whose general (asymptotic) solution is

$$\psi(x) \approx A\, J_{\frac{1}{2-\alpha}}\left(\frac{2\, x^{1-\alpha/2}}{2 - \alpha}\right) + B\, J_{-\frac{1}{2-\alpha}}\left(\frac{2\, x^{1-\alpha/2}}{2 - \alpha}\right) \tag{3.256}$$

where $J_\nu(z)$ is the Bessel function of first kind of index ν [9, 14, 25, 26]. For all non-zero values of the constants A, B, and all x_0, the function (3.256) has infinitely many zeros with $x > x_0$, which implies that there are infinitely many energy eigenstates with energy $E < 0$. We already know these states must be bound states.

3.8   One-Dimensional Systems III: $\mathcal{M} = \mathbb{R}_{\geq 0}$

When the particle is moving on the *half-line* $x \geq 0$, "physical continuity" imposes the Dirichlet boundary condition

$$\psi(0) = 0 \tag{3.257}$$

on finite energy states. Physically we may think of this system as a particle moving on the full real line subjected to a potential which is $+\infty$ for $x \leq 0$. Then we are reduced back to the analysis of the previous section by studying the potential

$$V(x) = \begin{cases} V(x) & x > 0 \\ W & x \leq 0, \end{cases} \tag{3.258}$$

and then taking the limit $W \to +\infty$ at the end. In this limit the half-line $x \leq 0$ is classically forbidden for all energies, and the wave eigenfunctions (3.209) vanish for $x \leq 0$.

The Wronskian argument shows that all energy eigenvalues, discrete or continuous are now *non-degenerate:* indeed the Wronskian of any two solutions is forced to be zero by the boundary condition. Physically: now a scattering state must consist of an incoming wave from $x = +\infty$ which gets reflected back at the hard wall in $x = 0$. The comparison and oscillation theorems are still valid, and the k-th bound state eigenfunction (when present) has $k + 1$ real zeros, i.e. $x = 0$ and other k positive zeros.

Example 3.8 (Free Motion in the Half-line) The Hamiltonian is $H = p^2/2m$, the energy spectrum is $E > 0$ with (non-normalized) eigenfunctions

$$\psi_E(x) = \sin(\sqrt{2mE}\,x) \qquad (\hbar = 1). \tag{3.259}$$

Example 3.9 (Harmonic Oscillator in the Half-line) We consider the particle moving in the half-line $\mathbb{R}_{\geq 0}$ subjected to a quadratic potential with Schrödinger equation and boundary condition ($m = \hbar = \omega = 1$)

$$-\frac{1}{2}\frac{d^2\psi_E}{dx^2} + \frac{1}{2}x^2\psi_E = E\psi_E, \qquad \psi_E(0) = 0. \tag{3.260}$$

As in Example 3.4, a simple integration by parts (cf. Problem 3.17) yields the identity

$$\left[\frac{d^2}{dx^2} - x^2 + 2E\right]\left(e^{x^2/2}\int_0^\infty s^{E-1/2}e^{-s^2/4}\cos\left(xs - \frac{\pi}{2}\left(E - \tfrac{1}{2}\right)\right)ds\right) = 0. \tag{3.261}$$

The solution

$$\psi_E(x) = e^{x^2/2}\int_0^\infty s^{E-1/2}e^{-s^2/4}\cos\left(xs - \frac{\pi}{2}\left(E - \tfrac{1}{2}\right)\right)ds \tag{3.262}$$

is easily seen to decay as $e^{-x^2/2}$ when $x \to +\infty$. To get valid solutions in $L^2(\mathbb{R}_{\geq 0})$ we have only to determine for which values of the parameter E the boundary condition at $x = 0$ is satisfied

$$0 \overset{?}{=} \psi_E(0) \equiv \cos\left[\frac{\pi}{2}\left(E - \tfrac{1}{2}\right)\right] \int_0^\infty s^{E - \frac{1}{2}} e^{-s^2/4} \, ds \equiv$$

$$\equiv 2^{E-1/2} \cos\left[\frac{\pi}{2}\left(E - \tfrac{1}{2}\right)\right] \Gamma\left(\frac{E}{2} + \frac{1}{4}\right) = \frac{2^{E-1/2}\,\pi}{\Gamma\left(\frac{1}{2} - \frac{1}{2}\left(E - \frac{1}{2}\right)\right)} \tag{3.263}$$

where we used the second functional equation for the Gamma function [14]

$$\Gamma(z)\,\Gamma(1-z) = \frac{\pi}{\sin(\pi z)}. \tag{3.264}$$

The entire function $1/\Gamma(z)$ has zeros for

$$z = 0, -1, -2, -3, \cdots \tag{3.265}$$

so we get a valid solution satisfying the boundary condition when

$$E = (2n + 1) + \tfrac{1}{2} \qquad n \in \mathbb{N}. \tag{3.266}$$

Of course, these eigenfunction are nothing else that the *odd* numbered energy eigenfunctions of the harmonic oscillator on the full line which—being odd—are automatically zero at the origin.

The Image Method
The last two examples illustrate a simple strategy to solve the Schrödinger equation in the half-line. One introduces the *mirror image* of the potential in the excluded region $x < 0$, and studies the system on the full line with the piece-wise continuous, locally bounded, potential

$$W(x) = \begin{cases} V(x) & \text{for } x > 0 \\ V(-x) & \text{for } x < 0. \end{cases} \tag{3.267}$$

By construction the potential $W(x)$ preserves parity and its eigenfunctions are alternatively even and odd. Keeping only the odd ones $\psi_{2k+1}(x)$, we get the solution to the original Schrödinger equation in $\mathbb{R}_{\geq 0}$ with the appropriate boundary condition $\psi_{2k+1}(0) = 0$.

Remark 3.9 If we are interested in the *Neumann* boundary condition at $x = 0$ i.e. $\psi'(0) = 0$—instead of the Dirichlet one $\psi(0) = 0$—we keep the *even* eigenfunctions.

3.9   One-Dimensional Systems IV: $\mathcal{M} = [0, L]$

A particle moving in the interval $[0, L]$ subjected to a (bounded below, piece-wise continuous) potential $V(x)$ with the standard-type Schrödinger Hamiltonian

$$H = -\frac{\hbar^2}{2m}\frac{d^2}{dx^2} + V(x) \tag{3.268}$$

may be seen as a particle moving in $\mathbb{R}$ with potential

$$W(x) = \begin{cases} V(x) & x \in [0, L] \\ +\infty & x \notin [0, L], \end{cases} \tag{3.269}$$

which, in particular, imposes the Dirichlet boundary conditions at the endpoints

$$\psi_E(0) = \psi_E(L) = 0 \tag{3.270}$$

on the solutions of the stationary Schrödinger equation

$$-\frac{\hbar^2}{2m}\frac{d^2\psi_E(x)}{dx^2} + V(x)\,\psi_E(x) = E\,\psi_E(x), \quad 0 \le x \le L. \tag{3.271}$$

All general results discussed for $\mathcal{M} = \mathbb{R}$ then carry over to $\mathcal{M} = [0, L]$:

(1)   the energy spectrum, $\{E_n\}_{n\in\mathbb{N}} \equiv \sigma(H)$, is purely discrete without accumulation points, and all energy eigenvalues are non-degenerate. Therefore all energy eigenfunctions $\{\psi_n(x)\}_{n\in\mathbb{N}}$ are normalizable and form a Hilbert basis of $L^2([0, L])$;

(2)   the n-th energy eigenfunction $\psi_n(x)$ has n nodes in the *interior* of the interval $[0, L]$. In particular the ground wave-function $\psi_0(x)$ vanishes only at the endpoints of the interval;

(3)   (parity): if $x \to L - x$ is a symmetry of the potential

$$V(L - x) = V(x) \tag{3.272}$$

the energy eigenfunctions satisfy

$$\psi_n(L - x) = (-1)^n\,\psi_n(x). \tag{3.273}$$

Example 3.10 (Free Particle Moving in $[0, L]$)   When $V(x) = 0$, the eigenvalue Eq. (3.271) with boundary conditions (3.270) has the obvious solution

$$\psi_n(x) = \sqrt{\frac{2}{L}}\,\sin\!\left((n + 1)\frac{\pi x}{L}\right), \tag{3.274}$$

Fig. 3.3 The potential of the
quantum model (3.277)

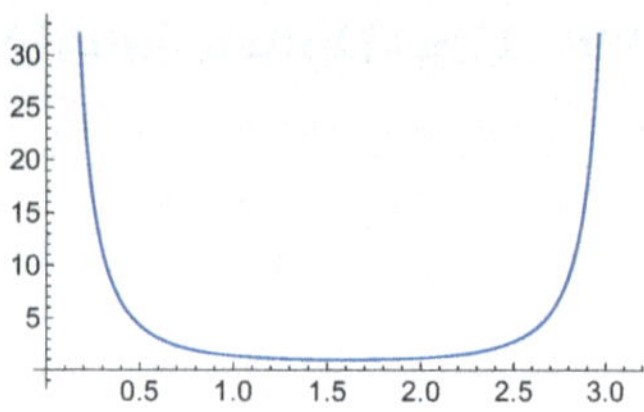

$$E_n = \frac{(n+1)^2 \pi^2 \hbar^2}{2mL^2} \qquad n = 0, 1, 2, \ldots, \tag{3.275}$$

It is immediate to check properties **(1)**, **(2)**, and **(3)** above.

Example 3.11 The solution to **Problem 10.2** in [1], that is, a classical particle moving in the interval $[0, \pi]$ with Hamiltonian ($m = 1$)

$$H_{\mathrm{cl}} = \frac{1}{2}\, p^2 + \frac{1}{\sin^2 x} - \frac{1}{2} \tag{3.276}$$

was extremely puzzling: as a function of the action canonical variable I, the Hamiltonian $H(I)$ had essentially the same form as for a free particle moving in the same interval, yet the two mechanical systems are not the same!

The same "puzzle" holds in the quantum setting. We consider a quantum particle moving in the interval $[0, \pi]$ with Schrödinger Hamiltonian[27] ($\hbar = 1$)

$$H = -\frac{1}{2}\frac{\mathrm{d}^2}{\mathrm{d}x^2} + \frac{1}{\sin^2 x} \tag{3.277}$$

see Fig. 3.3 for the shape of the potential. We start with a **Lemma** about polynomials.

Lemma 3.5 *Consider the sequence of polynomials of degree $\leq n$*

$$p_n(z) = U_n(z) - (n+1)\, T_n(z) \qquad n = 0, 1, 2, \ldots \tag{3.278}$$

where $T_n(z)$ (resp. $U_n(z)$) is the n-th Chebyshev polynomials of the first (resp. second) kind [9, 13]. Then

(a) $p_0(z) \equiv p_1(z) \equiv 0$ while $p_n(z)$ is a degree n polynomial for $n \geq 2$;
(b) $p_n(1) = p_n(-1) = 0$ while $p_n(z)$ ($n \geq 2$) has other $n - 2$ simple real zeros in the interval $-1 < z < +1$;
(c) the zeros of $p_n(z)$ interlace the zeros of $p_{n+1}(z)$ ($n \geq 2$);
(d) $p_n(-z) = (-1)^n\, p_n(z)$;

---

[27] The trick to solve this model is explained in the Appendix.

(e) we have the integrals

$$\int_{-1}^{+1} \frac{p_m(z)\, p_n(z)}{\sqrt{1 - z^2}}\, dz = (n^2 - 1)\frac{\pi}{2}\, \delta_{mn}, \qquad m, n \geq 2. \tag{3.279}$$

The sequence $p_n(z)$ is obviously complete (why?), so

Corollary 3.4 *The sequence of functions*

$$\psi_n(x) = \sqrt{\frac{2}{\pi}}\, \frac{p_{n+2}(\cos x)}{\sqrt{n^2 + 4n + 3}}, \qquad n = 0, 1, 2, \ldots \tag{3.280}$$

is an orthonormal Hilbert basis in $L^2([0, \pi])$.

It is easy to check that

$$\left(-\frac{1}{2}\frac{d^2}{dx^2} + \frac{1}{\sin^2 x} \right)\psi_n(x) = \frac{(n+2)^2}{2}\,\psi_n(x). \tag{3.281}$$

Since we have set $m = 1$, $\hbar = 1$ and $L = \pi$, this is *almost* the same energy spectrum as for the free system with the same parameters, Eq. (3.275). Each energy level of the present system is also an energy level for the free particle *except* that the free particle has an additional energy level $E_0 = 1/2$ at the bottom of the spectrum. This almost perfect match at the quantum level explains the classical puzzle in [1]. The phenomenon of two different quantum systems having the same spectrum up to some low-laying energy levels is not rare—in a sense it is generic. See Appendix.

3.9.1   Sturm-Liouville Problem

More generally, we may consider on the interval $[0, L]$ the ODE

$$-\frac{d}{dx}\left(A(x)\frac{d\psi}{dx} \right) + B(x)\,\psi(x) = \lambda\, C(x)\,\psi(x) \tag{3.282}$$

where $A(x)$ is positive in the open interval $(0, L)$. All ODEs can be put in this form where the linear operator in the LHS is formally self-adjoint. The study of this equation, with boundary conditions consistent with self-adjointness, is called the *Sturm-Liouville problem,* a classical topic in differential equation theory [15, 28–31]. Between the admissible boundary conditions we have the two physically relevant ones

$$\begin{aligned} \text{Dirichlet} \quad & \psi(0) = \psi(L) = 0 \\ \text{periodic} \quad & \psi(0) = \psi(L) \text{ and } \psi'(0) = \psi'(L). \end{aligned} \tag{3.283}$$

The Sturm comparison and oscillation theorems of Sects. 3.6 and 3.7, and all their consequences, are still valid (and usually stated) in this generality. There is only one additional issue to stress [15]

Fact 3.9 *If $A(x) > 0$ in the closed interval $[0, L]$, there is no accumulation point for the zeros of a solution, which then are finite in number, so the spectrum is purely discrete. The same holds if $A(x)$ vanishes at one end, say $A(0) = 0$, and the integral of $1/A(x)$ converges at this endpoint*

$$\int_0^a \frac{dx}{A(x)} < +\infty, \quad \textit{for } a > 0. \tag{3.284}$$

When the integral (3.284) *diverges, there is still no accumulation point in the* interior *of the interval, but the endpoint at which the integral* (3.284) *diverges may or may* not be *an accumulation point of zeros of the solution. If one (or both) endpoints are accumulation points of zeros, the eigenfunction belongs to the continuous spectrum.*

The statement follows from Eq. (3.176) and the Prüfer equation (3.171)

$$\#(\text{zeros in } [0, a]) \leq \int_0^a \max\left\{1/A(x), \max\{B(x), 0\}\right\} dx$$
$$\leq \int_0^a \frac{dx}{A(x)} + O(1). \tag{3.285}$$

Geometric Interpretation Let us explain the geometric meaning of this analytic statement. Consider the open interval $(0, L)$ equipped with a general Riemannian metric $ds^2 = g(x)\, dx^2$. The change of coordinates from x to the arc-length

$$x \mapsto s(x) \equiv \int_*^x \sqrt{g(x)}\, dx \tag{3.286}$$

transforms the metric into the usual flat one ds^2. The natural Schrödinger equation, when written in the original coordinate x, reads (cf. Eq. (2.459))

$$-\frac{d}{dx}\left(\sqrt{g}\, g^{-1}\frac{d\psi}{dx}\right) + \sqrt{g}\, V\psi = E\sqrt{g}\,\psi, \tag{3.287}$$

which is of the Sturm-Liouville form with

$$A(x) = 1/\sqrt{g(x)}, \quad B(x) = \sqrt{g(x)}\, V(x) \quad \text{and} \quad C(x) = \sqrt{g(x)}. \tag{3.288}$$

Now

$$\int_0^a \frac{dx}{A(x)} = \int_0^a \sqrt{g}(x)\, dx = \int_0^{s(a)} ds = s(a) - s(0), \tag{3.289}$$

so: the integral in the LHS diverges $\Leftrightarrow s(0) = -\infty \Leftrightarrow$ when written in the coordinate s the configuration space is the half-line $\mathbb{R}_{\leq s(L)}$ with the usual flat metric $\mathrm{d}s^2$. We are back to the situation in Sect. 3.8, and the spectrum may or may not be discrete depending on the behavior of the potential $V(x(s))$ as $s \to -\infty$. When the spectrum is continuous, the generic (generalized) eigenfunction has infinitely many zeros which accumulate at infinity in the Cartesian coordinate s, that is, at the endpoint $x = 0$ of the original interval $[0, L]$. This happens, for instance, when $V(x) = 0$.

3.10 One-Dimensional Systems V: $\mathcal{M} = S^1$

Let L be the length of the circumference of the circle S^1. We may see the Schrödinger equation on S^1

$$-\frac{\mathrm{d}^2\psi(x)}{\mathrm{d}x^2} + V(x)\,\psi(x) = E\,\psi(E) \tag{3.290}$$

as the equation in the interval $[0, L]$ with periodic boundary conditions

$$\psi(0) = \psi(L) \quad \text{and} \quad \psi'(0) = \psi'(L), \tag{3.291}$$

or, equivalently, on the line $\mathbb{R}$ subjected to the periodicity condition

$$\psi(x + L) = \psi(x) \quad \forall\, x \in \mathbb{R}. \tag{3.292}$$

We assume $V(x)$ to be piece-wise continuous and locally (hence globally) bounded.

Now the boundary conditions do not force the Wronskian of two solutions to vanish, as it was the case in Sect. 3.9; hence

Fact 3.10 **(1)** *An eigenvalue of the Schrödinger equation* (3.290) *on the circle S^1 may be either non-degenerate or doubly degenerate.* **(2)** *The spectrum of the Hamiltonian H is purely discrete, hence there is a ground state.* **(3)** *The ground state is non-degenerate and has no zero.* **(4)** *The zeros of a real solution of* (3.290) *are simple, and the zeros of two linearly-independent real solutions with the same energy interlace.* **(5)** *Let*

$$E_0 < E_1 \leq E_2 \leq E_3 \leq \cdots \tag{3.293}$$

be the non-decreasing *list of energy levels with repetitions whenever the eigenvalue is doubly degenerate. The number of zeros N_n of the n-th real eigenfunction satisfies*

$$N_n \in \{n - 1, n, n + 1\}. \tag{3.294}$$

(6) *If $V(a - x) = V(x)$ for some a, the real energy eigenfunctions may be chosen to be even or odd under the orientation reversing map $x \leftrightarrow a - x$.*

The existence of a ground state follows from the fact that the spectrum is discrete and bounded below. We may assume its wave-function to be real. By the comparison theorem, if a wave eigenfunction has two zeros we can find an eigenfunction with lower energy. Then the ground wave-functions must have either no zeros or a single simple zero. But a periodic real function cannot have *just one* simple zero, so the ground wave-function has no zero at all. If there was a second eigenfunction with the same energy, taking suitable linear combination of the two, we will construct a ground wave-function with some zeros, getting a contradiction. Hence the ground state is unique. More generally, periodicity of the solution implies periodicity of its Prüfer angle $\vartheta(x)$ up to a shift by an integral multiple of 2π; thus

$$\vartheta(L) - \vartheta(0) \in 2\pi\mathbb{Z}, \tag{3.295}$$

and the number of zeros is always even for all energy eigenstates (cf. Lemma 3.2). Now we check these predictions in the simplest possible example.

Example 3.12 (Free Particle Moving in S^1) The quantum Hamiltonian $H = p^2/2m$ commutes with the momentum operator p, so the energy eigenstates are just the momentum eigenstates

$$\frac{1}{\sqrt{L}} \exp(\mathrm{i}px/\hbar). \tag{3.296}$$

The periodicity condition (3.292) sets a constraint on the momentum eigenvalues p which are then discrete and quantized

$$p \equiv p_n = \frac{2\pi\hbar}{L}n, \quad n \in \mathbb{Z}, \tag{3.297}$$

so the energy level are

$$E_n = \frac{4\pi^2\hbar^2 n^2}{2mL^2}, \quad n = 0, 1, 2, \ldots, \tag{3.298}$$

all *doubly degenerate* except for the ground state $n = 0$ which is non degenerate with a constant wave function. For $n > 0$ the real wave eigenfunctions are

$$\sqrt{\frac{2}{L}} \cos(2\pi nx/L) \quad \text{and} \quad \sqrt{\frac{2}{L}} \sin(2\pi nx/L) \tag{3.299}$$

which are, respectively, even and odd under the parity operation $x \leftrightarrow -x$.

Number of States

Let us count the number $N(E)$ of states with energy $\leq E$. Since $V(x)$ is necessarily bounded in S^1, the asymptotic form of $N(E)$ for large E ($E \gg \max V(x)$) is independent of the potential, and may be obtained by looking to the free theory.[28] We have

$$N(E_n) = \#(\text{states with energy } E \leq E_n) = 2n + 1 \tag{3.300}$$

On the other hand, the volume in phase space of states with energy $\leq E_n$ is

$$\text{Vol}(H \leq E_n) = L \int_{-p_n}^{p_n} dp = 2n(2\pi\hbar), \tag{3.301}$$

so that we have

$$N(E) = \frac{\text{Vol}(H \leq E)}{2\pi\hbar} + O(1) = \frac{L\sqrt{2mE}}{\pi\hbar} + O(1) \quad \text{as } E \to \infty, \tag{3.302}$$

confirming our previous result in Sect. 2.13 that a region of the classical phase space contains a number of quantum states given (roughly) by its volume divided by $(2\pi\hbar)^n$, where n is the number of degrees of freedom.

3.10.1   General Periodic Conditions

We may consider the more general boundary conditions

$$\psi(x + L) = e^{2\pi i \alpha} \psi(x) \qquad 0 \leq \alpha < 1, \tag{3.303}$$

which are still consistent with the operator $-i\hbar \, d/dx$ being Hermitian.

In the free system with Hamiltonian $H = p^2/2m$, the eigenfunctions now become

$$\psi_n(x) = \frac{1}{\sqrt{L}} \exp\left(2\pi i(n + \alpha)\frac{x}{L}\right) \tag{3.304}$$

with momentum and energy

$$p = \frac{2\pi\hbar}{L}(n + \alpha), \tag{3.305}$$

---

[28] The math oriented reader may get the same conclusion using the fundamental theorems of Sturm-Liouville theory.

$$E = \frac{(2\pi\hbar)^2}{2mL^2}(n+\alpha)^2, \quad n \in \mathbb{Z}. \tag{3.306}$$

For generic α the boundary condition (3.303) breaks explicitly the parity symmetry $x \leftrightarrow -x$. The only exceptions are $\alpha = 0$ and $\alpha = 1/2$ where

$$\alpha = 0: \qquad \psi_n(-x) = \psi_{-n}(x) \tag{3.307}$$

$$\alpha = \tfrac{1}{2}: \qquad \psi_n(-x) = \psi_{-n-1}(x). \tag{3.308}$$

When $\alpha \neq 0, \tfrac{1}{2}$ all energy levels are non-degenerate. When $\alpha = 0$ all levels are doubly degenerated but for the ground state which is unique as per Fact 3.10. Finally, when $\alpha = 1/2$, *all* energy levels—including the ground one—are doubly degenerate:

$$E_n \equiv \frac{(2\pi\hbar)^2}{2mL^2}(n+\tfrac{1}{2})^2 = \frac{(2\pi\hbar)^2}{2mL^2}\big((-1-n)+\tfrac{1}{2}\big)^2 \tag{3.309}$$

The two ground states are obtained by setting $n = -1, 0$ in Eq. (3.304).

Spontaneous Breaking of Symmetry
When $\alpha = \tfrac{1}{2}$ there are *two ground states* which are interchanged by parity. This leads to the following definition:

Definition 3.3 A unitary symmetry U of the quantum Hamiltonian H

$$UU^\dagger = U^\dagger U = 1, \quad UHU^\dagger = H, \tag{3.310}$$

is *spontaneously broken* iff its ground states (of energy $E_0 \equiv \min \sigma(H)$) transform non-trivially under the symmetry, that is, if U restricted to the E_0-eigenspace of H is not multiplication by a common phase $e^{i\phi}$.[29]

In particular the symmetry is tautologically *unbroken* whenever the ground state is non-degenerate. The concept of spontaneous breaking of a symmetry is a central idea in Modern Physics. The free particle with the boundary condition $\alpha = \tfrac{1}{2}$ (called the *anti-periodic boundary condition*) yields a first example of this deep phenomenon:

Fact 3.11 *The free particle moving in S^1 with the anti-periodic boundary condition breaks explicitly the parity symmetry $\mathscr{P}$.*

Remark 3.10 The argument around (3.295) shows that the *real* energy eigenfunctions of any Schrödinger equation on the circle with the anti-periodic boundary

---

[29] When $e^{i\phi} \neq 1$, we just replace $U \rightsquigarrow U' \equiv e^{-i\phi}U$ which has the same effect as U on all observables and—if the symmetry U is unbroken—acts as the identity in the subspace of states of minimal energy $E_0 = \min \sigma(H)$.

condition (i.e. $\alpha = \frac{1}{2}$) have an *odd* number of zeros. For $\alpha \neq 0, \frac{1}{2}$ the wave-functions cannot be chosen to be real, and have no zeros since when its real part vanishes the imaginary part does not.

Spectral Flow

Varying continuously α from 0 to 1, the boundary condition (3.303) returns to its initial form, i.e. we get back the original quantum system. Consequently the $\alpha = 1$ energy spectrum (3.306) is the same as the $\alpha = 0$ one *as a set*. However each individual energy eigenvalue and eigenfunction may or may not return back to itself. For instance in the free model (3.306) they do not get back, but rather get shifted by one unit. This phenomenon is called the *spectral flow*. It is quite a useful feature: suppose we know just one solution of the Schrödinger equation (3.290), $\psi(x, \alpha)$, as an analytic function of the boundary condition parameter α, whose energy is $E(\alpha)$. The infinite sequence of functions

$$\psi_{\alpha,n}(x) = \psi(x, \alpha + n), \qquad n \in \mathbb{Z}. \tag{3.311}$$

are also eigenfunctions, now with energy $E(\alpha + n)$, which satisfy the same generalized boundary condition. If the analytic continuations of energy is not periodic in α

$$E(\alpha + n) \neq E(\alpha) \qquad n \in \mathbb{Z}, \tag{3.312}$$

that is, if the energy is a *multi-valued function* of the phase $e^{2\pi i\alpha}$, the state $\psi_{\alpha,n}(x)$ is linearly independent of $\psi(x, \alpha)$, and we get new energy eigenfunctions by simple analytic continuation of the original solution in the parameter α.

3.11 The Bohm-Aharonov Effect

The model studied in Sect. 3.10.1 presents a first instance of a general phenomenon of Quantum Physics which is of the utmost importance both conceptually and for several practical applications.

The redefinition of the wave-function (i.e. change of Schrödinger representation by a gauge transformation as discussed in detail in Sect. 2.12)

$$\psi(x) = e^{-2\pi i\alpha x/L} \, \psi(x)_{\text{old}} \tag{3.313}$$

transforms the generalized periodic condition (3.303) in the plain one

$$\psi(x + L) = \psi(x) \tag{3.314}$$

at the price of modifying the Schrödinger representation of momentum to

$$p = -i\hbar e^{-2\pi i\alpha x/L}\frac{d}{dx}e^{2\pi i\alpha x/L} = -i\hbar\frac{d}{dx} + \frac{2\pi\hbar\alpha}{L}. \tag{3.315}$$

We change notation and write $\tilde{p}$ for the Hermitian operator $-i\hbar d/dx$. The momentum now is

$$p = \tilde{p} + \frac{2\pi\hbar\alpha}{L}, \tag{3.316}$$

while the Hamiltonian reads

$$H = \frac{1}{2m}p^2 = \frac{1}{2m}\left(\tilde{p} + \frac{2\pi\hbar\alpha}{L}\right)^2 \equiv \frac{1}{2m}\left(\tilde{p} - A\right)^2, \tag{3.317}$$

so that the Hamiltonian is identically in form to the one for a particle of electric charge $+1$ moving in S^1 coupled to a background *magnetic potential*

$$A = -\frac{2\pi\hbar\alpha}{L}. \tag{3.318}$$

For the sake of comparison, let us write the classical Lagrangian of the corresponding classical system

$$L = \frac{m}{2}\dot{x}^2 - \frac{2\pi\hbar\alpha}{L}\dot{x} \equiv \frac{m}{2}\dot{x}^2 - \frac{d}{dt}\left(\frac{2\pi\hbar\alpha}{L}x\right). \tag{3.319}$$

Classically adding a total time derivative to the Lagrangian does not change the physics: the Lagrangian equations of motion will not be affected by the addition ([1] §. 3.3), and hence *no classical physical observable is sensitive to the value of α.* There is a good (classical) reason for this: the magnetic field dA is identically zero on the configuration manifold S^1 (dA is a 2-form, hence zero in a 1-dimensional manifold!), and a classical particle may feel a magnetic force only where the magnetic field is non-zero!

Now consider the physical set up of Fig. 3.4. There is no magnetic field along the circle where the particle is constrained to move, so *classically* the particle cannot feel the presence of the magnetic flux Φ which is concatenated to the circle. However, by Stokes theorem ([1] chap. 2),

$$\oint_{S^1} A = \int_{disk} dA = \Phi \neq 0, \tag{3.320}$$

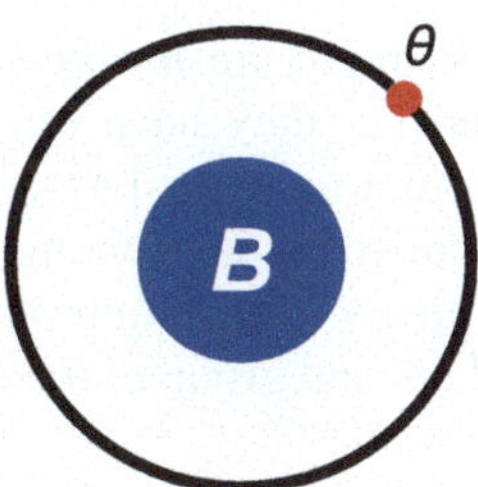

Fig. 3.4 A charged quantum particle is constrained to move in a circle laying in the x-y plane of $\mathbb{R}^3$ centered at the origin. The figure shows the x–y plane with the *circle* drawn in *black*. Along the vertical direction (normal to plane) there is a solenoid, centered on the z-axis: inside the solenoid there is a constant magnetic field B directed in the positive z-direction. The *blue disk* in the figure represents the intersection of the interior of the solenoid with the x–y plane. Outside the *blue disk* the magnetic field is *zero*. While the magnetic field vanishes along the *black circle*, there is a *magnetic flux* Φ concatenated with the *black circle* equal to BA (A is the area of the *blue disk*)

so the magnetic *potential A* is non-zero along the circle even if the magnetic *field* itself vanishes. Up to a gauge transformation we may choose A to be the constant $A = \Phi/L$. In the quantum set-up this leads to the previous system (3.317) with

$$\alpha = -\frac{\Phi}{2\pi\hbar}. \tag{3.321}$$

Now, contrary to the classical case, the value of α *matters:* indeed the value of the physical observable "energy" does depend non-trivially on α, see Eq. (3.306). More precisely:

Fact 3.12 *Only the class α mod 1 is observable: two α's differing by an integer are* physically indistinguishable—*no experiment may discriminate between them. This is already obvious from the boundary condition (3.303).*

This example shows that a quantum charged particle feels the effect of the magnetic field (in the solenoid) *even remaining away* from the region where the magnetic field is present! This quantum phenomenon is known as the *Bohm-Aharonov effect*, and is one of the most important features of Quantum Mechanics crucial for many of its practical applications.

Let us explain the phenomenon in the language of Differential Geometry ([1] chap. 2). Let $\mathcal{M}$ be a general configuration space. A magnetic potential on $\mathcal{M}$ is (roughly) a one-form $A \in \Omega^1(\mathcal{M})$ and its field strength—the *magnetic field*—is $B \equiv dA$. A gauge potential A is *flat* iff the magnetic field vanishes, $dA \equiv 0$, while it is *pure gauge,* or *exact,* ($\equiv$ physically trivial: no observable can detect it) iff $A = df$ for some $f \in \Omega^0(\mathcal{M})$. By definition the space of flat potentials modulo the exact ones is the first *de Rham cohomology group*

$$H^1_{\mathrm{DR}}(\mathcal{M}) = \{\xi \in \Omega^1(\mathcal{M}) : d\xi = 0\}\big/\{df : f \in \Omega^0(\mathcal{M})\}. \tag{3.322}$$

Whenever $H^1_{\mathrm{DR}}(\mathcal{M}) \neq 0$ on $\mathcal{M}$ there are *flat potentials* ($\equiv$ no magnetic field) which are physically non-trivial i.e. they have *observable effects*. The physical effects depend only on the cohomology class $[A] \in H^1_{\mathrm{DR}}(\mathcal{M})$ of the flat potential A not on the chosen representative in its cohomology class. Thus the Bohm-Aharonov effect is a general property of quantum physics which applies whenever the (Abelianization of the) fundamental group of the configuration space $\mathcal{M}$ is non-trivial. For instance, for the circle

$$H^1_{\mathrm{DR}}(S^1) = \mathbb{R}, \tag{3.323}$$

and we have a one-parameter family of non-trivial magnetic potentials on S^1 parametrized by α.

The actual story is slightly subtler since, geometrically speaking, the magnetic potential A is a $U(1)$-*connection on a complex line bundle on* $\mathcal{M}$ not a plain 1-form. This has the consequence that only the class mod 1 of the normalized flux $\Phi/(2\pi\hbar)$ concatenated with the circle may be detected by the charged particle moving outside the solenoid,[30] as indeed we found by analyzing the free particle with generalized boundary conditions, cf. Fact 3.12: the physically observable effects depend only on the class $[\alpha] \in \mathbb{R}/\mathbb{Z}$. For a general manifold $\mathcal{M}$ we have

$$\left(\begin{array}{c} \textbf{space of physically inequivalent} \\ \textbf{FLAT magnetic potentials } A \end{array} \right) \simeq H^1(\mathcal{M}, \mathbb{R})/H^1(\mathcal{M}, \mathbb{Z}). \tag{3.324}$$

We shall revisit the Bohm-Aharonov effect in Chap. 6 in terms of path integrals.

3.12   One-Dimensional Systems VI: Periodic Potentials

Finally we consider a particle moving in $\mathbb{R}$ subjected to a periodic potential

$$V(x + L) = V(x). \tag{3.325}$$

Now the limits $\lim_{x \to \pm x} V(x)$ do not exist, and the analysis of Sect. 3.7 is useless. In particular the spectrum of H cannot have the "usual" form, i.e. continuous above a certain threshold and discrete below. A new class of energy spectra needs to emerge.

The translation by L, implemented by the unitary operator

$$U = \exp(i\,\boldsymbol{p}L/\hbar), \tag{3.326}$$

is a symmetry of the Hamiltonian,

$$U H U^\dagger = H \quad \Leftrightarrow \quad U H = H U, \tag{3.327}$$

---

[30] This is the basis of Dirac's theory of the magnetic monopole and the quantization of electric charge [32]. We shall discuss more precisely these issues in the final sections of Chap. 5.

hence we can diagonalize simultaneously U and H. U is unitary, so its eigenvalues are phases $\mathrm{e}^{2\pi \mathrm{i}\alpha}$, while its spectrum is continuous, in facts $\sigma(U)$ is the unit circle $U(1)$. The Hilbert space, or rather the larger space $\mathcal{S}^\vee$, is then a direct integral[31] of generalized U-eigenspaces

$$\mathcal{S}^\vee = \bigoplus_\alpha \mathcal{H}_\alpha, \qquad U\big|_{\mathcal{H}_\alpha} = \mathrm{e}^{2\pi \mathrm{i}\alpha}, \tag{3.328}$$

and we may compute the eigenfunctions of H in each $\mathcal{H}_\alpha$ separately. The Schrödinger representation of a state in $\mathcal{H}_\alpha$ is a function with the periodicity property

$$\psi_\alpha(x + L) = \mathrm{e}^{2\pi \mathrm{i}\alpha} \psi_\alpha(x) \quad \Leftrightarrow \quad \psi_\alpha(x) \in \mathcal{H}_\alpha, \tag{3.329}$$

so, when restricted to a generalized direct integrand $\mathcal{H}_\alpha$, we get back the Schrödinger equation on S^1 with the generalized periodic boundary condition (3.303). The novelty is that while in Sect. 3.10 the parameter α was a given *fixed* number, part of the physical definition of the system (i.e. the effect of a specific background magnetic field *á la* Bohm-Aharonov), now α is merely the label of a subspace $\mathcal{H}_\alpha$ of vectors in $\mathcal{S}^\vee$ and hence takes all possible values $0 \le \alpha < 1$.

A differential equation of the form

$$-\frac{\mathrm{d}^2 \psi(x)}{\mathrm{d}x^2} + V(x)\,\psi(x) = E\psi(x) \tag{3.330}$$

where the potential function $V(x)$ is periodic, is called the *Hill equation,* the phase $\mathrm{e}^{2\pi \mathrm{i}\alpha}$ is known as the *Floquet phase,* and our discussion above is part of *Floquet theory,* see e.g. [33].

From Sect. 3.10 we known that *for fixed* α the Eq. (3.330) has a purely discrete spectrum. For *generic* α the eigenvalues are non-degenerate, so

$$E_0(\alpha) < E_1(\alpha) < E_2(\alpha) < \cdots \tag{3.331}$$

The energy level $E_k(\alpha)$ depends continuously on α; hence to a given integer k there corresponds a connected set of allowed energies

$$B_k = \{E_k(\alpha) \colon 0 \le \alpha < 1\} \subset \sigma(H) \tag{3.332}$$

called *a spectral band.* Thus the energy spectrum $\sigma(H)$ organizes itself in a sequence of bands $\{B_k\}$ of varying widths

$$\Delta_k = \sup_\alpha E_k(\alpha) - \inf_\alpha E_k(\alpha). \tag{3.333}$$

---

[31] We abuse notation and write $\bigoplus$ for the direct integral.

When

$$\delta_k \equiv \inf_{\alpha} E_k(\alpha) - \sup_{\beta} E_{k-1}(\beta) > 0 \tag{3.334}$$

the k-th band B_k is separated from the previous band B_{k-1} by a *gap* in the energy spectrum of size δ_k. Bands which are very narrow and widely gapped, $\Delta_k \ll \delta_k, \delta_{k+1}$, behave qualitatively as isolated eigenvalues.

On the contrary, when $\delta_k \leq 0$ the gap between the bands B_k, B_{k-1} *closes*, and the two bands may overlap: the eigenvalues in the intersection

$$E \in B_k \cap B_{k-1} \subset \sigma(H) \tag{3.335}$$

are *doubly degenerate* (generalized) energy eigenvalues. Triple or higher band intersections are empty since Eq. (3.330) cannot have more than 2 linearly independent solutions for a given E.

For large k, that is, for large energy $E \gg \max V(x)$, the spectrum is asymptotic to the spectrum of the free theory, i.e. continuous without gaps and doubly degenerate. Thus all gaps close as $k \to \infty$, and we have

$$\inf_{\alpha} E_{k+2}(\alpha) - \sup_{\alpha} E_k(\alpha) \to 0 \quad \text{for } k \to \infty. \tag{3.336}$$

All these properties will be illustrated in Sect. 3.13.6 in an easy-to-solve Hill equation. In that example $V(x)$ is not piece-wise continuous, but it is still integrable which is just as good for math (and phys) purposes.

Remark 3.11 The fact that the energy spectrum of a periodic system may organize itself in bands separated by gaps has an enormous relevance for our everyday life. All electric/electronic devices work because the energy spectrum of the electrons moving inside the periodic lattice of a conductor have a *conducting band* which is gapped with respect to the non-conducting ones.

Procedure to Compute the Spectral Bands We describe the algorithm to compute (in principle) the functions $E_k(\alpha)$. We write $\{S(x; E), C(x; E)\}$ for the basis of solutions to the 2nd order ODE

$$-\frac{\mathrm{d}^2 \phi(x)}{\mathrm{d}x^2} + V(x)\phi(x) = E\phi(x) \tag{3.337}$$

with the respective initial conditions

$$S(0; E) = 0 \quad S'(0; E) = 1, \qquad C(0; E) = 1 \quad C'(0; E) = 0, \tag{3.338}$$

which satisfy the Wronskian identity

$$C(x; E)S'(x; E) - C'(x; E)S(x; E) = 1 \qquad \forall\, x \in \mathbb{R}. \tag{3.339}$$

We fix α and take E to be in the discrete set $\{E_k(\alpha)\}_{k\in\mathbb{N}}$. The eigenfunction $\psi_\alpha(x; E) \in \mathcal{H}_\alpha$ is a linear combination of the above two solutions

$$\psi_\alpha(x; E) = a(\alpha)\, S(x; E) + b(\alpha)\, C(x; E). \tag{3.340}$$

Imposing the boundary conditions

$$\psi_\alpha(L) = e^{i\alpha}\,\psi_\alpha(0), \qquad \psi'_\alpha(L) = e^{i\alpha}\,\psi'_\alpha(0) \tag{3.341}$$

yields the linear equation

$$\begin{pmatrix} S(L; E) & C(L; E) - e^{i\alpha} \\ S'(L; E) - e^{i\alpha} & C'(L; E) \end{pmatrix} \begin{pmatrix} a(\alpha) \\ b(\alpha) \end{pmatrix} = 0. \tag{3.342}$$

A non-zero solution exists iff the determinant of the 2×2 matrix vanishes. Taking into account (3.339), we get the following implicit equation for the function $E(\alpha)$

$$C\big(L; E(\alpha)\big) + S'\big(L; E(\alpha)\big) = 2\cos\alpha. \tag{3.343}$$

The inverse function $E = E(\cos\alpha)$ is multivalued with countably many branches. The k-th branch solution of the transcendental Eq. (3.343) yields $E_k(\alpha)$. The symmetry $\alpha \leftrightarrow -\alpha$ of the spectrum reflects the time-reversal symmetry: the Hamiltonian in Eq. (3.330) is a real differential operator, while complex conjugation of the wave function flips the sign of α. See Sect. 3.13.6 for an example worked out in detail.

3.13 Important Examples

We consider some fundamental examples of one-dimensional Schrödinger equations mainly to illustrate some peculiar phenomena of Quantum Mechanics which are not present in the classical setting.

3.13.1 *One-Dimensional Potential Well*

We consider a particle moving on $\mathbb{R}$ subjected to the following potential

$$V(x) = \begin{cases} -\Lambda < 0 & \text{for } |x| < a \\ 0 & \text{for } |x| > a \end{cases} \tag{3.344}$$

We are interested in the number of bound states of this system. In particular we want to show that for all positive a and Λ there is *at least one* bound state, a result we anticipated and used when discussing the existence of bound states in Sect. 3.7.

From our general discussion in Sect. 3.7 we know that the bound states are the energy eigenstates with energy $E < 0$. We also know that the bound states are non-degenerate with real wave-functions, and that parity even and parity odd states alternate. A bound state of energy $-\Lambda < E < 0$ has a wave function $\psi(x)$ of the form

$$
\begin{cases}
\pm A\, e^{kx} & \text{for } x < -a \\[4pt]
B\, e^{iqx} \pm B\, e^{-iqx} & \text{for } -a < x < a \\[4pt]
A\, e^{-kx} & \text{for } x > a
\end{cases}
\tag{3.345}
$$

where

$$
k = \sqrt{2m|E|}\big/\hbar \quad \text{and} \quad q = \sqrt{2m(\Lambda - |E|)}\big/\hbar,
\tag{3.346}
$$

while the sign $\pm$ is $+$ (resp. $-$) for *even* (resp. *odd*) solutions. Note that

$$
k^2 + q^2 = \frac{2m\Lambda}{\hbar^2}.
\tag{3.347}
$$

To find the coefficients A, B, we have to impose that the function is C^1 at $x = \pm a$. By symmetry, it is enough to check the point $x = a$. For even eigenstates

$$
\psi(a^+) = \psi(a^-) \qquad \Rightarrow \qquad A e^{-ka} = 2B\cos(qa)
\tag{3.348}
$$

$$
\psi'(a^+) = \psi'(a^-) \qquad \Rightarrow \qquad -k\, A e^{-ka} = -2q B \sin(qa).
\tag{3.349}
$$

An *even* solution with $A, B \neq 0$ exists iff

$$
k = q \tan(qa).
\tag{3.350}
$$

Likewise, an *odd* solution exists for

$$
k = -q \cot(qa).
\tag{3.351}
$$

Writing $\xi = ka$, $\eta = qa$, we are led to solve the system

$$
\begin{array}{cc}
\textbf{even} & \textbf{odd} \\[6pt]
\begin{cases} \xi = \eta \tan\eta \\ \eta^2 + \xi^2 = R^2 \end{cases} &
\begin{cases} \xi = -\eta \cot\eta \\ \eta^2 + \xi^2 = R^2 \end{cases}
\end{array}
\tag{3.352}
$$

Fig. 3.5 Graphical solution to Eq. (3.352). The *blue curve* is $\xi = \eta \tan \eta$ while the *red curve* is $\xi = -\eta \cot \eta$. The *black circle* is the curve $\xi^2 + \eta^2 = R^2$ for $R = 6$. The intersection of the *blue* (resp. *red*) *curve* with the *circle* correspond to an even (resp. odd) bound state of energy $E = \frac{\hbar^2 \eta^2}{2ma^2} - \Lambda \equiv \Lambda\left(\frac{\eta^2}{R^2} - 1\right) < 0$

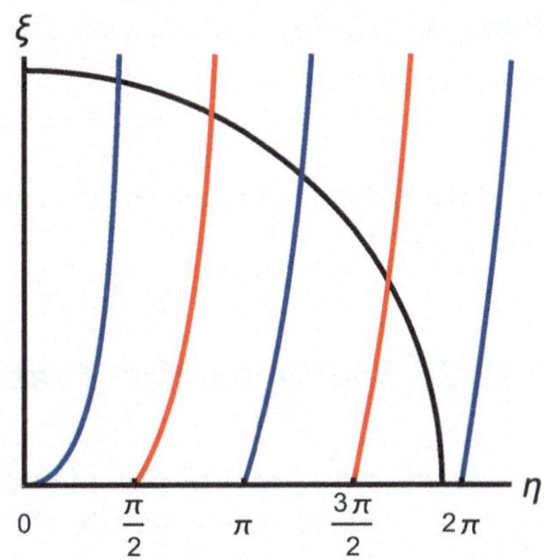

where we introduced the quantity

$$R^2 \equiv \frac{2m\Lambda a^2}{\hbar^2}. \tag{3.353}$$

The solutions to Eq. (3.352) are easily obtained graphically, see Fig. 3.5.

From the figure it is obvious that the number of bound states is

$$\#(\text{even bound states}) = 1 + \left[\frac{R}{\pi}\right] \tag{3.354}$$

$$\#(\text{odd bound states}) = \left[\frac{R}{\pi} - \frac{1}{2}\right] \tag{3.355}$$

where $[x]$ is the *integer part* of the real number x (the largest integer not greater than x). We see that *at least one* bound state (the ground state) is present for all values of R, i.e. for all values of a and Λ, as previously claimed. A different strategy to get this result is sketched in Problem 3.11.

The number of zeros of the bound state wave-function corresponding to an intersection with the circle at the point of abscissa η is

$$\#(\text{zeros of even wave-function}) = 2\left[\frac{\eta}{\pi}\right] \tag{3.356}$$

$$\#(\text{zeros of odd wave-function}) = 1 + 2\left[\frac{\eta}{\pi} - \frac{1}{2}\right] \tag{3.357}$$

as expected from the oscillation theorem and parity invariance.

For the scattering states at positive energy see Problem 3.12.

Remark 3.12 As we shall see in Chap. 5, higher dimensional potential wells

$$V(r) = -\Lambda\,\Theta(a - r) \tag{3.358}$$

(r is the radial coordinate in $\mathbb{R}^d$ with $d \geq 3$) do not have bound states for small positive a, Λ.

3.13.2 Square Barrier: Tunneling Effect

To illustrate the two dual phenomena of the *tunneling effect* and of *reflection over the barrier,* we invert the overall sign in the square-well potential (3.344) so that its profile now looks as

$$V(x) = \underline{\quad\quad\quad\lceil\;\rceil\quad\quad\quad} \tag{3.359}$$

We reparametrize it in the form

$$V(x) = \begin{cases} \Lambda > 0 & \text{for } 0 < x < L \\ 0 & \text{otherwise,} \end{cases} \tag{3.360}$$

and call it a *square barrier* of width L and height Λ. In this system there are no bound states, only scattering states propagating to infinity. In classical physics its dynamics is pretty obvious: particles incoming from infinity with energy $E < \Lambda$ will reach the barrier and be reflected back, with no chance of crossing to the other side of the barrier; dually, particles with $E > \Lambda$ will bypass the barrier and reach infinity on the other side, with no chance of being reflected back. Both statements are *false* in the quantum world:

Claim *At all energies $E > 0$, the scattering states have a non-zero probability of both being reflected and being transmitted on the other side of the barrier. However the probability of a classically forbidden process is* **exponentially small** *as*

$$\frac{m|\Lambda - E|L^2}{\hbar^2} \to \infty. \tag{3.361}$$

We consider a particle incoming from $x = -\infty$ with energy $0 < E < \Lambda$ which will be, with some probability, partly reflected and partly transmitted by the barrier. In other words, we look for a solution of the form

$$\psi(x) = \begin{cases} e^{\mathrm{i}px} + R\,e^{-\mathrm{i}px} & x < 0 \\ A\,e^{qx} + B\,e^{-qx} & 0 < x < L \\ T e^{\mathrm{i}p(x-L)} & x > L \end{cases} \tag{3.362}$$

where

$$p^2 = \frac{2mE}{\hbar^2} > 0, \qquad q^2 = \frac{2m(\Lambda - E)}{\hbar^2} > 0. \tag{3.363}$$

$R\,e^{-ipx}$ is the *reflected wave* and $T\,e^{ip(x-L)}$ is the wave *transmitted* on the other side of the obstacle. Conservation of the probability current (i.e. unitarity of the process) yields

$$1 - |R|^2 = |T|^2, \tag{3.364}$$

that is, the sum of the probabilities that particle is reflected and is transmitted is 1. C^1 regularity at $x = 0$ and $x = L$ requires

$$
\begin{aligned}
\psi(0^-) = \psi(0^+) &\quad\Rightarrow\quad & 1 + R = A + B \\
\psi'(0^-) = \psi'(0^+) &\quad\Rightarrow\quad & ip(1 - R) = q(A - B) \\
\psi(L^-) = \psi(L^+) &\quad\Rightarrow\quad & A\,e^{qL} + B\,e^{-qL} = T \\
\psi'(L^-) = \psi'(L^+) &\quad\Rightarrow\quad & q(A\,e^{qL} - B\,e^{qL}) = ip\,T
\end{aligned}
\tag{3.365}
$$

whose solution is

$$R = \frac{(p^2 + q^2)\sinh(Lq)}{(p^2 - q^2)\sinh(Lq) + 2ipq\cosh(Lq)} \tag{3.366}$$

$$T = \frac{(2ipq)}{(p^2 - q^2)\sinh(Lq) + 2ipq\cosh(Lq)} \tag{3.367}$$

$$A = -\frac{p(p - iq)e^{-Lq}\sinh(Lq)}{(p^2 - q^2)\sinh(Lq) + 2ipq\cosh(Lq)} \tag{3.368}$$

$$B = -\frac{p(p - iq)e^{-Lq}\sinh(Lq)}{(p^2 - q^2)\sinh(Lq) + 2ipq\cosh(Lq)} \tag{3.369}$$

Using $p^2 + q^2 = 2m\Lambda/\hbar^2$, the amplitudes of the reflected and transmitted waves may be written in a simple form

$$|R|^2 = \frac{(2m\Lambda)^2\sinh^2(Lq)}{(2m\Lambda)^2\sinh^2(Lq) + 4\hbar^2 p^2 q^2} \tag{3.370}$$

$$|T|^2 = \frac{4\hbar^2 p^2 q^2}{(2m\Lambda)^2\sinh^2(Lq) + 4\hbar^2 p^2 q^2} \tag{3.371}$$

which satisfy the unitary relation (3.364). As $\hbar \to 0$ or as $Lq \to \infty$, i.e. in the limit (3.361), $|R|^2 \to 1$ and $|T|^2 \to 0$ as in the classical theory.

For finite $\hbar$ and qL the transmitted wave is *non zero*, that is, the particle has a finite, but exponentially small probability to get on the other side of the barrier:

$$|T|^2 = O(e^{-2Lq\sqrt{2m(\Lambda-E)}/\hbar}). \tag{3.372}$$

One says that the quantum particle has "tunneled" through the barrier: this is the *tunneling effect,* one of the most crucial and characteristic aspects of Quantum Physics.

Dually, when the energy is larger than the height of the barrier, $E > \Lambda$, the quantum particle has a finite probability to get reflected back (*reflection over the barrier*). The amplitudes for $E > \Lambda$ are the analytic continuation of the ones for $E < \Lambda$ (the continuation from real to imaginary values of q). Writing $s \equiv |q|$ we get

$$|R|^2 = \frac{(2m\Lambda)^2 \sin^2(Ls)}{(2m\Lambda)^2 \sin^2(Ls) + 4\hbar^2 p^2 s^2} \tag{3.373}$$

$$|T|^2 = \frac{4\hbar^2 p^2 s^2}{(2m\Lambda)^2 \sin^2(Ls) + 4\hbar^2 p^2 s^2} \tag{3.374}$$

When the initial scattering state is in $L^2(\mathbb{R})$ (i.e. is a physically realizable state) the reflected amplitude goes to zero as $Ls \to \infty$.

3.13.3 δ-*Function Potential*

In Sect. 3.13.1 we found that the number of bounds states of a square potential well depends on the single quantity R. In particular in the limit $a \to 0$, $\Lambda \to \infty$, with $a\Lambda$ fixed, there is precisely one bound state. In this limit the potential gets proportional to the δ-function. Therefore we expect that the Schrödiger equation

$$-\frac{\hbar^2}{2m}\frac{d^2\psi}{dx^2} + g\,\delta(x)\,\psi(x) = E\,\psi(x) \tag{3.375}$$

has a single bound state for $g < 0$ and no bound state for $g > 0$. Let us confirm this expectation by an explicit computation. Since the potential diverges at the origin, the energy eigenfunction is continuous but its derivative jumps at $x = 0$: indeed integrating both sides of (3.375) from $-\epsilon$ to ϵ we see that the only effect of the interaction is to prescribe the boundary condition at $x = 0$

$$\psi(0^-) = \psi(0^+),$$
$$\psi'(0^+) - \psi'(0^-) = g\,\psi(0). \tag{3.376}$$

A negative energy solution in $L^2(\mathbb{R})$ has the form

$$\psi(x) = \begin{cases} A\,e^{-qx} & x > 0 \\ B\,e^{qx} & x < 0 \end{cases} \qquad q \equiv \frac{\sqrt{-2mE}}{\hbar} > 0. \tag{3.377}$$

The boundary conditions (3.376) give $A = B$ and

$$2Aq = -gA, \qquad A \neq 0 \tag{3.378}$$

which has no non-trivial solution with q positive when $g > 0$ and the single solution $q = |g|/2$ when $g < 0$, as expected.

3.13.4 Constant Field

We consider a charged particle moving on a line $\mathbb{R}$ subjected to a constant electric field (equivalently: a massive particle in a constant Newtonian gravitational field) whose Hamiltonian is

$$H = \frac{p^2}{2m} + g\,q \qquad g > 0. \tag{3.379}$$

This Hamiltonian is *not* bounded below but in a very mild way: the electric field is constant in space, so translations are symmetries, and the physics is the same at all $x \in \mathbb{R}$, and nothing particularly wild may happen anywhere. From our general discussion in Sect. 3.7 we expect the following properties:

(i) the generalized eigenvalues are non-degenerate, so we may take the energy eigenfunctions to be real;
(ii) the spectrum is continuous $-\infty < E < \infty$: indeed the energy level E shifts under spatial translation:

$$\psi_E(x + a) \equiv \psi_{E-ga}(x); \tag{3.380}$$

(iii) each energy eigenfunction $\psi_E(x)$ has infinitely many zeros, but there is a x_E such that for $x > x_E$ the function $\psi_E(x)$ has no zero.

Consider the Lie algebra $\mathfrak{f}$ generated by the three operators 1, p and $Q \equiv H/g$. The only non-trivial commutation relation in $\mathfrak{f}$ is

$$[Q, p] = i\hbar \tag{3.381}$$

i.e. the pair (Q, p) generates a canonical algebra. By the von Neumann theorem, all irreducible representations of the canonical algebra are unitary equivalent to the usual Schrödinger representation $(x, -i\hbar\, d/dx)$. Hence the solution to the spectral problem for the Hamiltonian (3.379) is trivial: we have a continuous of generalized eigenvalues of the Hamiltonian $H \equiv g\,Q$ ranging from $-\infty$ to $+\infty$, and all generalized eigenfunctions in the momentum representations are phases $|\langle p|\psi_E\rangle|^2 = 1$ while

$$\int_{\mathbb{R}} dx\ \psi_{E'}(x)^* \psi_E(x) = g\,\delta(E' - E). \tag{3.382}$$

Although this already settles the issue, let us solve the Schrödinger equation in the "dummy" way. We work in the p-representation: the Schrödinger equation becomes

$$H\phi_E(p) = \frac{p^2}{2m}\phi_E(p) + ig\hbar\,\frac{d\phi_E(p)}{dp} = E\phi_E(p), \tag{3.383}$$

where $\phi_E(p)$ is the Fourier transform of the q-representation wave function $\psi_E(x)$. The solution to the ODE (3.383) is

$$\phi_E(p) = \exp\left[\frac{i}{g\hbar}\left(\frac{p^3}{6m} - E\,p\right)\right] \tag{3.384}$$

which is readily seen to agree with the predictions of the von Neumann theorem. To get the Schrödinger-representation wave function we have only to take the inverse Fourier transform

$$\begin{aligned}
\psi_E(x) &= \int_{\mathbb{R}} \frac{dp}{2\pi\hbar}\ \exp\left[\frac{i}{g\hbar}\frac{p^3}{6m} + \frac{ip}{\hbar}\left(x - \frac{E}{g}\right)\right] \equiv \\
&= \int_0^{\infty} \frac{dp}{\pi\hbar}\ \cos\left[\frac{i}{g\hbar}\frac{p^3}{6m} + \frac{ip}{\hbar}\left(x - \frac{E}{g}\right)\right].
\end{aligned} \tag{3.385}$$

The wave-function $\psi_E(x)$ may be expressed in terms of a special function called the *Airy function* $\mathrm{Ai}(x)$ [34] (a special instance of a Bessel function, in turn a special case of Whittaker function [14]), which is the entire function defined by the integral representation

$$\mathrm{Ai}(X) = \int_0^{\infty} \frac{dt}{\pi}\ \cos\left(\frac{t^3}{3} + xt\right) \equiv \frac{1}{\pi}\sqrt{\frac{x}{3}}\,K_{\pm\frac{1}{3}}\left(\frac{2x^{3/2}}{3}\right) \tag{3.386}$$

where, as before, $K_\nu(z)$ is the modified Bessel function of the second kind of index ν [9, 34]. Since $K_\nu(z) \approx (\pi/2z)^{1/2}e^{-z}$ for large positive z, we see that for large positive x

$$\mathrm{Ai}(x) \approx \frac{\exp(-\frac{2}{3}x^{3/2})}{2\sqrt{\pi}\,x^{1/4}}\left(1 + O(x^{-2/3})\right) \tag{3.387}$$

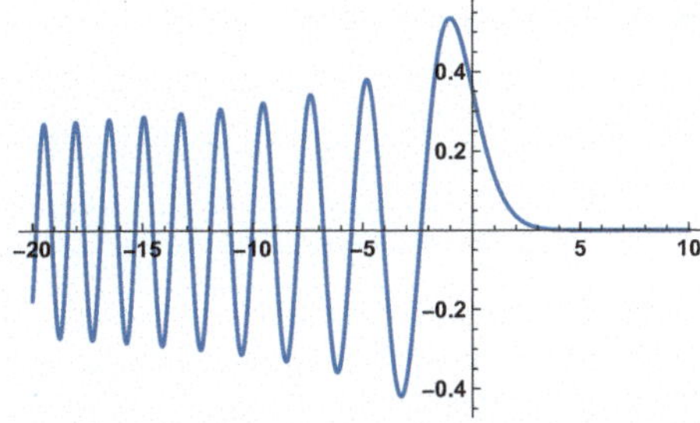

Fig. 3.6 Plot of the Airy function Ai(x)

while for $x \to -\infty$ the function oscillates. See Fig. 3.6 for the plot of Ai(x).

The entire functions Ai(x) and Ai$'(x)$ have infinitely many (simple) zeros all on the negative real axis

$$
\begin{aligned}
\mathrm{Ai}(a_k) &= 0 \quad \text{where} \quad 0 > a_1 > a_2 > a_3 > a_4 > \cdots \\
\mathrm{Ai}'(a_k') &= 0 \quad \text{where} \quad 0 > a_1' > a_2' > a_3' > a_4' > \cdots
\end{aligned}
\tag{3.388}
$$

See table 9.9.1 in [9] for the first few a_k, a_k' and other informations about the zeros of the Airy functions.

3.13.5   The Morse Potential

This is a simple one-dimensional potential with application to bi-atomic molecules [35, 36]. The Schrödinger equation is

$$
-\psi'' + \frac{2m}{\hbar^2}(A\,e^{2x} - 2B\,e^{x})\psi = \frac{2mE}{\hbar^2}\psi
\tag{3.389}
$$

where $A > 0$ to have a potential bounded below. We take $B > 0$ and look for bound states of the system, that is, for states with $E < 0$ (there are no bound states at threshold). The potential is represented in Fig. 3.7. Writing

$$
y = x + \frac{1}{2}\log\left(\frac{2mA}{\hbar^2}\right), \quad
\alpha = \sqrt{\frac{2m}{\hbar^2 A}}\,B, \quad
\lambda^2 = -\frac{2mE}{\hbar^2} > 0,
\tag{3.390}
$$

Eq. (3.389) becomes

$$
\frac{d^2\psi}{dy^2} - e^{2y}\psi + 2\alpha\,e^{y}\psi - \lambda^2\psi = 0.
\tag{3.391}
$$

Fig. 3.7 The Morse potential

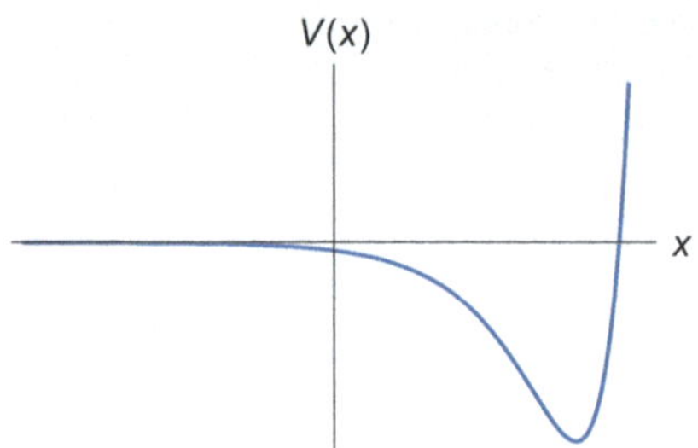

The change of variables ($\lambda > 0$)

$$\psi(y) = \exp[-e^y + \lambda y]\,\chi(2e^y), \qquad z = 2e^y, \tag{3.392}$$

transforms the ODE (3.391) into

$$z\,\chi''(z) + (2\lambda + 1 - z)\chi'(z) + (\alpha - \lambda - 1/2)\chi(z) = 0 \tag{3.393}$$

which is Kummer's equation (3.195). We shall present the theory of this equation in Chap. 5, see in particular the **BOX** on page 316. We refer there for a detailed discussion. The bottom line is that when

$$\alpha - \lambda - 1/2 = n \in \mathbb{N} \tag{3.394}$$

the ODE (3.393) is solved by the *Laguerre degree-n polynomial* $L_n^{(2\lambda)}(z)$. Thus when the parameter E is

$$E = E_n \equiv -\frac{1}{A}\left(B - \hbar\sqrt{\frac{A}{2m}}(n + 1/2)\right)^2 \qquad n = 0, 1, 2, \cdots \tag{3.395}$$

we have the solution

$$\psi_n(x) = \exp\Big[-e^x + (\alpha - n - 1/2)x\Big]\,L_n^{(2\lambda)}(2\,e^x). \tag{3.396}$$

However this solution is in $L^2(\mathbb{R})$ iff $n < \alpha - 1/2$, that is, we have only *finitely many bond states* $\psi_n(x)$ with n an integer in the range

$$n = 0, 1, \ldots, \left[\sqrt{\frac{2m}{A}}\frac{B}{\hbar} - \frac{1}{2} - \epsilon\right]. \tag{3.397}$$

In particular when $2\sqrt{2m}\,B < \sqrt{A}\hbar$ there are *no* bound states.

3.13.6 *The Poisson Periodic Potential*

We solve a model with a periodic potential in $\mathbb{R}$ to illustrate explicitly the *band* phenomenon. As periodic potential we take the kernel of the *Poisson summation* formula [37], that is, the mass distribution of the integers

$$\delta_{\mathbb{Z}}(x) \overset{\text{def}}{=} \sum_{n\in\mathbb{Z}} \delta(x-n) \tag{3.398}$$

which allows to convert sums into integrals

$$\sum_{n=-\infty}^{+\infty} F(n) = \int_{-\infty}^{+\infty} \delta_{\mathbb{Z}}(x)\, F(x)\, dx. \tag{3.399}$$

A simple but very fundamental result is the *Poisson summation formula* [37]:

Theorem 3.13 (Poisson) $\delta_{\mathbb{Z}}(x)$ *is its own Fourier transform*

$$\delta_{\mathbb{Z}}(x) = \int_{-\infty}^{+\infty} \mathrm{e}^{2\pi\mathrm{i}xy}\, \delta_{\mathbb{Z}}(y)\, \mathrm{d}y \equiv \sum_{n=-\infty}^{+\infty} \mathrm{e}^{2\pi\mathrm{i}nx}. \tag{3.400}$$

Proof See Problem 3.20.   □

We focus on the Schrödinger equation ($\hbar = 2m = 1$)

$$-\frac{\mathrm{d}^2}{\mathrm{d}x^2}\psi_E(x) + g\, \delta_{\mathbb{Z}}(x)\, \psi_E(x) = E\, \psi_E(x). \tag{3.401}$$

For brevity we take the parameter g to be positive, i.e. we consider the *repulsive* case where the energy E of all states is *positive*. For the attractive potential see Problem 3.22. We write $[x]$ for the *integer part* of x, i.e. largest integer not bigger than $x \in \mathbb{R}$, and $\{x\}$ for its *fractional part*, $\{x\} \equiv x - [x]$ with $0 \le \{x\} < 1$.

By the general theory of Sect. 3.12, and the fact that the motion is free in the interval between two integers, we infer that the wave-function with energy $E = k^2$ should have the general form

$$\psi_k(x) = \mathrm{e}^{\mathrm{i}\alpha[x]}\Big(A\mathrm{e}^{\mathrm{i}k\{x\}} + B\mathrm{e}^{-\mathrm{i}k\{x\}}\Big), \qquad -\pi < \alpha \le \pi, \ \ k \ge 0, \tag{3.402}$$

for some constants A and B with $B = 0$ when $k = 0$.

The parameter α is the *Floquet phase*. The connection formulae at $x = n$ are

$$\psi_k\big|_{x=n^+} - \psi_k\big|_{x=n^-} = 0 \tag{3.403}$$

$$\psi_k'\big|_{x=n^+} - \psi_k'\big|_{x=n^-} = g\, \psi_k\big|_{x=n} \tag{3.404}$$

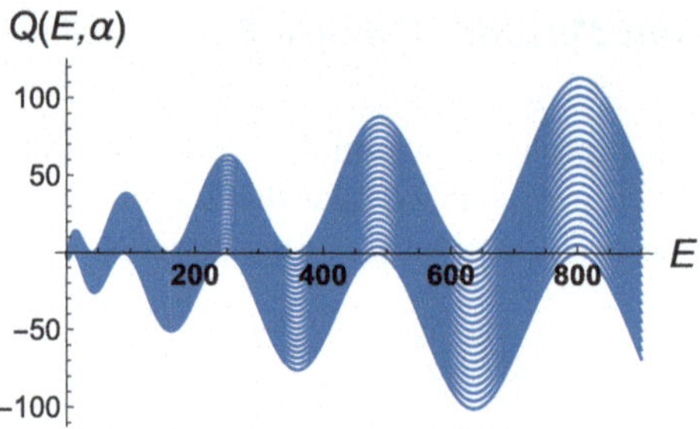

Fig. 3.8 The function $Q(E,\alpha)$ for $g = 1$ in the range $0 < E \leq 900$. The various blue curves refer to values of $\cos\alpha$ going from -1 to $+1$ in steps $\Delta\cos\alpha = 0.1$. A crossing of a curve with the positive real axis corresponds to a generalized energy eigenvalue with the corresponding energy and Floquet phase. A part for a tiny region near the origin (so small to be invisible in this large scale plot), for each point in the positive real axis there is a curve crossing it, i.e. the spectrum is continuous after some small E_0 and the gaps between the high energy bands are closed

which have no non-zero solution when $k = 0$, while for $k > 0$ they are equivalent to the linear equation

$$\begin{pmatrix} e^{i\alpha} - e^{ik} & e^{i\alpha} - e^{-ik} \\ (ik - g)e^{i\alpha} - ike^{ik} & (-ik - g)e^{i\alpha} + ike^{-ik} \end{pmatrix} \begin{pmatrix} A \\ B \end{pmatrix} \equiv M(\alpha, k)\begin{pmatrix} A \\ B \end{pmatrix} = 0$$

$$(3.405)$$

for the coefficients A, B. A non-zero solution exists iff the determinant of the matrix $M(\alpha, k)$ is zero:

$$\det M(\alpha, k) = -2\,i e^{i\alpha}\big(2k\cos\alpha - 2k\cos k - g\sin k\big) = 0. \tag{3.406}$$

Hence we have an eigenstate of energy $\sqrt{E}$ and Floquet phase α whenever

$$Q(E, \alpha) \stackrel{\text{def}}{=} \sqrt{E}\cos\alpha - \sqrt{E}\cos\sqrt{E} - g\,\sin\sqrt{E} = 0. \tag{3.407}$$

Note that the condition is even in α, so the energy is an even function of α: this reflects the fact that the potential preserves parity[32] $V(-x) = V(x)$. At high energy, when $E \gg g^2$ we may forget the last term in $Q(E, \alpha)$ and we get the energy levels

$$E_n(\alpha) \approx (2\pi n + \alpha)^2, \quad n \in \mathbb{Z}, \quad -\pi < \alpha \leq \pi \tag{3.408}$$

This is, of course, the spectrum of the free particle moving on $\mathbb{R}$ which has a doubly-degenerate continuous spectrum. In other words, as explained in Sect. 3.12, at high energy the gaps between the bands close, and we get a connected continuous spectrum. The situation in this large E regime is shown in Fig. 3.8.

---

[32] It is also a consequence of the time-reversal symmetry, as remarked at the end of Sect. 3.12.

The situation at low energy $E \ll g^2$ is more interesting. At $g = \infty$ the energy levels become $\sqrt{E} = n\pi$, $n \in \mathbb{N}$. Indeed, as $g \to \infty$ we must recover the result for a free particle constrained to move in the interval $[0, 1]$: in this limit the transmission across the δ-function barrier vanishes. To get the corrections to this asymptotic behavior, we expand the energy in inverse powers of g in the form

$$\sqrt{E} = n\pi \left(1 + \frac{a_1}{g} + O(g^{-2})\right). \tag{3.409}$$

To the first order in $1/g$ Eq. (3.407) becomes

$$a_1 = -1 + (-1)^n \cos\alpha \tag{3.410}$$

and

$$E_n(\alpha) = n^2\pi^2 \left(1 + \frac{2}{g}((-1)^n - \cos\alpha) + O\left(\frac{1}{g^2}\right)\right) \tag{3.411}$$

for $n = 1, 2, \ldots$. Thus, in the regime $g \gg n\pi$, the n-th band has a width

$$\Delta_n = \frac{4n^2\pi^2}{g}\left(1 + O\left(\frac{1}{g^2}\right)\right) \tag{3.412}$$

while the $(n + 1)$-th band is separated from the n-th one by a spectral gap of size

$$\delta_n = \min\left(E_{n+1}(\alpha) - E_n(\beta)\right) =$$
$$= (2n + 1)\pi^2 + \frac{2\pi^2}{g}(2n^2 + 2n + 1)[1 - (-1)^n] + O\left(\frac{1}{g^2}\right) \tag{3.413}$$

and $\delta_n \gg \Delta_{n+1}$. See Fig. 3.9.

Appendix: The SUSY Trick

To solve the Schrödinger equations of Examples 3.5 and 3.11 we used a trick that physically may be seen as an application of *supersymmetry* (SUSY). Supersymmetry will be discussed as a quantum symmetry in the last sections of Chap. 4, and then reinterpreted in the language of path integrals in the final sections of Chap. 6. Here we just limit to describe a "trick" to solve one-dimensional Schrödinger equations which is inspired by SUSY. For more details and generalizations of the method we refer to Chaps. 4 and 6.

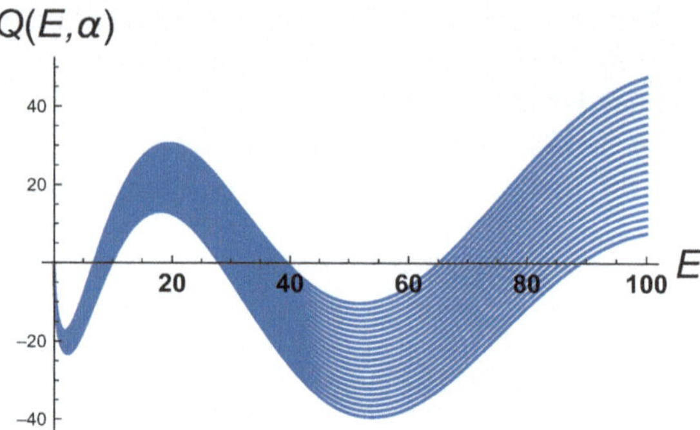

Fig. 3.9 The function $Q(E, \alpha)$ for $g = 20$ in the range $0 < E \le 100$. The various *blue curves* refer to values of $\cos \alpha$ going from -1 to $+1$ in steps $\Delta \cos \alpha = 0.1$. A crossing of a curve with the positive real axis corresponds to a generalized energy eigenvalue with the corresponding energy and Floquet phase. We see that for low n there are large gaps between the bands

In the language of [5] §. 6.4, where the Schrödinger equation of a Supersymmetric Quantum Mechanics (SQM) model is mapped to the Fokker-Planck equation which describe the probability distribution of a Langevin stochastic ODE, the trick amounts to comparing the forward and backward propagation in time for a diffusive process. Mathematically the trick was known to Riccati some four centuries ago.

Consider the differential operator

$$L = \frac{\mathrm{d}}{\mathrm{d}x} + F(x) \tag{3.414}$$

We can form two (formally) non-negative self-adjoint 2nd order differential operators

$$L^{\dagger} L = \left(-\frac{\mathrm{d}}{\mathrm{d}x} + F(x) \right) \left(\frac{\mathrm{d}}{\mathrm{d}x} + F(x) \right) = -\frac{\mathrm{d}^2}{\mathrm{d}x^2} + F^2(x) - \frac{\mathrm{d}F(x)}{\mathrm{d}x} \tag{3.415}$$

$$L L^{\dagger} = \left(\frac{\mathrm{d}}{\mathrm{d}x} + F(x) \right) \left(-\frac{\mathrm{d}}{\mathrm{d}x} + F(x) \right) = -\frac{\mathrm{d}^2}{\mathrm{d}x^2} + F^2(x) + \frac{\mathrm{d}F(x)}{\mathrm{d}x} \tag{3.416}$$

which are proportional to the Schrödinger Hamiltonians of two one-dimensional systems with $\hbar$ set to 1 and potentials

$$V_{\mp}(x) = \frac{1}{2m} \left(F^2(x) \mp \frac{\mathrm{d}F(x)}{\mathrm{d}x} \right). \tag{3.417}$$

Suppose we know the solution to the eigenvalue equation for the first potential

$$L^{\dagger} L \, \psi = 2m E \, \psi. \tag{3.418}$$

Multiplying by L both sides we get

$$LL^{\dagger}(L\psi) = 2mE(L\psi) \tag{3.419}$$

so either $L\psi = 0$, which by (3.418) may happen only when $E = 0$, or $L\psi$ is a solution of the eigenvalue equation for the second potential with the same energy $E \neq 0$. We note that

$$\left\| L|\psi\rangle \right\|^2 = \langle\psi|L^{\dagger}L|\psi\rangle = 2mE\langle\psi|\psi\rangle \equiv E\left\| |\psi\rangle \right\|^2 \tag{3.420}$$

so, when $E \neq 0$, if the solution ψ to the first eigenfunction equation is normalizable (in either the L^2 sense or the generalized continuum sense) so is the solution $L\psi$ to the second spectral problem. Thus if $M = \mathbb{R}$ we get an explicit map between the eigenvectors and eigenvalues which is one-to-one *except* for the zero-energy eigenstates. When $\partial M \neq \varnothing$ one has the check that $L\psi$ satisfies the correct boundary conditions: it does so when $V_-(x)$ blows up on ∂M, but not otherwise.

Example 3.13 $F(x) = -m\omega x$. $V_{\mp}(x) = \frac{1}{2}(m\omega^2 \pm \omega)$ and the two systems are both harmonic oscillators of the same frequency ω with a constant shift of the Hamiltonian $H_- = H_+ + \omega$. The map $\psi \mapsto L\psi$ then increases the energy by ω. Comparing with the algebraic approach to the harmonic spectrum we see that $L = a^{\dagger}, L^{\dagger} = a$.

Hence L and $L^{\dagger}$ may be seen as generalizations of the raising and lowering operators $a^{\dagger}, a$ of the algebraic approach to the harmonic spectrum.

Example 3.14 $F(x) = -\tanh x$. One has

$$2mV_-(x) = 1 \qquad 2mV_+(x) = 1 - \frac{2}{\cosh^2 x} \tag{3.421}$$

so the trick maps the solutions of the free Schrödinger equation (with the energy shifted by a constant) to the one for the system in Example 3.5, *except for the zero energy sector*. Since the free Hamiltonian has no bound states, the system in Example 3.5 has at most one bound state at zero energy

$$L^{\dagger}\psi_{\text{bound}}(x) = 0 \quad \Rightarrow \quad \frac{1}{\sqrt{2}}\frac{1}{\cosh^2 x} \tag{3.422}$$

The continuous spectrum has energies $E \geq 1/2m$, so there is an energy gap between the unique bound state and the scattering states.

Example 3.15 $F(x) = \tan x$ in the interval $(-\pi/2, \pi/2)$. One has

$$2mV_-(x) = 1, \qquad 2mV_+ = \frac{2}{\cos^2 x} - 1, \tag{3.423}$$

which maps the free system in the interval into the one in Example 3.11.

Problems

3.1 Check explicitly that the kernel (3.51) satisfies the Schrödinger PDE.

3.2 Write the Hermitian operators which generate the group of Bogoliubov transformations.

3.3 Show that Eq. (3.104) is an isomorphism of Hilbert spaces.

3.4 Write the Bargmann kernels of the group $Sp(2, \mathbb{R}) \simeq SL(2, \mathbb{R})$ of linear canonical transformations.

3.5 Prove Eq. (3.223).

3.6 Prove Lemma 3.4.

3.7 Prove the claim (3.254).

3.8 Consider the Schödinger equation $-\psi'' + V\psi = \lambda\psi$ in $[0, L]$ with the boundary conditions $\psi(0) = \psi(L)$. Parametrize the solution in the form

$$\psi(x) = r \sin \phi \qquad \psi' = \sqrt{\lambda}\, r \cos \phi.$$

Show that the number of zeros in $[0, L]$ is $(\phi(L) - \phi(0))/\pi + 1$. Use the differential equation for ϕ to get a precise estimate of this quantity.

3.9 Write the energy wave-functions for the states of the step potential in Example 3.2 with energies $E > V_0$.

3.10 Solve the Schrödinger equation for a particle moving in $\mathbb{R}$ subjected to the potential

$$V(x) = \begin{cases} \frac{m\omega^2}{2}x^2 & \text{for } x > 0 \\ 0 & \text{for } x < 0 \end{cases}$$

3.11 Show that the finite potential well in Sect. 3.13.1 has at least one bound state for all a, Λ by solving the corresponding Prüfer equation

$$\vartheta' = \cos^2\vartheta + (E - V)\sin^2\vartheta.$$

3.12 Find the generalized energy eigenfunctions for the scattering states of non-negative energy in the finite potential well of Sect. 3.13.1.

3.13 Solve the Schrödinger equation in $\mathbb{R}$ with the potential

$$V(x) = \begin{cases} 0 & \text{for } x < 0 \\ -\Lambda < 0 & \text{for } 0 < x < L \\ U > 0 & \text{for } x > L. \end{cases}$$

3.14 Solve the Schrödinger equation in $\mathbb{R}_{\geq 0}$ with the potential

$$V(x) = \begin{cases} -\Lambda < 0 & \text{for } 0 \leq x < L \\ 0 & \text{for } x > L. \end{cases}$$

3.15 Solve the Schrödinger equation in $\mathbb{R}_{\geq 0}$ with the potential

$$V(x) = \Lambda\,\Theta(x - L_1)\,\Theta(L_2 - x) \quad \Lambda > 0, \quad L_2 > L_1$$

3.16 Same as Exercise 3.15 but with $\Lambda < 0$.

3.17 Prove Eq. (3.261).

3.18 (Triangular Well) Solve the Schrödinger equation in the half-line $\mathbb{R}_{\geq 0}$ with potential

$$V(x) = g\,x, \qquad g > 0.$$

3.19 Solve the Schrödinger equation in $\mathbb{R}$ with potential $V(x) = g|x|$.

3.20 Prove the Poisson summation formula, Theorem 3.13.

3.21 Solve the Schrödinger equation for a particle moving in the half-line $\mathbb{R}_{\geq 0}$ with potential $V = g\,\delta(x - L)$, $L > 0$.

3.22 Study the states of negative energy for a particle moving in the Poisson periodic potential (3.398) with negative coupling constant $g < 0$ (the attractive case).

3.23 Solve the Poisson periodic model in the momentum representation.

3.24 Solve the Schrödinger equation for the periodic potential

$$V(x) = V(x + a)$$

where for $0 \leq x \leq a$

$$V(x) = \Lambda\,\Theta(b - x) \quad 0 < b < a, \quad 0 \leq x \leq a,$$

with $\Theta(z)$ the step function.

3.25 Solve the Schrödinger equation in $\mathbb{R}$ with the periodic potential

$$V(x) = \frac{2}{\pi^2}\sum_{k=1}^{\infty}(-1)^k\,\frac{\cos(k\pi x)}{k^2} + \frac{1}{6}.$$

References

1. S. Cecotti, *Analytic Mechanics. A Concise Textbook* (Springer, Berlin, 2024)
2. M. Reed, B. Simon, *Methods of Modern Mathematical Physics*, vols. I–IV (Academic Press, Cambridge, 1970)
3. M. van der Put, Symbolic analysis of differential equations, in *Some Tapas of Computer Algebra*, ed. by A.M. Cohen, H.Cuypers, H. Sterk (Springer, Berlin, 1999), pp. 208–236
4. S. Cecotti, *Supersymmetric Field Theories. Geometric Structures and Dualities* (Cambridge University Press, Cambridge, 2015)
5. S. Cecotti, *Statistical Mechanics. A Concise Advanced Textbook* (Springer, Berlin, 2024)
6. J. Humphreys, *Introduction to Lie Algebras and Representation Theory* (Springer, Berlin, 1980)
7. A.W. Knapp, *Lie Groups Beyond an Introduction*, 2nd edn. (Birkhäuser, Basel, 2002)
8. A. Erdèlyi et al., *Higher Transcendental Functions*, vol. II (McGraw Hill, New York, 1954)
9. F.W. Olver, D.M. Lozier, R.F. Boisvert, C.W. Clark, W. Charles W. (eds.), *NIST Handbook of Mathematical Functions* (Cambridge University Press, Cambridge, 2010). Available on-line at https://dlmf.nist.gov
10. J.S. Milne, *Algebraic Groups. The Theory of Group Schemes of Finite Type over a Field* (Cambridge University Press, Cambridge, 2017)
11. M. van der Put, M.F. Singer, *Galois Theory of Linear Differential Equations. A Series of Comprehensive Studies in Mathematics*, vol. 328 (Springer, Berlin, 2003)
12. Y. Haraoka, *Linear Differential Equations in the Complex Domain. From Classical Theory to Forefront. Lecture Notes in Mathematics*, vol. 2271 (Springer, Berlin, 2020)
13. G. Szegö, *Orthogonal Polynomials* (AMS, Providence, 1975)
14. G.E. Andrews, R. Askey, R. Roy, *Special Functions*. Encyclopedia of Mathematics and Its Applications, vol. 71 (Cambridge University Press, Cambridge, 2009)
15. C. Bennewitz, M. Brown, R. Weikard, *Spectral and Scattering Theory for Ordinary Differential Equations Vol. I: Sturm-Liouville Equations*. Universitext (Springer, Berlin, 2020)
16. R. Magnus, *Essential Ordinary Differential Equations*. Springer Undergraduate Mathematics Series (Springer, Berlin, 2023)
17. Q. Kong, *A Short Course in Ordinary Differential Equations*. Universitext (Springer, Berlin, 2014)
18. H. Năstase, *Quantum Mechanics. A Graduate Course* (Cambridge University Press, Cambridge, 2023)
19. P. Deligne, *Équations Différentielles à Points Singuliers Réguliers*. Lecture Notes in Mathematics, vol. 163 (Springer, Berlin, 1970)
20. M. Yoshida, *Fuchsian Differential Equations. With Special Emphasis on the Gauss-Schwarz Theory*. Aspects of Mathematics, vol. E11 (Springer, Berlin, 1987)
21. J.J. Gray, *Linear Differential Equations and Group Theory from Riemann to Poincaré*, 2nd edn. (Birkhäuser, Basel, 2000)
22. N.M. Katz, *Rigid Local Systems* (Princeton University Press, Princeton, 1996)
23. S. Cecotti, Fuchsian ODEs as Seiberg dualities. Adv. Theor. Math. Phys. **27**(8), 2429–2490 (2023). arXiv:2212.09370 [hep-th]
24. S.A. Rakityansky, *Jost Functions in Quantum Mechanics. A Unified Approach to Scattering, Bound, and Resonant State Problems* (Springer, Berlin, 2022)
25. E.T. Whittaker, G.N. Watson, *A Course of Modern Analysis*, 4th edn. (Cambridge University Press, Cambridge, 1927)
26. G.N. Watson, *A Treatise on the Theory of Bessel Functions*, 2nd edn. (Cambridge University Press, Cambridge, 1944)
27. S.J. Gustafson, I.M. Sigal, *Mathematical Concepts of Quantum Mechanics*, 3rd edn. Univesi-Text (Springer, Berlin, 2020)
28. A.N. Karapetyants, V.V. Kravchenko, *Methods of Mathematical Physics Classical and Modern* (Birkhäuser, Basel, 2022)

29. R. Courant, D. Hilbert, *Methods of Mathematical Physics*, vol. 1 (Wiley, Hoboken, 1989)
30. M. Masjed-Jamei, *Special Functions and Generalized Sturm-Liouville Problems* (Birkhäuser, Basel, 2020)
31. V.V. Kravchenko, *Direct and Inverse Sturm-Liouville Problems. A Method of Solution* (Birkhäuser, Basel, 2020)
32. S. Coleman, The magnetic monopole fifty years later, in *Gauge Theories of High Energy Physics*, Part 1, Les Houches 1881 (North-Hollands, Amsterdam, 1983), pp. 461–553
33. C. Chicone, *Ordinary Differential Equations with Applications*, 3rd edn. (Springer, Berlin, 2024)
34. O. Vallé, M. Soares, *Airy Functions and Applications to Physics*, 2nd edn. (World Scientific, Singapore, 2010)
35. G. Auletta, M. Fortunato, G. Parisi, *Quantum Mechanics* (Cambridge University Press, Cambridge, 2009)
36. P.M. Morse, Diatomic molecules according to the wave mechanics. II. Vibrational levels. Phys. Rev. **34**, 57–64 (1929)
37. H. Iwaniec, E. Kowalski, *Analytic Number Theory*. AMS Colloquium Publications, vol. 53 (AMS, Providence, 2004)

Chapter 4
Symmetry, Angular Momentum, Statistics

Symmetry and its diverse dynamical realizations are central concepts in Physics: the time-evolution itself is a one-parameter family of symmetries. Symmetry is an organizing principle of Physics, a structural constraint on the form of the dynamical laws, and also a valuable tool to get exact results quickly, bypassing long hard computations. In this chapter we introduce symmetry in the context of Quantum Mechanics, state a few general properties of it, and then focus on the two examples most needed in the concrete applications: *spatial rotations* and *permutations of identical particles*. The Galilei symmetry of non-relativistic Quantum Mechanics is also discussed. In the last sections of the chapter we introduce *supersymmetry* which, apart for its interest in its own right, is a valuable tool for computations in Quantum Mechanics as mentioned in the Appendix to Chap. 3 and further highlighted in Chap. 6. We end by a sketch of the Witten index for supersymmetric Quantum Mechanics and its relation to Algebraic and Differential Topology.

We have already seen some examples of symmetries in Quantum Mechanics: spatial translations in $\mathbb{R}^n$, parity $x \leftrightarrow -x$, and time reversal $t \leftrightarrow -t$. In Sect. 2.15 we have proven that the group of isometries of the configuration space M which preserve the various potentials are symmetries for the natural Hamiltonians, and that the *Noether theorem* holds at the full quantum level for this class of symmetries. Now we study symmetry in a more general and systematic way.

4.1 Generalities on Symmetries

In any physical theory—classical, quantum, or anything else might be formulated in the future—we have the

Definition 4.1 A *kinematical symmetry* S is an automorphism of the space $\mathfrak{W}$ of physical states which preserves all its defining structures.

S. Cecotti, *Quantum Mechanics*, UNITEXT for Physics,
https://doi.org/10.1007/978-3-031-98824-0_4

Since the Fundamental Laws of Physics are the same today as they will be tomorrow and the day after tomorrow (hopefully!), the space $\mathfrak{W}$ of possible states and its defining structures—which are dictated by the Fundamental Laws—should be preserved by the time-evolution, that is,

Fundamental Principle *The dynamical time-evolution of any physical system is a one-parameter group(oid[1]) $U(t) \subset \mathsf{Aut}(\mathfrak{W})$ of automorphisms of its state space $\mathfrak{W}$.*

Example 4.1 In Classical Mechanics the time-evolution starting from time t_0 is a one-parameter family of *canonical transformations,* i.e. symplectomorphisms $S(t, t_0) \colon \mathcal{W} \to \mathcal{W}$ of the phase space $\mathcal{W}$ ([1] chap. 8).

Definition 4.2 A *dynamical symmetry*[2] is an automorphism $S \in \mathsf{Aut}(\mathfrak{W})$ which commutes with the family $U(t) \subset \mathsf{Aut}(\mathfrak{W})$ of time-evolution automorphisms, i.e.

$$U(t)S = SU(t) \quad \text{in } \mathsf{Aut}(\mathfrak{W}) \text{ for all } t \in \mathbb{R}. \tag{4.1}$$

Wigner Theorem

Preliminarily we have to determine the group of automorphisms of the space of quantum states. The answer may be not fully obvious since the space of quantum states *is not* the Hilbert space $\mathcal{H}$ (as we often say by abuse of language) but rather its *projective* version $\mathbb{P}(\mathcal{H})$, see Chap. 1. The elements of $\mathsf{Aut}(\mathbb{P}(\mathcal{H}))$ are the *invertible maps* $S \colon |\psi\rangle \mapsto |S\psi\rangle$ which preserve the observable *quantum amplitudes*

$$\frac{\langle S\psi'|S\psi\rangle\langle S\psi|S\psi'\rangle}{\langle S\psi'|S\psi'\rangle\langle S\psi|S\psi\rangle} = \frac{\langle \psi'|\psi\rangle\langle \psi|\psi'\rangle}{\langle \psi'|\psi'\rangle\langle \psi|\psi\rangle} \in \mathbb{R}_{>0} \quad \text{for all } |\psi'\rangle, \, |\psi\rangle \neq 0. \tag{4.2}$$

We already know two *obvious* subgroups of $\mathsf{Aut}(\mathbb{P}(\mathcal{H}))$:

1st: we have the automorphisms of the Hilbert space $\mathcal{H}$, namely the *unitary transformations* $U \colon \mathcal{H} \to \mathcal{H}$, which are *a fortiori* automorphisms of $\mathbb{P}(\mathcal{H})$. The canonical map

$$\mathcal{U} \equiv \mathsf{Aut}(\mathcal{H}) \to \mathsf{Aut}(\mathbb{P}(\mathcal{H})) \tag{4.3}$$

has a non-trivial kernel given by the group $U(1) = \{e^{i\alpha} \cdot \mathbf{1}\}$ of unitary multiples of the identity $\mathbf{1}$;

2nd: the formalism of Quantum Physics is linear over the field $\mathbb{C}$, but all physical observables are *real,* so the Galois group

$$\mathsf{Gal}(\mathbb{C}/\mathbb{R}) \simeq \mathbb{Z}/2\mathbb{Z} \equiv \mathbb{Z}_2 \tag{4.4}$$

---

[1] The time-evolution forms a group $\simeq \mathbb{R}$ for a time-independent system, but only a groupoid for a time-dependent one.

[2] Sometimes the term "dynamical symmetry" is reserved for dynamical symmetries which are not of the Noether class i.e. they do not belong to the isometry group of the configuration space $\mathcal{M}$.

fixes all observables—in particular the quantum amplitudes (4.2)—and hence must be a kinematical symmetry.

We conclude that the group of *quantum kinematical symmetries* should—at least— contain the extension $\mathsf{Obvious}$ of the Galois group by the unitary transformations modulo phases

$$1 \to \mathsf{Aut}(\mathcal{H})/U(1) \to \mathsf{Obvious} \to \mathsf{Gal}(\mathbb{C}/\mathbb{R}) \to 1. \tag{4.5}$$

Elements $U \in \mathsf{Aut}(\mathcal{H})/U(1)$ are unitary transformation $U : |\psi\rangle \mapsto |U\psi\rangle$ with the defining property

$$\langle U\psi'|U\psi\rangle = \langle\psi'|\psi\rangle \quad \text{for all } |\psi'\rangle, \, |\psi\rangle, \tag{4.6}$$

identified up to overall phases which do not affect this equality. The elements of $\mathsf{Obvious}$ which map in the non-trivial element of $\mathsf{Gal}(\mathbb{C}/\mathbb{R})$, i.e. in the complex conjugation $(\cdot)^*$, are ***anti**-unitary* transformations, i.e. anti-linear maps

$$T(a_1|\psi_1\rangle + a_2|\psi_2\rangle) = a_1^*\langle T\psi_1| + a_2^*\langle T\psi_2| \tag{4.7}$$

sending kets to bras and viceversa, such that

$$\langle T\psi'|T\psi\rangle = \langle\psi|\psi'\rangle = \langle\psi'|\psi\rangle^* \quad \text{for all } |\psi'\rangle, \, |\psi\rangle, \tag{4.8}$$

again identified up to overall phases.

Example 4.2 We shown in Sect. 3.1 that, in absence of magnetic fields and spins, time-reversal is an *anti*-unitary dynamical symmetry.

The Wigner theorem is the pretty obvious statement:

Theorem 4.1 (Wigner) *The group* $\mathsf{Obvious}$ *is the full story:*

$$\mathsf{Aut}(\mathbb{P}(\mathcal{H})) \equiv \mathsf{Obvious}, \tag{4.9}$$

i.e. the kinematical symmetries are realized by either unitary or anti-unitary maps unique up to the overall phase.

Proof The usual algebraic proofs, even the supposedly smart ones (cf. chap. 7 of [2]), are long, boring, and non-illuminating. We give a *geometric* proof (less elementary, but providing deeper insights). First notice that it suffices to prove the **Theorem** for a Hilbert space of finite dimension since S preserves orthogonality and so maps (possibly non-linearly!) the states lying in a subspace $V \subset \mathcal{H}$ of dimension n into states lying in a subspace $W \subset \mathcal{H}$ of the same dimension. Hence we may assume the space of states to be $\mathbb{P}(\mathbb{C}^n)$ with no loss. As a Riemannian manifold,

$$\mathbb{P}(\mathbb{C}^n) \equiv \mathbb{P}^{n-1} \simeq U(n)/U(n-1) \tag{4.10}$$

is a compact *symmetric space* [3] with a unique (up to normalization) metric invariant under $U(n)$, which is both Kähler and Einstein with a positive curvature. This invariant metric is known as the *Fubini-Study metric* [4–7]. For instance, when $n = 2$, $\mathbb{P}(\mathbb{C}^2) = SO(3)/SO(2)$ is the sphere S^2 and the Fubini-Study metric is its usual "round" metric

$$ds^2 = d\theta^2 + \sin^2\theta\, d\phi^2. \tag{4.11}$$

Two quantum states are equivalence classes $[|\psi\rangle]$, $[|\psi'\rangle]$ which we identify with two points $p, p' \in \mathbb{P}(\mathbb{C}^n)$. A moment thought shows that the distance between these two points, measured with the invariant Fubini-Study metric, is

$$d(p, p') = \arccos \sqrt{\frac{\langle\psi'|\psi\rangle\langle\psi|\psi'\rangle}{\langle\psi'|\psi'\rangle\langle\psi|\psi\rangle}}. \tag{4.12}$$

We conclude that the maps that "preserve the quantum amplitudes" (4.2) are exactly the *isometries* of the projective space $\mathbb{P}(\mathbb{C}^n)$. We quote some well-known facts from Differential Geometry [8]:

Proposition 4.1 **(1)** *The group of isometries of a compact Riemannian manifold is a compact Lie group.* **(2)** *Let M be a Kähler manifold with a point $m \in M$ such that $\det R_{ij}|_m \neq 0$. Then all isometries of M are either holomorphic or anti-holomophic.* **(3)** *The group of complex automorphisms of $\mathbb{P}(\mathbb{C}^n)$ is $GL(n, \mathbb{C})/GL(1, \mathbb{C})$.*

By **(1)**–**(3)** the group of holomorphic isometries of $\mathbb{P}(\mathbb{C}^n) \equiv \mathbb{P}^{n-1}$ should be contained in the maximal compact subgroup

$$U(n)/U(1) \subset GL(n, \mathbb{C})/GL(1, \mathbb{C}) \tag{4.13}$$

of its complex automorphism group, and since the group of holomorphic isometries of the symmetric space $U(n)/U(n-1)$ contains $U(n)/U(1)$, it must coincide with $U(n)/U(1)$. Then the group of isometries of $\mathbb{P}(\mathbb{C}^n)$ is the extension of the $\mathbb{Z}_2$ group generated by complex conjugation by the (compact) Lie group

$$U(n)/U(1) \simeq SU(n)/\mathbb{Z}_n \overset{\text{def}}{=} PSU(n). \tag{4.14}$$

This completes the proof of Wigner's Theorem 4.1. $\square$

Example 4.3 When the Hilbert space is 2-dimensional, $\mathcal{H} \simeq \mathbb{C}^2$, the space of quantum states $\mathbb{P}^1 \equiv \mathbb{P}(\mathbb{C}^2)$ is isomorphic to the unit sphere

$$S^2 = \{x^2 + y^2 + z^2 = 1\} \subset \mathbb{R}^3 \tag{4.15}$$

with the induced metric from the flat ambient space. The isometry group is then $O(3) = SO(3) \times \mathbb{Z}_2$ where $\mathbb{Z}_2$ is generated by three-dimensional *parity*

$$\mathscr{P}: \quad \mathbf{x} \to -\mathbf{x}, \qquad \mathbf{x} \in \mathbb{R}^3 \tag{4.16}$$

which is an *anti-unitary* symmetry of the Riemann sphere $\mathbb{P}^1 \equiv \mathbb{P}(\mathbb{C}^2) \simeq S^2$. Equivalently,[3] $\mathbb{Z}_2$ flips the orientation of S^2: indeed it acts on the usual "geographic" coordinates (θ, ϕ) as $(\theta, \phi) \leftrightarrow (\pi - \theta, \phi + \pi)$, flipping the sign of the S^2 volume form $\sin \theta \, d\theta \wedge d\phi$.

Remark 4.1 When the Hilbert space is finite-dimensional, $\mathcal{H} \simeq \mathbb{C}^n$, all symmetries of the quantum system are realized either as holomorphic symplectomorphisms of the space of quantum states $\mathbb{P}(\mathbb{C}^n)$, equipped with its canonical Fubini-Study symplectic structure, or as anti-holomorphic morphisms which flip the sign of the symplectic form. In particular a connected one-parameter group $S(t)$ of quantum symmetries is the flow of a holomorphic Hamiltonian vector field on the space of states $\mathbb{P}(\mathbb{C}^n)$.

Remark 4.2 The time evolution is a particular instance of symmetry generated (in the time-independent case) by the unitary operator

$$U(t) = \exp(-iHt) \quad (\text{setting } \hbar = 1). \tag{4.17}$$

The fact that $U(t)$ is unique only up to an overall phase reflects the fact that the Hamiltonian H is unique only up to a shift by a constant: this non-uniqueness is a basic physical property of energy in the non-relativistic theory. In classical mechanics, the Hamiltonian H is (by definition) the momentum map of the action of the Lie group $\mathbb{R}$ on phase-space $\mathcal{W}$ given by the time-evolution ([1] chap. 7). A priori a momentum map is defined up to an additive constant. We see that the ambiguity by a phase of the unitary operator which represents a symmetry is the quantum counterpart to the ambiguity by an additive constant of the momentum map in symplectic geometry.

Dynamical Symmetries

In a time-independent quantum system, a unitary transformation S is a *dynamical symmetry* iff it commutes with the Hamiltonian H up to an inessential constant shift[4]

$$HS = S(H + C), \qquad C \in \mathbb{R}, \tag{4.18}$$

---

[3] Recall that the antiholomorphic isometries of a Riemannian surface Σ are the ones which reverse the orientation of Σ.

[4] An example of such a symmetry with a non-zero shift C is spatial translations in a constant electric field, see Sect. 3.13.4.

or, even more generally, up to a redefinition of the Schrödinger representation (i.e.
a change of gauge in the sense of Sect. 2.12.1)

$$S^{-1}HS = \mathrm{e}^{\mathrm{i}f(q)}H\mathrm{e}^{-\mathrm{i}f(q)} + C. \tag{4.19}$$

Indeed all physical observables are invariant under gauge transformations, so
changing the Schrödinger representation of the Hamiltonian does not modify the
physics, hence a unitary transformation S satisfying (4.19) is still a valid dynamical
symmetry. In Sect. 5.7 we shall see important examples of dynamical symmetries
realized in this general fashion.

Non-Invertible "Symmetries"
A crucial implicit assumption in the above discussion is that the map S is *invertible,*
otherwise there may exist *more general* non-invertible "symmetries" which, while
not symmetries in the strict sense of our definition, are quite useful to solve the
dynamical problem (when they are present). Here we limit to a baby example.

Example 4.4 Let $|\psi_n\rangle = |n\rangle/\sqrt{n!}$ be the normalized energy eigenstates of the
harmonic oscillator, cf. Sect. 3.5. The linear map acting on the orthonormal Hilbert
basis $\{|\psi_n\rangle\}_{n\in\mathbb{N}}$ as

$$S\colon |\psi_n\rangle \mapsto |\psi_{n+1}\rangle, \tag{4.20}$$

called the *shift*, does preserve the quantum amplitudes (4.2) and satisfies the
dynamical condition (4.18) (with $C = \hbar\omega$) so that

$$\left|\langle S\psi'|\mathrm{e}^{-\mathrm{i}Ht/\hbar}|S\psi\rangle\right|^2 = \left|\langle\psi'|\mathrm{e}^{-\mathrm{i}Ht/\hbar}|\psi\rangle\right|^2 \quad \text{for all } |\psi\rangle, |\psi'\rangle, \text{ and } t. \tag{4.21}$$

However S is a mere *isometry* of $\mathcal{H}$, not a unitary map, since it is *non-invertible*
because the ground state $|0\rangle$ is not in its image. Clearly we may exploit the semi-
group $\{S^n : n \in \mathbb{N}\}$ of non-invertible symmetries to solve the harmonic oscillator in
yet another way.

The Dual Action on Operators
In Chap. 2 we presented two different pictures for the particular symmetry "time
evolution": the symmetry acted on state vectors (resp. operators) in the Schrödinger
(resp. Heisenberg) picture. Correspondingly all unitary symmetry S has two dual
natural actions:

$$\bullet \text{ on states} \qquad |\psi\rangle \to |S\psi\rangle \equiv S|\psi\rangle \tag{4.22}$$

$$\bullet \text{ on operators} \qquad \mathcal{O} \to \mathcal{O}^S \equiv S^{-1}\mathcal{O}S, \tag{4.23}$$

which induce the same action on all physical quantities

$$\langle\psi'|\mathcal{O}^S|\psi\rangle = \langle S\psi'|\mathcal{O}|S\psi\rangle. \tag{4.24}$$

The action on operators is a *right* adjoint action, that is, composing the actions of S_1 and S_2 we get the action of $S_1 S_2$ not of $S_2 S_1$

$$(\mathcal{O}^{S_1})^{S_2} = S_2^{-1} S_1^{-1} \mathcal{O} S_1 S_2 = \mathcal{O}^{S_1 S_2}. \tag{4.25}$$

We stress that the ambiguity of S by a phase does not affect the action on operators (4.23), hence does not affect the transformation of any physical observable under the symmetry which is then free of ambiguities.

4.1.1 Unitary Symmetries

In view of the Wigner theorem, a *connected* Lie group G of symmetries of our quantum system is necessarily realized by a family of *unitary* transformations. Writing $\mathcal{U} \equiv \mathsf{Aut}(\mathcal{H})$ for the group of unitary transformations of the Hilbert space $\mathcal{H}$ (whose dimension is at most countable), we get a continuous group homomorphism

$$\varrho: G \to \mathcal{U}/U(1) \equiv \mathsf{Aut}^+ \, \mathbb{P}(\mathcal{H}). \tag{4.26}$$

Replacing G with $G/\ker \varrho$ we may assume ϱ injective. Let $\{g(t)\} \subset G$ be the one-parameter subgroup tangent to the vector $\ell \in T_e G \equiv \mathfrak{g}$ at the identity e of the Lie group G

$$\mathfrak{g} \ni \ell = \dot{g}|_{t=0} \qquad g(0) = e, \qquad g(t)g(s) = g(t+s). \tag{4.27}$$

($\mathfrak{g}$ is, by definition, the Lie algebra of G cf. [4] chap. 2). By the Stone theorem there is a (densely defined) Hermitian operator L such that

$$\exp(\mathrm{i}Lt) \in \mathcal{U} \tag{4.28}$$

is a representative of $\varrho(g(t)) \in \mathcal{U}/U(1)$. Shifting L by a real constant, $L \rightsquigarrow L + c$, changes the representative but not the image of $\exp(\mathrm{i}Lt)$ in $\mathcal{U}/U(1)$. Hence L is unique only up to an additive constant. Making a choice of representatives (i.e. of additive constants) of a basis of $\mathfrak{g}$, we get a linear map

$$\varrho_*: \mathfrak{g} \equiv T_e G \to \mathrm{i}\,\mathfrak{H} \qquad \varrho_*: \ell \mapsto \mathrm{i}L, \tag{4.29}$$

where $\mathfrak{H}$ stands for the $\mathbb{R}$-space of (densely defined) Hermitian operators acting on $\mathcal{H}$. Applying ϱ to the two sides of the Baker–Campbell–Hausdorff identity (Eq. (2.217)) and differentiating, one sees that ϱ_* is a Lie algebra homomorphism up to constant shifts[5] which arise from the ambiguity in the choice of representatives L of the elements ℓ of $\mathfrak{g}$:

$$[\varrho_*(\ell_1), \varrho_*(\ell_2)] = \varrho_*([\ell_1, \ell_2]) + i\, c(\ell_1, \ell_2), \qquad (4.30)$$

where in the LHS the bracket is the commutator of quantum operators while the bracket in the RHS is the Lie bracket in $\mathfrak{g}$. We **claim** that when G is *semisimple* we may choose the representatives $iL_a \equiv \varrho_*(\ell_a)$ so that the constants $c(\cdot, \cdot)$ in (4.30) vanish, i.e. we have a genuine *Lie algebra homeomorphism*

$$[\ell_a, \ell_b] = f_{ab}{}^c \ell_c \quad \Rightarrow \quad [iL_a, iL_b] = f_{ab}{}^c iL_c, \qquad (4.31)$$

where $\{\ell_a\}$ is a basis of $\mathfrak{g}$ and the real coefficients $f_{ab}{}^c$ are the structure constants of $\mathfrak{g}$. See **BOX** on page 215 for a proof of the **claim**. In this situation ϱ may be lifted to a continuous group homomorphism $\varrho\colon G \to \mathcal{U}$. A symmetry action $G \to \mathcal{U}/U(1)$ which *cannot be lifted* to a homomorphism $G \to \mathcal{U}$ is said to be *projective*.

Remark 4.3 The statement of the **claim** is *false* for Abelian Lie algebras. As a counter-example, consider a system with $\mathcal{M} = \mathbb{R}$. Acting on the space of states $\mathbb{P}(\mathcal{H})$ (or on operators), the Lie algebra of translations in both x and the dual momentum p is isomorphic to the Abelian Lie algebra $\mathbb{R}^2$. However, when we lift the action from $\mathbb{P}(\mathcal{H})$ to the Hilbert space $\mathcal{H}$, the Lie algebra $\mathbb{R}^2$ gets replaced by its Heisenberg central extension whose commutator

$$[q, p] = i\hbar \qquad (4.32)$$

has an extra term proportional to the identity which cannot be eliminated by any redefinition of the operators q, p. The action of the translations in phase space $\mathcal{W} \equiv \mathbb{R}^2$ is then a *projective* one. A physically more intriguing example of the projective action on the Hilbert space of this centrally extend Heisenberg Lie algebra will be discussed in Sect. 5.7.

Corollary 4.1 *When a symmetry group, preserving the quantum amplitudes* (4.2), *is a connected semisimple Lie group G, we may choose the additive constants in the Hermitian generators L_a in such a way that the map ϱ*

$$\varrho\colon G \to \mathcal{U}(\mathcal{H}) \qquad e^{t^a \ell_a} \mapsto e^{it^a L_a} \qquad (4.33)$$

is a genuine group homomorphism. Likewise $\ell_a \mapsto iL_a$ is an isomorphism of Lie algebras (onto its image).

From the quantum Noether theorem of Sect. 2.16 we also have

Corollary 4.2 *When a connected symmetry group G is a* closed subgroup *of the isometry group* $\mathsf{Iso}(\mathcal{M})$ *of the configuration space $\mathcal{M}$ which preserves the*

---

[5] Compare with the corresponding equation in classical mechanics, eq.(7.83) of [1].

potentials, i.e. G *is a group of Noether symmetries whose generators are Noether charges* L_a *which (as differential operators) kill the constants, the map* ϱ_* *is a homeomorphism of Lie algebras*[6]

$$\varrho_*(\ell_a) = -\mathrm{i}\,\hbar\,\ell_a \qquad \ell_a \in \mathfrak{g} \subset \mathfrak{iso}(\mathcal{M}), \tag{4.34}$$

and $\varrho\colon G \to \mathcal{U}(L^2(\mathcal{M}))$ *is a genuine group homomorphism for all Noether symmetry groups* G *which leave invariant the potentials.*

BOX: Proof of Claim

The map $c\colon \wedge^2 \mathfrak{g} \to \mathbb{R}$, $\ell_1 \wedge \ell_2 \mapsto c(\ell_1, \ell_2)$ is a *2-cochain* in the Chevalley-Eilenberg complex of the Lie algebra $\mathfrak{g}$ valued in the trivial module $\mathbb{R}$ [9]. The Jacobi identity yields

$$0 = -c([\ell_1, \ell_2], \ell_3) - c([\ell_2, \ell_3], \ell_1) - c([\ell_3, \ell_1], \ell_2) \equiv \mathrm{d}c(\ell_1, \ell_2, \ell_3)$$

where d is the Chevalley-Eilenberg differential [9]. Then c is a *2-cocycle*. We are free to redefine our generators by adding a constant, $L_a \rightsquigarrow L_a + b(\ell_a)$, which makes

$$c(\ell_1, \ell_2) \rightsquigarrow c(\ell_1, \ell_2) + b([\ell_1, \ell_2]) \quad \text{i.e.} \quad c \rightsquigarrow c + \mathrm{d}b.$$

We conclude that the obstruction to getting rid of the constant shift in (4.30) is the *class* of the cocycle $c \in H^2(\mathfrak{g}, \mathbb{R})$ in the 2nd cohomology group of $\mathfrak{g}$ valued in the trivial $\mathfrak{g}$-module $\mathbb{R}$. $H^2(\mathfrak{g}, \mathbb{R})$ vanishes for semisimple Lie algebras: see e.g. Chapter VII of [10] especially **Proposition 6.3**

4.1.2 Unitary Representations

Let G be a Lie group and V a complex Hilbert space (finite dimensional or separable). An *unitary*[7] *representation* of G on V is a continuous group homomorphism

$$\rho\colon G \to \mathcal{U}(V) \tag{4.35}$$

in the group $\mathcal{U}(V)$ of unitary automorphisms of V. Corollaries 4.1 and 4.2 say that a connected symmetry group G which is semisimple or Noether acts on the Hilbert space $\mathcal{H}$ by a unitary representation.

---

[6] We identify the Lie algebra of the isometry group of $\mathcal{M}$ with the Lie algebra of Killing vectors which we see as first-order differential operators acting on a dense subspace of functions in $L^2(\mathcal{M})$.

[7] All our representations are unitary.

The representation is *trivial* iff $\rho(g) = 1$ for all $g \in G$. V is the *(vector) space of the representation* ρ, and we abusively refer to V as the "representation" or, non-abusively, as a *G-module*. The *dimension* of the representation V is $\dim V$.

If V_1 and V_2 are two representations of G, a *G-map* is a continuous linear map

$$f : V_1 \to V_2 \tag{4.36}$$

such that

$$f\rho_1(g) = \rho_2(g)f \quad \text{for all } g \in G. \tag{4.37}$$

We also say that f is an *intertwiner* between V_1 and V_2. Two representations

$$\rho_1 : G \to \mathcal{U}(V_1) \quad \text{and} \quad \rho_2 : G \to \mathcal{U}(V_2) \tag{4.38}$$

are *isomorphic* iff there is an *invertible* G-map $f : V_1 \to V_2$. The *direct sum* of two representations, written $V_1 \oplus V_2$ for short, is the representation on the vector space $V_1 \oplus V_2$ given by

$$\rho(g)(v_1 \oplus v_2) = \rho_1(g)v_1 \oplus \rho_2(g)v_2 \qquad v_a \in V_a, \ g \in G. \tag{4.39}$$

A *sub-representation* $S \subset V$ is a closed linear subspace preserved by $\rho(G)$

$$\rho(G)S \subseteq S. \tag{4.40}$$

The restriction $\rho(G)|_S : S \to S$ makes S into a representation in its own right. A representation V is *irreducible* if it contains no non-trivial sub-representations besides $\{0\}$ and V. The equality

$$0 = \langle w|Uv\rangle = \langle U^{-1}w|v\rangle \quad \text{for all } |v\rangle \in S, \ |w\rangle \in S^{\perp}, \ U \in \rho(G), \tag{4.41}$$

shows that the orthogonal complement $S^{\perp}$ to S is also a sub-representation, and V is decomposed in the orthogonal direct sum of the two unitary representations

$$V = S \oplus S^{\perp}. \tag{4.42}$$

If S (or $S^{\perp}$) is not irreducible we can decompose V further. If V is finite dimensional, the decomposition process will stop after finitely many steps, and we get that V is the direct sum of irreducible representations $V = \oplus_a V_a$. If V is separable the situation may be subtler: for instance, for the group of unitary translations $\{\exp(i\boldsymbol{p}a)\}_{a\in\mathbb{R}}$ acting in $L^2(\mathbb{R})$; the Hilbert space $\mathcal{H} \equiv L^2(\mathbb{R})$ is dense in a larger space $\mathcal{S}^{\vee}$ which is the *direct integral* (not sum!)[8] of irreducible represen-

---

[8] For direct integral decompositions see e.g. [11] or for a simplified treatment [12].

tations (namely the momentum eigenspaces). $\mathcal{H}$ itself cannot be decomposed into irreducible representations of the translation group. This "bad" behavior is related to the fact that $\mathbb{R}$ is a *non-compact* group, see below for the "nice" compact case.

Lemma 4.1 (Schur Lemma)[9] **(1)** *Let V_1, V_2 be irreducible G-modules, of at most countable dimension, and $m : V_1 \to V_2$ a G-map. If V_1, V_2 are not isomorphic $m=0$.* **(2)** *Let V be a* unitary *irreducible G-module (of at most countable dimension) and $m : V \to V$ a G-map from the representation V to itself: then m is a multiple of the identity.*

Proof **(1)** The kernel and the cokernel of m are subrepresentations hence 0 or the full space. If the kernel is 0, so must be the cokernel, and m is an isomorphism. Otherwise the map is zero. **(2)** If $m : V \to V$ is a G-map,

$$m^\dagger \rho(g)^\dagger = \rho(g)^\dagger m^\dagger \quad \Rightarrow \quad m^\dagger \rho(g)^{-1} = \rho(g)^{-1} m^\dagger, \tag{4.43}$$

since the representation is unitary i.e. $\rho(g)^\dagger \rho(g) = 1$. Thus $m^\dagger$ is also a G-map and so are the Hermitian operators $m + m^\dagger$ and $\mathrm{i}(m^\dagger - m)$. The spectrum $\sigma(m + m^\dagger)$ is then *non-empty* by the spectral theorem (Chap. 2). Let

$$\mu \in \sigma(m + m^\dagger) \subset \mathbb{R}. \tag{4.44}$$

By definition of spectrum, $m + m^\dagger - \mu$ is a G-map which is not an isomorphism. Since the representation V is irreducible, $m + m^\dagger - \mu$ then must be the zero map, so $m + m^\dagger = \mu \in \mathbb{R}$. The same argument yields $\mathrm{i}(m^\dagger - m) = v \in \mathbb{R}$ and hence

$$m = \tfrac{1}{2}(\mu + \mathrm{i}v) \in \mathbb{C}. \tag{4.45}$$

$\square$

Operations on Representations
Given two G-modules V_1 and V_2, we may define two new G-modules, namely their *direct sum* and their *tensor product*

$$V_1 \oplus V_2 \simeq V_2 \oplus V_1, \qquad V_1 \otimes V_2 \simeq V_2 \otimes V_1 \tag{4.46}$$

by the obvious rules ($v_a \in V_a$, $a = 1, 2$)

$$\rho(g)(v_1 \oplus v_2) = \rho_1(g)v_1 \oplus \rho_2(g)v_2,$$
$$\rho(g)(v_1 \otimes v_2) = \rho_1(g)v_1 \otimes \rho_2(g)v_2. \tag{4.47}$$

---

[9] The Schur lemma takes the simple form stated in the main text because we are working over an *algebraically closed field* namely $\mathbb{C}$.

The set of representations of G—taken modulo isomorphism—then forms a commutative semi-ring under the operations $\oplus$ and $\otimes$. Its completion to a ring is called the *representation ring* $\mathscr{R}(G)$ *of* G and its elements *virtual representations* of G.

The decomposition

$$V \otimes V = V \odot V \oplus V \wedge V \tag{4.48}$$

into *symmetric* and *anti-symmetric* subspaces is preserved by the G-action, so from the representation V we can form two "quadratic" representations: its *symmetric square* $V \odot V$ and its *anti-symmetric square* $V \wedge V$. More generally we have a representation on the n-th tensor power of V

$$V^{\otimes n} \equiv \overbrace{V \otimes V \cdots \otimes V}^{n \text{ factors}}, \tag{4.49}$$

as well as on the symmetric and anti-symmetric tensor powers $V^{\odot n}$ and $V^{\wedge n}$. In particular the *tensor algebra*[10] *of* V

$$T^{\bullet}V = \bigoplus_{n=0}^{\infty} V^{\otimes n} \tag{4.50}$$

carries a canonical induced G-representation. If $R \subset T^{\bullet}V$ is a bilateral ideal of relations, which is invariant under the action of G, the quotient algebra $T^{\bullet}V/R$ also carries a natural linear G-action.

Example 4.5 Let $w_a = (q^i, p_j)$ $(i, j = 1, \ldots, n, a = 1, \ldots, 2n)$ be the $2n$-vector of canonical operators acting on $L^2(\mathbb{R}^n)$, and

$$R_{ab} \equiv w_a w_b - w_b w_a - i\hbar\,\Omega_{ab} = 0 \tag{4.51}$$

the canonical commutation relations. The vector space $W = (w_a)$ carries a representation of $Sp(2n, \mathbb{R}) \equiv \{M \in GL(n, \mathbb{R}) : M^t \Omega M = \Omega\}$,

$$w'_a = M_a{}^b w_b \qquad M \in Sp(2n, \mathbb{R}), \tag{4.52}$$

which leaves invariant the ideal (R_{ab}). Hence we have an induced representation of $Sp(2n, \mathbb{R})$ on the non-commutative algebra of "polynomial" quantum operators acting on $L^2(\mathbb{R}^n)$

$$\mathfrak{A}_{\text{poly}} = T^{\bullet}W/(R_{ab}) \tag{4.53}$$

which in each subspace $W^{\otimes m}$ reduces to a summand of the m-tensor representation.

---

[10] Recall that the tensor algebra of a $\mathbb{C}$-space V is the $\mathbb{C}$-algebra over the vector space $T^{\bullet}V = \bigoplus_{n=0}^{\infty} V^{\otimes n}$ (where $V^{\otimes 0} = \mathbb{C}$) with the product $x, y \mapsto x \otimes y$.

Example 4.6 Let $V \simeq \mathbb{C}^n$ be a representation of the group G. Let $z_1, \ldots, z_n \in V^\vee$ be the coordinates in V. The polynomial algebra in the z_i's is isomorphic to the symmetric tensor algebra

$$\mathbb{C}[z_1, \ldots, z_n] = \bigoplus_{n \geq 0} \odot^n V^\vee \equiv T^\bullet V^\vee / (z_i \otimes z_j - z_j \otimes z_i). \tag{4.54}$$

Thus the space of homogeneous polynomials of degree n, $\odot^n V^\vee$, carries a natural representation of G (typically reducible).

Semisimple Symmetries

We focus on connected semisimple Lie symmetries G which have genuine unitary representations in $\mathcal{H}$ by Corollary 4.1. We are particularly interested in the unitary irreducible representation of a semisimple Lie group G.

Theorem 4.2 *Assume the Lie group G to be connected and semisimple.*
(1) *If G is non-compact, all non-trivial unitary irreducible representations are necessarily infinite dimensional.* **(2)** *If G is compact, the irreducible unitary representations are all finite dimensional.*

Proof

(1) With no loss we may assume G essentially simple, i.e. its non-trivial normal subgroups are finite. A non-trivial unitary representation of finite dimension n is a group homomorphism

$$\rho : G \to U(n) \qquad \rho(G) \neq \{1\}. \tag{4.55}$$

The kernel of ρ is a normal subgroup hence a *finite* subgroup $F \triangleleft G$. Replacing G with G/F we may assume ρ injective, so an isomorphism on its image:

$$G \simeq \rho(G). \tag{4.56}$$

A deep theorem[11] states that the image $\rho(G) \subset U(n)$ of a connected semisimple Lie group G is *closed* in $U(n)$, hence *compact* because $U(n)$ is compact. Since G is non-compact, Eq. (4.56) gives a contradiction.

(2) We need the following result:

$$\square$$

Lemma 4.2 *G a compact Lie group. On G (seen as a manifold) there is a positive measure, i.e. a volume form μ, which is invariant under left and right translations*

$$L_h^* \mu = \mu, \qquad R_h^* \mu = \mu \quad \text{where} \quad L_h : g \mapsto hg, \quad R_h : g \mapsto gh, \tag{4.57}$$

---

[11] See **Theorem** (4.6.3) of [13].

and has total volume 1

$$\int_G \mu = 1,\tag{4.58}$$

while $\int_U \mu > 0$ for all open sets $U \subset G$. μ is called the (normalized) Haar measure *of G* [14].

Proof Let $m = \dim G$. Choose a faithful matrix representation[12] of G and let $\omega = g^{-1}\mathrm{d}g$ be the associated Maurer-Cartan form ([1] **Definition 2.11**). The m-form $\tilde{\mu} \equiv \mathrm{tr}(g^{-1}\mathrm{d}g)^m$ has constant sign in a neighborhood of the identity and multiplying it by ± 1 we get a locally positive m-form $\tilde{\mu}$ while

$$L_h^* \tilde{\mu} = \pm \mathrm{tr}[((hg)^{-1}\mathrm{d}(hg))^m] = \pm \mathrm{tr}[(g^{-1}h^{-1}h\mathrm{d}g)^m] = \tilde{\mu}\tag{4.59}$$

$$R_h^* \tilde{\mu} = \pm \mathrm{tr}[((gh)^{-1}\mathrm{d}(gh))^m] = \pm \mathrm{tr}[h^{-1}(g^{-1}\mathrm{d}g)^m h] = \tilde{\mu}\tag{4.60}$$

so that $\tilde{\mu}$ is positive everywhere in G. Then set

$$\mu = \frac{\tilde{\mu}}{\int_G \tilde{\mu}}.\tag{4.61}$$

$\square$

Henceforth we write the Haar measure on a compact group simply as $\mathrm{d}g$.

Cont. Proof of Part (2) of Theorem Let V be an irreducible unitary G-representation. For $|v\rangle \in V$ consider the operator

$$K_v = \int_G \mathrm{d}g \; g|v\rangle\langle v|g^{-1} : V \to V,\tag{4.62}$$

where we write $\rho(g)|v\rangle$ simply as $g|v\rangle$. By construction, K_v is a G-map for all $|v\rangle \in V$, so a multiple of the identity $K_v = \lambda_v \cdot 1$. Then

$$\lambda_v \||w\rangle\|^2 = \lambda_v \langle w|w\rangle = \langle w|K_v|w\rangle = \int_G \mathrm{d}g \; \langle w|g|v\rangle\langle v|g^{-1}|w\rangle =$$

$$= \int_G \mathrm{d}g \; \langle v|g^{-1}|w\rangle\langle w|g|v\rangle = \langle v|K_w|v\rangle = \lambda_w \||v\rangle\|^2\tag{4.63}$$

for all $|v\rangle, |w\rangle$ so $\lambda_v = C \||v\rangle\|^2$ i.e.

$$\int_G \mathrm{d}g \; g|v\rangle\langle v|g^{-1} = C \||v\rangle\|^2\tag{4.64}$$

---

[12] They exists by Ado's theorem [15].

for some constant C. We claim that $C > 0$. Indeed

$$\langle v|g|v\rangle\langle v|g^{-1}|v\rangle = |\langle v|g|v\rangle|^2 \tag{4.65}$$

is non-negative everywhere in G and non-zero in an open neighborhood of 1. Taking the trace of both sides of (4.64) we get

$$0 < C = \frac{1}{\dim V} \tag{4.66}$$

i.e. $\dim V$ is finite. $\qquad\square$

Statement **(2)** holds also when G is compact Abelian $G \simeq U(1)^k$. Indeed by Stone theorem a $U(1)$ group is represented by the one-parameter family of unitary operators $\exp(\mathrm{i}Lt)$ with $\sigma(L) \subseteq 2\pi\mathbb{Z}$ *discrete* (see Sect. 3.10), and each eigenspace of L yields an irreducible representation of $U(1)$. Hence **(2)** holds for all compact Lie groups.

Warning 5 On the contrary the assumption that G is semisimple is crucial in part **(1)**. Consider the non-compact Lie group $\mathbb{R}$ immersed in $U(1)^2$ as

$$\eta: t \mapsto (\mathrm{e}^{2\pi \mathrm{i}t}, \mathrm{e}^{2\pi \mathrm{i}\zeta t}) \in U(1)^2 \subset U(2). \tag{4.67}$$

where $\zeta \in \mathbb{R}$ is *irrational*. Note that the closure $\overline{\eta(\mathbb{R})}$ of the one-dimensional image is the full two-dimensional group $U(1)^2$, so the subgroup $\eta(\mathbb{R}) \subset U(1)^2$ is *not* closed (hence non-compact). We see $U(1)^2$ as the group of translations on the 2-torus with Hilbert space $L^2(S^1 \times S^1)$ and Hilbert basis $\{\mathrm{e}^{\mathrm{i}mx+\mathrm{i}ny}\}_{m,n\in\mathbb{Z}}$. The two $U(1)$'s are, respectively, the group of translations in x and y. The Hilbert space is the direct sum of the one-dimensional irreducible representations of the non-compact Lie group $\mathbb{R}$

$$V_{m,n} = \mathbb{C}\,e^{\mathrm{i}mx+\mathrm{i}ny} \quad \text{where} \quad t \mapsto e^{\mathrm{i}(m+n\zeta)t} \in U(1) \quad \text{and} \quad t \in \mathbb{R}. \tag{4.68}$$

The good new from Theorem 4.2 is that for *compact* Lie groups G we need only to understand the unitary representations in the *finite*-dimensional spaces $\mathbb{C}^n$. For future use we state what we learned from the proof of the **Theorem**:

Corollary 4.3 *Let the G-module V be a direct sum of finite-dimensional irreducible unitary representations* $V = \oplus_a V_a$ *and let* $|a, k\rangle$ $(k = 1, \ldots, \dim V_a)$ *be an orthonormal basis of* V_a. *Then*

$$\int_G \mathrm{d}g\,\langle a, i|g|b, j\rangle\langle c, k|g^{-1}|d, l\rangle =$$

$$= \begin{cases} 0 & \text{unless } a = b,\ c = d,\ \text{and } V_a \simeq V_c \\[2mm] \dfrac{\delta_{il}\,\delta_{jk}}{\dim V_a} & \text{if } a = b = c = d. \end{cases} \tag{4.69}$$

Finite-Dimensional Representations

In finite dimension three different algebraic objects share the same representation theory: the Lie group G, its Lie algebra $\mathfrak{g}$, and its universal enveloping algebra

$$\mathsf{U}_\mathfrak{g} \equiv T^\bullet\mathfrak{g}\big/\big([\ell_a, \ell_b] - f_{ab}{}^c \ell_c\big). \tag{4.70}$$

Let $\rho : \mathfrak{g} \to \mathsf{End}(V)$ be a unitary representation of $\mathfrak{g}$ in V with $\dim V = n$. Fixing an orthonormal basis in V, the representation ρ is concretely a set of $\dim \mathfrak{g}$ Hermitian $n \times n$ matrices L_a such that

$$[L_a, L_b] = -\mathrm{i} f_{ab}{}^c L_c, \qquad a, b, c = 1, \ldots, \dim \mathfrak{g}, \tag{4.71}$$

and the corresponding representation of $\mathsf{U}_\mathfrak{g}$ is given by polynomials in the matrices $\mathrm{i}L_a$, while the representation of the group G by the exponential of these matrices[13]

$$g(t^a) \equiv \exp_e(t^a \ell_a) \rightsquigarrow \rho(g(t^a)) = \exp(\mathrm{i} t^a L_a) \tag{4.72}$$

$$\exp(\mathrm{i} t^a L_a) \stackrel{\text{def}}{=} \sum_{n=0}^{\infty} \frac{(\mathrm{i} t^a L_a)^n}{n!} \in U(n), \tag{4.73}$$

where ℓ_a is a basis of the Lie algebra $\mathfrak{g}$, and t^a are local coordinates on G centered at the identity e. It is typically more convenient to study the Representation Theory of the Lie algebras (or their universal envelopes) than of the Lie groups. This study is our next task. We start from the physically most useful symmetry algebra which will provide insights on the general theory. We quote without proof (see e.g. [17]):

Theorem 4.3 *Let the Lie group G be semisimple and V a faithful representation. All finite-dimensional representations of G are sub-representations of some tensor power $\otimes^k V$ of V.*

Example 4.7 The group $SO(3)$ has the defining (hence faithful) three dimensional representation V given by its action on $\mathbb{R}^3$. The elements of $V \simeq \mathbb{R}^3 \otimes \mathbb{C}$ are written as 3-component vectors v_i ($i = 1, 2, 3$) which transform under a rotation $R \in SO(3)$ as $v_i \rightsquigarrow R_i{}^j v_j$ with

$$R \in GL(3, \mathbb{R}) \quad \text{such that} \quad RR^t = R^t R = 1 \quad \text{and} \quad \det R = 1. \tag{4.74}$$

---

[13] In Eq. (4.72) $\mathbf{exp}_e$ is the *exponential map* (in the sense of Differential Geometry [16]) from the tangent space at the identity $T_e G \equiv \mathfrak{g}$ to the Lie group G. We stress that $\mathbf{exp}_e : \mathfrak{g} \to G$, $\ell \mapsto \exp(\ell)$ is surjective but *not injective* in general. On the contrary $\exp(\cdot)$ is the usual exponential of matrices, given by the sum of the series in (4.73) which is easily seen to converge for all matrices $t^a L_a$.

The elements of $\otimes^k V$, called (contravariant) k-tensors, are written as $T_{i_1 i_2 \cdots i_k}$, where the k indices take the value $i_j = 1, 2, 3$; under $SO(3)$ they transform as

$$T_{i_1 i_2 \cdots i_k} \rightsquigarrow R_{i_1}{}^{j_1} R_{i_1}{}^{j_1} \cdots R_{i_1}{}^{j_1} T_{j_1 j_2 \cdots j_k}. \tag{4.75}$$

The k-tensor representation $\otimes^k V$ decomposes into trace parts and tensor spaces with different symmetry properties under its indices. All $SO(3)$ representations are obtained in this way: in facts they may all be represented by tensors which are totally symmetric in their k indices and traceless in the sense that $T_{iii_3 \cdots i_k} = 0$ (sum over repeated indices implied).

Characters. Representation Ring
We assume the symmetry group G to be semisimple and compact.

Definition 4.3 Let V be a finite-dimensional representation of G. The *character* of V is the function $G \to \mathbb{C}$ given by the trace of the group elements g in the space V

$$\chi_V : G \to \mathbb{C}, \qquad \chi_V(g) \overset{\text{def}}{=} \operatorname{tr}_V(g) \ g \in G. \tag{4.76}$$

The character is a *class function*, i.e. it depends only on the *conjugacy class* of g

$$\chi_V(hgh^{-1}) = \chi_V(g) \quad \forall \, h \in G. \tag{4.77}$$

We write $C(G)$ for the ring of (complex valued) class functions on G. Clearly two isomorphic representations have the same character.

Proposition 4.2 *The character* $\chi : \mathscr{R}(G) \to C(G)$ *is a ring homomorphism*

$$\chi_{V \oplus W} = \chi_V + \chi_W, \qquad \chi_{V \otimes W} = \chi_V \, \chi_W \tag{4.78}$$

while

$$\chi_V(1) = \dim V. \tag{4.79}$$

Let $V^\vee$ *be the dual representation of* V. *Then*

$$\chi_{V^\vee} = \chi_V^*. \tag{4.80}$$

Proof Only the last statement requires a proof. If the matrix g represents an element of G in the representation V, the matrix $(g^{-1})^t$ represents it in $V^\vee$. Since the representation is necessarily unitary, $g^{-1} = g^\dagger \equiv (g^*)^t$, and the matrix in the $V^\vee$ representation is g^* whose trace is $(\operatorname{tr} g)^*$. $\qquad\qquad\square$

Theorem 4.4 (Orthonormality of Characters) *Let $\{V_a\}$ be the set of pair-wise non-isomorphic, irreducible representations of the compact Lie group G. Then*

$$\int_G \chi_a(g)^* \, \chi_b(g) \, \mathrm{d}g = \delta_{ab} \tag{4.81}$$

where $\mathrm{d}g$ *is the Haar measure normalized to* 1. *A representation* V *is irreducible iff*

$$\int_G |\chi_V(g)|^2 \, \mathrm{d}g = 1. \tag{4.82}$$

Proof In (4.69) set $a = b = c = d$ and contract with $\delta^{ik}\delta^{jl}$. $\square$

In particular two representations V_1, V_2 are *isomorphic* if and only if they have the same character

$$V_1 \simeq V_2 \quad \Leftrightarrow \quad \chi_{V_1}(g) = \chi_{V_2}(g) \quad \forall\, g \in G. \tag{4.83}$$

Then we identify the representation ring $\mathscr{R}(G)$ with its image in $C(G)$ (the *character ring*). $\mathscr{R}(G)$ is generated as a $\mathbb{Z}$-module by the characters $\{\chi_a\}$ of the set $\{V_a\}$ of isoclasses of irreducible representations

$$\mathscr{R}(G) = \bigoplus_a \mathbb{Z}\chi_a. \tag{4.84}$$

The actual characters (as contrasted to *virtual* ones) are the sums of the χ_a's with non-negative coefficients.

Suppose the character $\chi_V(g)$ is *real for all* g. It follows from (4.80) that we have the isomorphism $V^\vee \simeq V$. An irreducible representation V_a is called *complex* iff $V_a^\vee \not\simeq V_a$, i.e. if $\chi_a(G) \not\subset \mathbb{R}$. When the character χ_a is real, $V_a \otimes V_a \simeq V_a^\vee \otimes V_a$ contains the trivial representation *exactly once*[14] in the form of the resolution of identity $1 = \sum_i |a, i\rangle\langle i, a|$. Since

$$V_a \otimes V_a = \odot^2 V_a \oplus \wedge^2 V_a, \tag{4.85}$$

the trivial representation is either symmetric or antisymmetric. In the first case we say that the representation V is *real* and in the second case that V is *quaternionic*.

---

[14] The number of the trivial representations in $V_a^\vee \otimes V_a$ is given, by orthogonality, by the integral

$$\int_G \mathrm{d}g\, \chi_0^* \, \chi_{V_a^\vee \otimes V_a} = \int_G \mathrm{d}g\, \chi_a^* \chi_a = 1, \qquad V_a \text{ irreducible,}$$

where $\chi_0 \equiv 1$ is the character of the trivial representation.

A simple computation yields the character formulae (see e.g. [18] Lemma 3.1)

$$\chi_{\odot^2 V_a}(g) = \frac{1}{2}\left(\chi_a(g) + \chi_a(g^2)\right) \tag{4.86}$$

$$\chi_{\wedge^2 V_a}(g) = \frac{1}{2}\left(\chi_a(g) - \chi_a(g^2)\right), \tag{4.87}$$

and we get the

Theorem 4.5 (Frobenius-Schur Indicator) *Let V_a be an irreducible representation of the compact group G: we have*

$$\left(\begin{smallmatrix} Frobenius-Schur \\ indicator \end{smallmatrix}\right) \overset{\text{def}}{=} \int_G dg\,\chi_a(g^2) = \begin{cases} 0 & complex \\ 1 & real \\ -1 & quaternionic. \end{cases} \tag{4.88}$$

When the unitary representation V_a (of dimension n) is *real* (resp. *quaternionic*) the image of the group $\rho(G) \subset U(n)$ is contained in the subgroup $O(n) \subset U(n)$ (resp. $Sp(n) \subset U(n)$) and we also say that the representation is *orthogonal* (resp. *symplectic*). In the symplectic case n must be *even*.

Let V and W be two representations. We wish to write their tensor product $V \otimes W$ as a direct sum of irreducible representations V_a

$$V \otimes W = \bigoplus_{a \in I(G)} V_a^{\oplus n_a} \quad \text{where} \quad V_a^{\oplus n_a} \overset{\text{def}}{=} \overbrace{V_a \oplus V_a \oplus \cdots \oplus V_a}^{n_a \text{ summands}} \tag{4.89}$$

($I(G)$ the set of isoclasses of irreducible representations of G). The non-negative integer n_a is the *multiplicity* of V_a in V. Up to isomorphism, the representation V is determined by its set of multiplicities $\{n_a\}_{a \in I(G)}$. By the orthogonality theorem they have the explicit expression

$$n_a = \int_G dg\,\chi_a(g)^* \chi_{V \otimes W}(g) \equiv \int_G dg\,\chi_a(g)^* \chi_V(g)\,\chi_W(g). \tag{4.90}$$

4.2 Angular Momentum: Spin

The following physically important situation will be studied in detail in Chap. 5: a quantum particle moves in the physical space, $\mathbb{R}^3$, subjected to a central potential with Hamiltonian[15]

---

[15] Sum over repeated indices implicit!

$$H = \frac{p_i p_i}{2m} + V(q^i q^i) \tag{4.91}$$

where $V(r^2)$ is a potential which depends only on the distance r from the origin of $\mathbb{R}^3$. The subgroup of isometries of $\mathbb{R}^3$ which fix the origin, $O(3)$, is obviously a symmetry of the system. The connected component of this symmetry group, $SO(3)$, is generated by three Killing vectors of the Euclidean metric on $\mathbb{R}^3$

$$k_i = x_j \frac{\partial}{\partial x_k} - x_k \frac{\partial}{\partial x_j} \tag{4.92}$$

(i, j, k) a cyclic permutation of $(1, 2, 3)$.

In classical mechanics the Noether theorem [1] allows us to transform these Killing symmetries in conserved quantities (Noether charges), namely the three components of the classical *angular momentum* vector

$$L_i^{cl} = x_j p_k - x_k p_k$$
$$(i, j, k) \text{ a cyclic permutation of } (1, 2, 3), \tag{4.93}$$

while the arguments in Sect. 2.16 show that we have the corresponding three conserved quantities also at the quantum level—they are the components L_i of the *quantum* angular momentum—which, when written in the Schrödinger representation as differential operators acting on $L^2(\mathbb{R}^3)$, are just the Killing vectors k_i themselves up to the overall factor $-i\hbar$

$$L_i \equiv -i\hbar\, k_i = -i\hbar \left(x_j \frac{\partial}{\partial x_k} - x_k \frac{\partial}{\partial x_j} \right)$$
$$(i, j, k) \text{ a cyclic permutation of } (1, 2, 3), \tag{4.94}$$

or, written in a representation independent form,

$$L_i = q_j p_k - q_k p_j \quad (i, j, k) \text{ cyclic permutation.} \tag{4.95}$$

The L_i's are Hermitian operators which generate rotations of the canonical operators q^i, p_i: this action is just the restriction to the subgroup

$$SO(3) \subset U(3) \subset Sp(6, \mathbb{R}) \tag{4.96}$$

of the linear unitary transformations discussed in Sect. 2.11.1 and also in Sect. 3.5.2. Indeed, in the notations of Sect. 3.5.2,

$$L_i = i\,\epsilon_{ijk}\, a_j^\dagger a_k. \tag{4.97}$$

This formula, obvious in view of the general theory developed in the previous chapters, is called the *Schwinger oscillator model* of the angular momentum. From Eq. (4.97) and the isomorphism (3.80), or by direction computation, one finds

$$[L_1, q_j] = [q_2 p_3 - q_3 p_2, q_j] = -i\hbar\,\delta_{3j}\,q_2 + i\hbar\,\delta_{2j}\,q_3, \qquad (4.98)$$

that is,

$$[L_i, q_j] = \begin{cases} i\hbar\,q_k & (i, j, k)\ \text{cyclic permutation} \\ 0 & i = j. \end{cases} \qquad (4.99)$$

That is, up to the factor $i\hbar$, we get the same formulae as for the Poisson bracket of L_i^{cl} and q_j given by the 3×3 antisymmetric matrices which represent an infinitesimal rotation in $\mathbb{R}^3$ around the i-th axis

$$(R_i)_{kl} = \epsilon_{ikl}. \qquad (4.100)$$

As it is obvious from Eq. (4.96) the same formula holds with the q_i's replaced by the p_i's

$$[L_i, q_j] = i\hbar (R_i)_{jk}\, q_k, \qquad [L_i, p_j] = i\hbar (R_i)_{jk}\, p_k. \qquad (4.101)$$

By construction the three 3×3 matrices R_i in Eq. (4.100) form the defining 3-dimensional representation $V_3 \equiv \mathbb{R}^3 \otimes \mathbb{C}$ of the Lie algebra $\mathfrak{so}(3)$ in the standard basis of $\mathbb{R}^3$; indeed

$$[R_i, R_j]_{kl} = \epsilon_{ikm}\epsilon_{jml} - i \leftrightarrow j = -\delta_{ij}\delta_{kl} + \delta_{il}\delta_{jk} - i \leftrightarrow j =$$
$$= \delta_{il}\delta_{jk} - \delta_{jl}\delta_{ik} = \epsilon_{ijm}\epsilon_{mlk} = -\epsilon_{ijm}(R_m)_{kl}, \qquad (4.102)$$

that is, the matrices R_i satisfy the $\mathfrak{so}(3)$ Lie algebra relations. Note that the R_i's form a basis of the $\mathbb{R}$-space of antisymmetric 3×3 matrices.

The argument in Example 4.5 shows that the quantum operators which are polynomials in the canonical operators q_i, p_i transform according to the corresponding tensor representation. The three operators L_1, L_2, L_3 span the 3-dimensional representation $V_3 \wedge V_3 \simeq V_3$ of $SO(3)$, that is, they are the three components of a vector, so

$$[L_i, L_j] = i\hbar (R_i)_{jk} L_k = i\hbar\,\epsilon_{ijk} L_k, \qquad (4.103)$$

that is, the Lie algebra $\mathfrak{so}(3)$ is isomorphic—as a module over itself—to the 3-dimensional vector representation V_3. Equations (4.103) are the commutation relations of the quantum angular momentum: they just say that L_i transforms as a vector under $SO(3)$ rotations. Equations (4.103) have the same form as the Poisson brackets of the classical angular momentum which also say that the angular momentum is a vector in $\mathbb{R}^3$. In particular, if $S \in \mathfrak{A}_{\text{pol}}$ is a (polynomial) quantum operator which is a scalar, like $p_i p_i$ or $q^i q^i$, we have

$$S \text{ scalar operator} \quad \Rightarrow \quad [L_i, S] = 0. \tag{4.104}$$

Hence for the Hamiltonian in Eq. (4.91),

$$[L_i, H] = 0 \qquad i = 1, 2, 3, \tag{4.105}$$

that is, in this case the three generators of the $SO(3)$ symmetry L_i ($i = 1, 2, 3$) are *conserved* observables ($\equiv$ the quantum angular momentum) as predicted by the quantum Noether theorem (cf. Sect. 2.16). The quantum angular momentum operator transforms under $SO(3)$ as a vector, Eq. (4.103), so its square-length

$$L^2 \equiv L_i L_i \in \mathfrak{A}_{\text{pol}} \tag{4.106}$$

transforms as a *scalar*, and hence

$$[L^2, L_i] = [L^2, H] = 0, \quad i = 1, 2, 3. \tag{4.107}$$

Note that the operator L^2 commutes with all functions $f(L_1, L_2, L_3)$, that is, it is a *central element* in the universal enveloping algebra $\mathsf{U}_{\mathfrak{so}(3)}$. Non-trivial central elements in the universal enveloping algebra $\mathsf{U}_{\mathfrak{g}}$ of a Lie algebra $\mathfrak{g}$ are called *Casimir operators* of the enveloping algebra $\mathsf{U}_{\mathfrak{g}}$. Schur's lemma implies

Corollary 4.4 *The Casimir operators of a Lie algebra $\mathfrak{g}$ act as multiples of the identity in each irreducible representation of $\mathfrak{g}$.*

L^2 is the basic Casimir operator of the Lie algebra $\mathfrak{so}(3)$. All other central elements of $\mathsf{U}_{\mathfrak{so}(3)}$ have the form $f(L^2)$ for some polynomial $f(\cdot)$.

Remark 4.4 L_i transforms as a vector under $SO(3)$ but *not* under the full isometry group fixing the origin $O(3)$. Indeed under parity $\mathscr{P}$

$$\mathscr{P} q_i \mathscr{P} = -q_i \quad \mathscr{P} p_i \mathscr{P} = -p_i \quad \text{but} \quad \mathscr{P} L_i \mathscr{P} = L_i \tag{4.108}$$

One says that the angular momentum is a *pseudo*-vector.

Orbital Angular Momentum vs. Spin

The operators (4.94) which generate the *geometric* $SO(3)$ rotations of the Hilbert space $L^2(\mathbb{R}^3)$ are the components of the *orbital angular momentum* which arises from the motion of the particle around the origin of $\mathbb{R}^3$. A quantum mechanical particle may have, in addition, another kind of angular momentum called its *spin* (or *intrinsic angular momentum*). It arises in the following way. Let $\mathscr{S}$ be a finite dimensional space carrying an *irreducible* representation of the Lie algebra $\mathfrak{so}(3)$ generated by three Hermitian matrices $S_i : \mathscr{S} \to \mathscr{S}$ ($i = 1, 2, 3$) which satisfy the

Lie algebra relation

$$[S_i, S_j] = i\hbar\, \epsilon_{ijk}\, S_k. \tag{4.109}$$

Consider the Hilbert space

$$\mathcal{H} = \mathcal{S} \otimes L^2(\mathbb{R}^3), \tag{4.110}$$

on which the operators S_i and L_i act (respectively on the first and second factor space[16]). The three operators

$$J_i = L_i + S_i, \quad i = 1, 2, 3 \tag{4.111}$$

called the *total angular momentum*, also represent the $\mathfrak{so}(3)$ Lie algebra

$$[J_i, J_j] = i\hbar\, \epsilon_{ijk}\, J_k \quad \text{acting on } \mathcal{H}. \tag{4.112}$$

In particular the operators J_i transform as the components of a vector $\mathbf{J} \in \mathbb{R}^3$. We say that our quantum particle *has spin s* iff

$$\frac{\dim \mathcal{S} - 1}{2} = s \in \frac{1}{2}\mathbb{N}. \tag{4.113}$$

It is a consequence of relativistic quantum physics that massive particles may have a non-zero spin. In the real world most elementary particles have spin. For instance: the electron and the proton have spin $s = 1/2$, while the W-boson has spin 1. In most quantum systems only the total angular momentum J_i is conserved as a consequence of the $SO(3)$ rotational symmetry, and not L_i and S_i separately.

As we shall see momentarily, the eigenvalues of S_i are bounded in absolute value by $s\hbar$, so in the classical limit $\hbar \to 0$ the total angular momentum reduces to the orbital one L_i, and the spin is a *purely quantum degree of freedom* without classical counterpart.

Following tradition, we shall reserve the notation L_i for the orbital angular momentum. The generators of angular momentum will be otherwise written J_i.

4.3   Unitary Representations of Angular Momentum

We assume that our "abstract" Hilbert space $\mathcal{H}$ carries an unitary representation of the Lie algebra $\mathfrak{so}(3)$ generated by the action of three quantum operators J_1, J_2, J_3 which satisfy the Lie algebra (4.112). To simplify the formulae we rescale

---

[16] Following the notation which is standard in Quantum Mechanics, we write simply S_i (resp. L_i) for $S_i \otimes 1$ (resp. $1 \otimes L_i$).

$J_i \rightsquigarrow J_i/\hbar$ or, equivalently, we set $\hbar = 1$ by a choice of units. When the Hilbert space $\mathcal{H}$ is separable, the representation is highly reducible, since all irreducible representations of the compact group $SO(3)$ are finite dimensional by Theorem 4.2.

Classification and Construction

Our first task is to classify and construct *all* the irreducible unitary representations of the Lie algebra $\mathfrak{so}(3)$, where we see isomorphic representations as the *same* representation. As the reader will surely notice, our method below is modeled on the algebraic construction of the energy spectrum for the harmonic oscillator in Sect. 3.5.3.

Recall that $SO(3) \simeq SU(2)/\mathbb{Z}_2$ ([1] chaps. 2, 5) so we have the Lie algebra isomorphism

$$\mathfrak{so}(3) \simeq \mathfrak{su}(2), \tag{4.114}$$

and we shall use the $\mathfrak{so}(3)$ and $\mathfrak{su}(2)$ languages interchangeably. The relation between the Lie groups $SU(2)$ and $SO(3)$ will be revisited in Sect. 4.4. See also the off-text **BOX** on page 236.

The Casimir operator

$$J^2 \equiv J_i J_i \equiv J_i^\dagger J_i, \tag{4.115}$$

being a non-negative operator, acts as multiplication by a non-negative real number $\lambda \geq 0$ in any irreducible $\mathfrak{so}(3)$-module V (cf. Corollary 4.4). We work in a basis $\{|m\rangle\}$ of V where the 3rd component J_3 is diagonal i.e.

$$J_3|m\rangle = m|m\rangle, \qquad J^2|m\rangle = \lambda|m\rangle. \tag{4.116}$$

We leave to the reader to check[17] that $\{J_3, J^2\}$ is a maximal system of commuting operators in the sense that any operator in $\mathsf{End}(V)$ which commutes with J_3 has the form $f(J_3, J^2)$ for some function f.

We diagonalize the action of J_3 on the Lie algebra $\mathfrak{so}(3)$ by setting

$$J_\pm = J_1 \pm iJ_2 \qquad J_+^\dagger = J_-, \tag{4.117}$$

that is, we use a *complex basis*[18] of $\mathfrak{so}(3) \otimes \mathbb{C} \simeq V_3$ instead of the usual real one described after Eq. (4.101). In the new basis[19] the Lie algebra $\mathfrak{so}(3)$ reads

$$[J_3, J_\pm] = \pm J_\pm \qquad [J_+, J_-] = 2J_3, \tag{4.118}$$

---

[17] The statement will be clear from the following construction anyhow.

[18] Below we shall write simply $\mathfrak{so}(3)$ for the *complexified* Lie algebra. The (complexified) usual tensors in three dimensions, $T \in T^\bullet\mathbb{R}^3 \otimes \mathbb{C}$, are called *spherical tensors* when written in this complex basis. See e.g. §. 107 of [19].

[19] In Representation Theory of semisimple Lie algebras this is the Serre-Chevalley basis [20].

which imply the identities

$$J_{\pm}J_{\mp} = (J_1 \pm iJ_2)(J_1 \mp iJ_2) = J_1^2 + J_2^2 \mp i[J_1, J_2] =$$
$$= J_1^2 + J_2^2 \pm J_3 = J^2 - J_3^2 \pm J_3 \equiv \lambda - J_3^2 \pm J_3 \quad \text{on } V. \tag{4.119}$$

Therefore

$$J_{\mp}^{\dagger} J_{\mp}|m\rangle \equiv J_{\pm}J_{\mp}|m\rangle = (\lambda - m^2 \pm m)|m\rangle. \tag{4.120}$$

Since $J_{\mp}^{\dagger} J_{\mp}$ are non-negative operators, we get the bounds

$$m(m \pm 1) \le \lambda \tag{4.121}$$

on the eigenvalues m of J_3 in the irreducible unitary representation V. Now

$$J_3(J_{\pm}|m\rangle) = [J_3, J_{\pm}]|m\rangle + J_{\pm}J_3|m\rangle =$$
$$= \pm J_{\pm}|m\rangle + mJ_{\pm}|m\rangle = (m \pm 1)J_{\pm}|m\rangle, \tag{4.122}$$

so the vector $J_{\pm}^k|m\rangle$—*if non-zero*—is an eigenvector of J_3 with eigenvalue $(m \pm k)$. In other words, J_- (resp. J_+) is a *lowering* (resp. *raising*) operator[20] by 1 for the eigenvalue of J_3. An eigenvalue of the form $(m+k)$ would violate the bound (4.121) for $k \gg 1$, so there is a minimal positive integer k such that $J_+^k|m\rangle = 0$. We write $|\ell\rangle \in V$ for a normalized vector proportional to $J_+^{k-1}|m\rangle \ne 0$: it has the properties

$$J_3|\ell\rangle = \ell|\ell\rangle, \quad J_+|\ell\rangle = 0, \quad \langle\ell|\ell\rangle = 1, \tag{4.123}$$

where ℓ is the *maximal eigenvalue* of J_3 in V. From Eq. (4.120) we have

$$0 = \|J_+|\ell\rangle\|^2 = \langle\ell|J_-J_+|\ell\rangle = (\lambda - \ell(\ell + 1))\||\ell\rangle\|^2, \tag{4.124}$$

and hence the maximal eigenvalue ℓ of J_3 in V and the eigenvalue λ of the Casimir operator J^2 are related by the formula

$$J^2\big|_V \equiv \lambda = \ell(\ell + 1). \tag{4.125}$$

Dually, the eigenvalue $(m - k)$ also violates the bound (4.121) for $k \gg 1$, so there is a minimal integer $k \ge 0$ such that $J_-^{k+1}|\ell\rangle = 0$ and then

$$0 = \|J_-^{k+1}|\ell\rangle\|^2 = \langle\ell|J_+^k J_+ J_- J_-^k|\ell\rangle =$$
$$= (\ell(\ell + 1) - (\ell - k)(\ell - k - 1))\|J_-^k|\ell\rangle\|^2, \tag{4.126}$$

---

[20] Cf. the algebraic discussion of the energy levels of the harmonic oscillator in Sect. 3.5.3.

which gives $\ell - k = -\ell$, or

$$2\,\ell = k = 0, 1, 2, \ldots \tag{4.127}$$

Thus ℓ is either a *non-negative integer* $\ell = 0, 1, 2 \ldots$ or a *positive half-integer* $\ell = \frac{1}{2}, \frac{3}{2}, \frac{5}{2}, \ldots$. We collect our findings:

SUMMARY: Irreducible Representations of $\mathfrak{so}(3) \equiv \mathfrak{su}(2)$
For each *positive integer* $2\ell + 1$ we have an irreducible $\mathfrak{so}(3)$-module V_ℓ, unique up to isomorphism, of dimension $2\ell + 1$

$$\dim V_\ell = 2\ell + 1 \tag{4.128}$$

which is spanned by the $2\ell + 1$ vectors

$$|\ell\rangle, \ J_-|\ell\rangle, \ J_-^2|\ell\rangle, \ \cdots, \ J_-^{2\ell-1}|\ell\rangle, \ J_-^{2\ell}|\ell\rangle \tag{4.129}$$

with Casimir eigenvalue

$$J^2|_{V_\ell} = \ell(\ell + 1) \tag{4.130}$$

and J_3 spectrum

$$\sigma(J_3) = \{-\ell, -\ell + 1, -\ell + 2, \cdots, \ell - 1, \ell\} \tag{4.131}$$

$$\text{with eigenvectors:} \quad J_3(J_-^k|\ell\rangle) = (\ell - k)(J_-^k|\ell\rangle). \tag{4.132}$$

All irreducible unitary representation of the Lie algebra $\mathfrak{so}(3) \simeq \mathfrak{su}(2)$ is isomorphic to precisely one V_ℓ. We call V_ℓ the *representation of spin ℓ*, where ℓ is either a non-negative integer or a positive half-integer.

Remark 4.5 The spectra in V_ℓ of the operators J_1, J_2 are identical to the spectrum of J_3 since the three components of the vector $\mathbf{J}$ are conjugated by $SO(3)$ rotations.

Remark 4.6 In the literature one encounters two ways of labelling the irreducible representations of $\mathfrak{su}(2)$: in terms of the highest eigenvalue ℓ of J_3 (the *spin* of the representation), written V_ℓ, or in term of its dimension written in boldface. So an alternative notation for V_ℓ is $V_{2\ell+1}$ or simply $\mathbf{2\ell + 1}$.

Matrix Elements

The three operators J_i are represented in V_ℓ by $(2\ell+1) \times (2\ell+1)$ Hermitian matrices which satisfy the commutation relations (4.103) with $\hbar = 1$ or, equivalently, the relations (4.118) when written in the basis where J_3 acts diagonally. We wish to write these matrices explicitly in a basis of V_ℓ where J_3 is diagonal. Following the standard convention, we write $|m, \ell\rangle$ for a normalized eigenvector of J_3 in V_ℓ with eigenvalue m where

$$m = \ell \bmod 1 \quad \text{with} \quad -\ell \le m \le \ell, \tag{4.133}$$

that is,

$$\langle m', \ell | J_3 | m, \ell\rangle = m\, \delta_{m,m'} \tag{4.134}$$

$$\langle m', \ell | J_\pm | m, \ell\rangle = 0 \quad \text{unless } m' = m \pm 1. \tag{4.135}$$

Then

$$
\begin{aligned}
\ell(\ell + 1) - m^2 \pm m &= \langle m, \ell | J_\pm J_\mp | m, \ell\rangle = \\
&= \sum_{m'} \langle m, \ell | J_\pm | m', \ell\rangle \langle m', \ell | J_\mp | m, \ell\rangle = \\
&= \langle m, \ell | J_\pm | m \mp 1, \ell\rangle \langle m \mp 1, \ell | J_\mp | m, \ell\rangle = \\
&= \left| \langle m \mp 1, \ell | J_\mp | m, \ell\rangle \right|^2,
\end{aligned}
\tag{4.136}
$$

and we may fix the arbitrary phases of the states $|m, \ell\rangle$ so that

$$\langle m \mp 1, \ell | J_\mp | m, \ell\rangle = \sqrt{\ell(\ell + 1) - m^2 \pm m}. \tag{4.137}$$

Using $J_1 = \frac{1}{2}(J_+ + J_-)$ and $J_2 = \frac{1}{2i}(J_+ - J_-)$, we get the matrix elements

$$
\begin{aligned}
\langle m', \ell | J_1 | m, \ell\rangle = {}&\frac{1}{2}\delta_{m',m+1}\sqrt{\ell(\ell + 1) - m^2 - m}\, + \\
&+ \frac{1}{2}\delta_{m',m-1}\sqrt{\ell(\ell + 1) - m^2 + m}
\end{aligned}
\tag{4.138}
$$

$$
\begin{aligned}
\langle m', \ell | J_2 | m, \ell\rangle = {}&\frac{1}{2i}\delta_{m',m+1}\sqrt{\ell(\ell + 1) - m^2 - m}\, - \\
&- \frac{1}{2i}\delta_{m',m-1}\sqrt{\ell(\ell + 1) - m^2 + m}
\end{aligned}
\tag{4.139}
$$

Characters

We adopt the $SU(2)$ viewpoint. All rotations in $SU(2)$ are conjugated to a rotation by an angle 2θ around the third axis generated by the unitary operator $\exp(2i\theta J_3)$

with $0 \leq \theta < 2\pi$. Its character in the irreducible representation V_ℓ is ($\ell \in \mathbb{N}/2$)

$$\chi_\ell(e^{2i\theta J_3}) = e^{2\ell i\theta} + e^{(2\ell-2)i\theta} + \cdots + e^{-2\ell i\theta} =$$
$$= \frac{\sin((2\ell + 1)\theta)}{\sin \theta} = U_{2\ell}(\cos \theta) \tag{4.140}$$

where $U_n(x)$ is a degree-n polynomial known as the *n-th Chebyshev polynomial of the second kind* [9]. The $U_n(x)$'s satisfy the obvious 3-term recursion

$$U_1(x) U_n(x) = U_{n+1}(x) + U_{n-1}(x) \qquad U_0 = 1, \ \ U_{-1} = 0. \tag{4.141}$$

Seen as a relation in the $SU(2)$-representation ring $\mathcal{R}(SU(2))$, the Chebyshev recursion relation (4.141) reads

$$V_{1/2} \otimes V_\ell = V_{\ell+1/2} \oplus V_{\ell-1/2} \qquad (V_\ell = 0 \text{ for } \ell < 0). \tag{4.142}$$

The dimension of the representation V_ℓ is given by

$$\dim V_\ell = \chi_\ell(1) = \lim_{\theta \to 0} U_{2\ell}(\cos \theta) = 2\ell + 1. \tag{4.143}$$

The orthonormality of the characters reads[21]

$$\frac{2}{\pi} \int_0^\pi U_{2\ell}(\cos \theta) \, U_{2\ell'}(\cos \theta) \, \sin^2\theta \, d\theta = \delta_{\ell,\ell'}. \tag{4.144}$$

The characters $U_{2\ell}(\cos \theta)$ are real, hence the irreducible representations of $SU(2)$ are either *real* or *quaternionic*; the Frobenius-Schur indicator is

$$\frac{2}{\pi} \int_0^\pi U_{2\ell}(\cos(2\theta)) \sin^2\theta \, d\theta = (-1)^{2\ell}, \tag{4.145}$$

hence the representations with ℓ integer are *real* ($\equiv$ *orthogonal*) and those with ℓ half-integral are *quaternionic* ($\equiv$ *symplectic*). We also say that the representation with ℓ integral are *tensor* representations while those with ℓ half-integral are *spinor(ial)*.

Reiterating the recursion relation (4.141) we get ($m \leq n$)

$$U_n(x) U_m(x) = \sum_{k=0}^{m} U_{n-m+2k}(x) \tag{4.146}$$

---

[21] In Eq. (4.144) we use the *Weyl integration formula* [21] to rewrite an integral with respect to the normalized Haar measure dg on $SU(2)$ as an integral over the maximal torus $U(1) \subset SU(2)$.

which yields the rule for the decomposition of the tensor product of two irreducible $SU(2)$-representations into irreducible ones

$$V_\ell \otimes V_{\ell'} = \bigoplus_{k=0}^{2\ell'} V_{\ell-\ell'+k} \qquad \text{for } \ell' \le \ell. \tag{4.147}$$

4.3.1   *SU(2)-Representations: Examples*

In the one-dimensional irreducible representation V_0 all operators J_i vanish, so V_0 is the trivial $\mathfrak{so}(3)$ representation which describes states of zero angular momentum invariant under arbitrary rotations. The first non-trivial case is the two-dimensional representation $V_{1/2}$.

The Spin Representation $V_{1/2}$
Following tradition we write S_i ($i = 1, 2, 3$) for the generators of $\mathfrak{su}(2)$ in the *spinorial* representation $V_{1/2}$. In the basis where S_3 is diagonal (with eigenvalues written in decreasing order) Eqs. (4.138) and (4.139) become

$$S_1 = \frac{1}{2}\begin{pmatrix} 0 & 1 \\ 1 & 0 \end{pmatrix}, \quad S_2 = \frac{1}{2}\begin{pmatrix} 0 & -i \\ i & 0 \end{pmatrix}, \quad S_3 = \frac{1}{2}\begin{pmatrix} 1 & 0 \\ 0 & -1 \end{pmatrix} \tag{4.148}$$

The S_i's satisfy the $\mathfrak{so}(3)$ Lie algebra relations

$$[S_i, S_j] = i\epsilon_{ijk} S_k. \tag{4.149}$$

The representation with $\ell = 1/2$ is called the *spin representation*. The three 2×2 matrices

$$\sigma_1 = \begin{pmatrix} 0 & 1 \\ 1 & 0 \end{pmatrix}, \quad \sigma_2 = \begin{pmatrix} 0 & -i \\ i & 0 \end{pmatrix}, \quad \sigma_3 = \begin{pmatrix} 1 & 0 \\ 0 & -1 \end{pmatrix} \tag{4.150}$$

are the *Pauli matrices*. They form a basis of the $\mathbb{R}$-space of traceless 2×2 Hermitian matrices, that is, a basis of the Lie algebra $\mathfrak{su}(2)$. This shows again the Lie algebra isomorphism

$$\mathfrak{su}(2) \simeq \mathfrak{so}(3) \tag{4.151}$$

BOX: Quaternions and the Spin Representation

The *quaternions* $\mathbb{H}$ are an associative non-commutative $\mathbb{R}$-algebra. An element $x \in \mathbb{H}$ has the form

$$x = x_0 + ix_1 + jx_2 + kx_3$$

where i, j, k are the three *imaginary units* with the multiplication table

$$i^2 = j^2 = k^2 = -1, \quad ij = -ji = k, \quad jk = -kj = i, \quad ki = -ik = j,$$

and x_0, x_1, x_2, x_3 are real coefficients. The quaternionic complex conjugate of x is

$$\bar{x} = x_0 - ix_1 - jx_2 - kx_3$$

and its absolute value squared is

$$|x|^2 \equiv \bar{x}x = x\bar{x} = x_0^2 + x_1^2 + x_2^2 + x_3^2 \geq 0$$

We have the basic identity

$$|xy| = |x|\,|y|$$

The quaternions of unit norm

$$U = \left\{ y \in \mathbb{H} \colon |y| \equiv y_0^2 + y_1^2 + y_2^2 + y_3^2 = 1 \right\} \tag{$\spadesuit$}$$

acting by multiplication on the left, resp. on the right, give two groups G_L, G_R which act by isometries on $\mathbb{H}$, i.e. they fix $|x|^2 \equiv x_0^2 + x_1^2 + x_2^2 + x_3^2$, by the basic identity. Hence

$$(G_L \times G_R)/\{\pm 1\} \subseteq SO(4)$$

and we must have equality by dimension considerations. From eq.($\spadesuit$) we see that the group $U \simeq S^3 \subset \mathbb{R}^4$ while $S^3 \simeq SU(2)$; then $SO(4)$ is $(SU(2)_L \times SU(2)_R)/\mathbb{Z}_2$ where *the two copies of $SU(2)$ act in the spin $\frac{1}{2}$ representation*. We may identify $\mathbb{H}$ with the $\mathbb{R}$-space of 2×2 matrices $\hat{x}$, with $\hat{x} + \hat{x}^\dagger$ a multiple of 1 and $\hat{x} - \hat{x}^\dagger$ traceless, via the correspondence

$$x \equiv x_0 + ix_1 + jx_2 + kx_3 \rightsquigarrow \hat{x} \equiv x_0 + i\sigma_1 x_1 + i\sigma_2 x_2 + i\sigma_3 x_3 \equiv \begin{pmatrix} x_0 + ix_3 & ix_1 + x_2 \\ ix_1 - x_2 & x_0 - ix_3 \end{pmatrix}$$

given by the *Pauli matrices* σ_i. We have $\bar{x}x = \det \hat{x}$, while the action of $SU(2)_L \times SU(2)_R$ is given by the product on the left and on the right by unit quaternions

$$\hat{x} \rightarrow \hat{u}_L \hat{x} \hat{u}_R \qquad \hat{u}_L, \hat{u}_L \in SU(2) \Leftrightarrow |u_L| = |u_R| = 1$$

A similar story holds for octonions leading to the triality property of $SO(8)$ [15]

that was exploited several times in [1] (see also Sect. 4.4 below). The connection of the Pauli matrices with the (associative) division $\mathbb{R}$-algebra $\mathbb{H}$ of *quaternions* (see **BOX** on page 236) yields the further isomorphism $\mathfrak{su}(2) \simeq \mathfrak{sp}(1)$.

The Pauli matrices σ_i give a representation of the Clifford algebra $\mathsf{Cl}_+(3)$ in dimension 3 [15], i.e. the algebra with 3 generators $\{\sigma_1, \sigma_2, \sigma_3\}$ and relations

$$\sigma_i\sigma_j + \sigma_j\sigma_i = 2\,\delta_{ij}, \qquad i, j = 1, 2, 3. \tag{4.152}$$

The σ_i's then have the multiplication table

$$\sigma_i\sigma_j = \delta_{ij} + i\epsilon_{ijk}\,\sigma_k. \tag{4.153}$$

Given a 3-vector $\mathbf{v} \equiv (v_i) \in \mathbb{R}^3$, we write $\hat{\mathbf{v}}$ for the traceless Hermitian 3×3 matrix

$$\hat{\mathbf{v}} \equiv v_i\sigma_i = \begin{pmatrix} v_3 & v_1 - iv_2 \\ v_1 + iv_2 & -v_3 \end{pmatrix}, \tag{4.154}$$

$$\det\hat{\mathbf{v}} = -(v_1^2 + v_2^2 + v_3^2) \equiv -v^2, \tag{4.155}$$

$\hat{\mathbf{v}}$ is the element of $\mathfrak{g}$ which corresponds to the vector $\mathbf{v} \equiv (v_i)$ *via* the isomorphism of (real) $\mathfrak{g}$-modules $\mathfrak{g} \simeq V_1^{\mathbb{R}} \equiv \mathbb{R}^3$. The matrix $\hat{\mathbf{v}}$ satisfies the algebraic equation

$$\hat{\mathbf{v}}^2 = v_i v_j \sigma_i\sigma_j = v_i v_j(\delta_{ij} + i\epsilon_{ijk}\sigma_k) = v_i v_i = \mathbf{v}^2, \tag{4.156}$$

and, more generally,

$$\hat{\mathbf{v}}^m = \begin{cases} v^m & \text{for } m \text{ even} \\ v^{m-1}\hat{\mathbf{v}} \equiv v^m\,\hat{\mathbf{n}} & \text{for } m \text{ odd} \end{cases} \tag{4.157}$$

where $v \equiv \sqrt{v_i v_i}$ is the length of the vector v_i and $\mathbf{v} \equiv v\mathbf{n}$ with $\mathbf{n} = (n_i)$ the unit vector pointing in the direction of $\mathbf{v}$. The multiplication rule (4.153) yields

$$\hat{\mathbf{v}}\hat{\mathbf{w}} = \mathbf{v} \cdot \mathbf{w} + i\widehat{\mathbf{v} \wedge \mathbf{w}} \tag{4.158}$$

where $\mathbf{v} \wedge \mathbf{w}$ is the cross product of vectors in $\mathbb{R}^3$. In Problem 4.1 the reader is invited to show the identity

$$\hat{\mathbf{w}}\hat{\mathbf{v}}\hat{\mathbf{w}} = -\mathbf{w}^2\,\hat{\mathbf{v}} + 2(\mathbf{w} \cdot \mathbf{v})\hat{\mathbf{w}}. \tag{4.159}$$

The elements of the Lie group $SU(2)$, acting in the defining 2-dimensional representation $V_{1/2}$, are obtained by exponentiating the elements $\hat{\mathbf{v}} \in \mathfrak{g}$

$$
\begin{aligned}
\exp(\mathrm{i}\hat{\mathbf{v}}) &= \sum_{k=0}^{\infty} \mathrm{i}^k \frac{\hat{\mathbf{v}}^k}{k!} = \sum_{k=0}^{\infty}(-1)^k \frac{\hat{\mathbf{v}}^{2k}}{(2k)!} + \mathrm{i}\sum_{k=0}^{\infty}(-1)^k \frac{\hat{\mathbf{v}}^{2k+1}}{(2k+1)!} = \\
&= \sum_{k=0}^{\infty}(-1)^k \frac{\hat{\mathbf{v}}^{2k}}{(2k)!} + \mathrm{i}\sum_{k=0}^{\infty}(-1)^k \frac{\hat{\mathbf{v}}^{2k+1}}{(2k+1)!} = \\
&= \cos v + \mathrm{i}\,\hat{\mathbf{n}}\sin v.
\end{aligned}
\tag{4.160}
$$

In particular a rotation around the third axis by an angle θ is given by

$$
\exp\!\left(\mathrm{i}\theta S_3\right) = \exp\!\left(\frac{\mathrm{i}}{2}\theta\sigma_3\right) = \cos(\theta/2) + \mathrm{i}\sigma_3\sin(\theta/2)
\tag{4.161}
$$

and, more generally, a rotation around the axis $\mathbf{n} = (n_i)$ by the unitary 2×2 matrix

$$
R_{\mathbf{n}}(\theta) = \exp(\mathrm{i}\theta\, n_i\sigma_i/2) \equiv \cos(\theta/2) + \mathrm{i}\,\hat{\mathbf{n}}\sin(\theta/2)
\tag{4.162}
$$

with the properties

$$
R_{\mathbf{n}}(\theta)R_{\mathbf{n}}(-\theta) = 1, \quad R_{\mathbf{n}}(\theta)^{\dagger} = R_{\mathbf{n}}(-\theta), \quad \det R_{\mathbf{n}}(\theta) = 1.
\tag{4.163}
$$

The dual action of $SO(3)$ on the operators $\hat{\mathbf{v}} \in \mathsf{End}(V_{1/2})$ (cf. Sect. 4.1) is

$$
\begin{aligned}
\hat{\mathbf{v}} &\rightsquigarrow R_{\mathbf{n}}(\theta)^{-1}\hat{\mathbf{v}}\,R_{\mathbf{n}}(\theta) = \\
&= \cos^2(\theta/2) + 2\cos(\theta/2)\sin(\theta/2)\,\widehat{\mathbf{n}\wedge\mathbf{v}} + \sin^2(\theta/2)\big(-\hat{\mathbf{v}} + 2(\mathbf{n}\cdot\mathbf{v})\hat{\mathbf{n}}\big) \\
&= \hat{\mathbf{v}}_{\parallel} + \cos\theta\,\hat{\mathbf{v}}_{\perp} + \sin\theta\,\widehat{\mathbf{n}\wedge\mathbf{v}_{\perp}}
\end{aligned}
\tag{4.164}
$$

where $\mathbf{v}_{\parallel}$ and $\mathbf{v}_{\perp}$ are, respectively, the components of the vector $\mathbf{v} = (v_i)$ parallel and orthogonal to $\mathbf{n}$, and we used the identities (4.158), (4.159).

3-Dimensional Vector Representation V_1

The complex basis of $\mathbb{R}^3 \otimes \mathbb{C}$ where J_3 is diagonal with decreasing eigenvalues, is

$$
e_1 = \frac{1}{\sqrt{2}}(1,\mathrm{i},0)^t, \quad e_2 = (0,0,1)^t, \quad e_3 = \frac{1}{\sqrt{2}}(1,-\mathrm{i},0)^t
\tag{4.165}
$$

In this basis

$$
J_1 = \frac{1}{\sqrt{2}}\begin{pmatrix} 0 & 1 & 0 \\ 1 & 0 & 1 \\ 0 & 1 & 0 \end{pmatrix}, \quad J_2 = \frac{1}{\sqrt{2}}\begin{pmatrix} 0 & -\mathrm{i} & 0 \\ \mathrm{i} & 0 & -\mathrm{i} \\ 0 & \mathrm{i} & 0 \end{pmatrix}, \quad J_3 = \begin{pmatrix} 1 & 0 & 0 \\ 0 & 0 & 0 \\ 0 & 0 & -1 \end{pmatrix}
\tag{4.166}
$$

We leave to the reader to check that these matrices satisfy the $\mathfrak{so}(3)$ commutation relations and that they generate the well known $SO(3)$ rotations when rewritten in the usual real basis of $\mathbb{R}^3$.

4.4 $SO(3)$ vs. $SU(2)$: Superselection Rules

We say that a state $|\psi\rangle$ which transforms in the irreducible representation V_j has *angular momentum j* or *spin j*, depending on the physical context. j may be an integer or a half-integer. Consider a rotation around the third axis by an angle $\theta = 2\pi$ which in V_j is implemented by the operator

$$\exp(2\pi i J_3) = \begin{cases} 1 & \text{if } j \in \mathbb{N} \\ -1 & \text{if } j \in \mathbb{N} + \frac{1}{2}, \end{cases} \tag{4.167}$$

where $\mathbf{J} = (J_i)$ is the total angular momentum (Eq. (4.111)). In $SO(3)$ a rotation by 2π is the identity, so only the representations of $\mathfrak{so}(3) \simeq \mathfrak{su}(2)$ with integral j exponentiate to representations[22] of $SO(3)$.

The elements of the matrix group $SU(2)$ are the exponential of linear combinations of the generators $S_i \equiv \sigma_i/2$ in the $j = 1/2$ representation, see Eq. (4.162). In the group $SU(2)$ the rotation by 2π is represented by the matrix

$$-1 \equiv \begin{pmatrix} -1 & 0 \\ 0 & -1 \end{pmatrix} \neq 1 \tag{4.168}$$

which is *not* the identity but merely *central*, i.e. $\exp(2\pi i S_3)$ commutes with all the S_i's. Indeed the center of $SU(2)$ is $\{\pm 1\}$, while $SO(3)$ has trivial center. Hence the map[23]

$$SU(2) \to SO(3), \qquad e^{i\theta_i S_i} \mapsto e^{i\theta_i R_i} \tag{4.169}$$

is a *double cover* with kernel $\{\pm 1\} \subset SU(2)$ which yields an isomorphism for the quotient group

$$SU(2)/\mathbb{Z}_2 \simeq SO(3). \tag{4.170}$$

In [1] (and in Sect. 4.1) this isomorphism was deduced from the fact that $\mathbb{P}^1 \equiv \mathbb{P}(\mathbb{C}^2)$ has a group of holomorphic isometries

$$(z_1, z_2) \mapsto A(z_1, z_2) \qquad A \in U(2), \tag{4.171}$$

---

[22] Note that the representations of $SO(3)$ are exactly the representations of $SU(2)$ which are orthogonal i.e. *real* (as contrasted to *quaternionic*).

[23] The R_i's are 3×3 matrices defined in Eq. (4.100).

while the multiples of the identity act trivially, so the group acting effectively on $\mathbb{P}^1$ is $U(2)/U(1) \equiv SU(2)/\{\pm 1\}$. On the other hand $\mathbb{P}^1$ is the Riemann sphere S^2 whose group of orientation preserving isometries is $SO(3)$. Hence we get

$$SO(3) \simeq SU(2)/\mathbb{Z}_2, \tag{4.172}$$

where $\mathbb{Z}_2 \vartriangleleft SU(2)$ is generated by the central rotation[24] of angle 2π.

Fact 4.6 *As a manifold $SU(2) \simeq S^3$ hence it is simply-connected. On the other hand $\pi_1(SO(3)) = \pi_1(SU(2)/\mathbb{Z}_2) = \mathbb{Z}_2$.*

Proof The general element of $SU(2)$ has the form

$$A = \begin{pmatrix} z_1 & z_2 \\ -z_2^* & z_1^* \end{pmatrix} \quad z_1, z_2 \in \mathbb{C} \text{ with } |z_1|^2 + |z_2|^2 = 1. \tag{4.173}$$

Writing $z_k = x_{2k-1} + \mathrm{i}x_{2k}$ $(k = 1, 2)$ the equation $|z_1|^2 + |z_2|^2 = 1$ becomes

$$\sum_{k=1}^{4} x_k^2 = 1 \tag{4.174}$$

which is the equation of the unit sphere $S^3 \subset \mathbb{R}^4$. $\qquad\qquad\square$

Definition 4.4 The irreducible representations V_s of $\mathfrak{su}(2)$ with half-integral s are called *spinor(ial) representation*. The representations V_s with integral s are called *orthogonal representations* or *tensor representation*. Their exponentials produce *faithful* representations of $SU(2)$ and, respectively, $SO(3)$. The quantum number s is called *spin*.

As seen before, for s integral (resp. half-integral) the representation V_s yields a group embedding $SO(3) \hookrightarrow SO(2s + 1)$ (resp. $SU(2) \hookrightarrow Sp(2s + 1)$) and hence is a *real* (resp. *quaternionic*) representation in the Frobenius-Schur sense. We shall return to the fundamental dichotomy integral/half-integral spin in Sect. 4.12 below.

The Spin Superselection Rule
Now it is time to dwell on a fine point that we glossed over when discussing the superposition principle in Chaps. 1 and 2.

Suppose that our quantum system—possibly very elaborate, with lots of tricky subsystems—is located in the ordinary space $\mathbb{R}^3$. Let J_i be the Hermitian operators which generate the overall spatial rotations of our system—they may be very involved operators, and may or may not be conserved in time—their only purpose is to describe the relation between the result of an experiment on our system and the result of the same experiment when we rotate our experimental apparatus by

---

[24] All rotations by 2π around any axis are the same, since the element is central.

arbitrary angles in the physical space $\mathbb{R}^3$ (keeping the system itself untouched). The operators J_i must obey the $\mathfrak{su}(2)$ Lie algebra

$$[J_i, J_j] = i\epsilon_{ijk} J_k, \tag{4.175}$$

but are otherwise arbitrary. The (bounded) Hermitian operator[25]

$$(-1)^F \stackrel{\text{def}}{=} e^{2\pi i J_3}, \qquad \left((-1)^F\right)^\dagger = (-1)^F \tag{4.176}$$

which generates the rotation by an angle 2π, satisfies the algebraic equation

$$\left((-1)^F\right)^2 = 1, \tag{4.177}$$

so the Hilbert space of our system—however involved—decomposes into the direct sum of the $+1$ and -1 eigenspaces

$$\mathcal{H} = \mathcal{H}_+ \oplus \mathcal{H}_- \qquad (-1)^F\big|_{\mathcal{H}_\pm} = \pm 1. \tag{4.178}$$

In the language of Definition 4.4 $\mathcal{H}_+$ is the subspace of the tensor states $|B\rangle$ and $\mathcal{H}_-$ is the subspace of the spinor ones $|F\rangle$.

Rotating the experimental apparatus by 2π means doing nothing: the outcome of the experiment cannot change. Since this holds for *all* possible experiments, we conclude that the physical state cannot change if we rotate it by a 2π angle. The vector $|\psi\rangle$ which *represents* the state may however get multiplied by an overall phase $|\psi\rangle \to e^{i\phi}|\psi\rangle$ since this does not change the physical state nor any measurable quantity. Now consider a general vector $|B\rangle + |F\rangle \in \mathcal{H}$, with $|B\rangle \in \mathcal{H}_+$ and $|F\rangle \in \mathcal{H}_-$, and rotate it by 2π

$$|B\rangle + |F\rangle \xrightarrow{\quad (-1)^F \quad} |B\rangle - |F\rangle. \tag{4.179}$$

The final vector is a multiple of the initial one iff $|F\rangle = 0$ or $|B\rangle = 0$ but not otherwise. We conclude that we cannot build physical states by superposing a tensor and a spinor state. The space of physical states is a *disjoint union*

$$\mathbb{P}(\mathcal{H}_+) \sqcup \mathbb{P}(\mathcal{H}_-). \tag{4.180}$$

A rule which forbids certain superpositions of states is called a *superselection rule*,[26] and the maximal subspaces whose elements may be freely superposed are the *superselection sectors*. The general story is as follows:

---

[25] The reason of the notation $(-1)^F$ will be clarified in Sect. 4.12.

[26] *Super*selection rule since there exist (totally innocuous) ordinary *selection rules*, cf. Chap. 8.

Superposition Principle 6 *Let S be a Hermitian operator with the property that it commutes with all physical observables $\mathcal{O}$:*

$$S\mathcal{O} = \mathcal{O}S, \tag{4.181}$$

that is, S belongs to the center $Z(\mathfrak{A}_{\text{obs}})$ of the algebra $\mathfrak{A}_{\text{obs}}$ of quantum observables. Let

$$\mathcal{H} = \bigoplus_{s \in \sigma(S)} \mathcal{H}_s \tag{4.182}$$

be the decomposition of the Hilbert space $\mathcal{H}$ in eigenspaces[27] of S. Then each eigenspace $\mathcal{H}_s$ is a superselection sector, *namely we are allowed to superpose only states belonging to the same eigenspace $\mathcal{H}_s$ of S.*

The *spin superselection rule* is this **Principle** applied to the rotation by 2π, $S \equiv (-1)^F$. Another example are the gauge transformations. They commute with all observables by construction. To be precise, in stating the above **Principle** we must restrict to the algebra $\mathfrak{A}_{\text{obs}}$ of compactly-supported observables $\mathcal{O}$: *alas,* all our experiments are performed using an apparatus which fits in some—possibly huge—compact $K \in \mathbb{R}^3$. All such observables will commute with electromagnetic gauge transformations which reduce to some constant α at spatial infinity. By Gauss' Law, the eigenvalue of these gauge transformations is $e^{i\alpha q}$ where q is the electric charge of our state. We conclude that states of different electric charges cannot be superimposed. The same story applies to the magnetic charge or any other charge that is measured by a flux at infinity.

4.5 Spherical Harmonics

We return to the representation of $SO(3)$ on $L^2(\mathbb{R}^3)$, that is, to the *orbital angular momentum* of a quantum particle moving in $\mathbb{R}^3$. We introduce the usual spherical coordinates (r, θ, ϕ) in $\mathbb{R}^3$

$$x_1 = r\sin\theta\cos\phi, \quad x_2 = r\sin\theta\sin\phi, \quad x_3 = r\cos\theta, \tag{4.183}$$

and rewrite the operators L_1, L_2, L_3 as differential operators in spherical coordinates: by the quantum Noether theorem of Chap. 2 the L_i's are $-i\hbar$ times the $\mathfrak{so}(3)$

---

[27] The reader may easily convince herself that the spectrum of S must be discrete.

Killing vectors on the unit sphere $S^2 \equiv \{r = 1\} \subset \mathbb{R}^3$ with the induced "round" metric

$$ds^2 = d\theta^2 + \sin^2\theta \, d\phi^2. \tag{4.184}$$

We set $\hbar = 1$ (to reinsert $\hbar$ just make $L_i \rightsquigarrow \hbar L_i$). An easy computation (left to the reader) gives:

$$L_1 = i\left(\sin\phi \frac{\partial}{\partial\theta} + \cos\phi \cot\theta \frac{\partial}{\partial\phi}\right) \tag{4.185}$$

$$L_2 = -i\left(\cos\phi \frac{\partial}{\partial\theta} - \sin\phi \cot\theta \frac{\partial}{\partial\phi}\right) \tag{4.186}$$

$$L_3 = -i\frac{\partial}{\partial\phi} \tag{4.187}$$

$$L_\pm = e^{\pm i\phi}\left(\pm\frac{\partial}{\partial\theta} + i\cot\theta \frac{\partial}{\partial\phi}\right) \tag{4.188}$$

The Casimir operator $L^2 \in \mathsf{U}_{\mathfrak{so}(3)}$ is a non-negative second-order differential operator acting on the functions $\psi \in \Omega^0(S^2)$ which is invariant under all isometries of S^2 and kills the constants. Up to overall normalization there is precisely one such object, namely the Laplacian of the round metric (4.184)

$$\Delta = -\frac{1}{\sin\theta}\frac{\partial}{\partial\theta}\left(\sin\theta \frac{\partial}{\partial\theta}\right) - \frac{1}{\sin^2\theta}\frac{\partial^2}{\partial\phi^2} \tag{4.189}$$

In facts the relative normalization is just 1

$$L^2 = \Delta, \tag{4.190}$$

as one may check applying both sides to a simple function like θ. In other words: The Casimir operator L^2—seen as a second-order differential operator acting on functions on the sphere S^2—is the Schrödinger-representation Hamiltonian for a free[28] particle constrained to move on the unit sphere of metric (4.184) (we use units where $2m = 1$ and $\hbar = 1$). In yet other words: L^2 is the quantum Hamiltonian obtained by quantization of the classical one $H_{\mathrm{cl}} = g^{ij} p_i p_j$—where g^{ij} is the

---

[28] "Free" here means that there is no potential, hence no force apart for the one implicitly produced by the constraint that the particle should remain on the surface of the sphere.

inverse of the metric (4.184)—with the prescription that the ground state, given by the constant function on S^2

$$\psi_0(\theta, \phi) = \frac{1}{\sqrt{4\pi}}, \tag{4.191}$$

has zero energy. We conclude:

Fact 4.7 *Studying the* Representation Theory *of $SO(3)$ on $L^2(\mathbb{R}^3)$ is equivalent to solving the Schrödinger equation for the free quantum particle moving on $M = S^2$ whose Hilbert space we identify with $L^2(S^2)$ through the Schrödinger representation* (cf. Sect. 2.15).

We adopt the quantum particle viewpoint. The two operators $H \equiv L^2$ and L_3 form a complete system of commuting conserved quantities for the quantum particle moving on S^2. In particular the angle ϕ—canonically conjugate to L_3— is a *cyclic* coordinate, so the free Schrödinger equation on S^2 is *separable* (as the corresponding classical Hamilton-Jacobi equation, [1] §. 9.3.2).

We look for simultaneous eigenvectors $|m, \ell\rangle$ of L^2 and L_3 which satisfy

$$L_3|m, \ell\rangle = m|m, \ell\rangle,$$
$$L^2|m, \ell\rangle = \ell(\ell + 1)|m, \ell\rangle, \tag{4.192}$$
$$m = \ell \bmod 1, \quad |m| \leq \ell,$$

where now ℓ must be a non-negative *integer* since the Hilbert space carries a representation of $SO(3)$. We wish to compute the wave-functions of the eigenstates $|m, \ell\rangle$ in the $L^2(S^2)$ Schrödinger representation i.e. the functions

$$Y_{m,\ell}(\theta, \phi) \overset{\text{def}}{=} \langle \theta, \phi | m, \ell\rangle, \tag{4.193}$$

normalized in the $L^2(S^2)$ norm

$$\int_{S^2} \sin\theta \, d\theta \, d\phi \, Y_{m',\ell'}(\theta, \phi)^* Y_{m,\ell}(\theta, \phi) = \delta_{m'm} \delta_{\ell'\ell}. \tag{4.194}$$

The eigenfunction $Y_{m,\ell}(\theta, \phi)$ are called *spherical harmonics*.

In the Schrödinger representation the eigenvalue Eq. (4.192) read

$$-i\frac{\partial}{\partial\phi} Y_{m,\ell}(\theta, \phi) = m \, Y_{m,\ell}(\theta, \phi), \tag{4.195}$$

$$-\left[\frac{1}{\sin\theta}\frac{\partial}{\partial\theta}\left(\sin\theta\frac{\partial}{\partial\theta}\right) + \frac{1}{\sin^2\theta}\frac{\partial^2}{\partial\phi^2}\right]Y_{m,\ell}(\theta, \phi) = \ell(\ell + 1)Y_{m,\ell}(\theta, \phi). \tag{4.196}$$

The separation of variables takes the obvious form

$$Y_{m,\ell}(\theta, \phi) = e^{im\phi}\, F_{m,\ell}(\theta). \tag{4.197}$$

We may state a priori that $F_{m,\ell}(\theta)$ is a polynomial in $\cos\theta$ and $\sin\theta$ of degree ℓ by the relation of the spherical harmonics with the *harmonic polynomials*, see next section. Hence the above differential equations can be solved in closed form. In particular from these equations we see that we can always choose[29]

$$F_{m,\ell}(\theta) \equiv F_{-m,\ell}(\theta). \tag{4.198}$$

Legendre Polynomials

We consider first the case $m = 0$ which is of independent interest.

We set $F_{0,\ell}(\theta) = P_\ell(\cos\theta)$. Using the identity

$$\left(\sin\theta\,\frac{\partial}{\partial\theta}\right)^2 P(\cos\theta) =$$
$$= (1 - \cos^2\theta)^2\, P''(\cos\theta) - 2(1 - \cos^2\theta)\cos\theta\, P'(\cos\theta), \tag{4.199}$$

we reduce Eq. (4.196) to ($x \equiv \cos\theta$)

$$\tilde{H} P_\ell(x) \equiv -\frac{\mathrm{d}}{\mathrm{d}x}\left[(1 - x^2)\frac{\mathrm{d}P_\ell(x)}{\mathrm{d}x}\right] = \ell(\ell + 1) P_\ell(x), \tag{4.200}$$

where $P_\ell(x)$ may be chosen to be real, while Eq. (4.194) reduces to the orthonormality condition in $L^2([-1, 1])$ (with free boundary conditions![30])

$$\int_{-1}^{+1} \mathrm{d}x\, P_{\ell'}(x)\, P_\ell(x) = \delta_{\ell,\ell'}. \tag{4.201}$$

Remark 4.7 Equation (4.200) is a Sturm-Liouville equation and also a Fuchs ODE with regular singularities at $\{-1, 1, \infty\}$. Since the relevant solutions are rational functions, in fact polynomials, its differential Galois group should be finite: indeed it is a cyclic group.

The differential operator $\tilde{H}$ sends polynomials of degree $\leq d$ into polynomials of degree $\leq d$, so we may diagonalize it recursively in the finite-dimensional spaces

---

[29] This choice involves, of course, a conventional choice of phases for the states.

[30] Note that the reduced Hamiltonian $\tilde{H} \equiv -\frac{\mathrm{d}}{\mathrm{d}x}(1 - x^2)\frac{\mathrm{d}}{\mathrm{d}x}$ is Hermitian in $L^2([-1, 1])$ without the need of imposing specific boundary conditions.

of polynomials of degree $d = 0, 1, 2, 3, \cdots$; moreover

$$\tilde{H} x^\ell = \ell(\ell + 1)x^\ell + \text{lower order},\tag{4.202}$$

so that $P_\ell(x)$ is a polynomial of degree ℓ, whose zeros are all in the interval $[-1, 1]$ by the oscillation theorem of Sturm-Liouville theory (Chap. 3), while parity invariance implies

$$P_\ell(-x) = (-1)^\ell P_\ell(x).\tag{4.203}$$

Hence, up to normalization, $P_0 \propto 1$, $P_1 \propto x$, and $P_2 \propto x^2 - c$. The constant c is uniquely determined by the orthogonality condition

$$0 = \int_{-1}^{+1} dx\, P_0(x) P_2(x) \propto \int_{-1}^{+1} dx (x^2 - c) = \frac{2}{3} - 2c \quad \Rightarrow \quad c = 1/3.\tag{4.204}$$

Likewise we can determine recursively $P_\ell(x)$ by imposing orthogonality with respect $P_0, P_1, \ldots, P_{\ell-1}$; in this way we construct all relevant solutions to the eigenvalue Eq. (4.200). We make the story shorter:

Proposition 4.3 *Define the sequence of degree-ℓ polynomials*

$$P_\ell(x) \stackrel{\text{def}}{=} \frac{1}{2^\ell \ell!} \frac{d^\ell}{dx^\ell} (x^2 - 1)^\ell,\tag{4.205}$$

called the Legendre polynomials. *The $P_\ell(x)$'s are solutions to the ODE (4.200). The polynomials*

$$\psi_\ell(x) = \sqrt{\ell + \frac{1}{2}}\, P_\ell(x)\tag{4.206}$$

are a complete orthonormal system in $L^2([-1, 1])$. In particular, when x, y are in the intervals $-1 \le x \le 1$ and $-1 \le y \le 1$ we have the resolution of the identity

$$\sum_{\ell=0}^{\infty} \left(\ell + \tfrac{1}{2}\right) P_\ell(x) P_\ell(y) = \delta(x - y).\tag{4.207}$$

Proof One easily checks by integration by parts that

$$\int_{-1}^{+1} P_\ell(x) P_{\ell'}(x)\, dx = \frac{\delta_{\ell\ell'}}{\ell + \tfrac{1}{2}}\tag{4.208}$$

the statement then follows by uniqueness of the solution to the recursive procedure described after Eq. (4.204). $\qquad\square$

As all sequence of orthogonal polynomials [22, 23] the Legendre polynomials satisfy a 3-term recursion in the degree ℓ

$$(\ell + 1) P_{\ell+1}(z) - (2\ell + 1)z\, P_{\ell}(z) + \ell\, P_{\ell-1}(z) = 0. \tag{4.209}$$

Generating Function for Legendre Polynomials Let us consider the following physical situation: in $\mathbb{R}^3$ we have an electric charge at the point $\mathbf{p} \equiv (1, 0, 0)$ which sources the Coulomb potential

$$\phi(\mathbf{r}) \equiv \frac{1}{|\mathbf{r} - \mathbf{p}|} = \frac{1}{(1 - 2r\cos\theta + r^2)^{1/2}} \equiv \sum_{n=0}^{\infty} r^n f_n(\cos\theta), \tag{4.210}$$

where θ is the angle between $\mathbf{r}$ and the x-axis, and the last equality is the definition of the sequence of functions $\{f_n(z)\}_n$. The Coulomb potential is a harmonic function for $\mathbf{r} \neq \mathbf{p}$. Thus

$$0 = -\Delta\phi(\mathbf{r}) = \left(\frac{1}{r^2}\frac{\partial}{\partial r}r^2\frac{\partial}{\partial r} + \frac{1}{r^2\sin\theta}\frac{\partial}{\partial\theta}\sin\theta\frac{\partial}{\partial\theta} \right) \sum_n r^n f_n(\cos\theta) =$$

$$= \frac{1}{r^2}\sum_n r^n \left[n(n+1)f_n(\cos\theta) + \frac{1}{\sin\theta}\frac{\partial}{\partial\theta}\left(\sin\theta\frac{\partial f_n(\cos\theta)}{\partial\theta} \right) \right] \tag{4.211}$$

Comparing with Eq. (4.196) we see that the functions $f_n(\cos\theta)$ must be equal to $c_n P_n(\cos\theta)$ for some coefficients c_n. To evaluate the c_n we exploit the fact that $P_n(1) = 1$ for all n (a property manifest from Eq. (4.205)); then

$$\sum_n r^n c_n = \sum_n r^n f_n(1) = \frac{1}{1-r} \quad\Rightarrow\quad c_n = 1 \;\forall\, n. \tag{4.212}$$

We got

Proposition 4.4 *The generating function of the Legendre polynomial is the Coulomb potential in $\mathbb{R}^3$*

$$\sum_{n=0}^{\infty} r^n P_n(\cos\theta) = \frac{1}{(1 - 2r\cos\theta + r^2)^{1/2}}. \tag{4.213}$$

Remark 4.8 (Ultraspherical Polynomials) We may do the corresponding exercise with the Coulomb potential in $\mathbb{R}^d$ for arbitrary d. We get

$$\frac{1}{(1 - 2r\cos\theta + r^2)^{(d-2)/2}} = \sum_{n=0}^{\infty} r^n \, C_n^{((d-2)/2)}(\cos\theta), \tag{4.214}$$

where $C_n^{(\lambda)}(z)$ are the *ultraspherical polynomials*, (a.k.a. *Gegenbauer polynomials*) of degree n [22–25]. The ultraspherical polynomials play in higher dimension the same role as the Legendre polynomials in $d = 3$. The LHS of (4.214) is the *generating function* of the Gegenbauer polynomials and may be taken as their definition; for alternate expressions see [22–25]. For $\lambda = 1$ (i.e. $d = 4$) we have

$$C_n^{(1)}(\cos\theta) \equiv U_n(\cos\theta) \tag{4.215}$$

i.e. the $d = 4$ ultraspherical polynomials coincide with the *Chebyshev polynomials of the second kind,* that is, with the characters of the irreducible representations of $SU(2)$ cf. Sect. 4.3. This, of course, reflects the fact that the unit sphere $S^3 \subset \mathbb{R}^4$ is the group $SU(2)$ itself.

Legendre polynomials have lots of others interesting and useful properties (recursion relations, integral representations, etc...) besides being solutions to the Schrödinger equation (4.200) which satisfy the orthogonality property (4.208). We refer the reader to [22–25].

Associated Legendre Polynomials
When $m > 0$ we set

$$F_{-m,\ell}(\theta) \equiv F_{m,\ell}(\theta) = P_\ell^m(\cos\theta), \tag{4.216}$$

and Eq. (4.196) reduces to

$$(1-x^2)\frac{\mathrm{d}^2 P_\ell^m(x)}{\mathrm{d}x^2} - 2x\frac{\mathrm{d} P_\ell^m(x)}{\mathrm{d}x} - \frac{m^2}{1-x^2}P_\ell^m(x) + \ell(\ell+1)P_\ell^m(x) = 0. \tag{4.217}$$

Luckily we don't need to solve this equation by hand. We know from Sect. 4.3 that $Y_{m+1,\ell}(\phi, \theta)$ is, up to normalization, just $L_+ Y_{m,\ell}(\phi, \theta)$. Using Eq. (4.188) this becomes a recursion relation for the $P_\ell^m(x)$ of the form

$$P_\ell^{m+1}(x) = c_{m+1}(1 - x^2)^{(m+1)/2}\frac{\mathrm{d}}{\mathrm{d}x}\left((1 - x^2)^{-m/2} P_\ell^m(x)\right) \tag{4.218}$$

where c_{m+1} are normalization coefficients to be chosen according to convenience. We use the formula (4.205) for $P_\ell^0(x)$ as the starting point of the recursion. Then, with a suitable choice of the c_m's, the solution becomes[31]

$$P_\ell^m(x) = \frac{(1-x^2)^{m/2}}{2^\ell\,\ell!}\,\frac{\mathrm{d}^{\ell+m}}{\mathrm{d}x^{\ell+m}}(x^2-1)^\ell, \qquad (4.219)$$

called the *associated Legendre polynomials*. For m even $P_\ell^m(x)$ is an actual polynomial in x, i.e. in $\cos\theta$, while when m odd it is $\sin\theta$ times a polynomial in $\cos\theta$. The associated Legendre polynomials satisfy the normalization condition

$$\int_{-1}^{+1} P_\ell^m(x)\,P_{\ell'}^m(x)\,\mathrm{d}x = \frac{2\,(\ell+m)!}{(\ell-m)!\,(2\ell+1)}\delta_{\ell,\ell'}, \qquad (4.220)$$

and the *spherical harmonics* are[32]

$$Y_{m,\ell}(\theta,\phi) = \frac{1}{\sqrt{4\pi}}\,\sqrt{\frac{(\ell-|m|)!\,(2\ell+1)}{(\ell+|m|)!}}\;\mathrm{e}^{\mathrm{i}m\phi}\,P_\ell^{|m|}(\cos\theta). \qquad (4.221)$$

The $Y_{m,\ell}(\theta,\phi)$'s are the wave-functions of our particle freely moving on S^2 in absence of potentials. They satisfy the proper $L^2(S^2)$ normalization condition

$$\int_{S^2} \sin\theta\,\mathrm{d}\theta\,\mathrm{d}\phi\;Y_{m,\ell}(\theta,\phi)^*\,Y_{m',\ell'}(\theta,\phi) = \delta_{\ell,\ell'}\,\delta_{m,m'}. \qquad (4.222)$$

An important property of the spherical harmonics is their transformation under parity $\mathscr{P}$: $(x_1,x_2,x_3) \to (-x_1,-x_2,-x_3)$ which written in polar coordinates becomes

$$\mathscr{P}\colon Y_\ell^m(\theta,\phi) \rightsquigarrow Y_\ell^m(\pi-\theta,\pi+\phi) = (-1)^\ell Y_\ell^m(\theta,\phi), \qquad (4.223)$$

a formula which easily follows from the properties of the associate Legendre polynomials.

Remark 4.9 The spherical harmonics are tabled in many books (see e.g. [26]) so you don't need to compute them by hand (but pay attention to the different conventions used by each author). They are also implemented in MATHEMATICA:

---

[31] Equation (4.219) is called the *Rodrigues formula* for the associate Legendre polynomials. We stress that the formula makes sense also for m negative, returning a function proportional to $P_\ell^m(x)$; indeed

$$\frac{(-1)^{-m}}{2^\ell\,\ell!}(1-x^2)^{-m/2}\frac{\mathrm{d}^{\ell-m}}{\mathrm{d}x^{\ell-m}}(x^2-1)^\ell = (-1)^m\frac{(\ell-m)!}{(\ell+m)!}P_\ell^m(x).$$

[32] The overall phases are, of course, conventional. In the literature the spherical harmonics are written with different phase conventions.

the command $\mathsf{SphericalHarmonicY}[\ell, m, \theta, \phi]$ calls the function $Y_{m,\ell}(\theta, \phi)$ and you may use it in your computations.

Remark 4.10 Wave-functions with small integral ℓ have got traditional names denoted by a lower-case Latin letter

$$
\begin{array}{c|ccccccccc}
\ell & 0 & 1 & 2 & 3 & 4 & 5 & 6 & 7 & \cdots \\
\hline
\text{name} & s & p & d & f & g & h & i & k & \cdots
\end{array}
\tag{4.224}
$$

So "s-wave" stands for a state of zero orbital angular momentum, "p-wave" for a state with $\ell = 1$, and so on.

4.6   Harmonic Polynomials vs. Spherical Harmonics in $\mathbb{R}^d$

To generalize the theory of spherical harmonics from the Representation Theory of $SO(3)$ in $L^2(\mathbb{R}^3)$ to the one of $SO(d)$ in $\mathbb{R}^d$, we use their relation with the more handy *harmonic polynomials* [24, 27, 28]. For the sake of comparison we start with the case $d = 3$ and then extend the results to general d.

Harmonic Polynomials in $\mathbb{R}^3$

Let x_1, x_2, x_3 be the Cartesian coordinates of $\mathbb{R}^3$. The homogeneous polynomials in x_1, x_2, x_3 of degree ℓ carry the (reducible) representation $\odot^\ell V_1$ of $SO(3)$ ($\equiv$ the *symmetric ℓ-th tensor power* of the 3-dimensional representation V_1—cf. Example 4.6) of dimension

$$
\dim \odot^\ell V_1 = \binom{\ell + 2}{2}.
\tag{4.225}
$$

The Laplacian $\Delta = -\partial_{x_i}\partial_{x_i}$ is $SO(3)$-invariant. Then the kernel of the map

$$
\Delta \colon \odot^\ell V_1 \to \odot^{\ell-2} V_1
\tag{4.226}
$$

is also a $SO(3)$-representation W_ℓ whose elements are called *harmonic polynomials* in $\mathbb{R}^3$ of degree ℓ. We **claim** that the map (4.226) is surjective: this is just the statement that the Poisson equation $\Delta \psi = p$ has a polynomial solution for all polynomial source p, a fact easily establish by induction on the degree of p. Therefore the representation W_ℓ has dimension

$$
\dim W_\ell = \binom{\ell + 2}{2} - \binom{\ell}{2} = 2\ell + 1
\tag{4.227}
$$

We **claim** that there is an isomorphism of $SO(3)$-modules

$$
W_\ell \simeq V_\ell,
\tag{4.228}
$$

i.e. W_ℓ is the irreducible $SO(3)$ representation of spin ℓ. To prove the **claim** it is enough to compute the eigenvalues of the Casimir operator L^2 acting on W_ℓ. Let $p(x_1, x_2, x_3) \in W_\ell$. Going to spherical coordinates

$$x_1 = r \cos\phi \sin\theta, \quad x_2 = r \sin\phi \sin\theta, \quad x_3 = r \cos\theta, \tag{4.229}$$

we have

$$p(x_1, x_2, p_3) = r^\ell f(\phi, \theta) \tag{4.230}$$

for a function $f(\phi, \theta) \in L^2(S^2)$ given by the restriction to the unit sphere of the polynomial function $p(x_i)$. We write L^2 for the differential operator (4.189) acting on (a dense domain in) $L^2(S^2)$. When written in spherical coordinates, the Laplacian Δ for the flat metric in $\mathbb{R}^3$ becomes

$$\Delta = -\frac{1}{r^2}\frac{\partial}{\partial r}r^2\frac{\partial}{\partial r} + \frac{1}{r^2}L^2. \tag{4.231}$$

Then, if p is a harmonic polynomial,

$$0 = \Delta p = -\frac{1}{r^2}\left(\frac{\partial}{\partial r}r^2\frac{\partial}{\partial r} - L^2\right)r^\ell f = -r^{\ell-2}\left(\ell(\ell+1)f - L^2 f\right), \tag{4.232}$$

so f is an eigenfunction of L^2 with eigenvalue $\ell(\ell+1)$, hence a spherical harmonic of angular momentum ℓ. Since the number of linearly-independent polynomials in the $SO(3)$-representation W_ℓ is equal to the dimension of the space of spherical harmonic $\{Y_{m,\ell}(\phi, \theta)\}$ with angular momentum ℓ, which span the irreducible $SO(3)$-module V_ℓ, the isomorphism (4.228) is established. We conclude:

> **Spherical Harmonics vs. Harmonic Polynomials**
> The $2\ell+1$ spherical harmonics $Y_{m,\ell}(\theta, \phi)$ are a basis of the restrictions to the unit sphere $S^2 \subset \mathbb{R}^3$ of the harmonic polynomials of degree ℓ

This also shows that the spherical harmonics have the form $e^{im\phi}$ times a polynomial in $\cos\theta$ and $\sin\theta$ as we found in the previous section by direct computation.

Harmonic Polynomials in $\mathbb{R}^d$

We wish to construct the eigenfunctions of the Laplacian $\Delta_{S^{d-1}}$ on the unit sphere $S^{d-1} \subset \mathbb{R}^d$—that is, to solve the Schrödinger equation for a free particle moving on the unit $(d-1)$-dimensional sphere S^{d-1} with its symmetric metric (cf. Fact 4.7).

From the previous paragraph it is evident that it is more convenient to study the algebra

$$\mathfrak{P}^{(d)} \equiv \mathbb{C}[x_1, \ldots, x_d] \tag{4.233}$$

of polynomial functions in the ambient space $\mathbb{R}^d$. We recall that the flat metric on $\mathbb{R}^d$ is related to the symmetric metric $\mathrm{ds}^2_{\mathrm{sphere}}$ on the unit sphere $S^{d-1} \subset \mathbb{R}^d$ by

$$\mathrm{ds}^2_{\mathrm{flat}} = \mathrm{d}r^2 + r^2 \mathrm{ds}^2_{\mathrm{sphere}}, \tag{4.234}$$

where $r^2 = x_i x_i$. The algebra $\mathfrak{P}^{(d)}$ is graded by the degree ℓ of the polynomials

$$\mathfrak{P}^{(d)} = \bigoplus_{\ell \geq 0} \mathfrak{P}_\ell^{(d)}, \tag{4.235}$$

and the dimensions of the homogeneous summands are

$$\dim \mathfrak{P}_\ell^{(d)} = \binom{d + \ell - 1}{\ell} \equiv \binom{\ell + d - 1}{d - 1}. \tag{4.236}$$

The homogeneous component $\mathfrak{P}_\ell^{(d)}$ of degree ℓ carries a representation of $SO(d)$ (cf. Example 4.6). Now consider the $SO(d)$-map

$$\Delta^{(d)} : \mathfrak{P}_\ell^{(d)} \to \mathfrak{P}_{\ell-2}^{(d)}, \qquad p \mapsto \Delta p, \tag{4.237}$$

with $\Delta^{(d)} = -\partial_{x_i} \partial_{x_i}$ the Laplacian in $\mathbb{R}^d$, and let $W_\ell^{(d)}$ be its kernel

$$W_\ell^{(d)} = \ker \Delta^{(d)} \subset \mathfrak{P}_\ell^{(d)}, \tag{4.238}$$

which is also a $SO(d)$ representation. Again the map is surjective, so[33]

$$N_\ell^{(d)} \stackrel{\mathrm{def}}{=} \dim W_\ell^{(d)} = \dim \mathfrak{P}_\ell^{(d)} - \dim \mathfrak{P}_{\ell-2}^{(d)} \equiv$$
$$\equiv \binom{\ell + d - 1}{d - 1} - \binom{\ell + d - 3}{d - 1}. \tag{4.239}$$

We **claim** that $W_\ell^{(d)}$ is an *irreducible representation* of $SO(d)$; in facts, as we are going to check, it is the irrepresentation of highest weight $[\ell, 0, \ldots, 0]$ (in the

---

[33] This easier-to-deduce formula agrees with the one given in §.9.3 of [24] as one sees using the recursion relations for the binomial coefficients.

Dynkin notation [20, 21, 29–31]). The dimension of this representation is $N_\ell^{(d)}$, while its quadratic Casimir[34] is

$$J^2\Big|_{W_\ell^{(d)}} \equiv \lambda_\ell = \ell(\ell + d - 2). \tag{4.240}$$

see e.g. [31]. Let us prove this formula. We use the fact that—just as it was the case for $\mathbb{R}^3$—the quadratic Casimir J^2 coincides with the Laplacian on the unit sphere $\Delta_{S^{d-1}}$ when acting on functions in $\mathbb{R}^d$. From the metric (4.234) we see that the Laplacians in $\mathbb{R}^d$ and S^{d-1} are related as

$$\Delta_{\mathbb{R}^d} = -\frac{1}{r^{d-1}} \frac{\partial}{\partial r} r^{d-1} \frac{\partial}{\partial r} + \frac{1}{r^2} \Delta_{S^{d-1}} \tag{4.241}$$

Applying this operator to a homogeneous polynomial $p \in W_\ell^{(d)}$ which has the form $p(x) = r^\ell Y(\theta_i)$, where the θ_i are coordinates on S^{d-1}, we get

$$0 = \Delta_{\mathbb{R}^d}\, p(x) = \left(-\frac{1}{r^{d-1}} \frac{\partial}{\partial r} r^{d-1} \frac{\partial}{\partial r} + \frac{1}{r^2} \Delta_{S^{d-1}} \right) r^\ell Y(\theta) =$$

$$= r^{\ell-2}\Big[-\ell(\ell + d - 2)Y + \Delta_{S^{d-1}} Y \Big]. \tag{4.242}$$

Definition 4.5 The elements $p \in W_\ell^{(d)}$ are called *harmonic polynomials of degree ℓ in $\mathbb{R}^d$*. They form the irreducible representation of $SO(d)$ with Dynkin label$[\ell, 0, \ldots, 0]$, Casimir eigenvalue $\ell(\ell + d - 2)$, and dimension $N_\ell^{(d)}$ (Eq. (4.239)).

The restriction of the harmonic polynomials to the unit sphere,

$$V^{(d)} = \bigoplus_{\ell=0}^{\infty} V_\ell^{(d)}, \qquad V_\ell^{(d)} \overset{\text{def}}{=} W_\ell^{(d)}\big|_{r=1}, \tag{4.243}$$

form an algebra of functions $Y \equiv p|_{r=1}$ on S^{d-1} which carry a $SO(d)$-representation, in facts denoting the irreducible $SO(d)$-modules by their Dynkin label

$$V^{(d)} = \bigoplus_{\ell=0}^{\infty} [\ell, 0, \ldots, 0] \tag{4.244}$$

---

[34] **Beware!** In the literature the quadratic Casimir is normalized in different ways. We take the normalization such that J^2 is equal to the Laplacian when acting on functions on S^{d-1}. For $d = 3$ this gives back the usual operator L^2. Other authors set the Casimir equal to the Hamiltonian of a particle with $m = \hbar = 1$ so their Casimir is $\frac{1}{2}\Delta$.

with quadratic Casimir

$$\ell(\ell + d - 2). \tag{4.245}$$

Definition 4.6 The elements of $V_\ell^{(d)}$ are the *d-dimensional spherical harmonics of angular momentum* ℓ. For $d = 3$ we recover the classical spherical harmonics in the usual three-dimensional space.

Proposition 4.5 *The space of spherical harmonic* $V^{(d)}$ *is dense in* $L^2(S^{d-1})$. *Hence* $V^{(d)}$ *contains a Hilbert basis of* $L^2(S^{d-1})$ *and we may see Eq.* (4.244) *as a decomposition of* $L^2(S^{d-1})$ *into finite-dimensional irreducible unitary representations of the symmetry group* $SO(d)$.

Proof By induction on degree the following isomorphism of $SO(d)$-modules is easily established

$$\mathbb{C}[r^2] \otimes \left(\bigoplus_{\ell \geq 0} W_\ell^{(d)} \right) \simeq \mathbb{C}[x_1, \ldots, x_d]. \tag{4.246}$$

(This is an elementary instance of **Theorem 6.3.3** of [32]). This means that the restrictions of harmonic polynomials on the surface $r = 1$ yield all the functions on the unit sphere which may be obtained by restricting arbitrary elements of the polynomial ring $\mathbb{C}[x_1, \ldots, x_d]$. The restriction $\mathbb{C}[x_1, \ldots, x_d]|_{r=1}$ is a ring of functions which separate the points of S^{d-1}, and since S^{d-1} is compact, the ring is dense in $L^2(S^{d-1})$ by the Stone-Weierstrass theorem. Since $\mathbb{C}[x_1, \ldots, x_d]|_{r=1} = V^{(d)}$, we get the claim. $\qquad\square$

Fact 4.8 *The* $SO(d)$ *representation* $V_\ell^{(d)}$ *decomposes into irreducible representations of the subgroup* $SO(d - 1) \subset SO(d)$ *as*

$$V_\ell^{(d)} = \bigoplus_{\ell'=0}^{\ell} V_{\ell'}^{(d-1)} \tag{4.247}$$

where all representations $V_{\ell'}^{(d-1)}$ *with* $\ell' \leq \ell$ *appear exactly once.*

Indeed

$$N_\ell^{(d)} = \sum_{\ell'=0}^{\ell} N_{\ell'}^{(d-1)}, \tag{4.248}$$

where we used the sum **0.15**.1 of [25].

4.6.1 Quantum Mechanics on Spheres S^n

The above Representation-Theoretical results have a clear physical meaning. Consider a quantum particle freely moving on a n-dimensional sphere, with the symmetric metric, in absence of potentials. The Hilbert space of the system is $L^2(S^n)$ and the Hamiltonian is $H \equiv \Delta$ (setting $\hbar = 2m = 1$ by a suitable choice of units). Then the discrete energy levels are (cf. Eq. (4.245))

$$E_\ell = \ell(\ell + n - 1) \qquad \ell = 0, 1, 2, \ldots, \tag{4.249}$$

with degeneracy

$$N_\ell^{(n+1)} = \binom{\ell + n}{n} - \binom{\ell + n - 2}{n}. \tag{4.250}$$

If one is interested in the explicit wave-functions in the Schrödinger representation, she has simply to compute the harmonic polynomials in $\mathbb{R}^{n+1}$, which is easily done by recursion in their degree ℓ, and then restrict them to the unit sphere.

Example 4.8 (Recovering the Ultraspherical Polynomials) To illustrate how one gets higher dimensional spherical harmonics from harmonic polynomials, let us consider the wave-functions of the states with angular momentum ℓ and "cylindrical symmetry" i.e. which are invariant under a subgroup

$$SO(n) \subset SO(n + 1) \tag{4.251}$$

of the isometry group. These cylindrically-symmetric wave-functions are the higher dimensional analogues of the $m = 0$ spherical harmonics on S^2 which are given by Legendre polynomials. We need to look for harmonic polynomials of the form $F(x_{n+1}, \sum_{i=1}^n x_i x_i)$ which are homogeneous of degree ℓ. They are of the form

$$F = (y + z^2)^{\ell/2} \, \Phi\left(\frac{z}{\sqrt{y + z^2}}\right) \qquad \text{where} \quad y = \sum_{i=1}^n x_i^2, \quad z = x_{n+1}. \tag{4.252}$$

The condition that F is a harmonic polynomial yields

$$0 = -\Delta F\Big|_{r=1} = \left(\frac{\partial^2 F}{\partial z^2} + 2n\frac{\partial F}{\partial y} + 4y\frac{\partial^2 F}{\partial y^2}\right)\Bigg|_{y=1-z^2} = \tag{4.253}$$

$$= (1 - z^2)\Phi''(z) - nz\,\Phi'(z) + \ell(\ell + n - 1)\Phi(z)$$

which is the ODE satisfied by the ultraspherical polynomials $C_\ell^{((n-1)/2)}(z)$ (here $z \equiv \cos\theta$ and $d \equiv n + 1$).

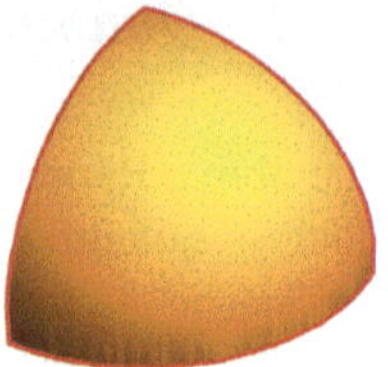

Fig. 4.1 The positive octant on the unit sphere, i.e. the equilateral spherical triangle whose angles and sides are all equal to $\pi/2$

4.6.2 Example: QM on a Spherical Triangle

The reader may prefer to skip this subsection in a first reading.

The method of the harmonic polynomials is quite powerful: it allows us to solve in an easy way many other systems, such as Quantum Mechanics on cosets of the form S^n/Γ where $\Gamma \subset O(n+1)$ is a discrete subgroup. As a further example we consider a quantum particle freely moving in an equilateral spherical triangle $T \subset S^2$ whose angle and sides are all equal to $\pi/2$, that is, our particle is constrained to move in the positive octant of the unit sphere (Fig. 4.1)

$$T = \left\{ (x, y, z) \in \mathbb{R}^3 : x^2 + y^2 + z^2 = 1, \ x \geq 0, \ y \geq 0, \ z \geq 0 \right\}. \tag{4.254}$$

The Hamiltonian $H = \Delta$ is the symmetric Laplacian on the sphere (equal to the $SO(3)$ quadratic Casimir operator) but now the wave-function satisfies the Dirichlet boundary condition

$$\psi(\phi, \theta)\Big|_{\partial T} = 0. \tag{4.255}$$

Of the $O(3)$ symmetries of the free Hamiltonian on the sphere, only the subgroup $\mathfrak{S}_3$ permuting the three vertices (and the three sides) of T is a symmetry. The spectrum of H is discrete and each energy eigenspace $\mathcal{H}_n$ decomposes into irreducible representations of $\mathfrak{S}_3$. We wish to compute the spectrum of H as well as the decomposition of each energy eigenspace into irreducible $\mathfrak{S}_3$-modules.

While the problem may be solved using more traditional methods, the technique of harmonic polynomials simplifies the task a lot. We may take care of the boundary condition (4.255) by the method of images, which amounts to look for eigenfunctions of Δ defined on the full S^2 with the property of having a simple zero along the three maximal circles

$$\theta = \pi/2, \quad \phi = 0, \quad \text{and} \quad \phi = \pi/2, \tag{4.256}$$

that is, we look for eigenfunctions of Δ on S^2 which are odd under the reflection with respect to these circles. The eigenfunctions of Δ are the restrictions to $r = 1$ of the harmonic polynomials. The image method then requires us to keep only the harmonic polynomials in x_1, x_2, x_3 which are *odd* under the flip $x_i \leftrightarrow -x_i$ of each one of the three variables, that is, that are the sum of monomials of odd degree in each variable. Hence the relevant polynomials have the structure

$$x_1 x_2 x_3 \, P(x_1^2, x_2^2, x_3^2) \tag{4.257}$$

for some polynomial $P(z_1, z_2, z_3)$. The number of linear independent polynomials of degree $\ell = 2k+3$ of the form (4.257) is equal to the number of linear independent polynomials in 3 variables of degree k, i.e.

$$\#\big(\text{degree } 2k+3 \text{ polynomials odd in } x_i \leftrightarrow -x_i \forall i\big) = \binom{k+2}{2} \tag{4.258}$$

The reflections are symmetries of the Laplacian Δ, so applying Δ to a polynomial of the form (4.257) we get a polynomial of the same form but with $k \rightsquigarrow k-1$. Since the map is surjective, the dimension of the *harmonic polynomials* of the form (4.257) is

$$\binom{k+2}{2} - \binom{k+1}{2} = k+1. \tag{4.259}$$

The harmonic polynomials of the form (4.257) are precisely the eigenfunction of Δ satisfying the boundary condition $\psi|_{\partial T} = 0$ with eigenvalue

$$E_k = \ell(\ell+1) = (2k+3)(2k+4), \quad k = 0, 1, 2, \ldots \tag{4.260}$$

We conclude that the spectrum of a quantum particle moving on the positive octant of the sphere (with $\hbar = 2m = 1$) is (4.260) while the k-th level has degeneracy $k+1$. In particular the ground state ($k=0$) is unique[35] with wave function

$$\psi_0 \propto x_1 x_2 x_3 = (\cos\phi \sin\theta)(\sin\phi \sin\theta)\cos\theta. \tag{4.261}$$

From this construction we see that the states in the k-th energy level transform under $\mathfrak{S}_3$ in the representation

$$\mathcal{H}_k \simeq V_{\mathrm{st}}^{\odot k} \ominus V_{\mathrm{st}}^{\odot(k-1)}, \quad \dim \mathcal{H}_k = k+1, \tag{4.262}$$

where V_{st} is the standard *reducible* 3-dimensional permutation representation. It is well known that the standard representation decomposes into irreducible representations as

$$V_{\mathrm{st}} = V_2 \oplus V_0 \tag{4.263}$$

---

[35] This result is consistent with the general theorem we shall prove in Sect. 6.13.

with V_0 the trivial representation. Let t be an indeterminate; for $g \in \mathfrak{S}_3$ we have the well-known character identity [18]

$$\sum_{n=0}^{\infty} t^n \operatorname{Tr}_{V_{\mathrm{st}}^{\odot n}}(g) = \det(1 - tg)^{-1} \tag{4.264}$$

where the determinant is computed in the 3-dimensional standard representation. We have 3 conjugacy classes, identity $\mathbf{1}$, cyclic permutation c, and 2-cycle s, with

$$\det(1 - t\mathbf{1})^{-1} = (1 - t)^{-3} \tag{4.265}$$

$$\det(1 - tc)^{-1} = (1 - t^3)^{-1} \tag{4.266}$$

$$\det(1 - ts)^{-1} = (1 - t)^{-1}(1 - t^2)^{-1} \tag{4.267}$$

Let $N_0(k)$, $N_1(k)$ and $N_2(k)$ be (respectively) the number of copies of the trivial, sign, and non-abelian irreducible representations of $\mathfrak{S}_3$ in the k-th energy level $\mathcal{H}_k$ ($k \geq 0$). The character table of $\mathfrak{S}_3$ then yields

$$\sum_{k \geq 0} t^k N_0(k) = \frac{1}{6}\left[\frac{1}{(1-t)^2} + \frac{2}{1+t+t^2} + \frac{3}{1-t^2}\right] \tag{4.268}$$

$$\sum_{k \geq 0} t^k N_1(k) = \frac{1}{6}\left[\frac{1}{(1-t)^2} + \frac{2}{1+t+t^2} - \frac{3}{1-t^2}\right] \tag{4.269}$$

$$\sum_{k \geq 0} t^k N_2(k) = \frac{1}{3}\left[\frac{1}{(1-t)^2} - \frac{1}{1+t+t^2}\right]. \tag{4.270}$$

4.7 Composition of Angular Momenta

In Quantum Mechanics the following situation is rather common. The system contains distinct sets of degrees of freedom and the rotation symmetry $SU(2)$ acts on each set separately. For instance a particle moving in $\mathbb{R}^3$ has both orbital and spin degrees of freedom, and the orbital angular momentum L_i generates rotations of the orbital degrees of freedom while the spin operators S_i rotate the spin ones. Another example is an interacting system composed of several particles moving in $\mathbb{R}^3$: each one has its own orbital angular momentum $L_i^{(\alpha)}$ and spin operator $S_i^{(\alpha)}$.

In these situations the Hilbert space is a tensor product of the Hilbert spaces (separable or finite-dimensional) associated to each set of degrees of freedom, and the a-th factor space carries its representation of $\mathfrak{su}(2)$ generated by its own angular momentum operators $J_i^{(a)}$ ($i = 1, 2, 3$)

$$J_i^{(a)}\left(|\psi_1\rangle \otimes \cdots \otimes |\psi_a\rangle \otimes \cdots \otimes |\psi_n\rangle\right) \equiv$$
$$\equiv |\psi_1\rangle \otimes \cdots \otimes \left(J_i^{(a)}|\psi_a\rangle\right) \otimes \cdots \otimes |\psi_n\rangle. \tag{4.271}$$

The operators $J_i^{(a)}$ have commutators ($\hbar = 1$)

$$[J_i^{(a)}, J_j^{(b)}] = i\delta^{ab}\epsilon_{ijk}J_k^{(a)}. \tag{4.272}$$

As far as the rotational degrees of freedom are concerned,[36] a state in the Hilbert space is labelled by the $2n$ quantum numbers (m_a, j_a) $(a = 1, \ldots, n)$ where m_a, is the eigenvalue of the a-th operator $J_3^{(a)}$ and $j_a(j_a + 1)$ is the eigenvalue of the a-th Casimir $(J^{(a)})^2 \equiv J_i^{(a)} J_i^{(a)}$ (not summed over a!). In facts the $2n$ operators

$$J_3^{(1)},\ (J^{(1)})^2,\ J_3^{(2)},\ (J^{(2)})^2,\ \cdots,\ J_3^{(n)},\ (J^{(n)})^2, \tag{4.273}$$

form a complete set of observables (relatively to the rotational degrees of freedom, cf. Footnote 36). In these situations, typically, only the *total* angular momentum

$$J_i = \sum_{a=1}^{n} J_i^{(a)}, \qquad [J_i, J_j] = i\epsilon_{ijk}J_k \tag{4.274}$$

is conserved by symmetry reasons. It is then natural to classify the states according to the representations of the conserved overall angular momentum, i.e. in terms of the eigenvalues of J_3 and $J^2 \equiv J_i J_i$, because the partial quantum numbers m_a change with time and hence are of little use in describing the dynamics of our intricate system. Mathematically this amounts to decomposing the tensor product $\mathfrak{su}(2)$-representation $V_{j_1} \otimes \cdots \otimes V_{j_n}$ into irreducible $\mathfrak{su}(2)$-representations V_j.

For simplicity we consider the case of just *two* angular momenta $J_i^{(1)}, J_i^{(2)}$; the general case of n angular momenta may be recovered by a recursive use of the formula for $n = 2$. The four observables $J_3^{(1)}, (J^{(1)})^2, J_3^{(2)}, (J^{(2)})^2$ form a complete set of commuting operators for the rotational degrees of freedom, and we can use their eigenvalues to label the elements of a basis of $V_{j_1} \otimes V_{j_2}$

$$|m_1, j_1; m_2, j_2\rangle \equiv |m_1, j_1\rangle \otimes |m_2, j_2\rangle \tag{4.275}$$

(we omit the non-rotational degrees of freedom from the notation). On the other hand the four operators $J_3, J^2, (J^{(1)})^2$ and $(J^{(2)})^2$ also form a complete set of

---

[36] Usually there are other degrees of freedom describing, say, translational and vibrational motions of the several subsystems.

commuting observables, so we may work in the orthonormal basis where they are diagonal

$$J_3|j_1, j_2; m, j\rangle = m|j_1, j_2; m, j\rangle, \tag{4.276}$$

$$J^2|j_1, j_2; m, j\rangle = j(j+1)|j_1, j_2; m, j\rangle, \tag{4.277}$$

$$(J^{(a)})^2|j_1, j_2; m, j\rangle = j_a(j_a+1)|j_1, j_2; m, j\rangle \quad a = 1, 2. \tag{4.278}$$

The two bases are related by a unitary transformation

$$|j_1, j_2; m, j\rangle = \sum_{m_1, m_2} |m_1, j_1; m_2, j_2\rangle\langle m_1, j_1; m_2, j_2|j_1, j_2; m, j\rangle \tag{4.279}$$

Definition 4.7 The matrix elements $\langle m_1, j_1; m_2, j_2|j_1, j_2; m, j\rangle$ of the unitary transformation connecting the above two bases of $V_{j_1} \otimes V_{j_2}$ are called *Clebsch-Gordan coefficients.*

A few properties of the Clebsch-Gordan coefficients are pretty obvious from their definition:

CG1 $\langle m_1, j_1; m_2, j_2|j_1, j_2; m, j\rangle$ vanishes unless $m = m_1 + m_2$;
CG2 by Eq. (4.147) it also vanishes unless

$$|j_1 - j_2| \le j \le j_1 + j_2 \quad \text{and} \quad j = j_1 + j_2 \bmod 1; \tag{4.280}$$

CG3 by a suitable choice of the phase of the basis vectors, we may take the Clebsch-Gordan coefficients to be *real;*
CG4 since the transformation is unitary (and real), we have:

$$\sum_{m_1, m_2} \langle m_1, j_1; m_2, j_2|j_1, j_2; m, j\rangle\langle m_1, j_1; m_2, j_2|j_1, j_2; m', j'\rangle = \delta_{m,m'}\delta_{j,j'} \tag{4.281}$$

$$\sum_{m,j} \langle m_1, j_1; m_2, j_2|j_1, j_2; m, j\rangle\langle m_1', j_1; m_2', j_2|j_1, j_2; m, j\rangle = \delta_{m_1,m_1'}\delta_{m_2,m_2'} \tag{4.282}$$

The identity

$$J_\pm|j_1, j_2; m, j\rangle =$$

$$= (J_\pm^{(1)} + J_\pm^{(2)}) \sum_{m_1, m_2} |m_1, j_1; m_2, j_2\rangle\langle m_1, j_1; m_2, j_2|j_1, j_2; m, j\rangle, \tag{4.283}$$

together with the known action of the lowering/raising operators $J_{\mp}$, $J_{\mp}^{(a)}$—cf. Eq. (4.137)—yields a recursion relation in m, m_1, m_2 for the Clebsch-Gordan coefficients which then can be computed explicitly. Their computation is a bit lenghty, but there is no need to carry out it by hand: there exist plenty of tables of Clebsch-Gordan coefficients. The coefficients are automatically produced by Mathematica with the instruction ClebschGordan[$\{j_1, m_1\}, \{j_2, m_2\}, \{j, m\}$].

Often the Clebsch-Gordan coefficients are written in terms of the *Wigner* $3j$ *coefficients:*

$$\begin{pmatrix} j_1 & j_2 & j \\ m_1 & m_2 & -m \end{pmatrix} \overset{\text{def}}{=} \frac{(-1)^{j_1-j_2+m}}{\sqrt{2j+1}} \langle m_1, j_1; m_2, j_2 | j_1, j_2; m, j \rangle. \qquad (4.284)$$

See [19]. There are similar formulae for the addition of 3 (resp. 4) angular momenta in terms of coefficients called $6j$-*symbols* (resp. $9j$-*symbols*). They may be computed using Mathematica when needed.

4.8   Rotation Operators and Quantum Mechanics on S^3

We wish to compute the matrix that represents the elements of the group $SU(2)$ in an irreducible representation V_j. The geometry of the group $SU(2)$ was discussed in [1] §.5.5 and is briefly reviewed in the **BOX** on page 262. We parametrize its elements by the Euler angles ϕ, θ, and ψ, that is, we write the general element of $SU(2)$ in the form

$$e^{i\phi J_3} e^{i\theta J_2} e^{i\psi J_3}, \qquad (4.285)$$

and we compute its matrix elements

$$R^j_{m,m'}(\phi, \theta, \psi) \overset{\text{def}}{=} \langle m, j | e^{i\phi J_3} e^{i\theta J_2} e^{i\psi J_3} | m', j \rangle \qquad (4.286)$$

that is,

$$e^{i\phi J_3} e^{i\theta J_2} e^{i\psi J_3} | m', j \rangle = \sum_{\substack{m=-j \\ m=j \bmod 1}}^{j} | m, j \rangle R^j_{m,m'}(\phi, \theta, \psi), \qquad (4.287)$$

in an arbitrary irreducible representation V_j, as a function of Euler's angles.

BOX: Geometry of S^3, Euler Angles, and All That

We we see $SU(2)$ as the group of unitary 2×2 matrices of determinant 1

$$u(z_1, z_2) \equiv \begin{pmatrix} z_1 & z_2 \\ -z_2^* & z_1^* \end{pmatrix} \in SU(2) \quad \text{with } (z_1, z_2) \in \mathbb{C}^2 \text{ and } |z_1|^2 + |z_2|^2 = 1$$

The last conditions shows that $SU(2) \simeq S^3$ as manifolds. We parametrize S^3 in terms of the three angles ϕ, θ, ψ by writing

$$z_1 = e^{i(\phi+\psi)/2} \cos(\theta/2), \qquad z_2 = e^{i(\phi-\psi)/2} \sin(\theta/2) \qquad (\spadesuit)$$

In terms of the spin generators $s_i = \sigma_i/2$ we then have

$$u(z_1, z_2) = e^{i\phi s_3} e^{i\theta s_2} e^{i\psi s_3}$$

a formula which identifies our three angles ϕ, θ, ψ with Euler's angles of the same name. Euler parametrized $SO(3) \simeq SU(2)/\mathbb{Z}_2$ and took all three angles to be periodic of period 2π

$$\phi \sim \phi + 2\pi, \quad \theta \sim \theta + 2\pi, \quad \psi \sim \psi + 2\pi.$$

However to cover the full $SU(2)$, which is a double cover of $SO(3)$, we must take one of the three angles, say ψ, to be periodic of period 4π. The rotation by 2π, $\exp(2\pi i s_3)$, is then the central element $-1 \in SU(2)$. The Maurer-Cartan form ([1] chap. 2) is then

$$u^{-1} du = \frac{i}{2} \begin{pmatrix} \cos\theta \, d\phi + d\psi & e^{-i\psi}(\sin\theta \, d\phi - i \, d\theta) \\ e^{i\psi}(\sin\theta \, d\phi + i \, d\theta) & -\cos\theta \, d\phi - d\psi \end{pmatrix}$$

and the bi-invariant metric is

$$ds^2 \equiv -2\,\mathrm{tr}(u^{-1} du \odot u^{-1} du) = d\theta^2 + \sin^2\theta \, d\phi^2 + (\cos\theta \, d\phi + d\psi)^2. \qquad (\ddagger)$$

Warning 6 The bi-invariant metric is unique only up to overall normalization. The metric ($\ddagger$) is written in the normalization which is natural as a Riemannian metric on the group manifold $SU(2)$ in the sense that the Laplacian Δ of ds^2 is equal to the Schrödinger representation of the *quadratic Casimir* of $\mathfrak{su}(2)$ acting on $L^2(SU(2))$, that is, the spectrum of Δ is $\ell(\ell+1)$ with $\ell \in \mathbb{N}/2$. However this is *not* the same normalization as the metric $ds_{S^3}^2$ on the unit sphere $S^3 \subset \mathbb{R}^4$ induced by the Euclidean metric in the ambient space. From eq. $\spadesuit$ we have

$$ds_{S^3}^2 \equiv |dz_1|^2 + |dz_2|^2 = \frac{1}{4} ds^2,$$

so the natural Laplacians on the unit sphere S^3 and on $SU(2)$ are related by

$$\Delta_{S^3} = 4\,\Delta_{SU(2)} \xrightarrow{\text{spectrum}} n(n+2) \text{ where } n = 2\ell \in \mathbb{N}$$

To compute the matrix elements (4.286) we change gears, and give an alternative physical interpretation to the functions $R^j_{m,m'}(\phi, \theta, \psi)$. As a Riemannian manifold, the group $SU(2)$ is the unit 3-sphere $S^3 \subset \mathbb{R}^4$ with the induced metric from the ambient Euclidean space $\mathbb{R}^4$ which is invariant for the ambient rotation group[37]

$$SO(4) \sim SU(2) \times SU(2). \tag{4.288}$$

The two $SU(2)$'s correspond, respectively, to the symmetry under left $g \mapsto hg$ and right $g \mapsto gh$ translations of the $SU(2)$ manifold. Changing normalization (cf. **BOX** on page 262), we write the bi-invariant metric in the form:

$$ds^2 = d\theta^2 + \sin^2\theta \, d\phi^2 + (\cos\theta \, d\phi + d\psi)^2 \tag{4.289}$$

which makes manifest that S^3 is a $U(1)$ principal bundle[38] over S^2 with non-trivial Chern class (*magnetic charge* in physics terms).

For the moment we put aside the problem of computing the matrix elements of the $SU(2)$ group elements, and study instead the Quantum Mechanics of a particle moving on S^3 with the "round" metric (4.289) in absence of external potentials. We already solved this dynamical problem, for spheres of arbitrary dimension n, in Sect. 4.6.1. What remains to be done is to specialize that result to the particular case $n = 3$ and then write the result in a slightly more explicit form. Notice that $n = 3$ is special because $SO(4)$ is the only orthogonal group which is *not* essentially simple, cf. Eq. (4.288).

The two $SU(2)$'s symmetries in (4.288) act by multiplication on the particle's Hilbert space $L^2(S^3)$, one on the left and one on the right:

$$\begin{aligned}
\text{left action:} \quad & \psi(g) \mapsto \psi(h^{-1}g) \\
\text{right action:} \quad & \psi(g) \mapsto \psi(gh).
\end{aligned} \tag{4.290}$$

We write L_i and R_i for the generators of the two copies of the $\mathfrak{su}(2)$ algebra. The two group actions commute and so do their respective Lie algebras

$$[L_i, L_j] = i\epsilon_{ijk}L_k, \quad [R_i, R_j] = i\epsilon_{ijk}R_k, \quad [L_i, R_j] = 0. \tag{4.291}$$

The $\mathfrak{su}(2)$ Casimir L^2 commutes with the group action, so its left and right actions should agree: then we have the constraint

$$L^2 = R^2. \tag{4.292}$$

Therefore a state of our particle in $L^2(S^3)$ is labelled by the *three* quantum numbers

$$m, \quad m', \quad j \tag{4.293}$$

---

[37] $\sim$ stands for *isogeny* i.e. equality up to finite subgroups.

[38] This non-trivial bundle $S^3 \to S^2$ with fiber S^1 is known as the *Hopf bundle*.

corresponding to the eigenvalues of the three operators L_3, R_3 and $L^2 \equiv R^2$ which form a complete system of commuting observables for the quantum particle on S^3. We work in a basis of wave-functions where these three operators are diagonal

$$\psi^j_{m,m'}(\phi, \theta, \psi) = \langle \phi, \theta, \psi | m, m', j \rangle. \tag{4.294}$$

Comparing the constructions, we see that the wave-functions $\psi^j_{m,m'}(\phi, \theta, \psi)$ and the matrix elements $R^j_{m,m'}(\phi, \theta, \psi)$ of the rotation operator are eigenfunctions of L_3, R_3 and L^2 with the same eigenvalues. Since these observables form a complete set, the eigenfunctions $\psi^j_{m,m'}(\phi, \theta, \psi)$ and $R^j_{m,m'}(\phi, \theta, \psi)$ should represent the same physical state. However their are normalized in different ways: their exact relation is

$$\psi^j_{m,m'}(\phi, \theta, \psi) = \sqrt{2j+1}\; R^j_{m,m'}(\phi, \theta, \psi). \tag{4.295}$$

This follows from Corollary 4.3 that we recast in the following

Lemma 4.3 *Let $\{|v_i\rangle\}_i$ be an orthonormal basis of an irreducible representation V of a compact Lie group G. Then*

$$\int_G \mathrm{d}g\, |\langle v_i | g | v_j \rangle|^2 = \frac{1}{\dim V} \quad \text{for all } i, j. \tag{4.296}$$

We know how to get the explicit form of the particle wave-functions $\psi^j_{m,m'}(\phi, \theta, \psi)$: just solve the Schrödinger equation. We also know that the Hamiltonian is the Casimir operator, whose eigenvalues are $j(j+1)$, and moreover we know that it coincides with the Laplacian Δ of the metric (4.289). Hence to get the matrix elements $R^j_{m_l,m_r}(\phi, \theta, \psi)$ we have only to solve the Schrödinger equation

$$\Delta \psi^j_{m_l,m_r} = j(j+1)\psi^j_{m_l,m_r} \tag{4.297}$$

in a basis where L_3 and R_3 are diagonal with eigenvalues m_l and m_r, respectively. The Laplacian can easily be written using the rules in Sect. 2.15. One gets

$$\Delta = -\frac{1}{\sin\theta}\frac{\partial}{\partial\theta}\sin\theta\frac{\partial}{\partial\theta} - \frac{1}{\sin^2\theta}\frac{\partial^2}{\partial\phi^2} - \frac{1}{\sin^2\theta}\frac{\partial^2}{\partial\psi^2} + 2\frac{\cos\theta}{\sin^2\theta}\frac{\partial^2}{\partial\psi\,\partial\phi}. \tag{4.298}$$

We separate variables in the Schrödinger equation (4.297) by writing the wave-function in the form

$$\begin{aligned}
\psi^j_{m_l,m_r}(\phi, \theta, \psi) &= \\
&= \mathrm{e}^{im_l\phi}\left(\cos\tfrac{\theta}{2}\right)^{m_l+m_r}\left(\sin\tfrac{\theta}{2}\right)^{m_l-m_r} Q^j_{m_l,m_r}(\cos\theta)\,\mathrm{e}^{im_r\psi}
\end{aligned} \tag{4.299}$$

in terms of functions $Q^j_{m_l,m_r}(\cos\theta)$ of the single variable $\cos\theta$ to be determined. The reader may easily verify the following differential identity

$$\left\{ -\frac{1}{\sin\theta}\frac{d}{d\theta}\sin\theta\frac{d}{d\theta} + \frac{m_l^2 + m_r^2}{\sin^2\theta} - 2\frac{m_l\,m_r\,\cos\theta}{\sin^2\theta} \right\}\left(\cos\tfrac{\theta}{2}\right)^{m_l+m_r}\left(\sin\tfrac{\theta}{2}\right)^{m_l-m_r} Q(\cos\theta) =$$

$$= -\left(\cos\tfrac{\theta}{2}\right)^{m_l+m_r}\left(\sin\tfrac{\theta}{2}\right)^{m_l-m_r}\left\{ \sin^2\theta\, Q''(\cos\theta) + 2\big(m_r - (m_l+1)\cos\theta\big)Q'(\cos\theta)+ \right.$$

$$\left. + \frac{1}{\sin^2\theta}\big(m_l^2 + m_r^2 - 2m_lm_r\cos\theta - m_l(m_l+1)\sin^2\theta\big)Q(\cos\theta)\right\} \qquad (4.300)$$

so that the Schrödinger equation (4.297) reduces to the following ODE for the function $Q(x) \equiv Q^j_{m_l,m_r}(x)$

$$(1 - x^2)Q''(x) + 2\big(m_r - (m_l+1)x\big)Q'(x) + (j - m_l)(j + m_l + 1)Q(x) = 0. \qquad (4.301)$$

This ODE is a special instance of the hypergeometric 2nd order ODE: setting

$$\alpha = m_l - m_r, \quad \beta = m_l + m_r, \quad n = j - m_l \qquad (4.302)$$

Eq. (4.301) becomes the celebrated *Jacobi equation* in its standard form

$$(1 - x^2)Q'' + [\beta - \alpha - (\alpha + \beta + 2)x]Q' + n(n + \alpha + \beta + 1)Q = 0, \qquad (4.303)$$

which has a solution in polynomials of degree n, the *Jacobi polynomials*

$$P_n^{(\alpha,\beta)}(x). \qquad (4.304)$$

See [22–24] for the magic properties of these hypergeometric polynomials. Here we only mention the *Rodrigues explicit formula* and their *orthogonality property* (which reflects the orthogonality of wave eigenfunctions in $L^2(S^3)$)

$$(1 - x)^\alpha (1 + x)^\beta P_n^{(\alpha,\beta)}(x) = \frac{(-1)^n}{2^n n!}\frac{d^n}{dx^n}\big[(1 - x)^{n+\alpha}(1 + x)^{n+\beta}\big] \qquad (4.305)$$

$$\int_{-\infty}^{+\infty} P_n^{(\alpha,\beta)}(x)\, P_m^{(\alpha,\beta)}(x)\,(1 - x)^\alpha (1 + x)^\beta dx =$$

$$\qquad (4.306)$$

$$= \frac{2^{\alpha+\beta+1}\Gamma(n + \alpha + 1)\Gamma(n + \beta + 1)}{(2n + \alpha + \beta + 1)\,\Gamma(n + \alpha + \beta + 1)\,n!}.$$

Using the identifications (4.302),(4.299) and (4.295) we get that the properly
normalized rotation matrices are

$$R^j_{m_l,m_r}(\phi, \theta, \psi) = \sqrt{\frac{(j+m_l)!\,(j-m_l)!}{(j+m_r)!(j-m_r)!}}\; \mathrm{e}^{\mathrm{i}m_l\phi+\mathrm{i}m_r\psi} \times$$

$$\times \left(\cos\tfrac{\theta}{2}\right)^{m_l+m_r} \left(\sin\tfrac{\theta}{2}\right)^{m_l-m_r} P^{(m_l-m_r,m_l+m_r)}_{j-m_l}(\cos\theta). \tag{4.307}$$

We summarize the Quantum Mechanics of a free quantum particle moving in
$S^3 \simeq SU(2)$ in the following

Theorem 4.9

(1) *The Hilbert space $L^2(SU(2))$ decomposes in irreducible representation of the
$SU(2)$ group acting on itself by left translations as*

$$L^2(SU(2)) = \bigoplus_{j=\frac{1}{2}\mathbb{N}} V_j^{\oplus \dim V_j}, \tag{4.308}$$

*that is, every irreducible representation V_j of $SU(2)$ appears in $L^2(SU(2))$
with a multiplicity equal to its dimension $2j + 1$.* **(2)** *The matrix elements
$R^j_{m_l m_r}(g)$ of $g \in SU(2)$ form a complete orthogonal system in $L^2(SU(2))$.*
(3) *In terms of the irreducible representations of $SU(2)_L \times SU(2)_R$*

$$L^2(SU(2)) = \bigoplus_{j=\frac{1}{2}\mathbb{N}} V_j^L \otimes V_j^R \tag{4.309}$$

where each irreducible representation $V_j^L \otimes V_j^R$ appears exactly once.

Remark 4.11 The wave-functions $\psi^j_{m_l,m_r}(\phi, \theta, \psi)$ are polynomials in $\cos(\phi/2)$,
$\sin(\phi/2)$, $\cos(\theta/2)$, $\sin(\theta/2)$, $\cos(\psi/2)$, and $\sin(\psi/2)$. This follows from their
relation with the harmonic polynomials in $\mathbb{R}^4$ as in Sect. 4.6.

4.9 QM on Lie Groups: The Peter-Weyl Theorem

In the discussion of previous section we may replace $SU(2)$ with any connected
compact Lie group G, and study the motion of a quantum particle on the group
manifold G equipped with the (unique up to scale) bi-invariant symmetric metric.
The Hilbert space is $L^2(G)$ and we have a unitary *left* action $G \curvearrowright L^2(G)$ given by

$$\psi(g) \to \psi(h^{-1}g) \qquad h \in G, \tag{4.310}$$

as well as an action by right translation

$$\psi(g) \rightarrow \psi(gh) \qquad h \in G. \tag{4.311}$$

We know that $L^2(G)$ is the direct sum of finite-dimensional irreducible representations of the left action (4.310). Let us show first that *all* irreducible unitary representations V_a of G appear in $L^2(G)$ with multiplicity at least $\dim V_a$.

Fix an arbitrary irreducible representation V_a of dimension d_a and let $\{|i; a\rangle\}$ $(i = 1, \ldots, \dim V_a)$ be an orthonormal basis. We consider the matrix elements of the group

$$R^a(g)_{ij} = \langle i; a|g|j; a\rangle \qquad g \in G \tag{4.312}$$

which satisfy the group law

$$R^a(hg)_{ij} = R^a(h)_{ik}\, R^a(g)_{kj} \qquad h, g \in G. \tag{4.313}$$

Now fix the second index j and consider the $\dim V_a$ functions on G

$$\psi^{a,j}(g)_i \equiv \langle i; a|g|j; a\rangle \quad i = 1, \ldots, d_a. \tag{4.314}$$

As functions on G they satisfy

$$h \cdot \psi^{a,j}(g)_i = \psi^{a,j}(h^{-1}g)_i = R^a_{ik}(h^{-1})\, \psi^{a,j}(g)_k \tag{4.315}$$

and then

$$h_1 h_2 \cdot \psi^{a,j}(g)_i = R^a_{ik}(h_2^{-1})h_1 \cdot \psi^{a,j}(g)_k = R^a_{ik}(h_2^{-1})R^a_{kl}(h_1^{-1})\psi^{a,j}(g)_l$$

$$= R^a_{ik}((h_1 h_2)^{-1})\psi^{a,j}(g)_k \tag{4.316}$$

i.e. for each fixed second index j the d_a functions $\psi^{a,j}(g)_i$ $(i = 1, \ldots, d_a)$ transform in the dual representation $V_a^\vee$. The d_a representations

$$\psi^{a,1}(g), \cdots, \psi^{a,d_a}(g) \tag{4.317}$$

associated to different second indices j are *linear independent*, otherwise there would be a non-trivial invariant subspace of V_a, while V_a is irreducible. The functions $\psi^{a,j}(g)_i$ are in $L^2(G)$: indeed they are bounded since

$$\sum_i |\psi^{a,j}(g)_i|^2 = 1 \tag{4.318}$$

by unitarity, while the volume of G is finite (normalized to 1). So $L^2(G)$ contains at least d_a copies of the representation $V_a^\vee$. Since V_a was arbitrary, this holds for all irreducible unitary representations of G. Then we can write

$$L^2(G) = \left(\widehat{\bigoplus_a} V_a^{\oplus d_a} \right) \oplus W \tag{4.319}$$

where W is the orthogonal complement. W also decomposes into irreducible unitary representations of G. Let $\psi(g)_i$ be a vector of functions in W transforming in the representation $V_a^\vee$

$$\psi(h^{-1}g)_i = R_{ij}^a(h^{-1})\,\psi(g)_j. \tag{4.320}$$

Then we have

$$\psi(g)_i = R_{ij}^a(g)\,\psi(1)_j \tag{4.321}$$

i.e. the function $\psi(g)_i$ is a linear combination of matrix elements of g and belongs to the first summand in the RHS of (4.319). Hence $W = 0$. These arguments lead us to the celebrated

Theorem 4.10 (Peter-Weyl: First Version) *G a compact Lie group.* **(1)** *The Hilbert space $L^2(G)$ decomposes in irreducible unitary representations of G*

$$L^2(G) = \widehat{\bigoplus_a} V_a^{\dim V_a} \tag{4.322}$$

where all (pairwise non-isomorphic) irreducible representations V_a appear with a multiplicity equal to their dimension $\dim V_a$. **(2)** *The rescaled matrix elements of the group action*

$$\sqrt{\dim V_a}\, R_{ij}^a(g) \tag{4.323}$$

for all irreducible representations V_a and all indices $i, j = 1, \ldots, \dim V_a$ form an orthonormal Hilbert basis for $L^2(G)$.

The last statement follows from Corollary 4.3.

Example 4.9 In the previous section we got the decomposition of $L^2(SU(2))$ into irreducible representations solving explicitly the Schrödinger equation by separation of variables. The results we got match the prediction of the Peter-Weyl theorem.

Remark 4.12 When G is simple, there is a unique (up to overall normalization) Hamiltonian quadratic in momenta which is bi-invariant under left and right translation. In the language of Representation Theory it is the quadratic Casimir operator of the Lie algebra $\mathfrak{g}$, while from the point of view of Riemannian geometry it is the Laplacian of the invariant metric. Representation Theory yields explicit formulae for the quadratic Casimirs of all finite-dimensional representations of compact Lie groups. These formulae, together with the Peter-Weyl theorem, yield a complete solution of the Schrödinger equation for a particle moving on the manifold G in absence of potentials. We refer the reader to textbooks on Lie groups and algebras for the details.

We may state the **Theorem** in a smarter way. Recall that the isometry group of G (with the symmetric metric) is $G_L \times G_R$ where G_L (resp. G_R) acts by left (resp. right) translation.

Theorem 4.11 (Peter-Weyl: Second Version) *$L^2(G)$ decomposes into irreducible representations of $G_L \times G_R$ in the form*

$$L^2(G) = \widehat{\bigoplus_a} V_a \otimes V_a^{\vee} \tag{4.324}$$

The sum is over all isoclasses of irreducible representations of G. The group G_L (resp. G_R) acts on the first (resp. second) index of the matrix elements of the irreducible representations of G.

We cannot resist quoting a few general consequences of the Peter-Weyl theorem which are frequently used in Theoretical Physics. We leave the proofs to the diligent reader as a nice exercise; less diligent readers may have a look to §. III.4 of [33].

Corollary 4.5

(1) *Every compact Lie group G admits a faithful unitary representation, that is, all compact Lie group G is a closed subgroup of $U(n)$ for some n.*

(2) *Let V a faithful representation of the compact Lie group G. Every irreducible G-representation W is a subrepresentation*

$$W \subset (\otimes^k V) \otimes (\otimes^l V^{\vee}) \tag{4.325}$$

for some k, l. **(3)** *Let $H \subset G$ be a closed subgroup of the compact Lie group G. Every irreducible representation of H is contained in the restriction to H of an irreducible representation of G.* **(4)** *Let $H \subset G$ be a closed subgroup. Then there exists a representation V of G and an element $v \in V$ such that the isotropy subgroup I_v of v*

$$I_v \stackrel{\text{def}}{=} \{g \in G : gv = v\} \subset G \tag{4.326}$$

is exactly H.

4.10 Quantum Mechanics on Compact Symmetric Spaces

The spheres and the compact Lie group are examples of compact symmetric spaces. There are two kinds of irreducible compact symmetric Riemannian manifolds [3, 4]:

Type I   are of the form G/K where G is a compact simple Lie group and $K \subset G$ is the Lie subgroup fixed by a Cartan involution $\theta : G \to G$;

Type II   are the compact simple Lie groups H seen as the coset

$$(H \times H)/H_{\text{diag}} \equiv G/K \qquad (4.327)$$

where

$$H_{\text{diag}} \equiv H \subset H \times H \qquad (4.328)$$

is the diagonal subgroup.

E.g. the d-sphere $S^d = SO(d+1)/SO(d)$ is of Type I. All symmetric spaces G/K have a G-invariant metric which is unique up to scale. Extending the analysis of the previous sections, we may study the Quantum Mechanics of a particle freely moving on any compact symmetric space G/K.

More generally, we may consider a Quantum Mechanical system whose configuration space $\mathcal{M}$ is an arbitrary G-homogeneous space G/H (G connected, compact, and semisimple) where the subgroup $H \subset G$ needs not be the fixed locus of a Cartan involution. Again on G/H we have (several) G-invariant Riemannian metrics and we write $L^2(G/H)$ for the Hilbert space of measurable functions on G/H equipped with the inner product defined by an invariant metric (cf. Chap. 2). As Hamiltonian we take the quadratic Casimir J^2 of the action of $\mathfrak{g}$ on $L^2(G/H)$.

We may see the functions on G/H as the functions on G which are invariant under the right action of $H \subset G$. Hence from the Peter-Weyl theorem we have [12]

$$L^2(G/H) = \widehat{\bigoplus_a} V_a \otimes (V_a^\vee)^H \qquad (4.329)$$

where for all G-module V the symbol V^H stands for the vector subspace $V^H \subset V$ which is left invariant by the subgroup H. We conclude that each irreducible G-representation V_a appears with a multiplicity equal dim V_a^H.

When the compact coset G/K is *symmetric*, one can show that dim V_a^K *is either zero or one* [12], so the Hilbert space of a particle freely moving on G/K is a direct sum of non-isomorphic irreducible G-representations all with multiplicity 1

$$L^2(G/K) = \widehat{\bigoplus_{V_a^K \neq 0}} V_a . \qquad (4.330)$$

Example 4.10 In the case of $S^n = SO(n+1)/SO(n)$ we saw that the Hilbert space was the sum of one copy of each irrepresentation with Dynkin label $[\ell, 0, \ldots, 0]$. Seeing $SO(n)$ as the group which rotates the first n components of the vector $(x_1, \ldots, x_{n+1})$ the $SO(n)$-invariant vector in $[\ell, 0, \ldots, 0]$ is the class of $(x_{n+1})^\ell$ in $\mathbb{C}[x_1, \ldots, x_{n+1}]/(x_i x_i)$.

Example 4.11 We see the compact Lie group G as the symmetric space $(G \times G)/G$. An irreducible $G \times G$-representation $V_a \otimes V_b$ contains a G-invariant vector if and only if $V_b \simeq V_a^\vee$, and in this case contains it with multiplicity 1 (cf. Footnote 14). This gives back our previous result about Quantum Mechanics on compact groups (cf. Theorem 4.11).

4.11 Galilei Symmetry

"Non-relativistic" mechanics is invariant under its own relativity principle, the *Galileian* one. The Noether conserved charge associated to the Galileian invariance was worked out, in the classical context, in *Example 3.10* of [1]

$$\mathbf{K} = t \sum_i \mathbf{p}_i - \sum_i m_i \mathbf{x}_i, \qquad (4.331)$$

where the sum is over all material points of mass m_i, position $\mathbf{x}_i \in \mathbb{R}^3$, and momentum $\mathbf{p}_i \in \mathbb{R}^3$. Conservation of $\mathbf{K}$ in time, $\dot{\mathbf{K}} = 0$, is equivalent to the theorem of the center-of-mass. The conserved vector $\mathbf{K} = (K_i)$ is called the *Galileian boost*.

The boost $\mathbf{K}$ is also conserved in non-relativistic closed *quantum* systems since the center-of-mass still moves at constant speed. More abstractly, we take the conservation of the boost $\mathbf{K}$ as the *definition* of a Galilei invariant quantum system.

As a generator of a quantum symmetry, the boost Hermitian operator

$$\mathbf{K} = t \sum_i \mathbf{p}_i - \sum_i m_i \mathbf{q}_i \qquad (4.332)$$

differs in two (related) ways from the ones considered in the previous sections:

(i) $\mathbf{K}$ is explicitly time-dependent,
(ii) $\mathbf{K}$ does not commute with the Hamiltonian H but rather (in units $\hbar = 1$)

$$0 = i\frac{d\mathbf{K}}{dt} = [\mathbf{K}, H] + i\frac{\partial \mathbf{K}}{\partial t} \quad \Rightarrow \quad [\mathbf{K}, H] = -i\mathbf{P} \equiv -i\sum_i \mathbf{p}_i. \qquad (4.333)$$

The Hamiltonian H, the boost $\mathbf{K} = (K_i)$, and the *total momentum* $\mathbf{P} = (P_i) \equiv \sum_i \mathbf{p}_i$ satisfy the commutator Lie algebra

$$[H, \mathbf{P}] = 0, \qquad\qquad [H, \mathbf{K}] = i\mathbf{P},$$
$$[P_i, P_j] = [K_i, K_j] = 0, \quad [K_i, P_j] = -i\delta_{ij} M, \tag{4.334}$$

where

$$M = \sum_i m_i \tag{4.335}$$

is the *total mass* of the system, which is a *central element* in the Lie algebra (4.334). We stress that the symmetry Lie algebra (4.334) is *not* semisimple, and then when $M \neq 0$ all its unitary representations are infinite-dimensional.

Galilei Boost of the Wave-Function We are interested in the transformation of the Schrödinger wave-function under a boost, i.e. in the relation between the wave function $\psi(\mathbf{x}_1, \ldots, \mathbf{x}_n)$ of a quantum state in our inertial frame and the wave function

$$\psi(\mathbf{x}_1, \ldots, \mathbf{x}_n)^{\mathbf{v}} \equiv \langle \mathbf{x}_1, \ldots, \mathbf{x}_n | \exp(i\mathbf{v}\cdot\mathbf{K}) | \psi, t \rangle \tag{4.336}$$

of the *same* state $|\psi, t\rangle$ as seen in an inertial frame which moves at relative velocity $\mathbf{v}$ with respect to our frame.

Recall from the BCH formula (Sect. 2.6) that, when $[A, B]$ is a c-number,

$$e^{i(A+B)} = e^{iA} e^{iB} e^{[A,B]/2}. \tag{4.337}$$

Hence, for all 3-vector $\mathbf{v}$

$$\exp(i\mathbf{v}\cdot\mathbf{K}) = e^{-i\sum_i m_i \mathbf{v}\cdot\mathbf{q}_i} e^{-iMt\, \mathbf{v}\cdot\mathbf{v}/2} e^{it\, \mathbf{v}\cdot\mathbf{P}}, \tag{4.338}$$

and the wave-function transforms under a change of inertial frame as

$$\psi(\mathbf{x}_1, \ldots, \mathbf{x}_n, t) \rightsquigarrow e^{i\mathbf{v}\cdot\mathbf{K}} \psi(\mathbf{x}_1, \ldots, \mathbf{x}_n, t) =$$
$$= \exp\left(-i\sum_i m_i \mathbf{v}\cdot\mathbf{x}_i - \frac{i}{2} Mt\, \mathbf{v}\cdot\mathbf{v} \right) \psi(\mathbf{x}_1 + t\mathbf{v}, \ldots, \mathbf{x}_n + t\mathbf{v}, t). \tag{4.339}$$

By construction $\exp(i\mathbf{v} \cdot \mathbf{K})$ is a time-dependent unitary transformation which is a dynamical symmetry iff the system is invariant under Galileian transformations. In particular the observable "probability density" transforms in the physically expected way

$$\left|\psi(\mathbf{x}_1,\ldots,\mathbf{x}_n,t)\right|^2\Big|_{\substack{\text{boosted}\\\text{frame}}} = \left|\psi(\mathbf{x}_1+t\mathbf{v},\ldots,\mathbf{x}_n+t\mathbf{v},t)\right|^2\Big|_{\substack{\text{original}\\\text{frame}}}. \qquad (4.340)$$

The Total Mass as a Superselected Central Charge In the above discussion we treated the total mass M as a numerical parameter. In actual fact M is a physical observable, and hence should be promoted to an *operator* acting on the Hilbert space $\mathcal{H}$. However, it should act as a c-number on each irreducible representation of the Galilei Lie algebra (4.334) since it is a central element.

We claim that the total mass M is a *superselected conserved charge*[39] in all Galilei invariant quantum systems. Indeed, consider a system with just one-particle and suppose we try to superimpose two states with wave-functions $\psi_1(\mathbf{x},t)$ and $\psi_2(\mathbf{x},t)$ where the particle has different masses $M_1 \neq M_2$. In the original frame we have a probability density

$$\left|c_1\,\psi_1(\mathbf{x},t) + c_2\,\psi_2(\mathbf{x},t)\right|^2, \qquad (4.341)$$

while in the boosted frame

$$\left|c_1\,\psi_1(\mathbf{x}+t\mathbf{v},t) + c_2\,e^{i(M_1-M_2)(\mathbf{v}\cdot\mathbf{x}-\frac{1}{2}t\mathbf{v}^2)}\,\psi_2(\mathbf{x}+t\mathbf{v},t)\right|^2 \qquad (4.342)$$

so that the experiments performed in the two frames will give physically contradictory results, unless $c_1 = 0$ or $c_2 = 0$.

More generally, Galileian relativity requires translations and boosts to commute *when acting on observables* in order not to produce conflicting measurements. This implies that all observables should commute with the total mass. For instance, the canonical operators $\mathbf{q}_i$, $\mathbf{p}_i$, and hence all operators acting on $L^2(\mathbb{R}^{3n})$, do commute with M.

Since all observables commute with M, from **Superposition Principle 6** we conclude that sectors of the Hilbert space with definite total mass are *superselection sectors,* and we are forbidden to superpose states of different total mass.

Center-of-Mass Theorem We introduce the *position of the center-of-mass vector*

$$\mathbf{R} \stackrel{\text{def}}{=} \frac{\sum_i m_i \mathbf{q}_i}{M} \qquad (4.343)$$

---

[39] M is conserved since it is the commutator of two conserved quantities, Eq. (4.334).

i.e. the operator which measures the mean position of the material points weighted with their masses. $\mathbf{R}$ is the canonically conjugate operator to the total momentum $\mathbf{P}$

$$[R_i, P_j] = i\hbar\,\delta_{ij}. \tag{4.344}$$

Theorem 4.12 (Quantum Theorem of the Center-of-mass) *Assume the symmetry algebra (4.334), with H time-independent and $\mathbf{K}$ conserved, Eq. (4.333), while the total mass M is a non-zero constant. Then*

(a) the total momentum $\mathbf{P}$ is conserved;
(b) the Hamiltonian has the form

$$H = \frac{\mathbf{P}^2}{2M} + H_{\mathrm{cm}} \tag{4.345}$$

where the center-of-mass reduced Hamiltonian H_{cm} contains only operators which commute with the center-of-mass operators $\mathbf{R}, \mathbf{P}$;
(c) the position of the center-of-mass moves along a straight line at constant speed

$$\mathbf{R} = \mathbf{v}\,t + \mathbf{R}_0 \quad where \quad \mathbf{v} = \frac{\mathbf{P}}{M}. \tag{4.346}$$

Proof $\mathbf{P}$ is the commutator of two conserved quantities, hence conserved. Then $[\mathbf{P}, H] = 0$ so H does not contain the conjugate variable $\mathbf{R}$. Then

$$-\,i\mathbf{P} = [\mathbf{K}, H] = -M[\mathbf{R}, H] = -iM\frac{\partial H}{\partial \mathbf{P}} \tag{4.347}$$

Integrating we get (4.345). Finally

$$0 = \dot{\mathbf{K}} = \mathbf{P} - M\dot{\mathbf{R}}, \tag{4.348}$$

together with the conservation of the total momentum P, yields (4.346). $\square$

4.12 Identical Particles: Bose and Fermi Statistics

We consider the following physically important situation: our quantum system is made of n *identical* particles, all with the same mass m and spin s (as well as any other charge they may carry), which move in the ordinary physical space $\mathbb{R}^3$. A priori the Hilbert space $\mathcal{H}_n$ of the n-particle system is the tensor product of n copies of the one-particle Hilbert space $\mathcal{H}_1$

$$\mathcal{H}_n = \otimes^n\mathcal{H}_1, \qquad \mathcal{H}_1 \simeq \mathbb{C}^{2s+1} \otimes L^2(\mathbb{R}^3), \tag{4.349}$$

and we write the Schrödinger wave-functions in the form

$$\psi(\boldsymbol{x}_1, \ldots, \boldsymbol{x}_n)_{m_1,\ldots,m_n} \tag{4.350}$$

where the index $-s \leq m_i \leq s$ labels the different spin states of the i-th particle in a basis where S_3 is diagonal. We assume that the Hamiltonian H is invariant under arbitrary permutations of the degrees of freedom of the n identical particles, so that the permutations $\pi \in \mathfrak{S}_n$ are symmetries of the Hamiltonian, and the Hilbert space $\mathcal{H}_n$ decomposes into irreducible unitary representations of the symmetric group $\mathfrak{S}_n$. We wish to motivate that only some very specific representations may appear in the physical Hilbert space.[40]

In classical physics the particles are always *distinguishable* (in principle): indeed their trajectories are well defined, so we may label them at some reference time t_0, then follow the motion of each one of them to keep track of their labels, and distinguish them at any other time t. In Quantum Physics the trajectories are not defined and the story is quite different: we have the

Fundamental Principle *Equal quantum particles are indistinguishable even in principle.*

The deep reason beyond this **Principle** lays in Relativistic Quantum Physics. Albeit we are working in the non-relativistic *approximation*—where the velocities are much smaller than the speed of light—all the real world systems at the end should be both *relativistic* and *quantum:* the compatibility conditions of these two fundamental principles lead to the above crucial **Principle**, which then holds for *all* real systems. Said in plain English: indistinguishability is needed in order to make the field-to-particle duality of Relativistic Quantum Physics to make sense. In Chap. 1 we exploited the duality between the Maxwell theory of light and its description in terms of photons to argue for the superposition principle. A moment thought shows that this duality would be untenable if we could distinguish between two photons with the same momentum and polarization. This point is especially transparent in the second quantized formulation of the field-particle duality (see [18] chap. 3). Thus, albeit a formal math proof of the **Fundamental Principle** is part of

---

[40] **Warning!** The discussion in the text is not fully general. Implicitly it assumes a *strong locality* property of our system which, roughly speaking, says that for a dense subspace of physical states we may take the wave-functions to be univalued and continuous everywhere in the physical space $\mathcal{M}$. We note that in the most general situation the Hilbert space carries a representation of the fundamental group $\pi_1(C_n)$ of the *configuration space*

$$C_n = \left\{(x_1, x_2, \ldots, x_n) \in \mathcal{M}^n : x_i \neq x_j \text{ for } i \neq j\right\}/\mathfrak{S}_n.$$

When the space $\mathcal{M}$ is contractible and $\dim \mathcal{M} \geq 3$, $\pi_1(C_n)$ is just the symmetric group $\mathfrak{S}_n$, but $\pi_1(C_n)$ is an infinite braid group when $\dim \mathcal{M} = 2$. Thus *exotic statistic* exists and the zoo is particularly reach in two-spatial dimensions. Quantum objects which obey an exotic statistic are called *anyons* and are important for quantum computations [34].

relativistic Quantum Field Theory, informally we may see the above statement as part of the superposition principle valid for general Quantum Physics.

Let the state of our system of n identical particles moving in $\mathbb{R}^3$ be described by the Schrödinger wave function

$$\psi(x_1, \ldots, x_n)_{m_1,\ldots,m_n} \in \otimes^n\big(\mathbb{C}^{2s+1} \otimes L^2(\mathbb{R}^3)\big). \tag{4.351}$$

Let $\pi \in \mathfrak{S}_n$ be any permutation. The wave-function obtained by permuting the identical particles ($\equiv$ permuting their orbital and spin degrees of freedom)

$$\psi(x_{\pi(1)}, \ldots, x_{\pi(n)})_{m_{\pi(1)},\ldots,m_{\pi(n)}} \tag{4.352}$$

should represent the *same state* since the particles are indistinguishable. Two (normalized) wave-functions describe the same state iff they differ only by a phase factor. Then, for all $\pi \in \mathfrak{S}_n$ we must have

$$\psi(x_{\pi(1)}, \ldots, x_{\pi(n)})_{m_{\pi(1)},\ldots,m_{\pi(n)}} = \lambda(\pi)\,\psi(x_1, \ldots, x_n)_{m_1,\ldots,m_n} \tag{4.353}$$

for some map $\lambda\colon \mathfrak{S}_n \to U(1)$. Consistency requires

$$\lambda(\pi_1\pi_2) = \lambda(\pi_1)\lambda(\pi_2) \quad \forall\, \pi_1, \pi_2 \in \mathfrak{S}_n, \tag{4.354}$$

and hence the map

$$\lambda\colon \mathfrak{S}_n \to U(1) \tag{4.355}$$

must be *a one-dimensional unitary representation of the permutation group* $\mathfrak{S}_n$. It is well known that $\mathfrak{S}_n$ has exactly *two* one-dimensional representations:[41]

(1) the *trivial* one $\lambda(\pi) = 1$ for all π;
(2) the *sign* one $\lambda(\pi) = (-1)^{|\pi|}$ which is one (minus one) is π is the product of an even (resp. odd) number of transpositions.

We conclude that the wave-function (4.351) is either *even* or *odd* under transposition of particles, that is,

$$\psi(x_1, \ldots, x_{i+1}, x_i, \ldots x_n)_{m_1,\ldots,m_{i+1},m_i,\ldots,m_n} = $$
$$= \pm\,\psi(x_1, \ldots, x_i, x_{i+1}, \ldots x_n)_{m_1,\ldots,m_i,m_{i+1},\ldots,m_n}. \tag{4.356}$$

---

[41] $\mathfrak{S}_n$ has a presentation in terms of $n-1$ generators $\sigma_1, \ldots, \sigma_{n-1}$ (transpositions) subjected to the relations

$$\sigma_i^2 = 1, \quad \sigma_i\sigma_{i+1}\sigma_i = \sigma_{i+1}\sigma_i\sigma_{i+1}, \quad \sigma_i\sigma_j = \sigma_j\sigma_i \text{ for } |i-j| > 1.$$

The map (4.355) sends each generator σ_i in a non-zero complex number λ_i. The second relation yields $\lambda_{i+1} = \lambda_i = \lambda$ for all i. Then the first relation yields $\lambda = \pm 1$.

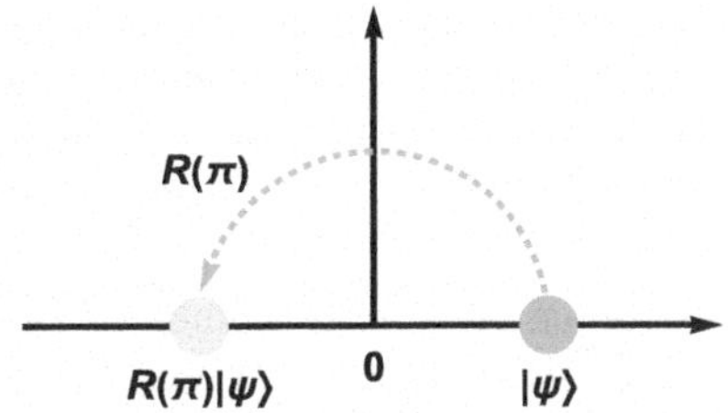

Fig. 4.2 The set-up of the heuristic argument in the text. $R(\pi)$ is a rotation by π around an axis orthogonal to the plane of the figure

When the sign is $+$ we say that the particles obey the *Bose-(Einstein) statistics* or simply that they are *bosons*; when the sign is $-$ we say that the particles obey the *Fermi-(Dirac) statistics* or that they are *fermions*. The use of the term "statistics" for the transformation properties of the system under permutations of its indistinguishable components, arises from the fact that the Statistical Mechanics of a gas of N equal quantum particles is determined by their symmetry under $\mathfrak{S}_N$, see e.g. [18] chap. 3.

A fundamental fact of Physics is the following result:

Spin & Statistics Theorem *Assume the space has dimension* ≥ 3 *or strong locality. Particles of integral (resp. half-integral) spin obey the Bose (resp. Fermi) statistics. Therefore the N identical particle Hilbert space* $\mathcal{H}_N$ *is*

$$
\mathcal{H}_N = \begin{cases} \odot^N \mathcal{H}_1 & \text{integral spin} \\ \wedge^N \mathcal{H}_1 & \text{half-integral spin.} \end{cases} \tag{4.357}
$$

This is a rigorous theorem[42] in the context of *relativistic* quantum theory (more precisely Quantum Field Theory). In this book we are studying *non-relativistic* quantum systems: however, as already stressed, all real physical systems are both quantum and relativistic, and we see non-relativistic Quantum Mechanics just as an approximation to the actual *relativistic* theory which is valid when the velocities are very small in comparison to the speed of light. In other words: we consider *only* non-relativistic quantum systems which *do arise* from fully-fledged relativistic systems in some appropriate limit, and we see non-relativistic Quantum Mechanics as a convenient—but approximate—computational scheme. Therefore all our quantum systems do obey the **Spin & Statistics theorem**.

We present a heuristic argument which makes the statement of the theorem to sound "natural". Consider a quantum state like the one in Fig. 4.2: its wave-function has support in some small region around the point $(a, 0, 0)$ on the 1st axis $(a > 0)$. We write $|\psi\rangle$ for its state vector. We apply to this state $|\psi\rangle$ the operator $R(\pi) = \exp(i\pi J_3)$ which generates a rotation by an angle π around the third axis which is normal to the plane of the figure. Next we consider the state with two identical

---

[42] The theorem was first proven by Pauli in [35] using physical methods; for a fully rigorous math proof (which also enlightens the deep meaning of the result) see e.g. [36].

copies of the same quantum object, one localized around the point $(a, 0, 0)$ on the x_1-axis and the other one around $(-a, 0, 0)$, whose state-vector is

$$|\psi\rangle \otimes R(\pi)|\psi\rangle \tag{4.358}$$

A rotation by π of the combined system around the third axis will interchange the two identical objects, so gives back the same state. We get

$$R(\pi)|\psi\rangle \otimes R(2\pi)|\psi\rangle = \pm R(\pi)|\psi\rangle \otimes |\psi\rangle \tag{4.359}$$

where the upper (lower) sign applies when the state $|\psi\rangle$ has integral (resp. half-integral) spin. We conclude that we get back the original vector (4.358) (i.e. the wave-function is univalued in $\mathbb{R}^3$) provided the rule

$$|\psi_1\rangle \otimes |\psi_2\rangle = \pm |\psi_2\rangle \otimes |\psi_1\rangle, \tag{4.360}$$

holds, i.e. iff particles with integral (resp. half-integral) spin obey the Bose (resp. Fermi) statistics.

A fundamental consequence of the **Spin & Statistics theorem** is the following principle:

Pauli Exclusion Principle *Two identical fermions cannot occupy the same one-particle state.*

Indeed, assume we have two fermions in the one-particle states $|\psi_1\rangle$ and $|\psi_2\rangle$. The state vector of the combined system is

$$|\psi_1\rangle \otimes |\psi_2\rangle - |\psi_2\rangle \otimes |\psi_1\rangle \in \mathcal{H}_2 \simeq \wedge^2 \mathcal{H}_1. \tag{4.361}$$

If $|\psi_1\rangle = |\psi_2\rangle$ this is the zero vector which does not represent any quantum state.

Example 4.12 Electrons (protons, neutrons, neutrinos, μ's, etc.) have spin $\frac{1}{2}$ and hence are fermions. Two electrons cannot be in the same quantum state. This fact is crucial for the stability of atomic matter.

Composite Systems Consider, say, a hydrogen atom. It is a bound state of two fermions, a proton and an electron, and its states have the form

$$\sum_{a,b} c_{ab}|\psi_a\rangle \otimes |\chi_b\rangle \in \mathcal{H}_p \otimes \mathcal{H}_e, \tag{4.362}$$

where $\mathcal{H}_p$ (resp. $\mathcal{H}_e$) is the Hilbert space of the proton (resp. electron) states. The total angular momentum, which is the sum of the orbital angular momentum of the electron around the proton and of the spins of the two particles, is *integral*. Hence the hydrogen atom as a whole is a boson. More generally,

Fact 4.13 *A composite system made of a number f fermions (and any number of bosons) is a boson for f even and a fermion for f odd.*

A subtle exception to this rule (*statistics transmutation*) will be presented in Fact 5.2.

We introduce an observable F, called the *Fermi number*, which counts the number of fermions is a quantum state. The eigenvalue f of F is the number of fermions in an eigenstate. While F may be non-conserved, or even not well-defined, the observable

$$(-1)^F \equiv \exp(i\pi F) \tag{4.363}$$

which counts the fermions mod 2 is always conserved and well-defined. Indeed by the **Spin & Statistics Theorem** $(-1)^F$ is just a rotation by 2π

$$(-1)^F = \exp(2\pi i J_3). \tag{4.364}$$

As explained in Sect. 4.4, the operator $(-1)^F$ commutes with *all observables*, hence *a fortiori* with all Hamiltonian H, and hence is always conserved.

4.12.1 $\mathbb{Z}_2$-Graded Hilbert Spaces ("Super-Mathematics")

In general a quantum system contains both bosons and fermions. The spin superselection rule introduces additional algebraic structures in the Hilbert space that is worth to formalize in the so-called "super-mathematics" where, informally speaking, one introduces a "super" version of all standard math notions.

In presence of fermions we have the non-trivial operator (4.364) which commutes with all observables and counts fermions mod 2. The operator $(-1)^F$ endows the Hilbert space with the structure of a *vector superspace*. A Hilbert *super*space is a Hilbert space $\mathcal{H}$ together with a $\mathbb{Z}_2$ grading

$$\mathcal{H} = \mathcal{H}_0 \oplus \mathcal{H}_1 \tag{4.365}$$

given by the grading operator $(-1)^F$

$$(-1)^F|_{\mathcal{H}_a} = (-1)^a, \qquad a = 0, 1 \bmod 2, \tag{4.366}$$

which decomposes $\mathcal{H}$ into the superselected spaces of *even* ($a = 0 \bmod 2$) and *odd* ($a = 1 \bmod 2$) states. The even states are bosonic, while the odd ones are fermionic, and we use the terminology even/odd and bosonic/fermionic interchangeably. The tensor product of two superspaces $\mathcal{H}_1$ and $\mathcal{H}_2$ is again a superspace with the $\mathbb{Z}_2$ grading $a = a_1 + a_2 \bmod 2$. It is convenient to define the isomorphism

$$\tau : \mathcal{H}_1 \otimes \mathcal{H}_2 \to \mathcal{H}_2 \otimes \mathcal{H}_1, \tag{4.367}$$

$$\tau : |A\rangle \otimes |B\rangle \mapsto (-1)^{ab}|B\rangle \otimes |A\rangle, \tag{4.368}$$

and define the super-symmetric tensor product as

$$\mathcal{H} \otimes_S \mathcal{H} = \mathcal{H} \otimes \mathcal{H}/\tau. \tag{4.369}$$

The grading extends to operators: the $\mathbb{Z}_2$ grading $a = 0, 1 \bmod 2$ of the operator A is defined by the rule

$$(-1)^F A = (-1)^a A(-1)^F \tag{4.370}$$

We also say that the even (resp. odd) operators are bosonic (resp. fermionic).

The spin superselection rule states that *only even operators are observable*. Nevertheless the odd operators may be very useful to study the system as we shall see in Sects. 4.13 and 4.14.

The natural bracket of two operators, which is their commutator in an ordinary (ungraded) Hilbert space, is now the *graded commutator* (or *supercommutator*)

$$[A, B] \overset{\text{def}}{=} AB - (-1)^{ab} BA, \tag{4.371}$$

where $a = 0, 1$ (resp. $b = 0, 1$) is the $\mathbb{Z}_2$-degree of A (resp. B). The supercommutator $[\cdot, \cdot]$ is the ordinary commutator $[A, B]$ if one of the two operators is even, and the *anticommutator*

$$\{A, B\} \overset{\text{def}}{=} AB + BA. \tag{4.372}$$

when they are both odd. The standing convention in theoretical physics is that a square bracket is antisymmetric in its two arguments, while a curly bracket is symmetric.

A *Lie super-algebra:* is a $\mathbb{Z}_2$-graded algebra $\mathfrak{L}^\bullet = \mathfrak{L}^0 \oplus \mathfrak{L}^1$, with both even and odd elements, which is equipped with a Lie *super*-bracket

$$[\cdot, \cdot]: \mathfrak{L}^\bullet \times \mathfrak{L}^\bullet \to \mathfrak{L}^\bullet, \qquad [\mathfrak{L}^a, \mathfrak{L}^b] \subseteq \mathfrak{L}^{a+b \bmod 2}. \tag{4.373}$$

$[A, B]$ is either the commutator or the anti-commutator of $A, B \in \mathfrak{L}$ depending on their $\mathbb{Z}_2$ degrees:

$$[A, B] = -(-1)^{ab}[B, A]. \tag{4.374}$$

The super-bracket of a Lie super-algebra satisfies the *super-Jacobi identity* (see [1] Chap. 2)

$$(-1)^{ac}[[A, B], C] + (-1)^{ba}[[B, C], A] + (-1)^{cb}[[C, A], B] = 0. \tag{4.375}$$

The simple Lie super-algebras are classified by Kac [37].

If $\mathcal{H} = \mathcal{H}_0 \oplus \mathcal{H}_1$ is a superspace, the *supertrace* of the operator A is

$$\mathrm{STr}_{\mathcal{H}}[A] \stackrel{\text{def}}{=} \mathrm{Tr}_{\mathcal{H}}[(-1)^F A] = \mathrm{Tr}_{\mathcal{H}_0}[A] - \mathrm{Tr}_{\mathcal{H}_1}[A]. \tag{4.376}$$

Just as the trace of an ordinary commutator is zero,[43] the supertrace of a supercommutator vanishes

$$\mathrm{STr}[A, B\} = 0. \tag{4.377}$$

Indeed, if A is even and B odd, then $[A, B\}$ is odd and has zero supertrace by definition. If both A, B have the same parity

$$\mathrm{Tr}\big((-1)^F AB\big) = \mathrm{Tr}\big(B(-1)^F A\big) = (-1)^{\deg B}\,\mathrm{Tr}\big((-1)^F BA\big). \tag{4.378}$$

The *superdeterminant* is defined as

$$\mathrm{SDet}(A) = \exp\big(\mathrm{STr}\log A\big). \tag{4.379}$$

4.12.2   *The Fermi Oscillator*

In a quantum system with Hilbert space $\mathbb{C}^2$, besides the identity $\mathbf{1}$, there are only three other linearly independent Hermitian operators, namely the three Pauli matrices σ_i, cf. Sect. 4.3.1. We form their complex linear combinations

$$b^\dagger \equiv \sigma_+ = \frac{\sigma_1 + i\sigma_2}{2} = \begin{pmatrix} 0 & 1 \\ 0 & 0 \end{pmatrix}, \qquad b \equiv \sigma_- = \frac{\sigma_1 + i\sigma_2}{2} = \begin{pmatrix} 0 & 0 \\ 1 & 0 \end{pmatrix}, \tag{4.380}$$

whose commutator is

$$[b^\dagger, b] \equiv \sigma_3 = \begin{pmatrix} 1 & 0 \\ 0 & -1 \end{pmatrix} \tag{4.381}$$

(cf. the complex basis $J_\pm$, J_3 for $\mathfrak{su}(2)$). The identification

$$(-1)^F \equiv \sigma_3, \tag{4.382}$$

gives to $\mathbb{C}^2$ the structure of a vector superspace with grading element σ_3. The operators b, $b^\dagger$ are *odd* with respect to this $\mathbb{Z}_2$ grading.

---

[43] For A, B Hilbert-Schmidt operators, or for one bounded and one of trace-class [38].

Shifting the Hamiltonian by a constant and changing basis, we may always write

$$H = \frac{\omega}{2}\sigma_3, \qquad \omega > 0 \tag{4.383}$$

with no loss of generality. The above operators satisfy the algebra

$$bb + bb = 0, \qquad b^\dagger b^\dagger + b^\dagger b^\dagger = 0, \qquad bb^\dagger + b^\dagger b = 1, \tag{4.384}$$

$$[H, b] = -\omega b, \qquad [H, b^\dagger] = \omega b^\dagger, \qquad H = \omega\left(b^\dagger b - \tfrac{1}{2}\right). \tag{4.385}$$

Eqs. (4.384) and (4.385) should be compared with Eq. (3.72) for the harmonic oscillator. The only difference between the pairs $a, a^\dagger$ and $b, b^\dagger$ are some flips of sign. Apart for the sign flip of the additive constant in H, the only modification is that the *minus* sign in the canonical commutation relations (written in units $\hbar = 1$)

$$aa - aa = 0, \quad a^\dagger a^\dagger - a^\dagger a^\dagger = 0, \quad aa^\dagger - a^\dagger a = 1, \tag{4.386}$$

is now replaced by a *plus,* that is, the canonical commutation relations are replaced by *anti-commutation* relations

$$\{b, b\} = \{b^\dagger, b^\dagger\} = 0, \quad \{b, b^\dagger\} = 1, \tag{4.387}$$

where the curly bracket $\{\cdot, \cdot\}$ is the *anticommutator* defined in Eq. (4.372). In plain English: when we replace the *even* operators $a, a^\dagger$ by their *odd* siblings $b, b^\dagger$ we have to change the commutator in the natural bracket for operators acting in a $\mathbb{Z}_2$-graded Hilbert space, which is the *anticommutator* for odd (fermionic) operators. In the light of "super-mathematics", we conclude that the quantum system with $\mathcal{H} = \mathbb{C}^2$ and Hamiltonian (4.383) is the fermionic counterpart of the usual (bosonic) harmonic oscillator. It is called the *Fermi oscillator.*

The ground state of the Fermi oscillator is

$$|0\rangle = \begin{pmatrix} 0 \\ 1 \end{pmatrix} \in \mathbb{C}^2, \qquad b|0\rangle = 0, \qquad E_0 = -\frac{\omega}{2}. \tag{4.388}$$

We stress that the zero point energy is now *negative*, the opposite of the zero point energy for the bosonic oscillator. The Fermi oscillator has just one excited state

$$|1\rangle \equiv b^\dagger|0\rangle = \begin{pmatrix} 1 \\ 0 \end{pmatrix} \in \mathbb{C}^2, \qquad b^\dagger|1\rangle = 0, \qquad E_1 = \frac{\omega}{2}. \tag{4.389}$$

The system is invariant under the interchange

$$b \leftrightarrow b^\dagger \qquad |0\rangle \leftrightarrow |1\rangle, \qquad H \leftrightarrow -H. \tag{4.390}$$

called the *particle-hole duality.*

The Fermi oscillator is a quantum system whose dynamics is fully captured by the spin-$\frac{1}{2}$ representation of $\mathfrak{su}(2)$: indeed, up to normalization, the only non-trivial Hermitian operators are the generators S_1, S_2 and S_3 of $\mathfrak{su}(2)$ in the spin-$\frac{1}{2}$ representation. We may play similar quantum games with other Lie groups and representations.

The Jordan-Wigner Construction

We consider the situation where we have several Fermi degrees of freedom $b_i^\dagger$, b_i ($i = 1, \ldots, N$). They satisfy the *canonical anti-commutation relation*

$$b_i b_j + b_j b_i = b_i^\dagger b_j^\dagger + b_j^\dagger b_i^\dagger = 0, \quad b_i b_j^\dagger + b_j^\dagger b_i = \delta_{ij}, \tag{4.391}$$

in perfect correspondence—up to the flip of sign from the $\mathbb{Z}_2$ grading—with the canonical commutation relations. The algebra generated by the b_i, $b_j^\dagger$ over $\mathbb{C}$ (resp. over $\mathbb{R}$) is called the *complex Clifford algebra in $2N$ generators* $\mathsf{Cl}(2N)_{\mathbb{C}}$ (resp. the positive real Clifford algebra $\mathsf{Cl}(2N)_+$) see e.g. [15]. For $N = 1$ it is isomorphic to the algebra generated by the Pauli matrices σ_i.

The operator algebra of N Fermi oscillators with Hamiltonian

$$H = \sum_{i=1}^{N} \omega_i \left(b_i^\dagger b_i - \frac{1}{2} \right) \tag{4.392}$$

is the complex *Clifford algebra* $\mathsf{Cl}(2N)_{\mathbb{C}}$. Therefore the Hilbert space is a Clifford module which is irreducible iff there are no other degrees of freedom (i.e. iff an operator which commutes with all b_i, $b_j^\dagger$ is a multiple of the identity). Working over $\mathbb{C}$ (up to isomorphism) there is a unique irreducible $\mathsf{Cl}(2N)_{\mathbb{C}}$-module $S_N \simeq \otimes^N \mathbb{C}^2$ [15] of dimension

$$\dim S_N = 2^N. \tag{4.393}$$

Clearly this unique module can be constructed out of N copies of the algebra of Pauli matrices which describes a single Fermi oscillator. We write $\sigma_+^{(a)}$, $\sigma_-^{(a)}$, $\sigma_3^{(a)}$ for the a-th copy of the Pauli matrices which act on the a-th factor space of the Hilbert space

$$\mathcal{H} \equiv \otimes_{a=1}^{N} (\mathbb{C}^2)^{(a)} \simeq \mathbb{C}^{2^N} \tag{4.394}$$

Clearly

$$[\sigma_\alpha^{(a)}, \sigma_\beta^{(b)}] = 0 \ \text{ for } a \neq b, \ \ \alpha, \beta = \pm, 3. \tag{4.395}$$

To construct the Fermi operators b_i, $b_i^\dagger$ we write

$$b_i = \left(\prod_{a=1}^{i-1} \sigma_3^{(a)}\right)\sigma_-^{(i)}, \qquad b_i^\dagger = \left(\prod_{a=1}^{i-1} \sigma_3^{(a)}\right)\sigma_+^{(i)}. \tag{4.396}$$

It is easy to see that these operators satisfy the Clifford algebra (alias canonical anticommutation relations) (4.391). This construction, known as the *Jordan-Wigner* one, is equivalent to the recursive construction of the universal irreducible complex Clifford module in math, see e.g. [15].

Remark 4.13 (The Fermionic Bogoliubov Group $\equiv$ Spin Group) The group of linear unitary transformations preserving the canonical commutation relations with N bosonic degrees of freedom is $Sp(2N, \mathbb{R})$. The corresponding group of linear unitary transformations preserving the fermionic anti-commutator relations for N Fermi degrees of freedom is $O(2N)$. The Hilbert superspace $S_N = S_{N,+} \oplus S_{N,-}$ decomposes an even and an odd superselected sectors and each of them carries a representation of the Lie algebra $\mathfrak{so}(2N)$. $S_{N,\pm}$ are irreducible $\mathfrak{so}(2N)$-representations of dimension 2^{N-1} known as the *spinorial representations s* and c [33]. They are representations of a double-cover of $SO(2N)$ called the *spin group* $\mathsf{Spin}(2N)$ [14, 33]. The relation between $SO(2N)$ and $\mathsf{Spin}(2N)$ is a generalization of the relation between $SO(3)$ and $SU(2) \equiv \mathsf{Spin}(3)$ that we discussed at length above.

For more on the Bose and Fermi statistics, and the related topic of second quantization, see [18] chap. 3.

4.13   Supersymmetric Quantum Mechanics

In Quantum Physics we have new discrete fermionic degrees of freedom b_i, $b_i^\dagger$ without classical analogues. This opens the possibility of a new kind of "Noether" symmetry which "rotates" ordinary *even* degrees of freedom into purely-quantum *odd* (fermionic) ones. Such symmetries are called *supersymmetries.* They act on Hilbert *superspaces* with a $\mathbb{Z}_2$-grading operator $(-1)^F$. While the supersymmetries are generated by conserved Noether charges Q^a—called *supercharges*—these new conserved quantities *are never* observables since they transform even states into odd ones (and viceversa)

$$Q^a(-1)^F = -(-1)^F Q^a, \tag{4.397}$$

thus violating the spin superselection rule of Sect. 4.4. The theory of quantum mechanical systems which admit some supersymmetry is called *Supersymmetric Quantum Mechanics* or SQM for short.

In the final sections of this chapter we describe some very simple examples of supersymmetric quantum systems from the Hamiltonian viewpoint of the Schrödinger representation. The deeper Lagrangian viewpoint will be discussed in Sect. 6.24 of Chap. 6. See also [18] chap. 6.

4.13.1 The Simplest Supersymmetric Model

We start with a system with just one even and one odd degree of freedom described by the four canonical operators q, p, b, $b^\dagger$ acting on the Hilbert space

$$\mathcal{H} = \mathbb{C}^2 \otimes L^2(\mathbb{R}). \tag{4.398}$$

$b, b^\dagger$ act on the factor $\mathbb{C}^2$ as in Eq. (4.380), while q, p act on $L^2(\mathbb{R})$ as described in Chap. 2. The Hilbert space $\mathcal{H}$ is graded in even an odd states with grading operator

$$(-1)^F = -\sigma_3 \tag{4.399}$$

which acts on the first factor space $\mathbb{C}^2$ of $\mathcal{H}$. We call a supersymmetric system with these degrees of freedom *a "superparticle" moving on the line* $\mathbb{R}$: it is the supersymmetric analog of the usual particle on $\mathbb{R}$; in addition to the usual even variables q, p it contains their odd "fermionic partners" b, $b^\dagger$.

Let Q_1 be a (non-zero) *Noether supercharge*: it is a Hermitian $(-1)^F$-*odd* operator which is a polynomial[44] in p of degree 1 and *conserved,* i.e. it commutes with the Hamiltonian H. We stress that a supersymmetric model has *at least two* linearly independent supercharges: indeed, if Q_1 is a Hermitian supercharge,

$$Q_2 \equiv iQ_1(-1)^F \tag{4.400}$$

is a second one. Then the complex supercharge

$$Q \equiv Q_1 + iQ_2 \equiv Q_1(1 - (-1)^F) \tag{4.401}$$

is non-zero and *nilpotent*

$$\begin{aligned}
Q^2 &= Q_1(1 - (-1)^F)Q_1(1 - (-1)^F) = \\
&= Q_1(1 - (-1)^F)(1 + (-1)^F)Q_1 = 0.
\end{aligned} \tag{4.402}$$

By construction Q kills the even states and maps the odd ones into even states. Its adjoint $Q^\dagger$ kills the odd states and sends the even ones into odd states.

---

[44] Recall that the Noether charges are polynomials in p of degree 1 [1].

The anti-commutator of Q with its adjoint $Q^\dagger$

$$\{Q, Q^\dagger\} \equiv QQ^\dagger + Q^\dagger Q \tag{4.403}$$

is an even, conserved, non-negative, Hermitian operator which is a polynomial in p of degree at most 2. In one-dimensional systems all conserved even operators are functions of the Hamiltonian H, which we assume to be quadratic in momenta of the form

$$H = \frac{1}{2}\,p^2 + V(q) + \mu(q)(b^\dagger b - bb^\dagger), \tag{4.404}$$

where the last term is the most general p-independent coupling between the Fermi and Bose degrees of freedom that we can write with just one q and one pair $b, b^\dagger$. In QFT this kind of interaction between even and odd degrees of freedom is called a *Yukawa coupling*, and we adopt this terminology also in the present mechanical context.

We conclude that the RHS of (4.403) must be linear in the Hamiltonian. Shifting H by a constant, and changing the overall normalization of Q, we may always set the algebra of our conserved (super)charges in the form

$$\{Q, Q\} = \{Q^\dagger, Q^\dagger\} = 0, \qquad \{Q, Q^\dagger\} = 2H,$$
$$[H, Q] = [H, Q^\dagger] = 0, \qquad \{(-1)^F, Q\} = \{(-1)^F, Q^\dagger\} = 0. \tag{4.405}$$

Equation (4.405) is the *supersymmetry algebra* (SUSY algebra for short) with *two* conserved Hermitian supercharges

$$Q_1 = \frac{1}{2}(Q + Q^\dagger) \quad \text{and} \quad Q_2 = \frac{1}{2i}(Q^\dagger - Q), \tag{4.406}$$

which satisfy the anti-commutation relation

$$\{Q_a, Q_b\} = \delta_{ab} H, \qquad Q_a^\dagger = Q_a, \qquad a, b = 1, 2. \tag{4.407}$$

Equation (4.405) is a first example of *Lie **super**-algebra* in the sense of Eqs. (4.373) and (4.375).[45] There are other SUSY algebras, with more than two supercharges, which satisfy the algebra (4.407) with $a, b = 1, \ldots, 2N$. They cannot be realized with just one bosonic and one fermionic degree of freedom. At the end of this section we shall briefly discuss a class of supersymmetric quantum systems with *four* supercharges ($N = 2$).

With only one fermionic degree of freedom, Q must be linear in $b, b^\dagger$ since all odd expressions of higher order in the $b, b^\dagger$ are equal to a linear combination of

---

[45] We stress that the Lie super-algebra (4.405) is *not* semi-simple.

them. We consider the complex nilpotent supercharges Q and $Q^\dagger$ (cf. Eq. (4.402));
they must have the form[46]

$$Q = b p + f(q)b + g(q)b^\dagger,$$
$$Q^\dagger = b^\dagger p + f(q)^* b^\dagger + g(q)^* b \tag{4.408}$$

for some complex functions f, g. Now

$$0 = Q^2 = (g p + g f)(b b^\dagger + b^\dagger b) - \mathrm{i} g' b b^\dagger = g(p + f) - \mathrm{i} g' b b^\dagger = 0, \tag{4.409}$$

hence we must have $g = 0$ and

$$Q = b(p + f(q)), \qquad Q^\dagger = b^\dagger(p + f(q)^*). \tag{4.410}$$

Then

$$QQ^\dagger + Q^\dagger Q = \left[p^2 + |f|^2 + (f + f^*)p - \mathrm{i}(f^*)' \right] b b^\dagger +$$
$$+ \left[p^2 + |f|^2 + (f + f^*)p - \mathrm{i} f' \right] b^\dagger b = \tag{4.411}$$
$$= p^2 + |f|^2 + (f + f^*)p - \mathrm{i}(f^*)' b b^\dagger - \mathrm{i} f' b^\dagger b.$$

Comparing with (4.404) we get $f(x) = \mathrm{i} F(x)$ with $F(x)$ a real function of x. We
write $F(x) = W'(x)$, so that the Schrödinger representation of the supercharges as
odd first-order differential operators acting on $\mathbb{C}^2 \otimes L^2(\mathbb{R})$ is ($\hbar = 1$)

$$Q = -\mathrm{i} \left(\begin{smallmatrix} 0 & 0 \\ 1 & 0 \end{smallmatrix}\right)\left(\frac{\mathrm{d}}{\mathrm{d}x} - \frac{\mathrm{d}W}{\mathrm{d}x}\right) \equiv \mathrm{e}^W\left[-\mathrm{i} \left(\begin{smallmatrix} 0 & 0 \\ 1 & 0 \end{smallmatrix}\right)\frac{\mathrm{d}}{\mathrm{d}x} \right]\mathrm{e}^{-W} \tag{4.412}$$

$$Q^\dagger = -\mathrm{i} \left(\begin{smallmatrix} 0 & 1 \\ 0 & 0 \end{smallmatrix}\right)\left(\frac{\mathrm{d}}{\mathrm{d}x} + \frac{\mathrm{d}W}{\mathrm{d}x}\right) \equiv \mathrm{e}^{-W}\left[-\mathrm{i} \left(\begin{smallmatrix} 0 & 1 \\ 0 & 0 \end{smallmatrix}\right)\frac{\mathrm{d}}{\mathrm{d}x} \right]\mathrm{e}^{W}. \tag{4.413}$$

The scalar potential and the Yukawa coupling are

$$V(x) = \frac{1}{2}\left(W'(x)\right)^2 \qquad \mu = \frac{1}{2}W''(x). \tag{4.414}$$

The function $W(x)$ is called the *superpotential*. Acting on (a dense domain of)
$\mathbb{C}^2 \otimes L^2(\mathbb{R}) \equiv \mathcal{H}$ the Hamiltonian takes the form

$$H = \frac{1}{2}\left[p^2 + (W')^2 + W''\sigma_3 \right]. \tag{4.415}$$

---

[46] Notice that we are free to redefine Q by an overall phase without modifying the SUSY super-
algebra.

H commutes with $(-1)^F \equiv -\sigma_3$, so the two operators can be diagonalized simultaneously. Acting on bosonic states with $(-1)^F = 1$ (resp. fermionic states with $(-1)^F = -1$) the Hamiltonian reduces to a standard one-dimensional Sturm-Liouville 2nd-order differential operator[47]

$$\text{on bosons} \qquad H_B = \frac{1}{2}\left(-\frac{\mathrm{d}^2}{\mathrm{d}x^2} + (W')^2 - W''\right) \tag{4.416}$$

$$\text{on fermions} \qquad H_F = \frac{1}{2}\left(-\frac{\mathrm{d}^2}{\mathrm{d}x^2} + (W')^2 + W''\right) \tag{4.417}$$

Suppose that $W''(x) > 0$ everywhere, so that $W(x)$ is *convex*. Then $H_B < H_F$, and the ground state of the system must be a bosonic state by the Sturm comparison theorem. The ground state is then unique by the general properties of the one-dimensional systems, see Chap. 3. The relation $\{Q, Q^\dagger\} = 2H$ implies that the spectrum of the Hamiltonian is non-negative, so if we find an eigenfunction of H_B with zero eigenvalue it should be the ground state. Consider the function

$$\psi_0(x) = \mathrm{e}^{-W(x)} \quad \Rightarrow \quad H_B \, \mathrm{e}^{-W(x)} = 0 \tag{4.418}$$

It is a function without zeros which—if normalizable in either the discrete or the continuum senses—describes a zero-energy ground state. We shall elaborate on such states in the next section.

Example 4.13 (The Supersymmetric Harmonic Oscillator)   We take

$$W(x) = \frac{1}{2}\omega x^2, \qquad \omega > 0, \tag{4.419}$$

The Hamiltonians on the bosonic/fermionic sectors are

$$H_B = \frac{1}{2}(p^2 + \omega^2 q^2) - \frac{\omega}{2}, \qquad H_F = \frac{1}{2}(p^2 + \omega^2 q^2) + \frac{\omega}{2} \tag{4.420}$$

i.e. they are the harmonic Hamiltonian of frequency ω shifted so that the ground state has zero-energy while the lowest energy fermionic state has energy ω. The energy levels are

$$\begin{aligned}
\text{bosonic energy levels:} &\quad 0,\ \omega,\ 2\omega,\ 3\omega,\ 4\omega,\ \cdots \\
\text{fermionic energy levels:} &\quad \omega,\ 2\omega,\ 3\omega,\ 4\omega,\ \cdots
\end{aligned} \tag{4.421}$$

All energy levels come in Bose/Fermi pairs except for the zero-energy ground state which in this model is a bosonic state. As we shall see in the next section, this is

---

[47] Recall that we are setting $\hbar = 1$.

the general pattern whenever supersymmetry is not spontaneously broken. Indeed the supercharges Q, $Q^\dagger$ commute with the Hamiltonian, hence they send energy eigenstates into eigenstates of the same energy (or to zero) while mapping bosonic states into fermionic ones and viceversa.

Remark 4.14 (The SUSY Trick) Now we can see that the trick used in the Appendix to Chap. 3 to solve certain one-dimensional Schrödinger equation was just to compare the eigenfunctions and eigenvectors for the Sturm-Liouville operator H_F to the ones for H_B using the fact that the two sets are related by supersymmetry, and hence their energy levels are equal with the possible exception of the zero-energy ground state. The idea is that if we know the spectrum of either one of the two operators H_B or H_F we can get spectrum and eigenfunctions of the other one by exploiting the supersymmetry relating them.

Generalization: The Superparticle in $\mathbb{R}^n$
The previous construction may be generalized to n bosonic degrees of freedom $(\boldsymbol{q}_i, \boldsymbol{p}_i)$ coupled to n fermionic ones $(b_i, b_i^\dagger)$ $(i = 1, \ldots, n)$ which satisfy, respectively, the canonical commutator and anti-commutator relations when acting on the Hilbert space

$$\mathcal{H} = \mathbb{C}^{2^n} \otimes L^2(\mathbb{R}^n). \tag{4.422}$$

The *superpotential* $W(x)$ is now a real smooth function on $\mathbb{R}^n$. Then the operators

$$Q = i\, b_i \left(-\frac{\partial}{\partial x^i} + \frac{\partial W}{\partial x_i}\right) \equiv e^W (b_i\, \boldsymbol{p}_i) e^{-W} \tag{4.423}$$

$$Q^\dagger = -i\, b_i^\dagger \left(\frac{\partial}{\partial x^i} + \frac{\partial W}{\partial x^i}\right) \equiv e^{-W} (b_i^\dagger\, \boldsymbol{p}_i) e^{W} \tag{4.424}$$

$$H = \frac{1}{2} \sum_i \left[-\frac{\partial^2}{\partial x^i \partial x^i} + \left(\frac{\partial W}{\partial x_i}\right)^2\right] + \frac{1}{2} \frac{\partial^2 W}{\partial x^i\, \partial x^j} [b_i^\dagger, b_j] \tag{4.425}$$

satisfies the Lie super-algebra (4.405). Therefore the system with Hamiltonian (4.425) is supersymmetric with conserved supercharges Q and $Q^\dagger$, cf. Eqs. (4.423) and (4.424).

4.13.2 Superparticle Moving in $\mathcal{M}$

Next we consider the supersymmetric extension of a particle moving on a Riemannian manifold $\mathcal{M}$ (which we assume connected and oriented) in absence of potentials. The usual "bosonic" Hilbert space $L^2(\mathcal{M})$ is promoted to the Hilbert superspace

$$\mathcal{H}^{\bullet} = \widehat{\bigoplus_{0 \le k \le m} \mathcal{H}^k}, \qquad \mathcal{H}^k \overset{\text{def}}{=} \overline{\Omega^k(\mathcal{M})_c}, \tag{4.426}$$

where $m = \dim \mathcal{M}$, and $\Omega^k(\mathcal{M})_c$ is the space of compactly-supported smooth complex-valued k-forms on $\mathcal{M}$. As usual, the overbar denotes closure in the Hilbert space norm defined by the Hodge Hermitian product on differential forms

$$\langle \alpha | \beta \rangle = \int_{\mathcal{M}} \bar{\alpha} \wedge *\beta \tag{4.427}$$

where $*$ is the Hodge dual defined by the Riemannian metric on $\mathcal{M}$, and the bar is complex conjugation. The $\mathbb{Z}_2$ grading of $\mathcal{H}^{\bullet}$ is given by the parity of the form degree

$$F\Big|_{\mathcal{H}^k} = k, \qquad (-1)^F\Big|_{\mathcal{H}^k} = (-1)^k, \tag{4.428}$$

that is, even (resp. odd) states are represented by even (resp. odd) differential forms. The Fermi operators $b_i^{\dagger}$, b_i act on the differential forms as[48]

$$b_i^{\dagger} \alpha = \mathrm{d}x_i \wedge \alpha, \qquad b_i \alpha = \iota_{\partial_{x_i}} \alpha, \tag{4.429}$$

an identification which is easily seen to be consistent with the canonical anti-commutator algebra [5, 39, 40]:

$$\mathrm{d}x_i \wedge \mathrm{d}x_j + \mathrm{d}x_j \wedge \mathrm{d}x_i = \iota_{\partial_{x_i}} \iota_{\partial_{x_j}} + \iota_{\partial_{x_j}} \iota_{\partial_{x_i}} = 0 \tag{4.430}$$

$$\iota_{\partial_{x_i}} \mathrm{d}x_j + \mathrm{d}x_j \wedge \iota_{\partial_{x_i}} = \delta_{ij}. \tag{4.431}$$

In the zero-form sector, $\mathcal{H}^0$, the natural Hamiltonian is $H = \frac{1}{2}\Delta$, with Δ the Laplacian (in unit where $m = \hbar = 1$). The natural extension to $\mathcal{H}^{\bullet}$ is

$$H = \tfrac{1}{2}\Delta \tag{4.432}$$

where now Δ is the Hodge Laplacian acting on differential forms. Let

$$\mathrm{d} : \Omega^{\bullet}(\mathcal{M})_c \to \Omega^{\bullet+1}(\mathcal{M})_c \tag{4.433}$$

---

[48] As in [1] chap. 2, ι_v is the (odd) operator $\iota_v : \Omega^{\bullet}(\mathcal{M})_c \to \Omega^{\bullet-1}(\mathcal{M})_c$ which contracts differential forms with the vector field v.

be the *exterior derivative* ([1] chap. 2), seen as an unbounded operator defined in a dense domain of $\mathcal{H}^\bullet$. The usual statements of Hodge theory [5, 39, 40] are equivalent to

Fact 4.14 *The four (densely defined) operators*

$$Q^\dagger \equiv \mathrm{d}, \qquad Q = \mathrm{d}^\dagger \equiv -*\,\mathrm{d}*,$$

$$H = \frac{1}{2}\Delta, \qquad (-1)^F = (-1)^k \tag{4.434}$$

satisfy the SUSY super-algebra (4.405) *and hence yield a Schrödinger representation of supersymmetry on the Hilbert space* $\mathcal{H}^\bullet$, *Eq.* (4.426), *which describes the free "superparticle" moving on* $\mathcal{M}$.

Adding a Superpotential

Now we may switch on a superpotential $W: \mathcal{M} \to \mathbb{R}$. Just as in the case $\mathcal{M} \equiv \mathbb{R}^n$, Eqs. (4.423) and (4.424), this amounts to conjugating the "free super-charges (4.434)" by the multiplication operators $\mathrm{e}^{\pm W}$:

$$Q^\dagger = \mathrm{e}^{-W}\mathrm{d}\,\mathrm{e}^{W}, \qquad Q = \mathrm{e}^{W}\mathrm{d}^\dagger\mathrm{e}^{-W}. \tag{4.435}$$

The algebra (4.405) still holds for an "interacting" Hamiltonian H_W which now contains potential and Yukawa terms

$$H_W = \frac{1}{2}\Big(\Delta + g^{ij}\partial_{x_i}W\partial_{x_j}W - \mathscr{L}_{\nabla W} - \mathscr{L}^\dagger_{\nabla W}\Big) \tag{4.436}$$

where ∇W is the gradient vector of the superpotential W and $\mathscr{L}_v$ the Lie derivative along the vector field v ([1] chap. 2).

There are more general supersymmetric quantum mechanical systems, but we shall not review them here for brevity.

Extended Supersymmetry

Suppose that the Riemannian manifold $\mathcal{M}$ is actually Kähler [4–7, 40]. The Hilbert space now has two gradings ($\equiv$ conserved charges)

$$\mathcal{H}^{\bullet,\bullet} = \widehat{\bigoplus_{p,q}} \overline{\Omega^{p,q}(\mathcal{M})_c} \tag{4.437}$$

where $\Omega^{p,q}(\mathcal{M})$ is the space of differential forms of type (p, q) with compact support [5, 6] and $\mathbb{Z}_2$ grading $(-1)^F = (-1)^{p+q}$. We have four well-known first order differential *odd* operators acting on a dense domain in $\mathcal{H}^{\bullet\bullet}$

$$\partial: \Omega^{p,q}(\mathcal{M})_c \to \Omega^{p+1,q}(\mathcal{M})_c \qquad \overline{\partial}: \Omega^{p,q}(\mathcal{M})_c \to \Omega^{p,q+1}(\mathcal{M})_c$$
$$\partial^\dagger: \Omega^{p,q}(\mathcal{M})_c \to \Omega^{p-1,q}(\mathcal{M})_c \qquad \overline{\partial}^\dagger: \Omega^{p,q}(\mathcal{M})_c \to \Omega^{p,q-1}(\mathcal{M})_c \tag{4.438}$$

We may interpreted these four operators as conserved supercharges for the super-particle on the Kähler manifold $\mathcal{M}$ in absence of superpotentials. We form four Hermitian supercharges

$$Q_1 = \partial + \partial^\dagger, \quad Q_2 = \mathrm{i}(\partial - \partial^\dagger), \quad Q_3 = \overline{\partial} + \overline{\partial}^\dagger, \quad Q_4 = \mathrm{i}(\overline{\partial} - \overline{\partial}^\dagger) \tag{4.439}$$

The well-known Kähler-Hodge identities [5, 6] are equivalent to the statement that these four supercharges and the Hamiltonian $H \equiv \Delta$ (the Laplacian acting on forms) satisfy the 4-supercharge SUSY algebra

$$\{Q_a, Q_b\} = \delta_{ab}H, \quad a, b, = 1, \dots, 4. \tag{4.440}$$

We may also switch on a superpotential while preserving four conserved super-charges, but now W should be a holomorphic function W on $\mathcal{M}$. Then one considers the two complex supercharges

$$\partial + \mathrm{d}\overline{W}\wedge, \qquad \overline{\partial} + \mathrm{d}W\wedge \tag{4.441}$$

and their Hermitian conjugates.

4.14 SUSY Representation Theory and Witten Index

We now consider the Representation Theory of the supersymmetry algebra with 2 supercharges (4.405). The Hamiltonian commutes with all supergenerators, so it is a c-number E in every irreducible representation of the SUSY superalgebra (called a *supermultipliet*). Then, for any normalizable state $|\psi\rangle$ in the irreducible representation,

$$2E = 2E\,\langle\psi|\psi\rangle = \langle\psi|2H|\psi\rangle =$$
$$= \langle\psi|QQ^\dagger|\psi\rangle + \langle\psi|Q^\dagger Q|\psi\rangle = \|Q^\dagger|\psi\rangle\|^2 + \|Q|\psi\rangle\|^2 \tag{4.442}$$

So a state is SUSY invariant iff it has zero energy. Otherwise $Q_1|\psi\rangle \neq 0$ is a state of opposite parity and the same energy. We summarize the situation:

SUMMARY: SUSY Representations

In a supersymmetric system the energy is non-negative $E \geq 0$ and a state $|\psi\rangle$ has zero energy if and only if

$$Q|\psi\rangle = Q^\dagger|\psi\rangle = 0, \tag{4.443}$$

that is, iff the state $|\psi\rangle$ is invariant under supersymmetry (i.e. iff $|\psi\rangle$ spans a one-dimensional irreducible representation of SUSY). The zero-energy states may be chosen to be eigenstates of the grading $(-1)^F$. All other irreducible representations have $E > 0$ and consists of a bosonic $|B\rangle$ and a fermionic state $|B\rangle$ with

$$|F\rangle = \sqrt{\frac{2}{E}}\,Q_1|B\rangle, \qquad |B\rangle = \sqrt{\frac{2}{E}}\,Q_1|F\rangle \tag{4.444}$$

That is, states with positive energy form Bose-Fermi pairs

The second sentence follows from the fact that if $|\psi\rangle$ is a zero energy state so are $(1 \mp (-1)^F)|\psi\rangle$ if non-zero.

Supersymmetry Breaking

Recall Definition 3.3: a symmetry is *unbroken* iff the ground state(s) are invariant under its action. The notion applies to supersymmetry as well. But a state is invariant under supersymmetry if and only if it has zero energy, in which case it is automatically a ground state since energy is non-negative. Therefore

---

SUSY Breaking

Supersymmetry is dynamically unbroken if and only if there are states of zero energy, i.e. if zero belongs to the spectrum of the Hamiltonian H. States of zero-energy are called *supersymmetric ground states*. If no state of zero-energy exist, supersymmetry is *spontaneously broken.*

Example 4.14 Consider a superparticle moving on $\mathbb{R}$ subjected to a linear superpotential $W(x) = -\mu x$ with $\mu \neq 0$. The Hamiltonian is

$$H = -\frac{1}{2}\frac{\mathrm{d}^2}{\mathrm{d}x^2} + \frac{1}{2}\mu^2, \tag{4.445}$$

whose spectrum $\sigma(H) \geq \mu^2$. There is no zero energy state, so supersymmetry is spontaneously broken.

We generalize the example.

Example 4.15 Consider a superparticle moving on $\mathbb{R}$ with a superpotential $W(x)$ which is a polynomial of degree n:

$$W(x) = \tfrac{1}{n}ax^n + \text{lower degree terms.} \tag{4.446}$$

The system of two linear ODEs

$$\begin{cases} \mathrm{i}\,\mathrm{e}^{-W}Q\Psi \equiv \dfrac{\mathrm{d}}{\mathrm{d}x}\left(\mathrm{e}^{-W}\left(\begin{smallmatrix} 0 & 0 \\ 1 & 0 \end{smallmatrix}\right)\Psi\right) = 0 \\[12pt] \mathrm{i}\,\mathrm{e}^{W}Q^{\dagger}\Psi \equiv \dfrac{\mathrm{d}}{\mathrm{d}x}\left(\mathrm{e}^{W}\left(\begin{smallmatrix} 0 & 1 \\ 0 & 0 \end{smallmatrix}\right)\Psi\right) = 0 \end{cases} \tag{4.447}$$

has the two solutions

$$\Psi_B = \left(\begin{smallmatrix} 0 \\ 1 \end{smallmatrix}\right)\mathrm{e}^{-W(x)}, \qquad \Psi_F = \left(\begin{smallmatrix} 1 \\ 0 \end{smallmatrix}\right)\mathrm{e}^{W(x)}, \tag{4.448}$$

where Ψ_B (resp. Ψ_F) is *bosonic* (resp. *fermionic*) i.e. an eigenvector of the grading operator $(-1)^F$ with eigenvalue $+1$ (resp. -1). Clearly the two solutions cannot both belong to $\mathcal{H} \equiv \mathbb{C}^2 \otimes L^2(\mathbb{R})$. We distinguish three cases:

(1) when $n = 2m$ is *even* and the leading coefficient is positive, $a > 0$, Ψ_B is asymptotically small as $x \to \pm\infty$,

$$\Psi_B \approx \begin{pmatrix} 0 \\ 1 \end{pmatrix} \exp\left(-\frac{ax^{2m}}{2m}\right), \tag{4.449}$$

and hence Ψ_B is normalizable, while Ψ_F blows up exponentially at infinity. Hence we have an unique zero-energy state which is *bosonic;*

(2) when $n = 2m$ is *even* and the leading coefficient is negative, $a < 0$, Ψ_B and Ψ_F interchange their roles: Ψ_F is exponentially small at infinity and belongs to $\mathcal{H}$, while Ψ_B is badly non-normalizable. We have again a unique supersymmetric ground state but this time is *fermionic;*

(3) when $n = 2m + 1$ is *odd*, both Ψ_B and Ψ_F are exponentially small at one of the two ends of the real line and exponentially large at the other end. Hence both solutions do not belong to $\mathcal{H}$ (nor $\mathcal{S}^{\vee}$), there are no zero-energy states, and supersymmetry is *spontaneously broken.*

Positive-Energy Representations

Next we consider an irreducible representation with $E > 0$. Acting on the irreducible representation, the two Hermitian supercharges Q_1, Q_2 satisfy the algebra

$$\{Q_a, Q_b\} = \delta_{ab}\, E \qquad a, b = 1, 2 \tag{4.450}$$

cf. Eq. (4.407). Consider the three Hermitian operators

$$\Gamma_1 = \frac{1}{\sqrt{2E}}(Q + Q^\dagger), \qquad \Gamma_2 = \frac{1}{i\sqrt{2E}}(Q - Q^\dagger), \qquad \Gamma_3 = (-1)^F. \qquad (4.451)$$

They satisfy the Clifford algebra $\mathsf{Cl}(3)_+$ with three Hermitian generators

$$\{\Gamma_a, \Gamma_b\} = 2\,\delta_{ab} \quad a, b = 1, 2, 3. \qquad (4.452)$$

Up to isomorphism, there is only one irreducible $\mathsf{Cl}(3)_+$-module which is 2-dimensional. The operators Γ_a are represented on $\mathbb{C}^2$ by the Pauli matrices

$$\Gamma_1 = \sigma_1, \quad \Gamma_2 = \sigma_2, \quad \Gamma_3 = \sigma_3, \qquad (4.453)$$

i.e. any irreducible representation of the algebra (4.450) with $E > 0$ has one bosonic and one fermionic state with the same energy. Any other positive-energy representation is the direct sum of copies of this 2-dimensional one. In particular in any positive-energy representation we have equal numbers of fermionic and bosonic states. Compare with our previous analysis of the supersymmetric harmonic oscillator Example 4.13.

Remark 4.15 More generally, when we have a supersymmetry algebra with $2N \geq 2$ supercharges

$$\{Q_a, Q_b\} = \delta_{ab}\, H, \quad Q_a^\dagger = Q_a, \quad a, b = 1, 2, \dots, 2N, \qquad (4.454)$$

a positive-energy irreducible representation is the irreducible $\mathsf{Cl}(2N + 1)_+$-module of dimension 2^N which contains 2^{N-1} bosonic and 2^{N-1} fermionic states.

Witten Index
We wish to compute the *supertrace* of the imaginary time evolution operator $\mathrm{e}^{-\beta H}$:

$$\mathrm{Tr}[(-1)^F \mathrm{e}^{-\beta H}] = \mathrm{Tr}_{\mathcal{H}^0}[\mathrm{e}^{-\beta H}] - \mathrm{Tr}_{\mathcal{H}^1}[\mathrm{e}^{-\beta H}]. \qquad (4.455)$$

The states with $E > 0$ come in pairs—one even and one odd—and they cancel each others pairwise. Therefore the supertrace get contributions only from the zero-energy states which belong to 1-dimensional representations and hence are not paired. Assuming H is gapped, so that the counting of zero-energy states in not ambiguous, we get

$$\Delta \overset{\text{def}}{=} \mathrm{Tr}[(-1)^F \mathrm{e}^{-\beta H}] =$$
$$= \#(\text{even zero-energy states}) - \#(\text{odd zero-energy states}) \qquad (4.456)$$

which, in particular, is independent of β. The supertrace of $\mathrm{e}^{-\beta H}$ is called the *Witten index* of the supersymmetric system [41, 42]. The fundamental importance of the Witten index stems from two basic facts:

WI1 the Witten index Δ is invariant under all continuous deformations of the model which are compatible with supersymmetry and vanish asymptotically at infinity in the configuration space $\mathcal{M}$. E.g. we can modify continuously the superpotential W and/or the Riemannian metric g_{ij} in an arbitrary way—provided we keep fixed their asymptotic behavior at infinity in $\mathcal{M}$—and Δ will remain invariant;

WI2 if the Witten index $\Delta \neq 0$ supersymmetry is necessarily unbroken.

WI1 implies that Δ may be computed in any limit (such as: weak coupling, classical limit, flat/symmetric metric, large mass, etc.) where the computation trivializes. Therefore Δ is usually easy to compute, while by **WI2** it may be used to control the subtle dynamical realization of supersymmetry and thus it yields precious informations on the exact physics of our system even in a strongly-coupled deeply quantum regime.

WI2 is obvious in view of Eq. (4.443). We justify **WI1**. Our supersymmetric Hamiltonian depends continuously on some parameters t_a

$$H = H(t_a). \tag{4.457}$$

The energy levels $E_n(t_a)$ then are continuous functions of these parameters. For some values of the t_a's some generically non-zero eigenvalues $E_n(t_a)$ may become zero, so the number of SUSY preserving states may jump for certain values t_a^0 of the parameters. However, the states with an energy

$$E(t_a) \to 0 \quad \text{as} \quad t_a \to t_a^0 \quad \text{and} \quad E(t_a) \neq 0 \quad \text{for} \quad t_a \neq t_a^0 \tag{4.458}$$

should form positive-energy representations of supersymmetry, and hence consist of even/odd pairs whose net contribution to the index Δ is *zero*. Then the index will not change even when the precise number of zero-energy states jumps. However this argument is fully correct only when the configuration space $\mathcal{M}$ is (effectively) *compact*. Otherwise subtler phenomena may occur: some zero-energy states may "escape at infinity" in configuration space or (conversely) "come in from infinity". We illustrate this issue in a prototypical example.

Example 4.16 Consider a superparticle moving in $\mathbb{R}$ subjected to the harmonic superpotential

$$W(x) = \frac{1}{2}\omega x^2 - \mu x, \tag{4.459}$$

which produces the harmonic potential

$$V(x) = \frac{1}{2}(\omega x - \mu)^2.$$

(4.460)

As discussed around Eq. (4.449), this system has a *unique* bosonic zero-energy state with the Gaussian wave-function

$$\psi_0(x) = \left(\frac{\omega}{\pi}\right)^{1/4} e^{-\mu^2/(2\omega)} \begin{pmatrix} 0 \\ 1 \end{pmatrix} e^{-\frac{1}{2}\omega x^2 + \mu x} \in \mathbb{C}^2 \otimes L^2(\mathbb{R})$$

(4.461)

picked at the minimum of potential $x_{\rm m} = \mu/\omega$. The Witten index is then $\Delta = 1$. Now we send $\omega \to 0$: in this limit the only zero-energy state "escapes to infinity", $x_{\rm m} \to +\infty$, and we remain with the superpotential $W(X) = -\mu x$ which has $\Delta = 0$ (cf. Example 4.14).

However this tricky phenomenon cannot happen if the variation of the Hamiltonian $\partial H / \partial t_a$ goes to zero at the ends of $\mathcal{M}$ (and *a fortiori* when $\mathcal{M}$ is closed). In this situation the Witten index is a truly invariant of the continuous family of supersymmetric Hamiltonians $\{H(t_a)\}$.

Cohomological Interpretation

The previous discussion may be reformulated in the language of cohomology, making some arguments more transparent. Let us focus on the nilpotent complex supercharge Q. Since $Q^2 = 0$, we may define a Q-cohomology valued in the Hilbert space $\mathcal{H}$. We start from the space $\mathcal{Z}$ of *Q-closed states*

$$\mathcal{Z} \overset{\text{def}}{=} \{|\psi\rangle \in \mathcal{H} : Q|\psi\rangle = 0\},$$

(4.462)

which is naturally decomposed into even and odd subspaces

$$\mathcal{Z} = \mathcal{Z}^0 \oplus \mathcal{Z}^1.$$

(4.463)

$\mathcal{Z}$ contains the Q-exact states i.e. the Q-image $Q\mathcal{H} \subset \mathcal{Z}$. The cohomology space is the quotient of the space $\mathcal{Z}$ of Q-closed vectors by the Q-exact ones. We define the bosonic (even) and fermionic (odd) Q-cohomology spaces as

$$H_Q^0 = \mathcal{Z}^0 / Q\mathcal{H}^1$$

(4.464)

$$H_Q^1 = \mathcal{Z}^1 / Q\mathcal{H}^0.$$

(4.465)

Proposition 4.6 *Suppose that the ground states are gapped. The eigenspace of even (resp. odd) zero-energy states is canonically isomorphic to the cohomology space H_Q^0 (resp. H_Q^1). In particular the number n_B of bosonic SUSY ground-states (resp.*

the number n_F of fermionic SUSY ground-states) is equal to $\dim H_Q^0$ *(resp.* H_Q^1*).
The Witten index*

$$\Delta = \dim H_Q^0 - \dim H_Q^1 \tag{4.466}$$

is then the Euler's characteristic of the Q-complex.

Proof Let $|\psi\rangle \in \mathcal{H}$ be a Q-closed state

$$Q|\psi\rangle = 0 \tag{4.467}$$

Its Q-cohomology class is the set of elements of $\mathcal{H}$ of the form

$$|\psi\rangle + Q|\eta\rangle. \tag{4.468}$$

In each class we look for the representative with the minimal norm. Let $|\psi_0\rangle$ be this
representative. For any state $|\eta\rangle \in \mathcal{H}$ the function

$$\begin{aligned}
\left\| |\psi_0\rangle + t\, Q|\eta\rangle \right\|^2 = \\
= \langle \psi_0|\psi_0\rangle + t^*\langle \eta|Q^\dagger|\psi_0\rangle + t\langle \psi_0|Q|\eta\rangle + tt^*\langle \eta|Q^\dagger Q|\eta\rangle, \quad t \in \mathbb{C}
\end{aligned} \tag{4.469}$$

has its minimum at $t = 0$. Taking the derivative with respect to t^* and setting $t = 0$
we get

$$\langle \eta|Q^\dagger|\psi_0\rangle = 0 \quad \text{for all } \langle \eta| \in \mathcal{H} \tag{4.470}$$

which, together with (4.467), implies

$$Q|\psi_0\rangle = Q^\dagger|\psi_0\rangle = 0 \quad \Rightarrow \quad 0 = (QQ^\dagger + Q^\dagger Q)|\psi_0\rangle = 2H|\psi_0\rangle, \tag{4.471}$$

that is, all non-zero cohomology classes are represented by a zero-energy eigenvector, namely the vector of smallest norm in the class. On the other hand any
zero-energy eigenvector is Q-closed, and hence represents a Q-cohomology class,
and is the vector of smallest norm in its class by Eq. (4.469).   □

Examples: Relative Cohomology and Morse Theory
We present a few elementary examples of the use of the Witten index which
also illustrate its deep relations with various topics in Algebraic and Differential
Topology.

Example 4.17 (Superparticle Moving on $\mathbb{R}$) We return to the system of Example 4.15. We take the superpotential $W(x)$ to be a *generic* real polynomial of
degree n. By the previous arguments, the Witten index Δ is invariant under all
continuous deformations of the coefficients of the polynomial $W(x)$ that do not

change the degree n of the polynomial.[49] We consider the family of superpotentials $W_t(x) = t\, W(x)$ $(t > 0)$ which leads to a family of Hamiltonians

$$H_t = \frac{1}{2}p^2 + \frac{t^2}{2}(W')^2 + \frac{t}{2}W''\sigma_3 \qquad (4.472)$$

Note that the rescaling $H_t \to t^{-2}H_t$ identifies t with $1/\hbar$, so that the limit $t \to \infty$ is just the classical limit. In this limit the zero energy configurations are the points $x_a \in \mathbb{R}$ where the potential vanishes, i.e. the real zeros x_a of the polynomial $W'(x)$ of degree $(n - 1)$:

$$W'(x_a) = 0. \qquad (4.473)$$

The number N_0 of real zeros of $W'(x)$ satisfies

$$N_0 \leq n - 1, \quad N_0 = n - 1 \bmod 2. \qquad (4.474)$$

In particular when $(n - 1)$ is odd, we have *at least one* real zero. As $t \to +\infty$ the wave-functions of the ground states localize at the minima of the potentials, and the Schrödinger equation reduces to local equations around these critical points. We expand the Hamiltonian near the zero x_a getting

$$H_t|_{x_a} \approx \frac{1}{2}p^2 + \frac{t^2}{2}W''(x_a)^2(x - x_a)^2 + \frac{t}{2}W''(x_a)\sigma_3 \qquad (4.475)$$

which is the supersymmetric harmonic oscillator (cf. Example 4.13) which has a unique zero-energy state which is bosonic (resp. fermionic) iff $W''(x_a) > 0$ (resp. $W''(x_a) < 0$). That is, the classical ground state x_a has a grading $(-1)^F$ equal to the sign of the Hessian of W at x_a, i.e. x_a is a bosonic (resp. fermionic) ground state iff it is a local minimum (resp. maximum) of the superpotential $W(x)$. Hence, the computation of the Witten index in the limit $t \to \infty$ (equivalently $\hbar \to 0$) gives

$$\Delta = \sum_{\substack{\text{real zeros} \\ x_a \text{ of } W'(x)}} \operatorname{sign}\big(W''(x_a)\big). \qquad (4.476)$$

Since local maxima and local minima alternate along $\mathbb{R}$ for all real function $W(x)$, Δ is zero when N_0 is even while $\Delta = \pm 1$ when N_0 is odd, the sign being $+1$ if the rightmost zero of W' is a local minimum and -1 if it is a local maximum. In particular Δ depends only on the degree of $W(x)$ and the sign of its leading coefficient, as required by **WI1**. This index computation agrees with the explicit solution of the Schrödinger equation in Example 4.15.

---

[49] The sign of the leading coefficient is invariant under such deformations.

Remark 4.16 (Relative Cohomology) The above discussion applies to all smooth superpotentials on the line $\mathbb{R}$ with a finite number of classical ground states i.e. with a finite number of critical points ($\equiv$ zeros of $W'(x)$). The number of bosonic/fermionic ground states depends only on the asymptotic behavior of $W(x)$ as $x \to \pm\infty$. Let Λ be some large real number, and $\mathbb{R}_{<-\Lambda} \subset \mathbb{R}$ be the domain in $\mathbb{R}$ where $W(x) < -\Lambda$. Then the spaces of even/odd ground states are isomorphic, respectively, to the even/odd *relative cohomology groups* [43]

$$H_Q^0 \simeq H^0(\mathbb{R}, \mathbb{R}_{<-\Lambda}) \tag{4.477}$$

$$H_Q^1 \simeq H^1(\mathbb{R}, \mathbb{R}_{<-\Lambda}). \tag{4.478}$$

This isomorphism between the Hilbert-space Q-cohomology and the relative cohomology of the pair $(\mathcal{M}, \mathcal{M}_{<-\Lambda})$, where $\mathcal{M}$ is the configuration space and $\mathcal{M}_{<-\Lambda} \subset \mathcal{M}$ the domain where the superpotential is $< -\Lambda$, is in fact quite general: it applies to all (connected, oriented) Riemannian manifold $\mathcal{M}$ and all smooth superpotentials with finitely many critical points:

$$H_Q^a \simeq \bigoplus_{k=a \bmod 2} H^k(\mathcal{M}, \mathcal{M}_{<-\Lambda}) \tag{4.479}$$

for Λ large enough.[50] See [44].

Example 4.18 (Free Superparticle Moving on a Compact $\mathcal{M}$) The Hamiltonian is proportional to the Hodge Laplacian acting on differential forms, so the wavefunctions of the zero-energy states with Fermi number $F = k$ are precisely the harmonic k-forms. The number of such states is then, by definition, the k-th Betti number $b_k \equiv \dim H^k(\mathcal{M})$ which is a topological invariant of $\mathcal{M}$. Thus, in this situation, the number of ground states of any given Fermi number k is invariant under all deformations of the Riemannian metric on $\mathcal{M}$. The Witten index is

$$\Delta = \sum_{\substack{\text{ground} \\ \text{states}}} (-1)^F \equiv \sum_{k=0}^{m} (-1)^k b_k \equiv \chi(\mathcal{M}) \tag{4.480}$$

where $\chi(\mathcal{M})$ is the Euler characteristic of the compact manifold $\mathcal{M}$, which is a basic topological invariant of the manifold.

Example 4.19 (Morse Theory) More generally our superparticle moving in the *compact* manifold $\mathcal{M}$ may be subjected to a superpotential $W(x)$ that we take to be a sufficiently generic smooth function $W: \mathcal{M} \to \mathbb{R}$. Again, we consider the family of superpotentials $W_t = tW$. The Witten index cannot depend on t. Taking $t \to 0$ we get back that $\Delta = \chi(\mathcal{M})$. Indeed, in this very simple set-up, we have the stronger result:

---

[50] That is, $\Lambda > \max_a |W(x_a)|$ where $W(x_a)$ are the critical values of $W(x)$.

Fact 4.15 *The number N_k of SUSY ground states with Fermi number k is independent of t for each k individually (and not just their signed sum $\Delta \equiv \sum_k (-1)^k N_k$). Setting $t = 0$ we get*

$$N_k = b_k \quad \text{the } k\text{-th Betti number.} \tag{4.481}$$

This follows from the fact that the supercharge $Q_t \equiv \mathrm{e}^{-tW} \mathrm{d}\, \mathrm{e}^{tW}$ is conjugate to the exterior derivative d and hence the dimension of its cohomology spaces are independent of t and equal to $\dim H^k_{\mathrm{DR}}(\mathcal{M})$.

More interesting is the classical limit $t \to \infty$. We get one classical zero-energy state for critical point $x_a \in \mathcal{M}$ where the differential of W vanishes:

$$\mathrm{d}W(x_a) = 0. \tag{4.482}$$

The argument after Eq. (4.475) gives that the Fermi number of the classical ground state at x_a is the number of negative eigenvalues[51] of the Hessian $\partial_{x_i} \partial_{x_j} W|_{x_a}$ called the *Morse index $i(x_a)$ of the critical point x_a*. Then the equality

$$\text{Witten index} \equiv \chi(\mathcal{M}) = \sum_{\substack{\text{critical} \\ \text{points}}} (-1)^{i(x_a)} \tag{4.483}$$

should hold for all (generic) smooth function $W : \mathcal{M} \to \mathbb{R}$. Indeed this is the basic result of Morse theory [45, 46]. For more details see Witten [47].

Remark 4.17 Example 4.19 may be generalized to $\mathcal{M}$ non-compact using the relative strategy in Remark 4.16. In the non-compact case the invariance of the Witten index reproduces all the basic results of the *relative Morse theory*.

Remark 4.18 More generally all theorems in *index theory* may be easily proven by applying the Witten index technique to suitable SQM systems [48]. In the Appendix to Chap. 5 these techniques are used to prove the Riemann-Roch theorem.

Problems

4.1 Show identity (4.159).

4.2 Write the wave-functions of a free particle moving on the unit sphere $S^3 \subset \mathbb{R}^4$ in terms of the harmonic polynomials in $\mathbb{R}^4$ restricted to the unit sphere.

---

[51] Since we are assuming that $W(x)$ is generic, the Hessian at the critical points x_a has no zero eigenvalue.

References

1. S. Cecotti, *Analytic Mechanics. A Concise Textbook* (Springer, Berlin, 2024)
2. P. Sharan, *Some unusual topics in Quantum Mechanics*. Lecture Notes in Physics, vol. 1020, 2nd edn. (Springer, Berlin, 2023)
3. S. Helgason, *Differential Geometry and Symmetric Spaces* (Academic Press New-York/London, 1962), 2nd edn. (1978)
4. A.L. Besse, *Einstein Manifolds* (Springer, Berlin, 1987)
5. P. Griffiths, J. Harris, *Principles of Algebraic Geometry* (Wiley, Hoboken, 1978)
6. D. Huybrechts, *Complex Geometry. An Introduction.* Universitext (Springer, Berlin, 2005)
7. J.-P. Demailly, *Complex Analytic and Differential Geometry.* Book on line https://www-fourier.ujf-grenoble.fr/demailly/manuscripts/agbook.pdf
8. S. Kobayashi, *Transformations Groups in Differential Geometry.* Classics in Mathematics (Springer, Berlin, 1995)
9. C. Chevalley, S. Eilenberg, Cohomology theory of Lie groups and Lie algebras. Trans. Am. Math. Soc. **63**, 85–124 (1948)
10. P.J. Hilton, U. Stammbach, *A Course in Homological Algebra*, 2nd edn. (Springer, Berlin, 1997)
11. N.R. Wallach, *Real Reductive Groups I* (Academic Press, Cambridge, 1988); N. R. Wallach, *Real Reductive Groups II* (Academic Press, Cambridge, 1992)
12. J.-P. Anker, B. Orsted (eds.), *Lie Theory. Harmonic Analysis on Symmetric Spaces – General Plancherel Theorems.* Progress in Mathematics, vol. 230 (Birkhaüser, Basel, 2005)
13. D.W. Morris, *Ratner's Theorems on Unipotent Flows.* Book available on-line at arXiv:math/0310402
14. M.M. Postnikov, *Geometry VI. Riemannian Geometry* (Springer, Berlin, 2001)
15. M.M. Postnikov, *Leçons de géometrie. Groupes et algébres de Lie* (Mir, Moscow, 1982)
16. S. Lang, *Fundamentals of Differential Geometry.* Graduate Texts in Mathematics, vol. 191 (Springer, Berlin, 1999)
17. J.S. Milne, *Algebraic Groups. The Theory of Group Schemes of Finite Type over a Field* (Cambridge University Press, Cambridge, 2017)
18. S. Cecotti, *Statistical Mechanics. A Concise Advanced Textbook* (Springer, Berlin, 2024)
19. L.D. Landau, E.M. Lifschitz, *Quantum Mechanics. Non-Relativistic Theory*, 3rd edn. (Butterworth-Heinemann, Oxford, 1981)
20. J.-P. Serre, *Complex Semisimple Lie Algebras* (Springer, Berlin, 2001)
21. D. Bump, *Lie Groups.* Graduate Texts in Mathematics, vol. 225, 2nd edn. (Springer, Berlin, 2013)
22. G. Szegö, *Orthogonal Polynomials* (AMS, Providence, 1975)
23. F.W. Olver, D.M. Lozier, R.F. Boisvert, C.W. Clark, W. Charles W. (eds.), *NIST Handbook of Mathematical Functions* (Cambridge University Press, Cambridge, 2010). Available on-line at https://dlmf.nist.gov
24. G.E. Andrews, R. Askey, R. Roy, *Special Functions.* Encyclopedia of Mathematics and Its Applications, vol. 71 (Cambridge University Press, Cambridge, 2009)
25. I.S. Gradshteyn, I.M. Ryzhik, *Table of Integrals, Series, and Products*, 7th edn. (Elsevier, Amsterdam, 2007)
26. K. Konishi, G. Paffuti, *Quantum Mechanics. A New Introduction* (Oxford University Press, Oxford, 2009)
27. S. Axler, P. Bourdon, W. Ramey, *Harmonic Function Theory.* Graduate Texts in Mathematics, vol. 137, 2nd edn. (Springer, Berlin, 2001)
28. J. Avery, *Hyperspherical Harmonics. Applications in Quantum Theory.* Reidel Texts in the Mathematical Sciences, vol. 5 (Kluwer Academic Publishers, Dordrecht, 1989)
29. J. Humphreys, *Introduction to Lie Algebras and Representation Theory* (Springer, Berlin, 1980)
30. A.W. Knapp, *Lie Groups Beyond an Introduction*, 2nd edn. (Birkhäuser, Basel, 2002)

31. B.C. Hill, *Lie Groups, Lie Algebras, and Representations. An Elementary Introduction* (Springer, Berlin, 2010)
32. N. Chriss, V. Ginzburg, *Representation Theory and Complex Geometry*. Modern Birkhaüser Classics (Birkhaüser, Basel, 2010)
33. T. Bröker, T. tom Dieck, *Representations of Compact Lie Groups*. Graduate Texts in Mathematics, vol. 98 (Springer, Berlin, 1985)
34. T.R. Govindarajan, P. Ramadevi, *Geometry and Topology of Low Dimensional Systems. Chern-Simons Theory with Applications* (Springer, Berlin, 2024)
35. W. Pauli, The connection between spin and statistics. Phys. Rev. **58**, 716–722 (1940)
36. R.F. Streater, A.S. Wightman, *PCT, Spin and Statistics, and All That*. Princeton Landmarks in Mathematics and Physics (Princeton University Press, Princeton, 2001)
37. V.G. Kac, Lie superalgebras. Adv. Math. **26**, 8 (1977)
38. V. Moretti, *Spectral Theory and Quantum Mechanics. Mathematical Foundations of Quantum Theories, Symmetries and Introduction to the Algebraic Formulation*, 2nd edn. (Springer, Berlin, 2017)
39. J. Jost, *Riemannian Geometry and Geometric Analysis*. UniversiText, 7th edn. (Springer, Berlin, 2017)
40. S.I. Goldberg, *Curvature and Homology* (Dover, Garden City, 1970)
41. E. Witten, Constraints on supersymmetry breaking. Nucl. Phys. **B202**, 253 (1982)
42. S. Cecotti, L. Girardello, Functional measure, topology and dynamical supersymmetry breaking. Phys. Lett. B **110**, 39 (1982)
43. R. Bott, L.W. Tu, *Differential Forms in Algebraic Topology* (Springer, Berlin, 1982)
44. K. Hori, A. Iqbal, C. Vafa, D-branes and mirror symmetry. arXiv:hep-th/0005247
45. J. Milnor, *Morse Theory* (Princeton University Press, Princeton, 1963)
46. M.W. Hirsch, *Differential Topology*. Graduate Texts in Mathematics, vol. 33 (Springer Berlin, 1976)
47. E. Witten, Supersymmetry and Morse theory. J. Differ. Geom. **17**, 661–692 (1982)
48. L. Alvarez-Gaumé, Supersymmetry and the Atiyah-Singer index theorem. Commun. Math. Phys. **90**, 161 (1983)

Chapter 5
Schrödinger Equation II

In this chapter we study the Schrödinger equation in some physically relevant systems with several degrees of freedom, including the hydrogen atom and the Landau levels of an electron moving in an uniform magnetic field in either the Euclidean or the Hyperbolic space. The central potential $-\alpha/r$ is studied in an arbitrary number of space dimensions to highlight its deep algebraic and geometric structures. The motion of a charged particle in the field of a magnetic monopole is also analyzed.

In all the systems we consider the Schrödinger equation is *separable* because of the existence of enough commuting conserved quantities. In some instance the system is *superseparable*: the dynamics of these lucky models may be solved by purely algebraic methods exploiting the Representation Theory of their higher symmetry. We also put to good use the Peter-Weyl theory developed in Chap. 4 to solve the quantum rotator and other quantum systems whose configuration space $\mathcal{M}$ is a compact Lie group or, more generally, a homogeneous Riemannian manifold.

5.1 Quantum Particles in a Central Potential

We have seen in Sect. 4.11 that the theorem of the center-of-mass holds in Quantum Mechanics in the same form as in the classical theory (for systems having a classical analogue). Therefore the *quantum two-body problems*—where the two bodies interact through a potential $V(r)$ which depends only on their distance r—reduce to single-body problems where the one body moves in the spherical symmetric potential $V(r)$, while the mass of the single body is the *reduced mass*

$$m = \frac{m_1 m_2}{m_1 + m_2} \tag{5.1}$$

of the masses of the two original bodies ([1] §. 5.2). For details see Problem 5.1.

© The Author(s), under exclusive license to Springer Nature Switzerland AG 2025
S. Cecotti, *Quantum Mechanics*, UNITEXT for Physics,
https://doi.org/10.1007/978-3-031-98824-0_5

We now study the reduced system of a quantum particle (without spin) moving in $\mathbb{R}^d$ ($d \geq 2$) subjected to a potential $V(r)$ which depends only on the distance from the origin. The Hamiltonian operator has the general form

$$H = \frac{1}{2m} p_i p_i + V\left(\sqrt{q^i q^i}\right) \tag{5.2}$$

($i = 1, 2, \ldots, d \geq 2$). The most relevant case is $d = 3$, the dimension of the ordinary physical space, but we shall be general in order to highlight the deep nature of the symmetries and their implications. To have a physically meaningful system, the Hamiltonian should be bounded below; in particular this requires that $V(r)$ is bounded below outside a small ball of radius r_0 centered at the origin of $\mathbb{R}^d$.

The Hamilton-Jacobi equation of the corresponding classical model is integrable by separation of variables for all d: in facts the Hamiltonian is invariant under arbitrary $O(d)$ rotations around the origin; together with conservation of energy, this $O(d)$ symmetry provides enough conserved quantities in involution to make the classical model integrable in the Liouville sense ([1] chaps. 9, 10). In the classical theory the motion takes place in a fixed plane, so d may be reduced to 2 without loss. This property does not hold in the quantum case (since the notion of "orbit" is no longer defined), however we still have the $O(d)$ symmetry and this suffices to guarantee integrability of the Schrödinger equation by separation of variables as we observed in Chap. 3.

In polar coordinates the (stationary) Schrödinger equation takes the form

$$\left(-\frac{\hbar^2}{2m} \frac{1}{r^{d-1}} \frac{\partial}{\partial r} r^{d-1} \frac{\partial}{\partial r} + \frac{\hbar^2}{2mr^2} \Delta_{S^{d-1}} + V(r) \right) \psi(r, \omega) = E\,\psi(r, \omega), \tag{5.3}$$

where $\Delta_{S^{d-1}}$ is the Laplacian of the round metric on the unit sphere $S^{d-1} \subset \mathbb{R}^d$, and ω are angular coordinates in S^{d-1}. To separate the variables, we look for solutions of the form

$$\psi(r, \omega) = R_E(r)\, Y_{\lambda,m}(\omega) \tag{5.4}$$

where $Y_{\lambda,m}(\omega)$ are *spherical harmonics in d dimensions*. The spherical harmonics satisfy the following equation and normalization condition

$$\Delta_{S^{d-1}} Y_{\lambda,m}(\omega) = \lambda\, Y_{\lambda,m}(\omega) \tag{5.5}$$

$$\int_{S^{d-1}} d\Omega(\omega)\, Y_{\lambda,m}(\omega)^*\, Y_{\lambda,m'}(\omega) = \delta_{m,m'}. \tag{5.6}$$

where $d\Omega(\omega)$ is the volume form on S^{d-1}. The function $R_E(r)$ satisfies the ODE

$$\left(-\frac{1}{r^{d-1}} \frac{d}{dr} r^{d-1} \frac{d}{dr} + \frac{\lambda}{r^2} + \frac{2m}{\hbar^2} V(r) \right) R_E(r) = \frac{2mE}{\hbar^2}\, R_E(r). \tag{5.7}$$

Depending on the physical situation, the energy E may belong to the *discrete* or to the *continuum* spectrum of the Hamiltonian H. As discussed in Sect. 3.2, the first case corresponds to a *bound state* where the particle bounds itself to the center of force or, more physically, the two components of the original two-body system attract each other so strongly that they cannot escape away from their partner, and form a *stable* composed object. The prototypical example is a proton and an electron forming a hydrogen atom. The continuum eigenstates describe physical processes where the particle comes in from infinity, scatters in the potential, and then escapes again to spatial infinity.[1] For discrete energy levels E_n the normalization condition reads

$$\int_0^\infty r^{d-1}\, dr\, |R_{E_n}(r)|^2 = 1. \tag{5.8}$$

As shown in Chap. 4, the Hilbert space $L^2(\mathbb{R}^d)$ decomposes in finite-dimensional unitary representations of the compact semisimple Lie group $SO(d)$. The eigenvalue λ in Eq. (5.5) is the *quadratic Casimir* of the $SO(d)$-representation, and the *d-dimensional spherical harmonics* $Y_{\lambda,m}(\omega)$ are constructed as in the Chap. 4 using the Representation Theory of the Lie group $SO(d)$. The index m labels the states forming a basis of the irreducible representation with Casimir eigenvalue λ. We have seen in Chap. 4 that for small d the spherical harmonics are[2]

d	Casimir		Spherical harmonic
$d = 2$	$\lambda = \ell^2$	$\ell \in \mathbb{N}$	$Y_{\ell,\pm}(\phi) = e^{\pm i\ell\phi}$
$d = 3$	$\lambda = \ell(\ell+1)$	$\ell \in \mathbb{N}$	$Y_{\ell,m}(\phi,\theta)$
$d = 4$	$\lambda = \ell(\ell+2)$	$\ell \in \mathbb{N}$	$\sqrt{\ell+1}\, R_{\ell/2,m,m'}(\phi,\theta,\psi)$

$$\tag{5.9}$$

More generally we can construct the d-dimensional *spherical harmonics* by restricting the *harmonic polynomials* in $\mathbb{R}^d$ to the unit sphere $S^{d-1} \subset \mathbb{R}^d$, see Sect. 4.6. There we saw that $L^2(S^{d-1})$ is a direct sum of pair-wise non-isomorphic irreducible representations

$$L^2(S^{d-1}) = \widehat{\bigoplus_{\ell \in \mathbb{N}}} V_\ell^{(d)} \tag{5.10}$$

where the irreducible representation $V_\ell^{(d)}$ has Casimir eigenvalue

---

[1] Under mild regularity conditions on $V(r)$ (say $r^2 V(r)$ is bounded below), it is not possible that an incoming particle gets "captured" by the center of force. Indeed, the differential operator in the LHS of (5.15) is real and has a unique (up to scale) non-zero solution satisfying the boundary condition (5.18) which may be chosen to be real. Then, for r large, $\chi_E(r) \approx C e^{i p_r r} + C^* e^{-i p_r r}$ and the amplitudes of the waves incoming from and outgoing to infinity are equal, that is, the net probability that the particle gets "trapped" in a bounded region is zero.

[2] For the case $d = 4$ compare with the **Warning** in the box on page 262.

$$\lambda_\ell = \ell(\ell + d - 2), \quad \ell = 0, 1, 2, 3, \cdots \tag{5.11}$$

and dimension

$$\dim V_\ell^{(d)} \equiv N_\ell^{(d)} = \binom{\ell + d - 1}{d - 1} - \binom{\ell + d - 3}{d - 1}. \tag{5.12}$$

The radial Eq. (5.7) may be simplified using the differential identity

$$\begin{aligned}
\frac{1}{r^{d-1}} \frac{d}{dr}\left[r^{d-1} \frac{d}{dr}\left(\frac{f(r)}{r^{(d-1)/2}} \right) \right] &= \\
&= \frac{1}{r^{(d-1)/2}}\left[f''(r) - \frac{(d-1)(d-3)}{4r^2} f(r) \right],
\end{aligned} \tag{5.13}$$

where the primes stand for derivatives with respect to r. Then, setting

$$\chi_E(r) = r^{(d-1)/2} R_E(r), \tag{5.14}$$

we get the linear ODE

$$-\chi_E'' + \frac{4\lambda + (d-1)(d-3)}{4r^2}\chi_E + \frac{2m}{\hbar^2}\left(V(r) - E \right)\chi_E(r) = 0, \tag{5.15}$$

while the normalization condition (5.8) also simplifies

$$\begin{aligned}
\int_0^\infty dr\, |\chi_{E_n}(r)|^2 = 1 && E_n \in \sigma_p(H) \\
\int_0^\infty dr\, \chi_{E'}(r)^* \chi_E = \delta(E - E') && E, E' \in \sigma_e(H).
\end{aligned} \tag{5.16}$$

The ODE (5.15) and the normalization (5.16) have exactly the same form as for the one-dimensional Schrödinger equation in the half-line $r \geq 0$ subjected to the effective potential

$$V_{\text{eff}}(r) = \frac{\hbar^2}{2m} \frac{(2\ell + d - 2)^2 - 1}{4\,r^2} + V(r), \tag{5.17}$$

where the non-negative integer ℓ is the *angular quantum number*. From Eq. (5.14) we also see that regularity of the wave-function $R_E(r)$ gives the boundary condition

$$\chi_E(0) = 0, \tag{5.18}$$

and the equivalence with the one-dimensional problem in the half-line $\mathbb{R}_{\geq 0}$ is then complete. We shall call Eq. (5.15) (with the Dirichlet condition at $r = 0$) the *radial*

Schrödinger equation. The reduction of the central-potential problem to the radial Schrödinger equation is the quantum counterpart of the classical reduction to a one-dimensional effective system describing the radial motion ([1] §. 9.3.2).

Having reduced the 2-body problem to a one-dimensional Schrödinger equation, all the results of Chap. 3 apply (in particular the Sturm-Liouville theorems):

(1) Let $V_\infty = \lim_{r \to \infty} V(r)$. States of energy $E < V_\infty$ are *bound states,* while states with energy $E > V_\infty$ are *scattering states.* States at the threshold $E = V_\infty$ require a case-by-case analysis.

(2) In the discrete spectrum (i.e. for bound states) the *oscillation theorem* holds: the n-th eigenfunction $\chi_{E_n}(r)$ has n nodes and the nodes of successive eigenfunctions interlace.

The wave-function of the ground state should not vanish anywhere.[3] Since the functions $Y_{\lambda,m}(\omega)$ have zeros unless they correspond to the trivial representation of $SO(d)$, we conclude that *the ground state belongs to the trivial representation* (in the traditional terminology: it is an *s*-wave). In other words: the $SO(d)$ symmetry is never spontaneously broken, a result that we shall prove in a much broader context in Chap. 6.

Behavior Near the Origin

We consider the asymptotic behavior of $R_E(r)$ as $r \to 0$. There are five possibilities: *(i)* as $r \to 0$ the function $r^2 V(r)$ may go to zero, or *(ii)* may go to a finite constant κ, or *(iii)* may go to $+\infty$, or *(iv)* may go to $-\infty$, or *(v)* the limit may not exist.

In case *(i)*, the ODE (5.7) reduces as $r \to 0$ to the asymptotic equation

$$-\frac{1}{r^{d-1}}\frac{\mathrm{d}}{\mathrm{d}r}r^{d-1}\frac{\mathrm{d}}{\mathrm{d}r}R + \frac{\lambda}{r^2}R = 0. \tag{5.19}$$

Setting $R = \mathrm{const.}\, r^\alpha$ we get a quadratic equation for the exponent α:

$$-\alpha(\alpha + d - 2) + \lambda = 0. \tag{5.20}$$

Using Eq. (5.11) for λ, we see that the unique solution which is regular as $r \to 0$ behaves as

$$R(r) \approx \mathrm{const.}\, r^\ell \quad \text{for } r \approx 0 \tag{5.21}$$

---

[3] The statement holds for systems with no magnetic fields. See *a contrario* Sect. 5.7.

for all spatial dimensions d. In case *(ii)* we get the same asymptotic Eq. (5.19) with the replacement

$$\lambda \rightsquigarrow \tilde{\lambda} \equiv \lambda + 2m\kappa/\hbar^2. \tag{5.22}$$

If $\tilde{\lambda} < 0$ the Eq. (5.20) has no real positive solution.[4] Then when κ is negative and $|\kappa|$ large enough, the particle falls in the center, see §. 35 of [3] for details.

In case *(iii)* the wave-function vanishes as $r \to 0$ more rapidly than any power, while in case *(iv)* it grows more rapidly than any power, so the particle falls in the center and there is no physically admissible solution.

5.2   The Free Particle in Polar Coordinates

In Sect. 3.4 we studied a quantum free particle moving in $\mathbb{R}^d$ using translational invariance or, equivalently, the separation in *Cartesian coordinates* of the free Schrödinger equation in the Hilbert space $L^2(\mathbb{R}^d)$. The free motion in $\mathbb{R}^d$ is classically *superseparable* ([1] chap. 9) and this property extends to the quantum setting. We already mentioned the separation of the free particle Schrödinger equation in *parabolic cylinder* coordinates in the **box** on page 140: in these coordinates we reduce to a pair of one-dimensional systems satisfying the parabolic-cylinder ODE ($\equiv$ biconfluent hypergeometric). Alternatively one may use *parabolic-spherical* coordinates which reduce the free Schrödinger equation to (a special case of) the confluent hypergeometric ODE, cf. Problem 5.2. In §§. 9.6.1, 9.6.2 of ref. [1] the separation of the free system was also discussed in *spherical* coordinates, *polar-cylindrical* coordinates, in the *elliptic-spherical* coordinates, *elliptic-cylindrical* coordinates, and any combinations thereof ([1] §§. 9.6.1, 9.6.2). The handbook [4] lists eleven inequivalent coordinate systems which separate the Schrödinger equation[5] for a free particle moving in $\mathbb{R}^3$. Here we limit ourselves to separation in spherical coordinates, leaving the elliptic ones to the diligent reader as Problem 5.3, and referring the reader to [4] for fancier coordinate systems.

To simplify the following expressions we set

$$k \equiv \sqrt{\frac{2mE}{\hbar^2}}, \qquad \nu \equiv \ell + \frac{d}{2} - 1. \tag{5.23}$$

---

[4] By Descartes' rule of signs [2].

[5] The Schrödinger equation for a quantum particle moving in $\mathbb{R}^3$ in absence of potentials is equivalent to the Helmholtz equation whose separation is discussed in [4].

The (reduced) radial one-dimensional Schrödinger equation then reads

$$\chi'' - \frac{\nu^2 - \frac{1}{4}}{r^2}\chi + k^2\chi = 0. \tag{5.24}$$

This is a well-known form of the Bessel ODE (see e.g. **8.491**.5 of [5]). A basis of its solutions is given by

$$\chi_1(r) = \sqrt{r}\, J_\nu(kr) \quad \text{and} \quad \chi_2(r) = \sqrt{r}\, Y_\nu(kr), \tag{5.25}$$

where $J_\nu(z)$ and $Y_\nu(z)$ are the *Bessel functions of index ν of the first* and, respectively, *second kind* [5–7]. We know quite a lot about these special functions: dozens of different integral representations, series expansions, functional relations, asymptotic series, etc., see e.g. [5–7] or the classical treatises [8, 9]. The behavior of these functions as $z \to 0$ is[6]

$$J_\nu(z) = \left(\frac{z}{2}\right)^\nu \frac{1}{\Gamma(\nu+1)}\left(1 + O(z^2)\right) \tag{5.26}$$

$$Y_\nu(z) \approx \begin{cases} \left(\frac{z}{2}\right)^{-\nu} \frac{1}{\Gamma(-\nu+1)} & \nu \neq 0 \\ 2\log\left(\frac{z}{2} + C\right) & \nu = 0. \end{cases} \tag{5.27}$$

The solution such that $R(r) \approx r^\ell$ as $r \to 0$ is then written in terms of the *Bessel function of the first kind*

$$R(r) = \frac{\chi(r)}{r^{(d-1)/2}} = r^{1-d/2} J_\nu(kr) \approx$$
$$\approx \text{const.}\, r^{1-d/2+\nu} \equiv \text{const.}\, r^\ell, \qquad r \approx 0, \tag{5.28}$$

hence the regular solution to Eq. (5.24) in d spatial dimensions is

$$\chi_{k,\nu}(r) = \sqrt{r}\, J_\nu(kr) \quad \text{i.e.} \quad R_{k,\ell}(r) = \frac{J_{\ell+d/2-1}(kr)}{r^{d/2-1}}. \tag{5.29}$$

We claim that the radial wave-functions (5.29) are properly normalized as generalized eigenfunction which belong to the continuum energy spectrum, that is,

$$\int_0^\infty r^{d-1}\, dr\, R_{k,\ell}(r)\, R_{k',\ell}(r) = \delta(k - k'). \tag{5.30}$$

---

[6] Here $C \approx 0.5772$ is Euler's constant.

This follows from the well-known asymptotics of the Bessel function $J_\nu(z)$ for large argument z [5, 6]

$$J_\nu(z) \approx \sqrt{\frac{2}{\pi z}} \cos\left(z - \frac{\pi \nu}{2} - \frac{\pi}{4}\right) + O(z^{-3/2}), \tag{5.31}$$

so that the *normalized* radial wave-function behaves for large r is

$$R_{k,\ell} \approx \sqrt{\frac{2}{\pi}} \frac{1}{r^{(d-1)/2}} \cos\left(kr - \frac{\pi(2\ell + d - 1)}{4}\right) \qquad kr \ggg 1 \tag{5.32}$$

which gives Eq. (5.30). The Bessel functions satisfy the recursion relations

$$z\frac{\mathrm{d}}{\mathrm{d}z} J_\nu(z) \pm \nu J_\nu(z) = \pm z J_{\nu \mp 1}(z) \tag{5.33}$$

$$2\frac{\mathrm{d}J_\nu}{\mathrm{d}z} = J_{\nu-1} - J_{\nu+1}, \tag{5.34}$$

$$2\nu J_\nu = z J_{\nu-1} + z J_{\nu+1}. \tag{5.35}$$

In the most relevant case, $d = 3$, the solutions to the ODE (5.24) are actually *elementary functions*

$$\sqrt{z}\, J_{\ell+1/2}(z) = (-1)^\ell z^{\ell+1} \sqrt{\frac{2}{\pi}} \frac{\mathrm{d}^\ell}{(z\,\mathrm{d}z)^\ell}\left(\frac{\sin z}{z}\right) \tag{5.36}$$

E.g. for $\ell = 0$ the function is simply $\sqrt{2/\pi}\,\sin z$.

The energy eigenfunctions of the free particle moving in $\mathbb{R}^3$ with fixed angular momentum quantum numbers ℓ, m

$$\psi_{k,\ell,m}(r, \phi, \theta) = R_{k,\ell}(r)\, Y_{m,\ell}(\theta, \phi) \tag{5.37}$$

are called *spherical waves*. Traditionally one says s-wave, p-wave etc., to mean a spherical wave with $\ell = 0$, resp. $\ell = 1$, etc., see table (4.224).

5.2.1 *Expansion of Plane Waves into Spherical Waves*

We have two complete systems of energy eigenfunctions for the free particle moving in $\mathbb{R}^d$: the *plane waves* studied in Sect. 3.4 by separating variables in the Cartesian coordinates, and the *spherical waves* defined above by separation of the stationary

Schrödinger equation in spherical coordinates. There are several others complete systems of waves: *parabolic-cylindrical waves* from the separation in parabolic-cylindrical coordinates (**box** on page 140), *elliptic waves*, and so on, but we focus on the two systems which are more commonly used in the applications.

Plane waves and spherical waves are two distinct complete systems in the (generalized) energy eigenspaces of the free particle moving in $\mathbb{R}^d$: therefore there are "change of basis" coefficients relating them. Concretely this means an expansion of the plane waves in spherical waves. This expansion is a major tool for scattering problems, S-matrix theory, bootstrap equations, etc., which is called the *expansion in partial waves*. For $d = 3$ it reads [5]

$$e^{i\boldsymbol{k}\cdot\boldsymbol{r}} \equiv e^{ikr\cos\theta} = \sqrt{\frac{\pi}{2}} \sum_{\ell=0}^{\infty} i^\ell (2\ell + 1) \frac{J_{\ell+1/2}(kr)}{\sqrt{kr}} P_\ell(\cos\theta) \tag{5.38}$$

where $P_n(z)$ are the Legendre polynomials. Equation (5.38) is an expansion of the plane wave in the LHS in spherical waves with $SO(3)$ quantum numbers ($\ell, m = 0$) with m the eigenvalue of L_3 where we choose the 3rd axis to be in the direction of propagation of the plane wave $\boldsymbol{k}/|\boldsymbol{k}|$.

Proof of Eq. (5.38) To prove the formula is straightforward. Indeed:

$$i\cos\theta \sum_n i^n (2n + 1) \frac{J_{n+1/2}(z)}{\sqrt{z}} P_n(\cos\theta) =$$

$$= \sum_n i^{n+1} (2n + 1) \frac{J_{n+1/2}(z)}{\sqrt{z}} \Big[(n + 1)P_{n+1}(\cos\theta) + nP_{n-1}(\cos\theta)\Big] =$$

$$= \sum_n i^n P_n(\cos\theta) \Big[n\frac{J_{n-1/2}(z)}{\sqrt{z}} - (n + 1)\frac{J_{n+3/2}(z)}{\sqrt{z}}\Big] =$$

$$= \sum_n i^n (2n + 1) P_n(\cos\theta) \frac{d}{dz}\frac{J_{n+1/2}(z)}{\sqrt{z}} = \frac{d}{dz} \sum_n i^n (2n + 1) \frac{J_{n+1/2}(z)}{\sqrt{z}} P_n(\cos\theta)$$

$$\tag{5.39}$$

where in the second line we used the recursion relation (4.209) for the Legendre polynomials, in the third line we rearranged the sum, and in the last line we used the two recursion relations (5.35) for Bessel functions. Equation (5.39) is the 1st order linear differential equation satisfied by the LHS of (5.38) (with $z \equiv kr$)

$$\frac{d}{dz}e^{iz\cos\theta} = i\cos\theta\, e^{iz\cos\theta}, \tag{5.40}$$

hence the two functions $e^{iz\cos\theta}$ and $\sum_n i^n (2n+1)\frac{J_{n+1/2}(z)}{\sqrt{z}} P_n(\cos\theta)$ are equal up to an overall integration constant which may depend only on $\cos\theta$. As $z \to 0$

$$\sum_n i^n (2n+1)\frac{J_{n+1/2}(z)}{\sqrt{z}} P_n(\cos\theta) \to \sqrt{\frac{2}{\pi}}, \qquad (z \to 0), \tag{5.41}$$

and this fixes the overall normalization to be the one in Eq. (5.38). $\square$

The corresponding formula for $d = 2$ is [5]

$$\exp(ikr\cos\phi) = \sum_{m=-\infty}^{+\infty} i^m J_m(kr)\, e^{im\phi}, \tag{5.42}$$

called the *Jacobi-Anger expansion*. We leave its easy proof to the reader.

The general formula,[7] which holds in $\mathbb{R}^d$ for all d, is (we set $\lambda \equiv d/2 - 1$ and assume $\lambda > 0$)

$$e^{iz\cos\phi} = 2^\lambda\, \Gamma(\lambda) \sum_{k=0}^{\infty} i^k (k+\lambda)\frac{J_{k+\lambda}(z)}{z^\lambda}\, C_k^{(\lambda)}(\cos\phi) \tag{5.43}$$

where $C_k^{(\lambda)}(\cos\phi)$ are the *ultraspherical polynomials* in $\mathbb{R}^{2(\lambda+1)}$ of degree k, cf. Remark 4.8 where the generating function of these polynomials was identified with the Coulomb potential in d dimensions. An alternative explicit expression is[8]

$$C_n^{(\lambda)}(x) = \frac{(-1)^n\,(2\lambda)_n}{2^n(\lambda+\frac{1}{2})_n\, n!}\frac{1}{(1-x^2)^{\lambda-1/2}}\frac{d^n}{dx^n}(1-x^2)^{n+\lambda-\frac{1}{2}}. \tag{5.44}$$

The ultraspherical ($\equiv$ *Gegenbauer*) polynomials satisfy the 3-term recursion relation

$$C_{n+1}^{(\lambda)}(x) = 2\frac{n+\lambda}{n+1} x\, C_n^{(\lambda)}(x) - \frac{n+2\lambda-1}{n+1} C_{n-1}^{(\lambda)}(x). \tag{5.45}$$

In Problem 5.4 the reader is asked to prove Eq. (5.43).

---

[7] See §. 10.23.9 of [7].

[8] The notation $(a)_n$ stands for the *Pochhammer symbol*, see **box** on page 316.

5.3   The Isotropic Harmonic Oscillator

In Sect. 3.5 we studied the isotropic harmonic oscillator in d-dimensions with Hamiltonian (in Birkhoff form)

$$H = \frac{\omega}{2} \sum_{i=1}^{d} (p_i p_i + q^i q^i) \tag{5.46}$$

and solved it in two ways: *(i)* algebraically noticing that it is invariant under a symmetry group $U(d)$, and *(ii)* by separation of variables in Cartesian coordinates. The states with energy

$$E = \hbar \omega \left(m + \frac{d}{2} \right) \qquad m \in \mathbb{N}, \tag{5.47}$$

are

$$a_{i_1}^\dagger a_{i_2}^\dagger \cdots a_{i_m}^\dagger |0\rangle, \quad 1 \le i_j \le d. \tag{5.48}$$

They form manifestly the irreducible totally symmetric m-indices representation $\odot^m F$ of $U(d)$ (F being the fundamental d-dimensional $U(d)$-representation) with degeneracy

$$\# \left\{ \text{states with energy } E = \hbar\omega \left(m + \frac{d}{2} \right) \right\} = \binom{m + d - 1}{d - 1}. \tag{5.49}$$

The very fact that the harmonic oscillator has a dynamical[9] symmetry $U(d)$ implies that it is *superseparable* i.e. separable in more than one equivalence class of coordinates (cf. [1] §. 9.7). In facts the model (5.46) is manifestly separable in *spherical coordinates* in addition to Cartesian ones. We now study its solution in spherical coordinates which is of independent interest.

---

[9] A symmetry is *dynamical* (in this sense!) if it is not *geometric*, i.e. its (conserved) Hermitian generators M_a are not all given by Killing vectors of the configuration manifold but rather by higher rank Killing tensors, see [1].

BOX: Kummer's Confluent Hypergeometric ODE

Kummer's form of the *confluent hypergeometric ODE* is (cf. [6] §. 4.1)

$$z \frac{d^2 y(z)}{dz^2} + (b - z) \frac{dy(z)}{dz} - a\, y(z) = 0. \qquad \text{(Kum)}$$

A standard solution, denoted $M(a, b, z)$, is the generalized hypergeometric sum $_1F_1(a, b; z)$

$$M(a, b, z) \equiv {}_1F_1(a, b; z) \stackrel{\text{def}}{=} \sum_{s=0}^{\infty} \frac{(a)_s}{(b)_s\, s!} z^s$$

where, as always, (for $s \in \mathbb{N}$)

$$(a)_s \stackrel{\text{def}}{=} a(a+1)(a+2)\cdots(a+s-1) \equiv \frac{\Gamma(a+s)}{\Gamma(a)}$$

is the *Pochhammer symbol*. See [5] **9.211** for integral representations and functional relations for the solutions of Kummer's ODE. A second linear-independent solution is $z^{1-b} {}_1F_1(a + 1 - b, 2 - b; z)$. *Kummer's first transformation* for the generalized hypergeometric sum $_1F_1(a, b; z)$ is

$$_1F_1(a, b; z) = e^z \, {}_1F_1(b - a, b; -z)$$

Another solution of (Kum), more important for Quantum Mechanics, is the *Kummer function* $U(a, b, z)$, which for large z goes as $U(a, b, z) \approx z^{-a}$ while all others solutions grow exponentially as $z \to \infty$. The *principal determination* of $U(a, b, z)$ is defined in the complex plane with a cut along the negative real axis. It is easy to check that is admits the integral representation

$$U(a, b, z) = \frac{1}{\Gamma(a)} \int_0^{\infty} e^{-zt} t^{a-1} (1 + t)^{b-a-1}\, dt \qquad (\spadesuit)$$

and the Poincaré asymptotic expansion for large z

$$U(a, b, z) \approx z^{-a} \sum_{s=0}^{\infty} (-1)^s \frac{(a)_s (a - b + 1)_s}{s!} z^{-s} \qquad (\blacktriangledown)$$

From ($\spadesuit$) we see that $U(a, b, z)$ diverges as $z \to 0$ unless $a = 0, -1, -2, -3, \cdots$ in which case the Poincaré asymptotic series ($\blacktriangledown$) becomes a finite sum, and $U(a, b, z)$ reduces to a polynomial in z of degree $-a$ proportional to a *Laguerre polynomial*

$$U(-n, \alpha + 1, z) = (-1)^n\, n!\, L_n^{(\alpha)}(z) \qquad n = 0, 1, 2, \cdots$$

The Kummer function $U(a, b, z)$ satisfies a number of functional/recursions relations [5, 6] e.g.

$$\frac{d}{dz} U(a, b, z) = -a\, U(a + 1, b + 1, z), \qquad U(a, b, z) = z^{1-b}\, U(a - b + 1, 2 - b, z)$$

5.3.1   Solution in Spherical Coordinates

After separation of variables, the reduced radial Schrödinger equation (5.15) takes
the form ($\hbar = 1$)

$$\chi'' - \frac{\nu^2 - \frac{1}{4}}{r^2}\chi - r^2\chi + \frac{2E}{\omega}\chi = 0. \tag{5.50}$$

where ν is as in Eq. (5.23). Comparing with the solution in Cartesian coordinates, we
know a priori that the solutions are of the form of $\exp(-r^2/2)$ times a polynomial
in r. The Sturm-Liouville spectral problems whose eigenfunctions have all the form
of a polynomial times a *fixed* function have been classified by Bochner [10, 11].
A quick look to the list of possibilities shows that the relevant polynomials must
belong to the family of *Laguerre polynomials*

$$L_n^{(\alpha)}(r) \overset{\text{def}}{=} \frac{1}{n!} r^{-\alpha} e^r \frac{d^n}{dr^n}(r^{n+\alpha}e^{-r}), \tag{5.51}$$

or, explicitly,

$$L_n^{(\alpha)}(r) = \sum_{k=0}^{n} (-1)^k \frac{(\alpha + k + 1)_{n-k}}{(n-k)!\,k!} r^k \tag{5.52}$$

where $(a)_s$ is the Pochhammer symbol (cf. **box** on page 316). The Laguerre
polynomials $L_n^{(\alpha)}(r)$ are particular solutions to *Kummer's confluent hypergeometric
ODE*

$$r\frac{d^2}{dr^2}L_n^{(\alpha)}(r) + (\alpha + 1 - r)\frac{d}{dr}L_n^{(\alpha)}(r) + n\,L_n^{(\alpha)}(r) = 0 \tag{5.53}$$

(cf. **box** on page 316) that satisfy the orthonormality condition

$$\int_0^\infty e^{-r} r^\alpha dr\, L_n^{(\alpha)}(r)\, L_m^{(\alpha)}(r) = \frac{\Gamma(n + \alpha + 1)}{n!}\delta_{m,n}, \tag{5.54}$$

which, in particular, entails that $L_n^{(\alpha)}(r)$ has n (distinct) real roots on the positive
real axis. We consider the function

$$\chi_n^{(\nu)}(r) = e^{-r^2/2}\, r^{\nu+1/2}\, L_n^{(\nu)}(r^2) \tag{5.55}$$

From Eq. (5.53) it is immediate that this function satisfies the ODE

$$\frac{d^2}{dr^2}\chi_n^{(\nu)}(r) - \frac{\nu^2 - \frac{1}{4}}{r^2}\chi_n^{(\nu)}(r) - r^2\,\chi_n^{(\nu)}(r) + (4n + 2\nu + 2)\chi_n^{(\nu)}(r) = 0. \tag{5.56}$$

We conclude that $\chi_n^{(\nu)}(r)$ are the solutions to the radial eigenvalue problem (5.50) while the energy eigenvalues are

$$E = \left(2n + \ell + \frac{d}{2}\right)\omega, \qquad n, \ell = 0, 1, 2, 3, \cdots \tag{5.57}$$

Note that states with *even* (resp. *odd*) energy level have even (resp. odd) angular momentum ℓ. The radial wave functions are

$$R_{n,\ell}(r) = \frac{e^{-r^2/2}\,r^{\ell+(d-1)/2}\,L_n^{(\ell+d/2-1)}(r^2)}{r^{(d-1)/2}} =$$
$$= e^{-r^2/2}\,r^\ell\,L_n^{(\ell+d/2-1)}(r^2). \tag{5.58}$$

Since

$$L_n^{(\ell+d/2-1)}(0) = \frac{\Gamma(n + \ell + d/2)}{n!\,\Gamma(\ell + d/2)} > 0, \tag{5.59}$$

formula (5.58) is in agreement with the general result that $R(r) \approx \text{const.}\,r^\ell$ as $r \to 0$. We leave to the reader to check that the number of eigenfunctions we found solving Kummer's ODE exactly matches the group theoretical expectation (5.49).

Remark 5.1 In Sect. 3.5.4 we considered the *one*-dimensional harmonic oscillator in the Bargmann quantization. The Bargmann energy eigenfunctions, seen as elements of $L^2(\mathbb{R}^2)$, have the form

$$(x + iy)^m e^{-r^2/2} = r^m e^{im\phi} e^{-r^2/2} \qquad r = \sqrt{x^2 + y^2}, \tag{5.60}$$

which are the energy eigenfunctions of the *two*-dimensional harmonic oscillator with $n = 0$, $m = \ell$ and energy $E = \omega\ell$ (after subtracting the zero-point energy ω). We have the inequality $E/\omega \geq \ell$ that says that the energy is not lower (when measured in natural units) than some conserved charge (here angular momentum). Inequalities of this sort are called *BPS bounds*. *BPS states* are states which saturate the bound, i.e. which satisfy the equality $E/\omega = \ell$. In the isotropic *two*-dimensional harmonic oscillator we have a BPS bound saying that the energy (normalized so that the ground state has energy zero) is not lower than the angular momentum. BPS states then have the particularity of having holomorphic wave-functions (up to the weight factor $e^{-r^2/2}$). In this instance the BPS states may be identified with states of the **one**-dimensional harmonic oscillator via Bargmann quantization.

5.4   The Spherical Potential Well

We consider a particle moving in $\mathbb{R}^d$ in presence of a spherically symmetric potential well

$$V(r) = \begin{cases} -\Lambda < 0 & r \le a \\ 0 & r > a \end{cases} \tag{5.61}$$

We want to study the bound states of this system, in particular understand for which range of the parameters Λ and a a bound state exists. Recall from Sect. 3.13.1 that, in a *one-dimensional* potential well, *at least one* bound state exists for all $\Lambda, a > 0$. We wish to show that when $d \ge 3$—on the contrary—there is *no bound state* for Λ, a small enough.

We start with the case $d = 3$. When a bound state with energy $E_B < 0$ exists, the ground state, which has energy

$$E_0 \le E_B < 0, \tag{5.62}$$

must be a bound state by the Sturm-Liouville comparison theorem. Hence to prove that no bound state is present it suffices to show that *the ground state is not a bound state*. We know that the ground state carries zero angular momentum, $\ell = 0$ (again by the comparison theorem). Setting $\ell = 0$, the $d = 3$ radial Eq. (5.15) becomes

$$-\frac{\hbar^2}{2m}\chi'' + V(r)\chi = E\chi \tag{5.63}$$

which coincides with the Schrödinger equation for the one-dimensional potential well of Chap. 3 *except* that now r takes value in $\mathbb{R}_{\ge 0}$ and we must impose the boundary condition $\chi(0) = 0$ for regularity. We know how to proceed: we double the system to the full line $\mathbb{R}$ by the image method, and then keep only the *odd* wave-functions (cf. Sect. 3.8). Now, while the one-dimensional well has always at least one *even* bound state, it may or may not have an *odd* bound state depending on the parameters Λ, a. In Fig. 3.5 the odd bound states are represented by the crossings of the red curve with the circles of radius $R = \sqrt{2m\Lambda a}/\hbar$. We see from the plot in that **Figure** that there are no odd bound states whenever

$$\frac{\sqrt{2m\Lambda a}}{\hbar} \equiv R < \frac{\pi}{2}. \tag{5.64}$$

Hence when this inequality is satisfied the spherical potential well in three dimensions has no bound state and all energy eigenstates are scattering states.

Going from $d = 3$ to $d > 3$ the effective potential of the radial Sturm-Liouville equation at zero angular momentum, $\ell = 0$, increases as

$$V(r) \rightsquigarrow V(r) + \frac{\hbar^2}{2m} \frac{(\frac{d}{2} - 1)^2 - \frac{1}{4}}{r^2} \tag{5.65}$$

and the energy of the ground state—which for small Λa was already non-negative when $d = 3$—can only increase (cf. Theorem 3.5(**2**)).

5.5   The Hydrogen Atom in Arbitrary Dimension

The hydrogen atom is the two-body quantum system composed of a light electron carrying electric charge -1 and a much heavier proton of charge $+1$ which interact through the Coulomb potential

$$-\frac{\alpha}{|\boldsymbol{r}_p - \boldsymbol{r}_e|}. \tag{5.66}$$

Separating the center-of-mass motion, we effectively remain with the electron moving in the central potential $-\alpha/r$ with Hamiltonian

$$H = \frac{1}{2m} \sum_{i=1}^{d} \boldsymbol{p}_i^2 - \frac{\alpha}{r} \qquad \text{where} \quad r^2 \equiv \sum_{i=1}^{d} x_i x_i \tag{5.67}$$

The real-world hydrogen atom corresponds, of course, to the physical case of three dimensions $d = 3$. However, as for the harmonic oscillator, studying the motion in an arbitrary number d of space dimensions makes the symmetries and the deep structures of the Schrödinger equation much more transparent.

In the classical case the motion takes place in a fixed plane determined by the conserved *angular momentum* antisymmetric 2-tensor

$$J_{ij}^{\mathrm{cl}} = -J_{ji}^{\mathrm{cl}} = x_i\,p_j - x_j\,p_i, \qquad i, j = 1, \ldots, d, \tag{5.68}$$

and the dynamical problem gets effectively reduced to the $d = 2$ case. Classically the only effect of increasing d is that the motion may now take place in several distinct planes differing by a $SO(d)$ rotation.

In the quantum set-up we have still an (operator-valued) conserved angular momentum 2-tensor

$$J_{ij} = -J_{ji} = x_i\,\boldsymbol{p}_j - x_j\,\boldsymbol{p}_i, \qquad i, j = 1, \ldots, d, \tag{5.69}$$

with commutation relations

$$[J_{ij}, J_{kl}] = \mathrm{i}\delta_{il}\, J_{kj} - \mathrm{i}\delta_{jk}\, J_{il} - \mathrm{i}\delta_{jl}\, J_{ki} + \mathrm{i}\delta_{ik}\, J_{jl}, \tag{5.70}$$

and we should expect only modest modifications when we make $d \rightsquigarrow d + 1$, except that the degeneracy of each energy level must increase rather rapidly with d to match the classical fact that the motion may take place in "several $SO(d)$-rotated planes in $\mathbb{R}^d$": in the quantum set-up the different orientations of the plane get reflected in the dimensions of the relevant irreducible unitary representations of $SO(d)$. These dimensions have the form

$$N_k^{(d)} \equiv \binom{k+d-1}{d-1} - \binom{k+d-3}{d-1} \qquad k = 0, 1, 2, 3, \cdots . \tag{5.71}$$

The Coulomb potential $-\alpha/r$ *vanishes at infinity.* From the general discussion in Sect. 3.2 we know that when $\alpha > 0$, that is, when the Coulomb potential is *attractive,* there may be three kinds of states:

B   *bound states* with energy $E < 0$ which belong to the discrete spectrum $\sigma_p(H)$ of the Hamiltonian H. Their classical counterparts are the *elliptic* orbits which are motions taking place in a bounded region of $\mathbb{R}^d$;

S   *scattering states* with $E > 0$ which belong to the continuum spectrum of energy $\sigma_e(H)$. They are the quantum counterpart to *hyperbolic* orbits;

T   energy eigenstates (in the generalized sense) at the *threshold energy $E = 0$,* i.e. the quantum counterpart to *parabolic* orbits.

The transition from the discrete to the continuous spectrum has the following feature: the threshold energy $E = 0$ is an *accumulation point* of discrete energy eigenlevels:

$$\lim_{n \to \infty} E_n = 0. \tag{5.72}$$

This already follows from the criterion (3.254), and will be confirmed by the explicit solution below. When $\alpha < 0$—i.e. the potential is repulsive—only positive-energy scattering states are present in the spectrum.

5.5.1   Algebraic Solution

To simplify the expressions, in this subsection we set $\hbar = m = 1$.

Classically in any dimension $d \geq 2$ the bounded trajectories of a particle in the potential $-\alpha/r$ are *closed,* that is, they take place in a 2-plane as a consequence of conservation of the angular momentum J_{ij} and, in addition, the *perihelion* in the fixed plane *does not precede* (see [1] §.5.3). This means that the position of the perihelion is a classically conserved quantity proportional to the *Runge-Lenz vector*

A_i ([1] §. 5.3). We may use the Runge-Lenz vector to completely solve the classical dynamics of a particle in the $-\alpha/r$ potential, see [1] §. 5.3 for details.

The Runge-Lenz vector is conserved also in the quantum theory and we can still use it to solve completely the dynamical problem, but now we have to pay attention to the order of operators when we write it:

Lemma 5.1 *The quantum Runge-Lenz vector*

$$A_i \equiv A_i^\dagger \overset{\text{def}}{=} \frac{1}{2}(J_{ij}p_j + p_j J_{ij}) - \frac{\alpha x_i}{r} = J_{ij}p_j + i\frac{d-1}{2}p_i - \frac{\alpha x_i}{r} \tag{5.73}$$

is conserved for all spatial dimension d

$$[H, A_i] = 0 \qquad [J_{ij}, A_k] = i(\delta_{ik}A_j - \delta_{jk}A_i) \tag{5.74}$$

The 2nd equation just says that A_k transforms as a vector under $SO(d)$ rotations.

Proof Problem 5.5.

The components of A_i do not commute; instead we have

$$[A_i, A_j] = -2i H J_{ij} \tag{5.75}$$

see Problem 5.6. The conserved quantities H, J_{ij}, A_i preserve the (generalized) energy eigenspaces, so we can study the representations of the symmetry they generate independently in each fixed energy eigenspace. We distinguish three situations: $E < 0$, $E = 0$, and $E > 0$.

(1) When $E < 0$ we define the new Hermitian operators

$$J_{d+1,i} \equiv -J_{i,d+1} \overset{\text{def}}{=} \frac{A_i}{\sqrt{-2E}} \tag{5.76}$$

then Eqs. (5.70), (5.74), and (5.75) unify into the commutators

$$[J_{ab}, J_{cd}] = i\delta_{ac}J_{bd} - i\delta_{bc}J_{ad} - i\delta_{ad}J_{bc} + i\delta_{bd}J_{ac} \qquad a, b = 1, \ldots, d+1, \tag{5.77}$$

which show that the Lie algebra of the symmetries, *when acting on states of negative energy*, is $\mathfrak{so}(d+1)$, that is, the Lie algebra $\mathfrak{iso}(S^d)$ of infinitesimal isometries of the d-sphere. The eigenspace $\mathcal{H}_E$ of a given negative energy $E < 0$ then decomposes into unitary irreducible representations of $SO(d+1)$. Since this group is semisimple and compact, these irreducible representations are *finite dimensional*.

(2) When $E = 0$ the Lie algebra of symmetries is (5.70), (5.74) and

$$[A_i, A_j] = 0. \tag{5.78}$$

Changing notation $A_i \rightsquigarrow P_i$ we identify the $E = 0$ symmetry algebra with the Lie algebra $\mathrm{iso}(\mathbb{R}^d)$ of infinitesimal isometries of the Euclidean space $\mathbb{R}^d$ (rotations and translations).

(3) When $E > 0$ the operators defined in Eq. (5.76) are *anti-Hermitian*. Writing them as $\mathrm{i} \equiv \sqrt{-1}$ times a Hermitian operator, we conclude that the scattering states form unitary representations of the *non-compact* semisimple Lie group $SO(d, 1)$ (the isometry group of the d-dimensional hyperbolic space). Since $SO(d, 1)$ is non-compact, these representations are necessarily *infinite-dimensional*.

Remark 5.2 The three symmetry groups we got for $E < 0$, $E = 0$, and $E > 0$ are the three maximal isometry groups for a d-dimensional Riemannian manifold with constant curvature (respectively) > 0, $= 0$, and < 0.

For the rest of this subsection we focus on the $E < 0$ case (bound states). We write $J^2 = J_{ij}J_{ij}/2$ for the quadratic Casimir of $SO(d)$. A simple calculation yields the identities[10]

$$ J_{ij}\,p_j\,J_{ik}\,p_j = J^2 p^2 \qquad\qquad J_{ij}\,p_i\,p_j = 0 \tag{5.79} $$

$$ p_i\,J_{ij} = -\mathrm{i}(d-1)\,p_j \qquad\qquad p_i\,J_{ij}\,p_j = -\mathrm{i}(d-1)\,p^2 \tag{5.80} $$

$$ J_{ij}\,p_j\,x_i = J^2 \qquad\qquad x_i\,J_{ij}\,p_j = J^2 - \mathrm{i}(d-1)x_i\,p_i \tag{5.81} $$

(see Problem 5.7) so that

$$ A_i A_i = \left(J_{ij}\,p_j + \mathrm{i}\frac{d-1}{2}\,p_i - \frac{\alpha x_i}{r} \right)^2 = $$

$$ = \left(2J^2 + \frac{(d-1)^2}{2} \right)\left(\frac{p^2}{2} - \frac{\alpha}{r} \right) + \alpha^2 \equiv \tag{5.82} $$

$$ \equiv 2H\left(J^2 + \tfrac{1}{4}(d-1)^2 \right) + \alpha^2, $$

or, equivalently,

$$ J_{i,d+1}J_{i,d+1} = -J^2 - \tfrac{1}{4}(d-1)^2 - \frac{\alpha^2}{2H}. \tag{5.83} $$

Therefore the $SO(d+1)$ quadratic Casimir is

$$ \frac{1}{2}J_{ab}J_{ab} \equiv J_{i,d+1}J_{i,d+1} + J^2 = -\tfrac{1}{4}(d-1)^2 + \frac{\alpha^2}{(-2H)}. \tag{5.84} $$

When the energy is negative, $E < 0$, the $SO(d+1)$ symmetry is realized unitarily on each energy eigenspace $\mathcal{H}_E$, which then decomposes into irreducible

---

[10] We write the canonical operators acting on $L^2(\mathbb{R}^3)$ as x_i and p_j with $[x_i, p_j] = \mathrm{i}\delta_{ij}$ ($\hbar \equiv 1$).

representations of $SO(d + 1)$. From (5.84) we see that the $SO(d + 1)$ Casimir acts as a multiple of the identity in each $\mathcal{H}_E$; since the enveloping algebra generated by the $SO(d + 1)$ generators $\{J_{ij}, A_i\}$ contains a complete set of commuting operators, we conclude[11] that $\mathcal{H}_E$ is an *irreducible* $SO(d + 1)$ representation (for $E < 0$).

Which irreducible representations of $SO(d + 1)$ do appear?

When seen as a representation of the subgroup $SO(d) \subset SO(d + 1)$ generated by the J_{ij}'s, $\mathcal{H}_E$ contains only the $SO(d)$ representations $V_\ell^{(d)}$ arising from the restriction of harmonic polynomials in $\mathbb{R}^d$ (the spherical harmonics) and each one with multiplicity at most one. These are precisely the properties of the irreducible representations $V_k^{(d+1)}$ which arise from the harmonic polynomials in $\mathbb{R}^{d+1}$ (cf. Fact 4.8), so we conclude that for all energy level E in the discrete spectrum there is a $k \in \mathbb{N}$ such that the energy eigenspace

$$\mathcal{H}_E \simeq V_k^{(d+1)} \tag{5.85}$$

is given by the space of spherical harmonics *in one more dimension* with quantum number k. We know from Chap. 4 that the values of the $SO(d+1)$ quadratic Casimir on the $V_k^{(d+1)}$'s are

$$k(k + d - 1) \qquad k = 0, 1, 2, 3, \cdots. \tag{5.86}$$

Plugging these values in the LHS of Eq. (5.84) (with $H \equiv E$), and solving for E, we extract the allowed discrete energy levels

$$E_k^{(d)} = -\frac{\alpha^2}{2} \frac{1}{(k + \frac{d-1}{2})^2} \qquad k = 0, 1, 2, 3, \cdots \tag{5.87}$$

The degeneracy $D_k^{(d)}$ of the k-th energy level is equal to the number of harmonic polynomials of degree k in one more dimension $d + 1$:

$$D_k^{(d)} \equiv N_k^{(d+1)} = \binom{k + d}{d} - \binom{k + d - 2}{d}. \tag{5.88}$$

For instance in the physical $d = 3$ dimension this gives

$$D_k^{(3)} = (k + 1)^2. \tag{5.89}$$

From Eq. (5.87) we see that $E = 0$ is an accumulation point of the discrete energy eigenvalues as advertised.

In the next subsection we solve the Schrödinger PDE for the hydrogen atom and confirm all the results of the present algebraic analysis.

---

[11] An alternative argument for the irreducibility uses the properties of the Sturm-Liouville radial equation proven in Chap. 3.

5.5.2   *Schrödinger Equation in Spherical Coordinates*

The actual hydrogen atom is three-dimensional, but—again—we may as well consider the central potential $-\alpha/r$ in $\mathbb{R}^d$. In facts, as we already remarked, the classical problem is essentially independent of d, and so the quantum problem should be nice for all d. Working in $\mathbb{R}^d$ the radial Schrödinger equation becomes

$$\chi'' - \frac{\nu^2 - \frac{1}{4}}{r^2}\chi + \frac{\beta}{r}\chi + \lambda\chi = 0 \tag{5.90}$$

where

$$\nu = \ell + \frac{d}{2} - 1, \quad \beta = \frac{2m}{\hbar^2}\alpha, \quad \lambda = \frac{2m}{\hbar^2}E. \tag{5.91}$$

Bound States

We consider first the discrete energy eigenstates with $E < 0$, i.e. we look for the bound states. We rescale the radial coordinate r by setting

$$\chi(r) = W(2\sqrt{-\lambda}\,r), \quad x = 2\sqrt{-\lambda}\,r, \quad \kappa = \frac{\beta}{2\sqrt{-\lambda}}. \tag{5.92}$$

The ODE (5.90) then becomes the *confluent hypergeometric equation* in its standard *Whittaker form*

$$\frac{d^2W}{dx^2} + \left(\frac{\frac{1}{4} - \nu^2}{x^2} + \frac{\kappa}{x} - \frac{1}{4}\right)W = 0, \tag{5.93}$$

whose solutions are the *Whittaker functions,* cf. **box** on page 327 (for more details see [5–7]). There is one Whittaker function, denoted as $W_{\kappa,\nu}(x)$, which goes exponentially to zero when $x \to \infty$, as it must be for a bound state wave function. All the other linearly-independent solutions grow exponentially at infinity and hence are not valid eigenfuctions (not even in a generalized sense). For generic k, ν the good function $W_{k,\nu}(x)$ has the integral representation

$$W_{\kappa,\nu}(x) =$$
$$= \frac{x^{\nu+1/2}\,e^{-x/2}}{2^{2\nu}\,\Gamma(\nu - \kappa + \frac{1}{2})} \int_0^\infty e^{-xt}\, t^{\nu-\kappa-\frac{1}{2}}\, (t + 1)^{\nu+\kappa-\frac{1}{2}}\, dt. \tag{5.94}$$

The diligent reader is invited to proof this formula by plugging it in the LHS of (5.93) and checking that it gives zero. The asymptotic expansion for large (positive) x of $W_{k,\nu}(x)$ reads

$$W_{\kappa,\nu}(x) \approx e^{-x/2}x^\kappa \sum_{s=0}^\infty (-1)^s \frac{(\frac{1}{2} + \nu - \kappa)_s\,(\frac{1}{2} - \nu - \kappa)_s}{s!}\left(\frac{1}{x}\right)^s, \tag{5.95}$$

where $(a)_s$ is the *Pochhammer symbol* (cf. **box** on page 316). The integral in the RHS of (5.94) *diverges* in the limit $x \to 0$. Hence the regularity condition

$$\chi(r) \approx \text{const. } r^{\nu+1/2} \quad \text{as } r \to 0 \tag{5.96}$$

is **not** satisfied, unless the pre-factor vanishes which happens for

$$\kappa - \nu - \tfrac{1}{2} \equiv n = 0, 1, 2, 3, \cdots \tag{5.97}$$

in which case the sum (5.95) terminates at $s = n$ giving a polynomial (times $x^{\nu+1/2}e^{-x/2}$). We know from Sect. 5.3.1 that the polynomial solutions to the confluent hypergeometric ODE are the *Laguerre polynomials* $L_n^{(\alpha)}(x)$: comparing Eqs. (5.52) and (5.95) it is immediate to see that

$$W_{n+\nu+\frac{1}{2},\nu}(x) = (-1)^n n!\, e^{-x/2} x^{\nu+1/2}\, L_n^{(2\nu)}(x). \tag{5.98}$$

Thus the bound states satisfy the condition

$$\kappa = \frac{\beta}{2\sqrt{-\lambda}} = n + \ell + \frac{d-1}{2}, \qquad n, \ell = 0, 1, 2, 3, \cdots \tag{5.99}$$

that is,

$$E_{n,\ell} = -\frac{m\alpha^2}{2\hbar^2} \frac{1}{(n + \ell + \frac{d-1}{2})^2} \tag{5.100}$$

in agreement with the algebraic result (5.87) (where we set $m = \hbar = 1$). We see that the only effect of changing the dimension $d \rightsquigarrow d + 2$ is to shift n by one unit and, of course, change the degeneracy of the eigenvalues since now for fixed n, ℓ we have

$$N_\ell^{(d)} = \binom{\ell + d - 1}{d - 1} - \binom{\ell + d - 3}{d - 1} \tag{5.101}$$

states. The total degeneracy $D_k^{(d)}$ of the k-th discrete energy level agrees with the algebraic prediction (5.88): indeed from Eq. (4.248)

$$D_k^{(d)} \equiv \sum_{\ell=0}^{k} N_\ell^{(d)} = N_k^{(d+1)}. \tag{5.102}$$

The bound-states radial wave-functions are

$$\chi_{n,\ell}(r) = W_{n+\ell+\frac{(d-1)}{2},\ell+\frac{d}{2}-1}\left(\frac{2m\alpha}{\hbar^2} \frac{r}{n + \ell + \frac{d-1}{2}}\right). \tag{5.103}$$

BOX: Whittaker's Confluent Hypergeometric ODE

In Kummer's confluent hypergeometric ODE (cf. **box** on page 316) we make the replacements

$$W(z) = e^{-z/2} z^{\mu+1/2} \Phi(z), \qquad \mu = \tfrac{1}{2}(b-1), \qquad \kappa = \tfrac{1}{2}b - a,$$

and we get the *Whittaker's standard form* of the confluent hypergeometric ODE

$$\frac{d^2 W}{dz^2} + \left(\frac{\kappa}{z} + \frac{\frac{1}{4} - \mu^2}{z^2} - \frac{1}{4} \right) W = 0$$

The interesting function which goes exponentially to zero at infinity is

$$W_{\kappa,\mu}(z) = e^{-z/2} z^{\frac{1}{2}+\mu} \, U(\tfrac{1}{2} + \mu - \kappa, 1 + 2\mu, z)$$

whose integral representation and large z asymptotic expansion may be read in (5.94) and (5.95)

Scattering States

The continuous spectrum takes all the positive values $\sigma_e(H) = [0, \infty)$. For each value of E the angular momentum quantum number ℓ takes all possible integral values from 0 to ∞. The radial Schrödinger equation is still the Whittaker ODE (5.93) but we see from Eq. (5.92) that now x and κ are imaginary

$$\chi(r) = W(2\,i\sqrt{\lambda}\,r), \quad x \equiv iy = 2\,i\sqrt{\lambda}\,r, \quad \kappa = \frac{\beta}{2\,i\sqrt{\lambda}} = -i\eta \qquad (5.104)$$

where $\eta > 0$ (resp. $\eta < 0$) for an attractive (resp. repulsive) Coulomb potential. Regularity and reality select the real Whittaker solution with

$$W(iy) \approx \text{const.} \, y^{\nu+1/2} \quad \text{as } y \approx 0. \qquad (5.105)$$

The particular solution

$$W(iy) = C \, y^{\nu+1/2} e^{-iy/2} \, {}_1F_1\big(\tfrac{1}{2} + \nu + i\eta, 1 + 2\nu, iy\big), \qquad (5.106)$$

where ${}_1F_1(a, b, c)$ is Kummer's hypergeometric sum (**box** on page 316) and C a real normalization constant, satisfies all the required conditions: in particular the reality of $W(iy)$ follows from Kummer's first transformation (**box** on page 316). The large y asymptotics is then

$$C' \cos\left(\frac{y}{2} + \eta \log y - \frac{\pi}{2}(\nu + \tfrac{1}{2}) - \arg \Gamma(\nu + \tfrac{1}{2} + i\eta) \right) + O(1/y) \qquad (5.107)$$

so the reduced one-dimensional wave function $\chi(r)$ is asymptotic to a plane wave ($\equiv$ a momentum eigenstate). Using the normalization convention for momentum eigenstates, Eq. (2.401), the large r asymptotics of the normalized radial wavefunctions becomes

$$R_{E,\ell}(r) =$$

$$= \frac{2}{r^{(d-1)/2}} \cos\left[\frac{\sqrt{2mE}}{\hbar} r + \sqrt{\frac{m}{2E}} \frac{\alpha}{\hbar} \log\left(\frac{2\sqrt{2mE}}{\hbar} r \right) - \delta_{E,\ell} \right] + \cdots \qquad (5.108)$$

where the *phase shift* $\delta_{E,\ell}$ is

$$\delta_{E,\ell} \equiv \frac{\pi}{2}\left(\ell + \tfrac{d-1}{2}\right) - \arg \Gamma\left(\ell + \tfrac{d-1}{2} + i\sqrt{\frac{m}{2E}} \frac{\alpha}{\hbar} \right). \qquad (5.109)$$

Energy Eigenfunctions at Threshold

At $E = 0$ the ODE (5.90) reduces to

$$\chi'' - \frac{\nu^2 - \tfrac{1}{4}}{r^2}\chi + \frac{\beta}{r}\chi = 0 \qquad (5.110)$$

which can be transformed in the Bessel equation (cf. [5] **8.491**,4) with solution

$$\chi(r) = \sqrt{r}\, J_{2\nu}(2\sqrt{\beta r}\,). \qquad (5.111)$$

5.5.3 *Solution in Parabolic Coordinates*

We know from [1] §. 9.7 that the motion of a particle in a central potential $-\alpha/r$ is separable in both the polar and the parabolic coordinates. Recall that the statement "the equation is separable in a particular system of coordinates" actually means that it is separable in a full equivalence class of coordinate systems, where two systems are in the same class iff they have the same level hypersurfaces for the coordinate functions. Hence, following [3], we take the liberty of using a non-conventional choice of coordinates in the parabolic class: in our "parabolic cylindrical" coordinates the flat metric in $\mathbb{R}^3$ reads

$$ds^2 = \frac{\xi + \eta}{4\xi} d\xi^2 + \frac{\xi + \eta}{4\eta} d\eta^2 + \xi\eta\, d\phi^2. \qquad (5.112)$$

These coordinates are related to the Cartesian ones x_i ($i = 1, 2, 3$) by

$$(x_1, x_2, x_3) = (\sqrt{\xi\eta}\cos\phi, \sqrt{\xi\eta}\sin\phi, \tfrac{1}{2}(\xi - \eta)), \tag{5.113}$$

$$r \equiv \sqrt{x_i x_i} = \tfrac{1}{2}(\xi + \eta). \tag{5.114}$$

The Laplacian becomes

$$\Delta = -\frac{4}{\eta + \xi}\left[\frac{\partial}{\partial\xi}\left(\xi\frac{\partial}{\partial\xi}\right) + \frac{\partial}{\partial\eta}\left(\eta\frac{\partial}{\partial\eta}\right)\right] - \frac{1}{\xi\eta}\frac{\partial^2}{\partial\phi^2} \tag{5.115}$$

while the Coulomb potential takes the form

$$V = -\frac{\alpha}{r} = -\frac{2\alpha}{\xi + \eta}. \tag{5.116}$$

We solve the stationary Schrödinger equation in three dimensions for $E < 0$ (we set $m = \hbar = 1$). We separate the variables by writing

$$\psi(\xi, \eta, \phi) = F(\xi)\,G(\eta)\,e^{im\phi}, \quad m \in \mathbb{Z}, \tag{5.117}$$

and we get the equations

$$\xi\frac{d^2}{d\xi^2}F + \frac{dF}{d\xi} - \frac{m^2}{4\,\xi}F + \frac{E}{2}\xi F = -\lambda F \tag{5.118}$$

$$\eta\frac{d^2}{d\eta^2}G + \frac{dG}{d\eta} - \frac{m^2}{4\,\eta}G + \frac{E}{2}\eta\,G = (\lambda - \alpha)\,G. \tag{5.119}$$

For $E < 0$ we set

$$F(\xi) = f\left(\sqrt{-2E}\,\xi\right) \qquad\qquad x = \sqrt{-2E}\,\xi, \tag{5.120}$$

$$G(\eta) = g\left(\sqrt{-2E}\,\eta\right) \qquad\qquad y = \sqrt{-2E}\,\eta, \tag{5.121}$$

and the above equations become

$$x\frac{d^2 f(x)}{dx^2} + \frac{df(x)}{dx} - \frac{m^2}{4}\frac{f(x)}{x} - \frac{1}{4}x\,f(x) = -\frac{\lambda}{\sqrt{-2E}}f(x) \tag{5.122}$$

$$y\frac{d^2 g(y)}{dy^2} + \frac{dg(y)}{dy} - \frac{m^2}{4}\frac{g(y)}{y} - \frac{1}{4}y\,g(y) = \frac{\lambda - \alpha}{\sqrt{-2E}}g(y) \tag{5.123}$$

which are yet another form of the confluent hypergeometric equation[12] whose normalizable solutions are[13]

$$f_{\lambda,|m|}(x) = e^{-x/2} x^{|m|/2} L_n^{(|m|)}(x)$$ (5.124)

with eigenvalue

$$\lambda = \sqrt{-2E} \left(n + \frac{|m|+1}{2} \right) \quad n = 0, 1, 2, 3, \ldots$$ (5.125)

and

$$g_{\lambda,\sigma}(y) = e^{-y/2} y^{|m|/2} L_{n'}^{(|m|)}(y)$$ (5.126)

with eigenvalue

$$\alpha - \lambda = \sqrt{-2E} \left(n' + \frac{|m|+1}{2} \right) \quad n' = 0, 1, 2, 3, \ldots$$ (5.127)

Hence

$$\sqrt{-2E}(n + n' + |m| + 1) = \alpha$$ (5.128)

and we get back the spectrum (5.87) ($d = 3$)

$$E = -\frac{\alpha^2}{2} \frac{1}{(n + n' + |m| + 1)^2}.$$ (5.129)

5.5.4 *The Stark Effect*

In [1] **Example 9.14**, we recalled that Lagrange in 1766 reduced to a quadrature the classical problem of a charged particle moving in the electric field which is the superposition of a Coulomb field produced by a fixed charge at the origin of $\mathbb{R}^3$ and an constant electric field $\mathcal{E}$ pointing in the direction of the third axis

$$V(x) = -\frac{\alpha}{|x|} + \mathcal{E}x_3, \qquad x = (x_1, x_2, x_3) \in \mathbb{R}^3.$$ (5.130)

---

[12] Cf. [7] Table 18.8.1 entry 10.

[13] In Eq. (5.124), as always, $L_n^{(\alpha)}(x)$ is the Laguerre polynomial of degree n and parameter α, cf. Eq. (5.51) for their definition.

Lagrange method was to separate variables in parabolic coordinates where the potential reads

$$V(\xi, \eta) = -\frac{2\alpha}{\xi + \eta} + \frac{\mathcal{E}}{2}(\xi - \eta) \tag{5.131}$$

The separation of variables extends to the quantum Schrödinger equation. One gets the ODEs

$$\xi \frac{\mathrm{d}^2}{\mathrm{d}\xi^2} F + \frac{\mathrm{d}F}{\mathrm{d}\xi} - \frac{m^2}{4\xi} F + \frac{E}{2}\xi F - \frac{\mathcal{E}}{4}\xi^2 F = -\lambda F \tag{5.132}$$

$$\eta \frac{\mathrm{d}^2}{\mathrm{d}\eta^2} G + \frac{\mathrm{d}G}{\mathrm{d}\eta} - \frac{m^2}{4\eta} G + \frac{E}{2}\eta\, G + \frac{\mathcal{E}}{4}\eta^2 F = (\lambda - \alpha)\, G. \tag{5.133}$$

Unfortunately these ODEs are not in the short list of linear ODEs of the second order whose general eigenvalues and eigenfunctions may be written in a closed form.[14] So, while the Schrödinger equation is "solvable", the solution is not fully explicit.

5.6   The Quantum Rigid Rotator

We now discuss the quantum system which is analogue to a classical free spin (a *spinning top*) i.e. a rigid body with one point kept fixed in absence of external forces (i.e. zero potential). Since the word "spin" is reserved in Quantum Mechanics to the intrinsic angular momentum of a particle, in the quantum context this system is called the (quantum) *rigid rotator* to avoid confusion.

A rigid rotator in $\mathbb{R}^3$ is a system with configuration manifold

$$\mathcal{M} \equiv SO(3) \tag{5.134}$$

where a point $R \in SO(3)$ in the configuration space $\mathcal{M}$ is identified with the rotation between a fixed inertial reference frame in $\mathbb{R}^3$ and a reference frame, co-moving with the rigid body, whose axes have origin at its fixed point and are oriented along the *principal inertia axes* of the rigid body ([1] §.5.4). Rotations of the inertial frame are tautologically a symmetry of the problem, so this system has (at least) a $SO(3)$ group of symmetries which act on $\mathcal{M}$ by right translation

$$R \mapsto Rh^{-1}, \qquad h \in SO(3). \tag{5.135}$$

---

[14] As Riemann stressed in 1851, the short list of ODEs with close-form general eigenvalues/eigenvectors essentially corresponds to Fuchsian ODEs whose monodromy representation is *rigid*.

Let

$$\omega \equiv (dR)R^{-1} \in \Omega^1(SO(3)) \otimes \mathfrak{so}(3) \tag{5.136}$$

be the *right-invariant* Maurer-Cartan form on the configuration manifold $SO(3)$ (see
[1] chap. 2). The general Riemannian metric on $SO(3)$ which is invariant under
right-translations is determined by a positive-definite quadratic form $I(\cdot, \cdot)$ on the
Lie algebra $\mathfrak{so}(3)$

$$ds^2 = I(\omega, \omega) \in \Omega^1(SO(3)) \odot \Omega^1(SO(3)). \tag{5.137}$$

We write I_1, I_2, I_3 for the eigenvalues of the quadratic form $I(\cdot, \cdot)$ which coincide
with the *principal momenta of inertia* of the physical rigid body ([1] §. 5.4). From
the formulae in the **box** on page 262 we see that, when written in Euler's angles, the
metric (5.137) takes the explicit form

$$
\begin{aligned}
ds^2 =& I_1(\cos\psi\,\sin\theta\,d\phi - \sin\psi\,d\theta)^2 + \\
&+ I_2(\sin\psi\,\sin\theta\,d\phi + \cos\psi\,d\theta)^2 + I_3(\cos\theta\,d\phi + d\psi)^2.
\end{aligned}
\tag{5.138}
$$

The quantum Hamiltonian (acting on $L^2(SO(3))$ in the Schrödinger representation)
is then

$$H = \frac{\hbar^2}{2}\Delta_I, \tag{5.139}$$

with Δ_I the Laplacian for the right-invariant metric (5.138).

Remark 5.3 In the above discussion we may replace $SO(3)$ by any other compact
Lie group G. A right-invariant metric is still given by a positive quadratic form
$I(\cdot, \cdot)$ on its Lie algebra $\mathfrak{g}$

$$ds^2 = I(\omega, \omega) \in \Omega^1(G) \odot \Omega^1(G), \tag{5.140}$$

with $\omega = (dg)g^{-1}$ the right-invariant Maurer-Cartan form on G. The quantum
Hamiltonian is proportional to the Laplacian for this Riemannian metric acting on
(a dense domain in) $L^2(G)$.

Let us study the symmetries of the quantum rotator. As we have seen in Chap. 4,
on the manifold $SO(3)$—hence on the Hilbert space $L^2(SO(3))$—we have *two*
commuting $SO(3)$ actions, namely left and right translations. We write $SO(3)_L$
(resp. $SO(3)_R$) for the copy of the group which acts on the left (resp. on the right).
Let L_a and J_a be the respective generators: we stress that *as* 3×3 *matrices* L_a and
J_a are identical copies, but they play different physical roles. The induced action of
J_a on $L^2(SO(3))$ represents the angular momentum *as defined in the inertial frame*,
so, *as operators acting on the Hilbert space*, the J_a's are conserved quantities of the

system in absence of external potentials. Now while, as matrices,

$$L_a R \neq R L_a \equiv R J_a \tag{5.141}$$

for a general rotation $R \in SO(3)$, the $\mathfrak{so}(3)_L$ Casimir element L^2 of the left algebra commutes with all group elements, so, as matrices

$$L^2 R = R L^2 \equiv R J^2, \tag{5.142}$$

and therefor the $\mathfrak{so}(3)_L$ Casimir L^2 gets identified with a central element of the universal enveloping algebra of $\mathfrak{so}(3)_R$. We conclude that

$$L^2 = J^2 \quad \text{equality as actions on } L^2(SO(3)) \tag{5.143}$$

We already proved this equality in Chap. 4 both from the point of view of Quantum Mechanics on the 3-sphere and from the Peter-Weyl perspective (cf. Example 4.11).

For a general inertia tensor I the only conserved quantities are the functions $f(J_1, J_2, J_3)$ of the inertial frame angular momentum. For special quadratic forms $I(\cdot, \cdot)$ the *symmetry enhances*. Writing I as a diagonal 3×3 matrix with positive,[15] eigenvalues I_1, I_2, I_3, the special cases are:

Sp $I_1 = I_2 = I_3$ (*spherical rotator*). In this case the symmetry is the full

$$SO(3)_L \times SO(3)_R, \tag{5.144}$$

the metric (5.137) is the unique (up to scale) bi-invariant metric, and all components of both vectors L_a, J_a are conserved. A maximal set of commuting conserved charges is $L_3, J_3, J^2 \equiv L^2$ and the system is integrable on the nose;

Sy $I_1 = I_2 \neq I_3$ (*symmetric rotator*). The symmetry now is

$$SO(2)_L \times SO(3)_R, \tag{5.145}$$

where $SO(2)_L$ are the rotation around the body's symmetry axis. We have still the three commuting conserved charges

$$L_3, J_3, J^2 \equiv L^2 \tag{5.146}$$

and the quantum system is again integrable on the nose.

In the *general case* where all I_a's are different, we still have three commuting conserved quantities, namely H, J^2, J_3, so the system is still integrable in the

---

[15] If one of the I_a vanishes the configuration space becomes $S^2 = SO(3)/SO(2)$ see [1], footnote 7 of chap. 5.

technical sense that the Schrödinger equation can be reduced to linear ODEs by separation of variables. However the resulting ODE is not in the short list of ODEs whose general eigenvalues and eigenfunctions can be written in a simple closed form.[16] As it will be clear below, in a finite time we can compute *in closed form* only finitely many eigenvalues/eigenfunctions.

We consider each case in turn.

Spherical Rotator

When $I_1 = I_2 = I_3 \equiv I$ the metric (5.138) is equal to I times the round metric (4.289) on $S^3/\mathbb{Z}_2$. This is the QM system on the group manifold $SO(3)$ with Hamiltonian the Laplacian Δ of the symmetric metric which is the quadratic Casimir

$$\Delta \equiv L^2 \equiv J^2 \tag{5.147}$$

that we already solved twice in Sects. 4.8 and 4.9. We know from the Peter-Weyl theorem that the Schödinger representation of the energy eigenstates is given (up to normalization) by the $SO(3)$ group matrix elements. Since here the group is $SO(3)$ instead of its double cover $SU(2)$, only the representations with integer ℓ will appear. Then the energy eigenstates are

$$\frac{E_\ell}{\hbar^2} = \frac{1}{2I}\ell(\ell+1) \quad \ell \in \mathbb{N} \tag{5.148}$$

while the energy level E_ℓ has degeneracy $(2\ell+1)^2$.

Symmetric Rotator

The Hamiltonian H can be written as

$$H = \frac{1}{2I_1}(L_1^2 + L_2^2) + \frac{1}{2I_3}L_3^2 \equiv \frac{1}{2I_1}L^2 + \left(\frac{1}{2I_3} - \frac{1}{2I_1}\right)L_3^2 \tag{5.149}$$

while the operators L_3, J_3 and $L^2 \equiv J^2$ are conserved. We write $\ell(\ell+1)$ for the eigenvalue of L^2 ($\ell \in \mathbb{N}$), m for the eigenvalue of L_3 with $-\ell \le m \le \ell$ ($m \in \mathbb{N}$), and k for the eigenvalue of J_3 with $-\ell \le k \le \ell$ ($k \in \mathbb{N}$). The energy levels $E_{\ell,m,k}$ then are

$$\frac{E_{\ell,m,k}}{\hbar^2} = \frac{\ell(\ell+1)}{2I_1} + \left(\frac{1}{2I_3} - \frac{1}{2I_1}\right)m^2 \tag{5.150}$$

and each energy level has a degeneracy $2\ell+1$ given by the $2\ell+1$ possible values of k. Up to normalization the wave eigenfunction is given by the matrix element of the rotation of Euler's angle ϕ, θ, ψ

$$\psi_{\ell,m,k} = \sqrt{2\ell+1}\, R_{k,m}^\ell(\phi, \theta, \psi), \tag{5.151}$$

see Chap. 4, especially Eq. (4.295).

---

[16] Cf. Footnote 14.

Asymmetric Rotator

Now the quantum Hamiltonian is

$$H = \frac{1}{2I_1}L_1^2 + \frac{1}{2I_2}L_2^2 + \frac{1}{2I_3}L_3^2 =$$
$$= a\,L^2 + bL_3^2 + c\,(L_1 + iL_2)^2 + c\,(L_1 - iL_2)^2 \tag{5.152}$$

where

$$a = \frac{1}{4I_1} + \frac{1}{4I_2}, \quad b = \frac{1}{2I_3} - \frac{1}{4I_1} - \frac{1}{4I_2}, \quad c = \frac{1}{8I_1} - \frac{1}{8I_2}. \tag{5.153}$$

$L^2(SO(3))$ decomposes into finite dimensional irreducible representations according to the Peter-Weyl theorem

$$L^2(SO(3)) = \bigoplus_{\ell \in \mathbb{N}} V_\ell^{\oplus(2\ell+1)} \tag{5.154}$$

where V_ℓ is the irreducible representation of spin ℓ. The Hamiltonian (5.152) preserves each $V_\ell \simeq \mathbb{C}^{2\ell+1}$; in fact it preserves the finer splitting

$$V_\ell = V_{\ell,+} \oplus V_{\ell,-} \tag{5.155}$$

where $V_{\ell,+}$ (resp. $V_{\ell,-}$) is the subspace of V_ℓ spanned by eigenvectors of L_3 with even (resp. odd) eigenvalue whose dimension is

$$d_{\ell,\pm} \equiv \dim V_{\ell,\pm} = \ell + \max\{\pm(-1)^\ell, 0\}. \tag{5.156}$$

Acting on each subspace $V_{\ell,\pm} \subset \mathcal{H}$ the Hamiltonian H reduces to a Hermitian *tridiagonal* $d_{\ell,\pm} \times d_{\ell,\pm}$ matrix $H_{\ell,\pm}$

$$H_{\ell,\pm} : V_{\ell,\pm} \to V_{\ell,\pm} \tag{5.157}$$
$$H_{\ell,\pm} = a\ell(\ell+1) + bL_3^2 + cL_+^2 + cL_-^2 \tag{5.158}$$

The energy levels with quantum numbers ℓ and $\pm$ are the eigenvalues of the finite-dimensional matrices $H_{\ell,\pm}$. Each such eigenvalue has a degeneracy $(2\ell+1)$ arising from the possible values of the quantum number k. We see that the computation of a finite set of low-lying energy levels is reduced to the diagonalizing of a few finite-dimensional matrices.

5.6.1 Rotators for General Compact Lie Groups

As explained in Remark 5.3, the above analysis extends to rotators whose config-uration spaces are arbitrary compact Lie groups G equipped with a right-invariant Riemannian metric $I(\omega, \omega)$. The symmetry group of the problem is

$$H_L \times G_R, \tag{5.159}$$

where G_R is the group acting on itself by right translation, while $H_L \subset G_L$ is the subgroup of left-translations which leave invariant the quadratic form $I(\cdot, \cdot)$ on the Lie algebra $\mathfrak{g}$ of G. If λ_a is an orthonormal basis of $\mathfrak{g}$, and $I_{ab} = I(\lambda_a, \lambda_b)$, the Hamiltonian is

$$H = \frac{1}{2} I^{ab} L_a L_b \tag{5.160}$$

where L_a are the Noether generators of the group G_L acting on the left written in the basis $\{\lambda_a\}$ of $\mathfrak{g}$ and the I^{ab} is the inverse of the matrix I_{ab}. Again H preserves each finite-dimensional subspace in the Peter-Weyl decomposition of $L^2(G)$ into irreducible G_L-modules

$$L^2(G) = \bigoplus_{\text{ir}} V_{\text{ir}}^{\oplus \dim V_{\text{ir}}}, \tag{5.161}$$

and hence computing the energy eigenvalues reduces to a sequence of diagonaliza-tions of finite dimensional matrices $H_{\text{ir}} \in \mathsf{End}(V_{\text{ir}})$. For a generic inertia matrix I_{ab} each eigenvalue of H_{ir} has degeneracy $\dim V_{\text{ir}}$ from the number of copies of each irreducible representation V_{ir} in $L^2(G)$. When I_{ab} is proportional to the identity matrix we get back the particle moving on the compact group G with the bi-invariant metric that we already solved in Sect. 4.9.

5.7 Motion in a Constant Magnetic Field

Quantum Mechanics in presence of magnetic fields presents very peculiar and interesting phenomena which are impossible in absence of magnetic fields such as *spontaneous symmetry breaking*.[17] We already encountered a baby instance of these dynamical subtleties when discussing the Bohm-Aharonov effect in Sect. 3.11. In this and the next two sections we study three prototypical examples of Quantum

---

[17] For the general proof of absence of spontaneous symmetry breaking in absence of magnetic fields see Sect. 6.13.

Mechanics in presence of non-zero magnetic fields both for their intrinsic physical interest and as an illustration of the new quantum phenomena.

Constant Magnetic Field

We consider a charged particle moving in $\mathbb{R}^3$ subjected to a constant magnetic field of intensity B pointing in the vertical direction $\hat{z}$. We may write the vector potential 1-form A in various gauges: the three most frequently used are

$$A^{(1)} = B\, x\, \mathrm{d}y \tag{5.162}$$

$$A^{(2)} = -B\, y\, \mathrm{d}x \tag{5.163}$$

$$A^{(3)} = \frac{B}{2}(x\, \mathrm{d}y - y\, \mathrm{d}x) \equiv \frac{B}{2} r^2 \mathrm{d}\phi \tag{5.164}$$

where the cylindrical coordinates (r, ϕ, z) are related to the Cartesian ones (x, y, z) as

$$(x, y, z) = (r\cos\phi, r\sin\phi, z). \tag{5.165}$$

We set $\hbar$ and the electric charge q of the particle to 1 by a choice of units. The quantum Hamiltonian has the same form as the classical one, except that now the canonical variables are operators[18] (densely) acting on $L^2(\mathbb{R}^3)$

$$H = \frac{1}{2m}\left((\boldsymbol{p}_x - A_x)^2 + (\boldsymbol{p}_y - A_y)^2 + \boldsymbol{p}_z^2 \right). \tag{5.166}$$

The coordinate z is cyclic, and we may look for energy eigenfunctions of the form $\exp(ikz)\,\Phi(x, y)$, so that the Schrödinger equation reduces to

$$\frac{1}{2m}\left((\boldsymbol{p}_x - A_x)^2 + (\boldsymbol{p}_y - A_y)^2 \right)\Phi(x, y) = \left(E - \frac{k^2}{2m} \right)\Phi(x, y), \tag{5.167}$$

i.e. to the particle moving on the x-y plane with the same magnetic potential A but with a shifted energy eigenvalue $E \rightsquigarrow E - k^2/2m$. From now on we work in the plane $\mathbb{R}^2$ and forget the third cyclic direction z altogether.

Review of the Classical Motion

For the sake of comparison, let us first recall the situation in the classical case. The trajectories of the charged particle in the constant magnetic field are *circles* in the plane around a center $(x_0, y_0) \in \mathbb{R}^2$ and the frequency of the circular motion is the

---

[18] As before we write the components of momentum and coordinates in boldface to emphasize that they are operators.

cycloton one

$$\omega_c = \frac{B}{m} \tag{5.168}$$

which is a constant independent of the energy E. The fact that the frequency is independent of energy means that the classical energy has the form

$$E(I) = \omega_c I, \tag{5.169}$$

linear in the *action variable* I conjugate to the rotation angle θ, just as for the harmonic oscillator whose energy has also the form (5.169) because the frequency is independent of the initial conditions for the oscillator too, cf. [1] eq.(10.48). The coordinates (x_0, y_0) of the center of the orbit are then *constants of motion* akin to the position of the perihelion in the Coulomb potential (cf. the discussion at the beginning of Sect. 5.5.1). Then, classically, the system has three functionally-independent conserved charges: x_0, y_0 and I. This is the maximal number for a non-trivial system with two degrees of freedom,[19] and the system is therefore *superintegrable* (hence *superseparable*). The origin of the three conservation laws is physically obvious: since the magnetic field is everywhere the same in the $\mathbb{R}^2$ plane, translations and rotations are symmetries, and hence their three generators are conserved. However the Hamiltonian is *not* invariant under the *naive* isometries of the plane: the physical symmetries are combinations of these isometries with suitable gauge transformations, see [1] *Example 3.16*. On the other hand we know[20] that a classical energy $E(I) = \omega_c I$ implies, after quantization, energy levels of the form

$$E_n = \omega_c(n + \text{const.}) \equiv \frac{B}{m}(n + \text{const.}), \qquad n \in \mathbb{N}, \tag{5.170}$$

exactly as it happens in the harmonic oscillator which also has $E(I) = \omega I$ classically. We stress that the three functionally independent conserved quantities x_0, y_0 and I *cannot be in involution:* by Liouville theorem the maximal number of functional independent conserved quantities which are in involution is the number of degrees of freedom which is *two* for a particle moving in the plane.

The Algebra of Quantum Conserved Quantities
Since the magnetic field $B \equiv dA$ is constant, it is invariant for all isometries of the plane. Under an infinitesimal isometry generated by a Killing vector $k \in \mathfrak{iso}(\mathbb{R}^2)$

$$\mathscr{L}_k B = 0 \quad \Rightarrow \quad \mathscr{L}_k A = d\alpha_k. \tag{5.171}$$

---

[19] See [1] **Theorem 6.2**.

[20] This follows from the Born-Sommerfeld quantization to be discussed in Chap. 8. It also follows from the **Historical Motivation** in §. 10.6 of [1].

Thus the quantum symmetry is the combination of the infinitesimal isometry k of the plane with the gauge transformation of parameter $-\alpha_k$ needed to transform back the potential in the original chosen gauge.

α_k depends on both $k \in \mathfrak{iso}(\mathbb{R}^2)$ and the chosen gauge for A. The coordinates y, x, and ϕ are cyclic, respectively, in the first, second, and third gauge in Eqs. (5.162)–(5.164). The shifts in the cyclic variables are infinitesimal isometries with $\alpha_k = 0$, and the conserved quantity is simply the canonical momentum dual to the cyclic coordinate. The symmetries which are not just shifts of a cyclic variable include a non-trivial gauge transformation $A \rightsquigarrow A - \mathrm{d}\alpha_k$.

In the first gauge $A^{(1)}$ (Eq. (5.162)) the generators T_x, T_y of the translations in the x-, y-directions and the generator J of rotations read

$$T_x = p_x - By, \qquad T_y = p_y, \qquad J = x p_y - y p_x + \frac{B}{2}(y^2 - x^2), \qquad (5.172)$$

and, while their analytic expressions will look different in the other gauges, these Noether charges are gauge invariant *as physical observables*, in the sense that their expectation values on physical states are independent of all choices, cf. Problem 5.8.[21] It is easy to check that these operators commute with the Hamiltonian (Problem 5.9), so T_x, T_y, J are conserved quantities also in the quantum set-up. In the classical theory the coordinates of the center of the circular orbit are

$$(x_0, y_0) = \left(\frac{T_y}{B}, -\frac{T_x}{B} \right) \qquad (5.173)$$

(cf. Problem 5.10). In the quantum set-up x_0, y_0 get promoted to operators $\mathbf{x}_0$, $\mathbf{y}_0$ defined in terms of the Noether charges by Eq. (5.173). The algebra of symmetries is gauge-independent and takes the form ($\hbar = 1$)

$$[T_x, T_y] = -\mathrm{i}B, \qquad [J, T_x] = \mathrm{i}T_y, \qquad [J, T_y] = -\mathrm{i}T_x. \qquad (5.174)$$

From these equations we see that the effect of the constant magnetic field B is to produce a *central extension* of the Abelian Lie algebra of translations generated by T_x and T_y. We stress that the resulting translation algebra is isomorphic to the canonical algebra $[p, q] = -\mathrm{i}$ under the map

$$T_x \mapsto \sqrt{B}\, p, \qquad T_y \mapsto \sqrt{B}\, q, \qquad (5.175)$$

---

[21] The reader may have a look to the corresponding story in the classical set-up in §. 6.5 of [1], and in particular to eq.(6.59) therein which is the classical version of Eq. (5.174) below.

or equivalently, in terms of the coordinates of the center,

$$[x_0, y_0] = -\frac{i}{B},$$ (5.176)

that is, the *two coordinates of the center* (x_0, y_0) *do not commute* and cannot be measured simultaneously with certainty. Indeed, since the algebra (5.176) is isomorphic to the canonical one, the Heisenberg indetermination principle applies to the position (x_0, y_0) of the center with the identification

$$\hbar \rightsquigarrow 1/B.$$ (5.177)

Since the irreducible unitary representation of the canonical commutator algebra is unique and well-known (Chap. 2), this simple algebraic observation yields precious informations about the present quantum system (see below).

We stress that when the particle moves in the *infinite* plane $\mathbb{R}^2$ the isometry Lie group generated by T_x, T_y, J is *non-compact*, hence all its non-trivial unitary representations are *infinite dimensional,* in facts copies of $L^2(\mathbb{R})$ by the von Neumann theorem. Since the energy eigenspaces carry a non-trivial unitary representation of the Heisenberg group, each energy eigenspace should have infinite dimension, i.e. *all energy eigenvalues are infinitely degenerate.* In particular the vector space of ground states $\mathcal{H}_0 \subset \mathcal{H}$ is naturally isomorphic to $L^2(\mathbb{R})$.

Solution of the Schrödinger Equation
We set $\hbar = 1$. In the gauge (5.162) the two-dimensional Hamiltonian reads

$$H = \frac{1}{2m}\left(p_x^2 + (p_y - Bx)^2\right).$$ (5.178)

We separate variables by looking to solution with a fixed eigenvalue k of the conserved momentum p_y

$$\phi(x, y) = e^{iky} F(x),$$ (5.179)

so that the Schrödinger equation reduces to the ODE

$$-\frac{1}{2m}\frac{d^2 F}{dx^2} + \frac{B^2}{2m}(x - x_0)^2 F = E F, \qquad x_0 = \frac{k}{B}.$$ (5.180)

Translating $x \rightsquigarrow x - x_0$ we identify the linear differential operator in the LHS with the Hamiltonian of the one-dimensional harmonic oscillator with the cycloton frequency $\omega_c = B/m$. The discrete energy eigenvalues then are

$$E = \frac{B}{m}\left(n + \frac{1}{2}\right) \quad n = 0, 1, 2, 3, \cdots$$ (5.181)

with eigenfunction $F(x)$ given by the n-th Hermite polynomial times a Gaussian as in Chap. 3. This result agrees with the expectations from classical physics considerations, cf. Eq. (5.170).

The space of quantum states with a definite n is called the $(n+1)$-*th Landau level.* E.g. states with $n = 0$ are said to belong to the *first Landau level:* they are the states of lowest possible energy, that is, the *ground states.* In absence of magnetic fields (and spins) the ground state must be unique and invariant under all symmetries (see Sect. 6.13): in the present *magnetic* system, on the contrary, the ground energy level has a *huge degeneration.* The degeneration arises because the energy is invariant under translations of the center (x_0, y_0).

The Number of Ground States
Let us count the number of states in the first Landau level, i.e. in the energy eigenspace of energy $E_0 = B/2m$. Each eigenspace should carry a unitary representation of the symmetry Lie algebra (5.174). To make the count well defined we compactify $\mathbb{R}^2$ on a two-torus T^2 by imposing periodic boundary conditions on the wave-functions

$$\psi(x + L_1, y) = \psi(x, y + L_2) = \psi(x, y), \qquad (5.182)$$

where L_1, L_2 are the sides of T^2. The isometry group of a compact manifold is compact, so now the quantum symmetry group is compact, and all its unitary representations are finite-dimensional: in particular the number of ground states, while it may be huge, is finite.

Let us first consider the motion of a *classical* particle in the torus T^2 subjected to a constant magnetic field B. The center (x_0, y_0) is now a point in T^2. Since the Poisson bracket of the coordinates of T^2 is *non zero*

$$[x_0, y_0]_{\mathrm{PB}} = -1/B \neq 0, \qquad (5.183)$$

T^2 is now naturally a *symplectic manifold* that we may see as a phase space with symplectic 2-form

$$\Omega = B\, dy_0 \wedge dx_0. \qquad (5.184)$$

Let us now count the ground states of the quantum system. The operators x_0, y_0 commute with H, so the zero-energy eigenspace is the quantum system obtained by quantizing the phase space T^2 with symplectic form Ω. By Fact 2.10 the asymptotic number of quantum states for large $L_1 L_2$ is the symplectic volume of this phase space T^2 divided by $(2\pi\hbar)$ (in our units $\hbar = 1$). Therefore (for $L_1 L_2 |B| \gg 1$) we have:

$$\#(\text{ground states in the first Landau level}) = \frac{1}{2\pi} \int_{T^2} B =$$

$$= \frac{\text{magnetic flux through } T^2}{2\pi} = \frac{L_1 L_2 |B|}{2\pi}. \tag{5.185}$$

The educated reader has certainly recognized the RHS to be the *first Chern class* of the gauge connection A. Indeed the above formula—which is just the *Heisenberg indetermination principle* (Sect. 2.13) applied to the canonical pair of operators x_0, $B y_0$—is a fundamental result in Algebraic and Analytic Geometry known as the *Riemann-Roch theorem* [12–14]. Let us revisit that story.

5.7.1 *Complex Geometric Interpretation*

We work on the torus T^2 with possibly generalized periodic boundary conditions of the form[22]

$$\psi(x + L_1, y) = e^{i\alpha_1(y)} \psi(x, y),$$

$$\psi(x, y + L_2) = e^{i\alpha_2(x)} \psi(x, y), \tag{5.186}$$

in presence of a gauge field $A = A_x dx + A_y dy$. The torus T^2 may be replaced by any oriented compact surface Σ with minor modifications which we leave to the reader as an exercise.

The Schrödinger wave functions $\psi(x, y)$ are sections of a complex line bundle $\mathcal{L} \to T^2$ on which A is a $U(1)$ connection.[23] Mathematically speaking, the generalized boundary condition (5.186) specifies the $U(1)$ cocycle which specifies the line bundle $\mathcal{L}$ [13, 15–17]. The Schrödinger representation of the gauge-invariant (observable) velocities v_i

$$v_i = -\frac{i}{m}\left(\frac{\partial}{\partial x^i} - i A_i(x)\right) \equiv -\frac{i}{m} D_i \tag{5.187}$$

are now proportional to the *covariant derivatives* D_i of the connection A, while the canonical momenta are given by the "naive" non-covariant derivatives $-i\partial_i$.

We see the torus T^2 as a complex manifold of dimension 1, and use the complex coordinate $z = x + iy$ with complex conjugate $\bar{z} = x - iy$. We also set $m = 1$ to

---

[22] $\alpha_1(y)$ and $\alpha_2(x)$ are *real* functions.

[23] As always, there is a discrepancy by an overall $i \equiv \sqrt{-1}$ between the math and physics conventions. Be careful to use your preferred convention in a consistent manner.

make the geometry more transparent. In complex dimension 1 all smooth bundles with connection are holomorphic bundles,[24] so we may choose a *complex* gauge such that the covariant derivatives ($\equiv$ velocities) read

$$D_{\bar{z}} = \partial_{\bar{z}}, \qquad D_z = e^V \partial_z e^{-V} \equiv \partial_z - \partial_z V,$$
$$\text{i.e.} \quad A_{\bar{z}} = 0, \qquad iA_z = \partial_z V, \tag{5.188}$$

with V a real function on T^2. In the Hilbert space $\mathcal{H}(\mathcal{L})$ of sections ψ of $\mathcal{L} \to T^2$ with finite norm

$$\|\psi\|^2 \overset{\text{def}}{=} \int_{T^2} dz\, d\bar{z}\, e^{-V(z,\bar{z})} |\psi(z, \bar{z})|^2 < +\infty, \tag{5.189}$$

we have

$$D_z = -(D_{\bar{z}})^\dagger. \tag{5.190}$$

Equations (5.188) and (5.189) yield the Schrödinger representation of a quantum particle moving in the torus T^2 and subjected to a magnetic field

$$iF_{z\bar{z}} = [D_z, D_{\bar{z}}] = \partial_{\bar{z}} \partial_z V. \tag{5.191}$$

The relation with the magnetic field written in the usual real notation is

$$iF_{z\bar{z}}\, dz \wedge d\bar{z} = iF_{z\bar{z}}(dx + idy) \wedge (dx - idy) =$$
$$= 2\, F_{z\bar{z}}\, dx \wedge dy \equiv B_{xy}\, dx \wedge dx, \tag{5.192}$$

so the function V which corresponds to an uniform magnetic field $B > 0$ is

$$V = \frac{B}{2} z\bar{z}. \tag{5.193}$$

The second-order differential operator which is natural from the viewpoint of complex geometry [13, 15–17] is

$$H_{\text{CG}} = \frac{1}{2} v_i v_i = -2\, D_z D_{\bar{z}} \equiv -\frac{1}{2} e^V (\partial_x - i\partial_y) e^{-V} (\partial_x + i\partial_y) \tag{5.194}$$

Let us see how this natural complex-geometric operator compares with the original Schrödinger-representation Hamiltonian H_{SC} in Eq. (5.178). It is obvious from

---

[24] Recall that a bundle with connection A over a complex manifold M is *holomorphic* iff its curvature 2-form $F = dA + A^2$ has type $(1, 1)$ [13, 15]. *All* 2-forms have type $(1, 1)$ in complex dimension 1.

(5.194) that the relation between the Schrödinger wave-functions ψ_{SC} and the sections ψ_{CG} which are natural from a complex-geometric viewpoint is

$$\psi_{\mathrm{CG}} = \mathrm{e}^{V/2}\psi_{\mathrm{SC}}. \tag{5.195}$$

Indeed this identifications yields an isometry of Hilbert spaces

$$L^2(\mathbb{R}^2) \to \mathcal{H}(\mathcal{L}). \tag{5.196}$$

Under the identification (5.195) the "geometric" Hamiltonian H_{CG} becomes

$$-\frac{1}{2}\mathrm{e}^{V/2}(\partial_x - \mathrm{i}\partial_y)\mathrm{e}^{-V}(\partial_x + \mathrm{i}\partial_y)\mathrm{e}^{V/2} =$$

$$= -\frac{1}{2}\left\{\left(\partial_x + \frac{\mathrm{i}}{2}\partial_y V\right) - \mathrm{i}\left(\partial_y - \frac{\mathrm{i}}{2}\partial_x V\right)\right\}\left\{\left(\partial_x + \frac{\mathrm{i}}{2}\partial_y V\right) + \mathrm{i}\left(\partial_y - \frac{\mathrm{i}}{2}\partial_x V\right)\right\}$$

$$= -\frac{1}{2}\left\{\left(\partial_x + \frac{\mathrm{i}}{2}\partial_y V\right)^2 + \left(\partial_y - \frac{\mathrm{i}}{2}\partial_x V\right)^2 + \mathrm{i}\left[\partial_x + \frac{\mathrm{i}}{2}\partial_y V, \partial_y - \frac{\mathrm{i}}{2}\partial_x V\right]\right\} =$$

$$= H_{\mathrm{SC}} - \frac{1}{2}B_{xy} \tag{5.197}$$

where H_{SC} is the usual Schrödinger Hamiltonian (5.167) (with $m = \hbar = 1$) in presence of the magnetic potential

$$A_x = -\frac{1}{2}\partial_y V, \qquad A_y = \frac{1}{2}\partial_x V. \tag{5.198}$$

Note that all two-dimensional gauge potentials (A_x, A_y) may be set in the form (5.198) by a gauge transformation: indeed the only condition on the vector potential for the existence of a "prepotential" V is that (A_x, A_y) must be written in the standard Coulomb gauge

$$\partial_x A_x + \partial_y A_y = 0. \tag{5.199}$$

The last term in the RHS of (5.197) is the magnetic field in the plane:

$$B_{xy} \equiv \partial_x A_y - \partial_y A_x \equiv \frac{1}{2}\left(\frac{\partial^2 V}{\partial x^2} + \frac{\partial^2 V}{\partial y^2}\right). \tag{5.200}$$

There are physically natural situations where this second term is actually present in the physical Hamiltonian which then agrees with the standard geometric operator H_{CG}. In our simplified set-up, spinless charged particles, this term is not present *but* when the magnetic field B_{xy} is the constant B, the additional term amounts to a harmless constant shift of the Hamiltonian, and the complex-geometric operator

H_{CG} is just as good a Hamiltonian as the original one

$$H_{\mathrm{CG}} = H_{\mathrm{SC}} - E_0 \tag{5.201}$$

where $E_0 = B/2$ is the ground state energy. In this situation the geometric operator H_{CG} is just the physical Hamiltonian normalized in the natural way, i.e. shifted to make the energy E of the ground state equal zero. In the complex notation the Schrödinger equation becomes

$$- D_z \partial_{\bar{z}} \, \psi_{\mathrm{CG}} = \frac{(E - B/2)}{2} \, \psi_{\mathrm{CG}}. \tag{5.202}$$

Comparing with the previous solution of the Schrödinger equation, we conclude:

Fact 5.1 *The states of the first Landau level, i.e. the ground states on the 2-torus T^2 in presence of a constant magnetic field B with $E = 0$, are* exactly *the states whose geometrical wave-functions $\psi_{\mathrm{CG}}(z)$ are* holomorphic sections *of the corresponding complex line bundle* $\mathcal{L}$

$$(zero \ energy \ eigenspace) \equiv \Gamma(T^2, \mathcal{L}) \subset \mathcal{H}(\mathcal{L}). \tag{5.203}$$

Remark 5.4 Note that the ground states of a particle moving in $\mathbb{R}^2 \simeq \mathbb{C}$ in presence of a constant magnetic field and the states of a harmonic oscillator in the Bargmann quantization (Sect. 3.5.4) have the same wave-functions (for $\omega = \omega_c \equiv B/m$) but quite different Hamiltonians.

Relation to the Riemann-Roch Theorem

The *Riemann-Roch theorem* specialized to a 2-torus T^2 says that the dimension of the vector space $\Gamma(T^2, \mathcal{L})$ of holomorphic sections of a line bundle $\mathcal{L}$ is [12–14]

$$\dim \Gamma(T^2, \mathcal{L}) = \deg \mathcal{L} \equiv c_1(\mathcal{L}) \equiv$$
$$\equiv \int_{T_2} \frac{B}{2\pi} \, \mathrm{d}x \wedge \mathrm{d}y = \frac{\text{magnetic flux}}{2\pi} \tag{5.204}$$

which is the result (5.185) we got from the Representation Theory of the canonical commutation relations *via* the Heisenberg indetermination principle.

Equation (5.204) is the special instance of Riemann-Roch in the 2-torus ($\equiv$ genus 1). In the Appendix to this chapter we address the general case and prove the Riemann-Roch theorem for a general complex manifold of arbitrary dimension.

Remark 5.5 A quantum charged particle in a general magnetic field $B(x, y) \, \mathrm{d}x \wedge \mathrm{d}y$ on a manifold of real dimension 2 is a simple example of the geometric aspects of the Schrödinger representation briefly discussed in Sect. 2.15. In particular the extra term $- B_{xy}$ has exactly the form of the quantum ambiguity by a coupling linear in the curvatures discussed there.

5.8 Charged Particle in the Field of a Magnetic Monopole

We consider a quantum charged particle moving in $\mathbb{R}^3$ in presence of a heavy point-like magnetic monopole sitting at the origin.[25] Physically it would be natural to take the particle to be an electron, which has spin $\frac{1}{2}$, but for simplicity we assume it to be spinless. We normalize the particle's electric charge to be 1.

Geometry of the Magnetic Monopole
The flux of the magnetic field through any sphere S_r^2 centered at the origin is $2\pi q$ where $q \in \mathbb{Z}$ is the (quantized) magnetic charge of the monopole. Written in spherical coordinates, the magnetic field 2-form then reads

$$B = \frac{q}{2} \sin\theta \, d\theta \wedge d\phi, \tag{5.205}$$

that is, B is $q/2$ times the volume form of the unit sphere $S^2 \subset \mathbb{R}^3$. Hence the restriction $B|_{S^2}$ represents a non-trivial cohomology class in $H^2(S^2)$ and cannot be written as dA for a global vector potential A which is everywhere regular. This reflects the fact that, in this case, the gauge potential A is a connection on a topologically non-trivial Hermitian line bundle $\mathcal{L}_q$ over the base

$$X \equiv S^2 \times \mathbb{R}_{>0} \simeq \mathbb{R}^3 \setminus \{0\}, \tag{5.206}$$

whose Chern class $[B/2\pi]$ is q times the generator of

$$H^2(S^2 \times \mathbb{R}_{>0}, \mathbb{Z}) \simeq \mathbb{Z}. \tag{5.207}$$

The charged particle's wave "functions" $\Psi(r, \theta, \phi)$ are not functions at all, but rather (square-integrable) *sections of the complex line bundle*

$$\pi : \mathcal{L}_q \to X \equiv S^2 \times \mathbb{R}_{>0}, \tag{5.208}$$

in other words: the Hilbert space $\mathcal{H} \equiv \overline{\mathcal{V}}$ is the closure of the space

$$\mathcal{V} = \left\{ \Psi : X \to \mathcal{L}_q \;\middle|\; \pi \circ \Psi = \mathrm{Id}_X, \;\; \int_X d\mu \, |\Psi|^2 < \infty \right\}. \tag{5.209}$$

Let $\mathcal{L} = \mathcal{L}_1$ be the line bundle for a monopole of unit magnetic charge. Its circle sub-bundle[26] $U(\mathcal{L})$ is the principal $U(1)$-bundle with Chern class 1; the gauge field A of a unit monopole is a principal connection on $U(\mathcal{L})$. The line bundle $\mathcal{L}_q$ is

---

[25] For background on magnetic monopoles see [18].
[26] The circle sub-bundle $U(L)$ of a Hermitian complex line bundle L is the bundle over the same base whose fibers are the elements of the fibers of L of unit norm.

the complex vector bundle associated to the principal $U(1)$-bundle $U(\mathcal{L})$ through the one-dimensional $U(1)$ representation $e^{i\theta} \to e^{iq\theta}$, equivalently, $\mathcal{L}_q$ is the q-th power $\mathcal{L}^q$ of the line bundle $\mathcal{L}$.

$U(\mathcal{L})$ is the pull-back to $S^2 \times \mathbb{R}_{>0}$ of a most celebrated principal $U(1)$-bundle over S^2: the *Hopf bundle*

$$\varpi : S^3 \to S^2. \tag{5.210}$$

The restriction of the monopole gauge field A to the unit sphere $S^2 \subset \mathbb{R}^3$ is then a principal connection on the Hopf bundle. This fact is already obvious from the symmetric metric of S^3 ($\equiv$ the total space of the Hopf bundle) which we wrote in Eq. (4.289):[27]

$$\mathrm{d}s^2 = \overbrace{\mathrm{d}\theta^2 + \sin^2\theta\, \mathrm{d}\phi^2}^{\text{metric on base } S^2} + 4 \overbrace{\left(\tfrac{1}{2}\cos\theta\, \mathrm{d}\phi + \mathrm{d}\tilde{\psi}\right)^2}^{U(1) \text{ connection form}} \tag{5.211}$$

where $\tilde{\psi} = \psi/2$ is the coordinate along the fiber rescaled to be a periodic angle of period 2π. ψ is the usual Euler angle which is periodic of period 4π in the double cover $SU(2) \equiv S^3$ of $SO(3)$.

The Angular Schrödinger Equation
The system is invariant by rotations around the origin, hence the Schrödinger equation should separate into a radial and an angular equation. However now the angular equation is a bit subtle because—as it was the case for a constant magnetic field in Sect. 5.7—the actual symmetries are non-trivial combinations of geometric isometries and gauge transformations.

We start by considering the angular Schrödinger equation or, equivalently, the Quantum Mechanics of a (spinless) charged particle constrained to move on the unit sphere $S^2 \subset \mathbb{R}^3$ in the magnetic field produced by a magnetic monopole of charge q sitting at the origin of $\mathbb{R}^3$.

We adopt the viewpoint of a charged particle moving on S^2. To solve this quantum mechanical problem, we may proceed in two ways: the *dummy* way and the *smart* one. The dummy procedure is to write A in some specific gauge: our A will be necessarily singular somewhere in S^2. For instance we may take

$$A = -\frac{q}{2}\cos\theta\, \mathrm{d}\phi \quad \Rightarrow \quad \mathrm{d}A = \frac{q}{2}\sin\theta\, \mathrm{d}\theta \wedge \mathrm{d}\phi \equiv B, \tag{5.212}$$

which is singular at the two poles $\theta = 0, \pi$. However the singularities are mere gauge artifacts (Dirac half-strings [18]) and hence have no physical consequence except for the technical nuisance of having to work with singular expressions.

---

[27] For more elaboration, see e.g. §. 6.1 of [19].

The *smart move* is to pull back the angular part of the Schrödinger equation from S^2 to the total space of the Hopf bundle $S^3 \equiv SU(2)$. We read the Laplacian on $SU(2)$ in Eq. (4.298) which we reproduce here

$$\Delta_{SU(2)} = -\frac{1}{\sin\theta}\frac{\partial}{\partial\theta}\sin\theta\frac{\partial}{\partial\theta} - \frac{1}{\sin^2\theta}\frac{\partial^2}{\partial\phi^2} - \frac{1}{\sin^2\theta}\frac{\partial^2}{\partial\psi^2} + \frac{2\cos\theta}{\sin^2\theta}\frac{\partial^2}{\partial\psi\,\partial\phi}$$

$$(5.213)$$

The Euler angle ψ is a coordinate along the Hopf fiber of ϖ periodic of period 4π. Acting on functions f which are constant along the fibers, $\partial_\psi f = 0$, we have

$$\Delta_{SU(2)} f(\phi,\theta) = \Delta_{S^2} f(\phi,\theta), \tag{5.214}$$

that is, the Laplacian on the total space $SU(2)$ reduces to the Laplacian on the base S^2 (with its standard metric) as we saw in Chap. 4. Consider now the action of $\Delta_{SU(2)}$ on a function Φ such that

$$\partial_\psi\Phi = i\frac{q}{2}\Phi \quad\Rightarrow\quad \Phi(\phi,\theta,\psi) = e^{iq\psi/2}\,\Psi(\phi,\theta), \tag{5.215}$$

for some function $\Psi(\phi,\theta)$ on the sphere S^2. One has the identity

$$\Delta_{SU(2)}\Phi = -\frac{1}{\sin\theta}\frac{\partial}{\partial\theta}\sin\theta\frac{\partial}{\partial\theta}\Phi + \frac{1}{\sin^2\theta}\left(-i\frac{\partial}{\partial\phi} - \frac{q}{2}\cos\theta\right)^2\Phi. \tag{5.216}$$

From this formula we see that the differential operator

$$\frac{\hbar^2}{2m}\left(-\frac{1}{r^2}\frac{\partial}{\partial r}r^2\frac{\partial}{\partial r} + \frac{1}{r^2}\Delta_{SU(2)}\right) \tag{5.217}$$

acting on functions of the form $e^{iq\psi/2}\,\Psi(r,\phi,\theta)$ is equal to

$$e^{iq\psi/2}\,H_q\Psi(r,\phi,\theta) \tag{5.218}$$

where H_q is the Hamiltonian of our particle of charge 1 in the field of a magnetic monopole with magnetic charge $q \in \mathbb{Z}$. We already know the eigenvalues and eigenfunctions of $\Delta_{SU(2)}$ from the *Peter-Weyl theorem:* the eigenfunctions satisfying (5.215) are, up to normalization, the $SU(2)$ matrix elements with $m' = q/2$. Using Eq. (4.295), we have

$$e^{iq\psi/2}\,\Psi^\ell_{m,q}(\phi,\theta) \equiv \sqrt{2\ell+1}\,R^\ell_{m,q/2}(\phi,\theta,\psi)$$

$$m = -\ell, -\ell+1, \cdots, \ell-1, \ell, \tag{5.219}$$

with eigenvalues

$$\ell(\ell+1) \qquad \text{and degeneracy } 2\ell+1, \tag{5.220}$$

where

$$\ell = \frac{q}{2} \mod 1 \quad \text{and} \quad \ell \geq \frac{|q|}{2}. \tag{5.221}$$

Note that the *orbital angular momentum* ℓ of our spinless charged particle moving in S^2 cannot be lower than $|q|/2$ and

$$\ell = \ell_{\min} \equiv \frac{|q|}{2} \quad \text{iff the state is a ground state on } S^2. \tag{5.222}$$

Therefore the number of ground states is

$$2\ell_{\min} + 1 = |q| + 1, \tag{5.223}$$

and the $SU(2)$ rotational symmetry is *spontaneously broken* (completely for q odd, to $\{\pm 1\}$ for q even). Moreover, when the magnetic charge q is *odd* the orbital angular momentum ℓ is *half-integral*.

The functions $\Psi^\ell_{m,q}(\phi,\theta)$ in Eq. (5.219), for *fixed* q, are called the *monopole spherical harmonics* for magnetic charge q. Their explicit expressions are given by Jacobi polynomials, see Eq. (4.307). For $q = 0$ (no monopole) they reduce to the usual spherical harmonics $Y_{m,\ell}(\phi,\theta)$ studied in Chap. 4.

Remark 5.6 (Riemann-Roch Again) The monopole spherical harmonics are the energy eigenfunctions of a charged particle constrained to move on the unit sphere S^2 in presence of a constant magnetic flux through the surface. The Riemann-Roch theorem applies to this situation too. For the sphere (i.e. for a Riemann surface of genus 0) the theorem states [12]

$$\#(\text{ground states}) = 1 + \frac{(\text{magnetic flux})}{2\pi} = 1 + \frac{4\pi\frac{|q|}{2}}{2\pi} = \tag{5.224}$$

$$= 1 + |q| = 1 + 2\ell_{\min},$$

which agrees with the result (5.222) we got from the explicit solution of the Schrödinger equation. It is amusing that the dimensions of the irreducible $SU(2)$-representations can be recovered from the Riemann-Roch theorem!

Remark 5.7 Contrary to the case of a constant magnetic field in $\mathbb{R}^3$, the Lie algebra of symmetries for a charged particle moving in the field of a monopole is isomorphic to the symmetry algebra $\mathfrak{iso}(S^2)$ without central extensions. This agrees with the general theorems of Chap. 4 since now the Lie algebra $\mathfrak{iso}(S^2) \simeq \mathfrak{so}(3)$ is semisimple and has no central extensions.

Physical Interpretation

If we consider a particle of charge e moving in the field of a magnetic monopole of charge q, we get

$$\ell_{\min} = \frac{|eq|}{2}. \tag{5.225}$$

This result has an alternative physical interpretation.[28] The charged particle and the monopole source, respectively, a static electric $\mathbf{E}$ and a static magnetic field $\mathbf{B}$. According to the Poynting theorem, the electromagnetic field has an angular momentum

$$\mathbf{J}_{\mathrm{EM}} = \int \mathbf{r} \wedge (\mathbf{E} \wedge \mathbf{B}) \mathrm{d}^3 x. \tag{5.226}$$

By symmetry, $\mathbf{J}_{\mathrm{EM}}$ points in the direction of the vector $\mathbf{r}_e - \mathbf{r}_m$ connecting the two point sources. Replacing in this formula the Coulomb field of the charged particle and the static magnetic field of the monopole

$$\mathbf{B} = \frac{q}{2} \frac{\mathbf{r}}{r^3}, \tag{5.227}$$

we get (see **box** on page 351)

$$|\mathbf{J}_{\mathrm{EM}}| = \frac{|eq|}{2} \tag{5.228}$$

independently of the distance between the point charge and the monopole. We conclude that the minimal angular momentum $\ell_{\min}$ in the monopole spherical harmonics is nothing else than the part of the total angular momentum stored in the electromagnetic fields. In addition we have the "mechanical" angular momentum due to the motion of the charge.

---

[28] See §. **6.13** of [20] and references therein.

BOX: $\mathbf{J}_{\mathrm{EM}}$ in the Monopole-Charge System

The angular momentum stored in the electromagnetic fields sourced by a magnetic monopole of charge q at the origin and a point charge e at $\mathbf{r} = \mathbf{r}_e$ is

$$\mathbf{J}_{\mathrm{EM}} \equiv \int \mathbf{r} \wedge (\mathbf{E} \wedge \mathbf{B}) \mathrm{d}^3 x = \frac{q}{2} \int \frac{\mathbf{r}}{r^3} \wedge (\mathbf{E} \wedge \mathbf{r}) \mathrm{d}^3 x \qquad \text{for} \quad \mathbf{B} = \frac{q}{2} \frac{\mathbf{r}}{r^3}$$

The i-th component of the integrand is

$$\left[\frac{\mathbf{r} \wedge (\mathbf{E} \wedge \mathbf{r})}{r^3} \right]_i = \left(\frac{\delta_{ij}}{r} - \frac{r_i r_j}{r^3} \right) E_j = E_j \partial_j \frac{r_i}{r} = \partial_j \left(E_j \frac{r_i}{r} \right) - \frac{r_i}{r} \partial_j E_j$$

The total derivative term does not contribute to the integral, and

$$\mathbf{J}_{\mathrm{EM}} = -\frac{q}{2} \int \frac{\mathbf{r}}{r} \nabla \cdot \mathbf{E} \, d^3 x$$

Since $\nabla \cdot \mathbf{E} = e \, \delta(\mathbf{r} - \mathbf{r}_e)$,

$$\mathbf{J}_{\mathrm{EM}} = -\frac{eq}{2} \int \frac{\mathbf{r}}{r} \delta(\mathbf{r} - \mathbf{r}_e) \, d^3 x = -\frac{eq}{2} \frac{\mathbf{r}_e}{r_e}$$

Dirac Quantization of Charge

In view of the quantization of angular momentum in half-integer units (Chap. 4), the above observation leads to the fundamental

Dirac Quantization of Charge *The product of the electric and magnetic charges of any two objects is always an integer. In particular, if a monopole of non-zero charge exists, the values of all electric charges should be an integer (in suitable units). Dually, if a charged object exists, the magnetic charges are integrally quantized.*

Dirac proposed the existence of the monopole to explain the empirical fact that all electric charges in Nature are integral multiples of the charge of the electron. All Grand Unified Theories predict the existence of monopoles. See [18] for more.

Remark 5.8 An alternative proof the Dirac quantization of charge is purely geometric. By construction the wave-functions of a particle with electric charge e are sections of the line bundle $\mathcal{L}^e$, whose Chern class (a.k.a. *degree*) is

$$c_1(\mathcal{L}^e) = eq. \tag{5.229}$$

The Chern class of a complex line bundle is always an integer, so $eq \in \mathbb{Z}$.

Statistics Transmutation

Now we add to the Hamiltonian an attractive radial potential, say $-\alpha/r$, to produce bound states of our charged particle with the massive dyonic[29] body sitting at the origin of $\mathbb{R}^3$. In view of the **Spin & Statistics theorem** we have:

Fact 5.2 (Statistics Transmutation) *The Fermi parity operator ($\equiv \mathbb{Z}_2$ grading) $(-1)^F$ of the bound states of a particle with electric charge e (and no magnetic charge) and a dyon of magnetic charge q is*

$$(-1)^F\Big|_{\substack{bound \\ state}} = (-1)^F\Big|_{\substack{charged \\ particle}} (-1)^F\Big|_{dyon} (-1)^{eq} \tag{5.230}$$

In particular whenever eq is odd *the bound states of two bosons is a fermion and the bound states of a boson and a fermion is a boson.*

This transmutation from the Bose to the Fermi statistics (and viceversa) is a fundamental property of magnetic monopoles with odd magnetic charge.

Radial Schrödinger Equation

We return to our original problem, a quantum particle moving in $\mathbb{R}^3$ in presence of a magnetic monopole. We may separate the Schrödinger equation

$$H\Phi(r,\theta,\phi) = E\,\Phi(r,\theta,\phi) \tag{5.231}$$

by looking for solutions of the form

$$\Phi(r,\theta,\phi) = F(r)\,R^\ell_{m,q/2}(\phi,\theta,\psi(\theta,\phi)) \tag{5.232}$$

where $\psi(\theta,\phi)$ is *any* local section of the Hopf bundle (5.210). Different choices of this section correspond to different gauge choices for the magnetic potential. A good gauge choice is, say, the zero function $\psi(\theta,\phi) \equiv 0$. The radial Schrödinger equation becomes

$$-\frac{1}{r^2}\frac{\mathrm{d}}{\mathrm{d}r}r^2\frac{\mathrm{d}}{\mathrm{d}r}F + \frac{\ell(\ell+1)}{r^2}F = \frac{2mE}{\hbar^2}\,F \tag{5.233}$$

which is *identical* to the radial equation for the free particle moving in $\mathbb{R}^3$ (cf. Sect. 5.2), and therefore has the same solutions, *except* that now the allowed values of ℓ are different, being restricted by Eq. (5.221).

The fact that the radial equation is not affected by the presence of the monopole has a simple physical explanation. A particle moving radially will feel no magnetic force since the Lorentz force $\mathbf{v}\wedge\mathbf{B}$ vanishes in this situation. Hence the radial motion is not affected by the magnetic monopole.

---

[29] A *dyon* is a particle carrying both electric and magnetic charges.

More generally we may add a radial potential $V(r)$ to the problem. The radial equation will be the same one as in absence of the magnetic field, except for the restriction on the allowed ℓ's. This comment applies, for instance, to the dyonic system in Fact 5.2.

5.9 Hyperbolic Landau Levels

In the previous two sections we studied the quantum mechanics of a non-relativistic particle moving on, respectively, the plane and the sphere, subjected to a constant magnetic field. The sphere and the plane are maximally symmetric Riemannian surfaces[30] of, respectively, positive and zero curvature. There is a third symmetric surface with negative curvature, namely the Poincaré upper half-plane (the *hyperbolic* plane)

$$\mathfrak{H} \overset{\text{def}}{=} \{z \in \mathbb{C} : \operatorname{Im} z > 0\} \equiv \{(x, y) \in \mathbb{R}^2 : y > 0\} \tag{5.234}$$

with the $SL(2, \mathbb{R})$-invariant metric

$$ds^2 = \frac{dx^2 + dy^2}{y^2}. \tag{5.235}$$

To make the story complete, it is natural to study also the Schrödinger equation for a charged particle moving on the upper half-plane $\mathfrak{H}$ subjected to a uniform magnetic field whose field-strength 2-form is proportional to the Poincaré volume form

$$dA = B \frac{dx \wedge dy}{y^2} \tag{5.236}$$

(B a real constant). We set the electric charge $e = 1$ and $\hbar = 1$.

According to the rules of Chap. 2, in the gauge

$$A = B \frac{dx}{y}, \tag{5.237}$$

the Hamiltonian is the differential operator

$$H = \frac{1}{2m} y^2 \left(-\frac{\partial^2}{\partial y^2} + \left(-i\frac{\partial}{\partial x} - \frac{B}{y} \right)^2 \right) \tag{5.238}$$

acting on wave-functions $\psi(x, y)$ with Hilbert norm

$$\|\psi\|^2 = \int_{-\infty}^{+\infty} dx \int_0^\infty dy \, \frac{|\psi(x, y)|^2}{y^2}. \tag{5.239}$$

---

[30] In the present context by a "surface" we mean a Riemannian manifold of real dimension 2.

Since the coordinate x is cyclic, we may separate variables by writing the wave-function in the form

$$\psi(x, y; k) = \exp(ikx)\,\phi(y; k) \tag{5.240}$$

with k a real constant. The Schrödinger equation then reduces to

$$-y^2 \frac{d^2\phi}{dy^2} + (B - yk)^2\phi = 2mE\phi \tag{5.241}$$

or, in the more canonical form,

$$\frac{d^2\phi}{dy^2} - \left(\frac{B^2 - 2mE}{y^2} - \frac{2kB}{y} + k^2\right)\phi = 0. \tag{5.242}$$

We note that flipping the sign of k is equivalent to flipping the sign of B. Thus we may assume $k \geq 0$ with no loss (but we must take into account the solutions with B changed of sign). Abusing language, we shall speak of "bound" states (resp. "scattering" states) to refer to bound (resp. scattering) states of the reduced one-dimensional Schrödinger equation (5.242). In particular for a "bound" state

$$\int_0^\infty \frac{|\phi(y)|^2}{y^2}\,dy < \infty. \tag{5.243}$$

For $k = 0$ the solutions to Eq. (5.241) are

$$\phi(y; 0) = y^{1/2} \exp\left[\pm\sqrt{1 + 4B^2 - 8mE}\,\log y\right] \tag{5.244}$$

which are "scattering" states when the argument of the exponential is imaginary[31] i.e. for

$$E \geq \frac{1 + 4B^2}{8m}, \tag{5.245}$$

or otherwise they do not represent physical states. When $k > 0$, we introduce the rescaled variable $z = 2ky$ and put the Schrödinger equation in the form

$$\frac{d^2\phi}{dz^2} + \left(-\frac{1}{4} + \frac{2mE - B^2}{z^2} + \frac{B}{z}\right)\phi = 0 \tag{5.246}$$

---

[31] When the argument vanishes, i.e. (5.245) is an equality, we have a "scattering" state at threshold.

which is the Whittaker equation (cf. **box** on page 327) with $\kappa = B$ and

$$\frac{1}{4} - \mu^2 = 2mE - B^2 \quad \Rightarrow \quad \mu = \sqrt{B^2 + \frac{1}{4} - 2mE}. \tag{5.247}$$

The "bound" state solutions are real, so the argument of the square root must be non-negative

$$E \le \frac{1 + 4B^2}{8m}, \tag{5.248}$$

The "bound" states then correspond to the Whittaker functions

$$\phi(y; k) = C \, W_{B,\mu}(2ky) \tag{5.249}$$

with $\mu > 0$ and $k > 0$ (when $k < 0$ we flip the sign of B to reduce to $k > 0$). These functions are not normalizable unless

$$\frac{1}{2} + \sqrt{B^2 + \frac{1}{4} - 2mE} - B \equiv \frac{1}{2} + \mu - \kappa = -n \equiv 0, -1, -2, \dots, \tag{5.250}$$

or,

$$\sqrt{B^2 + \frac{1}{4} - 2mE} = B - \left(n + \frac{1}{2}\right) \tag{5.251}$$

when the Whittaker functions may be written in terms of degree-n Laguerre polynomials

$$W_{B, B-n-\frac{1}{2}}(2ky) = C' \, e^{-ky} y^{B-n} \, L_n^{(2B-2n-1)}(2ky). \tag{5.252}$$

When the RHS of (5.251) vanishes for some integer n the solution becomes

$$C' \, e^{-ky} y^{1/2} L_n^{(0)}(2ky) \approx C' \, y^{1/2} \text{ as } y \to 0, \tag{5.253}$$

which is non-normalizable in the sense (5.243). We conclude that in order to have a "bound" state the two sides of (5.251) must be *strictly* positive with n a non-negative integer. Thus we get one "bound" state solution for all $k > 0$ and all integer n such that

$$0 \le n < B - \frac{1}{2} \tag{5.254}$$

all $k < 0$ and integers n such that

$$0 \le n < -B - \frac{1}{2} \tag{5.255}$$

with wave-functions

$$\psi(x, y; n, k) = C(n)\, e^{ikx}\, e^{-ky}\, (ky)^{B-n}\, L_n^{(2B-2n-1)}(2ky), \tag{5.256}$$

where $C(n)$ is a normalization constant which is independent of k. In particular, when $|B| \le 1/2$ there is no "bound" state, while for $B > 1/2$ (resp. $B < -1/2$) all bound states have $k > 0$ (resp. $k < 0$). The energy of the bound states are

$$E = \frac{|B|}{m}\left(n + \frac{1}{2}\right) - \frac{n(n-1)}{2m} \tag{5.257}$$

the first term being the Landau formula for the energy levels in the plane and the second one the contribution from the curvature of the manifold. We refer to the "bound" states with quantum number n as the *n-th hyperbolic Landau level.* Eq. (5.257) is the formula for the hyperbolic Landau levels while Eq. (5.256) yields their explicit wave-functions.

From (5.251) it is clear that the lowest hyperbolic Landau level corresponds to the lowest n, i.e. $n = 0$

$$E_{\text{ground states}} = \frac{|B|}{2m} \tag{5.258}$$

which is the same value as in the plane except that now $|B|$ must be larger than $\frac{1}{2}$. There are infinitely many ground states, and the translation symmetry in x is spontaneously broken. In fact the broken group is much bigger, as we are going to show.

Noether Symmetries

The isometry group of the Poincaré half-plane is $SL(2, \mathbb{R}) \rtimes \mathbb{Z}_2$ where the inversion of orientation $\mathbb{Z}_2$ acts as $x \leftrightarrow -x$ (and hence $k \leftrightarrow -k$) In presence of a magnetic field $\mathbb{Z}_2$ is *not* a symmetry since the magnetic field, being a 2-form, is odd under inversion of orientation. The three Killing vectors of $SL(2, \mathbb{R})$ may be written in the form

$$K_- = \frac{\partial}{\partial x} \tag{5.259}$$

$$K_0 = x\frac{\partial}{\partial x} + y\frac{\partial}{\partial y} \tag{5.260}$$

$$K_+ = (x^2 - y^2)\frac{\partial}{\partial x} + 2xy\frac{\partial}{\partial y} \tag{5.261}$$

They generate the $\mathfrak{sl}(2)$ Lie algebra

$$[K_0, K_\pm] = \pm K_\pm, \qquad [K_-, K_+] = 2K_0. \tag{5.262}$$

For the gauge potential (5.237)

$$\mathscr{L}_{K_-} A = \mathscr{L}_{K_0} A = 0, \quad \mathscr{L}_{K_+} A = -2B\,\mathrm{d}y. \tag{5.263}$$

Therefore, applying the quantum Noether theorem (Chap. 2) to K_- (translation in x) and K_0 (dilatation $(x, y) \rightarrow \lambda(x, y)$ with $\lambda > 0$) we get that the two operators

$$M_- = -\mathrm{i}\frac{\partial}{\partial x} \tag{5.264}$$

$$M_0 = -\mathrm{i}x\frac{\partial}{\partial x} - \mathrm{i}y\frac{\partial}{\partial y} \tag{5.265}$$

are conserved Noether charges. The third conserved operator is modified by adding a term of order zero in the derivatives which compensates the gauge transformation $A \rightarrow A - 2B\,\mathrm{d}y$ induced by the action of K_+

$$M_+ = -\mathrm{i}(x^2 - y^2)\frac{\partial}{\partial x} - 2\mathrm{i}xy\frac{\partial}{\partial y} + 2By \tag{5.266}$$

The three Noether charges generate the $\mathfrak{sl}(2, \mathbb{R})$ Lie algebra

$$[M_0, M_\pm] = \mp\mathrm{i}M_\pm, \qquad [M_-, M_+] = -2\mathrm{i}M_0 \tag{5.267}$$

without modifications. While in the case of the plane the presence of a homogeneous magnetic field modifies the Lie algebra of Noether charges, this does not happen for the hyperbolic plane and the sphere. Again, this is a manifestation of the theorem proven in the **box** on page 215. The Lie algebras $\mathfrak{so}(3)$ and $\mathfrak{sl}(2, \mathbb{R})$ are semisimple and hence do not admit any non-trivial central extension. On the contrary the Lie algebra of isometries of the plane is solvable, and, when we switch on a magnetic field, it gets modified by a non-trivial central extension.

Each energy eigenspace should be stable under the action of $\mathfrak{sl}(2, \mathbb{R})$; this applies in particular to the hyperbolic n-th Landau level which must form an unitary representation of $SL(2, \mathbb{R})$. Since this group is non-compact, by Theorem 4.2, the representations—if non-trivial—must be infinitely dimensional, and this yields a Representation Theoretic explanation of the fact that all hyperbolic Landau levels have infinite degeneration. The actions of M_- and M_0 on the hyperbolic Landau levels are obvious

$$M_- \, \psi(x, y; n, k) = k \, \psi(x, y; n, k) \tag{5.268}$$

$$M_0 \, \psi(x, y; n, k) = -\mathrm{i}k \, \frac{\partial}{\partial k}\Big(\psi(x, y; n, k)\Big). \tag{5.269}$$

However, to be more systematic, it is convenient to identify the states in a given hyperbolic Landau level with the functions $f(k) \in L^2(\mathbb{R}_{>0})$ (with boundary condition $f(0) = 0$) through the correspondence

$$f \rightsquigarrow \int_0^\infty dk \, f(k) \, \psi(x, y; k, n), \tag{5.270}$$

so that

$$M_- f = kf, \qquad M_0 f = \mathrm{i}\frac{\partial}{\partial k}kf. \tag{5.271}$$

Passing to the Laplace transform

$$\hat{f}(z) = \int_0^\infty \mathrm{e}^{-zk} f(k) \, dk, \tag{5.272}$$

the action takes the standard form

$$M_- \hat{f}(z) = -\frac{\partial}{\partial z}\hat{f}(z), \qquad M_0 \hat{f}(z) = -\mathrm{i}z\frac{\partial}{\partial z}\hat{f}(z) \tag{5.273}$$

of the corresponding generators of $\mathfrak{sl}(2, \mathbb{R})$ identified with the isometry Lie algebra of the unit disk $\mathfrak{D} = \{|z| < 1\} \subset \mathbb{C}$ with the Poicaré metric

$$\mathrm{d}s^2 = \frac{\mathrm{d}z \, \mathrm{d}\bar{z}}{(1 - |z|^2)^2}, \tag{5.274}$$

which is well-known to be globally isometric to the upper half-plane $\mathfrak{H}$ with the metric (5.235). This fixes the generator M_+ to be the third Killing vector in $\mathfrak{iso}(\mathfrak{D})$:

$$M_+ \hat{f}(z) = z^2 \frac{\partial}{\partial z}\hat{f}(z). \tag{5.275}$$

We conclude that the full symmetry group $SL(2, \mathbb{R})$ is spontaneously broken.

Number of Ground States

In all three symmetric surfaces, the sphere, the plane, and the hyperbolic plane, the number of ground states per unit area is proportional to

$$\max\left\{|B| + \frac{K}{2}, 0\right\} \tag{5.276}$$

where $K = +1$, 0, and $K = -1$ for a symmetric surface whose curvature is (respectively) positive, zero, or negative. The interpretation of this result is clear in view of the Gauss-Bonnet and Riemann-Roch theorems. The integral of the above expression over a compact surface (divided by 2π) is

$$\max\left\{c_1(\mathcal{L}) + \frac{1}{2}\chi, 0\right\} \tag{5.277}$$

where $c_1(\mathcal{L})$ is the Chern class of the line bundle over which A is a connection and $\chi = 2 - 2g$ is the Euler class. Equation (5.277) then is the dimension of the space of holomorphic sections of $\mathcal{L}$.

The Poincaré Cylinder

We may identify the coordinate x periodically[32]

$$x \sim x + 2\pi \qquad \psi(x + 2\pi, y) = \psi(x, y) \tag{5.278}$$

getting a half-cylinder with the Poincaré metric. Now k should be an *integer*. The "bound" states now become genuine bound states with normalizable wavefunctions. Then, in addition to the scattering states, we have $n = [B - 1/2]$ bound states for each positive integer k.

Appendix: General Riemann-Roch Theorem and SUSY

In the last three sections we commented on the relevance of the Riemann-Roch theorem for Riemann surfaces (complex dimension 1) for the physics of a charged particle moving on a surface in presence of a (general) magnetic field. The Riemann-Roch theorem has been generalized in *all* complex dimensions [21]. In this appendix we explain the Quantum Mechanical meaning of the general theorem.

Let the configuration space $\mathcal{M}$ be a compact Kähler manifold of complex dimension n, and $\mathcal{L} \to \mathcal{M}$ a holomorphic line bundle with a Hermitian fiber metric h. This corresponds to a magnetic field on $\mathcal{M}$ given by the global $(1, 1)$ form

$$B \equiv i\bar{\partial}\partial \log h. \tag{5.279}$$

---

[32] More general periodic conditions may be also considered.

As Hilbert space $\mathcal{H}$ we take the closure of the space of differential forms of types $(0, \bullet)$ with coefficients in $\mathcal{L}$

$$\mathcal{H} = \widehat{\bigoplus_q \overline{\Omega^{(0,q)}(\mathcal{L})}} \tag{5.280}$$

This is a Hilbert superspace with grading operator $(-1)^F \equiv (-1)^q$ and Hermitian product

$$\langle \psi_1 | \psi_2 \rangle = \int_{\mathcal{M}} \overline{\psi}_1 \wedge *(h\psi_2) \tag{5.281}$$

(the overbar stands for complex conjugation). On this superspace we realize the $N = 1$ SUSY algebra with operators

$$Q = \overline{\partial}, \quad Q^\dagger = \overline{\partial}^\dagger, \quad 2H = \overline{\partial}\,\overline{\partial}^\dagger + \overline{\partial}^\dagger\overline{\partial}. \tag{5.282}$$

Let $\mathcal{V}_q \subset \mathcal{H}$ be the space of zero energy states with degree ($\equiv$ Fermi number) q. The Witten index is

$$\Delta \equiv \sum_q (-1)^q \dim \mathcal{V}_q \equiv \chi(\mathcal{L}) \tag{5.283}$$

the Euler characteristic of the line bundle $\mathcal{L}$. The Riemann-Roch theorem is an explicit formula for $\chi(\mathcal{L})$ in terms of characteristic classes, i.e. in terms of the magnetic 2-form B and the curvature 2-form (with coefficients in $\mathfrak{u}(n)$)

$$\chi(\mathcal{L}) = \int_{\mathcal{M}} e^{B/2\pi} \, \mathsf{Td}(R) \tag{5.284}$$

where B is the magnetic field seen as a 2-form, R is the Riemann tensor seen as a 2-form with coefficients in $\mathfrak{u}(n)$ and $\mathsf{Td}(\cdot)$ is the Todd "polynomial" [21]. The prescription of the integral in the RHS is to select the component of the integrand which is a form of degree $2n$ and forget the rest.

For simplicity we prove the theorem for a *flat* manifold, $R = 0$, which also gives the large magnetic field behavior for any compact manifold. We leave the general case $R \neq 0$ to the reader as an instructive exercise. $\mathsf{Td}(R)$ is just 1 for a flat manifold. The SUSY Hamiltonian may be written (schematically) as

$$H = \frac{1}{2}(p_i - A_i)^2 + B_{ij}\, b_i^\dagger b_j \tag{5.285}$$

and the Witten index is

$$\Delta = \mathrm{Tr}[(-1)^F \, e^{-\beta H}]. \tag{5.286}$$

Since the RHS is independent of β, we are free to compute it for any value we please, for instance, asymptotically as $\beta \to 0$ where the Trotter formula yields

$$e^{-\beta H} \approx e^{-\beta(p_i - A_i)^2/2}\, e^{-\beta B_{ij}\, b_i^\dagger b_j}. \tag{5.287}$$

The diagonal matrix elements

$$\langle x, \alpha | (-1)^F e^{-\beta(p_i - A_i)^2/2}\, e^{-\beta B_{ij}\, b_i^\dagger b_j} | x, \alpha \rangle \tag{5.288}$$

(the label α refers to the Fermi degrees of freedom) are *gauge invariant*, and we are free to choose a gauge such that $A_i(x) = 0$ at the particular point x where we compute the amplitude. Hence, at the leading order as $\beta \to 0$

$$\langle x, \alpha | (-1)^F e^{-\beta(p_i - A_i)^2/2}\, e^{-\beta B_{ij}\, b_i^\dagger b_j} | x, \alpha \rangle \approx$$
$$\approx \frac{1}{(2\pi\beta)^n} \langle \alpha | (-1)^F e^{-\beta B_{ij}(x)\, b_i^\dagger b_j} | \alpha \rangle \tag{5.289}$$

At the point x we may diagonalize the $n \times n$ matrix $B_{ij}(x)$ so that the last factor becomes the product of single Fermi oscillator factors. For a single Fermi oscillator we may identify the Fermi operators b, $b^\dagger$ with Pauli matrices, so that the single oscillator factor becomes

$$\mathrm{tr}\left[\sigma_3\, e^{-\beta B \sigma_+ \sigma_-}\right]\Big|_{\beta \to 0} = \beta B + O(\beta^2). \tag{5.290}$$

Putting everything together, and keeping into account the combinatorics of the differential forms, we get the final expression for the Witten index

$$\Delta = \frac{1}{n!} \int_M \frac{B^n}{(2\pi)^n} \tag{5.291}$$

where B^n is the n-th exterior power of the 2-form B. In view of Eq. (5.283) this proves the Riemann-Roch theorem (5.284) in the case $R = 0$.

Problems

5.1 Reduce a two-body quantum problem to a single-body problem with the reduced mass.

5.2 Solve the Schrödinger equation for the free motion in $\mathbb{R}^d$ with Hamiltonian $H = p_i p_i / 2m$ by separation of variables in *parabolic spherical* coordinates.

5.3 Separate the Schrödinger equation for a free particle moving in $\mathbb{R}^2$ in *elliptic* coordinates.

5.4 Prove Eq. (5.43).

5.5 Prove Lemma 5.1.

5.6 Prove the commutator relation (5.75).

5.7 Check the identities (5.79)–(5.81).

5.8 Show that the conserved Noether charges of a particle moving in a magnetic field, while they form depend on the chosen gauge, are in fact gauge invariant in value as required for physical observables.

5.9 Prove that the operators (5.172) commute with the Hamiltonian of a charged particle in a constant magnetic field.

5.10 Prove the classical formulae (5.173).

5.11 Prove the formula (5.228) for the angular momentum of the electromagnetic field sourced by a monopole and an electric charge.

References

1. S. Cecotti, *Analytic Mechanics. A Concise Textbook* (Springer, Berlim, 2024)
2. M. Bensimhoun, Historical account and ultra-simple proofs of Descartes's rule of signs, De Gua, Fourier, and Budan's rule. arXiv:1309.6664
3. L.D. Landau, E.M. Lifschitz, *Quantum Mechanics. Non-Relativistic Theory*, 3rd edn. (Butterworth-Heinemann, Oxford, 1981)
4. P. Moon, D.E. Spencer, *Field Theory Handbook. Including Coordinate Systems, Differential Equations and Their Solutions*, 3rd reprinting (Springer, Berlin, 1988)
5. I.S. Gradshteyn, I.M. Ryzhik, *Table of Integrals, Series, and Products*, 7th edn. (Elsevier, Amsterdam, 2007)
6. G.E. Andrews, R. Askey, R. Roy, *Special Functions.* Encyclopedia of Mathematics and its Applications, vol. 71 (Cambridge University Press, Cambridge, 2009)
7. F.W. Olver, D.M. Lozier, R.F. Boisvert, C.W. Clark, W. Charles (eds.), *NIST Handbook of Mathematical Functions* (Cambridge University Press, Cambridge, 2010). Available on-line at https://dlmf.nist.gov
8. E.T. Whittaker, G.N. Watson, *A Course of Modern Analysis*, 4th edn. (Cambridge University Press, Cambridge, 1927)
9. G.N. Watson, *A Treatise on the Theory of Bessel Functions*, 2nd edn. (Cambridge University Press, Cambridge, 1944)
10. S. Bochner, Uber Sturm-Liouville polynomsysteme. Math. Z. **29**, 730–736 (1929)
11. W. Hahn, Uber die Jacobiscehn Polynome und zwei werwandte Polynomklassen. Math. Z. **39**, 634–638 (1935)
12. H.M. Farkas, I. Kra, *Riemann Surfaces.* Graduate Texts in Mathematics, vol. 71 (Springer, Berlin, 1995)
13. P. Griffiths, J. Harris, *Principles of Algebraic Geometry* (Wiley, Hoboken, 1978)
14. R. Friedman, *Algebraic Surfaces and Holomorphic Vector Bundles.* Universitext (Springer, Berlin, 1998)

15. S. Kobayashi, *Differential Geometry of Complex Vector Bundles* (Princeton University Press, Princeton, 1987)
16. D. Huybrechts, *Complex Geometry. An Introduction.* Universitext (Springer, Berlin, 2005)
17. J.-P. Demailly, *Complex Analytic and Differential Geometry.* Book on line https://www-fourier.ujf-grenoble.fr/demailly/manuscripts/agbook.pdf
18. S. Coleman, The magnetic monopole fifty years later, in *Gauge Theories of High Energy Physics*, Part 1, Les Houches 1881 (North-Hollands, Amsterdam, 1983), pp. 461–553
19. S. Cecotti, *Introduction to String Theory* (Springer, Berlin, 2023)
20. J.D. Jackson, *Classical Electrodynamics*, 2nd edn. (Wiley, Hoboken, 1975)
21. F. Hirzebruch, *Topological Methods in Algebraic Geometry* (Springer, Berlin, 1978)

Chapter 6
Path Integrals

In this chapter we describe a formulation of Quantum Mechanics which is alternative, but equivalent, to the Hilbert space one outlined in Chap. 2. In this formulation the quantum amplitudes are written as infinite-dimensional integrals over function spaces called *path integrals* (or *functional integrals*). The path integral formulation of Quantum Mechanics is more intuitive and often more handy for concrete computations than the Hilbert space one. Most of the material presented in this chapter is not covered in textbooks, and some of it is actually novel.

The five Sects. 6.17–6.21 discuss various *techniques* for the explicit computation of path integrals, some of which are a bit sophisticated. A reader only interested in the general theory, may prefer to skip these sections.

This chapter is dedicated to *exact* results and techniques for path integrals. For a survey of the main *approximate* methods to compute path integrals, see Chap. 8.

6.1 The Feynman-Kac Formula

Arguably the most important quantity in Quantum Mechanics is the *time-evolution kernel*

$$U(x, y; t) \equiv \langle x | e^{-iHt/\hbar} | y \rangle \tag{6.1}$$

where x, y are points in the configuration space $\mathcal{M}$ and $|x\rangle, |y\rangle \in \mathcal{S}(\mathcal{M})^\vee$ are generalized position eigenvectors. For concreteness we focus on a spinless system with configuration manifold $\mathcal{M} \equiv \mathbb{R}^d$ and refer to the system as a "particle" moving in $\mathbb{R}^d$. For the moment we assume the Hamiltonian to have the simple form

$$H = \sum_i \frac{p_i p_i}{2m} + V(x) \tag{6.2}$$

© The Author(s), under exclusive license to Springer Nature Switzerland AG 2025 365
S. Cecotti, *Quantum Mechanics*, UNITEXT for Physics,
https://doi.org/10.1007/978-3-031-98824-0_6

for some time-independent potential $V : \mathbb{R}^d \to \mathbb{R}$. Spins and magnetic fields will be added in later sections. We assume the Hamiltonian H to be bounded below. Adding a constant to H, we set the energy of the ground state to zero.[1]

The time-evolution kernel $U(x, y; t)$ satisfies the Schrödinger differential equation

$$\mathrm{i}\hbar \frac{\partial}{\partial t} U(x, y; t) = H_x\, U(x, y; t), \tag{6.3}$$

where H_x stands for the Schrödinger Hamiltonian seen as a differential operator acting on the first argument $x \in \mathbb{R}^n$, that is,

$$H_x = -\frac{\hbar^2}{2m} \frac{\partial^2}{\partial x^i\, \partial x^i} + V(x). \tag{6.4}$$

Using the evolution kernel $U(x, y; t)$, we solve the Schrödinger equation, with initial state $\psi(x, t_0)$ at time t_0, in the form

$$\begin{aligned}
\psi(x, t) &= \int_{\mathbb{R}^d} \mathrm{d}^d y \, \langle x | e^{-\mathrm{i}(t - t_0)H/\hbar} | y \rangle \langle y | \psi, t_0 \rangle = \\
&= \int_{\mathbb{R}^d} \mathrm{d}^d y \, U(x, y; t - t_0)\, \psi(y, t_0).
\end{aligned} \tag{6.5}$$

The Heat Kernel

Technically (and also conceptually) it is more convenient to work with the analytic continuation of the evolution kernel $U(x, y; t)$ to *imaginary time* by replacing

$$\mathrm{i}t \rightsquigarrow s \in \mathbb{R}. \tag{6.6}$$

We shall therefore mostly be concerned with the imaginary time kernel

$$K(x, y; s) = \langle x | e^{-sH/\hbar} | y \rangle \quad x, y \in \mathcal{M}, \;\; s \geq 0, \tag{6.7}$$

where, for the moment, $\mathcal{M}$ is any Riemannian manifold with metric g_{ij}. From $K(x, y; s)$ we may always recover the real time evolution kernel $U(x, y; t)$ by a (careful) analytic continuation. The real time evolution kernels form a *group* of *unitary* operators, while the imaginary time ones form a *semi-group*[2]

$$\begin{aligned}
K(x, y; t + s) &= \int_{\mathcal{M}} \sqrt{g}\, \mathrm{d}^d z \, \langle x | e^{-sH/\hbar} | z \rangle \langle z | e^{-tH/\hbar} | y \rangle \equiv \\
&\equiv \int_{\mathcal{M}} \sqrt{g}\, \mathrm{d}^d z \, K(x, z; s)\, K(z, y; t)
\end{aligned} \tag{6.8}$$

---

[1] More precisely: we assume $\inf \sigma(H) = 0$.

[2] The convention in footnote 1 is in force.

$(t, s > 0)$ of Hermitian *contractions*

$$\left\| \int_M \sqrt{g}\, d^d y\, K(x, y; s)\psi(y) \right\| \leq \|\psi(x)\| \qquad \forall\, s > 0. \tag{6.9}$$

Suppose the system is *gapped,* i.e. the ground state(s) belongs to the discrete spectrum and there is a finite energy gap between the ground state(s) and the first excited states (as it happens, say, for the Landau levels, cf. Sect. 5.7). Then we have

$$\lim_{s \to +\infty} \langle x|e^{-sH}|y\rangle = \langle x|P_H(0)|y\rangle =$$

$$= \sum_k \langle x|0, k\rangle\langle k, 0|y\rangle = \sum_k \psi_k^0(x)\psi_k^0(y)^*, \tag{6.10}$$

where $P_H(0)$ is the spectral projector on the zero-energy eigenspace, and

$$\psi_k^0(x) \equiv \langle x|0, k\rangle, \qquad H|0, k\rangle = 0, \qquad \langle 0, h|0, k\rangle = \delta_{kh}, \tag{6.11}$$

is an orthonormal basis of zero-energy eigenstates. The kernel $K(x, y; s)$ satisfies the imaginary time Schrödinger equation

$$\frac{\partial}{\partial s} K(x, y; s) + H_x K(x, y; s) = 0. \tag{6.12}$$

For a free system moving in the Riemannian manifold M, with the geometric natural Hamiltonian given by the Laplacian Δ acting (densely) on $L^2(M)$, the imaginary time Schrödinger PDE (6.12) is the *heat equation* on M, and the function

$$K(x, y; s)\colon M \times M \times \mathbb{R} \to \mathbb{R} \tag{6.13}$$

is the *heat kernel of* M. The kernel $K(x, y; s)$ is a fundamental quantity in Geometric Analysis on Riemannian manifolds [1]. When M is compact, the heat kernel $K(x, y; s)$ is a *compact operator* acting on the Hilbert space $L^2(M)$. We shall use the terms "heat kernel" and "imaginary time evolution kernel" interchangeably.

The Formula

For simplicity of notation we specialize to the case $M = \mathbb{R}^d$. We split the (imaginary) time interval $[0, t]$ in N small intervals of equal length that we call ϵ

$$0 \equiv t_0 < t_1 < t_2 < \cdots < t_N \equiv t, \qquad t_{k+1} - t_k = \epsilon, \tag{6.14}$$

and write $x(t_k) \equiv (x(t_k)^1, \ldots, x(t_k)^d)$ for the eigenvalues of the coordinate operators $q \equiv (q^1, \ldots, q^d)$ at (imaginary) time $t_k = k\epsilon$. The reiteration of Eq. (6.8) yields[3]

$$\langle x(t)|e^{-(t-t_0)H/\hbar}|x(t_0)\rangle =$$

$$= \int dx(t_{N-1}) \cdots dx(t_1) \, \langle x(t)|e^{-\epsilon H/\hbar}|x(t_{N-1})\rangle \langle x(t_{N-1})|e^{-\epsilon H/\hbar}|x(t_{N-2})\rangle \times$$

$$\times \langle x(t_{N-2})|e^{-\epsilon H/\hbar}|x(t_{N-3})\rangle \cdots \langle x(t_2)|e^{-\epsilon H/\hbar}|x(t_1)\rangle \langle x(t_1)|e^{-\epsilon H/\hbar}|x(t_0)\rangle.$$
$$(6.15)$$

We are interested in taking the limit $\epsilon \to 0$ of this equality while keeping $N\epsilon = (t - t_0)$ fixed. The Trotter formula says that for a pair of operators A and B satisfying the appropriate conditions (see Theorem 2.5)

$$e^{-(A+B)} = \lim_{N \to \infty} \left(e^{-A/N} e^{-B/N} \right)^N.$$
$$(6.16)$$

In the present instance the operators

$$A \equiv \frac{t - t_0}{\hbar} \frac{p^2}{2m} \quad \text{and} \quad B \equiv \frac{t - t_0}{\hbar} V$$
$$(6.17)$$

satisfy the condition of Theorem 2.5(2) *provided* $(t - t_0) > 0$, and then the formula (6.16) holds in the sense of convergence in the strong topology. Hence in the limit

$$\epsilon \equiv \frac{t - t_0}{N} \to 0$$
$$(6.18)$$

we have

$$\langle x(t+\epsilon)|e^{-\epsilon H/\hbar}|x(t)\rangle \approx \langle x(t+\epsilon)|e^{-\epsilon \, p^2/(2m\hbar)}|x(t)\rangle \exp[-\epsilon \, V(x(t))/\hbar]$$

$$= \int \frac{d^d p}{(2\pi\hbar)^d} \langle x(t+\epsilon)|p\rangle \, e^{-\epsilon \, p^2/(2m\hbar)} \langle p|x(t)\rangle \exp[-\epsilon \, V(x(t))/\hbar]$$

$$= \int \frac{d^d p}{(2\pi\hbar)^d} \, e^{i p^j (x_j(t+\epsilon) - x_j(t))/\hbar - \epsilon \, p^2/(2m\hbar)} \exp[-\epsilon \, V(x(t))/\hbar]$$

$$= \left(\frac{m}{2\pi\hbar\epsilon} \right)^{d/2} \exp\left[-\frac{\epsilon}{\hbar} \left(\sum_{i=1}^{d} \frac{m}{2} \left(\frac{x(t+\epsilon)^i - x(t)^i}{\epsilon} \right)^2 + V(x(t)) \right) \right]$$
$$(6.19)$$

---

[3] We write $d^d x(t)$ simply as $dx(t)$ to avoid cluttering.

By definition, in the limit $\epsilon \to 0$

$$\frac{x(t+\epsilon)^i - x(t)^i}{\epsilon} \to \dot{x}(t)^i \quad \text{velocity at time } t, \tag{6.20}$$

and then

$$\langle x(t+\epsilon)|e^{-\epsilon H/\hbar}|x(t)\rangle \approx \left(\frac{m}{2\pi\hbar\epsilon}\right)^{d/2} \exp\left[-\frac{\epsilon}{\hbar} L_E(\dot{x}, x)\right] \tag{6.21}$$

where $-L_E(\dot{x}, x)$ is the analytic continuation of the Lagrangian

$$L = \frac{m}{2}\dot{x}^i\dot{x}^i - V(x) \tag{6.22}$$

to imaginary time $t \rightsquigarrow it$ and $\dot{x} \rightsquigarrow -i\dot{x}$. L_E is called the *Euclidean Lagrangian* by analogy with the corresponding object in QFT. The heat kernel is then

$$\begin{aligned}
\langle x(t)|e^{-tH/\hbar}|x(0)\rangle &= \lim_{\substack{\epsilon\to 0 \\ \epsilon N=t}} \int \prod_{i=1}^{N} \frac{d^d x(t_i)}{\sqrt{2\pi\hbar\epsilon/m}} \exp\left[-\frac{1}{\hbar}\sum_{i=1}^{N} \epsilon\, L_E(k\epsilon)\right] = \\
&= \lim_{\substack{\epsilon\to 0 \\ \epsilon N=t}} \int \prod_{i=1}^{N} \frac{d^d x(t_i)}{\sqrt{2\pi\hbar/m}} \exp\left[-\frac{1}{\hbar}\int_0^t ds\, L_E(\dot{x}, x)\right] \\
&= \lim_{\substack{\epsilon\to 0 \\ \epsilon N=t}} \int \prod_{i=1}^{N} \frac{d^d x(t_i)}{\sqrt{2\pi\hbar/m}}\, e^{-S_E[x]/\hbar},
\end{aligned} \tag{6.23}$$

where $S_E[x]$ is the *Euclidean action*, i.e. $-i$ times the classical action S analytically continued to imaginary times

$$S_E[x] = \int_0^t ds \left(\frac{m}{2}\dot{x}^i\dot{x}^i + V(x)\right), \tag{6.24}$$

evaluated on the piecewise-linear trajectory ($\equiv$ *path*) $x(s)$ where the particle at the time $s = t_k$ is in the position $x(t_k)^i \in \mathbb{R}^d$ and travels with constant velocity

$$v(t_k)^i = \frac{x(t_{k+1})^i - x(t_k)^i}{t_{k+1} - t_k}, \tag{6.25}$$

during the small time intervals $[t_k, t_{k+1}]$ of length ϵ, from $x(t_k)^i$ to $x(t_{k+1})^i$. Before taking the limit $\epsilon \to 0$, i.e. for ϵ small but finite, in the RHS of (6.23) we integrate with a certain positive measure over the positions $x(t_k)^i$ of the particle at the intermediate times $t_k = k\epsilon$ ($1 \le k \le N-1$) while keeping *fixed* the positions at the initial and final times, $x(0)^i$ and $x(k\epsilon)^i \equiv x(t)^i$. As $\epsilon \to 0$ the piecewise-linear path becomes a general curve $x(t)$ with those fixed endpoints, and the measure converges[4] to a positive measure on the space of such paths which we denote as $[dx]$ and call the *functional measure*. An integral with respect to the measure $[dx]$ is then a *functional integral*. This yields the

Theorem 6.1 (The Feynman-Kac Formula) *One has*

$$\langle x|e^{-Ht/\hbar}|y\rangle = \int_{(y,0)}^{(x,t)} [dx]\, \exp\left(-\frac{S_E[x]}{\hbar}\right) \tag{6.26}$$

where the functional integral is over all paths with $x(0) = y$ and $x(t) = x$. The functional integral is defined as the limit for $\epsilon \to 0$ of the RHS *of* (6.23) *(in the strong topology).*

We shall comment later on the class of paths on which the functional measure is supported, in particular with regard to their regularity properties. Mathematically the positive measure

$$[dx]\, \exp\left(-\frac{S_E[x]}{\hbar}\right) \tag{6.27}$$

is well-defined: as we shall mention later, it is strictly related to the *Wiener measure* which describes the Brownian motion.[5] Following the physics parlance, we refer to the functional integral over the space of paths (6.26) as a *path integral*, more precisely as an *Euclidean* path integral, meaning that the integration is carried out in imaginary time. Had we worked in *real* time, we would have got the *real-time path integral*

$$\langle x|e^{-iHt/\hbar}|y\rangle = \int_{(y,0)}^{(x,t)} [dx]\, \exp\left(\frac{i}{\hbar} S[x]\right), \tag{6.28}$$

where now

$$S[x] = \int_0^t ds\, L(\dot{x}, x) = \int_0^t ds \left(\frac{m}{2}\dot{x}^i\dot{x}^i - V(x)\right) \tag{6.29}$$

is the physical (i.e. real-time) action computed on real-time paths with fixed endpoints at $x(0)$ and $x(t)$. However the real-time path-integral (6.28) is a formal expression which can be rigorously defined only via the analytic continuation of

---

[4] For a proof of convergence see e.g. [2].

[5] For background on the Brownian motion see chap. 6 of [3].

its Euclidean counterpart (6.26). The real-time functional measure is not mathematically well-defined per se, and also from the physical side there are rather deep reasons to define it via analytic continuation of the more fundamental Euclidean path integral.

We may think of the path integral as providing an explicit "integral representation" of the solutions to the Schrödinger equation. Indeed, the path integral satisfies by construction the Schrödinger equation (or rather the *heat* equation)

$$-\hbar \frac{\partial}{\partial t} \int_{(y,0)}^{(x,t)} [dx]\, e^{-\frac{1}{\hbar} S_E} = H_x \int_{(y,0)}^{(x,t)} [dx]\, e^{-\frac{1}{\hbar} S_E}, \tag{6.30}$$

where H_x is the Hamiltonian seen as a differential operator acting on the position x at the final time t.

We stress that the functional measure $[dx]$ has all the usual formal properties of the Lebesgue measure in $\mathbb{R}^d$, although some extra care is needed when manipulating infinite-dimensional integrals, and one should pay due attention to the proper prescriptions of each computation.

6.1.1 Adding Discrete Degrees of Freedom

We consider systems which, in addition to continuous degrees of freedom taking values in a manifold $\mathcal{M}$, have discrete degrees of freedom taking n values, so that the Hilbert space is conveniently written as $L^2(\mathcal{M}) \otimes \mathbb{C}^n$, and the observables as $n \times n$ matrices whose entries are operators acting on $L^2(\mathcal{M})$. The Hamiltonian is typically an $n \times n$ matrix of the form

$$H_{ab} = \delta_{ab}\left(\frac{\hbar^2}{2m}\Delta + V(x)\right) + \hbar\, M(x)_{ab} \tag{6.31}$$

where $M(x)_{ab}$ is an $n \times n$ matrix of functions on $\mathcal{M}$. From the Trotter formula we get the generalized Feynman-Kac formula

$$\langle a, x | e^{-tH/\hbar} | b, y \rangle =$$

$$= \int_{(y,0)}^{(x,t)} [dx]\, e^{-\frac{1}{\hbar} \int_0^t ds\,(m\dot{x}^2/2 + V)} \left(\mathbf{Pe}^{-\int_{x(s)} M(x(s))ds}\right)_{ab} \tag{6.32}$$

where the $n \times n$ matrix

$$\mathbf{Pe}^{-\int_{x(s)} M(x(s))ds} \tag{6.33}$$

is the *path-ordered* exponential in which the matrices $M(x)$ for different x's are ordered following the path. Parametrizing the path with imaginary time, $x = x(s)$, the path-ordered exponential reduces to the time-ordered one, cf. Sect. 2.9.4:

$$
\mathbf{P}e^{-\int_{x(s)} M(x(s))ds} \equiv
$$

$$
\equiv 1 - \int_0^t M(x(s))\,ds + \int_0^t M(x(s_1))\,ds_1 \int_0^{s_1} M(x(s_2))\,ds_2 + \cdots \tag{6.34}
$$

The version (6.32) of the Feynman-Kac formula is seldom used, since it is unpractical for concrete computations. It is however conceptually quite important. A more handy approach to discrete degrees of freedom will be sketched in the last two sections of this chapter.

6.2 The Canonical Path Integral

There is a more canonical version of the path integral where one integrates on paths in the *phase space* $\mathcal{W}$ instead of trajectories in the configuration space $\mathcal{M}$. The quantum measure of this approach is the functional version of the canonical Liouville measure ([4] §. 6.8). To get the new formulation, we go back to Eq. (6.19) which we rewrite in the form[6]

$$
\langle q(t_{k+\epsilon})|e^{-\epsilon H/\hbar}|q(t)\rangle =
$$

$$
= \int \frac{d^d p}{(2\pi\hbar)^d}\, e^{\mathrm{i} p \cdot (q(t+\epsilon) - q(t))/\hbar - \epsilon\, p^2/(2m\hbar)}\, \exp[-\epsilon\, V(q(t))/\hbar], \tag{6.35}
$$

so that $(q_k \equiv q(t_k))$

$$
\langle q(t)|e^{-(t-t_0)H/\hbar}|q(t_0)\rangle =
$$

$$
= \int \frac{d^d p_0}{(2\pi\hbar)^d} \prod_{k=1}^{N-1} \frac{d^d p_k\, d^d q_k}{(2\pi\hbar)^d}
$$

$$
\times \exp\left[\epsilon \sum_k \left(\frac{\mathrm{i} p_k (q(t_{k+1}) - q(t_k))}{\hbar\epsilon} - \frac{p(t_k)^2}{2m\hbar} - \frac{V(q(t_k))}{\hbar} \right) \right] \tag{6.36}
$$

$$
= \int \frac{d^d p_0}{(2\pi\hbar)^d} \prod_{k=1}^{N-1} \frac{d^d p_k\, d^d q_k}{(2\pi\hbar)^d}
$$

$$
\times \exp\left[\sum_k \left(\frac{\mathrm{i} p_k (q(t_{k+1}) - q(t_k))}{\hbar} - \frac{\epsilon\, H(p(t_k), q(t_k))}{\hbar} \right) \right]
$$

---

[6] To avoid cluttering, we omit the d-vector indices in the q^i's and p_i's.

where $H(p, q)$ is the physical classical Hamiltonian. Taking the limit $\epsilon \to 0$ we get

$$\langle x(t)|e^{-(t-t_0)H/\hbar}|x(t_0)\rangle =$$

$$= \int [dp\, dq]\, \exp\left(\frac{i}{\hbar} \int_{x(t_0)}^{x(t)} p\, dq - \frac{1}{\hbar} \int_{t_0}^{t} H(p(t'), q(t'))\, dt'\right) \tag{6.37}$$

which is an integral over paths in the phase space $\mathcal{W}$ with fixed boundary conditions

$$q(t) = x(t), \qquad q(t_0) = x(t_0) \tag{6.38}$$

for the q's and *free* boundary conditions for the p's. The canonical functional measure

$$[dp\, dq] = \lim_{N \to \infty} \prod_{k=1}^{N} \frac{d^d p_k\, d^d q_k}{(2\pi\hbar)^d} \tag{6.39}$$

is the "obvious" functional version of the Liouville volume normalized by the volume of the quantum unit cell which properly counts the number of quantum states at the intermediate times t_k (cf. Eq. (2.435)).

Remark 6.1 Note that the integration space for the Lagrangian and the Hamiltonian path integrals obey exactly the same boundary conditions as the ensembles of paths where we *classically* look for extrema of the functional

$$\int_{t_0}^{t} L\, dt' \quad \text{resp.} \quad \int_{0}^{t} [p\dot{q} - H]dt' \tag{6.40}$$

for the Lagrangian ([4] chap. 4) and, respectively, Hamiltonian ([4] §. 6.3) classical variational problems. This is crucial for the quantum-to-classical correspondence.

6.3 Physical Interpretations

The path integral approach has the advantage over the Schrödinger equation of being much more physically intuitive and inspiring. Let us explain its physical meaning. When discussing the relation of path integrals with physical reality we are forced to use the real-time functional integral, since physical phenomena happen in real time. The main difference of the real-time path integral with respect to its Euclidean brother is that now the "measure" is complex instead of real positive.

We return to the interference experiments we briefly discussed in Chap. 1 to motivate the superposition principle. In Fig. 6.1a we see the set-up of the original interference experiment. We know that the total amplitude in a point x on the screen

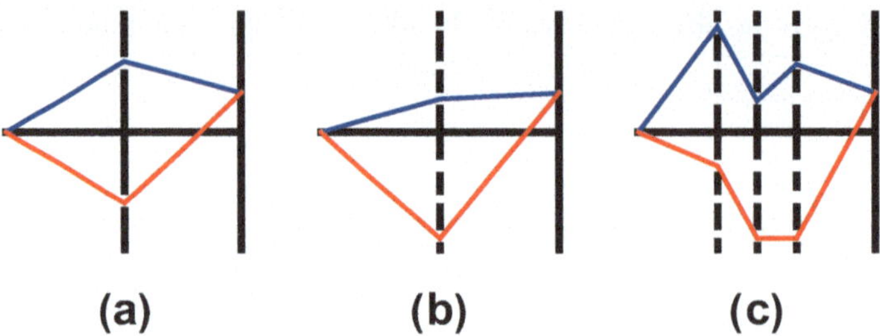

Fig. 6.1 Interference experiment with multiple screens and slits. (**a**) The total amplitude is the sum over the amplitudes of propagation through the two slits associated to the blue and the red path, respectively. (**b**) When we have several slits we have to sum over *all* slits. (**c**) When the particle passes through several screens, each one with a lot of slits, we have to sum over the amplitudes associated to the piecewise-linear path propagating through a series of slits whose positions at intermediate times is given by the position of the slit

is the sum $\psi_1(x) + \psi_2(x)$ of the amplitudes associated, respectively, to the particle going through the first and the second slit. If the intermediate panel has several slits, as in Fig. 6.1b, the amplitude becomes a sum $\sum_k \psi_k(x)$ over all the possible positions x_k of the slit in the vertical panel through which the particle may pass at time t_1. Consider now the situation where we have n parallel panels each one with m slits as in Fig. 6.1c. The amplitude in the final point x will be the sum over all possible amplitudes of the form

$$\psi_{i_1, i_2, \dots, i_n}(x) \qquad i_s = \in \{1, \dots, m\} \tag{6.41}$$

which corresponds to a *history* where the particle passed the first panel at the slit in position x_{i_1}, the second panel at the slit x_{i_2}, and so on, until the n-th panel which was passed at x_{i_n}. The total amplitude is then

$$\psi(x) = \sum_{\substack{\text{histories} \\ \{i_1,\dots,i_n\}}} \psi_{i_1,\dots,i_n}(x). \tag{6.42}$$

In other words: we must sum the amplitudes over *all possible time-histories of our particle,* taking into account all allowed *paths.* As the numbers of intermediate screens and slits go to infinity, $m, n \to \infty$, the sum becomes an integral over all paths between the fixed initial and final points. The amplitude for the propagation of the particle for an infinitesimal time is given[7] by $e^{iS/\hbar}$ and hence at the end we get

$$\psi(x) = \int_{\text{paths}} [dx]\, e^{iS/\hbar} \tag{6.43}$$

---

[7] This follows e.g. using the *eikonal equation* of geometric optics [4].

which is the real-time path integral. The path integral thus makes explicit the physical idea that the quantum particle *does not* follow a determinate trajectory but rather follows simultaneously all possible trajectories, each trajectory being "weighted" with a complex amplitude given by the exponential of $i/\hbar$ times the classical action of that trajectory. We stress that the (real time) path integral is a sum over *complex amplitudes* of the paths *not* of their probabilities (which are non-negative real numbers). Indeed, as already stressed in Chap. 1, the observed phenomenon of destructive interference could not happen if the probability of finding the particle in x was the sum of the (positive) probabilities of each path. We need the sum to contain terms with different phases in order to observe *destructive* interference (i.e. dark fringes). The path integral achieves this by summing paths with the correct *complex* weights. Again: Quantum Physics is not about classical probabilities. In Chap. 7 we will make this statement a lot sharper. However, if we analytically continue the complex amplitudes back to the Euclidean path integral, we do get a positive probability measure (in absence of magnetic fields and spins !) to which standard probability arguments and techniques apply.

States as Space-Time Histories

Path integrals also yield a new interpretation of *quantum states* which was especially emphasized by Feynman [5]. Consider an integral over paths

$$x : (-\infty, t_0] \to \mathbb{R}^d \tag{6.44}$$

which end up at time t_0 at the position $x_0 \in \mathbb{R}^d$ and have some sort of *fixed* boundary condition (or asymptotic behavior) $\bigstar$ as $t \to -\infty$, with possibly a time-dependent Hamiltonian in the far past, say for $t < t_* \ll t_0$, while the Hamiltonian H is time-independent for $t > t_*$. The value of the (real time) path integral is then a function

$$\psi(x_0, t_0) = \int_{\bigstar}^{(x_0, t_0)} [\mathrm{d}x]\, e^{iS[x]/\hbar} \tag{6.45}$$

of the boundary condition x_0 at time t_0. This function may be identified with a wave-function of the system at time t_0, hence with a physical state $|\psi, t_0\rangle \in \mathcal{H}$. This identification is consistent in the sense that it reproduces itself under time evolution: indeed, for all $t_1 > t_0$,

$$\psi(x, t_1) = \int \mathrm{d}^n y\, U(x, y; t_1 - t_0)\, \psi(x, t_0) \equiv \int_{\bigstar}^{(x, t_1)} [\mathrm{d}x]\, e^{iS[x]/\hbar} \tag{6.46}$$

by the real-time Feynman-Kac formula. In particular the state $|\psi, t\rangle$ so constructed is automatically a solution to the Schrödinger equation

$$i\hbar \frac{\mathrm{d}}{\mathrm{d}t}|\psi, t\rangle = H|\psi, t\rangle. \tag{6.47}$$

This leads to Feynman's notion of quantum state: *a quantum state is the path integral over its past space-time history,* that is:

> *A quantum state is the history of how it was produced* (either by natural time-evolution, or by the conscious action of a physicist who prepared the state for her experiment, or by the action of other external agents).

Hamiltonian Path Integrals

The interpretation of the canonical path integral (6.37) is similar, except that we integrate over the history of the system in its phase-space $\mathcal{W}$. The integral over the position $(q_k, p_k) \in \mathcal{W}$ at time t_k in Eq. (6.36) is done with the measure which precisely reproduces the quantum trace over the Hilbert space (cf. Eq. (2.432)), i.e. which counts the intermediate states at the time t_k weighted with the proper complex amplitude. The physical meaning of a sum over path histories is even more transparent than in the Lagrangian language.

6.4 $\pi_1(\mathcal{M}) \neq \{1\}$: Bohm-Aharonov Revisited

The path integral gives a more transparent and geometrical interpretation of the Bohm-Aharonov effect we studied in Sect. 3.11. To simplify the expressions, in this section we set $\hbar = 1$ by a choice of units.

We start from some general consideration. Suppose that our configuration space $\mathcal{M}$ is connected but not simply-connected. Then the space of continuous paths with fixed endpoints $x_0, x_1 \in \mathcal{M}$,

$$\mathsf{Path}(x_0, x_1) \equiv \left\{ x : [0, T] \to \mathcal{M} : x(0) = x_0, \ x(T) = x_1 \right\}, \tag{6.48}$$

decomposes in topological distinct classes under homotopy. We write

$$\mathsf{Hom}(x_0, x_1) \tag{6.49}$$

for the set of homotopy classes of paths from x_0 to x_1:

$$\mathsf{Path}(x_0, x_1) = \bigcup_{\gamma \in \mathsf{Hom}(x_0, x_1)} \mathsf{Path}(x_0, x_1; \gamma), \tag{6.50}$$

$$\mathsf{Path}(x_0, x_1; \gamma) \stackrel{\mathrm{def}}{=} \left\{ \ell \in \mathsf{Path}(x_0, x_1) : [\ell] = \gamma \right\} \tag{6.51}$$

Fixing a reference path $\ell_0 \in \mathsf{Path}(x_0, x_1)$ we see that

$$\mathsf{Hom}(x_0, x_1) \simeq \pi_1(\mathcal{M}, x_0) \quad \text{as sets via} \quad [\ell] \mapsto [\ell_0^{-1}\ell] \in \pi_1(\mathcal{M}, x_0). \tag{6.52}$$

The path integral (6.26), which represents the heat kernel, splits in a discrete sum of contributions, one from each topological class of paths:

$$\int_{(x_0,0)}^{(x_1,T)} [dx]\, e^{-S_E[x]} = \sum_{\gamma \in \mathsf{Hom}(x_0,x_1)} \int_\gamma [dx]\, e^{-S_E[x]} \tag{6.53}$$

where $\int_\gamma [dx]$ stands for the functional integral over paths $\ell \in \mathsf{Path}(x_0, x_1; \gamma)$.

Figure 6.1c shows an example of this situation: paths with the same endpoints x_0, x_1 which pass through different slits of a screen belong to different homotopy classes in $\mathsf{Hom}(x_0, x_1)$. We may think of modifying the system of screens by putting some special physical devices in their several slits, which interact with the passing particle. The particle-slit interaction adds a term μ_a to the action of a particle whenever it goes through the a-th slit. Writing $\mu(\gamma)$ for the sum of all μ_a's associated to the sequence of slits crossed by a particle when following a path of homotopy class γ, we get the path integral for the modified system

$$\sum_{\gamma \in \mathsf{Hom}(x_0,x_1)} e^{-\mu(\gamma)} \int_\gamma [dx]\, e^{-S_E[x]}. \tag{6.54}$$

This rough physical argument shows that, in general, the contributions to the path integral from the different topological sectors $\mathsf{Path}(x_0, x_1; \gamma)$ may be weighted with some extra weights $w(\gamma)$

$$\langle x_1|e^{-\beta H}|x_0\rangle = \sum_\gamma w(\gamma) \int_\gamma [dx]\, e^{-S_E[x]}. \tag{6.55}$$

We conclude that the set of homotopy weights $\{w(\gamma)\}$ is a necessary part of the definition of the quantum system whenever the configuration space (which we assume connected by default) is *non simply-connected*.

However the assignment $\gamma \mapsto w(\gamma)$ cannot be arbitrary. The heat kernel should be self-adjoint and satisfy the semi-group law. The first condition gives

$$\left(\sum_\gamma w(\gamma) \int_\gamma [dx]\, e^{-S_E[x]}\right)^* = \langle x_1|e^{-\beta H}|x_0\rangle^* =$$

$$= \langle x_0|e^{-\beta H}|x_1\rangle = \sum_\gamma w(\gamma^{-1}) \int_{\gamma^{-1}} [dx]\, e^{-S_E[x]} \tag{6.56}$$

that is,

$$w(\ell)^* = w(\ell^{-1}), \tag{6.57}$$

where we write $w(\ell)$ for $w([\ell])$. Likewise the semi-group property

$$\langle x_2|e^{-(\beta_1+\beta_2)H}|x_0\rangle = \int_M \sqrt{g}\, d^n x_1 \langle x_2|e^{-\beta_2 H}|x_1\rangle\langle x_1|e^{-\beta_1 H}|x_0\rangle \tag{6.58}$$

requires

$$w(\ell_2\ell_1) = w(\ell_2)\, w(\ell_1) \tag{6.59}$$

whenever the end point of ℓ_1 coincides with the starting point of ℓ_2, i.e. whenever the two paths ℓ_1, ℓ_2 may be concatenated to form the path $\ell_2\ell_1$. In particular the trivial path should have weight $w = 1$, and hence

$$1 = w(\ell)\, w(\ell^{-1}) = w(\ell)\, w(\ell)^*. \tag{6.60}$$

The weights $w(\gamma)$ then must be *phases*, and the phase associated to a concatenation of paths should be equal to the product of the phases of the several paths. Writing the path ℓ as the concatenation of a large number of paths ℓ_k of length ϵ, and writing $w(\ell_k) = \exp(i\epsilon A_k)$ with A_k *real*, in the limit $\epsilon \to 0$ we end up with

$$w(\ell) = \exp\left(i\int_\ell A\right), \tag{6.61}$$

where A is a real one-form on M. More precisely A is a $U(1)$ connection on a complex line bundle $\mathcal{L}$ over the configuration space M, and $w(\ell)$ is the *holonomy element* of the connection A along the path ℓ (known as the *Wilson line* in the QFT context). In the present set-up the weight $w(\ell)$ depends only on the class $[\ell]$ of the path. The integrals of A along two homotopic paths should be equal, and this forces the connection A to be *flat* i.e. to have zero *curvature* (called *field strength* in field theory)

$$F \equiv dA = 0. \tag{6.62}$$

The curvature 2-form F of a general $U(1)$ connection has the physical interpretation of a magnetic field. Here it vanishes only because we are assuming (for the moment !) that no magnetic field is present: in general the connection A *is not* flat and $F \neq 0$. But even when the connection *is* flat, and the magnetic field *is* zero, the presence of the connection A has a physically observable effect because the different topological sectors in the path integral are summed with the non-trivial weights $w(\gamma)$ given by the holonomy elements (6.61).

In this way we recover the situation described by the Bohm-Aharonov effect (cf. Sect. 3.11): a magnetic potential A with zero magnetic field, $F \equiv dA = 0$, has observable effects in Quantum Mechanics when the configuration space is not simply-connected.

> **Bohm-Aharonov Effect: Path Integral Interpretation**
> From the path integral viewpoint, the Bohm-Aharonov effect arises because we may give different weights to paths in different topological classes. Consistency requires the extra topological weight to be the holonomy element (6.61) of a flat $U(1)$ connection A along the path.

We stress that the imaginary-time action of a particle in the presence of a magnetic potential A ($\equiv U(1)$ connection on $\mathcal{M}$), not necessarily flat, has the form

$$S_E[x] = \int_0^T \left(\frac{1}{2} g_{ij} \dot{x}^i \dot{x}^j + V(x) \right) dt - i \int_0^T A_i \, \dot{x}^i \, dt, \tag{6.63}$$

as one sees by making the analytic continuation $it \rightsquigarrow s$ of the real time action

$$i \int \left(\frac{1}{2} g_{ij} \frac{dx^i}{dt} \frac{dx^j}{dt} + V(x) + A_i \frac{dx^i}{dt} \right) dt \rightsquigarrow$$

$$\rightsquigarrow - \int \left(\frac{1}{2} g_{ij} \frac{dx^i}{ds} \frac{dx^j}{ds} - V(x) - i A_i \frac{dx^i}{ds} \right) ds \equiv -S_E[x]. \tag{6.64}$$

In particular,

Fact 6.2 *The imaginary time functional measure $[dx] \exp(-S_E[x])$ is no longer real positive in presence of a background magnetic potential A.*

This fact has dramatic implications for the uniqueness of the ground state and symmetry breaking, see Sect. 6.13.

Remark 6.2 Around Eq. (6.60) we proved that the homotopy weights $\{w(\gamma)\}$ in Eq. (6.55) are *phases* using the semi-group law. A more direct argument goes as follows: the weights $\{w(\gamma)\}$ are *topological* features, hence, in particular, time independent. Hence they remain invariant under analytic continuation from imaginary to real time. The real time action

$$S_0 + \frac{1}{i} \log w(\ell) \tag{6.65}$$

should be real on physical grounds, and this implies that $w(\ell)$ is a phase for all ℓ.

Gauge Invariance While a connection A with zero field strength, $F \equiv dA = 0$, has observable quantum effects, these effects depend only the class of A modulo gauge equivalence, as we saw in Chap. 3. Indeed, consider two gauge-equivalent connections

$$A^{(2)} = A^{(1)} + df, \tag{6.66}$$

where f is a function which is globally defined on $\mathcal{M}$ or, more properly, which is multivalued in $\mathcal{M}$ with the differences between two determinations a multiple of 2π, so that $\exp(if)$ is a genuine *global function* $\mathcal{M} \to U(1)$ which yields the change of trivialization $\psi^{(1)} \rightsquigarrow e^{if}\psi^{(2)}$ of the associated line bundle. The holonomy elements of the two connections $A^{(1)}$, $A^{(2)}$ are related as

$$\begin{aligned} w(\ell)_2 &\equiv \exp\left(i\int_\ell A^{(2)}\right) = \\ &= \exp\left(i\int_\ell A^{(1)}\right)\exp\left(i\int_\ell df\right) = e^{if(x_1)}\, w(\ell)_1\, e^{-if(x_0)}, \end{aligned} \tag{6.67}$$

where x_0, x_1 are the fixed endpoints of ℓ. Since e^{if} is a global function on $\mathcal{M}$

$$\langle x_1 | e^{-\beta H^{(2)}} | x_0 \rangle = e^{if(x_1)} \langle x_1 | e^{-\beta H^{(1)}} | x_0 \rangle\, e^{-if(x_0)}, \tag{6.68}$$

where $H^{(a)}$ is the Hamiltonian computed with the gauge potential $A^{(a)}$. We see that the change of gauge (6.66) may be undone by redefining the Schrödinger representation by irrelevant globally-defined phases

$$|x\rangle \rightsquigarrow e^{-if(x)}|x\rangle \equiv |x\rangle_{\text{new}}, \tag{6.69}$$

which do not affect any physical observable. The *Riemann-Hilbert correspondence* states that there is an equivalence between flat $U(1)$ connections A modulo gauge equivalence and conjugacy classes of *monodromy representations*, i.e. of group homomorphisms

$$\varrho : \pi_1(\mathcal{M}) \to U(1). \tag{6.70}$$

Since $U(1)$ is Abelian, the map ϱ factors through the *Abelianization* of $\pi_1(\mathcal{M})$, which is the first homology group $H_1(\mathcal{M}, \mathbb{Z})$ [6], via the map

$$[\ell] \mapsto \exp\left(i\int_\ell A\right). \tag{6.71}$$

> **Summary: Topological Phases**
> To fully specify a quantum system with configuration space $\mathcal{M}$ and no magnetic fields (nor discrete degrees of freedom) we need to specify, in addition to the local physics on $\mathcal{M}$, an unitary representation $\varrho\colon H_1(\mathcal{M}, \mathbb{Z}) \to U(1)$. By the Riemann-Hilbert correspondence the last datum is equivalent to specifying a flat $U(1)$ connection modulo gauge equivalence

Non-Abelian Generalization

In the set-up of Sect. 6.1.1, where we have also discrete degrees of freedom acting on $\mathbb{C}^n$, the weights $w([\ell])$ of the different topological sectors take values in $U(n)$ and the complete definition of the system requires the specification of a non-Abelian monodromy representation

$$\varrho\colon \pi_1(\mathcal{M}) \to U(n), \tag{6.72}$$

given by the $U(n)$-valued holonomy elements

$$w(\ell) = \mathbf{P}\exp\left(\mathrm{i}\int_\ell A\right) \tag{6.73}$$

where now $\mathrm{i}A$ is a *flat* $U(n)$ connection which locally is a 1-form with coefficients in the Lie algebra $\mathfrak{u}(n)$. $\mathbf{P}$ is the path-order prescription, as in Sect. 6.1.1. Let $\{\mathrm{i}\lambda_a\}$ $(a = 1, \ldots, n^2)$ be $n \times n$ matrices making a basis of of $\mathfrak{u}(n)$. We see the λ_a's as *quantum operators* acting on the factor $\mathbb{C}^n$ of the Hilbert space $\mathcal{H} \equiv \mathbb{C}^n \otimes L^2(\mathcal{M})$. We expand the connection A in this basis,[8]

$$\mathrm{i}A = \mathrm{i}A^a(x)\lambda_a \tag{6.74}$$

where the A^a's are (locally) *real* one forms on $\mathcal{M}$. Comparing with the discussion in Sect. 6.1.1 we see that the imaginary time action has the form

$$S_E = S_E{}^0 - \mathrm{i}\int_0^T A^a(x)\lambda_a \, \mathrm{d}t. \tag{6.75}$$

The second term gives the coupling between the "orbital" degrees of freedom x^i's and the discrete ones λ_a. When the quantum system represents an elementary particle with spin which moves in $\mathcal{M}$, and the λ_a's are its spin operators $s_a \equiv \sigma_a/2$, the last term in Eq. (6.75) is called the *spin-orbit interaction,* a fundamental effect for real systems of particles (say electrons inside atoms and molecules).

---

[8] Sum over repeated indices implied!

6.5   Alternative Views on the Functional Measure

When computing path integrals explicitly, one seldom uses the definition of the functional measure as the limit $\epsilon \to 0$ in Eq. (6.23). There exist more user-friendly expressions for the path integral. Before going to that, we make a general observation whose validity will be even more evident in Sect. 6.8 below.

General Observation For most practical purposes it suffices to know the path integral up to overall normalization: often computing directly the integral with the proper absolute normalization may be tedious, while it is not worth the effort since, once we know the result up to an overall constant, we may easily recover the proper normalization in other ways, for instance by imposing that the heat kernel satisfies the appropriate semi-group law. Therefore in the literature one usually writes the path integral in the form

$$\mathcal{N}\int [\mathrm{d}x]\, e^{-S[x]/\hbar} \tag{6.76}$$

where $\mathcal{N}$ stands for an "overall normalization constant" in which we absorb all the annoying constant factors that we encounter along the computation without bothering to evaluate them.

We go back to the computation of the path integral (6.26). The integral is over all paths $x: [t_0, t_1] \to \mathbb{R}^d$ (in imaginary time) which satisfy the boundary conditions

$$x(t_0) = y \in \mathbb{R}^d, \qquad x(t_1) = x \in \mathbb{R}^d. \tag{6.77}$$

The integration variable $x(t)$ may be written as

$$x(t)^i = x_0(t)^i + h(t)^i \qquad i = 1, \dots, d, \tag{6.78}$$

where $x_0(t)$ is a *particular* path satisfying the boundary conditions (6.77), while $h(t)$ is a general path $h: [t_0, t_1] \to \mathbb{R}^d$ with

$$h(t_0)^i = h(t_1)^i = 0. \tag{6.79}$$

The functional measure is invariant under translations by fixed paths, that is,

$$[\mathrm{d}x] = [\mathrm{d}(x_0 + h)] = [\mathrm{d}h]. \tag{6.80}$$

We may expand $h(t)$ in a Fourier series of sines

$$h(t)^i = \sqrt{\frac{2}{(t_1 - t_0)}} \sum_{k=1}^{\infty} a_k^i \, \sin\left(\frac{k\pi(t - t_0)}{t_1 - t_0}\right), \tag{6.81}$$

where a_k^i are real coefficients. More generally, we may expand $h(t)$ in any other *orthonormal* Hilbert basis of (real) functions $\{\phi_k(t)\}$ in the interval $[t_0, t_1]$ which vanish at the endpoints:

$$h(t)^i = \sum_{k=1}^{\infty} b_k^i \, \phi_k(t). \tag{6.82}$$

We **claim** that the following formula holds

$$[dx] \equiv [dh] = \mathcal{N} \prod_{k \geq 1} \frac{d^d b_k}{(2\pi \hbar)^{d/2}}. \tag{6.83}$$

We stress that if this formula is valid for *one* Hilbert basis $\{\phi_k(t)\}$, it is valid for *all* of them with the same overall normalization $\mathcal{N}$. Indeed, the two bases are related by a linear real unitary transformation, and the functional Jacobian determinant

$$\det \left[\frac{\delta b_k}{\delta b'_j} \right] = 1. \tag{6.84}$$

The functional Jacobian determinant $\mathcal{J}$ from the original version of the functional measure

$$[dx] = {}^{``} \lim_{N \to \infty} \prod_{i=1}^{N} \frac{d^d x(t_i)}{\sqrt{2\pi \hbar / m}} {}^{"} \tag{6.85}$$

to the measure (6.83) is a constant which we may absorb in $\mathcal{N}$ and hence is inconsequential by the **General Observation**. Formally $\mathcal{J}$ looks like $m^{-\infty/2}$ and one may wonder how we make sense of similar quantities. Below we shall give concrete prescriptions to compute determinants of "infinite-dimensional matrices", i.e. determinants of unbounded operators acting in Hilbert spaces. There we shall see that, when properly defined, $\mathcal{J}$ is actually 1: see Remark 6.8.

6.6   Example: The Mehler Formula Again

In Sect. 3.5.7 we computed the heat kernel for the harmonic oscillator (the Mehler formula) using the Schrödinger representation and the properties of the Hermite polynomials. Now we give another proof of the formula using the Feynman-Kac formula. We set $\hbar = m = 1$ as before.

We split the path as

$$x(t) = y(t) + h(t), \tag{6.86}$$

where $y(t)$ is the solution to the imaginary time equations of motion

$$\ddot{y} = \omega^2 y, \tag{6.87}$$

which satisfies the boundary condition

$$y(0) = x_1, \qquad y(T) = x_2, \tag{6.88}$$

and $h(t)$ is a quantum fluctuation which vanishes at the endpoints: $h(0) = h(T) = 0$. The Euclidean action is

$$
\begin{aligned}
S_E[y + h] &= \frac{1}{2} \int_0^T \left((\dot{y} + \dot{h})^2 + \omega^2 (y + h)^2 \right) dt = \\
&= S_E[y] + S_E[h] + \int_0^T \left(\dot{h}\dot{y} + \omega^2 h y \right) dt = \\
&= S_E[y] + S_E[h] + \int_0^T \frac{d}{dt}(h\dot{y})\, dt + \int_0^T h\left(-\ddot{y} + \omega y \right) dt
\end{aligned}
\tag{6.89}
$$

The first integral in the last line is a boundary term which vanishes since $h(0) = h(T) = 0$, while the second integral is zero because $y(t)$ is a solution to the equations of motion (6.87). Then

$$
\langle x_2 | e^{-TH} | x_1 \rangle = \int_{(x_1,0)}^{(x_2,T)} [dx]\, e^{-S_E[x]} = e^{-S_E[y]} \int_{(0,0)}^{(0,T)} [dh]\, e^{-S_E[h]}. \tag{6.90}
$$

The path integral in the RHS is a factor independent of x_1, x_2 which may depend on the imaginary time T only through the combination ωT by dimensional analysis. This path integral is computed in Sect. 6.19

$$
\int_{(0,0)}^{(0,T)} [dh]\, e^{-S_E[h]} = \frac{1}{\sqrt{2\pi \sinh(\omega T)}}. \tag{6.91}
$$

This normalization factor may also be obtained algebraically by imposing the semigroup law on the kernel, or observing that its square must be a meromorphic function of ωT, of period $2\pi i$, with known poles and residues. The essential part of the formula (6.90) is the pre-factor $\exp(-S_E[y])$ which is purely classical (the exponential of Hamilton's principal function [4]). The general solution to Eq. (6.87) is

$$y(t) = A \sinh(\omega t + C) \tag{6.92}$$

with A, C constants which are determined by the boundary conditions as

$$A \sinh c = x_1, \quad A \sinh(\omega T + c) = x_2,$$

$$\Rightarrow \quad A \cosh c = \frac{x_2 - x_1 \cosh(\omega T)}{\sinh(\omega T)}. \tag{6.93}$$

The Hamilton principal function of the solution (6.92) is then

$$
\begin{aligned}
S_E[y] &= \frac{1}{2} \int_0^T (\dot{y}^2 + \omega^2 y^2)\mathrm{d}t = \frac{A^2 \omega^2}{2}\left(\sinh(2\omega T + 2c) - \sinh(2c) \right) \\
&= \frac{\omega}{2} \frac{(x_1^2 + x_2^2)\cosh(\omega T) - 2x_1 x_2}{\sinh(\omega T)},
\end{aligned}
\tag{6.94}
$$

and the heat kernel of the harmonic oscillator (with $m = \hbar = 1$) is

$$\langle x_2 | e^{-TH} | x_1 \rangle = \frac{1}{\sqrt{2\pi \sinh(\omega T)}} \exp\left[-\frac{\omega}{2} \frac{(x_1^2 + x_2^2)\cosh(\omega T) - 2x_1 x_2}{\sinh(\omega T)} \right] \tag{6.95}$$

which is the Mehler formula (3.153).[9]

6.7 Canonical Density Matrix, Traces, Free Energy

The imaginary time path integral computes the kernel

$$\langle x | e^{-\beta H} | y \rangle \tag{6.96}$$

of the imaginary time evolution operator in the Schrödinger representation $L^2(\mathbb{R}^d)$. This quantity has a physical interpretation of its own, besides the mathematical one of being the analytic continuation of the Schrödinger evolution kernel.

Canonical Density Matrix. Partition Function

Recall from Sect. 2.10 that a density matrix ρ of an ensemble of quantum states is a non-negative, trace-class, (bounded) Hermitian operator $\rho \in \mathcal{B}(\mathcal{H})$ normalized so that its trace is 1

$$\rho^\dagger = \rho, \qquad \rho \geq 0, \qquad \mathrm{Tr}_{\mathcal{H}}(\rho) = 1. \tag{6.97}$$

---

[9] In Sect. 3.5.7 the Hamiltonian is written in the normalized form (3.148), while in this section we used the standard form $H = (p^2 + \omega^2 q^2)/2$. The formula (6.95) is then obtained from (3.153) by the replacement $x_a \rightsquigarrow \sqrt{\omega} x_a$ ($a = 1, 2$).

We shall study density matrices more in depth in Chap. 7. Here we focus on their path integral representations.

For the class of Hamiltonians in Eq. (6.2) (with $V(x)$ bounded below), and $\beta > 0$, the operator $\mathrm{e}^{-\beta H}$ is positive, Hermitian, and trace-class. We write

$$\varrho = \frac{1}{Z(\beta)}\,\mathrm{e}^{-\beta H}, \tag{6.98}$$

for the imaginary time evolution operator $\mathrm{e}^{-\beta H}$ normalized so that it is the density matrix of an ensemble of quantum states

$$\mathrm{Tr}(\varrho) = 1 \quad \Rightarrow \quad Z(\beta) = \mathrm{Tr}(\mathrm{e}^{-\beta H}), \tag{6.99}$$

The operator ϱ has the physical interpretation of being the density matrix of the *canonical ensemble* at thermal equilibrium at the fixed temperature [3]

$$T = \frac{1}{k\beta} \tag{6.100}$$

where k is Boltzmann's constant that we usually set equal 1 by a choice of units [3]. The normalizing factor $Z(\beta)$ is the *canonical partition function* [3]. From Statistical Mechanics we know that the *free energy F* is [3]

$$F = E - TS = -T\,\log Z(T^{-1}), \tag{6.101}$$

where E is the *energy* of the system, and S its *thermal entropy* in equilibrium at the fixed temperature T. Then we have the following formula for the thermodynamical free energy F of our quantum system

$$\mathrm{e}^{-\beta F} = \mathrm{Tr}(\mathrm{e}^{-\beta H}) = \int \mathrm{d}^d x\,\langle x|\mathrm{e}^{-\beta H}|x\rangle. \tag{6.102}$$

The diagonal matrix element $\langle x|\mathrm{e}^{-\beta H}|x\rangle$ is given by the integral over *closed* paths that start at $x \in \mathbb{R}^d$ at imaginary time $s = 0$ and return to $x \in \mathbb{R}^d$ at imaginary time $s = \beta$. Since in Eq. (6.102) we also integrate over the position of the initial point x, we end up integrating over *all* closed paths which return to their starting point after an (imaginary) time β, i.e. we integrate over all paths periodic in imaginary time of period $\beta = T^{-1}$. We conclude:

> **Non-zero Temperature** $\equiv$ **Periodic in Imaginary Time**
>
> *The canonical partition function $Z \equiv e^{-\beta F}$ of a quantum mechanical system at temperature $T = \beta^{-1} > 0$ (in units where $k = 1$) is given by the path integral over all paths*
>
> $$\ell \colon S^1 \to \mathcal{M} \tag{6.103}$$
>
> *periodic in the imaginary time s of period $\beta = 1/T$.*

We demonstrated this fundamental fact only for spinless systems, but it is easy to see that it holds in full generality using the methods of Sect. 6.1.1.

Remark 6.3 If we change the overall normalization of the path measure by making $\mathcal{N} \rightsquigarrow \mathcal{N}'$, the free energy F will change by a term linear in T

$$F' = F - T \, \log(\mathcal{N}'/\mathcal{N}) \tag{6.104}$$

From Eq. (6.101) we see that this is equivalent to shifting the entropy S by an additive constant. It is a basic fact of thermodynamics that shifting S by a constant does not change the equations of state [3], i.e. the choice of $\mathcal{N}$ does not affect the physical observables.

Remark 6.4 When $\pi_1(\mathcal{M})$ is non-trivial, the periodic path integral decomposes into a sum over contributions from each homotopy class weighted with appropriate weights, cf. Sect. 6.4. This leads to thermal analogues of the Bohm-Aharonov effect.

Grand **Canonical Partition Functions**
Let $\{H, M_1, \ldots, M_s\}$ be a maximal set of independent, pair-wise commuting, Hermitian operators for our quantum system which includes the Hamiltonian H. We know from Chap. 2 that the M_a's are all conserved quantities, that they generate a group of unitary symmetries of the quantum system, and that they can be simultaneously diagonalized together with the Hamiltonian:

$$H|E, m_1, \ldots, m_k\rangle = E|E, m_1, \ldots, m_k\rangle, \tag{6.105}$$

$$M_a|E, m_1, \ldots, m_k\rangle = m_a|E, m_1, \ldots, m_k\rangle. \tag{6.106}$$

For simplicity we assume that the M_a are *Noether charges*,[10] i.e. associated to a Lie group of isometries of the configuration space $\mathcal{M}$ (see Sect. 2.16). Hence

$$e^{is_a M_a} q_j e^{-is_a M_a} = f(q; s_a)_j \tag{6.107}$$

---

[10] If one is interested in more general conserved quantities, which generate symmetries which mix q and p, one should work with the canonical path integral of Sect. 6.2 instead of the Lagrangian one we use in this section. Apart for this minor point, the discussion in the text applies in general.

for some Abelian group of isometries $\mathbb{R}^k \curvearrowright \mathcal{M}$ acting as

$$x_j \mapsto f(x, s_a)_j \equiv x_j^{(s)} \qquad x_j^{(0)} \equiv x_j, \quad s_a \in \mathbb{R}^k. \tag{6.108}$$

In particular

$$\mathrm{e}^{-\mathrm{i}s_a M_a}\big|x\big\rangle = \big|x^{(s)}\big\rangle, \qquad \big\langle x\big|\mathrm{e}^{\mathrm{i}s_a M_a} = \big\langle x^{(s)}\big|. \tag{6.109}$$

We may define a generalized partition function, called the *Grand Canonical partition function* [3],

$$\Xi(\beta, \mu_b) \overset{\text{def}}{=} \mathrm{Tr}[\mathrm{e}^{-\beta(H-\mu_a M_a)}] = \sum_{E, m_b} \mathrm{e}^{-\beta E - \mu_a m_a} \langle E, m_b | E, m_b \rangle \tag{6.110}$$

The parameter μ_a is called the *chemical potential* associated to the conserved quantity (charge) M_a.

We look for a path integral representing the general Grand Canonical partition function of a quantum system. We analytically continue the chemical potential by setting $\beta\mu_a = \mathrm{i}s_a$ with s_a real, so that the positive Hermitian operator $\exp(\beta\mu_a M_a)$ gets replaced by the *unitary symmetry* $U \equiv \exp(\mathrm{i}s_a M_a) \in \mathcal{U}(\mathcal{H})$. One has

$$\Xi(\beta, \mathrm{i}s_a/\beta) = \mathrm{Tr}\big[\mathrm{e}^{\mathrm{i}s_a M_a}\,\mathrm{e}^{-\beta H}\big] = \int \mathrm{d}^d x \,\big\langle x\big|\mathrm{e}^{\mathrm{i}s_a M_a}\,\mathrm{e}^{-\beta H}\big|x\big\rangle =$$

$$= \int \mathrm{d}^d x \,\big\langle x^{(s)}\big|\mathrm{e}^{-\beta H}\big|x\big\rangle = \int \mathrm{d}^d x \int_{(x,0)}^{(x^{(s)},\,\beta)} [\mathrm{d}x]\,\mathrm{e}^{-S_E[x]}. \tag{6.111}$$

Since $x \to x^{(s)}$ is a symmetry commuting with the imaginary-time evolution, we may restate this result in the form

Fact 6.3 *The* Grand Canonical *partition function* $\Xi(\beta, m u_a)$, *analytically continued to imaginary chemical potentials* $\mu_a = \mathrm{i}s_a/\beta$, *is given by an integral over all paths in* $\mathcal{M}$ *which satisfy the* generalized *periodic condition in imaginary time*

$$x(t + \beta)_j = x(t)_j^{(s)}, \tag{6.112}$$

that is, over paths which are periodic in imaginary time (of period β) up to the action of the symmetry $U : x_j \mapsto x_j^{(s)}$ where $U = \exp(\mathrm{i}s_a M_a)$.

6.8 Time-Ordered Correlations and Generating Functionals

To get a complete formalism of Quantum Mechanics, we need to be able to compute other quantities besides the evolution kernel. In this section we adopt the Heisenberg picture and define the Heisenberg operators at time t as (cf. Chap. 2)

$$A(t) = \mathrm{e}^{\mathrm{i}Ht} A\, \mathrm{e}^{-\mathrm{i}Ht} \tag{6.113}$$

where, for simplicity, we assume the Hamiltonian H to be time independent and set $\hbar = 1$. We shall generalize the story to arbitrary time-dependent Hamiltonians momentarily.

Let

$$0 < t_1 < t_2 < \cdots < t_k < t \tag{6.114}$$

be a sequence of intermediate time instants in increasing order. We are interested in computing the multi-time amplitude

$$\langle x | \mathrm{e}^{-\mathrm{i}Ht} A_k(t_k) A_{k-1}(t_{k-1}) \cdots A_2(t_2) A_1(t_1) | y \rangle \tag{6.115}$$

which generalizes the evolution kernel which is the special case of (6.115) where all Heisenberg operators A_j's are 1. To keep formulae short, we take $k = 2$: the generalization to arbitrary k is totally straightforward. We first assume that the A_j's commute with all q^i's, that is,

$$A_j = A_j(q^i) \quad \Rightarrow \quad \langle x_1 | A_j | x_2 \rangle = A(x_1)\, \delta(x_1 - x_2). \tag{6.116}$$

(cf. Corollary 2.6). Then

$$\langle x | \mathrm{e}^{-\mathrm{i}Ht} A_2(t_2) A_1(t_1) | y \rangle = \langle x | \mathrm{e}^{-\mathrm{i}H(t-t_2)} A_2\, \mathrm{e}^{-\mathrm{i}H(t_2-t_1)} A_1\, \mathrm{e}^{-\mathrm{i}Ht_1} | y \rangle =$$

$$= \int \mathrm{d}x_1\, \mathrm{d}x_2\, \mathrm{d}x_3\, \mathrm{d}x_4\, \langle x | \mathrm{e}^{-\mathrm{i}H(t-t_2)} | x_1 \rangle \langle x_1 | A_2 | x_2 \rangle \times$$

$$\times \langle x_2 | \mathrm{e}^{-\mathrm{i}H(t_2-t_1)} | x_3 \rangle \langle x_3 | A_1 | x_4 \rangle \langle x_4 | \mathrm{e}^{-\mathrm{i}Ht_1} | y \rangle =$$

$$= \int \mathrm{d}x_1\, \mathrm{d}x_3 \langle x | \mathrm{e}^{-\mathrm{i}H(t-t_2)} | x_1 \rangle A_2(x_1) \langle x_1 | \mathrm{e}^{-\mathrm{i}H(t_2-t_1)} | x_3 \rangle A_1(x_3) \langle x_3 | \mathrm{e}^{-\mathrm{i}Ht_1} | y \rangle =$$

$$= \int \mathrm{d}x_1\, \mathrm{d}x_3 \int_{(x_1,t_2)}^{(x,t)} [\mathrm{d}x]\, \mathrm{e}^{\mathrm{i}S[x]} A_2(x_1) \int_{(x_3,t_1)}^{(x_1,t_2)} [\mathrm{d}x]\, \mathrm{e}^{\mathrm{i}S[x]} A_1(x_3) \int_{(y,0)}^{(x_3,t_1)} [\mathrm{d}x]\, \mathrm{e}^{\mathrm{i}S[x]} =$$

$$= \int_{(y,0)}^{(x,t)} [\mathrm{d}x]\, \mathrm{e}^{\mathrm{i}S[x]} A_2(x(t_2))\, A_1(x(t_1))$$

$$\tag{6.117}$$

where the path integral in the last line is over all (real time) paths with fixed endpoints y, x at times 0 and t of the functional integrand

$$\mathcal{F}[x] = e^{iS[x]} A_2(x(t_2)) A_1(x(t_1)).$$
(6.118)

We summarize our findings

Summary: Path Integral with Operator Insertions
The real time path integral with *operator insertions*

$$\mathcal{N} \int_{(y,0)}^{(x,t)} [dx]\, e^{iS[x]} \prod_{j=1}^{k} A_j(x(t_j)) \qquad t_j\text{'s ordered as in (6.114),}$$
(6.119)

computes the matrix element

$$\langle x | e^{-iHt}\, \mathbf{T}\Big\{ \prod_j A_j(x(t_j)) \Big\} | y \rangle$$
(6.120)

where $\mathbf{T}$ is the *time-order* operation defined in Eq. (2.291). *In other words:* The path integral with the operator insertions $A_j(x(t_j))$ computes the matrix elements of the *time-ordered* product of the inserted operators

Ground State Correlations
We suppose that our quantum system has a normalizable ground state $|0\rangle$ (necessarily unique in absence of spins and magnetic fields, see Sect. 6.13). Then

$$\lim_{s \to +\infty} e^{-iHs} = |0\rangle\langle 0|$$
(6.121)

as one sees by defining the LHS as the analytic continuation of the imaginary time evolution operator. We rewrite (6.120) in the form

$$\langle x | e^{-iH(t-s)}\, \mathbf{T}\Big\{ \prod_j A_j(x(t_j - s)) \Big\} e^{-iHs} | y \rangle$$
(6.122)

and then take the limit $s \to +\infty$, $t - s \to +\infty$ while keeping $t_j - s \equiv \tau_j$ fixed. In the limit we get

$$\langle x|0\rangle\langle 0| \mathbf{T}\Big\{ \prod_j A_j(x(\tau_j)) \Big\} |0\rangle\langle 0|y\rangle,$$
(6.123)

that is, up to a simple factor, we get the expectation value of the time-ordered
product of operators in the ground state $|0\rangle$

$$\langle 0|\mathbf{T}\Big\{\prod_j A_j(x(\tau_j))\Big\}|0\rangle, \tag{6.124}$$

also called the (time-ordered) *correlation function* of the operators $A_j(\tau_j)$. The
proper normalization condition of these correlations is such that when the inserted
operator is the identity 1,

$$\langle 0|\mathbf{T}\{1\}|0\rangle \equiv \langle 0|1|0\rangle = \langle 0|0\rangle \tag{6.125}$$

we must get 1 for the normalized ground state $|0\rangle$. Hence the properly normalized
time-ordered correlation functions on the ground state are

$$\langle 0|\mathbf{T}\Big\{\prod_j A_j(x(t_j))\Big\}|0\rangle = \lim_{s\to+\infty} \frac{\int_{(y,-s)}^{(x,s)} [dx]\,e^{iS[x]} \prod_j A_j(x(t_j))}{\int_{(y,-s)}^{(x,s)} [dx]\,e^{iS[x]}} \tag{6.126}$$

where the annoying overall factor N and the boundary factors $\langle x|0\rangle\langle 0|y\rangle$ cancel
between numerator and denominator. We see from (6.126) that (assuming a *unique*
ground state) the functional integral over paths extending from time $-\infty$ to time
$+\infty$ does not depend on the initial conditions x and y (modulo factors which we
absorb in N). This "infinite time" path integral is usually written simply as

$$\int [dx]\,e^{iS[x]} \prod_j A_j(x(t_j)), \tag{6.127}$$

so that Eq. (6.126) may be written as

$$\langle 0|\mathbf{T}\Big\{\prod_j A_j(x(t_j))\Big\}|0\rangle = \frac{\int [dx]\,e^{iS[x]} \prod_j A_j(x(t_j))}{\int [dx]\,e^{iS[x]}} \tag{6.128}$$

We stress again that the overall factor N in the functional measure cancels out in the
ratio, that is, when computing a path integral we may disregard all constant factors,
as previously claimed. The LHS of Eq. (6.126) is usually written in the simplified
form

$$\mathbf{T}\Big\langle \prod_j A_j(x(t_j))\Big\rangle \tag{6.129}$$

and often the time-order prescription $\mathbf{T}$ is also left implicit. The functional integral with operator insertions (6.127) is computed (in facts *defined*) as the analytic continuation of the corresponding imaginary time integral

$$\int [\mathrm{d}x]\, \mathrm{e}^{-S_E[x]} \prod_j A_j(x(t_j)). \tag{6.130}$$

As we saw in Sect. 6.7, the inverse normalization factor

$$Z \equiv \int [\mathrm{d}x]\, \mathrm{e}^{-S_E[x]} \tag{6.131}$$

is the *partition function*. More precisely, the infinite imaginary time $\beta \to \infty$ path integral gives the canonical partition function at zero temperature. Now Z does not depend on time, so its analytic continuation to real time is equal to its value as computed by the imaginary-time path integral. This is a general rule for all static quantities.

General Operator Insertions   Up to now we assumed that the inserted operators commute with the q^i. Let us lift this unnatural restriction. For a system with Hamiltonian $H = p_i^2/2m + V$ we have the Heisenberg equations

$$p_i = m\, \dot{q}^i \tag{6.132}$$

so that, say,

$$\mathbf{T}\Big\langle p_i(t) \prod_j A_j(q(s_j)) \Big\rangle \tag{6.133}$$

(with $t \neq s_j$ for all j) is given by

$$m\frac{\mathrm{d}}{\mathrm{d}t}\mathbf{T}\Big\langle q(t)_i \prod_j A_j(q(s_j)) \Big\rangle = \frac{1}{Z}\int [\mathrm{d}x]\, \mathrm{e}^{\mathrm{i}S[x]}\, m\, \dot{x}(t)_i \prod_j A_j(x(s_j)). \tag{6.134}$$

We see that arbitrary correlations of canonical operators q^i, p_j are obtained by inserting in the path integral expressions which depend on both the position $x(t)$ and the velocity $\dot{x}(t)$ of the path at various intermediate times. The expression (6.134) is tricky when the insertion times coincide, i.e. when $t = s_j$ for some j, since velocities and positions *do not* commute. We shall analyze the issue of coincident times below, see in particular Eq. (6.166).

6.8.1 Adding Sources

Given a functional $F[x]$ of the path

$$x: \mathbb{R} \to \mathbb{R}^d \tag{6.135}$$

we define its functional derivative

$$\frac{\delta F[x]}{\delta x(t)^i} \qquad i = 1, \ldots, n, \tag{6.136}$$

as the coefficient of the first variation of the functional $F[x]$ in the sense of the Variational Calculus (see [4] chap. 4)

$$\delta F[x] = \int dt \, \frac{\delta F[x]}{\delta x(t)^i} \, \delta x^i(t) + \text{higher order in } \delta x^i(t). \tag{6.137}$$

Clearly

$$\frac{\delta x(s)^j}{\delta x(t)^i} = \delta^j{}_i \, \delta(s - t) \tag{6.138}$$

a formula which, together with the functional Leibniz rule, uniquely determines the functional derivative (6.136). Formally speaking: we treat the continuous variables t, s as additional "indices" labeling the functional variables, $x_{t,i} \equiv x(t)^i$, of our integral/differential functional calculus, and use the fact that the Dirac δ-function is the continuous counterpart to Kronecker's δ-symbol, $\delta_{s,t} \equiv \delta(s - t)$ cf. Sect. 2.4.2

$$\frac{\delta x(s)^j}{\delta x(t)^i} \rightsquigarrow \frac{\partial x_{s,j}}{\partial x_{t,i}} = \delta_{s,t} \, \delta_{j,i}. \tag{6.139}$$

For each fundamental degree of freedom $x(t)^i$ we integrate over in the path integral—which we call a *field* adopting the QFT language—we introduce a time-dependent *external driving force* $f(t)_i$, that is, we add to the Lagrangian a time-dependent extra potential linear in the "field" $x(t)^i$. In the imaginary-time formalism this is

$$L_E \rightsquigarrow L_E + x^i f_i(t) \tag{6.140}$$

$$\text{i.e.} \quad S_E[x] \rightsquigarrow S_E[x, f] = S_E[x] + \int dt \, x^i(t) \, f_i(t), \tag{6.141}$$

where now the action $S_E[x, f]$ is a functional depending on the two maps

$$x: \mathbb{R} \to \mathbb{R}^d \quad \text{and} \quad f: \mathbb{R} \to \mathbb{R}^d. \tag{6.142}$$

We require the map f to vanish rapidly as $t \to \pm\infty$, so that the path integral still represents an expectation value on the ground state $|0\rangle$ of the original system (cf. the argument around Eq. (6.123)). In the QFT context f is called the *source* of the field x and we shall use this term in our mechanical set-up as a synonym of "external driving force". The functional integral over the paths $x(t)$ now produces a functional of the sources

$$Z[f] = \int [dx]\, e^{-S_E[x,f]} \equiv \int [dx]\, e^{-S_E[x]} \exp\left(-\int dt\, x(t)^i\, f(t)_i\right) \qquad (6.143)$$

namely the (zero-temperature) partition function of the system coupled to the sources.

Now the n-point correlation function is simply[11]

$$\langle x(t_1)^{i_1} x(t_2)^{i_2} \cdots x(t_n)^{i_n}\rangle = (-1)^n\, \frac{1}{Z[0]}\, \frac{\delta^n Z[f]}{\delta f(t_1)_{i_1} \cdots \delta f(t_n)_{i_n}}\bigg|_{f\equiv 0} \qquad (6.144)$$

Definition 6.1 In view of the property (6.144), the partition function in presence of general sources, $Z[f]$, is called the *generating functional of the correlation functions*.

The correlations in presence of non-zero sources are given by the same formula (6.144) but setting f equal to their background value $f(t)_B$ instead of zero

$$\langle x(t_1)^{i_1} x(t_2)^{i_2} \cdots x(t_n)^{i_n}\rangle_{f(t)_B} =$$
$$= (-1)^n\, \frac{1}{Z[f(t)_B]}\, \frac{\delta^n Z[f]}{\delta f(t_1)_{i_1} \cdots \delta f(t_n)_{i_n}}\bigg|_{f\equiv f(t)_B}. \qquad (6.145)$$

In particular the 1-point function is

$$\langle x^i(t)\rangle = -\frac{\delta W[f]}{\delta f(t)_i}, \qquad (6.146)$$

where $W[f]$ is the functional

$$W[f] \stackrel{\text{def}}{=} \log Z[f], \qquad (6.147)$$

called the *generating functional of the **connected** correlation functions* which are

$$\langle x(t_1)^{i_1} x(t_2)^{i_2} \cdots x(t_n)^{i_n}\rangle_{\text{conn}} \stackrel{\text{def}}{=} (-1)^n\, \frac{\delta^n W[f]}{\delta f(t_1)_{i_1} \cdots \delta f(t_n)_{i_n}}\bigg|_{f\equiv 0} \qquad (6.148)$$

---

[11] As advertised, in the LHS we leave the symbol $\mathbf{T}$ implicit.

For the combinatoric aspects of the *connected* correlations and their statistical meaning see [3] §. 5.5. We shall return to the properties of the connected correlations in Chap. 8 in relation with modern perturbation theory.

Remark 6.5 We may see the generating functional

$$
Z[-\mathrm{i} f] = \int [\mathrm{d}x]\, e^{-S_E[x,-\mathrm{i} f]} \equiv \int [\mathrm{d}x]\, e^{-S_E[x]}\, \exp\!\left(\mathrm{i}\int \mathrm{d}t\; x(t)^i f(t)_i\right)
$$

$$(6.149)$$

as a *functional Fourier transform* of the functional $\exp(-S_E[x])$ in the path space, which returns another functional on the same space. This functional transform satisfies (formally) all the usual properties of the Fourier transform in $\mathbb{R}^n$. For a compilation of useful formulae for the path integrals reinterpreted as functional Fourier transforms see §. 9.1 of [2].

6.9   Integration by Parts: Schwinger-Dyson Equations

The *Ehrenfest theorem* of Quantum Mechanics states that the classical equations of motion are satisfied "in expectation value", that is, the matrix elements $\langle \psi_1 | q(t) | \psi_1 \rangle$ and $\langle \psi_1 | p(t) | \psi_1 \rangle$ of the Heisenberg canonical operators satisfy

$$
m \frac{\mathrm{d}}{\mathrm{d}t} \langle \psi_1 | q(t) | \psi_1 \rangle = \langle \psi_1 | p(t) | \psi_1 \rangle,
$$

$$(6.150)$$

$$
\frac{\mathrm{d}}{\mathrm{d}t} \langle \psi_1 | p(t) | \psi_1 \rangle = -\langle \psi_1 | V'(q(t)) | \psi_1 \rangle
$$

$$(6.151)$$

This is an immediate consequence of the Heisenberg equations of motion (in facts these relations holds for all matrix elements, not just for the diagonal ones). From the path integral perspective the theorem follows from a functional *integration by parts*. Indeed we have

$$
\int_{(y,0)}^{(x,t)} [\mathrm{d}x]\, \frac{\delta\, e^{\mathrm{i} S[x]}}{\delta x(s)^i} = 0 \qquad s \neq 0, t
$$

$$(6.152)$$

for all reasonable action functionals $S[x]$ since the integral of a total variation reduces to a boundary term *in function space* and this boundary contribution typically vanishes for the functionals $S[x]$ which arise from sound quantum systems. Computing the functional derivative in the LHS of (6.152) we get

$$
\left\langle x, t \left| \frac{\delta S[x]}{\delta x(s)^i} \right| y, 0 \right\rangle = 0, \qquad s \neq 0, t
$$

$$(6.153)$$

which is the matrix element of the classical equations of motion

$$\frac{\delta S[x]}{\delta x(t)^i} = 0 \tag{6.154}$$

between arbitrary states. This proves the Ehrenfest theorem.

Schwinger-Dyson Equations More generally, for any functional $F[x]$ we have the identity

$$\int [dx] \, \frac{\delta}{\delta x(t)^i} \left(F[x] \, e^{-S_E[x]} \right) = 0, \tag{6.155}$$

where now we integrate over paths from $t = -\infty$ to $t = +\infty$ (so we do not need to specify a boundary condition, as long as there is a unique ground state—see the argument around Eq. (6.123)). The above identity gives the equation

$$\left\langle \frac{\delta F[x]}{\delta x(t)^i} \right\rangle - \left\langle \frac{\delta S_E[x]}{\delta x(t)^i} \, F[x] \right\rangle = 0, \tag{6.156}$$

obeyed by all the imaginary time correlations. For instance

$$\left\langle \frac{\delta S_E[x]}{\delta x(t)^i} \, x(s_1)^{j_1} \cdots x(s_n)^{j_n} \right\rangle =$$

$$= \sum_{\ell=1}^{n} \delta^{j_\ell}{}_i \, \delta(t - j_\ell) \langle x(s_1)^{j_1} \cdots \widehat{x(s_\ell)^{j_\ell}} \cdots x(s_n)^{j_n} \rangle \tag{6.157}$$

where an overhat means that the factor is omitted. Equation (6.157) shows that:

> *The classical equation of motion are obeyed inside the correlation functions away from the times s_k where an operator is inserted*

The equalities (6.157) are known as the *Schwinger-Dyson equations*.

6.10 Free Particle in $\mathbb{R}^d$. Commutation Relations

The functional measure for the imaginary time path integral of a free quantum particle coincides with the Wiener measure in the sense of stochastic analysis. In particular it is a mathematically well-defined measure [2].

To simplify the notation we start with the case $d = 1$; the straightforward extension to arbitrary d is given at the end of the subsection. The free particle moving in $\mathbb{R}$ has the imaginary time action (we set $m = \hbar = 1$)

$$S_E[x] = \int \frac{1}{2}\dot{x}^2 \, dt. \tag{6.158}$$

We add a source f for the time derivative $\dot{x}$ of the field x

$$\begin{aligned} S_E[x, f] &\equiv \int dt \left(\frac{1}{2}\dot{x}^2 + \dot{x} f(t) \right) = \\ &= -\frac{1}{2} \int dt \, f(t)^2 + \frac{1}{2} \int dt \left(\frac{d}{dt}(x + F) \right)^2 \end{aligned} \tag{6.159}$$

where $\dot{F} \equiv f$. Since the functional measure is translationally invariant

$$[dx] = [d(x + F)], \tag{6.160}$$

we have

$$\begin{aligned} Z[f] &= \int [dx] \, e^{-S_E[x,f]} = e^{\frac{1}{2}\int dt \, f(t)^2} \int [d(x+F)] \, e^{-\frac{1}{2}\int dt \, (\dot{x}+\dot{F})^2} \\ &= e^{\frac{1}{2}\int dt \, f(t)^2} \int [dy] e^{-\frac{1}{2}\int dt \, \dot{y}^2} = \mathcal{N} e^{\frac{1}{2}\int dt \, f(t)^2}, \end{aligned} \tag{6.161}$$

where $\mathcal{N}$ is a constant independent of the source f. Then

$$\langle \dot{x}(t_1) \, \dot{x}(t_2) \rangle \equiv \frac{\delta^2 \exp\left[\frac{1}{2}\int dt \, f(t)^2 \right]}{\delta f(t_1) \, \delta f(t_2)} \Bigg|_{f=0} = \delta(t_1 - t_2), \tag{6.162}$$

where we wrote $x(t)$ in boldface to emphasize that it is a Heisenberg operator. We integrate this equality to get

$$\langle x(t_1) \, \dot{x}(t_2) \rangle = \Theta(t_1 - t_2) + \text{const.} \tag{6.163}$$

where $\Theta(z)$ is the Heaviside step function (2.164). In imaginary time the Heisenberg equations read[12]

$$i m \dot{x} = p. \tag{6.164}$$

---

[12] The Euclidean action is $\int (i p\dot{x} - p^2/2m - V) dt$, see Eq. (6.39). Its variation with respect to p yields $i\dot{x} - p/m = 0$.

For $\hbar = m = 1$ Eq. (6.163) yields

$$\langle \boldsymbol{x}(t_1)\,\boldsymbol{p}(t_2)\rangle \equiv \langle 0|\,\mathbf{T}\big(\boldsymbol{x}(t_1)\,\boldsymbol{p}(t_2)\big)|0\rangle = i\,\Theta\,(t_1 - t_2) + \text{const.} \tag{6.165}$$

Then

$$\langle 0|[\boldsymbol{x},\,\boldsymbol{p}]|0\rangle = \lim_{t_1 \to t_2^+}\Big(\langle \boldsymbol{x}(t_1)\,\boldsymbol{p}(t_2)\rangle - \langle \boldsymbol{x}(t_2)\,\boldsymbol{p}(t_1)\rangle\Big) = i, \tag{6.166}$$

which reproduces the canonical commutation relation. Integrating again (and using the symmetry $t_1 \leftrightarrow t_2$), we get

$$\langle \boldsymbol{x}(t_1)\,\boldsymbol{x}(t_2)\rangle = C - \frac{1}{2}|t_1 - t_2|, \tag{6.167}$$

where the integration constant $C = \langle \boldsymbol{x}^2\rangle$ is divergent. This is a consequence of translational invariance: the correlation of operators invariant under translation—such as $\dot{\boldsymbol{x}}$—are well defined, but not those of $\boldsymbol{x}$ itself. Indeed, we know that the translation symmetry is unbroken, and the only way that the equality

$$\langle (\boldsymbol{x} + a)^2\rangle = \langle \boldsymbol{x}^2\rangle \quad \forall\, a \in \mathbb{R} \tag{6.168}$$

may be *not wrong* is that both sides diverge. In the following subsection we discuss the meaning of the result (6.167) which is equivalent to the canonical commutation relations (cf. (6.166)). For future reference we note that the two-time correlation (6.167) satisfies the inhomogeneous linear differential equation

$$-\frac{d^2}{dt_1^2}\langle \boldsymbol{x}(t_1)\,\boldsymbol{x}(t_2)\rangle = \delta(t_1 - t_2). \tag{6.169}$$

For the free motion in $\mathbb{R}^d$ simply replace Eq. (6.162) with

$$\langle \dot{\boldsymbol{x}}^i(t_1)\,\dot{\boldsymbol{x}}^j(t_2)\rangle = \delta^{ij}\,\delta(t_1 - t_2), \qquad i,\, j = 1,\ldots,d. \tag{6.170}$$

6.11 More on the Functional Measure

The Support of the Path Measure

We continue the analysis of the example in Sect. 6.10. We have (for $t > 0$)

$$\langle \exp[\lambda(\boldsymbol{x}(t) - \boldsymbol{x}(0))]\rangle = \frac{1}{Z}\int [dx]\,e^{-\frac{m}{2\hbar}\int_{\mathbb{R}}\dot{x}^2\,ds + \lambda\int_0^t \dot{x}(s)\,ds} =$$

$$= \frac{1}{Z}\int [dx]\,e^{-\frac{m}{2\hbar}\int_{\mathbb{R}}(\dot{x} + \frac{\hbar\lambda}{m}\chi_t)^2\,ds + \frac{\hbar\lambda^2}{2m}t} = \exp\left[\frac{\hbar\lambda^2}{2m}t\right], \tag{6.171}$$

where λ is an indeterminate and χ_t the characteristic function of the interval $[0, t]$. In particular, expanding both sides in powers of λ to second order, we get

$$\langle (x(t) - x(0))^2 \rangle = \frac{\hbar}{m} t \tag{6.172}$$

which shows that the support of the functional measure (or Wiener measure) is on paths which are *continuous but everywhere non-differentiable* since the variance of the mean velocity

$$\left\langle \left(\frac{x(\epsilon) - x(0)}{\epsilon} \right)^2 \right\rangle = \frac{\hbar}{m\,\epsilon} \tag{6.173}$$

diverges as $\epsilon \to 0$. This is yet another manifestation of the canonical commutation relations, that is, of the fact that the trajectory is not a good concept in Quantum Mechanics.

In facts, the typical trajectory of the free motion in $\mathbb{R}^d$ with $d \geq 2$ has Hausdorff dimension 2. Another way of seeing the same fact is to compute the functional measure of the set of paths which have finite action (this set contains all paths which are piecewise of class C^1). We write the path integration variable $x(s)$ as a free motion solution plus a Fourier sum

$$x(s) = \frac{x(t) - x(0)}{t} s + \sqrt{\frac{2}{t}} \sum_{k \geq 1} a_k \sin(\pi k s / t). \tag{6.174}$$

So that

$$S_E = \frac{m}{2} \int_0^t \dot{x}^2 \, ds = \frac{m}{2t} (x(t) - x(0))^2 + \frac{m\pi^2}{2t^2} \sum_{k \geq 1} k^2 a_k^2. \tag{6.175}$$

Consider the measure $\mathsf{Mes}(\Lambda)$ of the paths with action

$$S_E \leq \frac{m}{2t} (x(t) - x(0))^2 + \Lambda^2. \tag{6.176}$$

We have the obvious bound

$$\mathsf{Mes}(\Lambda) \leq \mathsf{Mes}\left(\left\{ \frac{\sqrt{m}\,\pi}{\sqrt{2\hbar}\,t} k \, |a_k| \leq \Lambda, \ \forall k \right\} \right) =$$

$$= \prod_{k \geq 1} \frac{\int_{\frac{\sqrt{m}\,\pi}{\sqrt{2\hbar}\,t} k |a_k| \leq \Lambda} \exp[-\frac{m\pi^2}{2\hbar t^2} k^2 a_k^2] \, da_k}{\int_{\mathbb{R}} \exp[-\frac{m\pi^2}{2\hbar t^2} k^2 a_k^2] \, da_k} = \prod_{k \geq 1} \frac{\int_{|b_k| \leq \Lambda} \exp[-b_k^2] \, db_k}{\int_{\mathbb{R}} \exp[-b_k^2] \, db_k} \tag{6.177}$$

since the constant Jacobian of the change of variables $a_k \rightsquigarrow b_k \equiv \sqrt{\frac{m\pi^2}{2\hbar t^2}} k a_k$ cancels between numerator and denominator. The RHS of (6.177) is

$$\lim_{n\to\infty} \operatorname{erf}(\Lambda)^n \to 0 \quad \text{for } \Lambda < \infty, \tag{6.178}$$

where $\operatorname{erf}(z)$ is the *error function* [7].

Reflection Positivity

We have seen in Sect. 6.3 that a ket state $|\psi\rangle$ may be identified with the integral over all space-time past histories that lead to it: its wave-function $\langle x|\psi\rangle$ is a path integral over imaginary-time paths $(-\infty, 0) \to \mathbb{R}^d$ ending at x for $t = 0$ computed with a measure $d\mu[x]_\psi$ which reflects its *past* history (in imaginary time). Dually, the wave-function of a bra state $\langle\phi|y\rangle$ is an integral over its imaginary time *future* history.

Suppose we have a functional $F[x(t)]$ which depends only on the values of $x(t)$ for $t < 0$: we say that $F[x(t)]$ has *support on the half-line $t < 0$*. For instance, $F[x(t)]$ may be a product of operator insertions at times $t_k < 0$. The functional $F[x(t)]$ may be identified with a state $|F\rangle$ via

$$\langle x|F\rangle = \int^{(x,0)} [dx]\, e^{-S_E[x]}\, F[x(t)] \equiv \langle x|F[x(t)]|0\rangle, \tag{6.179}$$

where the **T**-order is implicit. The Hermitian conjugate bra state $\langle F|$ is then produced by the insertion of the time-reflected functional

$$\langle F|x\rangle = \int_{(x,0)} [dx]\, e^{-S_E[x]}\, F[x(t)]^\sharp \equiv \langle 0|F[x(t)]^\sharp|x\rangle \tag{6.180}$$

where the *time-reflected functional* is

$$F[x(t)]^\sharp \overset{\text{def}}{=} F[x(-t)]^*. \tag{6.181}$$

The functional $F[x(t)]^\sharp$ has support on the half-line $t > 0$. Now

$$0 < \langle F|F\rangle = \int d^d x\, \langle F|x\rangle\langle x|F\rangle =$$

$$= \int d^d x \int_{(x,0)} [dx]\, e^{-S_E[x]} F[x]^\sharp \int^{(x,0)} [dx]\, e^{-S_E[x]} F[x] \tag{6.182}$$

$$= \int [dx]\, e^{-S_E[x]}\, F[x]^\sharp\, F[x] = \left\langle F[x]^\sharp F[x] \right\rangle.$$

We conclude that the imaginary-time correlations of the time-ordered product of any operator insertions times its time-reflection is positive. This is a fundamental property of imaginary-time expectation values known as *reflection positivity*. It is one of the basic axioms in a rigorous approach to quantum physics [2]: it replaces in the functional approach the assumption that the Hilbert space norm is non-negative. Clearly only imaginary-time actions $S_E[x]$ consistent with reflection positivity may lead to physically consistent path integrals. From the obvious equality

$$Z = \int [\mathrm{d}x]\, \mathrm{e}^{-\int_0^{+\infty} S_E[x]\,\mathrm{d}t}\, \mathrm{e}^{-\int_{-\infty}^0 S_E[x]\,\mathrm{d}t} \tag{6.183}$$

we infer the condition on the Euclidean Lagrangian

$$L_E(x, -\dot{x}) = L_E(x, \dot{x})^*, \tag{6.184}$$

which implies, in particular, that the real-time Lagrangian $L(x, \dot{x}) = L_E(x, -i\dot{x})$ should be real.

6.12 Symmetries and Ward Identities

Symmetry is a central aspect of Quantum Mechanics. Let us see how it works in the path integral formulation.

The Classical Noether Theorem Revisited

We start with the reformulation of the *classical* Noether theorem ([4] Sect. 3.6) in the functional language. We use the imaginary-time Lagrangian L_E: in absence of spins and magnetic fields, L_E is equal to the real-time Lagrangian for the inverted potential $V \rightsquigarrow -V$. Suppose our classical system has a connected Lie group G of Noether symmetries, and let $\mathfrak{g}$ be its Lie algebra. We write $\alpha \cdot \delta x$ for the action of $\alpha \in \mathfrak{g}$ on the Lagrangian degrees of freedom that we collectively denote as $x \equiv (x^i)$. Since G is a symmetry of the Lagrangian L_E, to the linear order in $\alpha \in \mathfrak{g}$ we have

$$L_E[x + \alpha \cdot \delta x] + \alpha \cdot \frac{\mathrm{d}}{\mathrm{d}t} F(x, t) = L_E[x] \tag{6.185}$$

for a certain function $F(x, t)$ taking values in $\mathfrak{g}^\vee$ [4]. This equality holds for α *time-independent*. If we make α time-*dependent*, at the linear order we must get an extra term $\dot{\alpha} \cdot M$, proportional to the time derivative $\dot{\alpha}$ of α:

$$L_E[x + \alpha \cdot \delta x] - L_E[x] + \alpha \cdot \frac{\mathrm{d}}{\mathrm{d}t} F(x, t) = \dot{\alpha} \cdot M \stackrel{\mathrm{def}}{=} \dot{\alpha} \cdot \delta x^i \frac{\partial L_E}{\partial \dot{x}^i}. \tag{6.186}$$

Integrating from 0 to T we get

$$\int_0^T \left(L_E[x + \alpha \cdot \delta x] - L_E[x] \right) dt + \int_0^T \alpha \cdot \frac{dF}{dt} \, dt = \int_0^T \dot{\alpha} \cdot M \, dt, \qquad (6.187)$$

that is, we have the identity (to the linear order in α)

$$S_E[x + \alpha \cdot x] - S_E[x] + \alpha(T) \cdot F(T) - \alpha(0) \cdot F(0) = - \int_0^T \dot{\alpha} \cdot J \, dt \qquad (6.188)$$

where

$$- J \equiv M + F = \delta x^i \frac{\partial L_E}{\partial \dot{x}^i} + F \qquad (6.189)$$

is a physical observable, valued in the dual $\mathfrak{g}^\vee$ of the Lie algebra, that we call the *Noether charge* of the symmetry G. In the traditional proof of the Noether theorem one takes α to be time-independent. In the functional argument one takes $\alpha(t)$ to be a generic smooth map $\alpha : [0, T] \rightarrow \mathfrak{g}$ with the boundary conditions

$$\alpha(0) = \alpha(T) = 0. \qquad (6.190)$$

Then $S_E[x + \alpha \cdot \delta x] - S_E[x]$ is a variation of the functional $S_E[x]$ with fixed endpoints, which is zero (to first order in α) when evaluated on a solution of the (imaginary time) equations of motion: this follows directly from the variational characterization of the classical solutions ([4] chap. 4). Therefore, evaluating both sides of the identity (6.188) on a solution, we conclude that classically *on shell*

$$\int_0^T \dot{\alpha} \cdot J \, dt = 0 \qquad (6.191)$$

for all $\alpha : [0, T] \rightarrow \mathfrak{g}$ which vanishes at the endpoints, Eq. (6.190), that is, integrating by parts and using the fundamental lemma of variational calculus ([4] chap. 4)

$$\frac{d}{dt} J = 0, \qquad (6.192)$$

and the Noether charge J is a *conserved quantity* along the physical classical time-evolution. This is the Noether theorem of Classical Mechanics.

For our purposes we are instead interested in the identity (6.188) evaluated on a *generic* path $x : [0, T] \rightarrow M$ which is not necessarily a classical solution. Then

$$\frac{d}{dt} J(t) = \left. \frac{\delta S_E[g \cdot x]}{\delta \alpha(t)} \right|_{g=1} \qquad (6.193)$$

where

$$g(t) \cdot x \equiv \exp(\alpha(t)) \cdot x \tag{6.194}$$

is the path obtained by applying to the path $x(t)$ the time-dependent transformation $g(t) \equiv \exp(\alpha(t))$ satisfying $g(0) = g(T) = 1$.

Ward Identities

Let G be a connected symmetry group of our quantum system with a classical analogue. For all maps $g : [0, T] \to G$ we have

$$[d(g \cdot x)] = [dx], \tag{6.195}$$

and hence for all integrable functionals $F[x]$

$$\int_{(y,t)}^{(x,T)} [dx]\, F[g \cdot x] = \int_{(y,0)}^{(x,T)} [d(g \cdot x)]\, F[g \cdot x] = \int_{(g^{-1}(0)\cdot y,0)}^{(g^{-1}(T)\cdot x,T)} [dx]\, F[x]. \tag{6.196}$$

In particular, extending the time interval from $-\infty$ to $+\infty$, we may ignore the boundary conditions to get:

$$\int [dx]\, \frac{\delta}{\delta\alpha(t)} F[g \cdot x] = 0. \tag{6.197}$$

Taking $F[x] = G[x] \exp(-S_E[x])$ this gives

$$\left\langle \frac{\delta S_E[g \cdot x]}{\delta\alpha(t)}\Big|_{g=1} G[x] \right\rangle = \left\langle \frac{\delta G[g \cdot x]}{\delta\alpha(t)}\Big|_{g=1} \right\rangle \tag{6.198}$$

or, in view of Eq. (6.193)

$$\frac{d}{dt}\left\langle J(t)\, G[x] \right\rangle = \left\langle \frac{\delta G[g \cdot x]}{\delta\alpha(t)}\Big|_{g=1} \right\rangle. \tag{6.199}$$

A typical instance of this general identity is

$$\frac{d}{dt}\left\langle J(t)\, x^{i_1}(s_1) \cdots x^{i_k}(s_k) \right\rangle =$$

$$= \sum_{j=1}^{k} \delta(t - s_j) \left\langle x^{i_1}(s_1) \cdots \delta x^{i_j}(s_j) \cdots x^{i_k}(s_k) \right\rangle, \tag{6.200}$$

where $\delta x(s)$ is the infinitesimal variation of $x(s)$ generated by the Noether charge J. In the operator formulation δx is defined by the commutator

$$[J, x^j] = i\delta x^j. \tag{6.201}$$

The equalities of the form (6.199) and (6.200) are called *Ward identities* in QFT.

Physical Interpretation   The meaning of Eq. (6.200) is transparent. Let us write the real time analytic continuation of the equation

$$\frac{\mathrm{d}}{\mathrm{d}t}\mathbf{T}\Big\langle J(t)\,x^{i_1}(s_1)\cdots x^{i_k}(s_k)\Big\rangle =$$
$$= i\sum_{j=1}^{k}\delta(t-s_j)\,\mathbf{T}\langle x^{i_1}(s_1)\cdots \delta x^{i_j}(s_j)\cdots x^{i_k}(s_k)\rangle. \tag{6.202}$$

where we wrote explicitly the time-order operation $\mathbf{T}$ for emphasis. We assume $s_1 > s_2 \cdots > s_k$. We see from (6.202) that the Heisenberg picture quantum Noether charge $J(t)$ is constant in the time intervals

$$s_{j+1} < t < s_j \tag{6.203}$$

between two operator insertions. This just reflects the fact $J(t)$ is a conserved quantity. The time derivative of $J(t)$ has to vanish only away from the insertions of other operators, see the general principle in the gray box of Sect. 6.9. Next consider the limits $t \to s_j^{\pm}$: because of the time-order prescription, the limits from the left and the right are different

$$\lim_{t \to s_j^+}\mathbf{T}\langle\cdots J(t)\,x^{i_j}(s_j)\cdots\rangle - \lim_{t \to s_j^-}\mathbf{T}\langle\cdots J(t)\,x^{i_j}(s_j)\cdots\rangle =$$
$$= \mathbf{T}\langle\cdots [J, x^{i_j}(s_j)]\cdots\rangle = i\,\mathbf{T}\langle\cdots \delta x^{i_j}(s_j)\cdots\rangle \tag{6.204}$$

where in the last equality we used that J is the generator of the symmetry action and Eq. (6.201). We conclude that the time derivative in the LHS of (6.202) is a sum of δ-functions with coefficients as in Eq. (6.204).

6.13   Symmetry (Non-)Breaking

Let G be a symmetry group of the Hamiltonian. According to the definition in Sect. 3.10.1, the symmetry G is *unbroken* iff G acts on the energy eigenspace $V_0 \subset \mathcal{H}$ of ground states (lower energy eigenstates) as multiplication by scalars. In the important special case that G is a semisimple Lie group, this is equivalent to saying that V_0 is a direct sum of trivial representations (cf. **BOX** on page 215).

All symmetries G are automatically unbroken when the ground state is *unique,* i.e. the lowest energy eigenvalue E_0 is non-degenerate. In Chap. 3 we proved that the spinless one-dimensional systems have a unique ground state using the Sturm comparison theorem. However the story looks very different in presence of a magnetic potential A (even when the magnetic *field* $B \equiv dA$ vanishes):

SB1 in the model studied in Sect. 3.10.1, a particle moving in a circle S^1 with anti-periodic boundary conditions, the ground state is doubly degenerate and parity $\mathscr{P}$ is spontaneously broken;

SB2 in the systems studied in the last three sections of Chap. 5—a charged particle moving in a maximally symmetric surface M in presence of a uniform magnetic field—the ground state has an arbitrarily large degeneracy and the symmetry group $\mathsf{Iso}(M)$ is fully broken.

We need a simple criterion saying *when* symmetry breaking is possible. Path integrals give a short and elegant proof of the following results:

Theorem 6.4 *Let M be a* connected *Riemannian manifold. A spinless system with Hilbert space $L^2(M)$ and Hamiltonian*[13]

$$H = \frac{1}{2m}\Delta + V, \tag{6.205}$$

where Δ is the Laplacian and $V : M \to \mathbb{R}$ is continuous and bounded below, has a unique *ground state, hence all its symmetries are* unbroken. *Moreover the ground state wave function may be chosen to be everywhere positive (i.e. has no zeros).*

The proof is based on the following

Lemma 6.1 *Under the condition of the **Theorem**, the imaginary-time evolution kernel*

$$\langle m' | e^{-\beta H} | m \rangle \tag{6.206}$$

is strictly positive for all $\beta > 0$ and all pairs of points $m, m' \in M$.

Indeed the kernel (6.206) is given by a path integral of the form

$$\int_{(m,0)}^{(m',\beta)} [dx]\, e^{-S_E[x]} \tag{6.207}$$

---

[13] The more general Hamiltonian $H = (\Delta + \xi R)/2m + V$ (cf. Sect. 2.15) reduces to the case studied in the text with the replacement $V \rightsquigarrow V + \xi R/2m$.

which is easily seen to be positive for all m, m'. Here the crucial condition is that M is *connected*, so any two points are joined up by an open family of paths which has a positive functional measure.

The Perron-Frobenius theorem [8] states that the largest eigenvalue of a matrix with positive entries is non-degenerate and the associated eigenvector has positive components. The proof applies verbatim to the case of positive kernels. Hence the highest eigenvalue of $\exp[-\beta H]$, i.e. the lowest eigenvalue of H, is non-degenerate and the associated eigenfunction may be chosen to be positive. This is equivalent to saying that the lowest eigenvalue of H is non-degenerate and the corresponding wave-function has no nodes.

Remark 6.6 In presence of a background magnetic potential A the imaginary time path integral becomes[14]

$$\int\limits_{(m,0)}^{(m',\beta)} [dx]\, e^{-S_E[x]^0}\, e^{i\int_{x(t)} A} \tag{6.208}$$

where $S_E[x]^0$ is the Euclidean action in absence of magnetic fields and

$$\int_{x(t)} A \equiv \int_0^\beta A_i(x(t))\, \dot{x}^i(t)\, dt \tag{6.209}$$

is the integral of the 1-form A on the path $x(t)$ we are integrating over. Now the (imaginary time) functional measure is manifestly non-positive, indeed complex. If A is *flat*, $dA \equiv 0$, the phase

$$\exp\left(i\int_{x(t)} A\right) \tag{6.210}$$

depends only on the homotopy class of the path $x(t)$ (cf. Sect. 6.4); when M is non simply connected we may have a plurality of ground states as in **SB1**. Note that the Euclidean action in (6.208), while non-real, satisfies the condition (6.184) from reflection positivity.

6.14 Adding Magnetic Fields: Diamagnetic Inequalities

The real physical systems present all sort of magnetic effects—ferromagnetism, diamagnetism, paramagnetism, etc.—all of which are impossible according to the laws of classical physics. Indeed the *Bohr-Van Leeuwen theorem* of classical

---

[14] We take the electric charge to be 1.

Statistics Mechanics[15] states that the free energy of a classical system does not change if we switch on a magnetic field however strong. Quantum Mechanics has elegant explanations for all the fancy magnetic phenomena we observe in real materials. Ferromagnetism and paramagnetism arise from the coupling of the magnetic field to non-classical degrees of freedom such as the electrons' *spin*; hence in this section we limit ourselves to the phenomenon of *diamagnetism* which arises from the orbital motion of electrically charged particles in presence of a magnetic field. That the effect of the magnetic field on the thermodynamical quantities cannot be trivial is already clear from our discussion of the Landau levels in Sect. 5.7: in particular the number of ground states (hence the entropy at zero temperature) depends on the magnetic field through the Riemann-Roch theorem.

Let us recall the classical Bohr-Van Leeuwen argument [3]. The classical free energy F at temperature β^{-1} is given by

$$e^{-\beta F}\Big|_{\text{classical}} = \int \frac{\mathrm{d}^d p \, \mathrm{d}^d q}{(2\pi\hbar)^d} \exp\big[-\beta\, H(p_i - A_i, q^j) \big] \tag{6.211}$$

where $A \equiv A_i dq^i$ is the magnetic potential. The change of integration variables $p_i \rightsquigarrow p_i - A_i$, whose Jacobian is 1, eliminates the magnetic field completely, without changing F.

The corresponding formula in the quantum set-up is

$$e^{-\beta F}\Big|_{\text{quantum}} = \operatorname{Tr} e^{-\beta H} =$$

$$= \int_{\mathfrak{P}} [\mathrm{d}p\,\mathrm{d}q] \exp\left[i\int_0^\beta p_i\, \dot{q}^i \, \mathrm{d}t - \int_0^\beta H(p_i - A_i, q^j)\,\mathrm{d}t \right],$$

$$\tag{6.212}$$

where the integral is over the periodic paths in phase space of period β. Performing the same change of variables as in the classical argument, now gives

$$e^{-\beta F}\Big|_{\text{quantum}} = \int_{\mathfrak{P}} [\mathrm{d}p\,\mathrm{d}q] \exp\left[i\int_0^\beta (p_i + A_i)\,\dot{q}\,\mathrm{d}t - \int_0^\beta H(p_i, q^j)\,\mathrm{d}t \right] =$$

$$= \int_{\mathfrak{P}} [\mathrm{d}q]\, e^{-S_E[q]^0} \exp\left(i \oint q^* A \right), \tag{6.213}$$

where $S_E[q]^0$ is the imaginary-time action of the path $q(t)^i$ computed in absence of magnetic field and

$$\oint q^* A \tag{6.214}$$

---

[15] See **Theorem 2.5** in [3].

is the integral of the magnetic potential 1-form A along the closed ($\equiv$ periodic) path[16] $q(t)$. In particular the functional measure is now *complex*, not real positive as in absence of magnetic fields. Nevertheless the free energy F is still real, as it is obvious from the first line of Eq. (6.212), since H is Hermitian. We have

$$e^{-\beta F} = \left| \int_{\mathfrak{P}} [dq]\, e^{-S_E[q]^0} \exp\left(i \oint q^* A \right) \right| \le$$
$$\le \int_{\mathfrak{P}} [dq]\, \left| e^{-S_E[q]^0} \exp\left(i \oint q^* A \right) \right| = \int_{\mathfrak{P}} [dq]\, e^{-S_E[q]^0} = e^{-\beta F(B=0)} \tag{6.215}$$

We conclude

Fact 6.5 (Diamagnetic Inequalities) *In all spinless quantum systems, switching on a magnetic field increases the free energy.*

6.15   The δ-Functional

The simplest path integral corresponds to an action functional $S_E[x]$ which is linear in the field $x(t)$. Convergence requires the Euclidean action to be purely imaginary. We consider path integrals of the form

$$\mathcal{N} \int [dx] \exp\left(\frac{i}{\hbar} \int_{\mathbb{R}} dt\, x(t) f(t) \right) \tag{6.216}$$

for a certain given (real) source $f(t)$. Expanding both the field and the source in some (real) orthonormal Hilbert basis $\{\psi_k\}$,

$$x(t) = \sum_k x_k\, \psi_k(t), \qquad f(t) = \sum_k f_k\, \psi_k(t), \tag{6.217}$$

we get the integral[17]

$$\mathcal{N}' \prod_{k=1}^{\infty} \int dx_k \exp\left(\frac{i}{\hbar} x_k f_k \right) = \mathcal{N}'' \prod_{k=1}^{\infty} \delta(f_k) \equiv \mathcal{N}'' \delta[f(t)] \tag{6.218}$$

---

[16] More formally: $q^* A$ is the pull-back of the 1-form A to the imaginary time circle of length β.

[17] $\mathcal{N}$, $\mathcal{N}'$, $\mathcal{N}''$ and $\mathcal{N}'''$ stand for various overall constants that we don't need to evaluate by the **General Observation** in Sect. 6.5.

where $\delta[f(t)]$ is the δ-*functional,* namely the reproducing functional kernel for the functional measure defined by the property

$$F[x(t)] = \int [\mathrm{d}y]\,\delta[y(t) - x(t)]\,F[y(t)] \tag{6.219}$$

for all functionals $F[x]$. In the chain of equalities (6.218) we absorbed all constant Jacobians in the overall constant that we didn't bother to compute. In the rare situations where the precise overall constant is needed, one can compute the functional Jacobians by the ζ-regulation method to be introduced in Sect. 6.18.1.

6.16 Gaussian Functional Integrals

Next we consider path integrals where the Euclidean action S_E is a quadratic functional of the paths

$$\int_{\mathfrak{P}} [\mathrm{d}x]\,\exp\left[-\frac{1}{2}\int_M x^i D_{ij}\,x^j\,\mathrm{d}t\right]. \tag{6.220}$$

Path integrals of this form are called *Gaussian.* In this class of integrals the imaginary time action takes the form

$$S_E[x] = \frac{1}{2}\int_M x^i D_{ij}\,x^j\,\mathrm{d}t \tag{6.221}$$

for a linear operator $D = (D_{ij})$ acting on the space $\mathfrak{P}$ of paths $x^i : M \to \mathbb{R}^n$ over which we are integrating. For the moment we assume $D = (D_{ij})$ to be a *positive operator*; this condition will be lifted in Sect. 6.16.1. Here M is one of the four connected 1-dimensional manifolds: an interval, the half-line, the line $\mathbb{R}$, or a circle S^1 of length β. When $\partial M \neq \varnothing$ the paths in $\mathfrak{P}$ are required to satisfy an appropriate boundary condition (typically the Dirichlet one), while, when $\pi_1(M)$ is non-trivial, we need to specify the homotopy weights of paths. $D = (D_{ij})$ is called the *co-variance* operator of the Gaussian functional measure. In the typical applications to Quantum Mechanics D_{ij} is a differential operator which usually[18] is self-adjoint. Some examples are in order.

Example 6.1 The path integral representing the imaginary time evolution kernel (6.26) for the free motion in $\mathbb{R}^d$ with Lagrangian $L_E = \frac{m}{2}\dot{x}^2$ has the form (6.220) with

$$D_{ij} = -m\,\delta_{ij}\frac{\mathrm{d}^2}{\mathrm{d}t^2}, \qquad i,j = 1,\ldots,d, \tag{6.222}$$

---

[18] But not always!

and $\mathfrak{P}$ the space of maps $[0, T] \rightarrow \mathbb{R}^d$ with Dirichlet boundary conditions $x(T)^i = x^i, x(0)^i = y^i$. Indeed

$$S_E[x] = \frac{m}{2} \int_0^T dt\, \dot{x}^i \dot{x}^i = \frac{m}{2} \int_0^T dt\, \frac{d}{dt}(x^i \dot{x}^i) - \frac{m}{2} \int_0^T dt\, x^i \frac{d^2}{dt^2} x^i, \tag{6.223}$$

while the first term in the RHS is a surface contribution which vanishes by the Dirichlet condition $\dot{x}^i|_{\text{endpoint}} = 0$. In this example D_{ij} is self-adjoint and positive.

Example 6.2 For the harmonic oscillator with Hamiltonian

$$H = \frac{p^2}{2m} + \frac{m\omega^2}{2} q^2 \tag{6.224}$$

the operator D takes the form

$$D = -m \frac{d^2}{dt^2} + m\omega^2 \tag{6.225}$$

acting on maps $\mathcal{M} \rightarrow \mathbb{R}$. D is self-adjoint when acting on the natural spaces $\mathfrak{P}$ of paths.

Example 6.3 More generally, when the imaginary time Lagrangian with n degrees of freedom $x^1, \ldots x^n$ has the form

$$L_E = \frac{1}{2} A(t)_{ij}\, \dot{x}^i \dot{x}^j + \frac{1}{2} B(t)_{ij}\, x^i x^j, \tag{6.226}$$

where $A(t)_{ij}\, B(t)_{ij}$ are symmetric $n \times n$ real matrices whose entries are arbitrary functions of time with $A(t)_{ij}$ positive definite, the operator D_{ij} is

$$D_{ij} = -\frac{d}{dt}\left(A(t)_{ij} \frac{d}{dt} \right) + B(t)_{ij} \tag{6.227}$$

which is also self-adjoint (when $n = 1$ it is a Sturm-Liouville operator in normal form (3.157)).

Example 6.4 We consider the canonical path integral in phase space, Eq. (6.37), for a system with a quadratic Hamiltonian which may be time-dependent. The imaginary time action is

$$S_E[p, q] = \int_{\mathcal{M}} dt \left(-i p\dot{q} + \frac{A(t)}{2} p^2 + \frac{B(t)}{2} q^2 \right) \tag{6.228}$$

which may be written as

$$\text{boundary term} + \frac{1}{2} \int_M \mathrm{d}t \, \big(p \ \ q \big) \boldsymbol{D}\!\left(\begin{smallmatrix} p \\ q \end{smallmatrix}\right) \tag{6.229}$$

where

$$\boldsymbol{D} = \begin{pmatrix} A(t) & -\mathrm{i}\frac{\mathrm{d}}{\mathrm{d}t} \\ \mathrm{i}\frac{\mathrm{d}}{\mathrm{d}t} & B(t) \end{pmatrix} \tag{6.230}$$

$\boldsymbol{D}$ is *not* Hermitian. However we may replace $\boldsymbol{D}$ by a self-adjoint operator $\boldsymbol{D}'$ by analytic continuation of the field q to imaginary values

$$\text{boundary term} + \frac{1}{2} \int_M \mathrm{d}t \, \big(p \ \ \mathrm{i}q \big) \begin{pmatrix} A & -\frac{\mathrm{d}}{\mathrm{d}t} \\ \frac{\mathrm{d}}{\mathrm{d}t} & -B \end{pmatrix}\!\left(\begin{smallmatrix} p \\ \mathrm{i}q \end{smallmatrix}\right). \tag{6.231}$$

Computing a Gaussian Path Integral Going back to the general Gaussian path integral (6.220), we assume $\boldsymbol{D} \equiv (D_{ij})$ to be *self-adjoint* (after analytic continuation, if necessary) and real. We then decompose the path $x \in \mathfrak{P}$ in orthonormal (real) eigenfunctions of $\boldsymbol{D}$ acting on $\mathfrak{P}$

$$x(t) = \sum_k a_k \, \psi_k(t) \tag{6.232}$$

where the $\{\psi_k\}$'s satisfy

$$\boldsymbol{D}\psi_k(t) = \lambda_k \, \psi_k(t), \qquad \int_M \psi_k(t) \, \psi_l(t) \, \mathrm{d}t = \delta_{kl}, \tag{6.233}$$

so that the path integral becomes

$$\mathcal{N}\int_{\mathfrak{P}} [\mathrm{d}x] \, \mathrm{e}^{-S_E[x]} = \mathcal{N}' \prod_k \int \mathrm{d}a_k \, \mathrm{e}^{-\lambda_k a_k^2/2} =$$
$$= \mathcal{N}'' \prod_k \lambda_k^{-1/2} = \mathcal{N}''(\operatorname{Det} \boldsymbol{D})^{-1/2} \tag{6.234}$$

where

$$\operatorname{Det} \boldsymbol{D} = \prod_{\lambda_k \in \sigma(\boldsymbol{D})} \lambda_k \tag{6.235}$$

is the *functional determinant of the linear operator* D. The overall normalization constant $\mathcal{N}''$ is independent of the operator D (and hence can be consistently taken to be 1). The final result is

$$\mathcal{N} \int_{\mathfrak{P}} [\mathrm{d}x] \exp\left[-\frac{1}{2} \int_{M} x^{i} D_{ij} \, x^{j} \, \mathrm{d}t \right] = (\mathrm{Det}\, D)_{\mathfrak{P}}^{-1/2}, \tag{6.236}$$

where the functional determinant is computed in the class of functions $\mathfrak{P}$ we are integrating over. Formally one has

$$\mathrm{Det}\, D = \exp(\mathrm{Tr}\log D) = \exp\left(\int_{\mathbb{R}} \log \lambda \, \mathrm{d}\mathrm{Tr}\, P_{D}(\lambda) \right) \tag{6.237}$$

where $P_{D}(\lambda)$ is the *spectral family* of D (cf. Sect. 2.4.1). However the integral in the RHS is typically badly divergent, and our task is to give a proper definition/prescription of what we mean by the determinant $\mathrm{Det}\, D$ of the self-adjoint differential operator D. There are several approaches and we shall review some of them in Sects. 6.18 and 6.19 below.

Complex Gaussian Integrals

Consider the action of the harmonic oscillator, written in canonical variables, evaluated in the space of periodic phase-space paths

$$\begin{aligned}
S_{E}[q, p] &= \frac{1}{2} \int_{0}^{\beta} (iq\dot{p} - ip\dot{q} + \omega q^{2} + \omega p^{2})\mathrm{d}t = \\
&= \int_{0}^{\beta} \left[\frac{q - ip}{\sqrt{2}} \frac{\mathrm{d}}{\mathrm{d}t} \frac{q + ip}{\sqrt{2}} + \omega \frac{q - ip}{\sqrt{2}} \frac{q + ip}{\sqrt{2}} \right]\mathrm{d}t + \int_{0}^{\beta} \mathrm{d}t \frac{\mathrm{d}}{\mathrm{d}t} \frac{p^{2} + q^{2}}{4} \\
&= \int_{0}^{\beta} \left[\bar{a}\left(\frac{\mathrm{d}}{\mathrm{d}t} + \omega \right)a \right]\mathrm{d}t,
\end{aligned} \tag{6.238}$$

where a and $\bar{a}$ are the complex fields ($\equiv$ integration variables in path integrals) which correspond to the lowering/raising quantum operators a and $a^{\dagger}$ of the harmonic oscillator, cf. Eq. (3.71) with $\hbar = 1$. In the last line of Eq. (6.238) we omitted the surface terms which vanish for periodic orbits. Then the partition function of the harmonic oscillator can be written as

$$Z(\beta) = \int [\mathrm{d}a \, \mathrm{d}\bar{a}] \, \exp\left(-\int_{0}^{\beta} \left[\bar{a}\left(\frac{\mathrm{d}}{\mathrm{d}t} + \omega \right)a \right]\mathrm{d}t \right). \tag{6.239}$$

This is a prototypical example of *complex Gaussian integrals* of the form

$$\int [\mathrm{d}a \, \mathrm{d}\bar{a}] \, e^{-S_{E}[a,\bar{a}]} = \int [\mathrm{d}a \, \mathrm{d}\bar{a}] \exp\left(-\int \bar{a}^{i} D_{ij} a^{j} \, \mathrm{d}t \right), \tag{6.240}$$

where $D \equiv (D_{ij})$ is a differential operator acting on a suitable space of *complex* functions $\mathbb{R} \to \mathcal{M}$ and $\bar{a}^i$ is the complex conjugate field to a^i.

By decomposing a^i into its real and imaginary parts we reduce back to a product of two copies of the real Gaussian integral of the form (6.220); hence

$$\int [\mathrm{d}a \, \mathrm{d}\bar{a}] \, \exp\left(- \int \bar{a}^i D_{ij} a^j \, \mathrm{d}t \right) = (\mathrm{Det}\, D)^{-1}. \tag{6.241}$$

Again the determinant should be computed in the appropriate space of complex functions.

"Completing the Square"
More generally we are interested in path integrals whose action is the sum of a quadratic and a linear form in the fields x^i

$$S_E[x] = \frac{1}{2} \int_{\mathcal{M}} \mathrm{d}t \, x^i D_{ij} x^j + \int_{\mathcal{M}} \mathrm{d}t \, J_i(t) x^i. \tag{6.242}$$

As in Sect. 6.8.1 we call $J_i(t)$ the *source*. The path integral with linear sources may be seen as the functional Fourier transform of the Gaussian measure with covariance D_{ij} (cf. Remark 6.5). One can write

$$\int [\mathrm{d}x] \, \mathrm{e}^{-S_E[x]} =$$
$$= \mathrm{e}^{\frac{1}{2} \int \mathrm{d}t \, J_i (D^{-1})_{ij} J_j} \int [\mathrm{d}x] \, \mathrm{e}^{-\frac{1}{2} \int \mathrm{d}t \, (x+D^{-1}J)^i D_{ij} (x+D^{-1}J)^j}, \tag{6.243}$$

where D^{-1} is the inverse of the linear operator $D = (D_{ij})$. The transformation from the LHS to the RHS is dubbed "completing the square". Since the functional measure is invariant under the translation $x \rightsquigarrow x + D^{-1}J$, the integral in the RHS is independent of the source J and using Eq. (6.234) (with the overall constant set to 1)

$$\int [\mathrm{d}x] \, \exp\left(-\frac{1}{2} \int_{\mathcal{M}} \mathrm{d}t \, x^i D_{ij} x^j - \int_{\mathcal{M}} \mathrm{d}t \, J_i(t) x^i \right) =$$
$$= (\mathrm{Det}\, D)^{-1/2} \, \exp\left(\frac{1}{2} \int \mathrm{d}t \, J D^{-1} J \right). \tag{6.244}$$

It is convenient to introduce the integral kernel $G(t, s)$ of the inverse operator D^{-1}. This object is so central in the formalism that it has got several names: *Green's function, resolvent kernel, Euclidean propagator, fundamental solution.* We shall call it the *propagator* (or *Green's function*): it satisfies the differential equation

$$D_{ij} G(t, s)_{jk} = \delta_{ik} \, \delta(t - s), \tag{6.245}$$

where the differential operator $D \equiv (D_{ij})$ acts on the first argument t. See Eq. (6.169) for a simple example. It is convenient to think of D_{ij} as a (unbounded) operator acting on the Hilbert space $\mathbb{C}^n \otimes L^2(\mathscr{T})$ where $\mathscr{T}$ is the relevant 1-dimensional imaginary-time manifold (an interval, the line, or the circle). In the Dirac notation we have

$$G(t, s)_{ij} = \langle t, i | D^{-1} | s, j \rangle. \tag{6.246}$$

Hence the adjoint kernel

$$G^\dagger(t, s)_{ij} \equiv G(s, t)^*_{ji} \tag{6.247}$$

is the Green's function of the adjoint operator

$$D^\dagger G^\dagger(t, s) = \delta(t - s). \tag{6.248}$$

In particular when D is self-adjoint, so is $G(t, s) = (G(t, s)_{ij})$, and when D is real symmetric (as it is often the case)

$$G_{ij}(t, s) = G_{ji}(s, t). \tag{6.249}$$

Going back to Eq. (6.244), the final expression of our Gaussian integral with sources is

$$\int [dx] \exp\left(-\frac{1}{2} \int_{\mathcal{M}} dt \, x^i D_{ij} x^j - \int_{\mathcal{M}} dt \, J_i(t) x^i(t)\right) =$$
$$= (\mathrm{Det}\, D)^{-1/2} \exp\left(\frac{1}{2} \int dt \, ds \, J_i(t) G(t, s)_{ij} J_j(s)\right). \tag{6.250}$$

As in Sect. 6.8.1, taking functional derivatives with respect to the sources $J_i(t)$ we get arbitrary correlation functions for the Gaussian theories.

Connected Correlation Functions. Wick Theorem
From (6.250) we read the generating functional $W[J(t)]$ of the connected correlation functions (see their definition in (6.147)) for a model with quadratic action

$$W[J(t)] = \frac{1}{2} \int dt \, ds \, J_i(t) G(t, s)_{ij} J_j(s) + \text{const.} \tag{6.251}$$

which is quadratic in the sources $J_i(t)$. Then Eq. (6.148) gives:

Fact 6.6 *The **connected** correlation functions of a quantum system with the quadratic action (6.221), are:*

$$\langle x_i(t)\rangle_c = -\int ds\, G(t,s)_{ij} J_j(s) \qquad = 0 \text{ in absence of sources} \tag{6.252}$$

$$\langle x_i(t)\, x_j(s)\rangle_c = G(t,s)_{ij} \tag{6.253}$$

$$\langle x_{i_1}(t_1)\cdots x_{i_s}(t_s)\rangle_c = 0 \quad \text{for } s \geq 3. \tag{6.254}$$

The ordinary correlation functions are obtained by taking functional derivatives of the generating functional

$$Z[\boldsymbol{J}] \equiv \exp\big(W[\boldsymbol{J}]\big), \tag{6.255}$$

hence are given by the combinatorics of the Leibniz rule as sums of products of connected 2-time functions $\langle x_i(t)\, x_j(s)\rangle_c$. The *Wick theorem* gives an explicit combinatorial algorithm to write all correlation functions in terms of the propagator. To state the theorem, we introduce some notation (see Chap. 8 for more). A *pairing* p of the set $\{1, 2, \ldots, 2m\}$ is a decomposition into the disjoint union of m *pairs* ($\equiv$ sets with two elements) of the form:

$$\{1, \ldots, 2m\} = \coprod_{k=1}^{m} \{a_k, b_k\}. \tag{6.256}$$

The set $\mathfrak{P}_{2m}$ of all pairings of $2m$ objects contains $(2m!)/2^m m!$ elements. Setting the source to zero, an elementary recursion gives

Theorem 6.7 (Wick) *The n-time correlation functions*

$$\big\langle x_{i_1}(t_1)\, x_{i_2}(t_2)\cdots x_{i_n}(t_n)\big\rangle \tag{6.257}$$

of a quantum system with the quadratic action (6.221), and zero sources, are:

WT1   *zero when n is* odd. *In other words the $\mathbb{Z}_2$ symmetry $x_i \leftrightarrow -x_i$ is unbroken as required by Theorem 6.4;*

WT2   *when n is even the correlation (6.257) is given by a sum over the*

$$(n-1)!! \equiv \frac{n!}{2^{n/2}(n/2)!} \tag{6.258}$$

pairings of products of propagators, that is,

$$\big\langle x_{i_1}(t_1)\cdots x_{i_n}(t_n)\big\rangle = \sum_{p\in\mathfrak{P}_n}\ \prod_{\{a_k, b_k\}\in p} G(t_{a_k}, t_{b_{j_k}})_{i_{a_k} j_{b_k}}. \tag{6.259}$$

6.16.1 Zero-Modes and All That

A priori Eq. (6.245) determines $G(t, s)$ only up to a solution of the homogeneous equation

$$DG = 0. \tag{6.260}$$

However if the homogeneous equation has a non-zero solution, it means that D is not invertible, and this cannot happen when D is a positive operator, that is, when $\sigma(D) \subset (0, +\infty)$. Let us relax this condition, and allow for a non-zero kernel[19] of D: the field configurations ($\equiv$ paths) in the kernel of D are called *zero-modes* (of D). When $\ker D \neq 0$ Eq. (6.245) has no solution: the LHS is orthogonal to all zero-modes: consistency requires the RHS to have the same property which is not the case for the kernel of the identity. Let P be the projector on the kernel of D; Eq. (6.245) should be replaced by

$$DG = 1 - P. \tag{6.261}$$

We decompose the source as $J = PJ + (1 - P)J$, and the RHS of (6.250) becomes

$$\delta[PJ] \, V_0 \, (\mathrm{Det}^* D)^{-1/2} \exp\left(\frac{1}{2} \int dt \, ds \, J_i(t)(1 - P)G(t, s)_{ij}(1 - P)J_j(s) \right) \equiv$$

$$\equiv \delta[PJ] \, V_0 \, (\mathrm{Det}^* D)^{-1/2} \exp\left(\frac{1}{2} \int dt \, ds \, J_i(t)G(t, s)_{ij}J_j(s) \right) \tag{6.262}$$

where $\mathrm{Det}^* D$ stands for the product of all *non-zero* eigenvalues of D, and V_0 is the volume of the zero-modes, that is, the result of the integration over the finite-dimensional subspace of paths in the kernel of D. Note that the expression (6.262) is *well-defined*, that is, it remains invariant when we add to $G(t, s)$ any solution of the homogeneous equation (6.260).

Remark 6.7 In addition to zero eigenvalues, D may have negative (or complex) eigenvalues. In this case the path integral is computed by analytic continuation in the fields, getting the same expression (6.241) in terms of the determinant of the operator D. However in this situation, while the absolute value of the determinant $|\mathrm{Det}^* D|$ is easy to define and compute by the techniques to be illustrated in Sect. 6.18, its sign (more generally phase) $\mathrm{Det}^* D / |\mathrm{Det}^* D|$ is rather

---

[19] D may be non-invertible while $\ker D = 0$ since the would-be inverse is not bounded (cf. Chap. 2). In this case one has to supplement the Green's equation (6.245) with appropriate boundary conditions. This situation is common in real-time path integrals but never occurs in physically sound imaginary-time functional integrals. The correct boundary condition for the real-time Green's function is the one obtained by analytic continuation of the non-ambiguous imaginary-time one.

tricky and requires special care, except when there are only finitely many non-positive eigenvalues.

6.16.2 $\pi(\mathcal{M}) \neq 1$: A First Example

Up to now we consider Gaussian integrals over paths in $\mathbb{R}^n$. It is interesting to consider the case where the configuration space $\mathcal{M}$ has a more complicate topology. In this situation we have to sum over the homotopy classes of maps.

As a first example we take $\mathcal{M}$ to be a circle S^1 of length L and consider a free particle moving in the circle with Euclidean action

$$S_E[x] = \frac{1}{2} \int_0^\beta \dot{x}^2 \, dt. \tag{6.263}$$

This system was studied from the Schrödinger viewpoint in Sect. 3.11. We now compute the integral over paths periodic in imaginary time of period β. For the sake of illustrating a larger variety of techniques, we compute it in two different ways.

Free Motion in S^1: Lagrangian Approach   As discussed in Sect. 6.4, the novelty is that we have to sum over topological sectors of paths characterized by their *winding number w*:

$$x(t + \beta) = x(t) + wL \qquad w \in \mathbb{Z}. \tag{6.264}$$

Thus

$$\mathrm{Tr}\!\left[e^{-\beta H}\right]_{S^1} = \sum_{w \in \mathbb{Z}} \int_{\mathfrak{P}, w} [dx]\, e^{-S_E[x]} \tag{6.265}$$

In the sector of winding number w we write

$$x(t) = \frac{wLt}{\beta} + y(t), \tag{6.266}$$

where $y(t)$ is a periodic function of period β. In this topological sector

$$S_E[x]\big|_w = \frac{w^2 L^2}{2\beta} + \frac{1}{2} \int_0^\infty \dot{y}^2 dt \tag{6.267}$$

so that

$$\mathrm{Tr}\big[\mathrm{e}^{-\beta H}\big]_{S^1} = \sum_{w\in\mathbb{Z}} \mathrm{e}^{-w^2 L^2/2\beta} \cdot \int_{\mathfrak{P}} [\mathrm{d}y]\,\mathrm{e}^{-S_E[y]} =$$
$$= V_0 \big[\mathrm{Det}^*_\beta(-\tfrac{\mathrm{d}^2}{\mathrm{d}t^2})\big]^{-1/2} \sum_{w\in\mathbb{Z}} \mathrm{e}^{-w^2 L^2/2\beta}, \tag{6.268}$$

where $\mathrm{Det}^*_\beta(-\tfrac{\mathrm{d}^2}{\mathrm{d}t^2})$ is the determinant of the operator $-\tfrac{\mathrm{d}^2}{\mathrm{d}t^2}$ in the space of periodic functions of period β *orthogonal* to the constants (i.e. to the zero-modes). The most tricky point is to compute the zero-mode volume V_0. Expanding in Fourier series

$$y(t) = \frac{1}{\sqrt{\beta}}\Big(a_0 + \sum_{k\geq 1}(a_k\,\mathrm{e}^{2\pi i t/\beta} + a_k^*\,\mathrm{e}^{-2\pi i t/\beta})\Big) \tag{6.269}$$

the integration measure becomes

$$[\mathrm{d}y] = \frac{\mathrm{d}a_0}{\sqrt{2\pi}}\prod_{k\geq 1}\frac{\mathrm{d}a_k\,\mathrm{d}a_k^*}{2\pi}, \tag{6.270}$$

so in the zero-mode sector the measure is

$$\frac{\mathrm{d}a_0}{\sqrt{2\pi}} = \sqrt{\frac{\beta}{2\pi}}\,\mathrm{d}y_0, \tag{6.271}$$

where $y_0 \equiv a_0/\sqrt{\beta}$ is the constant term in the Fourier expansion (6.269). Integrating the constant mode y_0 along the circle of length L we get the zero-mode volume V_0

$$V_0 = \sqrt{\frac{\beta}{2\pi}}L. \tag{6.272}$$

The functional determinant in the RHS of (6.268) is computed in Example 6.13 below:

$$\mathrm{Det}^*_\beta(-\tfrac{\mathrm{d}^2}{\mathrm{d}t^2}) = \beta^2. \tag{6.273}$$

Hence

$$\mathrm{Tr}\big[\mathrm{e}^{-\beta H}\big]_{S^1} = \frac{L}{\sqrt{2\pi\beta}}\sum_{w\in\mathbb{Z}}\mathrm{e}^{-w^2 L^2/2\beta}. \tag{6.274}$$

We recall a special case of the *Poisson summation formula* (cf. Theorem 3.13):

$$\sqrt{a}\sum_{w\in\mathbb{Z}} e^{-\pi a w^2 + 2\pi i b w} = \sum_{n\in\mathbb{Z}} e^{-\pi(n-b)^2/a}. \tag{6.275}$$

Specializing this identity to the parameters

$$a = \frac{L^2}{2\pi\beta}, \qquad b = 0, \tag{6.276}$$

we get

$$\mathrm{Tr}\!\left[e^{-\beta H}\right]_{S^1} = \sum_{n\in\mathbb{Z}} e^{-\frac{1}{2}\left(\frac{2\pi n}{L}\right)^2\beta}. \tag{6.277}$$

i.e. the energy levels of the quantum system defined by the action (6.263) are

$$E_n = \frac{1}{2}\left(\frac{2\pi n}{L}\right)^2 \quad n \in \mathbb{Z} \tag{6.278}$$

which is the energy spectrum we got from the Schrödinger equation in Chap. 3.

Canonical Approach: Use of Cyclic Variables
The unnecessary lengthy computation in the previous example was performed for the sake of illustrating the Lagrangian Gaussian technique in a simple model. However, for the purpose of computing the path integral in the RHS of (6.265), there is a shorter procedure which exploits the fact that the coordinate x is a *cyclic degree of freedom*, and hence its dual momentum p is *conserved* by the Noether theorem.

The following method may be applied whenever one functional variable in the path integral is cyclic. It is quite a convenient technique.

Free Motion on S^1: Canonical Approach   Exploiting the cyclicity of the coordinate, the canonical path integral becomes essentially elementary

$$\mathrm{Tr}\!\left[e^{-\beta H}\right]_{S^1} = \int_{\mathfrak{P}} [dp\,dx]\, e^{i\int_0^\beta p\dot{x}\,dt - \frac{1}{2}\int_0^\beta p^2\,dt} =$$

$$= \int_{\mathfrak{P}} [dp\,dx]\, e^{i p(\beta)[x(\beta)-x(0)]+i[(p(\beta)-p(0)]x(0)}\, e^{-i\int_0^\beta x\dot{p}\,dt - \frac{1}{2}\int_0^\beta p^2\,dt} = \tag{6.279}$$

$$= \sum_{w\in\mathbb{Z}} \int_{\mathfrak{P}} [dp\,dx]\, e^{i p(\beta)Lw}\, e^{-i\int_0^\beta x\dot{p}\,dt - \frac{1}{2}\int_0^\beta p^2\,dt}$$

The action is now linear in the cyclic variable $x(t)$, so the integral in $[\mathrm{d}x]$ yields a δ-functional as in Sect. 6.15. The RHS of the above equation then becomes

$$
L \sum_{w \in \mathbb{Z}} \int_{\mathfrak{P}} [\mathrm{d}p]\, \delta[\dot{p}]\, \mathrm{e}^{\mathrm{i}pLw - \frac{1}{2}\int_0^\beta p^2\, \mathrm{d}t}.
\tag{6.280}
$$

The δ-functional $\delta[\dot{p}]$ expresses the fact that the conjugate momentum p is conserved. The functional integral $[\mathrm{d}p]$ then reduces to the ordinary integral $\mathrm{d}p$ with respect to the constant value of p. The above expression becomes simply

$$
L \sum_{w \in \mathbb{Z}} \int \frac{\mathrm{d}p}{2\pi}\, \mathrm{e}^{\mathrm{i}pLw - \frac{1}{2}p^2 \beta} =
$$

$$
= \sum_{n \in \mathbb{Z}} \int \frac{L\,\mathrm{d}p}{2\pi}\, \delta\!\left(\frac{pL}{2\pi} - n\right) \mathrm{e}^{-\frac{p^2}{2}\beta} = \sum_{n \in \mathbb{Z}} \mathrm{e}^{-\frac{1}{2}\left(\frac{2\pi n}{L}\right)^2 \beta}
\tag{6.281}
$$

without need of computing any functional determinant. In the second line we used again the Poisson summation formula, Theorem 3.13.

We close this section by summing up our findings:

Gaussian Path Integrals
From Eqs. (6.244) and (6.262) we conclude that computing a Gaussian integral with general sources amounts to:

(a) finding the propagator D^{-1} ($\equiv$ Green's function);
(b) computing the functional determinant $\mathrm{Det}^* D$.

These are our next two tasks which will keep us busy for three sections

6.17   Computing Green's Functions: Four Methods

There are several techniques to compute Green's functions of ordinary differential operators. We review the basic ones.

First Method
The most obvious method is to solve the non-homogeneous differential equation

$$
D G(t, s) = \delta(t - s) - P_0(t, s),
\tag{6.282}
$$

where $P_0(t, s)$ is the kernel of the projector on the zero-modes of D. The solution of this equation is unique up to the addition of a solution of the homogeneous equation.

We are interested in the specific solution which also solves the adjoint equation (6.248) in the second variable s.

Example 6.5 (1st Order Operator in $\mathbb{R}$) Consider the (non-Hermitian!) operator

$$\frac{d}{dt} + B(t) \tag{6.283}$$

acting on function defined on the full real line $\mathbb{R}$. Then the solutions which also solve the adjoint equation are

$$G(t, s) = \left(\Theta(t - s) + K\right) \exp\left[-\int_s^t B(u)\, du\right] \tag{6.284}$$

where $\Theta(x)$ is the step function and K a constant to be fixed so that $G(t, s)$ belongs to the right space of distributions for the application at hand.

Second Method
Let $\psi_n(t)$ and λ_n be, respectively, the eigenfunctions and eigenvalues of a *self-adjoint* differential operator D with purely discrete spectrum

$$D\psi_n = \lambda_n \psi_n. \tag{6.285}$$

Then the Green's function is

$$G(t, s) = \sum_n^\sharp \frac{\psi_n(t)\psi_n^*(s)}{\lambda_n}, \tag{6.286}$$

where the superscript $\sharp$ means that the terms with $\lambda_n = 0$ are omitted from the sum. When the spectrum is continuous the sum is replaced by an integral, and in general

$$G(t, s) = \sum_n^\sharp \frac{\psi_n(t)\psi_n^*(s)}{\lambda_n} + \int_{\sigma_e(D)} \frac{dP_D(\lambda)}{\lambda}. \tag{6.287}$$

where $\{P_D(\lambda)\}$ is the spectral family of D (see Chap. 2). Indeed,

$$D\sum_n^\sharp \frac{\psi_n(t)\psi_n(s)^*}{\lambda_n} = \sum_{n:\lambda_n\neq 0} \psi_n(t)\psi_n(s)^* = $$
$$= 1 - \sum_{n:\lambda_n=0} \psi_n(t)\psi_n(s)^* = 1 - P_0, \tag{6.288}$$

where we used the resolution of identity (2.189). This expression makes obvious that also the adjoint equation is satisfied and

$$G(t, s)_{ij} = G(s, t)_{ji}^*. \tag{6.289}$$

Example 6.6 (Constant Coefficients, $\mathcal{M} = \mathbb{R}$) We consider the Green's function
of a second order differential operator with constant coefficients, acting on functions
$f : \mathbb{R} \to \mathbb{R}$, of the form

$$-\frac{d^2}{dt^2} + \rho^2, \qquad \rho > 0. \tag{6.290}$$

Using Eq. (6.287) we get

$$G(t, s) = \int \frac{d\omega}{2\pi} \frac{e^{i\omega(t-s)}}{\omega^2 + \rho^2} = \frac{1}{2\rho}\, e^{-\rho|t-s|}. \tag{6.291}$$

Example 6.7 (Constant Coefficients: $\mathcal{M} = S^1$) We compute the Green's function
of the same operator (6.290) but with $\mathcal{M}$ a circle of length L. Now the Green's
function $G(t, s)_{S^1}$ is a periodic function of $x \equiv t - s$ of period L which satisfies the
ODE

$$\left(-\frac{d^2}{dx^2} + \rho^2\right) G(x)_{S^1} = \sum_{n \in \mathbb{Z}} \delta(x - nL). \tag{6.292}$$

Using the **image method**, we learn that the solution is a sum of translations of the
solution with source a single δ-function, given by (6.291). Thus for $0 \le x < 1$

$$\begin{aligned}
G(x)_{S^1} &= \frac{1}{2\rho} \sum_{n=-\infty}^{+\infty} e^{-\rho|x-nL|} = \\
&= \frac{1}{2\rho} \sum_{n \ge 0} e^{-\rho(x+nL)} + \frac{1}{2\rho} \sum_{n \ge 1} e^{-\rho(nL-x)} \\
&= \frac{1}{2\rho} \frac{\cosh(\rho(x - L/2))}{\sin(\rho L/2)}.
\end{aligned} \tag{6.293}$$

Alternatively, one starts from the representation in Eq. (6.286)

$$G(x)_{S^1} = \sum_{n=-\infty}^{+\infty} \frac{e^{2\pi i n x}}{\frac{4\pi^2 n^2}{L^2} + \rho^2} \tag{6.294}$$

and sums the known Fourier series in the RHS using formula **1.445**(2) of [9].

Example 6.8 (Harmonic Hamiltonian) We compute the Green's function of the
differential operator

$$D = \frac{1}{2}\left(-\frac{d^2}{dt^2} + t^2 + 1\right) \tag{6.295}$$

i.e. of the Hamiltonian of the one-dimensional harmonic oscillator with

$$\hbar = m = \omega = 1, \tag{6.296}$$

where we shifted the energy by $1/2$. Using the wave eigenfunctions computed in Sect. 3.5.5 we have

$$
\begin{aligned}
G(t, s) &= \frac{1}{\sqrt{\pi}} \sum_{n \geq 0} \frac{H_n(t)e^{-t^2/2} \, H_n(s)e^{-s^2/2}}{2^n n!(n+1)} \equiv \\
&\equiv \frac{1}{\sqrt{\pi}} e^{-(t^2+s^2)/2} \int_0^1 dz \sum_{n \geq 0} \frac{H_n(t) \, H_n(s)}{2^n n!} z^n
\end{aligned}
\tag{6.297}
$$

where $H_n(t)$ is the n-th Hermite polynomial. The sum inside the integral is easily computed with the help of the Meyer formula (3.149).

Third Method

The second line of Eq. (6.297) in Example 6.8 illustrates another way to get the Green function when we know the heat kernel of the operator $\boldsymbol{D}$

$$\langle t|e^{-\beta \boldsymbol{D}}|s\rangle, \tag{6.298}$$

as it is the case for the harmonic Hamiltonian. We first assume $\boldsymbol{D}$ to be a positive-definite operator with no zero-modes. In this case we may compute its Green's function by the formula

$$G(t, s) = \int_0^\infty d\beta \, \langle t|e^{-\beta \boldsymbol{D}}|s\rangle. \tag{6.299}$$

When zero modes are present (and $\boldsymbol{D}$ is non-negative)

$$G(t, s) = \int_0^\infty d\beta \, \langle t|\left(e^{-\beta \boldsymbol{D}} - P_0\right)|s\rangle, \tag{6.300}$$

where

$$P_0 = \lim_{\beta \to \infty} e^{-\beta \boldsymbol{D}} \tag{6.301}$$

is the projector to the zero-mode eigenspace. Indeed

$$
\begin{aligned}
\boldsymbol{D}G(t, s) &= \int_0^\infty d\beta \, \langle t|\boldsymbol{D}e^{-\beta \boldsymbol{D}}|s\rangle = -\int_0^\infty d\beta \frac{\partial}{\partial \beta} \langle t|e^{-\beta \boldsymbol{D}}|s\rangle = \\
&= \langle t|s\rangle - \lim_{\beta \to \infty} \langle t|e^{-\beta \boldsymbol{D}}|s\rangle \equiv \delta(t-s) - P_0(t, s).
\end{aligned}
\tag{6.302}
$$

Fourth Method

When D is a scalar second-order differential operator, that can always be put in the normal form

$$D = -\frac{\mathrm{d}}{\mathrm{d}t}\left(A(t)\frac{\mathrm{d}}{\mathrm{d}t}\right) + B(t), \qquad A(t) > 0, \ B(t) \text{ real}, \tag{6.303}$$

D is automatically self-adjoint. When we know ***any*** *one particular* solution ψ_0 of the homogeneous equation, $D\psi_0 = 0$, we can easily construct the Green's function of D. Indeed we may get a second solution ψ_1 with Wronskian

$$W[\psi_0, \psi_1] \equiv A(t)\left[\psi_0(t)\psi_1'(t) - \psi_0'(t)\psi_1(t)\right] = 1, \tag{6.304}$$

by a simple quadrature (cf. Eq. (3.167))

$$\psi_1(t) = \psi_0(t)\int^t \frac{\mathrm{d}y}{A(y)\,\psi_0(y)^2}. \tag{6.305}$$

Then a *particular* C^0 solution of the inhomogeneous equation

$$-\frac{\mathrm{d}}{\mathrm{d}t}\left(A(t)\frac{\mathrm{d}}{\mathrm{d}t}G(t,s)\right) + B(t)G(t,s) = \delta(t-s) \tag{6.306}$$

is

$$\begin{aligned}
G(t,s)_0 &= \Theta(s-t)[\psi_0(s)\,\psi_1(t) - \psi_0(t)\,\psi_1(s)] \equiv \\
&\equiv \Theta(s-t)\,\psi_0(s)\,\psi_0(t)\int_s^t \frac{\mathrm{d}y}{A(y)\,\psi_0(y)^2},
\end{aligned} \tag{6.307}$$

where, as always, $\Theta(x)$ is the step function. Then the Green's function must have the form

$$G(t,s) = G(t,s)_0 + a(s)\,\psi_0(t) + b(s)\,\psi_0(t)\int_0^t \frac{\mathrm{d}y}{A(y)\,\psi_0(y)^2} \tag{6.308}$$

where the functions $a(s)$, $b(s)$ are chosen so that $G(t,s)$ solves the adjoint ODE and it belongs to the proper space of functions to be the propagator for the physical problem at hand.

Example 6.9 We return to Example 6.6. For the operator in Eq. (6.290) we take $\psi_0(t) = \mathrm{e}^{-\rho t}$. Then

$$\psi_1(t) = \mathrm{e}^{-\rho t}\int_{-\infty}^t \frac{\mathrm{d}y}{\mathrm{e}^{-2\rho t}} = \frac{\mathrm{e}^{\rho t}}{2\rho}, \tag{6.309}$$

and then

$$G(t, s)_0 = \Theta(s - t)\frac{1}{2\rho}\left[e^{-\rho(s-t)} - e^{-\rho(t-s)}\right] \tag{6.310}$$

which is not good on *two counts*: *(1)* it grows exponentially as $t \to +\infty$ at fixed s, and *(2)* is not self-adjoint i.e. not symmetric in $s \leftrightarrow t$. However adding a suitable solution of the homogeneous equation

$$G(t, s)_0 + \frac{1}{2\rho}e^{-\rho(t-s)} \equiv \frac{1}{2\rho}e^{-\rho|t-s|} \tag{6.311}$$

to fix the first issue, we automatically fix also the second one, and we get the genuine Green's function, cf. (6.291).

A General Formula The last *Example* gives a strategy to get the Green's function for *all* 2nd order operators D of the form (6.303) acting in $L^2(\mathbb{R})$, where we suppose $A(t)$ to be time-independent[20] and

$$\lim_{t\pm\infty} B(t) > 0, \tag{6.312}$$

while D has no zero-modes. After the construction of the second solution, by taking linear combinations we form a special basis of solutions $\{\psi_-, \psi_+\}$ such that $\psi_\pm(t)$ goes to zero as $t \to \pm\infty$. The two solutions are linearly independent since there are no zero-modes. We can normalize them so that their Wronskian is 1. Then

$$G(t, s) = \begin{cases} \psi_-(s)\,\psi_+(t) & t > s \\ \psi_-(t)\,\psi_+(s) & t < s. \end{cases} \tag{6.313}$$

6.18 Functional Determinants I: General Methods

Our next task is to compute the functional determinants of differential operators. In the context of the path integrals which arise in Quantum Mechanics, the differential operators are *ordinary* ones, i.e. they act on a space of functions in *one* dimension. Some general facts hold in arbitrary dimension, i.e. for determinants of both ordinary and partial differential operators. In this section we focus on general properties which apply to differential operators D in any number n of variables (that we see as local coordinates in a manifold $\mathcal{M}$). In the next section we focus on the special techniques which may be used for ordinary differential operators. All

---

[20] We can always reduce to this case by the change of variables $t \rightsquigarrow s(t) \equiv \int^t \mathrm{d}x/A(x)$, see Eq. (3.289) and the discussion around it.

our examples will be in one dimension; for simple higher dimensional examples see
e.g. [3] chap. 5.

Resolvent and Fredholm Determinants

Consider the equalities

$$\frac{\partial}{\partial z}\log\mathrm{Det}[\boldsymbol{D}-z] = \frac{\partial}{\partial z}\mathrm{Tr}\log(\boldsymbol{D}-z) = -\mathrm{Tr}\left(\frac{1}{\boldsymbol{D}-z}\right). \qquad (6.314)$$

Thus, if we wish to compute the functional determinant $\mathrm{Det}[\boldsymbol{D}-z]$ modulo an
overall constant (that we may absorb into $\mathcal{N}$), it suffices to compute the trace of the
operator

$$\boldsymbol{R}(z) = \frac{1}{\boldsymbol{D}-z} \qquad (6.315)$$

called the *resolvent* of $\boldsymbol{D}$. By definition $\boldsymbol{R}(z)$ is singular when $z \in \sigma(\boldsymbol{D})$. To
compute the trace one may use the resolvent kernel

$$R(x, y; z) = \langle x|(\boldsymbol{D}-z)^{-1}|y\rangle, \qquad x, y \in \mathcal{M}, \qquad (6.316)$$

by writing

$$\mathrm{Tr}\left(\frac{1}{\boldsymbol{D}-z}\right) = \int_{\mathcal{M}} \mathrm{d}^n x\, R(x, x; z). \qquad (6.317)$$

Then

$$\mathrm{Det}(\boldsymbol{D}-z) = \mathrm{Det}(\boldsymbol{D})\exp\left(-\int_0^z \mathrm{d}w \int_{\mathcal{M}} \mathrm{d}^n x\, R(x, x; w)\right). \qquad (6.318)$$

Let us see how we can make the RHS more explicit. We start from the identity

$$(\boldsymbol{D}-z)^{-1} = \boldsymbol{D}^{-1}(1 - z\,\boldsymbol{D}^{-1})^{-1} = \sum_{k=0}^{\infty} z^k(\boldsymbol{D}^{-1})^{k+1} \qquad (6.319)$$

for all z's where the sum converges i.e. such that there are no eigenvalues of $\boldsymbol{D}$ in
the disk of radius[21] $|z|$. The integral kernel of $\boldsymbol{D}^{-1}$ is the Green's function $G(x, y)$,

---

[21] The method works in much more general situations. Suppose that the number of eigenvectors
with eigenvalues in the disk is finite—as it is the case for *all* reasonable differential operators $\boldsymbol{D}$.
We can still use the formulae in the main text for the determinant restricted to the orthogonal
complement of the finite dimensional space spanned by the finitely-many eigenvectors, and
write the original determinant as a product of the determinant of a matrix acting on this finite-
dimensional space times a functional determinant in the orthogonal complement whose power
series in z now converges.

so we get

$$R(x, y; z) = G(x, y) + z \int d^n x_1 G(x, x_1) G(x_1, y) +$$

$$+ z^2 \int d^n x_1 \, d^n x_2 \, G(x, x_1) G(x_1, x_2) G(x_2, y) + \tag{6.320}$$

$$+ z^3 \int d^n x_1 \, d^n x_2 \, d^n x_3 \, G(x, x_1) G(x_1, x_2) G(x_2, x_3) G(x_3, y) + \cdots$$

and the functional determinant takes the form

$$\frac{\mathrm{Det}(\boldsymbol{D} - z)}{\mathrm{Det}(\boldsymbol{D})} = \exp\Bigg(-z \int d^n x_0 \, G(x_0, x_0) +$$

$$- \frac{z^2}{2} \int d^n x_0 \, d^n x_1 \, G(x_0, x_1) G(x_1, x_0) + \cdots \Bigg). \tag{6.321}$$

The functional determinant defined in this way in terms of reiterated integral kernels is called the *Frobenius determinant* and is a central object in Frobenius' theory of linear integral equations.

Heat Kernel Methods

It is clear that, while the infinite determinant $\mathrm{Det}(\boldsymbol{D} - z)$ is formally divergent and requires an appropriate regularization to be well-defined (see below), the ratio of the two determinants in the LHS of (6.321) is well-defined under some mild conditions which are typically satisfied in the relevant applications. Integrating Eq. (6.314) we get an expression for the ratio of the two functional determinants in terms of the trace of the heat kernel of the operator $\boldsymbol{D}$

$$\log \frac{\mathrm{Det}(\boldsymbol{D} - z)}{\mathrm{Det}(\boldsymbol{D})} = \int_0^\infty \frac{ds}{s} (1 - e^{zs}) \, \mathrm{Tr}(e^{-\boldsymbol{D}s}) \equiv$$

$$\equiv \int_0^\infty \frac{ds}{s} (1 - e^{zs}) \int_{\mathcal{M}} d^n x \, \langle x | e^{-\boldsymbol{D}s} | x \rangle. \tag{6.322}$$

The integral over $\mathcal{M}$ in the RHS is convergent[22] for all positive-definite differential operator $\boldsymbol{D}$. The integral in s converges when $\mathrm{Re}\, z < \inf \sigma(\boldsymbol{D})$. The ratio

$$\frac{\mathrm{Det}(\boldsymbol{D} - z)}{\mathrm{Det}(\boldsymbol{D})} \tag{6.323}$$

---

[22] The statement is strictly true for $\mathcal{M}$ compact; otherwise there is a trivial linear divergence with the volume of $\mathcal{M}$. Indeed the logarithm of the functional determinant is proportional to the free energy F, and the free energy is an extensive thermodynamical quantity. Physically it is the free energy *per unit volume* F/V which is finite, and hence $\log[\mathrm{Det}(\boldsymbol{D} - z)/\mathrm{Det}(\boldsymbol{D})] = \mathrm{Vol}(\mathcal{M}) \cdot \text{(finite)}$. We abuse language and say that $\log \mathrm{Det}\, \boldsymbol{D}$ is *finite* when it is finite per unit volume of the configuration space $\mathcal{M}$.

is an *entire* function of z with zeros of multiplicity m when z is an eigenvalue of $\boldsymbol{D}$ of degeneracy m. Hence we may define it by analytic continuation of the expression (6.322) computed in the convergence half-plane $\mathrm{Re}\, z < \inf \sigma(\boldsymbol{D})$. We conclude that whenever we know the diagonal heat kernel $\langle x | e^{-\boldsymbol{D}s} | x \rangle$ we can compute the functional determinants by quadratures.

Let z_0 be an eigenvalue of $\boldsymbol{D}$ of multiplicity m so that (6.323) has a zero of order m at $z = z_0$. Then

$$\frac{\mathrm{Det}(\boldsymbol{D} - z)}{\mathrm{Det}(\boldsymbol{D})} = (z_0 - z)^m \frac{\mathrm{Det}^*(\boldsymbol{D} - z_0)}{\mathrm{Det}(\boldsymbol{D})} + O((z - z_0)^{m+1}) \tag{6.324}$$

so $\mathrm{Det}^*(\boldsymbol{D} - z_0)$ is computed by taking m derivatives of the ratio (6.323).

Zero Modes
Suppose the operator $\boldsymbol{D}$ is non-negative with a non-trivial kernel of dimension m. Then Eq. (6.322) is replaced by

$$\begin{aligned}
\log \frac{\mathrm{Det}(\boldsymbol{D} - z)}{(-z)^m\, \mathrm{Det}^*(\boldsymbol{D})} &= \int_0^\infty \frac{ds}{s}(1 - e^{zs})\mathrm{Tr}(e^{-\boldsymbol{D}s} - P_0) \equiv \\
&\equiv \int_0^\infty \frac{ds}{s}(1 - e^{zs}) \int_M d^n x\, \langle x | (e^{-\boldsymbol{D}s} - P_0) | x \rangle.
\end{aligned} \tag{6.325}$$

where P_0 is the projector on $\ker \boldsymbol{D}$.

6.18.1 ζ-*Determinants*

One may wish to compute directly determinants, instead of ratios of them. At the face level they look to be ill-defined divergent expressions, and one needs a detailed prescription to give a precise meaning to these expressions and evaluate them. ζ-*regularization* is the most convenient and widely used such prescription.

Let $\{\lambda_k\}$ be the spectrum of $\boldsymbol{D}$. One starts from the elementary identities

$$\begin{aligned}
\log \mathrm{Det}\, \boldsymbol{D} = \mathrm{Tr} \log \boldsymbol{D} &\equiv \sum_k{}' \log \lambda_k = \\
&= -\frac{\partial}{\partial s} \sum_k \lambda_k^{-s} \Big|_{s=0} \equiv -\frac{\partial}{\partial s} \mathrm{Tr}[\boldsymbol{D}^{-s}] \Big|_{s=0}
\end{aligned} \tag{6.326}$$

(the sum is replaced by an integral for a continuous spectrum). The function

$$\zeta_{\boldsymbol{D}}(s) \overset{\mathrm{def}}{=} \mathrm{Tr}\left[\frac{1}{\boldsymbol{D}^s} \right] = \sum_k \frac{1}{\lambda_k^s} \tag{6.327}$$

is the ζ-*function of the operator* D. For reasonable operators the trace converges in some half-plane $\mathrm{Re}\, s > s_0$. When D is an elliptic differential operator, its ζ-function $\zeta_D(s)$ can be analytically continued to a meromorphic function in the full complex s-plane. In this case the ζ-function is analytic around $s = 0$,[23] so its derivative at $s = 0$ exists and defines the ζ-regularized determinant through the formula (6.326)

$$\zeta\text{-determinant of } D \overset{\text{def}}{=} \exp\!\left(-\zeta_D'(0)\right). \tag{6.328}$$

The ζ-function of an operator D is mostly conveniently written in terms of the trace of its heat kernel (i.e. of the *partition function* if we see D as a Hamiltonian)

$$\mathrm{Tr}[D^{-s}] = \frac{1}{\Gamma(s)} \int_0^\infty dt\, t^{s-1}\, \mathrm{Tr}(e^{-Dt}) =$$
$$= \frac{1}{\Gamma(s)} \int_0^\infty dt\, t^{s-1} \int \sqrt{g}\, d^d x\, \langle x | e^{-Dt} | x \rangle, \tag{6.329}$$

where $\Gamma(s)$ is Euler's Gamma-function

$$\Gamma(s) = \int_0^\infty dt\, t^{s-1}\, e^{-s}. \tag{6.330}$$

Examples   We present some basic examples of the technique.

Example 6.10   The harmonic Hamiltonian (with a shift $+\omega/2$ in energy)

$$H = \frac{1}{2}\left(-\frac{d^2}{dx^2} + \omega^2 x^2 + \omega\right) \tag{6.331}$$

acting on $L^2(\mathbb{R})$ has eigenvalues $\lambda_n = n\omega$, $n = 1, 2, \ldots$, and its ζ-function is

$$\zeta_H(s) = \sum_{n \geq 1} \frac{1}{\omega^s n^s} = \omega^{-s} \zeta(s) \tag{6.332}$$

where $\zeta(s)$ is Riemann's ζ-function. Hence the ζ-determinant of H (densely) acting on $L^2(\mathbb{R})$ is

$$\mathrm{Det}_\zeta[H] = \exp\!\left(-\zeta'(0) + \log \omega\, \zeta(0)\right) = \sqrt{\frac{2\pi}{\omega}} \tag{6.333}$$

where we used [7]

$$\zeta(0) = -\frac{1}{2}, \qquad \zeta'(0) = -\frac{1}{2}\log(2\pi). \tag{6.334}$$

---

[23] For examples in d dimensions see appendix 2 to chapter 5 of [3].

Example 6.11 In many applications we need the determinant of the free particle Hamiltonian

$$D = -\frac{\mathrm{d}^2}{\mathrm{d}t^2} \tag{6.335}$$

acting on the space of functions in the interval $[0, L]$ with Dirichlet boundary conditions $\psi(0) = \psi(L) = 0$. Its eigenvalues and eigenfunctions are

$$\lambda_n = \frac{\pi^2 n^2}{L^2}, \quad \psi_n(t) = \sqrt{\frac{2}{L}} \sin(\pi n t/L), \quad n = 1, 2, 3, \ldots \tag{6.336}$$

and

$$\zeta_D(s) = \frac{L^{2s}}{\pi^{2s}} \zeta(2s) \tag{6.337}$$

Then

$$\log \mathrm{Det} \left(-\frac{\mathrm{d}^2}{\mathrm{d}t^2}\right)_{\substack{[0,L] \\ \mathrm{Dirichlet}}} = -\zeta_D'(0) = -2\log(L/2\pi)\zeta(0) - 2\zeta'(0) = \tag{6.338}$$

$$= \log(L).$$

This formula says that the determinant is linear in L. However the result suffers from a minor ambiguity since it depends on the units we use to measure lengths. Thus

$$\mathrm{Det} \left(-\frac{\mathrm{d}^2}{\mathrm{d}t^2}\right)_{[0,L]} = \Lambda L \tag{6.339}$$

for a constant Λ with dimension (length)$^{-1}$. The constant Λ depends on the physical context.[24]

Example 6.12 We compute the determinant of the operator

$$D_B = -\frac{\mathrm{d}^2}{\mathrm{d}x^2} + B^2, \tag{6.340}$$

---

[24] The expression for the determinant may also appear with different factors of $\sqrt{2\pi}$ arising from the change of normalization of the measure in position and momentum spaces from the symmetric one $\mathrm{d}x/\sqrt{2\pi\hbar}$ and $\mathrm{d}p/\sqrt{2\pi\hbar}$, implicitly used in most path integral computations, and the conventional one $\mathrm{d}x$ and $\mathrm{d}p/(2\pi\hbar)$. These conventional factors don't affect the observables.

with B constant, acting on $L^2(\mathbb{R})$. This operator has a continuous spectrum, and, as explained in footnote 22, only $\log D_B$ per unit length makes sense. Correspondingly we compute the ζ-function per unit length:

$$\frac{\zeta_B(s)}{L} = \frac{L^{-1}}{\Gamma(s)} \int_0^\infty dt\, t^{s-1} \operatorname{Tr}(e^{-D_B t}) = \frac{1}{\Gamma(s)} \int_0^\infty dt\, t^{s-1} e^{-B^2 t} \int \frac{dp}{2\pi} e^{-p^2 t}$$

$$= \frac{1}{\sqrt{4\pi}} \frac{1}{\Gamma(s)} \int_0^\infty dt\, t^{s-3/2} e^{-B^2 t} = \frac{1}{\sqrt{4\pi}} \frac{\Gamma(s-1/2)}{\Gamma(s)} B^{1-2s}. \tag{6.341}$$

We have

$$\frac{\Gamma(s-1/2)}{\Gamma(s)} = -\sqrt{4\pi}\, s + O(s^2). \tag{6.342}$$

Hence the ζ-determinant of D_B for a large interval of length $L \to \infty$ is

$$\mathrm{Det}_\zeta[D_B] = \exp(BL). \tag{6.343}$$

Remark 6.8   We have still to justify our claim at the end of Sect. 6.5 that

$$\mathrm{Det}\left(-m\frac{d^2}{dt^2} + m\,B^2\right) = \mathrm{Det}\left(-\frac{d^2}{dt^2} + B^2\right). \tag{6.344}$$

Now we show that this equality holds in the sense of ζ-determinants. The ζ-function of the operator on the left is

$$\tilde{\zeta}_B(s) = \frac{L}{\sqrt{4\pi}} m^{-s} \frac{\Gamma(s-1/2)}{\Gamma(s)} B^{1-2s} \tag{6.345}$$

and

$$\frac{d}{ds}\tilde{\zeta}_B(s)\bigg|_{s=0} = \frac{d}{ds}\zeta_B(s)\bigg|_{s=0} \tag{6.346}$$

by Eq. (6.342).

Example 6.13 (Determinant in Eq. (6.273))   We have to compute the determinant

$$\mathrm{Det}_\beta^*\left(-\frac{d^2}{dt^2}\right) \tag{6.347}$$

acting on periodic functions of period β and zero-mean. The eigenvalues are $(2\pi n/\beta)^2$ with $n \in \mathbb{Z} \setminus \{0\}$, hence

$$\zeta^*_{-\frac{d^2}{dt^2}}(s) = \left(\frac{\beta}{2\pi}\right)^{2s} \sum_{n \neq 0} \frac{1}{(n^2)^s} = 2\left(\frac{\beta}{2\pi}\right)^{2s} \zeta(2s). \tag{6.348}$$

and

$$\frac{d}{ds}\zeta^*_{-\frac{d^2}{dt^2}}(s)\Big|_{s=0} = 4\log\left(\frac{\beta}{2\pi}\right)\zeta(0) + 4\,\zeta'(0) = -2\log\beta \tag{6.349}$$

cf. Eq. (6.334). Then

$$\mathrm{Det}^*_\beta\left(-\frac{d^2}{dt^2}\right) = \exp\left[-\frac{d}{ds}\zeta^*_{-\frac{d^2}{dt^2}}(s)\Big|_{s=0}\right] = \beta^2. \tag{6.350}$$

6.19　Functional Determinants II: Special Methods

For ordinary differential operators things simplify and we have several special methods to evaluate their functional determinants. In this section we review some of these techniques. A fancier exact formula for the functional determinant is presented in Sect. 8.6.1 in the context of large-N expansions.

6.19.1　Determinants Over Periodic Paths: Monodromy Method

Often we have to compute integrals over periodic paths of fixed period β in imaginary time. This is the case, for instance, when we wish to study our quantum system at finite temperature $T = 1/\beta$, cf. Sect. 6.7. In this situation we may use the *monodromy method*.

1st Order Linear Differential Operators
Let $B(t)$ be a periodic function of period β. We consider the differential operator

$$D = \frac{d}{dt} + B(t) + \mu \tag{6.351}$$

acting on the space $\mathfrak{P}_\beta$ of periodic complex functions of period β. The eigenfunction equation

$$D\psi_\lambda = \lambda\,\psi_\lambda \qquad \lambda \in \mathbb{C} \tag{6.352}$$

has the general solution

$$\psi_\lambda(t) = \exp\left[(\lambda - \mu)t - \int_0^t B(s)\,\mathrm{d}s\right]\psi_\lambda(0) \tag{6.353}$$

which is periodic of period β iff

$$(\lambda - \mu)\beta - \int_0^\beta B(s)\,\mathrm{d}s = 2\pi\mathrm{i}n, \quad n \in \mathbb{Z}. \tag{6.354}$$

Then the eigenvalues λ_n of D acting in $\mathfrak{P}_\beta$ are

$$\lambda_n = \mu + \frac{2\pi\mathrm{i}}{\beta}n + \frac{1}{\beta}\int_0^\beta B(s)\,\mathrm{d}s \qquad n \in \mathbb{Z}, \tag{6.355}$$

and therefore

$$\frac{\partial}{\partial\mu}\log\mathrm{Det}(D)_\beta = \frac{\partial}{\partial\mu}\sum_n \log\lambda_n =$$

$$= \sum_{n\in\mathbb{Z}} \frac{1}{\mu + \frac{2\pi\mathrm{i}n}{\beta} + \frac{1}{\beta}\int_0^\beta B(s)\,\mathrm{d}s} = \tag{6.356}$$

$$= \frac{\beta}{\beta\mu + \int_0^\beta B(s)\,\mathrm{d}s} + 2\beta \sum_{n\geq 1} \frac{\beta\mu + \int_0^\beta B(s)\,\mathrm{d}s}{(\beta\mu + \int_0^\beta B(s)\,\mathrm{d}s)^2 + 4\pi^2 n^2}$$

Now recall the well-known identity (see e.g. **4.36.2** of [7])

$$\frac{d}{dz}\log\sinh z \equiv \coth z = \frac{1}{z} + 2\sum_{n\geq 1}\frac{z}{z^2 + \pi^2 n^2}. \tag{6.357}$$

Comparing with Eq. (6.356) we get

$$\frac{\partial}{\partial\mu}\log\mathrm{Det}(D)_\beta = \frac{\beta}{2}\coth\left(\frac{\beta\mu + \int_0^\beta B(s)\,\mathrm{d}s}{2}\right), \tag{6.358}$$

that is,

$$\mathrm{Det}(D)_\beta = 2\sinh\left(\frac{\beta\mu}{2} + \frac{1}{2}\int_0^\beta B(s)\,\mathrm{d}s\right) =$$

$$= e^{\frac{1}{2}(\beta\mu + \int_0^\beta B(s)\,\mathrm{d}s)}\left(1 - e^{-\beta\mu - \int_0^\beta B(s)\,\mathrm{d}s}\right). \tag{6.359}$$

We fixed the overall integration constant (which may be absorbed in $\mathcal{N}$) using some physical insight, see Example 6.14 and Sect. 6.23.2. For the benefit of the skeptical reader we compute the determinant in the ζ-regularization:

We set $w = \mu + \beta^{-1} \int_0^\beta B(s)\mathrm{d}s$. The ζ-function of D is

$$
\begin{aligned}
\zeta_D(s) &= \frac{1}{w^s} + \left(\frac{\beta}{2\pi}\right)^s \left[\sum_{n\geq 1} \frac{1}{(n - \frac{i\beta w}{2\pi})^s} + \sum_{n\geq 1} \frac{1}{(n + \frac{i\beta w}{2\pi})^s} \right] = \\
&= \frac{1}{w^s} + \left(\frac{\beta}{2\pi}\right)^s \left[\zeta\left(s, 1 - \frac{i\beta w}{2\pi}\right) + \zeta\left(s, 1 + \frac{i\beta w}{2\pi}\right) \right]
\end{aligned}
\tag{6.360}
$$

where $\zeta(s, a) \equiv \sum_{n\geq 0}(n + a)^{-s}$ is the Hurwitz ζ-function ([7] §. 25.11). One has

$$
\zeta(0, a) = \frac{1}{2} - a, \qquad \zeta'(0, a) = \log \Gamma(a) - \frac{1}{2} \log(2\pi),
\tag{6.361}
$$

cf. 25.11.13 and 25.11.18 in [7]. Hence

$$
\log \mathrm{Det}\, D \equiv -\zeta_D'(0) = \log(2\pi) + \log \left[\frac{\frac{\beta w}{2\pi}}{\Gamma\left(1 - \frac{i\beta w}{2\pi}\right) \Gamma\left(1 + \frac{i\beta w}{2\pi}\right)} \right]
$$

$$
= \log\left(2 \sinh \frac{\beta w}{2}\right)
\tag{6.362}
$$

where we used the Γ-function identity **8.332**.3 of [9]. The ζ-determinant agrees with (6.359).

Example 6.14 (Harmonic Oscillator in Complex Form) The partition function of the quantum harmonic oscillator at temperature $T = 1/\beta$ is (see Eq. (6.239))

$$
\mathrm{Tr}(\mathrm{e}^{-\beta H}) = \int [\mathrm{d}a\, \mathrm{d}\bar{a}]\, \mathrm{e}^{- \int_0^\beta \bar{a}(\mathrm{d}_t + \omega)a\, \mathrm{d}t} = \frac{1}{\mathrm{Det}(\mathrm{d}_t + \omega)_\beta},
\tag{6.363}
$$

where $\mathrm{d}_t \equiv \mathrm{d}/\mathrm{d}t$, and the determinant is computed over the space $\mathfrak{P}_\beta$ of periodic functions of period β (in imaginary time). Hence, from Eq. (6.359),

$$
\mathrm{Tr}(\mathrm{e}^{-\beta H}) = \frac{1}{2 \sinh(\beta\omega/2)} = \frac{\mathrm{e}^{-\beta\omega/2}}{1 - \mathrm{e}^{-\beta\omega}} \equiv \sum_{n\geq 0} \mathrm{e}^{-\beta\omega(n+1/2)}
\tag{6.364}
$$

which agrees with the direct computation of the spectrum in Chap. 3 including the zero-point energy. It also explains our insight on the overall normalization.

Example 6.15 (Generalized Periodicity) More generally we may compute the functional determinant over the space $\mathfrak{P}_{\beta,\xi}$ of functions satisfying the generalized periodicity condition

$$\psi(x + \beta) = e^{2\xi}\,\psi(x) \qquad \xi \in \mathbb{C}. \tag{6.365}$$

Equation (6.355) gets replaced by

$$\lambda_n = \frac{2\xi}{\beta} + \mu + \frac{2\pi i}{\beta}n + \frac{1}{\beta}\int_0^\beta B(s)\,\mathrm{d}s \qquad n \in \mathbb{Z}, \tag{6.366}$$

and

$$\mathrm{Det}\left[\frac{\mathrm{d}}{\mathrm{d}t} + B(t)\right]_{\beta,\xi} = 2\,\sinh\!\left(\xi + \frac{1}{2}\int_0^\beta B(s)\,\mathrm{d}s\right) \tag{6.367}$$

From the viewpoint of Statistical Mechanics, 2ξ plays the role of the chemical potential, cf. Fact 6.3. For the thermodynamical implications see [3].

k^{th} Order Linear Differential Operators

More generally we may consider the first operator

$$(\boldsymbol{D})_{ij} = \delta_{ij}\frac{\mathrm{d}}{\mathrm{d}t} + B(t)_{ij} \qquad i,j = 1,\ldots,k, \tag{6.368}$$

acting on k-vectors of functions satisfying the generalized periodic condition

$$\boldsymbol{\psi}(t + \beta) = \varXi\,\boldsymbol{\psi}(t), \quad \varXi \in GL(k,\mathbb{C}). \tag{6.369}$$

We write $\boldsymbol{\Phi}(t)$ for the $k \times k$ matrix which is the solution to the initial value problem

$$\boldsymbol{D}\boldsymbol{\Phi} = 0 \qquad \boldsymbol{\Phi}(0) = \mathbf{1}. \tag{6.370}$$

We know from Sect. 2.9.4 that $\boldsymbol{\Phi}(t)$ can be written as the $\mathbf{T}$-ordered exponential

$$\boldsymbol{\Phi}(t) = \mathbf{T}\exp\!\left(-\int_0^t B(s)\,\mathrm{d}s\right). \tag{6.371}$$

The $k \times k$ matrix $\boldsymbol{\Phi}(\beta)$ is the *monodromy* of the ODE (6.370).

The eigenfunctions of $\boldsymbol{D}$ have the form

$$\boldsymbol{\psi}_\lambda(t) = e^{\lambda t}\,\boldsymbol{\Phi}(t)\,\boldsymbol{\psi}_\lambda(0), \tag{6.372}$$

and the generalized periodicity condition (6.369) yields the equation

$$e^{\lambda\beta}\,\boldsymbol{\psi}_\lambda(0) = \boldsymbol{\Phi}(\beta)^{-1}\,\varXi\,\boldsymbol{\psi}_\lambda(0). \tag{6.373}$$

Let η_i be the principal determination of the logarithm of the i-th eigenvalue of the matrix $\boldsymbol{\Phi}(\beta)^{-1}\, \boldsymbol{\varXi}$ ($i = 1, \ldots, k$). The eigenvalues of $\boldsymbol{D}$ are

$$\lambda_{n,i} = \frac{1}{\beta}(2\pi i n + \eta_i), \qquad n \in \mathbb{Z}, \quad i = 1, \ldots, k. \tag{6.374}$$

Hence, from Eq. (6.359) we have

$$\mathrm{Det}[\boldsymbol{D}]_{\beta,\boldsymbol{\varXi}} = \prod_i \left(2\sinh(\eta_i/2)\right) = \pm\frac{\det(\boldsymbol{\varXi} - \boldsymbol{\Phi}(\beta))}{\sqrt{\det(\boldsymbol{\varXi})\,\det(\boldsymbol{\Phi}(\beta))}}. \tag{6.375}$$

Example 6.16 (2nd Order Differential Operator) Consider the differential operator

$$D = -\frac{\mathrm{d}^2}{\mathrm{d}t^2} + B(t), \qquad B(t + \beta) = B(t), \tag{6.376}$$

Generically there are two linear independent solutions $\psi_1(t)$ and $\psi_2(t)$ of the *Hill equation*[25] [10]

$$D\psi = 0 \tag{6.377}$$

which diagonalize the monodromy $\boldsymbol{\Phi}(\beta)$

$$\psi_1(t + \beta) = \mathrm{e}^{\eta_1}\psi_1(t), \qquad \psi_2(t + \beta) = \mathrm{e}^{\eta_2}\psi_2(t) \tag{6.378}$$

where we assume $\eta_1 \neq \eta_2 \bmod 2\pi i$ for simplicity. Then the determinant over the space of periodic functions is

$$\mathrm{Det}[D] = 4\sinh(\eta_1/2)\sinh(\eta_2/2). \tag{6.379}$$

More generally, the determinant over the space of functions satisfying the generalized periodicity condition

$$\psi(t + \beta) = \mathrm{e}^{2\pi i x}\psi(t) \quad x \in \mathbb{R} \tag{6.380}$$

is

$$\mathrm{Det}[D]_x = 4\sinh\left(\frac{\eta_1}{2} - \pi i x\right)\sinh\left(\frac{\eta_2}{2} - \pi i x\right). \tag{6.381}$$

For operators of the form (6.376) the Wronskian of two linearly-independent solutions is a constant. This is equivalent to saying that

$$\boldsymbol{\Phi}(t) \in SL(2, \mathbb{R}) \quad \text{in particular } \boldsymbol{\Phi}(\beta) \in SL(2, \mathbb{R}). \tag{6.382}$$

---

[25] See Eq. (3.330).

Thus $\eta_2 = -\eta_1 + 2\pi i k$. Then for generic η

$$\mathrm{Det}[D]_x = 4(-1)^{k+1}\left|\sinh\left(\frac{\eta}{2} - i\pi x\right)\right|^2 =$$
$$= (-1)^k\left[-4\sinh^2\frac{\eta}{2} - 4\sin^2(\pi x)\right]. \tag{6.383}$$

This is an example of the subtlety we mentioned in Remark 6.7: whereas the absolute value of a functional determinant is typically unambiguous, its sign (or phase) may depend on the prescription, here from the determination of the logarithm of the eigenvalues of the monodromy (see also the $\pm$ sign in front of Eq. (6.375)). This ambiguity is one of the effects of the *spectral flow* of the monodromy eigenvalues (cf. Sect. 3.10.1). The principal determination is the one with $k = 0$; however in the physical applications one observes that making $B(t) \rightsquigarrow B(t) + \mu$ and taking μ large enough, all eigenvalues of D becomes positive, so the physically natural determination of the determinant is the positive one (k odd).

However when $e^{\eta_1} = e^{\eta_2} = \pm 1$ we have two different possibilities: the Jordan normal form of the monodromy $\Phi(\beta)$ may be diagonal or consist of a 2×2 block. When $e^{2\pi i x} = e^{\eta_1} = e^{\eta_2}$ in the first case we have *two* zero modes, while in the second case only *one*.

Example 6.17   We specialize the previous example to a time independent coefficient $B = M^2$ and boundary conditions (6.380). Then formula (6.381) (k odd) yields

$$\mathrm{Det}[D]_x = 4\left|\sinh\left(\frac{M\beta}{2} - \pi i x\right)\right|^2 = 2\cosh(M\beta) - 2\cos(2\pi x). \tag{6.384}$$

This result may be recovered by a direct computation. The eigenvalues and eigenfunctions of D are

$$\lambda_k = 4\pi^2\frac{(x+k)^2}{\beta^2} + M^2, \quad \psi_k(t) = e^{2\pi i(x+k)t/\beta}, \quad k \in \mathbb{Z}, \tag{6.385}$$

and hence

$$\frac{1}{\beta}\frac{\partial}{\partial M}\log\mathrm{Det}_x(D) = \frac{1}{\beta}\frac{\partial}{\partial M}\sum_{k\in\mathbb{Z}}\log\lambda_k = \sum_{k\in\mathbb{Z}}\frac{2M\beta}{4\pi^2(x+k)^2 + M^2\beta^2} \tag{6.386}$$

which is equivalent to (6.384) in view of the easily established identity

$$\frac{1}{\pi}\sum_{k\in\mathbb{Z}}\frac{2y}{(k+x)^2 + y^2} = \coth[\pi(y + ix)] + \coth[\pi(y - ix)]. \tag{6.387}$$

6.19.2   The Analytic Method

Let D be, say, a differential operator of the form

$$D = -\frac{d^2}{dt^2} + B(t)^2 \tag{6.388}$$

acting on the space of functions in the interval $[0, L]$ with Dirichlet boundary conditions (b.c.) at the endpoints

$$\psi(0) = \psi(L) = 0. \tag{6.389}$$

We write D^0 for the operator $-\frac{d^2}{dt^2}$ acting on the same space.

Let $\Psi(t)$ be the unique solution to the initial value problem

$$-\frac{d^2}{dt^2}\Psi + B(t)^2\Psi = 0, \qquad \Psi(0) = 0, \quad \Psi'(0) = 1. \tag{6.390}$$

We **claim** that the determinant of the operator D in the interval with Dirichlet b.c. is

$$\frac{\mathrm{Det}\left[-\frac{d^2}{dt^2} + B(t)^2\right]_{[0,L]}}{\mathrm{Det}\left[-\frac{d^2}{dt^2}\right]_{[0,L]}} = \frac{\Psi(L)}{\Psi_0(L)}, \tag{6.391}$$

where $\Psi_0(t)$ is the solution to the initial value problem (6.390) for $B(t) \equiv 0$. Then we may choose the overall normalization $\mathcal{N}$ of the functional measure so that

$$\mathrm{Det}[D]_{[0,L]} = \Psi(L). \tag{6.392}$$

For instance, for the determinant computed in (6.338) we have $\Psi(t) = t$ and hence the determinant is just L, which is the result we got using the ζ-definition of the determinant.

The above formula may be rewritten in a more elegant way:

Corollary 6.1 *Let $\psi_1(t)$, $\psi_2(t)$ be two solutions of the ODE (6.390) normalized so that their Wronskian is 1*

$$W[\psi_1, \psi_2] \equiv \psi_1(t)\psi_2'(t) - \psi_1'(t)\psi_2(t) = 1. \tag{6.393}$$

Then the solution which vanishes at $t = t_0$ with $\psi'(t_0) = 1$ is

$$\psi(t) = \psi_1(t_0)\,\psi_2(t) - \psi_2(t_0)\,\psi_1(t), \tag{6.394}$$

*and the determinant of the operator D acting on the functions in the segment $[t_0, t_1]$
which vanish at the endpoints is*

$$\text{Det}[D]_{[t_0,t_1]} = \psi_1(t_0)\,\psi_2(t_1) - \psi_2(t_0)\,\psi_1(t_1). \tag{6.395}$$

Before going to the proof of our claim (6.391) we present an example.

Example 6.18 We take B constant. In this case the eigenvalues and eigenfunctions
are simply

$$\lambda_n = \frac{\pi^2}{L^2}n^2 + B^2, \qquad \psi_n = \sin(\pi n t/L), \qquad n = 1, 2, 3, \ldots \tag{6.396}$$

and

$$\frac{\partial}{\partial B}\log\left[\frac{\text{Det}\,D_{[0,L]}}{\text{Det}\,D^0_{[0,L]}}\right] = \frac{\partial}{\partial B}\log\prod_{n\geq 1}\left(\frac{\pi^2}{L^2}n^2 + B^2\right) =$$
$$= \sum_{n\geq 1}\frac{2BL^2}{\pi^2 n^2 + B^2 L^2}. \tag{6.397}$$

Using again the identity (6.357) now written in the form

$$\sum_{n\geq 1}\frac{2z}{z^2 + n^2\pi^2} = \frac{\mathrm{d}}{\mathrm{d}z}\log\left(\frac{\sinh z}{z}\right), \tag{6.398}$$

we get the general solution to Eq. (6.397)

$$\frac{\text{Det}\,D_{[0,L]}}{\text{Det}\,D^0_{[0,L]}} = f(L)\,\frac{\sinh(BL)}{BL}, \tag{6.399}$$

for some function $f(L)$. Taking $B \to 0$ the LHS must go to 1, so $f(L) = 1$. On the
other hand

$$\Psi(t) = \frac{1}{B}\sinh(Bt) \quad \text{and} \quad \Psi_0(t) = t \tag{6.400}$$

so that

$$\frac{\Psi(L)}{\Psi_0(L)} = \frac{\sinh(BL)}{BL}. \tag{6.401}$$

As $L \to \infty$ the determinant $\log\text{Det}[D]_{[0,L]}$ agrees asymptotically with the ζ-
determinant computed on the full line $\mathbb{R}$, cf. Eq. (6.343).

Proof of Claim Let D be the differential operator in (6.388). Consider the two meromorphic functions of $z \in \mathbb{C}$

$$\frac{\text{Det}[D - z]_{[0,L]}}{\text{Det}[D_0 - z]_{[0,L]}} \quad \text{and} \quad \frac{\Psi_z(L)}{\Psi_{0,z}(L)} \tag{6.402}$$

where

$$\begin{aligned}
(D - z)\Psi_z = 0, & \quad \Psi_z(0) = 0, & \quad \Psi_z'(0) = 1, \\
(D_0 - z)\Psi_{0,z} = 0, & \quad \Psi_{0,z}(0) = 0, & \quad \Psi_{0,z}'(0) = 1.
\end{aligned} \tag{6.403}$$

Both functions in (6.402) have zeros (resp. poles) when z is an eigenvalue of the operator D (resp. of D_0) acting on functions $f : [0, L] \to \mathbb{C}$ with Dirichlet b.c. $f(0) = f(L) = 0$. The ratio of the two functions in Eq. (6.402) is then an *entire function of $z \in \mathbb{C}$ without zeros*; since it goes to 1 at infinity, the two functions are equal.

Suppose that we know a particular solution $\psi_1(t)$ to Eq. (6.390). As seen in Chap. 3, a second solution $\psi_2(t)$ with Wronskian $W[\psi_1(t), \psi_2(t)] = 1$ is

$$\psi_2(t) = \psi_1(t) \int_{t_0}^{t} \frac{\mathrm{d}s}{\psi_1(s)^2}. \tag{6.404}$$

Plugging this solution in Eq. (6.395) we get

Corollary 6.2 *Let $\psi_1(t)$ be any particular solution to Eq. (6.390) with $\psi_1(t_0) \neq 0$. Then*

$$\text{Det}[D]_{[t_0,t_1]} = \psi_1(t_0)\, \psi_1(t_1) \int_{t_0}^{t_1} \frac{\mathrm{d}s}{\psi_1(t)^2}. \tag{6.405}$$

Maslov Index

The *Maslov index* μ of a second order differential operator

$$D \equiv -\frac{\mathrm{d}^2}{\mathrm{d}x^2} + V(x) \tag{6.406}$$

acting on the functions $\psi : [0, L] \to \mathbb{R}$, with Dirichlet boundary conditions at the endpoints $\psi(0) = \psi(L) = 0$, is the number μ of its negative eigenvalues. For $V(x)$ regular this number is finite.

We see the index $\mu(L)$ as a function of the upper endpoint L for a fixed function $V : \mathbb{R} \to \mathbb{R}$ bounded below. Since the lowest eigenvalue λ_0 satisfies the inequality

$$\lambda_0 \geq \min_x V(x) + \inf_{\|\psi\|=1} \int_0^L \mathrm{d}x \left(\frac{\mathrm{d}\psi}{\mathrm{d}x}\right)^2 = \min_x V(x) + \frac{\pi^2}{L^2}, \tag{6.407}$$

we have $\lim_{L\to 0}\mu(L) = 0$. This means that for L sufficiently small positive, all eigenvalues are positive, $\lambda_k > 0$ ($k = 0, 1, \dots$), and

$$\text{Det}[D]_{[0,L]} = C\,\Psi(L) > 0 \quad \text{for } L \approx 0, \tag{6.408}$$

where $\Psi(x)$ is the solution

$$D\Psi = 0 \quad \text{with} \quad \Psi(0) = 0, \quad \Psi'(0) = 1. \tag{6.409}$$

Now we increase L. When L reaches the first positive (simple) zero L_1 of $\Psi(x)$,

$$\Psi(L_1) = 0, \quad \Psi'(L_1) < 0, \quad L_1 > 0, \tag{6.410}$$

the determinant vanishes since $\Psi(x)$ is now a zero mode for D in $[0, L_1]$. For L slightly larger than L_1 the determinant is negative, and we have precisely one negative eigenvalue λ_0. When we reach the second positive zero L_2 of $\Psi(x)$, the second eigenvalue λ_1 becomes zero. Now $\Psi(x)$ is an eigenfunction of D in $[0, L_2]$ with eigenvalue zero and one internal node at $L_1 \in [0, L_2]$, which, in view of the oscillation theorem (Theorem 3.7), means that there is precisely one negative eigenvalue. For $L = L_2 + \epsilon$ the function $\Psi(L)$ is positive and hence we have two negative eigenvalues and no zero eigenvalue. As we increase L the third eigenvalue λ_2 decreases until L reaches the third zero L_3 of $\Psi(x)$ where $\lambda_2 = 0$ and $\Psi(x)$ is a zero mode with 2 internal zeros in $[0, L_3]$. Repeating the argument, we conclude

Fact 6.8 *The Maslov index $\mu(L)$ of the operator D acting on the functions $\psi: [0, L] \to 0$, with Dirichlet boundary conditions, is equal to the number of zeros of the solution $\psi(x)$ in Eq. (6.409) for $0 < x < L$.*

This result is equivalent to the oscillation theorem of Sturm-Liouville theory.

Matrix-Valued Differential Operators

For future applications in Chap. 8 we need the determinant of differential operators of the form

$$\boldsymbol{D}_{ij} = -\delta_{ij}\frac{\mathrm{d}^2}{\mathrm{d}t^2} + B(t)_{ij}, \quad i, j = 1, \dots, n \tag{6.411}$$

acting on column vectors

$$\boldsymbol{\psi}(t) = (\psi_1(t), \dots, \psi_n(t))^t \tag{6.412}$$

of functions in the interval $[0, L]$ satisfying the Dirichlet boundary condition

$$\boldsymbol{\psi}(0) = \boldsymbol{\psi}(L) = 0. \tag{6.413}$$

The same logics as before yields

Corollary 6.3 *Let $\boldsymbol{\Phi}(t)$ be an $n \times n$ matrix of functions satisfying*

$$D\boldsymbol{\Phi} = 0, \qquad \boldsymbol{\Phi}(0) = 0, \qquad \dot{\boldsymbol{\Phi}}(0) = \mathbf{1}_n. \tag{6.414}$$

Then

$$\mathrm{Det}[D]_{[0,L]} = \det \boldsymbol{\Phi}(L). \tag{6.415}$$

6.19.3 Computing $\mathrm{Det}^* D$

When zero modes are present, we need to compute the functional determinant in the space of functions *orthogonal to the zero modes* instead of the functional determinant over the full functional space which vanishes trivially in presence of zero eigenvalues. For definiteness we focus on the situations more frequently encountered in the applications, namely the second order differential operator (6.388) acting either on the space of functions in the interval $[0, L]$ which satisfy the boundary condition (6.389), or in the space of generalized periodic function (6.380). For brevity we limit ourselves to the case of the interval leaving the (similar) periodic problem to the diligent reader as a nice exercise. As discussed in Chap. 3, whenever $\ker D \neq 0$ the operator D has a *single* zero mode $\psi_0(t)$

$$D\psi_0 = 0, \qquad \psi_0(0) = \psi_0(L) = 0, \tag{6.416}$$

that we suppose to be known and normalized so that $\dot{\psi}_0(0) = 1$. Then we get a second linearly-independent solution $\psi_1(x)$ by a quadrature, see Eq. (3.167)

$$\psi_1(t) = \psi_0(t) \int_*^t \frac{\mathrm{d}s}{\psi_0(s)^2}, \qquad \psi_1(0) = -1, \tag{6.417}$$

and the Wronskian $W[\psi_0, \psi_1]$ of the two solutions is 1.

Consider now the functional determinant

$$\mathrm{Det}[D + z] = z \prod_k^* (\lambda_k + z), \tag{6.418}$$

where the superscript $*$ means that the product is taken over the eigenvalues in the space orthogonal to $\psi_0(x)$. Then

$$\mathrm{Det}^*[D] \equiv \prod_k^* \lambda_k = \frac{\mathrm{d}}{\mathrm{d}z}\mathrm{Det}(D + z)\Big|_{z=0}. \tag{6.419}$$

We write $\Psi(t, z)$ for the solution of the initial value problem

$$(D + z)\Psi(t, z) = 0, \quad \Psi(0, z) = 0, \quad \dot{\Psi}(0, z) = 1 \text{ for all } z. \tag{6.420}$$

We have

$$\Psi(t, z) = \psi_0(t) + z\,\Psi'(t) + O(z^2), \qquad \Psi'(t) \equiv \left.\frac{\partial\Psi(t, z)}{\partial z}\right|_{z=0} \tag{6.421}$$

where $\Psi'(t)$ is the solution to the non-homogeneous linear ODE

$$D\Psi' = -\psi_0, \qquad \Psi'(0) = \dot{\Psi}'(0) = 0, \tag{6.422}$$

which is conveniently computed by the method of *variation of the parameters* [11]

$$\Psi'(t) = -\int_0^t \left[\psi_0(t)\psi_1(s) - \psi_1(t)\psi_0(s)\right]\psi_0(s)\,\mathrm{d}s. \tag{6.423}$$

Using Eq. (6.392) the functional determinant in the $[0, L]$ interval is

$$\begin{aligned}
\mathrm{Det}(D + z) &= \Psi(L, z) = z\,\Psi'(L) + O(z^2) = \\
&= z\,\psi_1(L)\int_0^L \psi_0(s)^2\,\mathrm{d}s + O(z^2)
\end{aligned} \tag{6.424}$$

and then from (6.419)

$$\mathrm{Det}^* D = \psi_1(L)\int_0^L \psi_0(s)^2\,\mathrm{d}s = -\frac{1}{\dot{\psi}_0(L)}\int_0^L \psi_0(s)^2\,\mathrm{d}s. \tag{6.425}$$

Thus $\mathrm{Det}^* D$ is given by a simple quadrature in terms of the known function $\psi_0(t)$.

6.20   Non-Gaussian Path Integrals I: Methods

There are methods [12] that allow to compute explicitly "some" non-Gaussian path integrals. A part traditional "ad hoc" methods [13], there is a family of techniques inspired by *Witten index* techniques in supersymmetry [14, 15]. They may be interpreted physically in two ways:

(1)   One may repackage the imaginary-time Schrödinger equation as a Fokker-Planck ODE describing a dissipative phenomenon. The dissipation-fluctuation theorem may then be interpreted as a functional identity for the path integral which represents the solution of the Schrödinger equation. For this "thermo-dynamic/stochastic" interpretation of the method see [3] chap. 6;

(2)   One may relate the path integral under investigation to one in an auxiliary supersymmetric system, and then use the supersymmetry Ward identities to manipulate it in simpler expressions. The Schrödinger equation version of this approach was briefly described in the appendix to Chap. 3. A path integral treatment was mentioned in [3] chap. 3 and will be further considered in Sect. 6.23 below.

One may wonder what "some" means, in other words: *which non-Gaussian path integrals may be computed in closed form?* As stressed by Witten [16] the path integrals are best understood as contour integrals in a complexification of the space of paths. The path integrals along general complex "contours" give integral representations of the solutions of the Schrödinger equation in the complex domain. The object that controls the "complexity" of the path integral is then expected to be the monodromy group of the complex solutions or, more specifically, the differential Galois group of the Schrödinger equation (cf. **Box** on page 145). Roughly speaking, we should expect to be able to compute "explicitly" the path integral only when this Galois group is isogenous to a solvable algebraic group for all energy eigenvalues. Our examples below belong to this lucky class.

The Stochastic Form of a Path Integral

Suppose we have to compute a path integral (with operator insertions) for the one-dimensional system with imaginary time action

$$S_E[x] = \frac{1}{2}\int_0^\beta \mathrm{d}t \left(\dot{x}^2 + F(x)^2\right) \tag{6.426}$$

over the space $\mathfrak{P}$ of periodic paths with period β. Equation (6.426) is the general time-independent one-dimensional system: we can always reparametrize the configuration space to write the kinetic terms in the above form (cf. Eq. (3.289)) while by adding a constant to the bounded below potential $V(x)$ we may always assume $V(x) \geq 0$ so that, setting $F(x) = \sqrt{2V(x)}$ the Euclidean action takes the form (6.426). To fully specify the model it remains to say if the configuration space $\mathcal{M}$ where x takes values is $\mathbb{R}$, $\mathbb{R}_{>0}$, $[0, L]$, or S^1 (cf. Chap. 3). For brevity we focus on the first two cases.

More generally, we may consider a system with N degrees of freedom with an imaginary-time action of the *special kind*

$$S_E[x_i] = \int_0^\beta \mathrm{d}t \left(\frac{1}{2}\sum_{i=1}^N \dot{x}_i^2 + V(x_i)\right) \tag{6.427}$$

where the potential has the particular form

$$V(x_i) = \sum_{i=1}^N \left(\frac{\partial W(x)}{\partial x_i}\right)^2 \tag{6.428}$$

for some globally defined function $W(x)$ on the configuration space $\mathcal{M}$. When evaluated on a periodic path the action takes the simpler form

$$
\begin{aligned}
S_E[x_i] &= \frac{1}{2}\int_0^\beta dt \sum_i \left(\dot{x}_i^2 + \left(\frac{\partial W}{\partial x_i}\right)^2 \right) = \\
&= \frac{1}{2}\int_0^\beta dt \sum_i \left(\dot{x}_i + \frac{\partial W}{\partial x_i}\right)^2 - \int_0^\beta dt \sum_i \dot{x}_i \frac{\partial W}{\partial x_i} = \\
&= \frac{1}{2}\int_0^\beta dt \sum_i \left(\dot{x}_i + \frac{\partial W}{\partial x_i}\right)^2 - \int_0^\beta dt \, \frac{d}{dt} W = \\
&= \frac{1}{2}\int_0^\beta dt \sum_i \left(\dot{x}_i + \frac{\partial W}{\partial x_i}\right)^2 ,
\end{aligned}
\tag{6.429}
$$

where in the last step we used that the path is periodic of period β

$$
x_i(t) = x_i(t+\beta).
\tag{6.430}
$$

Now we make the functional change of variables

$$
x_i \rightsquigarrow h_i = \dot{x}_i + \frac{\partial W}{\partial x_i},
\tag{6.431}
$$

whose functional Jacobian is

$$
\left| \mathrm{Det}_\beta\left[\frac{\delta x_i}{\delta h_j}\right] \right| = \left| \mathrm{Det}_\beta\left[\frac{\delta h_i}{\delta x_j}\right]^{-1} \right| = \left| \mathrm{Det}_\beta\left[\delta_{ij}\frac{d}{dt} + \frac{\partial^2 W}{\partial x_i \, \partial x_j}\right]^{-1} \right| = \\
= \left| \frac{\det\left[\mathbf{T}e^{\int_0^\beta \partial_i \partial_j W dt}\right]^{1/2}}{\det\left[1 - \mathbf{T}e^{\int_0^\beta \partial_i \partial_j W dt}\right]} \right|
\tag{6.432}
$$

where $\mathrm{Det}_\beta[\cdot]$ is the functional determinant computed over the periodic functions of period β and in the second line we used Eqs. (6.371) and (6.375). Note that the RHS now contains only finite $N \times N$ determinants. Therefore

$$
\int_{\mathfrak{P}} [dx_i]\,e^{-S_E[x_i]}\, G[x_i] = \\
= \sum_{x[h]} \int \frac{[dh_i]\,e^{-\frac{1}{2}\int_0^\beta \sum_i h_i^2 \, dt}\,\left|\det\left[\mathbf{T}e^{\int_0^\beta \partial_i \partial_j W dt}\right]\right|^{1/2} G[x[h]]}{\left|\det\left[1 - \mathbf{T}e^{\int_0^\beta \partial_i \partial_j W dt}\right]\right|}
\tag{6.433}
$$

where $x_i = x[h]_i$ is the inverse of the functional change of variables (6.431), that is, a *periodic* solution to the ODE

$$\dot{x}_i + \frac{\partial W}{\partial x_i} = h(t)_i, \tag{6.434}$$

seen as a functional of the "noise" $h(t)_i$. The sum in (6.433) is over all periodic solutions, if more than one. We shall call the expression (6.433) the *stochastic form of the path integral*: indeed Eq. (6.434) is a *Langevin stochastic equation* with drift force $\partial_i W$ and *white noise* $h(t)_i$. A "white noise" is (by definition) a field whose correlations are the momenta of a Gaussian distribution with zero mean and variance

$$\langle h(t)_i\, h(s)_j \rangle = \delta_{ij}\, \delta(t - s). \tag{6.435}$$

In other words a *white noise* is a field whose correlation functions

$$\left\langle h(t_1)_{i_1} \cdots h(t_s)_{i_s} \right\rangle = \int [\mathrm{d}h]\mathrm{e}^{-\frac{1}{2}\int_0^\beta h_i h_i\, \mathrm{d}t}\, h(t_1)_{i_1} \cdots h(t_s)_{i_s}, \tag{6.436}$$

are given by Wick products (Theorem 6.7) of propagators (6.435) which are just (the kernel of) the *identity operator* 1, Eq. (6.435). For more about the stochastic interpretation of the imaginary-time path integrals see [3] chap. 6 or [14, 15].

The stochastic form of the path integral presents it as the expectation of the functional

$$\mathfrak{F}[h] = \frac{\left| \det\!\left[\mathbf{T}\,\mathrm{e}^{\int_0^\beta \partial_i \partial_j W \mathrm{d}t} \right] \right|^{1/2} G[x[h]]}{\left| \det\!\left[1 - \mathbf{T}\,\mathrm{e}^{\int_0^\beta \partial_i \partial_j W \mathrm{d}t} \right] \right|} \tag{6.437}$$

in the Gaussian measure of covariance 1

$$\int [\mathrm{d}x_i]\,\mathrm{e}^{-S_E[x]}\, G[x] = \left\langle \mathfrak{F}[h] \right\rangle_{\mathrm{Gauss}} \equiv \int [\mathrm{d}h_i]\,\mathrm{e}^{-\frac{1}{2}\int_0^\beta \sum_i h_i^2\, \mathrm{d}t}\, \mathfrak{F}[h]. \tag{6.438}$$

When $\mathfrak{F}[h]$ is reasonably simple, this gives an efficient method to compute the path integral as the examples in this and the next sections will illustrate (see also Sect. 8.6.1). The functional $\mathfrak{F}[h]$ may have a rather intricate functional form, yet be *reasonably simple* for the present purposes, if it satisfies a convenient functional differential equation or difference equation ($\equiv$ a functional recursion relation). The simplest $\mathfrak{F}[h]$'s arise when computing SUSY protected operator insertions in the supersymmetric completion of the quantum system (6.427): see Sect. 6.24. In this and the following sections we are interested in exploiting the technique for the ***non-supersymmetric*** model with action (6.427) where some extra ingenuity is required.

6.20.1   *Example: The Morse Path Integral*

As a first example, let us consider a one-dimensional quantum particle subjected to the Morse potential. The imaginary time action reads

$$S_E[x] = \frac{1}{2}\int_0^\beta ds\left(\dot{x}^2 + e^{2x} - 2\alpha\, e^x\right) \qquad \alpha > 0 \tag{6.439}$$

This is the same system we solved in Sect. 3.13.5 using the Schrödinger equation; here we have redefined the field x so that

$$m/\hbar^2 = 1, \quad A = 1/2, \quad B = \alpha/2, \tag{6.440}$$

while the parameter α has the same meaning as in Chap. 3. We wish to solve the model by a direct computation of its periodic path integral. In particular we wish to recover the spectrum of its bound states, i.e. the energy levels with $E < 0$, cf. Eq. (3.395). While in the literature there are ad hoc methods [13] to do that, we shall use the stochastic representation of the path integral.

We compute the integral over paths periodic in time, of period β of the form

$$\int_\mathfrak{P} [dx]\, e^{-S_E[x]}\, G[x] = \mathrm{Tr}\!\left(e^{-\beta H} G[x]\right) \tag{6.441}$$

where the operator insertion $G[x]$ is chosen to make the trace over the Hilbert space convergent for positive β. When $G[x]$ is a multiple of 1 the trace diverges due to the presence of scattering states with energy $E > 0$. To get a convergent trace we may, for instance, take $G[x] = \exp(x(t_0))$. For periodic paths we have

$$S_E[x] = -\frac{1}{2}\alpha^2\beta + \frac{1}{2}\int_0^\beta dt\,(\dot{x} + e^x - \alpha)^2, \tag{6.442}$$

cf. Eq. (6.429). We write $x[h]$ for the unique *periodic* solution to the associated Langevin equation, which in the present case is the Bernoulli ODE [11]

$$\frac{dx[h]}{dt} + e^{x[h]} - \alpha = h(t). \tag{6.443}$$

Explicitly

$$e^{-x[h](t)} = e^{-\int_0^t (h+\alpha)ds}\left[\int_0^t e^{\int_0^s (h+\alpha)du}\, ds + \frac{\int_0^\beta e^{\int_0^s (h+\alpha)du}\, ds}{e^{\int_0^\beta (h+\alpha)ds} - 1}\right]. \tag{6.444}$$

The function $x[h](t)$ is *real* iff the RHS is positive. This holds for all t iff the following constraint $\mathfrak{C}$ is obeyed

$$\mathfrak{C}: \qquad 0 < \int_0^\beta ds\,(h+\alpha) \equiv \int_0^\beta e^{x[h]}\,ds. \tag{6.445}$$

In the present situation Eq. (6.433) reduces to[26]

$$\beta\,\mathrm{Tr}\big[e^{-\beta H}e^{x(t_0)}\big] = e^{\alpha^2\beta/2}\int [dh]\,\frac{e^{-\frac{1}{2}\int_0^\beta h^2 dt}\int_0^\beta e^{x[h]}\,ds}{2\sinh\big(\frac{1}{2}\int_0^\beta e^{x[h]}\,ds\big)} \equiv$$
$$\equiv e^{\alpha^2\beta/2}\int [dh]\,\frac{e^{-\frac{1}{2}\int_0^\beta h^2 dt}\int_0^\beta (h+\alpha)ds}{2\sinh\big(\frac{1}{2}\int_0^\beta (h+\alpha)ds\big)}. \tag{6.446}$$

Now we decompose the periodic function $h(t)+\alpha$ in the constant mode y plus a fluctuation $\xi(t)$ with zero-mean

$$h(t)+\alpha = y+\xi(t), \qquad \int_0^\beta \xi(t)\,dt = 0. \tag{6.447}$$

A part for the Gaussian exponential, the integrand in (6.446) depends only on the zero mode y. Then the integrals over the non-zero modes $\xi(t)$ are Gaussian of covariance 1, and the integrals over them give just a factor 1. We remain with an ordinary integral in the domain $y \geq 0$ (cf. Eq. (6.445))[27]

$$\mathrm{Tr}\big[e^{-\beta H}e^{x(t_0)}\big] = \sqrt{\frac{2\beta}{\pi}}\int_0^\infty y\,dy\,\frac{e^{-\frac{1}{2}y^2\beta+\alpha y\beta}}{2\sinh(\frac{1}{2}y\beta)} =$$
$$= \sum_{n\geq 0}\sqrt{\frac{2\beta}{\pi}}\int_0^\infty y\,dy\,e^{-\frac{1}{2}y^2\beta+y\beta[\alpha-(n+\frac{1}{2})]}. \tag{6.448}$$

---

[26] We use that by translational invariance in time $\mathrm{Tr}[e^{-\beta H}\int_0^\beta A(t)\,dt] = \beta\,\mathrm{Tr}[e^{-\beta H}A(t_0)]$ for all insertions $A(t)$ and t_0.

[27] The overall factor $\sqrt{\beta}$ arises in the following way. We expand $h(t)+\alpha$ in the orthonormal system given by the Fourier series

$$h(t)+\alpha = \frac{1}{\sqrt{\beta}}\Big[a_0 + \sum_{k\geq 1}\big(a_k\,e^{2\pi ikt/\beta} + a_k^*\,e^{-2\pi ikt/\beta}\big)\Big]$$

and the functional measure is $[dh] = da_0\prod_{k\geq 1}da_k\,da_k^*$. The integrals over the non-zero modes a_k, a_k^* ($k\geq 1$) yield 1, while $da_0 = \sqrt{\beta}\,dy$. The factor $\sqrt{2/\pi}$ is just the normalization of the Gaussian of covariance 1 in a half-line.

Without the operator insertion, which yields the factor y in the integrand, we would get the partition function $\mathrm{Tr}[e^{-\beta H}]$ written as a formal series

$$\sum_{n\geq 0}\sqrt{\frac{2\beta}{\pi}}\int_0^\infty dy\, e^{-\frac{1}{2}y^2\beta+y\beta[\alpha-(n+\frac{1}{2})]}. \tag{6.449}$$

Although this series does not converge, the divergence is entirely due to the scattering states with $E > 0$, in facts to the "infra-red" states with $E \approx 0$. If we truncate the sum to some large $n \gg 1$, we can read the exact bound state spectrum from its large β behavior. Equivalently, we may use the convergent series in (6.448).

When studying the large β asymptotics of each term in the sum (6.449) by the saddle point method [17], we need to distinguish two cases:

(i) $\alpha - (n + \frac{1}{2}) > 0$;
(ii) $\alpha - (n + \frac{1}{2}) < 0$.

In case *(i)* the integrand has a saddle point inside the domain of integration $y > 0$, while in case *(ii)* there is no positive saddle point, and the integral is dominated by the region of small y (which is where the integral of the sum diverges). The terms with $\alpha - (n + \frac{1}{2}) > 0$ produce contributions to the partition function which *grow exponentially* as $\beta \to \infty$ while the others ones give contributions $O(\beta^{-1/2})$ in the large β limit. Indeed for $\alpha - (n + \frac{1}{2}) < 0$ the large β asymptotics is

$$\sqrt{\frac{2\beta}{\pi}}\int_0^\infty dy\, e^{-\frac{1}{2}y^2\beta+y\beta[\alpha-(n+\frac{1}{2})]} \approx \sqrt{\frac{2}{\pi\,\beta}}\frac{1}{(n+\frac{1}{2})-\alpha}+O(\beta^{-3/2}). \tag{6.450}$$

Since the partition function is a sum of terms of the form $e^{-\beta E}$, the exponentially large contributions are precisely the ones from energy eigenstates with $E < 0$, i.e. from the bound states. Therefore as $\beta \to \infty$ one has

$$\mathrm{Tr}[P(E < 0)\, e^{-\beta H}] =$$

$$= \sum_{0\leq n\leq \alpha-1/2}\sqrt{\frac{2\beta}{\pi}}\int_0^\infty dy\, e^{-y^2\beta/2+\beta y(\alpha-n-1/2)}+O(\beta^{-1/2}) = \tag{6.451}$$

$$= \sum_{0\leq n\leq \alpha-1/2}\exp\left[\tfrac{1}{2}\beta\left(\alpha-n-1/2\right)^2\right]$$

where $P(E < 0)$ is the spectral projector to states with negative energy, i.e. to the bound states. Since the bound states are normalizable and non-degenerate, there is no $O(\beta^{-1/2})$ correction to the formula in the last line which is then *exact*.

We conclude that the Morse potential has a bound state for each non-negative integer n strictly smaller than $\alpha - 1/2$ with precise energy level

$$E_n = -\frac{1}{2}\left(\alpha - (n + \tfrac{1}{2})\right)^2. \tag{6.452}$$

This is exactly what we found in Chap. 3 by solving the relevant Schrödinger equation ($\equiv$ confluent hypergeometric ODE) in terms of Laguerre polynomials.

6.21 Non-Gaussian Path Integrals II: Induction Method

An important application of the stochastic form of the path integrals is to produce recursion relations for sequences of path integrals which then may be computed by induction. As it will be clear in Sect. 6.24, these recursion relations are nothing else than the *supersymmetric Ward identities* of an auxiliary SUSY quantum system. We start with a few important examples. The first one is a bit formal because of convergence issues (similar to the ones we encountered in the Morse potential).

6.21.1 Free Radial Motion

Suppose that we want to compute the periodic path integral corresponding to the effective radial Schrödinger equation for the free motion in $\mathbb{R}^3$ which we studied from the Schrödinger viewpoint in Sect. 5.2. This is the path integral with Euclidean action

$$S[r]_\ell = \frac{1}{2} \int_0^\beta \left(\dot{r}^2 + \frac{\ell(\ell+1)}{r^2}\right) dt \tag{6.453}$$

where ℓ is a non-positive integer and $r > 0$. In absence of insertions the partition function $\mathrm{Tr}[e^{-\beta H_\ell}]$ diverges because of the scattering states of low energy. Hence we choose a convenient regulating operator $\mathcal{O} = \mathcal{O}[r]$ and consider

$$\mathrm{Tr}[e^{-\beta H_\ell}\mathcal{O}] = \int_{\mathfrak{P}} [dr]\,\mathcal{O}[r]\,e^{-S[r]_\ell} \tag{6.454}$$

as a function of the quantum number $\ell \in \mathbb{N}$. A convenient choice is, say,

$$\mathcal{O}[r] = \left(\int_0^\beta \frac{dt}{r(t)}\right)^2. \tag{6.455}$$

We have

$$
\mathrm{Tr}[e^{-\beta H_\ell}\mathcal{O}] - \mathrm{Tr}[e^{-\beta H_{\ell-1}}\mathcal{O}] =
$$

$$
= \int_{\mathfrak{P}} [dr]\,\mathcal{O}[r]\, e^{-\frac{1}{2}\int_0^\beta (\dot r^2 + \ell^2/r^2)\,dt} \left(e^{-\frac{1}{2}\int_0^\beta dt\,\ell/r^2} - e^{+\frac{1}{2}\int_0^\beta dt\,\ell/r^2} \right)
\tag{6.456}
$$

The idea of the *induction method* is to take advantage of the stochastic version of the path integral to evaluate explicitly the path integral in the RHS, and then use the result to compute the $\mathrm{Tr}[e^{-\beta H_\ell}\mathcal{O}]$'s recursively in the integer ℓ. We use the stochastic form of the path integral with the change of variable ($\equiv$ stochastic Langevin equation)

$$
r \rightsquigarrow h \equiv \dot r + \frac{\ell}{r},
\tag{6.457}
$$

to get

$$
\mathrm{Tr}[e^{-\beta H_\ell}\mathcal{O}] - \mathrm{Tr}[e^{-\beta H_{\ell-1}}\mathcal{O}] =
$$

$$
= \frac{1}{\ell^2} \int_{\mathfrak{P}} [dh]\, \frac{\left(\int_0^\beta h(t)\,dt\right)^2 e^{-\frac{1}{2}\int_0^\beta h^2 dt}}{2\,\sinh(-\frac{1}{2}\int_0^\beta dt\,\ell/r^2)} \left(e^{-\frac{1}{2}\int_0^\beta dt\,\ell/r^2} - e^{+\frac{1}{2}\int_0^\beta dt\,\ell/r^2} \right)
\tag{6.458}
$$

$$
= \frac{1}{\ell^2} \int_{\mathfrak{P}} [dh] \left(\int_0^\beta h(t)\,dt\right)^2 e^{-\frac{1}{2}\int_0^\beta h^2 dt} = \sqrt{\frac{2\beta^5}{\pi}}\,\frac{1}{\ell^2}\int_0^\infty dy\, y^2\, e^{-\beta y^2/2}
$$

$$
= \frac{\beta}{\ell^2},
$$

where we assumed that there is a unique positive periodic solution to the ODE (6.457) when $\int_0^\beta h(t)\,dt > 0$. This recursion relates recursively the path integral for general ℓ to the one for $\ell = 0$ which is Gaussian and easy to compute.

Example 6.19 (Radial Motion in $\mathbb{R}^d$) More generally, we may consider the free radial motion in $\mathbb{R}^d$ which is a one-dimensional system with effective potential (cf. Eq. (5.24))

$$
V_{\mathrm{eff}}(r) = \frac{\nu^2 - \frac{1}{4}}{2r^2}, \qquad \nu = \ell + \frac{d}{2} - 1, \quad \ell \in \mathbb{N}.
\tag{6.459}
$$

To get the recursion relation in the angular momentum ℓ, generalize the change of variables (6.457) to the Langevin equation

$$
r \rightsquigarrow h \equiv \dot r + \frac{\nu - \frac{1}{2}}{r},
\tag{6.460}
$$

and perform the same manipulations as in the $d = 3$ case, see Problem 6.2.

6.21.2   Harmonic Radial Motion

To get a well-defined path integral we consider the model with Euclidean action

$$S[x]_\ell = \frac{1}{2} \int_0^\beta \left(\dot{r}^2 + \frac{\ell(\ell+1)}{r^2} + \omega^2 r^2 \right) dt \tag{6.461}$$

with the constraint $r \geq 0$, whose Schrödinger equation is the radial equation for the isotropic harmonic oscillator in three dimensions with angular momentum ℓ, see Sect. 5.3.1. We write

$$Z_\ell \equiv \int_{\mathfrak{P}} [dr]\, e^{-S[r]_\ell} \equiv \mathrm{Tr}[e^{-\beta H_\ell}]. \tag{6.462}$$

We make the change of functional variables

$$\dot{r} + \frac{\ell}{r} + \omega r = h. \tag{6.463}$$

The Jacobian determinant

$$\mathfrak{J}[h] \equiv \mathrm{Det}_\beta\left[\frac{d}{dt} - \frac{\ell}{r^2} + \omega\right] = 2\sinh\left[\frac{\beta}{2}\omega - \frac{1}{2}\int_0^\beta \frac{\ell}{r[h]^2}\, dt\right] \tag{6.464}$$

now does not have a definite sign—$\mathfrak{J}[h]$ is positive for some functions $h(t)$ and negative for others. Using the explicit form of the functional determinant (6.464) we get the identity

$$e^{\beta\omega\ell} \int_{\mathfrak{P}} [dr]\, e^{-\frac{1}{2}\int_0^\beta (\dot{r}+\ell/r+\omega r)^2 dt}\, \mathrm{Det}_\beta\left[\frac{d}{dt} - \frac{\ell}{r^2} + \omega\right] =$$

$$= e^{\frac{1}{2}\beta\omega} \int_{\mathfrak{P}} [dr]\, e^{-S[r]_\ell} - e^{-\frac{1}{2}\beta\omega} \int_{\mathfrak{P}} [dr]\, e^{-S[r]_\ell} \equiv \tag{6.465}$$

$$\equiv e^{\frac{1}{2}\beta\omega} Z_\ell - e^{-\frac{1}{2}\beta\omega} Z_{\ell-1}.$$

One shows [14] that, for a gapped system like the present one, the quantity

$$\int_{\mathfrak{P}} [dr]\, e^{-\frac{1}{2}\int_0^\beta (\dot{r}+\ell/r+\omega r)^2 dt}\, \mathrm{Det}_\beta\left[\frac{d}{dt} - \frac{\ell}{r^2} + \omega\right] =$$

$$= \int_{\mathfrak{P}} [dh]\, e^{-\frac{1}{2}\int_0^\beta h^2 dt}\, \mathrm{sign}(\mathfrak{J}[h]) \tag{6.466}$$

is an *integer*, namely the *Witten index* Δ_ℓ of the underlying supersymmetric theory (see Sect. 6.24 below). In the present case the index is *zero* (for $\ell \geq 1$): the quickest way to see this is that the energy of the eigenstates cannot decrease when we increase ℓ, as a non-zero value of the RHS of (6.465) will imply.[28] Notice that this is possible since $\Im[h]$ has not a constant sign, contrary to the previous examples. We remain with the recursion relation

$$Z_\ell = e^{-\omega\beta} Z_{\ell-1}. \tag{6.467}$$

that is, the energy levels $E_n(\ell)$ for different values of ℓ are related by

$$E_n(\ell) = E_n(\ell - 1) + \omega \quad n = 0, 1, 2, \cdots. \tag{6.468}$$

Solving this elementary recursion we get

$$E_n(\ell) = \omega\ell + E_n(0) = \omega\ell + (2n + 1)\omega + \frac{\omega}{2}, \tag{6.469}$$

where we used the fact that the model (6.461) reduces for $\ell = 0$ to the one dimensional model in the half-line $\mathbb{R}_{\geq 0}$ which can be seen, via the image method, as the one-dimensional harmonic oscillator truncated to the odd sector, i.e. to the *odd* energy eigenstates. Of course this is the same result we got from the explicit solution of the Schrödinger equation, see Eq. (5.57).

Example 6.20 (Generalization to the Isotropic Harmonic Oscillator in d Dimensions) As in Example 6.19 one makes the redefinition

$$\ell \rightsquigarrow \nu - \frac{1}{2} \equiv \ell - \frac{d - 3}{2} \tag{6.470}$$

and goes through the same steps as in the $d = 3$ case.

6.21.3   Motion on the Sphere

We next consider the problem of computing the periodic path integral for a particle moving on a sphere S^2 of radius 1. The Schrödinger Hamiltonian is

$$H = -\frac{1}{\sin\theta}\frac{\partial}{\partial\theta}\left(\sin\theta\frac{\partial}{\partial\theta}\right) - \frac{1}{\sin^2\theta}\frac{\partial^2}{\partial\phi^2} \tag{6.471}$$

---

[28] A more detailed argument is as follows: the superpotential of the auxiliary SUSY model is $W(r) = \ell\log r + \frac{1}{2}\omega r^2$ in the half-line $\mathbb{R}_{\geq 0}$. The solutions of the Schrödinger equation with $E = 0$ are as in Eq. (4.448) i.e. $\exp(\pm W(r))$. For $\ell > 0$ neither is in $L^2(\mathbb{R}_{\geq 0})$, so the auxiliary model has no SUSY ground state and $\Delta = 0$.

acting on the Hilbert space $L^2(S^2, \sin\theta \, d\theta \, d\phi)$. One has

$$(\sin\theta)^{1/2} H (\sin\theta)^{-1/2} = -\frac{\partial^2}{\partial\theta^2} - \frac{1}{4}\frac{1}{\sin^2\theta} - \frac{1}{4} - \frac{1}{\sin^2\theta}\frac{\partial^2}{\partial\phi^2}, \tag{6.472}$$

acting on $L^2(d\theta \, d\phi)$. Hence we may compute the path integral using the usual "flat space" functional measure but with the effective Lagrangian

$$L_{\text{eff}} = \frac{1}{2}\left(\dot\theta^2 + \sin^2\theta \, \dot\phi^2 - \frac{1}{4}\frac{1}{\sin^2\theta} - \frac{1}{4}\right). \tag{6.473}$$

The path integral in the cyclic coordinate ϕ is easily performed by the canonical method illustrated in the second part of Sect. 6.16.2; then for all operator insertions $\mathcal{O}[\theta]$ independent of the cyclic variable ϕ:

$$\int_{\mathfrak{P}} [d\theta \, d\phi]\, e^{-S_E[\theta,\phi]}\mathcal{O}[\theta] = e^{\beta/8}\sum_{m\in\mathbb{Z}}\int_{\mathfrak{P}}[d\theta]\,e^{-\frac{1}{2}\int_0^\beta dt\left(\dot\theta^2 + \frac{m^2-\frac{1}{4}}{\sin^2\theta}\right)}\mathcal{O}[\theta] \tag{6.474}$$

where each path integral in the sum in the RHS is nothing else than the path integral for the effective one-dimensional system resulting from the separation of the cyclic variable. We are reduced to computing integrals on periodic paths $\theta: [0,\beta] \to [0,\pi]$ which compute $\text{Tr}[e^{-\beta H_m}\mathcal{O}]$ where H_m is the effective one-dimensional quantum Hamiltonian with the effective potential

$$V_m(\theta) = \frac{m^2 - \frac{1}{4}}{\sin^2\theta}. \tag{6.475}$$

The following identity is obvious

$$\text{Tr}[e^{-\beta H_{m-1}}\mathcal{O}] - \text{Tr}[e^{-\beta H_m}\mathcal{O}] =$$

$$= 2\,e^{\beta/8}\int_{\mathfrak{P}}[dx]\,e^{-\frac{1}{2}\int_0^\beta\left(\dot\theta^2 + \frac{(m-\frac{1}{2})^2}{\sin^2\theta}\right)dt}\,\sinh\left(\frac{1}{2}\int_0^\beta\frac{m-\frac{1}{2}}{\sin^2\theta}\,dt\right)\mathcal{O}[\theta]. \tag{6.476}$$

This equality allows us to compute the trace $\text{Tr}[e^{-\beta H_m}\mathcal{O}]$ by induction on the integer m using the stochastic representation of the path integral. The change of periodic functional variables

$$\theta(t) \rightsquigarrow h(\theta) \equiv \dot\theta - \left(m-\frac{1}{2}\right)\frac{\cos\theta}{\sin\theta}, \qquad 0 \le \theta \le \pi, \tag{6.477}$$

whose functional Jacobian determinant is

$$\left| \mathrm{Det}_\beta \left[\frac{d}{dt} + \left(m - \frac{1}{2} \right) \frac{1}{\sin^2 \theta(t)} \right] \right|^{-1} = \frac{1}{2} \left| \sinh \left(\frac{2m-1}{4} \int_0^\beta \frac{dt}{\sin^2 \theta(t)} \right) \right|^{-1},$$

(6.478)

gives the equality

$$2\, e^{\frac{1}{2}(m-\frac{1}{2})^2 \beta} \int_{\mathfrak{P}} [dx]\, e^{-\frac{1}{2} \int_0^\beta \left(\dot{\theta}^2 + \frac{(m-\frac{1}{2})^2}{\sin^2 \theta} \right) dt}\, \sinh \left(\frac{1}{2} \int_0^\beta \frac{m - \frac{1}{2}}{\sin^2 \theta}\, dt \right) =$$

(6.479)

$$= \int_{\mathfrak{I}} [dh]\, e^{-\frac{1}{2} \int_0^\beta h(t)^2\, dt}$$

where $\mathfrak{I} \subset \mathfrak{P}$ is the image of the functional map $x \mapsto h$ in the space of periodic functions. In the interval $0 < \theta < \pi$ the function

$$W'(\theta) \equiv -\frac{\cos \theta}{\sin \theta}$$

(6.480)

is monotonically increasing from $-\infty$ to $+\infty$. Then, by the Witten index arguments in Sect. 6.24 below, the Gaussian path integral in the RHS of Eq. (6.479) is just 1. Hence

$$2 \int_{\mathfrak{P}} [dx]\, e^{-\frac{1}{2} \int_0^\beta \left(\dot{\theta}^2 + \frac{(m-\frac{1}{2})^2}{\sin^2 \theta} \right) dt}\, \sinh \left(\frac{1}{2} \int_0^\beta \frac{m - \frac{1}{2}}{\sin^2 \theta}\, dt \right) = e^{-\frac{1}{2}(m-\frac{1}{2})^2 \beta}.$$

(6.481)

The recursion (6.476) (with $\mathcal{O} = 1$) then becomes

$$\mathrm{Tr}[e^{-\beta H_{m-1}}] - \mathrm{Tr}[e^{-\beta H_m}] = e^{\beta/8}\, e^{-\frac{1}{2}(m-\frac{1}{2})^2 \beta} \equiv e^{-\frac{1}{2}m(m-1)\beta},$$

(6.482)

whose solution is

$$\mathrm{Tr}[e^{-\beta H_m}] = \sum_{\ell \geq |m|} e^{-\frac{1}{2}\ell(\ell+1)\beta}$$

(6.483)

which is the result we obtained from the Representation Theory of Lie groups in Chap. 4.

Example 6.21 The same stochastic strategy allows to compute by induction on the integer m the path integrals (with suitable insertions) for the potentials

$$V_m(x) = \frac{\frac{1}{4} - m^2}{\cosh^2 x}.$$

(6.484)

The Langevin equation now is

$$\dot{x} + \left(m + \frac{1}{2}\right)\tanh x = h. \tag{6.485}$$

6.21.4   The Hydrogen Atom

The path integral for the hydrogen atom, i.e. a quantum particle moving in $\mathbb{R}^3$ subjected to the central potential $-\gamma/r$, in spherical coordinates (r, θ, ϕ) can be computed following three steps which parallel the separation of variables for the Schrödinger equation. In the first step one performs the integral in the cyclic field ϕ using the canonical method of Sect. 6.16.2. In the second step one integrates over the field θ: one gets an effective path integral in $[d\theta]$ with Euclidean Lagrangian of the form

$$L_{\text{eff}} = \frac{1}{2}r^2\dot{\theta}^2 + \frac{m^2 - \frac{1}{4}}{r^2 \sin^2\theta} \tag{6.486}$$

which is easily computed by recursion in m, getting the formula (6.483) with

$$\beta \rightsquigarrow \int_0^\beta \frac{dt}{r^2}. \tag{6.487}$$

In the third step one remains with a path integral for the radial motion of the hydrogen atom, i.e. the motion on the half-line $r \geq 0$ with effective potential

$$V(r)_{\text{eff}} = \frac{1}{2}\frac{\ell(\ell + 1)}{r^2} - \frac{\gamma}{r}. \tag{6.488}$$

We write H_ℓ for the Hamiltonian in the half-line $r \geq 0$ with the effective potential.

Again the partition function does not converge because of the scattering states with $E > 0$. We know from the analysis of the Morse path integral in Sect. 6.20.1 what we have to do: we compute the path integral with the insertion of the projector operator over negative energy states,

$$\text{Tr}\!\left[P(E < 0)\,e^{-\beta H_\ell}\right], \tag{6.489}$$

which gets rid of the continuous spectrum states and hence is convergent. We can read the bound states from this convergent path integral.

Just as the Morse path integral was obtained from the Liouville one by shifting the noise

$$h(t) \rightsquigarrow h(t) + \alpha, \tag{6.490}$$

in the stochastic representation (cf. Eq. (6.447)), we get the hydrogen atom by making the same shift in the stochastic form of the path integral for the free radial motion (Sect. 6.21.1), that is, we write

$$h + \alpha = \dot{r} + \frac{\ell}{r}.$$
(6.491)

Going through the same manipulations as in previous examples, we get the equality

$$e^{\alpha^2 \beta / 2 - \frac{1}{2} \int_0^\beta h^2 dt} \, \mathrm{Det}_\beta \left[\frac{\mathrm{d}}{\mathrm{d}t} - \frac{\ell}{r^2} \right] =$$
$$= \left(e^{\frac{1}{2} \int_0^\beta \left(\dot{r}^2 + \ell(\ell+1)/r^2 - 2\alpha\ell/r \right) dt} - e^{-\frac{1}{2} \int_0^\beta \left(\dot{r}^2 + (\ell-1)\ell/r^2 - 2\alpha\ell/r \right) dt} \right).$$
(6.492)

Following the same logic as before, we transform this equality in the following functional identity for the radial motion partition functions $Z_\ell(\gamma)$

$$\sqrt{\frac{2\beta}{\pi}} \int_0^\infty e^{-\frac{1}{2}\beta y^2 + \beta \alpha y} \, \mathrm{d}y = Z_{\ell-1}(\gamma) \Big|_{\gamma = \alpha\ell} - Z_\ell(\gamma) \Big|_{\gamma = \alpha\ell}.$$
(6.493)

We stress again that, while $Z_\ell(\gamma)$ is divergent before introducing a suitable operator insertion, the difference in the RHS is *finite* without insertions. Extracting the contributions to the partition function from the bound states, we get the recursion in the momentum quantum number ℓ

$$\mathrm{Tr}\left[P(E < 0)\, e^{-\beta H_{\ell-1}} \right] - \mathrm{Tr}\left[P(E < 0)\, e^{-\beta H_\ell} \right] = e^{\frac{\beta\gamma^2}{2\ell^2}},$$
(6.494)

whose solution is

$$\mathrm{Tr}\left[P(E < 0)\, e^{-\beta H_\ell} \right] = \sum_{n \geq 0} \exp\left[\frac{\beta\gamma^2}{2(n + \ell + 1)^2} \right],$$
(6.495)

that is, the discrete energy levels of the bound states are

$$E_{n,\ell} = -\frac{\gamma^2}{2} \frac{1}{(n + \ell + 1)^2}$$
(6.496)

which is, of course, the correct result for the discrete energy levels of the hydrogen atom.

Example 6.22 (Generalization to $V = -\gamma/r$ in d Dimensions) As in Example 6.19 just make the redefinition

$$\ell \rightsquigarrow \nu - \frac{1}{2} \equiv \ell - \frac{d-3}{2}.$$
(6.497)

6.22   The Semi-classical Limit of Path Integrals

One nice feature of the path integrals is their simple behavior in the semi-classical limit $\hbar \to 0$, which makes the relation between a classical mechanical systems and its quantum version particularly transparent. In the functional approach "quantization" is a rather intuitive procedure. We shall elaborate more on the semi-classical limit in Chap. 8 where it is used as an *approximation scheme.*

Review of Lagrange's Asymptotic Method

We recall Lagrange's saddle-point technique to evaluate ordinary integrals of the form[29]

$$\int_{\mathbb{R}^n} e^{-\lambda V(x)} \, d^n x \tag{6.498}$$

in the asymptotic limit $\lambda \to +\infty$. Here $V(x)$ is a smooth function bounded below. Suppose first that $V(x)$ has a unique (absolute) minimum at x_0 with value $V(x_0) = V_0$. We write

$$\int_{\mathbb{R}^n} e^{-\lambda V(x)} \, d^n x = e^{-\lambda V_0} \int_{\mathbb{R}^n} e^{-\lambda(V(x)-V_0)} \, d^n x. \tag{6.499}$$

The integrand in the RHS is exponentially small away from x_0, hence in the limit $\lambda \to +\infty$ the integral gets non-zero contributions only from an infinitesimal neighborhood of x_0, and we can replace $V(x) - V_0$ by its Taylor expansion around x_0

$$V(x) - V_0 = \frac{1}{2} a_{ij} (x - x_0)_i (x - x_0)_j + \text{higher order} \tag{6.500}$$

For large λ we can neglect the higher order terms, and compute the resulting Gaussian integral

$$\int e^{-\lambda V(x)} \, d^n x = d^{-\lambda V_0} (2\pi \lambda^{-1})^{n/2} (\det a)^{-1/2} \Big(1 + O(1/\lambda) \Big). \tag{6.501}$$

When there are more than one absolute minimum, we have to sum (or integrate) over the contributions from all the minima (also called *saddle points*).

The rules for the asymptotic evaluation of integrals of the form (6.498) then are:

SP1   find the critical points $(x_i^a) \in S$ of $V(x)$ where the gradient of the function $V(x)$ vanishes: $\partial_i V(x) = 0$ at $x = x^a$;

SP2   Compute the Hessian $h(x^a)_{ij} = \partial^2 V / \partial x_i \partial x_j |_{x=x^a}$ at the critical points. When the critical points form a manifold S of positive dimension, split

---

[29] For a more precise discussion see [3] §. 2.5.

(locally) the coordinates into coordinates y_m along S and normal coordinates x_i, with

$$h(y^a, x^a)_{ij} = \frac{\partial^2 V}{\partial x_i \partial x_j}\bigg|_{x=x^a, y=y^a} \tag{6.502}$$

the Hessian in the normal directions;

SP3   The asymptotic value of the integral is

$$\int e^{-\lambda V(x)}\, d^n x \approx (2\pi\lambda^{-1})^{n/2} \sum_a e^{-\lambda V(x_a)} (\det h(x^a))^{-1/2} \tag{6.503}$$

When $\dim S > 0$ the sum is replaced by an integral over the submanifold $S \subset \mathbb{R}^n$ with respect to the induced metric;

SP4   As $\lambda \to +\infty$ we need to keep only the contribution from the *dominating saddle-points* associated to absolute minima of $V(x_a)$.

Saddle-Point Methods for Path Integrals

We extend the above procedure from the finite dimensional integrals (6.498) to integrals over paths $\mathcal{T} \to \mathbb{R}^n$ where we first take $\mathcal{T}$ to be an interval $\mathcal{T} \equiv [0, T]$ and we fix the endpoints $x_0, x_T \in \mathbb{R}^n$ of the paths.

We wish to evaluate (imaginary time) path integrals of the typical form

$$\int_{(x_0,0)}^{(x_T,T)} [dx]\, e^{-\frac{1}{\hbar} S_E[x]}\, G[x] \tag{6.504}$$

in the asymptotic semi-classical limit $\hbar \to 0$ (i.e. $1/\hbar \to +\infty$). $G[x]$ is a functional ($\equiv$ operator insertion) with a smooth limit as $\hbar \to 0$.

The first step of the procedure requires to find the critical points of the functional $S_E[x]$ in the space of paths satisfying the appropriate boundary conditions

$$\frac{\delta S_E[x]}{\delta x(t)} = 0. \tag{6.505}$$

The saddle-point condition (6.505) is nothing else than the classical equations. Thus:

Fact 6.9 *As $\hbar \to 0$ the functional measure concentrates on a small functional neighborhood of the classical solutions.*

This **Fact** is the precise statement of the physical expectation that the limit $\hbar \to 0$ of Quantum Mechanics gives back Classical Mechanics. It clarifies how the macroscopic classical world emerges from the microscopic quantum reality: a system behaves classically whenever the effective support of the functional measure is along sharp trajectories.

Let $x(t)_{\mathrm{cl}}$ be a classical solution with minimal action and write

$$x(t) = x(t)_{\mathrm{cl}} + y(t) \qquad y(0) = y(t) = 0, \tag{6.506}$$

where $y(t)$ is the "quantum fluctuation" around the classical trajectory. The measure has support on nowhere smooth paths (Sect. 6.11), and this is irregularity arises from the tiny but non-smooth fluctuation $y(t)$. Then

$$S_E[x(t)_{\mathrm{cl}} + y(t)] = S_E[x(t)_{\mathrm{cl}}] +$$

$$+ \frac{1}{2} \int_0^T \mathrm{d}t \, y^i \, D[x_{\mathrm{cl}}]_{ij} \, y^j + \text{higher order in the } y\text{'s} \tag{6.507}$$

where the ordinary differential operator $D[x_{\mathrm{cl}}] = (D[x_{\mathrm{cl}}]_{ij})$ is the second variation operator around the classical solution x_{cl}, see [4] §. 4.1.

As $\hbar \to 0$ the quantum fluctuations $y(t)$ are of order $\hbar^{1/2}$ and we may neglect the higher order terms just as Lagrange did for the finite-dimensional integral (6.498). We remain with

$$\int_{(x_0,0)}^{(x_T,T)} [\mathrm{d}x] \, e^{-\frac{1}{\hbar} S_E[x]} \, G[x] \approx$$

$$\approx \sum_{x_{\mathrm{cl}}} e^{-\frac{1}{\hbar} S_E[x(t)_{\mathrm{cl}}]} \, G[x(t)_{\mathrm{cl}}] \int [\mathrm{d}y] e^{-\frac{1}{2\hbar} \int_0^T \mathrm{d}t \, y \, D[x_{\mathrm{cl}}] y} \tag{6.508}$$

$$\equiv \sum_{x_{\mathrm{cl}}} e^{-\frac{1}{\hbar} S_E[x(t)_{\mathrm{cl}}]} \, G[x(t)_{\mathrm{cl}}] \left(\mathrm{Det}[D[x_{\mathrm{cl}}]]\right)^{-1/2}$$

where the sum is over the classical solutions (if there are more than one with the given endpoints). When the different classical solutions belong to different homotopy classes we may insert topological factors in the sum, as we did in Sect. 6.4. In case we have a continuous family of classical solutions with the prescribed endpoints, the sum is replaced by an integral. The ratio of the exact path integral in the LHS and its leading asymptotic expression in the RHS is an asymptotic series of the form

$$1 + a_1\,\hbar + a_2\,\hbar^2 + \cdots \tag{6.509}$$

which is typically non-convergent.

Definition 6.2 The expression in the RHS of Eq. (6.508) is called the *semi-classical path integral*.

6.23 Spins *and All That*

We now consider the degrees of freedom which are classically "invisible". The simplest set-up is a two-state system with $\mathbb{Z}_2$-graded Hilbert space $\mathbb{C}^2$ called the *Fermi oscillator* which was studied in Sect. 4.12.2. As shown there, the Fermi oscillator is the *odd* (fermionic) analogue of the harmonic oscillator. For the benefit of the reader we recall some basic formula from Chap. 4.

The Hilbert space of the Fermi oscillator is bidimensional $\mathcal{H} \simeq \mathbb{C}^2$ and may be identified with the representation space $V_{1/2}$ of the spin-$\frac{1}{2}$ representation, cf. Sect. 4.3.1. The algebra of operators is spanned by the four operators 1 (identity), $b^\dagger$ (raising), b (lowering), and H (Hamiltonian) satisfying the supercommutator relations

$$\{b, b\} = 0, \qquad \{b^\dagger, b^\dagger\} = 0, \qquad \{b, b^\dagger\} = 1, \tag{6.510}$$

$$[H, b] = -\omega b, \qquad [H, b^\dagger] = \omega b^\dagger, \qquad H = \omega\big(b^\dagger b - \tfrac{1}{2}\big). \tag{6.511}$$

The ground state $|0\rangle$ of the Fermi oscillator is unique and satisfies

$$b|0\rangle = 0, \qquad E_0 = -\frac{\omega}{2} \tag{6.512}$$

and there is just one excited state

$$|1\rangle \equiv b^\dagger|0\rangle, \qquad b^\dagger|1\rangle = 0, \qquad E_1 = \frac{\omega}{2}. \tag{6.513}$$

6.23.1 *Grassmann Variables and Integrals*

If we reinsert $\hbar$ back in the expressions, the canonical commutation relation becomes $[a, a^\dagger] = \hbar$, reflecting the fact that in the classical limit the canonical variables *commute*. The corresponding statement for the b's is that they classically *anti*-commute. That is, the "classical" b's—which we denote as ξ to avoid all misunderstanding—are *odd elements* of a Grassmanian algebra which anticommute with everything *odd*. In particular they classically anticommute with themselves, and their "conjugates" i.e. they are *nilpotent:*

$$\lim_{\hbar \to 0} (b + b^\dagger)^2 = 0. \tag{6.514}$$

Hence in the classical limit all their physical effects disappear, and these degrees of freedom become "invisible". Following the QFT terminology, we shall call *fermionic* the degrees of freedom which classically anticommute with all odd operators.

We are led to the conclusion that, in order to describe the quantum dynamics of degrees of freedom like spin, we have to perform a functional integral over "classical fields" which are *odd* variables ξ^i in a *Grassmann algebra*, meaning that they satisfy the classical anticommutation relations

$$\xi^i \xi^j + \xi^j \xi^i = \bar{\xi}^i \bar{\xi}^j + \bar{\xi}^j \bar{\xi}^i = \bar{\xi}^i \xi^j + \xi^j \bar{\xi}^i = 0 \quad i, j = 1, \ldots, N, \tag{6.515}$$

where $\bar{\xi}^j$ is the odd classical variable which corresponds to $b^\dagger$. The Grassmann field $\bar{\xi}^j$ is an independent odd variable, although for notational convenience we write it as "it was" the complex conjugate of ξ^j: this makes the analogy with the harmonic oscillator more transparent.

Since a Grassmann odd variable ξ is nilpotent, $\xi^2 = 0$, all function of a single odd variable ξ is a polynomial of first order

$$f(\xi) = f_0 + \xi f_1, \tag{6.516}$$

where f_0, f_1 are constants of opposite Grassmann parity. The only consistent non-trivial definition of the integral of $f(\xi)$ is

$$\int d\xi \, f(\xi) = f_1, \tag{6.517}$$

see **BOX** on page 463. In particular the integral of a constant vanishes. Informally, the odd integral is the same as the derivative

$$\partial_\xi f(\xi) = f_1. \tag{6.518}$$

BOX: Integrals over an Odd Grassmann Variable

For the applications to Quantum Physics it is crucial that the integrals over an odd variable ξ still have the two basic properties of the ordinary integral $\int_{\mathbb{R}} dx\, f(x)$:

(i) to be linear in the function f;
(ii) to be translationally invariant $\int_{\mathbb{R}} dx\, f(x+a) = \int_{\mathbb{R}} dx\, f(x)$.

Thus we require

$$\int d\xi\, f(\xi + \eta) = \int d\xi\, f(\xi)$$

for all ξ-independent odd variable η. Then

$$\int d\xi\, f_0 - \eta \int d\xi\, f_1 + \int d\xi\, \xi\, f_1 = \int d\xi\, f_0 + \int d\xi\, \xi\, f_1,$$

showing that the integral of a constant vanishes. $\int d\xi\, \xi$ should be non-zero, otherwise the integral of *any* function will vanish, and the integral would be trivial. We may set $\int d\xi\, \xi = 1$ as a choice of normalization of the measure.

Let us now consider a Gaussian integral in finitely many odd variables

$$\int \exp\left(\sum_{i,j=1}^{N} \bar{\xi}^j A_{ij}\, \xi^j \right) \prod_{k=1}^{N} d\xi^k d\bar{\xi}^k \tag{6.519}$$

where A_{ij} is a numerical $N \times N$ matrix. Expanding out the exponential, we get a finite polynomial in the ξ^i, $\bar{\xi}^j$. By the rule (6.517), only the terms in the polynomial which contain precisely N ξ^i's and N $\bar{\xi}^j$'s contribute to the integral

$$\int e^{\bar{\xi} A \xi}\, d^N \xi\, d^N \bar{\xi} = \int \frac{1}{N!} \left(\sum_{i,j=1}^{N} \bar{\xi}^j A_{ij} \xi^j \right)^N \prod_{k=1}^{N} d\xi^k d\bar{\xi}^k \tag{6.520}$$

Each non-zero term in the expansion of the sum in RHS has the form $A_{i_1 j_1} \ldots A_{i_N j_N}$, where $\{i_1, \ldots, i_N\}$ and $\{j_1, \ldots, j_N\}$ are permutations of $\{1, \ldots, N\}$, times a sign produced by the permutation of the odd variables. The sum takes the form

$$\sum_{\pi,\sigma \in \mathfrak{S}_N} \frac{1}{N!}\, \text{sign}(\pi)\text{sign}(\sigma) A_{\pi(1)\sigma(1)} A_{\pi(2)\sigma(2)} \cdots A_{\pi(N)\sigma(N)} = \det A. \tag{6.521}$$

The same result may be obtained by changing the integration variables

$$\xi^i \to \eta_i \equiv A_{ij}\,\xi^j \tag{6.522}$$

without touching the independent variables $\bar\xi^i$. From the rule (6.517) (that is, from the fact that the integral over odd variables behaves as a derivative) we see that the Jacobian for a linear change of odd coordinates is the *inverse* of the usual one:

$$\prod_i \mathrm{d}(A_{ij}\xi^j)] = \frac{1}{\det A}\prod_j \mathrm{d}\xi^j \tag{6.523}$$

and hence

$$
\begin{aligned}
1 &= \int \exp\!\left[\sum_i \bar\xi^i \eta_i\right] \prod \mathrm{d}\bar\xi^i\,\mathrm{d}\eta_i = \\
&= \frac{1}{\det A}\int \exp\!\left[\sum_{i,j} \bar\xi^i A_{ij}\,\xi^j\right] \prod \mathrm{d}\bar\xi^i \mathrm{d}\xi^i
\end{aligned}
\tag{6.524}
$$

which yields back the result we got by expanding the exponential

$$\int \mathrm{e}^{\bar\xi A \xi}\, \mathrm{d}^N \xi\, \mathrm{d}^N \bar\xi = \det A. \tag{6.525}$$

6.23.2   *Path Integral for the Fermi Oscillator*

We now consider the path integral obtained from the harmonic oscillator functional integral in its complex form, Eq. (6.239), by replacing the commuting field $a(t), \bar a(t)$ with their anti-commuting cousins $\xi(t), \bar\xi(t)$. The formal arguments in the previous subsection suggest that this is the appropriate path integral to compute the quantum amplitudes of the *Fermi oscillator* of Sect. 4.12.2, since this is the "odd version" of the usual harmonic oscillator. To make the formulae look nicer, we introduce some notation. Let $f_i(t)$ be odd ($\equiv$ fermionic) Heisenberg operators. The time-order prescription for *odd* operators is

$$\mathbf{T}\big(f_{i_1}(t_1)\cdots f_{i_s}(t_s)\big) = (-1)^{|\pi|}\, f_{i_{\pi(1)}}(t_{\pi(1)})\cdots f_{i_{\pi(s)}}(t_{\pi(s)}) \tag{6.526}$$

where $\pi \in \mathfrak{S}_s$ is the permutation such that

$$t_{\pi(1)} > t_{\pi(2)} > \cdots > t_{\pi(s)} \tag{6.527}$$

and $(-1)^{|\pi|}$ is its sign. Now

Claim *We have the following identity between quantum amplitudes for the Fermi oscillator, as computed in the Hilbert space approach of Sect. 4.12.2, and the correlation functions computed by the Gaussian path integral over odd Grassmanian fields:*

$$\mathbf{T}\langle 0|\sigma_-(s_1)\cdots\sigma_-(s_n)\sigma_+(t_1)\cdots\sigma_+(t_m)|0\rangle =$$

$$= \frac{\int[d\xi\,d\bar\xi]\exp\left[-\int dt\,\bar\xi(d_t+\omega)\xi\right]\xi(s_1)\cdots\xi(s_n)\,\bar\xi(t_1)\cdots\bar\xi(t_m)}{\int[d\xi\,d\bar\xi]\exp\left[-\int dt\,\bar\xi(d_t+\omega)\xi\right]} \qquad (6.528)$$

where $\sigma_\pm(t) = e^{tH}\sigma_\pm e^{-tH}$ *are the Heisenberg spin operators of the Fermi oscillator, and* d_t *stands for the time derivative.*

Let us check that our **Claim** is correct. As for the usual Gaussian integrals, we may introduce in the action sources (which are now Grassmann odd background fields), and get a generating functional for the correlations of the $\xi(t)$, $\bar\xi(t')$ which is again the exponential of a quadratic form in the sources. Then the correlations with n operator insertions are determined by the quadratic form through the same combinatorics as before (the Wick theorem) except that now one should be careful with the signs arising from reordering the Fermi fields. Thus to check the **Claim** it suffices to check it for the correlation of one ξ with one $\bar\xi$ that is, in the operator notation,

$$\mathbf{T}\langle 0|\sigma_-(t)\sigma_+(0)|0\rangle. \qquad (6.529)$$

We first compute this amplitude using the Hilbert space formulation, and then compare the result with the corresponding path integral in the RHS of (6.528). One has

$$\mathbf{T}\langle 0|\sigma_-(t)\sigma_+(0)|0\rangle = \langle 0|e^{tH}\sigma_-e^{-tH}\sigma_+|0\rangle =$$

- $t > 0$

$$= (0\ 1)\begin{pmatrix} e^{t\omega/2} & 0 \\ 0 & e^{-t\omega/2} \end{pmatrix}\begin{pmatrix} 0 & 0 \\ 1 & 0 \end{pmatrix}\begin{pmatrix} e^{-t\omega/2} & 0 \\ 0 & e^{t\omega/2} \end{pmatrix}\begin{pmatrix} 0 \\ 1 \end{pmatrix} = e^{-t\omega} \qquad (6.530)$$

- $t < 0$ $\qquad \mathbf{T}\langle 0|\sigma_-(t)\sigma_+(0)|0\rangle = -\langle 0|\sigma_+e^{tH}\sigma_-e^{-tH}|0\rangle = 0. \qquad (6.531)$

On the other hand, by the general facts about Gaussian integrals, the two point function is the Green's function ($\equiv$ the kernel of the inverse of the operator in the quadratic action)

$$\langle\xi(t)\,\bar\xi(0)\rangle = \langle t|(d_t+\omega)^{-1}|0\rangle \equiv G(t). \qquad (6.532)$$

To check the **Claim** we have to prove the equality

$$G(t) = \begin{cases} e^{-\omega t} & t > 0 \\ 0 & t < 0. \end{cases} \tag{6.533}$$

Indeed Eq. (6.533) is the only everywhere bounded solution to the Green's equation

$$\left(\frac{\mathrm{d}}{\mathrm{d}t} + \omega\right) G(t) = \delta(t), \tag{6.534}$$

and the **Claim** follows.

6.23.3 Traces and Fermi Partition Functions

Replacing ordinary paths by paths valued in Grassmann odd variables has also another consequence. We **Claim** that to represent a trace in the ($\mathbb{Z}_2$ graded) Hilbert space of fermionic degrees of freedom, we need to integrate over Grassmanian fields $\xi(t)^i$, $\bar{\xi}(t)^i$ which are *anti*-periodic instead of periodic in imaginary time. The difference with respect to the bosonic case is that we get an extra sign when rearranging the order of operators along the imaginary-time circle S_β^1: indeed suppose we insert two fields $\xi_1(\epsilon)\,\xi_2(0)$ on the time circle, and then transport the field ξ_1 to the point $\beta - \epsilon$. For a periodic field we would end up to

$$\xi_1(-\epsilon)\xi_2(0) = -\mathbf{T}(\xi_1(-\epsilon)\,\xi_2(0)) \tag{6.535}$$

To get a covariance $\mathbf{T}\langle \xi_1(t)\,\xi_2(0)\rangle_\beta$ which is univalued on the circle, we should integrate over *anti-periodic* Grassmann-valued paths.

Therefore the partition function of the Fermi oscillator at inverse temperature β is given by the functional determinant

$$\mathrm{Det}[\mathrm{d}_t + \omega]_{\mathrm{anti}} \tag{6.536}$$

now computed over the *anti-periodic functions* of imaginary time. The determinant of first order differential operators with generalized periodic conditions was computed in Eqs. (6.365) and (6.367). To get the one with anti-periodic conditions we need to set $\xi = i\pi/2$ in Eq. (6.367):

$$\mathrm{Det}\left[\frac{\mathrm{d}}{\mathrm{d}t} + B(t)\right]_{\mathrm{anti}} = 2\sinh\left(\frac{\pi i}{2} + \frac{1}{2}\int_0^\beta B(s)\,\mathrm{d}s\right)$$
$$= 2\cosh\left(\frac{1}{2}\int_0^\beta B(s)\,\mathrm{d}s\right). \tag{6.537}$$

For the Fermi oscillator this gives

$$\mathrm{Tr}(\mathrm{e}^{-\beta H}) = 2\cosh(\beta\omega/2) = \mathrm{e}^{\beta\omega/2} + \mathrm{e}^{-\beta\omega/2} \tag{6.538}$$

which is the correct trace for a two-state system with energy levels $E_1 = \frac{1}{2}\omega$ and $E_0 = -\frac{1}{2}\omega$. From this example we see that the anti-periodic prescription is needed to produce a sum of contributions $\mathrm{e}^{-\beta E_n}$ with *positive* integral coefficients, as required for the physical partition function

More generally Fact 6.3 has a fermionic counterpart:

Fact 6.10 (Fermi Traces with Chemical Potentials) *The fermionic Grand Canonical partition function with imaginary chemical potentials* $\mu_a = \mathrm{i}s_a/\beta$

$$\mathrm{Tr}[\mathrm{e}^{\mathrm{i}s_a J_a}\,\mathrm{e}^{-\beta H}], \tag{6.539}$$

where J_a are conserved charges of the quantum system which generate a symmetry acting on the fermionic degrees of freedom as

$$\begin{aligned}
b_i &\rightsquigarrow \mathrm{e}^{\mathrm{i}s_a J_a} b_i \mathrm{e}^{-\mathrm{i}s_a J_a} \equiv R(s_a)_{ij} b_j \\
b_i^\dagger &\rightsquigarrow \mathrm{e}^{\mathrm{i}s_a J_a} b_i^\dagger \mathrm{e}^{-\mathrm{i}s_a J_a} \equiv R(s_a)_{ij}^* b_j^\dagger, \qquad (R_{ij}) \in U(N),
\end{aligned} \tag{6.540}$$

is given by a path integral over Grassmanian fields ξ_i, $\bar{\xi}_i$ ($i = 1, \dots, N$) which satisfy the generalized *anti-periodic condition in imaginary time*

$$\xi(t+\beta)_i = -R(s_a)_{ij}\xi(t)_j, \qquad \bar{\xi}(t+\beta)_i = -R(s_a)_{ij}^*\bar{\xi}(t)_j, \tag{6.541}$$

that is, over paths which are anti-periodic in imaginary time up to the action of the unitary symmetry $\exp(\mathrm{i}s_a J_a)$.

This functional integral approach to the spin degrees of freedom may look not very interesting per se (alas, it is the linear algebra of 2×2 matrices!) but it is the starting point of the second quantization for fermions ([3] chap. 3) and is a foundational element of Field Theory.

6.24 Supersymmetry and Path Integrals

The path integrals of the supersymmetric quantum systems have spectacular properties [14, 15], related to the invariance of the Witten index (cf. Sect. 4.14), which allow to compute them explicitly in many cases. All our tricks to compute non-Gaussian path integrals, such as the stochastic representation and the recursion method, consist in reinterpreting the path integral to be computed in terms of an auxiliary supersymmetric system, and then using the properties of the supersymmetric functional measure. The recursion relations we used in Sect. 6.21 to solve several problems may be interpreted as supersymmetric Ward identities for the auxiliary system. For brevity we focus on systems whose even ($\equiv$ bosonic) degrees of freedom take value in $\mathbb{R}^n$. The results are valid (with the obvious modifications) for general supersymmetric quantum models.

Let $W(x_i)$ be a superpotential depending on the bosonic variables $x_1, \ldots, x_n$ which are coordinates in $\mathbb{R}^n$. We write $\xi_i, \bar{\xi}_i$ ($i = 1, \ldots, n$) for the odd Grassmann fields corresponding to the fermionic annihilator/creator operators b_i, $b_i^\dagger$ which satisfy the Clifford algebra

$$b_i b_j + b_j b_i = b_i^\dagger b_j^\dagger + b_j^\dagger b_i^\dagger = 0, \quad b_i^\dagger b_j + b_j b_i^\dagger = \delta_{ij}. \tag{6.542}$$

The Euclidean ($\equiv$ imaginary time) Lagrangian of the supersymmetric system with superpotential $W(x)$ is

$$L_E = \frac{1}{2} \sum_i \left(\frac{\mathrm{d}x^i}{\mathrm{d}t} \right)^2 + \frac{1}{2} \sum_i \left(\frac{\partial W}{\partial x^i} \right)^2 + \sum_i \bar{\xi}_i \frac{\mathrm{d}}{\mathrm{d}t} \xi_i + \frac{\partial^2 W}{\partial x^i \partial x^j} \bar{\xi}_i \xi_j. \tag{6.543}$$

The Hamiltonian operator for this quantum model is then

$$H = -\frac{1}{2} \frac{\partial^2}{\partial x_i \, \partial x_i} + \frac{1}{2} \sum_i \left(\frac{\partial W}{\partial x^i} \right)^2 + \frac{1}{2} \frac{\partial^2 W}{\partial x^i \partial x^j} (b_i^\dagger b_j - b_j b_i^\dagger), \tag{6.544}$$

acting (densely) on $\mathbb{C}^{2^n} \otimes L^2(\mathbb{R}^n)$. The imaginary-time Lagrangian L_E in Eq. (6.543) is invariant, up to surface terms, under two supersymmetries with Grassmann odd constant parameters $\bar{\epsilon}$ and ϵ, respectively. The first one is

$$\delta x^i = \bar{\epsilon} \xi^i \qquad\qquad \delta \xi^i = 0$$
$$\delta \bar{\xi}^i = \bar{\epsilon} \left(-\dot{x}^i - \frac{\partial W}{\partial x^i} \right) \qquad \delta L_E = -\frac{\mathrm{d}}{\mathrm{d}t} \left(\bar{\epsilon} \frac{\partial W}{\partial x^i} \xi^i \right). \tag{6.545}$$

The second one is

$$\delta x^i = \bar{\xi}^i \epsilon \qquad\qquad \delta \bar{\xi}^i = 0$$
$$\delta \xi^i = \left(\dot{x}^i - \frac{\partial W}{\partial x^i} \right) \epsilon \qquad \delta L_E = \frac{\mathrm{d}}{\mathrm{d}t} \left(\bar{\xi}^i \dot{x}^i \epsilon \right). \tag{6.546}$$

These supersymmetries are generated by two complex supercharges Q and $\bar{Q}$:

$$[\bar{\epsilon} Q, \mathcal{O}] = \bar{\epsilon} [Q, \mathcal{O}\} = \mathrm{i} \delta_{\bar{\epsilon}} \mathcal{O}, \tag{6.547}$$

$$[\bar{Q} \epsilon, \mathcal{O}] = [\bar{Q}, \mathcal{O}\} \epsilon = \mathrm{i} \delta_\epsilon \mathcal{O} \tag{6.548}$$

where $\mathcal{O}$ is any operator (even or odd), and $[\cdot, \cdot\}$ is the *supercommutator* introduced in Sect. 4.12.1.

From these supersymmetries we can deduce Ward identities for the quantum amplitudes following the methods of Sect. 6.12 (with some extra care for the signs arising from the $\mathbb{Z}_2$ grading). In particular, whenever supersymmetry is *unbroken*— i.e. the ground state is annihilated by the supercharges, $Q^a|0\rangle = 0$— for all operator $\mathcal{O}$ we have the identities

$$\langle \delta_\epsilon \mathcal{O}\rangle = \langle \delta_{\bar\epsilon}\, \mathcal{O}\rangle = 0, \tag{6.549}$$

where $\delta_\epsilon \mathcal{O}$, $\delta_{\bar\epsilon}\mathcal{O}$ are the variations of $\mathcal{O}$ with respect to the above supersymmetries, as defined in Eqs. (6.547) and (6.548).

Finite Temperature vs. Witten Index

Consider the partition function at finite temperature

$$\mathrm{Tr}[e^{-\beta H}], \tag{6.550}$$

according to the rules developed in the previous sections, this Hilbert-space trace is given by a functional integral

$$\int_{\mathfrak{P}} [\mathrm{d}x] \int_{\mathfrak{A}} [\mathrm{d}\xi\, \mathrm{d}\bar\xi]\, e^{-\int_0^\beta L_E[x,\xi,\bar\xi]\, \mathrm{d}t} \tag{6.551}$$

over *periodic even fields* x_i and **anti**-*periodic odd fields* ξ_i, $\bar\xi_i$. Since the odd parameters ϵ, $\bar\epsilon$ in Eqs. (6.545) and (6.546) are time-independent, the two supersymmetries map periodic even fields into periodic odd ones, and hence the boundary conditions which produce the partition function (6.550) *break explicitly* supersymmetry at any non-zero temperature. This is to be expected, since—by definition—even and odd degrees of freedom have different statistics, hence quite different thermodynamics (see [5] chap. 3), and no symmetry of any sort may connect them at generic temperature.

To restore supersymmetry we should modify the boundary condition of the path integral in such a way that even and odd fields are either *both periodic* or both anti-periodic. The second choice is allowed only when the superpotential is even and it is a bit *ad hoc*, while the first one is intrinsic and canonical.

Fact 6.10 states that the path integral over *periodic* odd fields ξ_i, $\bar\xi_i$ computes the trace over the Hilbert space with the additional insertion of a conserved unitary operator U such that

$$U b_i U^{-1} = -b_i, \qquad U b_i^\dagger U^{-1} = -b_i^\dagger, \tag{6.552}$$

and, in general

$$U \mathcal{O}_{\mathrm{even}} U^{-1} = \mathcal{O}_{\mathrm{even}}, \qquad U \mathcal{O}_{\mathrm{odd}} U^{-1} = -\mathcal{O}_{\mathrm{odd}} \tag{6.553}$$

for all even (resp. odd) operators $\mathcal{O}_{\text{even}}$ (resp. $\mathcal{O}_{\text{odd}}$). In any system with fermionic degrees of freedom there is a canonical universal unitary operator with the properties (6.553), namely the $\mathbb{Z}_2$-*grading operator*

$$(-1)^F \equiv \exp\left[i\pi \sum_k b_k^\dagger b_k\right], \tag{6.554}$$

which counts fermions mod 2 and is $+1$ (resp. -1) on even (resp. odd) states.[30]

To simplify the discussion of the supersymmetric path integrals, we assume that the superpotential $W(x)$ in (6.543) is such that the spectrum of H is discrete (so that the traces over the Hilbert space converge). Then the periodic path integral is

$$\int_{\mathfrak{P}} [\mathrm{d}x\, \mathrm{d}\xi\, \mathrm{d}\bar{\xi}]\, e^{-S_E[x,\xi,\bar{\xi}]} = \mathrm{Tr}[(-1)^F e^{-\beta H}] =$$

$$= \sum_{\substack{n:\, \mathbf{even} \\ \text{state}}} e^{-\beta E_n} - \sum_{\substack{m:\, \mathbf{odd} \\ \text{state}}} e^{-\beta E_m} \tag{6.555}$$

We know from the representation theory of supersymmetry, Sect. 4.14, that all energy levels E_n—but the zero-energy ground states—are paired between even and odd states. Therefore all excited states cancel in pairs in the RHS of (6.555) and only the contribution from the zero-energy ($\equiv$ supersymmetric) ground states survives

$$\int_{\mathfrak{P}} [\mathrm{d}x\, \mathrm{d}\xi\, \mathrm{d}\bar{\xi}]\, e^{-S_E[x,\xi,\bar{\xi}]} = n_{\text{ev}}^0 - n_{\text{od}}^0 \equiv \Delta \in \mathbb{Z} \text{ the } \textbf{Witten index}, \tag{6.556}$$

where n_{ev}^0 (resp. n_{od}^0) is the number of *even* (resp. *odd*) ground states. We conclude:

Supersymmetric Traces and Witten Index

The path integral over *periodic* even an odd fields is the trace over the Hilbert space with the insertion of the $\mathbb{Z}_2$-grading operator $(-1)^F$ see Eq. (6.555). When the quantum system is supersymmetric and the spectrum of H is discrete (or, more generally, the ground states are *gapped*) $\mathrm{Tr}[(-1)^F e^{-\beta H}]$ is the Witten index $\Delta \in \mathbb{Z}$.

Remark 6.9 The result we got from physical considerations is very natural from the point of view of the "super-mathematics" that was briefly outlined in Sect. 4.12.1. The supersymmetric trace (6.555) is just the *supertrace* of the

---

[30] In all physically realizable systems this operator is identified with a 2π rotation (via the Spin and Statistic Theorem, cf. Sect. 4.12): in the present discussion we allow for more general quantum systems which however admits a grading operator $(-1)^F$ which satisfies (6.553).

imaginary-time evolution operator in the Hilbert *superspace*

$$\mathrm{Tr}[(-1)^F e^{-\beta H}] \equiv \mathrm{STr}_{\mathcal{H}}[e^{-\beta H}].$$
(6.557)

Its main property is that

$$i\,\mathrm{Tr}[(-1)^F e^{-\beta H} \delta_\epsilon \mathcal{O}_{\mathrm{odd}}] \equiv \mathrm{Tr}[(-1)^F e^{-\beta H} \{\bar{Q}, \mathcal{O}_{\mathrm{odd}}\}]\epsilon = 0,$$
(6.558)

which just expresses the fact that the path integral with periodic boundary condition is SUSY invariant.

We stress that, when the periodic path integral is convergent (i.e. a well defined function of β), it is automatically independent of β since

$$-2\frac{\partial}{\partial\beta}\mathrm{Tr}[(-1)^F e^{-\beta H}] =$$

$$= 2\mathrm{Tr}[(-1)^F e^{-\beta H} H] = \mathrm{Tr}[(-1)^F e^{-\beta H}\{Q, Q^\dagger\}]$$

$$= \mathrm{Tr}[(-1)^F e^{-\beta H} QQ^\dagger] + \mathrm{Tr}[(-1)^F e^{-\beta H} Q^\dagger Q] =$$

$$= \mathrm{Tr}[(-1)^F e^{-\beta H} QQ^\dagger] + \mathrm{Tr}[Q(-1)^F e^{-\beta H} Q^\dagger] =$$

$$= \mathrm{Tr}[(-1)^F e^{-\beta H} QQ^\dagger] - \mathrm{Tr}[(-1)^F Q\, e^{-\beta H} Q^\dagger] =$$

$$= \mathrm{Tr}[(-1)^F e^{-\beta H} QQ^\dagger] + \mathrm{Tr}[(-1)^F e^{-\beta H} QQ^\dagger] = 0$$
(6.559)

A similar argument shows that the periodic path integral is independent of all couplings λ_a entering in the Hamiltonian H as long as these couplings are consistent with supersymmetry (and the functional trace is defined for all λ_a), cf. Problem 6.3.

6.24.1 *Stochastic Interpretation*

Now we compute the path integral of the supersymmetric quantum system with Lagrangian (6.543). We consider first the case $n = 1$ with just one even and one odd degree of freedom. We take the even field $x(t)$ to be periodic of period β, while for the odd ones we use the generalized boundary condition

$$\mathfrak{P}(\alpha): \quad \xi(t + \beta) = e^{i\alpha}\,\xi(t), \qquad \bar{\xi}(t + \beta) = e^{-i\alpha}\,\bar{\xi}(t)$$
(6.560)

with $0 \le \alpha < 2\pi$. For $\alpha = \pi$ the path integral gives back the trace on the Hilbert space of the imaginary-time evolution operator $e^{-\beta H}$, while for general α we get

$$\int_{\mathfrak{P}(\alpha)} [dx\, d\xi\, d\bar{\xi}]\, e^{-S_E[x,\xi,\bar{\xi}]} = \mathrm{Tr}\!\left[e^{i(\alpha-\pi)F}\, e^{-\beta H}\right].$$
(6.561)

In particular for $\alpha = 0$ (periodic boundary condition) we get back the Witten index Δ. Integrating over the fermions, using the determinant formulae of Sect. 6.19, we get

$$\int_{\mathfrak{P}(\alpha)} [dx\, d\xi\, d\bar{\xi}]\, e^{-S_E[x,\xi,\bar{\xi}]} =$$

$$= \int_{\mathfrak{P}} [dx]\, e^{-\frac{1}{2} \int_0^\beta \left(\dot{x}^2 + (W')^2 \right) dt}\; \mathrm{Det}_{\beta,\alpha} \left(\frac{d}{dt} + W''(x(t)) \right) = \qquad (6.562)$$

$$= \int_{\mathfrak{P}} [dx]\, e^{-\frac{1}{2} \int_0^\beta \left(\dot{x}^2 + (W')^2 \right) dt}\; 2 \sinh \left(\frac{i\alpha}{2} + \frac{1}{2} \int_0^\beta W'' dt \right).$$

We now perform the functional change of variable

$$x \mapsto h = \dot{x} + W' \qquad (6.563)$$

to write the path integral in the stochastic form

$$\int_{\mathfrak{P}(\alpha)} [dx\, d\xi\, d\bar{\xi}]\, e^{-S_E[x,\xi,\bar{\xi}]} = \int_{\mathfrak{Z}} [dh]\, e^{-\frac{1}{2} \int_0^\beta h^2 dt}\; \frac{\sinh \left(\frac{i\alpha}{2} + \frac{1}{2} \int_0^\beta W'' dt \right)}{\left| \sinh \left(\frac{1}{2} \int_0^\beta W'' dt \right) \right|}$$

$$\qquad (6.564)$$

For $\alpha = 0$ this gives back the Witten index. When the fermionic determinant has a definite sign, say plus, the Witten index is the number of times the functional covering map $x \mapsto h$ covers the space of periodic functions, that is, the number of *periodic* solutions to the Langevin ODE (6.563) for a sufficiently generic periodic function $h(t)$. In the general case Δ is the *signed* number of the coverings, where covers which invert the orientation of the functional space count negative.[31] Hence we may interpret the Witten index as the "topological degree" of the functional "covering map" $x \mapsto h$ of the space of periodic function of period β.

The field $h(t)$ is a Gaussian "white noise" with two-point function (variance)

$$\langle h(t)\, h(s) \rangle = \delta(t - s), \qquad (6.565)$$

and Eq. (6.563) is the *Langevin stochastic differential equation* with drift force W' in presence of a stochastic (white) noise $h(t)$. We defined the "white noise" in

---

[31] An elegant example is worked out explicitly in Sect. 8.6.1 where $W'(x) = x^2$. In that case each periodic $h(t)$ with a positive mean is the image of precisely two periodic functions $x(t)$ with Jacobian determinants equal and opposite.

Eq. (6.436): its *connected* correlation functions are

$$\langle h(t) \rangle_{\text{conn}} = 0 \tag{6.566}$$

$$\langle h(t_1) h(t_2) \rangle_{\text{conn}} = \delta(t_1 - t_2) \tag{6.567}$$

$$\langle h(t_1) h(t_2) \cdots h(t_s) \rangle_{\text{conn}} = 0 \quad \text{for } s \geq 3. \tag{6.568}$$

For more details about the stochastic equations viewpoint on supersymmetry and it relation with Fokker-Planck diffusive processes see chap. 6 of [3]. There the relation between the path integrals and the Schrödinger equation is explained in the language of stochastic differential equations and diffusion processes in view of the quantum diffusion-fluctuation theorem.

Example 6.23 Suppose $W''(x) > 0$ for all $x \in \mathbb{R}$, i.e. the superpotential is *convex*. A typical example is the even superpotential

$$W(x) = \sum_{k=0}^{n} \lambda_k x^{2(n-k)} \quad \text{with} \quad \lambda_k > 0. \tag{6.569}$$

Then the periodic functional determinant $\text{Det}[d_t + W'']$ is everywhere positive in path space. Formally this means that the functional cover map $x(t) \mapsto h(s)$ is "unbranched", so that the differential equation

$$\dot{x} + W' = h \tag{6.570}$$

has one periodic solution for generic periodic h. Then this system has precisely one supersymmetric ground state which is even (because the functional measure is positive after integrating away the fermions). Let us check this conclusion by a direct computation in the Schrödinger representation. The Hilbert space is $\mathbb{C}^2 \otimes L^2(\mathbb{R})$ and the Hamiltonian is the 2×2 matrix of differential operators

$$H = -\frac{1}{2}\frac{d^2}{dx^2} + \frac{1}{2}(W')^2 - \frac{1}{2}W''\sigma_3 \tag{6.571}$$

where $\sigma_3 \equiv (-1)^F$. The Hamiltonian commutes with σ_3, i.e. the energy eigenstates have a definite Fermi parity (as required by the superselection rule). Since $W'' > 0$ for all x, the ground states should be even i.e. they satisfy $\sigma_3|\text{ground}\rangle = |\text{ground}\rangle$. A supersymmetric ground state has zero energy, so its wave-function satisfies the ODE

$$0 = -\frac{d^2\psi}{dx^2} + (W')^2\psi - W''\psi = 0 = -\left(\frac{d}{dx} - W'\right)\left(\frac{d}{dx} + W'\right)\psi. \tag{6.572}$$

A solution in $L^2(\mathbb{R})$ is $\psi(x) = \exp(-W(x))$ which is a bound state. This is the only ground state since we know that in the one-dimensional Schrödinger equation bound states are never degenerate (cf. Chap. 3).

Problems

6.1 Generalize the computation of the path integral in Sect. 6.16.2 by adding a background magnetic potential $A(x)$ along the circle.

6.2 Perform the periodic path integral for the radial motion of a particle in $\mathbb{R}^d$ for general d.

6.3 Show that in a supersymmetric model of the form (6.543) the periodic path integral is independent of the couplings in the Lagrangian as long as they preserve supersymmetry.

References

1. R. Schoen, S.-T. Yau, *Lectures on Differential Geometry* (International Press of Boston, Boston, 2010)
2. J. Glimm, A. Jaffe, *Quantum Physics. A Functional Integral Point of View* (Springer, New York, 1981)
3. S. Cecotti, *Statistical Mechanics. A Concise Advanced Textbook* (Springer, Berlin, 2024)
4. S. Cecotti, *Analytic Mechanics. A Concise Textbook* (Springer, Cham, 2024)
5. R.P. Feynman, The space-time approach to non-relativistic quantum mechanics. Rev. Mod. Phys. **20**, 367–387 (1948)
6. W. Fulton, *Algebraic Topology: A First Course* (Springer, New York, 1995)
7. F.W. Olver, D.M. Lozier, R.F. Boisvert, C.W. Clark, W. Charles (eds.), *NIST Handbook of Mathematical Functions* (Cambridge University Press, Cambridge, 2010); available on-line at https://dlmf.nist.gov
8. F. Gantmacher, *The Theory of Matrices*, vol. 2 (AMS, Providence, 2000)
9. I.S. Gradshteyn, I.M. Ryzhik, *Table of Integrals, Series, and Products*, 7th edn. (Elsevier, Amsterdam, 2007)
10. E.T. Whittaker, G.N. Watson, *A Course of Modern Analysis*, 4th edn. (Cambridge University Press, Cambridge, 1927)
11. E.L. Ince, *Ordinary Differential Equations* (Dover, New York, 1956)
12. S. Cecotti, A trick for the computation of some non-gaussian path integrals. Lett. Nuovo Cim. **42**, 63–66 (1985)
13. H. Kleinert, *Path Integrals in Quantum Mechanics, Statistics, Polymer Physics, and Financial Markets*, 5th edn. (World Scientific, Singapore, 2009)
14. S. Cecotti, L. Girardello, Functional measure, topology and dynamical supersymmetry breaking. Phys. Lett. B **110**, 39 (1982)
15. S. Cecotti, L. Girardello, Stochastic and parastochastic aspects if supersymmetric functional measures: anew nonperturbative approach to supersymmetry. Ann. Phys. **145**, 81–99 (1983)
16. E. Witten, A new look at the path integral of Quantum Mechanics, arXiv:1009.6032 [hep-th]
17. J.D. Murray, *Asymptotic Analysis* (Springer, New York, 1984)

Chapter 7
Open Systems and Quantum Entanglement

*Entanglement is not **one** but rather **the** characteristic trait of quantum mechanics*

E. Schrödinger [1]

In the previous chapters we have seen several aspects of Quantum Mechanics which strikingly deviate from the classical paradigm: quantum interference, the Bohm-Aharonov effect, the tunneling effect, etc. However the most dramatic contrast between classical and quantum physics is given by the phenomenon of *entanglement* between different quantum systems, in particular quantum systems which are well separated in space and do not interact with each other in any usual sense. Closely related to quantum entanglement are the theories of Quantum Information and Quantum Computing which recently have attracted a lot of attention both for their potential application to technological breakthroughs and for their relevance to Quantum Gravity. In this chapter we give a general overview of quantum entanglement and related quantum effects, together with some hints to the theory of Quantum Information.

General references for this chapter are, among many others, [2–9]. We recommend in particular the Lecture Notes by Preskill [10]. The Quantum Mechanics textbooks [11, 12] also contain introductory treatments of entanglement and Quantum Information.

We start with a discussion of *open quantum systems*.

7.1 States of Open Systems and Density Matrices

The mathematical framework of Quantum Physics presented in Chap. 2 was formulated for *closed* quantum systems whose states we described as elements of $\mathbb{P}(\mathcal{H})$, where $\mathcal{H}$ is a complex Hilbert space. In Sect. 2.10 we briefly touched upon the situation in which our system is composed by two subsystems A and B, in

© The Author(s), under exclusive license to Springer Nature Switzerland AG 2025
S. Cecotti, *Quantum Mechanics*, UNITEXT for Physics,
https://doi.org/10.1007/978-3-031-98824-0_7

which case the Hilbert space is the *tensor product* of the Hilbert spaces of the two subsystems[1]

$$\mathcal{H}_{AB} = \mathcal{H}_A \widehat{\otimes} \mathcal{H}_B. \tag{7.1}$$

A vector in $\mathcal{H}_{AB}$ has the general form

$$|\psi\rangle_{AB} = \sum_{i,j} C_{ij} |\phi_i\rangle_A \otimes |\chi_j\rangle_B. \tag{7.2}$$

where $\{|\psi_i\rangle\}$ (resp. $\{|\chi_j\rangle\}$) is an orthonormal basis of $\mathcal{H}_A$ (resp. $\mathcal{H}_B$) and C_{ij} is a matrix of complex coefficients. States of the special form

$$|\phi\rangle_A \otimes |\chi\rangle_B \tag{7.3}$$

are called *product states* (or *separable states*). The typical state in $\mathcal{H}_{AB}$ cannot be set in the product form (7.3).

In Sect. 2.10 we considered the situation where we can perform experiments (measures) only on the subsystem A, while the subsystem B is inaccessible to us. This is indeed the typical situation in the real world: all actual physical systems are in contact with other systems—called collectively their *environment E*. We carry out our experiments only on a partial subsystem A, neglecting the fact that it is part of a *larger quantum system* which includes our experimental apparatus, we the experimentalists, our laboratory, the campus of the university where our laboratory is located, our city, the Country, the planet Earth, etc. etc. We measure only observables of the subsystem A which, as operators acting on the full Hilbert space $\mathcal{H}_{AE} = \mathcal{H}_A \otimes \mathcal{H}_E$, have the form

$$\mathcal{O}_A \otimes \mathbf{1}_E, \tag{7.4}$$

where $\mathcal{O}_A$ stands for the self-adjoint operator $\mathcal{O}$ acting on $\mathcal{H}_A$ and

$$\mathbf{1}_E = \sum_i |i\rangle_E \, {}_E\langle i| \tag{7.5}$$

is the identity operator acting on the environment Hilbert space $\mathcal{H}_E$. The expectation value of the subsystem-A observable $\mathcal{O}_A$ in the quantum state

$$|\psi\rangle_{AE} = \sum_{i,j} C_{ij} |\phi_i\rangle \otimes |\chi_j\rangle \in \mathcal{H}_A \otimes \mathcal{H}_E \tag{7.6}$$

---

[1] In the sequel we often omit the hat over $\otimes$. We adopt the **standing convention** that all spaces are meant to be completed in the norm topology without mentioning this condition all the time.

is

$$_{AE}\langle\psi|(\mathcal{O}_A \otimes \mathbf{1}_E)|\psi\rangle_{AE} = \sum_{ij}(CC^\dagger)_{ij}\,\langle\phi_j|\mathcal{O}_A|\phi_i\rangle = \mathrm{Tr}_A(\mathcal{O}_A\,\varrho), \tag{7.7}$$

where

$$\varrho = \mathrm{Tr}_{\mathcal{H}_E}\big(|\psi\rangle_{AE}\,_{AE}\langle\psi|\big) =$$
$$= \sum_{ij,kl} C_{ij}C_{kl}^* \,\mathrm{Tr}_{\mathcal{H}_E}\big(|\phi_i\rangle \otimes |\chi_j\rangle\langle\chi_l| \otimes \langle\phi_k|\big) = \sum_{ij}|\phi_j\rangle(CC^\dagger)_{ij}\langle\phi_i| \tag{7.8}$$

is the *density matrix* produced by the partial trace over the "environment" degrees of freedom in $\mathcal{H}_E$, i.e. on the degrees of freedom to which we have no experimental access (cf. Sect. 2.10).

Open quantum systems differ from closed ones in three crucial aspects:

Op1 States are not elements of $\mathbb{P}(\mathcal{H})$ but density matrices $\varrho \in \mathcal{B}(\mathcal{H})$;
Op2 The time-evolution is not given by a unitary transformation;
Op3 Measurements are not given by orthogonal projections.

Let us start by considering quantum states in *open systems*.

7.1.1 States of an Open System

In an open system, we are forced to work with density matrices ϱ. In this context they are called *mixed states* or just *states,* reserving the term *pure state* for the special ones which are represented by a vector $|\phi\rangle \in \mathcal{H}$. For the purpose of this chapter the term *state* is defined as follows:

Definition 7.1 A *state* $\varrho \in \mathcal{B}(\mathcal{H})$ is an operator acting on the Hilbert space $\mathcal{H}$ of the open quantum system such that:

St1 the operator $\varrho \in \mathcal{B}(\mathcal{H})$ is *self-adjoint* $\varrho^\dagger = \varrho$;
St2 ϱ is *non-negative*, $\varrho \geq 0$, i.e. its spectrum $\sigma(\varrho) \subseteq \mathbb{R}_{\geq 0}$;
St3 ϱ is trace-class and *normalized* to 1

$$\mathrm{Tr}_{\mathcal{H}}\varrho = 1. \tag{7.9}$$

Then all eigenvalues λ of ϱ satisfy the bounds

$$0 \leq \lambda \leq 1, \tag{7.10}$$

that is, $\sigma(\varrho) \subset [0, 1]$. We write $\mathcal{S} \subset \mathcal{B}(\mathcal{H})$ for the *space of states* i.e. the space[2] of operators satisfying **St1**, **St2**, and **St3**.

Definition 7.2 A state ϱ is called *pure* iff it is an *orthogonal projector* (necessarily of rank 1), i.e. iff in addition to **St1**, **St2**, **St3** it satisfies the algebraic equation

$$\varrho^2 = \varrho, \tag{7.11}$$

in which case its eigenvalues are all 0 or 1, and 1 must have multiplicity precisely 1 by **St3**. Then $\varrho = |\psi\rangle\langle\psi|$ where $|\psi\rangle \in \mathcal{H}$ is a normalized eigenvector of ϱ associated to the eigenvalue 1.

The space of pure states in isomorphic to $\mathbb{P}(\mathcal{H})$ through the map that associates to a rank-1 projector $|\psi\rangle\langle\psi|$ the dimension-1 subspace $\mathbb{C}|\psi\rangle \subset \mathcal{H}$.

From **St1**, **St2**, and **St3** we get:

Fact 7.1 *The space* $\mathcal{S} \subset \mathcal{B}(\mathcal{H})$ *of states is* convex *i.e. if* ϱ_1, ϱ_2 *are states so is*

$$t\varrho_1 + (1 - t)\varrho_2 \quad \textit{for all } 0 \leq t \leq 1. \tag{7.12}$$

The points of a convex set C which cannot be written as non-trivial convex combinations of two distinct elements of C are called *extremal points* (of C).

Lemma 7.1 *The extremal points of* $\mathcal{S} \subset \mathcal{B}(\mathcal{H})$ *are the pure states.*

Proof All state ϱ may be diagonalized in some orthonormal basis $\{|e_i\rangle\}$

$$\varrho = \sum_i \lambda_i |e_i\rangle\langle e_i|, \qquad 0 \leq \lambda_i \leq 1, \qquad \sum_i \lambda_i = 1, \tag{7.13}$$

so any ϱ is a convex combination of pure states $|e_i\rangle\langle e_i|$. It remains to show that a pure state is automatically extremal. Assume that

$$|\psi\rangle\langle\psi| = t\varrho_1 + (1 - t)\varrho_2 \tag{7.14}$$

with $0 < t < 1$ and $\varrho_1 \neq \varrho_2$ both non-negative. The two operators

$$\hat{\varrho}_a \equiv (1 - |\psi\rangle\langle\psi|)\varrho_a(1 - |\psi\rangle\langle\psi|), \quad a = 1, 2 \tag{7.15}$$

are non-negative while

$$t\hat{\varrho}_1 + (1 - t)\hat{\varrho}_2 = 0 \tag{7.16}$$

so they both vanish, and hence $\varrho_1 = \varrho_2 = |\psi\rangle\langle\psi|$. $\qquad\square$

---

[2] "Space" here in meant in the sense of *point space*.

All states satisfy $\varrho^\dagger = \varrho$ and $\varrho^2 \leq \varrho$ (with equality iff ϱ is pure). Hence

$$\mathrm{Tr}(\varrho^\dagger \varrho) = \mathrm{Tr}(\varrho^2) \leq \mathrm{Tr}(\varrho) = 1, \tag{7.17}$$

so $\mathcal{S}$ is contained in the unit ball of the *Hilbert-Schmidt Hilbert space*

$$\mathcal{H}_{\mathrm{HS}} \simeq \mathcal{H} \widehat{\otimes} \mathcal{H}^\vee \tag{7.18}$$

of operators $A, B \in \mathcal{B}(\mathcal{H})$ with finite norm with respect to the Hermitian form $\langle \cdot, \cdot \rangle_{\mathrm{HS}}$ induced by the inner product on $\mathcal{H}$, that is,

$$\langle A, B \rangle_{\mathrm{HS}} \overset{\mathrm{def}}{=} \mathrm{Tr}(A^\dagger B). \tag{7.19}$$

This allows us to define a natural distance between two states

$$D(\varrho_1, \varrho_2)^2 = 2 \left\| (\varrho_1 - \varrho_2) \right\|_{\mathrm{HS}}^2 \equiv 2\,\mathrm{Tr}(\varrho_1 - \varrho_2)^2. \tag{7.20}$$

A simple computation gives that for two pure states $|\psi_1\rangle, |\psi_2\rangle$

$$D(|\psi_1\rangle, |\psi_2\rangle) = 2 \sin d(|\psi_1\rangle, |\psi_2\rangle)_{\mathrm{FS}}, \tag{7.21}$$

where $d(\cdot, \cdot)_{\mathrm{FS}}$ is the *Fubini-Study distance* between pure states defined in Eq. (4.12). We conclude that an isometry of the metric $D(\cdot, \cdot)$ on $\mathcal{S}$, which preserves the subspace of pure states, restricts in $\mathbb{P}(\mathcal{H}) \subset \mathcal{S}$ to an isometry of $d(\cdot, \cdot)_{\mathrm{FS}}$ which is either a *unitary* or an *anti-unitary* map by Wigner theorem.

States in Finite Dimension
The typical Hilbert space of Quantum Information theory is finite dimensional and for most of this chapter we focus on this case. When the Hilbert space $\mathcal{H}$ has finite dimension n, the space of *pure states*, i.e. of orthogonal projectors of rank 1, is

$$\mathbb{P}(\mathbb{C}^n) \equiv \mathbf{PC}^{n-1}, \tag{7.22}$$

while the general *states* ϱ form the manifold $\mathcal{S}_n$ of non-negative Hermitian $n \times n$ matrices with trace equal 1. One way to parametrize the space $\mathcal{S}_n$ is in terms of elements h of the $\mathbb{R}$-vector space of traceless Hermitian matrices—i.e. the matrices h giving the n-dimensional representation of the Lie algebra $\mathfrak{su}(n)$ in the physical convention that the generators are Hermitian. Then

$$\varrho = \frac{e^{-h}}{\mathrm{tr}(e^{-h})} \quad \text{where } h \in \mathfrak{su}(n) \quad h = h^\dagger, \; \mathrm{tr}(h) = 0. \tag{7.23}$$

The RHS of (7.23) is called the *exponential parametrization* of $\mathscr{S}_n$ which is especially suggestive from the Statistic Mechanics viewpoint.[3]

Bloch Vectors

Let λ^a ($a = 1, \ldots, n^2 - 1$) be traceless Hermitian $n \times n$ matrices which form an orthonormal basis of $\mathfrak{su}(n)$ normalized so that

$$\mathrm{tr}(\lambda^a \lambda^b) = \frac{1}{2}\delta_{ab}. \tag{7.24}$$

The multiplication table of the λ^a's has the form

$$\lambda^a \lambda^b = \frac{1}{2n}\delta_{ab} + d_{abc}\,\lambda^c + \mathrm{i}\,f_{abc}\,\lambda^c, \tag{7.25}$$

where the real coefficients f_{abc} are the totally antisymmetric structure constants of the Lie algebra $\mathfrak{su}(n)$, while d_{abc} is the totally symmetric invariant tensor associated to the cubic Casimir of $\mathfrak{su}(n)$

$$d_{abc} \equiv \mathrm{tr}\big(\{\lambda^a, \lambda^b\}\lambda^c\big). \tag{7.26}$$

(d_{abc} vanishes identically for $n = 2$). Then we write the elements of $\mathscr{S}_n$ in the form

$$\varrho(t) = \frac{1}{n}\mathbf{1} + \sum_a t_a \lambda^a, \tag{7.27}$$

where the real vector $t = (t_1, \ldots, t_{n^2-1})$ takes values in the open domain

$$B_n = \left\{ t \in \mathbb{R}^{n^2-1} : \sigma(t_a \lambda^a) \subset \left[-\tfrac{1}{n}, \tfrac{n-1}{n} \right] \right\} \subset \mathbb{R}^{n^2-1}. \tag{7.28}$$

t is called the *Bloch vector* of ϱ. The distance-square (7.20) between two states is

$$\mathrm{D}\big(\varrho(t), \varrho(s)\big)^2 = 2\sum_{a,b}(t_a - s_a)(t_b - s_b)\,\mathrm{tr}[\lambda^a \lambda^b] = (t - s)^2, \tag{7.29}$$

i.e. $\mathrm{D}\big(\varrho(t), \varrho(s)\big)$ is just the Euclidean distance $\sqrt{(t - s)^2}$ in the Bloch space $\mathbb{R}^{n^2-1}$.

Definition 7.3 The density matrix at the origin of the Bloch space

$$\varrho_\star = \frac{1}{n}\mathbf{1} \tag{7.30}$$

---

[3] Identifying h with βH one gets the canonical ensemble.

is called the *maximally mixed state* (or the *ignorance state*[4]).

Example 7.1 (The Bloch Sphere) We consider $\mathcal{H} \simeq \mathbb{C}^2$, the simplest non-trivial Hilbert space. The general state is a 2×2 matrix of the form

$$\varrho(\mathbf{t}) = \frac{1}{2}\big(1 + \mathbf{t} \cdot \boldsymbol{\sigma}\big), \tag{7.31}$$

where $\boldsymbol{\sigma} = (\sigma_1, \sigma_2, \sigma_3)$ are the three Pauli matrices (Sect. 4.3.1). Now

$$0 \le 4 \det \varrho(\mathbf{t}) = 1 - \mathbf{t} \cdot \mathbf{t} \tag{7.32}$$

implies that the space of states $\mathscr{S}_2$ is the *unit ball in* $\mathbb{R}^3$

$$\mathscr{S}_2 = \{\mathbf{t} \in \mathbb{R}^3 : |\mathbf{t}| \le 1\}, \tag{7.33}$$

which is indeed a *convex* domain in $\mathbb{R}^3$. The boundary of $\mathscr{S}_2$ is the *unit sphere*

$$\partial \mathscr{S}_2 \equiv S^2 \subset \mathbb{R}^3, \tag{7.34}$$

which is a copy of $\mathbb{P}(\mathbb{C}^2)$, the space of *pure* states. Indeed when $\mathbf{t} \equiv \mathbf{n}$ is a unit vector

$$\varrho(\mathbf{n})^2 = \frac{1}{4}\big(1 + 2\mathbf{n} \cdot \boldsymbol{\sigma} + (\mathbf{n} \cdot \boldsymbol{\sigma})^2\big) = \frac{1}{2}(1 + \mathbf{n} \cdot \boldsymbol{\sigma}) = \varrho(\mathbf{n}), \tag{7.35}$$

and the state $\varrho(\mathbf{n})$ is a projector, hence a pure state.

Remark 7.1 In finite dimension n a state $\varrho(\mathbf{t}) \in \mathscr{S}_n \subset \mathcal{B}(\mathbb{C}^n)$ satisfies the pure state condition

$$\varrho(\mathbf{t})^2 = \varrho(\mathbf{t}) \tag{7.36}$$

iff

$$|\mathbf{t}|^2 = \frac{2(n-1)}{n} \quad \text{and} \quad d_{abc}\, t_b t_c = \frac{n-2}{n} t_a, \tag{7.37}$$

so when $n > 2$ the pure states form a *proper subset*

$$\mathcal{P}u \subsetneqq S^{n^2 - 2} \tag{7.38}$$

---

[4] The second name will be motivated in Sect. 7.5.

of the sphere of radius $\sqrt{2(n-1)/2}$ in the Bloch space $\mathbb{R}^{n^2-1}$: $\mathcal{P}u$ is the real affine variety given by the zero locus of the n^2 quadratic polynomials (7.37). The affine variety $\mathcal{P}u$ is isomorphic (as a real manifold) to the complex projective space

$$\mathcal{P}u \simeq_{\text{smooth}} \mathbb{P}(\mathbb{C}^n) \equiv \mathbf{P}\mathbb{C}^{n-1}. \tag{7.39}$$

Qubits

In Quantum Information one usually works with *qubits* (or *q*-bits), the quantum counterpart of classical bits. A classical bit is a variable which can take two values: 0 and 1. A *qubit* is a state of a quantum system with the two-dimensional Hilbert space $\mathcal{H} \equiv \mathbb{C}^2$ which we may identify physically with the *Fermi oscillator* (see Sect. 4.12.2) or, equivalently, with the spin-$\frac{1}{2}$ representation of $\mathfrak{su}(2)$ (Sect. 4.3.1):

$$\mathcal{H} \simeq V_{1/2}. \tag{7.40}$$

The two basis vectors of $\mathcal{H} \simeq \mathbb{C}^2$ are written, interchangeably, as one of the following traditional symbols

$$|0\rangle = |\downarrow\rangle \equiv \begin{pmatrix} 0 \\ 1 \end{pmatrix} \qquad |1\rangle = |\uparrow\rangle \equiv \begin{pmatrix} 1 \\ 0 \end{pmatrix} \tag{7.41}$$

where the arrows stand, respectively, for spin "down" and spin "up" i.e. for the sign of the eigenvalue of $S_3 \equiv \frac{1}{2}\sigma_3$.

The Hilbert space of an n qubits system is the n-th tensor power of $V_{1/2}$

$$\bigoplus_{a=1}^{n} V_{1/2}^{(a)} \simeq \mathbb{C}^{2^n}. \tag{7.42}$$

Sometimes instead of qubits one uses *qutrits* whose Hilbert space $\simeq \mathbb{C}^3$.

7.1.2 Symmetries of Open Systems

Recall from Chap. 4 that a *kinematical symmetry* is an automorphism of the space of physical states which preserves all its defining structures. In the present set-up an automorphism of the convex space $\mathcal{S}$ should be an invertible map $m \colon \mathcal{S} \to \mathcal{S}$ compatible with the convex structure i.e. such that

$$m(t\varrho_1 + (1-t)\varrho_2) = t\, m(\varrho_1) + (1-t)\, m(\varrho_2) \tag{7.43}$$

for all $0 \leq t \leq 1$. This implies that m extends to an affine map

$$m \colon \mathbb{R}^{n^2-1} \to \mathbb{R}^{n^2-1} \tag{7.44}$$

of the Bloch space $\mathbb{R}^{n^2-1}$. It also implies that the map m sends the real affine variety of pure states $\mathcal{P}u \subset \mathbb{R}^{n^2-1}$ to itself because pure states are extremal (Lemma 7.1). It would be natural to ask also that the map m preserves the Hilbert-Schmidt distance

$$D(m(\varrho_1), m(\varrho_2)) = D(\varrho_1, \varrho_2) \tag{7.45}$$

between states. The Kadison theorem [13] states that this second condition is unnecessary, because it is an automatic consequence of (7.43). An automorphism of the space of states $\mathcal{S}$ then restricts to an isometry of $\mathbb{P}(\mathcal{H})$ by the argument around Eq. (7.21). Comparing with Wigner's theorem (Chap. 4) we get

Fact 7.2 *A kinematical symmetry of the space $\mathcal{S}$ of quantum states is given by either the adjoint action of a unitary transformation*

$$\varrho \rightsquigarrow U \varrho\, U^{-1} \qquad U \in \mathcal{U}(\mathcal{H}), \tag{7.46}$$

or by the adjoint action of an anti-unitary transformation.

7.1.3 Singular Value Decomposition

$\mathcal{S}$ is a convex subspace of the convex cone $\mathcal{P} \subset \mathcal{B}(\mathcal{H})$ of all non-negative (Hermitian) operators. This subsection contains a short digression about the properties of *general* non-negative operators.

From Chap. 2 we know that all *non-negative* operator $N \in \mathcal{P}$ admits a unique non-negative Hermitian *square root operator* $\sqrt{N}$ given explicitly by

$$\sqrt{N} = \int_0^\infty \sqrt{\lambda}\, dP_N(\lambda) \tag{7.47}$$

where $P_N(\lambda)$ is the *spectral family of projectors of the operator* N. Informally, $\sqrt{N}$ is the self-adjoint operator with the same eigenvectors of N whose eigenvalues are the non-negative square-roots of the eigenvalues of N.

Let $\mathcal{H}$ be a Hilbert space, $A \in \mathcal{B}(\mathcal{H})$ an arbitrary (bounded) operator, and $A^\dagger$ its adjoint. The operator $AA^\dagger : \mathcal{H} \to \mathcal{H}$ is non-negative, hence has a non-negative square-root

$$|A| \overset{\text{def}}{=} \sqrt{AA^\dagger}. \tag{7.48}$$

We decompose

$$\mathcal{H} = \ker A^\dagger \oplus (\ker A^\dagger)^\perp \tag{7.49}$$

One has

$$|A|(\ker A^\dagger)^\perp \subseteq (\ker A^\dagger)^\perp, \tag{7.50}$$

while, by construction, $|A|$ is invertible in $(\ker A^\dagger)^\perp$. Hence in $(\ker A^\dagger)^\perp$ we have

$$(|A|^{-1}A)(|A|^{-1}A)^\dagger = |A|^{-1}AA^\dagger|A| = |A|^{-1}|A|^2|A|^{-1} = 1 \tag{7.51}$$

so $\tilde{U} \equiv |A|^{-1}A$ is unitary in $(\ker A^\dagger)^\perp$. We may extend $\tilde{U}$ to a unitary operator

$$U = \tilde{U} + U' \tag{7.52}$$

in the full Hilbert space $\mathcal{H}$ by adding a unitary operator $U' \colon \ker A^\dagger \to \ker A^\dagger$. We get

Proposition 7.1 *An arbitrary operator $A \in \mathcal{B}(\mathcal{H})$ admits the* polar decomposition

$$A = |A|\,U = \sqrt{AA^\dagger}\,U, \tag{7.53}$$

where $|A|$ is non-negative and U is an unitary operator which is unique if and only if A is invertible.

Definition 7.4 The eigenvalues of the non-negative matrix $|A|$ are called the *singular values of A.*

One may generalize the **Proposition** to linear maps

$$A \colon \mathcal{H}_1 \to \mathcal{H}_2 \tag{7.54}$$

between two Hilbert spaces which may have different dimensions. $AA^\dagger$ and $A^\dagger A$ are non-negative self-adjoint operators acting, respectively, on $\mathcal{H}_2$ and $\mathcal{H}_1$.

Claim $AA^\dagger$ *and $A^\dagger A$ have the same non-zero eigenvalues λ_k.*

Proof Indeed suppose $|\psi\rangle \in \mathcal{H}_2$ is an eigenvector of $AA^\dagger$ with eigenvalue $\lambda \neq 0$

$$AA^\dagger|\psi\rangle = \lambda|\psi\rangle. \tag{7.55}$$

$A^\dagger|\psi\rangle$ is non-zero and an eigenvector of $A^\dagger A$ with the same non-zero eigenvalue λ

$$A^\dagger A(A^\dagger|\psi\rangle) = A^\dagger(AA^\dagger|\psi\rangle) = \lambda\,A^\dagger|\psi\rangle. \tag{7.56}$$

$$\square$$

It is convenient to make the two matrices $AA^\dagger$ and $A^\dagger A$ of the same size by adding zero eigenvalues to the smaller one. In this way we may always reduce to the case $\mathcal{H}_1 \simeq \mathcal{H}_2$.

Let $U \in \mathcal{U}(\mathcal{H}_1)$ and $V \in \mathcal{U}(\mathcal{H}_2)$ be the unitary transformations which diagonalize, respectively, $A^\dagger A$ and $AA^\dagger$ (with the common non-zero eigenvalues λ_k written in the same order for the two operators):

$$(A^\dagger A)_{ij} = U_{ik}\lambda_k U_{kj}^{-1}, \qquad (AA^\dagger)_{ij} = V_{ik}\lambda_k V_{kj}^{-1}. \tag{7.57}$$

Then the *singular value decomposition* of the linear map $A : \mathcal{H}_1 \to \mathcal{H}_2$ is

$$A_{ij} = V_{ik}\sqrt{\lambda_k}\, U_{kj}^{-1}, \tag{7.58}$$

that is,

Corollary 7.1 (Singular Value Decomposition) *Given a linear map $A : \mathcal{H}_1 \to \mathcal{H}_2$ between any two Hilbert spaces, we may write*

$$A = VDU^{-1} \tag{7.59}$$

where $V \in \mathcal{U}(\mathcal{H}_2)$, $U \in \mathcal{U}(\mathcal{H}_1)$ are unitary and $D : \mathcal{H}_1 \to \mathcal{H}_2$ is diagonal[5] with non-negative entries.

7.2 Schmidt Decomposition

We return to our physical discussion of open quantum systems. All real quantum systems interact with some "environment" E. The system of interest S and its environment E together form a *closed system* (at least ideally) to which the formalism of Chap. 2 may be applied. The pure states of the combined closed quantum system are represented by vectors in the total Hilbert space

$$\mathcal{H}_S \otimes \mathcal{H}_E, \tag{7.60}$$

where $\mathcal{H}_S$ is our system's Hilbert space and $\mathcal{H}_E$ is the environment's one. States in the tensor product $\mathcal{H}_S \otimes \mathcal{H}_E$ of two Hilbert spaces are called *bi-partite*. More generally, states in the Hilbert space

$$\mathcal{H}_1 \otimes \mathcal{H}_2 \otimes \cdots \otimes \mathcal{H}_k \quad (k \text{ factor spaces}) \tag{7.61}$$

are called *k-partite states*. In this chapter we shall be mainly concerned with bi-partite systems and states. A useful fact about bi-partite states is

---

[5] A rectangular matrix is *diagonal* iff it is the direct sum of a diagonal square matrix and the zero matrix. It becomes a diagonal matrix in the usual sense by adding zeros to make $\mathcal{H}_1 \simeq \mathcal{H}_2$.

Theorem 7.3 (Schmidt Decomposition) *Every bi-partite pure state*

$$|\Psi\rangle \in \mathcal{H}_S \otimes \mathcal{H}_E \tag{7.62}$$

can be expressed in the form

$$|\Psi\rangle = \sum_{i=1}^{N} \sqrt{\lambda_i}\, |e_i\rangle \otimes |f_i\rangle \tag{7.63}$$

where $\{|e_i\rangle\}_{i=1}^{N_1}$ *is an orthonormal basis for* $\mathcal{H}_S$, *and* $\{|f_i\rangle\}_{i=1}^{N_2}$ *is an orthonormal basis for* $\mathcal{H}_E$, *and the* λ_i's *are the (positive) eigenvalues of the reduced density matrix*

$$\varrho_S \equiv \mathrm{Tr}_{\mathcal{H}_E} |\Psi\rangle\langle\Psi| = \sum_{i=1}^{N_1} \lambda_i\, |e_i\rangle\langle e_i| \tag{7.64}$$

Here $N \leq \min\{N_1, N_2\}$. *(As always,* $N_1, N_2 \in \mathbb{N} \cup \{\aleph_0\}$*).*

We stress that the two orthonormal bases

$$\{|e_i\rangle\}_{i=1}^{N_1}, \qquad \{|f_i\rangle\}_{i=1}^{N_2} \tag{7.65}$$

depend on the particular state $|\Psi\rangle$. In a *fixed* pair of bases

$$\{|E_i\rangle\}_{i=1}^{N_1}, \qquad \{|F_i\rangle\}_{i=1}^{N_2} \tag{7.66}$$

the general bi-partite pure state has the form

$$|\Psi\rangle = \sum_{i=1}^{N_1} \sum_{j=1}^{N_2} C_{ij} |E_i\rangle \otimes |F_j\rangle \tag{7.67}$$

for a complex $N_1 \times N_2$ matrix C_{ij}. We give two proofs of the **Theorem**.

First Proof We may assume $N_1 \leq N_2$ with no loss. Let $|\Psi\rangle$ be the arbitrary pure state (7.67). We write

$$|\Phi_i\rangle \equiv \sum_{j} C_{ij} |F_j\rangle \quad \Rightarrow \quad |\Psi\rangle = \sum_{i=1}^{N_1} |E_i\rangle \otimes |\Phi_i\rangle. \tag{7.68}$$

We take the partial trace of $|\Psi\rangle\langle\Psi|$ over the environment Hilbert space $\mathcal{H}_E$

$$\varrho_S = \mathrm{Tr}_E (|\Psi\rangle\langle\Psi|) = \sum_{i=1}^{N_1} \sum_{j=1}^{N_2} |E_i\rangle\langle\Phi_j|\Phi_i\rangle\langle E_j| \in \mathcal{B}(\mathcal{H}_S). \tag{7.69}$$

Now we write ϱ_S in an orthonormal basis $\{|e_i\rangle\}$ of $\mathcal{H}_S$ where it is diagonal

$$\varrho_S = \sum_{i=1}^{N_1} \lambda_i \, |e_i\rangle\langle e_i|, \tag{7.70}$$

the eigenvalues λ_i being non-negative. Next we replace in Eq. (7.67) the generic basis $\{|E_i\rangle\}$ with the eigenvector basis $\{|e_i\rangle\}$ we have constructed in Eq. (7.70), and go through the same steps ending up with

$$|\Psi\rangle = \sum_{i=1}^{N_1} |e_i\rangle \otimes |\phi_i\rangle \tag{7.71}$$

for some states $|\phi_i\rangle \in \mathcal{H}_E$ (use Eq. (7.68) with $|\Phi_i\rangle \rightsquigarrow |\phi_i\rangle$). Taking the partial trace over $\mathcal{H}_E$ we get

$$\varrho_S = \sum_{i=1}^{N_1}\sum_{j=1}^{N_1} |e_i\rangle\langle\phi_j|\phi_i\rangle\langle e_j| \tag{7.72}$$

Comparing with (7.70), we learn that

$$\langle\phi_i|\phi_j\rangle = \lambda_i \, \delta_{ij} \quad \lambda_i \geq 0. \tag{7.73}$$

Setting

$$|f_i\rangle = \begin{cases} \dfrac{1}{\sqrt{\lambda_i}}|\phi_i\rangle & \text{if } \lambda_i \neq 0 \\ 0 & \text{otherwise} \end{cases} \tag{7.74}$$

we get the **Theorem**. $\qquad\qquad\square$

Second Proof We apply the singular value decomposition to the matrix C_{ij} in (7.67) seen as a map $\mathcal{H}_E \to \mathcal{H}_S$. The diagonal matrix D in Eq. (7.59) is

$$D = \mathrm{diag}(\sqrt{\lambda_1}, \cdots, \sqrt{\lambda_n}, 0, \cdots, 0). \tag{7.75}$$

$\qquad\qquad\square$

Consider now the reduced density matrix ϱ_E produced on the *second* Hilbert space $\mathcal{H}_E$ by the pure state $|\Psi\rangle \in \mathcal{H}_S \otimes \mathcal{H}_E$. Taking the trace over $\mathcal{H}_S$ we get

$$\varrho_E = \mathrm{Tr}_S(|\Psi\rangle\langle\Psi|) = \sum_i \lambda_i |f_i\rangle\langle f_i| \tag{7.76}$$

with the same *non-zero* coefficients λ_i as in ϱ_S.

Definition 7.5 The common non-zero coefficients $\lambda_i \neq 0$ in ϱ_S and ϱ_E (Eqs. (7.70) and (7.76)) are called the *Schmidt coefficients*. They satisfy

$$\lambda_i > 0, \qquad \sum_i \lambda_i = 1. \tag{7.77}$$

The number r of non-zero coefficients λ_i is called the *Schmidt number* (or *rank*) of the pure state $|\Psi\rangle$. r is the rank of both reduced matrices ϱ_S, ϱ_E.

The reduced matrix ϱ_S is a pure state iff its Schmidt number $r = 1$.

Definition 7.6 A bi-partite state $|\Psi\rangle \in \mathcal{H}_S \otimes \mathcal{H}_E$ is *entangled* iff $r > 1$.

A fundamental property of the Schmidt coefficients λ_i (and number r) is that they are invariant under all *local unitary transformations*, namely they do not change under the action of the group

$$\mathcal{U}(\mathcal{H}_S) \times \mathcal{U}(\mathcal{H}_E) \tag{7.78}$$

of unitary transformations which act separately on the two factor Hilbert spaces i.e. that act only on the degrees of freedom of one subsystem. Of course, r will change under a general unitary transformation $U \in \mathcal{U}(\mathcal{H}_S \otimes \mathcal{H}_E)$. We conclude this section by summarizing our results:

Corollary 7.2 (Reduction Lemma) *Let $|\Psi\rangle \in \mathcal{H}_S \otimes \mathcal{H}_E$ be a pure state and ϱ_S, ϱ_E be the reduced density matrices obtained by performing the partial trace on $\mathcal{H}_E$ and, respectively, $\mathcal{H}_S$. Then the spectra of ϱ_S and ϱ_E are identical except for the degeneracy of the zero eigenvalue.*

7.3 Purification and the Geometry of States

An important consequence of the Schmidt decomposition is

Corollary 7.3 (Purification) *Given a (mixed) state $\varrho \in \mathcal{B}(\mathcal{H})$, there exists a Hilbert space $\widetilde{\mathcal{H}}$ and a pure state $|\psi\rangle \in \mathcal{H} \otimes \widetilde{\mathcal{H}}$ such that*

$$\varrho = \mathrm{Tr}_{\widetilde{\mathcal{H}}}(|\psi\rangle\langle\psi|). \tag{7.79}$$

In other words, we may always replace our mixed state $\varrho \in \mathcal{B}(\mathcal{H})$ by a pure state in a larger Hilbert space $\mathcal{H} \otimes \widetilde{\mathcal{H}}$.

Definition 7.7 A pure state $|\psi\rangle \in \mathcal{H} \otimes \widetilde{\mathcal{H}}$ with the property (7.79) is called a *purification* of the (mixed) state $\varrho \in \mathcal{B}(\mathcal{H})$.

A purification of a state $\varrho \in \mathcal{B}(\mathcal{H})$ of Schmidt number r exists in the Hilbert space $\mathcal{H} \otimes \widetilde{\mathcal{H}}$ iff $\dim \widetilde{\mathcal{H}} \geq r$. However it is more convenient to fix once and for all the auxiliary Hilbert space $\widetilde{\mathcal{H}}$ independently of the particular state ϱ. Since ϱ is an element of the Hilbert-Schmidt space

$$\mathcal{H}_{\mathrm{HS}} \equiv \mathcal{H} \widehat{\otimes} \mathcal{H}^{\vee}, \tag{7.80}$$

while the Hilbert space satisfies $\mathcal{H}^{\vee} \simeq \mathcal{H}$, the canonical choice of the auxiliary Hilbert space $\widetilde{\mathcal{H}}$ is $\mathcal{H}$ itself. For instance, writing ϱ in some orthonormal basis $\{|E_i\rangle\}$ of $\mathcal{H}$, we may write

$$\varrho = \sum_{ij} C_{ij}|E_i\rangle\langle E_j| \rightsquigarrow |\psi\rangle = \sum_{ij}(\sqrt{C})_{ij}|E_i\rangle \otimes |E_j\rangle \in \mathcal{H} \widehat{\otimes} \mathcal{H} \tag{7.81}$$

where $\sqrt{C}$ is the *square root* of the non-negative matrix C_{ij} (cf. (7.47)).

We stress that the purification $|\psi\rangle$ is *highly non-unique*. The HJW[6] theorem (to be discussed in Sect. 7.3.3 below) states that two purifications $|\psi_1\rangle$, $|\psi_2\rangle$ of ϱ in $\mathcal{H} \otimes \mathcal{H}$ differ by

$$|\psi_2\rangle = (1 \otimes U)|\psi_1\rangle \quad \text{with} \ U \in \mathcal{U}(\mathcal{H}), \tag{7.82}$$

i.e. differ by an unitary transformation acting only on the degrees of freedom of the auxiliary system whose Hilbert space is the second copy of $\mathcal{H}$.

Corollary 7.4 (Purification Formula) *Seeing both the state ϱ and its purification ψ as elements of the Hilbert space $\mathcal{H}_{\mathrm{HS}} \equiv \mathcal{H} \widehat{\otimes} \mathcal{H}^{\vee}$, we have the* purification formula

$$\varrho = \psi \psi^{\dagger} \quad \text{with the equivalence } \psi \sim \psi U \text{ for } U \in \mathcal{U}, \tag{7.83}$$

while the polar decomposition (7.53) of the purification ψ is

$$\psi = \sqrt{\varrho}\, U. \tag{7.84}$$

The eigenvalues of ϱ are the squares of the singular values of ψ.

7.3.1 The Bures Geometry of States

The purification formula (7.83) has an elegant geometric interpretation which we briefly describe. For comparison sake we start by reviewing Cartan's geometry.

---

[6] HJW stands for Hughston, Jozsa, and Wootters.

Vielbeins, Symmetric Spaces, and All that

In Riemannian geometry the metric $g = (g_{\mu\nu})$ is a positive-definite symmetric $n \times n$ matrix. It is often convenient to write the metric in terms of the *vielbein* $e = (e^i_\mu) \in GL(n, \mathbb{R})$ in the form

$$g = ee^t \quad \text{or, writing the indices,} \quad g_{\mu\nu} = e^i_\mu e^i_\nu. \tag{7.85}$$

The vielbein e is unique up to multiplication on the right by an element of $O(n)$

$$e \rightsquigarrow e\,O, \quad O \in O(n). \tag{7.86}$$

Hence[7]

$$\begin{matrix} \text{convex cone of} \\ \text{positive } symmetric \\ n \times n \text{ matrices} \end{matrix} = GL^+(n, \mathbb{R})/SO(n) \simeq \mathbb{R}_{>0} \times SL(n, \mathbb{R})/SO(n), \tag{7.87}$$

i.e. the space of positive definite symmetric $n \times n$ matrices is the product of two symmetric spaces: $\mathbb{R}_{>0}$ (the overall normalization factor) and the Type III symmetric space

$$SL(n, \mathbb{R})/SO(n). \tag{7.88}$$

The same story applies to a *Hermitian* metric h: we introduce a vielbein $a \in GL(n, \mathbb{C})$ and write

$$h = aa^\dagger. \tag{7.89}$$

a is unique up to multiplication on the right by an element of $U(n)$

$$a \rightsquigarrow a\,U, \quad U \in U(n). \tag{7.90}$$

We conclude that

$$\mathcal{P}_H(n) \equiv \begin{matrix} \text{convex cone of} \\ \text{positive } Hermitian \\ n \times n \text{ matrices} \end{matrix} = \mathbb{R}_{>0} \times SL(n, \mathbb{C})/SU(n). \tag{7.91}$$

So, again, the space of positive-definite Hermitian matrices is a *symmetric space* which is the product of the two irreducible spaces $\mathbb{R}_{>0}$ and the Type IV symmetric space

$$SL(n, \mathbb{C})/SU(n) \simeq \begin{matrix} \text{space of positive Hermitian} \\ n \times n \text{ matrices of determinant} 1 \end{matrix} \tag{7.92}$$

via the identification $a \mapsto aa^\dagger$ (the *Cartan isomorphism*).

---

[7] $GL^+(n, \mathbb{R})$ is the group of $n \times n$ matrices with positive determinant.

Let G/K be a general symmetric space, where G is a connected real Lie group with trivial center and $K \subset G$ is the subgroup fixed by a Cartan involution $\theta \colon G \to G$ [14, 15]. G/K is the base of the canonical *principal K-bundle*

$$\mathsf{pr} \colon G \to G/K \tag{7.93}$$

which is equipped with a canonical G-homogeneous K-connection. To construct this connection, we decompose the Lie algebra $\mathfrak{g}$ of G in the Lie algebra $\mathfrak{k}$ of the compact subgroup $K \subset G$ and the orthogonal complement $\mathfrak{p}$

$$\mathfrak{g} = \mathfrak{k} \oplus \mathfrak{p} \tag{7.94}$$

Let $\mu \in \Omega^1(G) \otimes \mathfrak{g}$ be the left-invariant Maurer-Cartan form ([16] chap. 2), and $g \colon G/K \to G$ a section of pr. We may decompose the Maurer-Cartan form in the two direct summands of (7.94):

$$g^* \mu = g^{-1} dg = \omega + e, \quad \omega \overset{\text{def}}{=} (g^{-1} dg)_{\mathfrak{k}}, \quad e \overset{\text{def}}{=} (g^{-1} dg)_{\mathfrak{p}} \tag{7.95}$$

It is immediate to see that ω transforms as K-connection under the K-gauge transformation $g \rightsquigarrow gu, u \in K$

$$\omega \to u^{-1} \omega u + u^{-1} du. \tag{7.96}$$

ω is our canonical K-connection. One shows[8] that the one-form e with values in $\mathfrak{p}$ is a vielbein for the symmetric metric on G/K ($\equiv$ *Cartan metric*)[9]

$$ds^2_{\text{Car}} = e \otimes e^t \equiv \langle (g^{-1} dg)_{\mathfrak{p}}, (g^{-1} dg)_{\mathfrak{p}} \rangle. \tag{7.97}$$

Left-translations by elements of G are isometries for the Cartan metric. ω is the Levi-Civita (torsionless) connection for the Cartan metric. As we shall see in a moment, on the symmetric convex cone $\mathcal{P}_H(n)$ there are other interesting metrics besides the Cartan one.

Geometry of Quantum States

As it is customary in the present context, we assume the Hilbert space $\mathcal{H}$ to have finite dimension n, although most of the story may be easily extended to separable Hilbert spaces.

---

[8] A readable account may be found in chap. 5 of [17].
[9] $\langle \cdot, \cdot \rangle$ is the invariant metric on $\mathfrak{p}$.

The only differences between the purification formula (7.83) and Cartan's equations (7.89),(7.90) are:

(a)   the state ϱ is, in general, merely *non-negative* instead of *positive*;
(b)   the state ϱ is trace-normalized: $\operatorname{Tr}\varrho = 1$.

The closed convex cone of non-negative Hermitian $n \times n$ matrices is the closure $\overline{\mathcal{P}}_H(n)$ of the open convex cone $\mathcal{P}_H(n)$ of positive matrices, so item (a) requires that we replace the symmetric space $\mathcal{P}_H(n)$ by its closure $\overline{\mathcal{P}}_H(n)$. The space of non-negative matrices $\overline{\mathcal{P}}_H(n)$ is stratified in strata of matrices with fixed rank r ($\equiv$ the Schmidt number), and each irreducible component of a rank-r stratum is a copy of $\mathcal{P}_H(r)$. The geometry of the closed cone $\overline{\mathcal{P}}_H(n)$ may be recovered from the one of the open cone $\mathcal{P}_H$ by continuity, and we shall henceforth work in the "big" generic stratum of positive-definite Hermitian matrices $\mathcal{P}_H(n)$ which is a $U(n)$-principal bundle

$$\pi : GL(n, \mathbb{C}) \to GL(n, \mathbb{C})/U(n) \equiv \mathcal{P}_H(n) \tag{7.98}$$

$$\pi : \psi \mapsto \psi\psi^{\dagger} \equiv \varrho \tag{7.99}$$

whose fiber over the state ϱ consists of its purifications in the standard Hilbert space $\mathcal{H} \otimes \mathcal{H}^{\vee}$. In particular on $\mathcal{P}_H(n)$ we have Cartan's $U(n)$-connection $(\psi^{-1}d\psi)_{\mathfrak{u}(n)}$.

However, for the purpose of the geometry of quantum states, Cartan's symmetric metric is *not* natural: since the Cartan metric is already complete in the "big" stratum $\mathcal{P}_H(n)$ of states with Schmidt number $n = \dim\mathcal{H}$, all states with $r < n$ (say, the pure states with $r = 1$) are at *infinite* Cartan distance. On the contrary, the natural metric for the quantum states should extend nicely on the closed cone $\overline{\mathcal{P}}_H(n)$ and reduce on the pure state submanifold to the usual overlap distance[10]

$$D(|\psi_1\rangle, |\psi_2\rangle)^2 = 1 - |\langle\psi_1|\psi_2\rangle|^2, \tag{7.100}$$

or to the related Fubini-Study distance D_{FS} on $\mathbb{P}(\mathbb{C}^n)$

$$\cos D_{\mathrm{FS}}(|\psi_1\rangle, |\psi_2\rangle) = |\langle\psi_1|\psi_2\rangle|. \tag{7.101}$$

The appropriate distance in the closed cone $\overline{\mathcal{P}}_H(n)$ is given by the *Bures distance* which extends the overlap distance (7.100), while the *Bures angle* extends the Fubini-Study distance (7.101). Both Bures distance and angle arise from a Riemannian metric on the closed convex cone $\overline{\mathcal{P}}_H(n)$.

To construct the Bures metric on $\overline{\mathcal{P}}_H(n)$ we start from the space of $n \times n$ complex matrices

$$\overline{GL(n, \mathbb{C})} \equiv \mathcal{H} \otimes \mathcal{H}^{\vee} \tag{7.102}$$

---

[10] All pure states are normalized, unless explicitly stated otherwise.

which comes with the natural Euclidean distance

$$D(a, b)^2 = \|a - b\|^2 \equiv \mathrm{Tr}((a - b)(a - b)^\dagger) =$$
$$= \mathrm{Tr}(aa^\dagger) + \mathrm{Tr}(bb^\dagger) - \mathrm{Tr}(ab^\dagger) - \mathrm{Tr}(ba^\dagger) \tag{7.103}$$

defined by the flat Riemannian metric

$$\mathrm{d}s^2 = \mathrm{Tr}(\mathrm{d}a\,\mathrm{d}a^\dagger). \tag{7.104}$$

$\mathrm{d}s^2$ restricts to a metric on the total space $GL(n, \mathbb{C})$ of the bundle π (cf. (7.98)). The natural distance between two points ϱ_1, ϱ_2 in the base $\mathcal{P}_H(n)$ is given by the minimal Euclidean distance between two points in their respective fibers

$$\min\{D(\psi_1, \psi_2)\colon \pi(\psi_a) = \varrho_a,\ a = 1, 2\}. \tag{7.105}$$

This leads to the Bures distance on $\overline{\mathcal{P}}_H(n)$:

$$D_B(\psi_1\psi_1^\dagger, \psi_2\psi_2^\dagger)^2 \overset{\mathrm{def}}{=} \min_{U_1, U_2} \|\psi_1 U_1 - \psi_2 U_2\|^2, \tag{7.106}$$

whose geometry we study in the rest of this subsection and in the following one.

The Riemannian Bures Metric
The Bures distance $D_B(\cdot, \cdot)$ on $\overline{\mathcal{P}}_H(n)$ arises from a Riemannian metric $\mathrm{d}s_B^2$. The Bures length of a curve γ in the base $\mathcal{P}_H(n)$ is the Euclidean length of a lift of γ in the total space $GL(n, \mathbb{C})$ whose tangent is *everywhere horizontal,* i.e. orthogonal to the fibers. Let $\psi \in GL(n, \mathbb{C})$. The fiber through ψ is given by

$$F_\psi \equiv \{\psi U : U \in \mathcal{U}\}, \tag{7.107}$$

so that the tangent space to the fiber at ψ is

$$T_\psi F_\psi = \{i\psi H : H \text{ Hermitian}\}, \tag{7.108}$$

and then

$$(T_\psi F_\psi)^\perp = \{X \in \mathbb{C}(n)\colon X^\dagger\psi - \psi^\dagger X = 0\}. \tag{7.109}$$

Thus $\psi(t)$ is a curve with horizontal tangents iff

$$\dot\psi^\dagger\psi - \psi^\dagger\dot\psi = 0 \quad \Rightarrow \quad (\psi^{-1})^\dagger\dot\psi^\dagger = \dot\psi\psi^{-1}, \tag{7.110}$$

that is,

$$\dot\psi(t) = K(t)\psi(t) \quad \text{with } K \text{ Hermitian.} \tag{7.111}$$

The Bures metric is then the square-length of the horizontal tangent form

$$\dot{\psi}\, dt = dK\,\psi \qquad dK \equiv K\, dt, \tag{7.112}$$

that is,

$$ds_B^2 = \mathrm{Tr}(d\psi\, d\psi^\dagger) = \mathrm{Tr}(dK\,\psi\psi^\dagger dK) = \mathrm{Tr}(dK\varrho\, dK). \tag{7.113}$$

To make the expression more explicit, we first notice

$$d\varrho = d\psi\,\psi^\dagger + \psi\, d\psi^\dagger = dK\,\psi\psi^\dagger + \psi\psi^\dagger dK = dK\,\varrho + \varrho\, dK, \tag{7.114}$$

so that

$$ds_B^2 = \frac{1}{2}\,\mathrm{Tr}(dK\, d\varrho), \tag{7.115}$$

then we diagonalize ϱ in the form

$$\varrho = U\lambda U^{-1}, \tag{7.116}$$

where U is unitary and λ diagonal with entries equal to the eigenvalues of ϱ. Then for a horizontal differential $d\varrho$

$$\begin{aligned}
[U^{-1}d\varrho\, U]_{ij} &= [U^{-1}(U\lambda U^{-1}dK + dK\, U\lambda U^{-1})U]_{ij} = \\
&= [\lambda(U^{-1}dK\, U) + (U^{-1}dK\, U)\lambda]_{ij} = (\lambda_i + \lambda_j)(U^{-1}dK\, U)_{ij}
\end{aligned} \tag{7.117}$$

and hence

$$\begin{aligned}
ds_B^2 &= \frac{1}{2}\mathrm{Tr}\big[d\varrho\, dK\big] = \frac{1}{2}\mathrm{Tr}\big[(U^{-1}d\varrho\, U)(U^{-1}dK\, U)\big] = \\
&= \frac{1}{2}\sum_{i,j}\frac{[U^{-1}d\varrho\, U]_{ij}[U^{-1}d\varrho\, U]_{ji}}{\lambda_i + \lambda_j}.
\end{aligned} \tag{7.118}$$

Now

$$\begin{aligned}
\left[U^{-1}d\varrho\, U\right]_{ij} &= [U^{-1}d(U\lambda U^{-1})U^{-1})U]_{ij} = \\
&= \delta_{ij}\, d\lambda_i - (\lambda_i - \lambda_j)(U^{-1}dU)_{ij}
\end{aligned} \tag{7.119}$$

and the Bures metric on the closed convex cone of non-negative Hermitian matrix $\overline{\mathcal{P}}_H(n)$ is

$$ds_B^2 = \frac{1}{4}\sum_i \frac{d\lambda_i^2}{\lambda_i} + \sum_{i<j}\frac{(\lambda_i - \lambda_j)^2}{\lambda_i + \lambda_j}\,\big|(U^{-1}dU)_{ij}\big|^2. \tag{7.120}$$

The Hilbert-Schmidt Unit Sphere

Up to now ϱ was just a non-negative Hermitian matrix in $\overline{\mathcal{P}}_H(n)$. In particular Eq. (7.120) is the Bures metric on the full closed convex cone $\overline{\mathcal{P}}_H(n)$. To be a *quantum state,* ϱ must satisfy, in addition, the normalization condition **St3**, that is,

$$1 = \mathrm{Tr}(\varrho) = \mathrm{Tr}(\psi\psi^\dagger) = \|\psi\|^2 \tag{7.121}$$

i.e. the purification ψ must belong to the *Hilbert-Schmidt unit sphere*[11]

$$S_{\mathrm{HS}} \equiv \{a \in \mathcal{H} \otimes \mathcal{H}^\vee : \mathrm{Tr}(aa^\dagger) = 1\} \subset \mathcal{H} \otimes \mathcal{H}^\vee. \tag{7.122}$$

Restricting the bundle (7.98) to the space $\mathcal{S}(n) \subset \mathcal{H} \otimes \mathcal{H}^\vee$ ($\dim \mathcal{H} = n$) of *normalized* $n \times n$ density matrices (states) we get a projection

$$\pi : S_{\mathrm{HS}} \to \mathcal{S}(n), \qquad a \mapsto aa^\dagger, \tag{7.123}$$

whose total space is the Hilbert-Schmidt unit sphere S_{HS} which becomes a principal $U(n)$ bundle when restricted to the open dense domain

$$\mathcal{S}(n) \cap \mathcal{P}_H(n) \subset \mathcal{S}(n) \tag{7.124}$$

of *positive-definite* density matrices. Computing the minimum of the distance along the unit sphere S_{HS}—instead than in the ambient Hilbert-Schmidt Euclidean space as we did in the previous paragraph to end up with Eq. (7.120)—yields the *second Bures distance* $D_A(\cdot, \cdot)$. The two distances are related by the cord-length/arc-length relation in Euclidean geometry:

$$\cos D_A(\varrho_1, \varrho_2) = 1 - \frac{1}{2} D_B(\varrho_1, \varrho_2)^2 \tag{7.125}$$

D_A is called the *Bures angular distance* (or simply *Bures angle*). The Riemannian metric $\mathrm{d}s_A^2$ which yields the distance function $D_A(\cdot, \cdot)$ is the one induced by $\mathrm{d}s_B^2$ on the submanifold $\mathcal{S}(n) \subset \overline{\mathcal{P}}_H(n)$ of normalized states. Explicitly: now the eigenvalues λ_i are restricted by the linear constraint

$$\lambda_1 + \lambda_2 + \cdots + \lambda_n = 1, \tag{7.126}$$

and we plug-in this constraint in Eq. (7.120) i.e.

$$\mathrm{d}s_A^2 = \mathrm{d}s_B^2\Big|_{\lambda_n = 1 - \lambda_1 - \cdots - \lambda_{n-1}}. \tag{7.127}$$

---

[11] I.e. the sphere defined by the Hilbert-Schmidt Euclidean metric in $\mathcal{H} \otimes \mathcal{H}^\vee \simeq \mathbb{R}^{2n^2}$.

Angular Bures Geodesics We can easily describe the *geodesics* of the angular Bures metric $\mathrm{d}s_A^2$. An angular Bures geodesic is the projection to $\mathcal{S}(n)$ of a geodesic $\tilde{\gamma}$ on the unit sphere $S_{\mathrm{HS}} \subset \mathcal{P}_H(n)$ with the property that the tangent to $\tilde{\gamma}$ is everywhere orthogonal to the fibers of π. A geodesic on the sphere looks like

$$\psi(t) = \psi(0)\cos t + \dot{\psi}(0)\sin t \tag{7.128}$$

with

$$\mathrm{Tr}\big(\psi(0)\psi(0)^\dagger\big) = \mathrm{Tr}\big(\dot{\psi}(0)\dot{\psi}(0)^\dagger\big) = 1, \tag{7.129}$$

$$\mathrm{Tr}\big(\psi(0)\dot{\psi}(0)^\dagger + \dot{\psi}(0)\psi(0)^\dagger\big) = 0. \tag{7.130}$$

The condition of orthogonality with the fibers, Eq. (7.109), then yields

$$\dot{\psi}(0)^\dagger\psi(0) = \psi(0)^\dagger\dot{\psi}(0). \tag{7.131}$$

Given a density matrix $\psi(0)$ and any Hermitian matrix $\dot{\psi}(0)$ satisfying Eqs. (7.129), (7.130), and (7.131) we get a geodesic of the Bures angular metric $\mathrm{d}s_A^2$. It is easy to see from these expressions that the density matrices with a given block structure form a totally geodesic subspace of $\mathcal{S}(n)$.

7.3.2 Fidelity and Uhlmann's Theorem

We need a measure of how different two states ϱ_1, ϱ_2 are. For the pure states this measure is given by the deviation from 1 of the overlap between the two (normalized) states

$$1 - |\langle\psi_1|\psi_2\rangle|^2, \tag{7.132}$$

or, equivalently by the Fubini-Study metric $D_{\mathrm{FS}}(\cdot,\cdot)$

$$\cos D_{\mathrm{FS}}(|\psi_1\rangle, |\psi_2\rangle) = |\langle\psi_1|\psi_2\rangle|. \tag{7.133}$$

The overlap ($\equiv$ transition amplitude) is a measure of how "near" two pure states are. For two general states $\varrho_1, \varrho_2 \in \mathcal{B}(\mathcal{H})$ we replace the overlap by their *fidelity*.

Definition 7.8 The *fidelity* of the two states ϱ_1 and ϱ_2 is

$$F(\varrho_1, \varrho_2) = \max \big|\mathrm{Tr}(\psi_1\psi_2^\dagger)\big|^2 \tag{7.134}$$

where the maximum is taken over all possible purifications ψ_1, ψ_2 of the two states ϱ_1, ϱ_2.

The fidelity $F(\varrho_1, \varrho_2)$ is non-negative, vanishes when ϱ_1 and ϱ_2 have support on mutually orthogonal subspaces, and attains its maximaal value 1 if and only if the two states are identical. Moreover for pure states the fidelity reduces to the square-overlap

$$F\big(|\psi_1\rangle\langle\psi_1|, |\psi_2\rangle\langle\psi_2|\big) = \max_{U_1, U_2} \Big|\mathrm{Tr}\big(|\psi_1\rangle\langle\psi_1|U_1 U_2^{\dagger}|\psi_2\rangle\langle\psi_2|\big)\Big|^2 = $$
$$= \big|\langle\psi_2|\psi_1\rangle\big|^2 \, \max_{U} \Big|\langle\psi_1|U|\psi_2\rangle\Big|^2 = \big|\langle\psi_1|\psi_2\rangle\big|^2, \tag{7.135}$$

and more generally for a pure state $|\psi\rangle$ and a general one ϱ

$$F\big(|\psi\rangle\langle\psi|, \varrho\big) = \langle\psi|\varrho|\psi\rangle. \tag{7.136}$$

The quantity (7.134) is called *fidelity* for the following reason: suppose we have a general state ϱ and we wish to approximate it by a pure state $|\psi\rangle$. The accuracy ($\equiv$ "fidelity") of the approximation is clearly given by

$$F = \langle\psi|\varrho|\psi\rangle \tag{7.137}$$

($F = 1$ is 100% accuracy, $F = 0.98$ is 98% accuracy, etc.).

The Bures distance in the convex cone of non-negative Hermitian matrices $\overline{\mathcal{P}}_H(n)$ is also a measure of how different two states are. It is natural to look for its relation with fidelity. Comparing with Eq. (7.106) we get

Lemma 7.2 *The square distance in $\overline{\mathcal{P}}_H(n)$ is given by*

$$D_B(\varrho_1, \varrho_2)^2 = \mathrm{Tr}\, \varrho_1 + \mathrm{Tr}\, \varrho_2 - 2\sqrt{F(\varrho_1, \varrho_2)}. \tag{7.138}$$

The corresponding formula for the angular distance D_A in $\mathcal{S}(n)$ is

$$\cos D_A(\varrho_1, \varrho_2) = \sqrt{F(\varrho_1, \varrho_2)}. \tag{7.139}$$

The fidelity may be written directly in terms of the two states ϱ_1, ϱ_2:

Theorem 7.4 (Uhlmann's Theorem)   *One has*

$$F(\varrho_1, \varrho_2) = \big(\mathrm{Tr}|\sqrt{\varrho_2}\sqrt{\varrho_1}|\big)^2 = \left(\mathrm{Tr}\sqrt{\sqrt{\varrho_2}\,\varrho_1\sqrt{\varrho_2}}\right)^2. \tag{7.140}$$

Proof Let ψ_1, ψ_2 be purifications of (respectively) ϱ_1, ϱ_2. The polar decomposition (7.84) allows to write

$$\psi_1 = \sqrt{\varrho_1}\, U_1, \qquad \psi_2 = \sqrt{\varrho_2}\, U_2. \tag{7.141}$$

Then

$$\mathrm{Tr}(\psi_1\psi_2^\dagger) = \mathrm{Tr}(\sqrt{\varrho_1}\,U_1 U_2^\dagger\sqrt{\varrho_2}) = \mathrm{Tr}(\sqrt{\varrho_2}\sqrt{\varrho_1}\,U_1 U_2^\dagger). \tag{7.142}$$

We perform another polar decomposition

$$\sqrt{\varrho_2}\sqrt{\varrho_1} = \left|\sqrt{\varrho_2}\sqrt{\varrho_1}\right| V, \tag{7.143}$$

and write $U = V U_1 U_2^\dagger \in \mathcal{U}$. Then

$$F(\varrho_1,\varrho_2) = \max_{U\in\mathcal{U}} \left|\mathrm{Tr}\left(\left|\sqrt{\varrho_2}\sqrt{\varrho_1}\right| U\right)\right|^2. \tag{7.144}$$

Since $\left|\sqrt{\varrho_2}\sqrt{\varrho_1}\right|$ is positive, the maximum is obtained for $U = 1$. This yields
Eq. (7.140). $\qquad\qquad\square$

7.3.3　Schrödinger Mixing Theorem

The following result is sometimes called the *Schrödinger mixing theorem* (an
equivalent result is the already mentioned HJW theorem[12]). To formulate it properly,
we start by recalling some general notions of Matrix Analysis [2, 19].

Vector Majorization　Given a vector $x = (x_1,\dots,x_n) \in \mathbb{R}^n$ we write $x^\downarrow$ for the
vector obtained by re-arranging the components of x in a *non-increasing* order

$$x^\downarrow \equiv (x_1^\downarrow,\dots,x_n^\downarrow) \overset{\mathrm{def}}{=} (x_{\pi(1)}, x_{\pi(2)}, \cdots, x_{\pi(n)}) \tag{7.145}$$

where $\pi \in \mathfrak{S}_n$ is such that $x_i^\downarrow \equiv x_{\pi(i)} \geq x_{\pi(i+1)} \equiv x_{i+1}^\downarrow \ \forall\, i$.

We introduce a *partial order* $\prec$ in $\mathbb{R}^n$, written $x \prec y$ (x *precedes* y, or y *majorizes*
x), by

$$x \prec y \quad\Leftrightarrow\quad
\begin{cases}
\displaystyle\sum_{i=1}^{k} x_i^\downarrow \leq \sum_{i=1}^{k} y_i^\downarrow & k = 1, 2, \dots, n-1 \\[2mm]
\displaystyle\sum_{i=1}^{n} x_i^\downarrow = \sum_{i=1}^{n} y_i^\downarrow &
\end{cases} \tag{7.146}$$

---

[12] For Hughston, Jozsa, and Wootters [18].

Definition 7.9 A $n \times n$ matrix A_{ij} is called *bistochastic* (or *doubly stochastic*) iff the following two conditions hold

$$1) \quad A_{ij} \geq 0 \quad \text{for all } i, j \tag{7.147}$$

$$2) \quad \sum_{i=1}^{n} A_{ij} = \sum_{i=1}^{n} A_{ji} = 1 \ \text{ for all } j. \tag{7.148}$$

Lemma 7.3 (Schur) *(1) Let q, $p \in \mathbb{R}^n$. $q \prec p$ if and only if there is a bistochastic $n \times n$ matrix A such that $q = Ap$. (2) The matrix A may be chosen in the form $A_{ij} = |O_{ij}|^2$ for O_{ij} an orthogonal matrix.*

For a proof see e.g. **Theorem II.1.10** in [19]. *Birkhoff's theorem* states that the space of bistochastic matrices is convex with the permutation matrices as its extremal points [19].

Schrödinger Mixing Theorem We start with an ensemble of quantum pure states $\{|\psi_i\rangle\}_{i \in I}$, which are normalized *but not* necessarily pairwise orthogonal. The corresponding density matrix is

$$\varrho = \sum_i p_i |\psi_i\rangle\langle\psi_i|, \tag{7.149}$$

where $p_i \geq 0$ is the probability of the state $|\psi_i\rangle$ and $\sum_i p_i = 1$. Our task is to characterize the set of such *pure state ensembles*

$$\{|\psi_i\rangle, p_i\}_{i \in I} \tag{7.150}$$

which share the *same* density matrix ϱ. One also says that the ensemble $\{|\psi_i\rangle, p_i\}_{i \in I}$ *realizes* the state ϱ.

Suppose we have a second such ensemble[13] $\{|\phi_i\rangle, q_i\}_{i \in I}$ with probabilities q_i

$$\varrho = \sum_i q_i |\phi_i\rangle\langle\phi_i|. \tag{7.151}$$

We equate the purifications of ϱ in $\mathcal{B}(\mathcal{H} \otimes \mathbb{C}^{|I|})$ as computed from the two ensembles

$$\sum_i \sqrt{p_i}\, |\psi_i\rangle \otimes |\alpha_i\rangle = \sum_i \sqrt{q_i}\, |\phi_i\rangle \otimes |\beta_i\rangle, \tag{7.152}$$

---

[13] We take the cardinality $|I|$ of I to be equal to the maximum of the number of pure states in the two ensembles, adding, if necessary, extra states with probability zero to the smaller ensemble to make the two ensembles of equal cardinality.

where $|\alpha_i\rangle$, $|\beta_i\rangle$ are suitable orthonormal bases in $\mathbb{C}^{|I|}$. Any two orthonormal bases are related by a unitary transformation; let U be the unitary transformation such that $|\beta_i\rangle = U|\alpha_i\rangle$. Writing $U_{ji} = \langle\alpha_j|U|\alpha_i\rangle$, we have

$$|\beta_i\rangle = \sum_j |\alpha_j\rangle U_{ji}. \tag{7.153}$$

Plugging this expression in (7.152) we get

$$\sqrt{p_i}\,|\psi_i\rangle = \sum_j \sqrt{q_j}\,U_{ij}|\phi_j\rangle. \tag{7.154}$$

We have proven item **(1)** in the following

Theorem 7.5 (Mixing Theorem)

(1) *The two ensembles of pure states $|\{|\psi_i\rangle, p_i\}_{i\in I}$ and $|\{|\phi_i\rangle, q_i\}_{i\in I}$ have the same density matrix ϱ if and only if we can find a $|I| \times |I|$ unitary matrix U_{ij} such that Eq. (7.154) holds.*
(2) *Given a density matrix ϱ and a probability vector $\boldsymbol{p} = (p_i)_{i\in I}$, an ensemble $|\{|\psi_i\rangle, p_i\}_{i\in I}$ with*

$$\varrho = \sum_i p_i|\psi_i\rangle\langle\psi_i| \tag{7.155}$$

exists if and only if

$$\boldsymbol{p} \prec \boldsymbol{\lambda} \tag{7.156}$$

where $\boldsymbol{\lambda} = (\lambda_1, \ldots, \lambda_n)$ is the vector of eigenvalues of ϱ. That is,

$$\varrho = \sum_i \lambda_i\,|e_i\rangle\langle e_i| \tag{7.157}$$

with $\{|e_i\rangle\}$ the orthonormal *system of eigenvectors of ϱ.*

Proof It remains to prove **(2)**. The eigenvector orthonormal ensemble $\{|e_i\rangle, \lambda_i\}_{i\in I}$ realizes ϱ. Then from Eq. (7.154) we have

$$\sqrt{p_i}\,|\psi_i\rangle = \sum_j \sqrt{\lambda_j}\,U_{ij}|e_j\rangle, \tag{7.158}$$

for some unitary matrix U_{ij}. Multiplying this equation by $\langle \psi_i |$, and respectively $\langle e_j |$, we get (no summation over repeated indices unless explicitly indicated):

$$\sqrt{p_i} = \sum_j \sqrt{\lambda_j}\, U_{ij}\, \langle \psi_i | e_j \rangle \tag{7.159}$$

$$\sqrt{p_i}\, \langle e_j | \psi_i \rangle = \sqrt{\lambda_j}\, U_{ij}. \tag{7.160}$$

Then

$$p_i = \sqrt{p_i} \cdot \sqrt{p_i} = \sum_j \sqrt{\lambda_j}\, U_{ij}\, \sqrt{p_i}\, \langle \psi_i | e_j \rangle =$$
$$= \sum_j \sqrt{\lambda_j}\, U_{ij}\, \sqrt{\lambda_j}\, U_{ij}^* = \sum_j U_{ij} U_{ij}^*\, \lambda_j = \sum_j D_{ij} \lambda_j \tag{7.161}$$

where the matrix $D_{ij} = |U_{ij}|^2$ is bistochastic. The **Theorem** now follows from Schur's Lemma 7.3. $\qquad\square$

7.4 Dynamical Evolution and TPCP Maps

Having discussed the *states* of an open quantum system, we now examine its *time evolution* which, starting from a generic state at time t_0, that is, from an initial density matrix $\varrho(t_0) \in \mathcal{S} \subset \mathcal{B}(\mathcal{H})$, returns the evolved state $\varrho(t)$ at time t. The time evolution is a map

$$\mathcal{E}_t : \mathcal{S} \to \mathcal{S} \tag{7.162}$$

which should preserve all defining structures of the space of states $\mathcal{S}$. Such maps are usually called *quantum channels,* a name borrowed from Quantum Information theory. The map $\mathcal{E}_t$ is *not unitary* (in general), and our first task is to characterize the maps which *may be* time-evolutions of open systems, i.e. we need to understand the property that generalizes to open systems the unitary condition which applies to the closed ones.

To motivate the following math definition, we start at time $t = 0$ from a product pure state of the closed system composed by our system S and the environment E

$$|\psi_0\rangle \otimes |a_0\rangle \in \mathcal{H}_S \otimes \mathcal{H}_E. \tag{7.163}$$

We choose an orthonormal basis $\{|a\rangle\}$ for the environment Hilbert space $\mathcal{H}_E$. The combined system is closed, so its time evolution is given by a unitary operator

$$U(t) \in \mathcal{U}(\mathcal{H}_S \otimes \mathcal{H}_E). \tag{7.164}$$

Expanding $U(t)$ in the basis $\{|a\rangle\}$ of $\mathcal{H}_E$, we may write the state at time t in the form

$$U(t)(|\psi_0\rangle \otimes |a_0\rangle) = \sum_a M_a(t)|\psi_0\rangle \otimes |a\rangle \tag{7.165}$$

for certain linear operators $M_a \equiv M_a(t) \in \mathcal{B}(\mathcal{H}_S)$. Since $U(t)$ is unitary in the big space $\mathcal{H}_S \otimes \mathcal{H}_E$, the norm of any initial state is preserved

$$
\begin{aligned}
1 &= \||\psi_0\rangle \otimes |a_0\rangle\|^2 = \|U(t)(|\psi_0\rangle \otimes |a_0\rangle)\|^2 = \\
&= \sum_{a,b} \langle a| \otimes ((\langle\psi_0|M_a^\dagger)(M_b|\psi_0\rangle)) \otimes |b\rangle = \sum_{a,b} \langle a|b\rangle \ \langle\psi_0|M_a^\dagger M_b|\psi_0\rangle \\
&= \sum_a \langle\psi_0|M_a^\dagger M_a|\psi_0\rangle,
\end{aligned}
\tag{7.166}
$$

an equality which is true for all normalized initial states $|\psi_0\rangle$ if and only if

$$\sum_a M_a^\dagger M_a = 1. \tag{7.167}$$

The initial state

$$\varrho_0 = |\psi_0\rangle\langle\psi_0| \in \mathcal{B}(\mathcal{H}_S) \tag{7.168}$$

will then evolve in the state given by the partial trace over the environment of the evolved state (7.165)

$$\mathrm{Tr}_{\mathcal{H}_E}\left(\sum_{a,b} M_a(t)|\psi_0\rangle \otimes |a\rangle\langle b| \otimes \langle\psi_0|M_b^\dagger\right) = \sum_a M_a(t)\varrho_0 M_a(t)^\dagger. \tag{7.169}$$

While we assumed that the initial state was pure, the map

$$\mathcal{E}_t : \varrho \mapsto \sum_a M_a(t)\varrho\, M_a(t)^\dagger \tag{7.170}$$

is linear in ϱ, so it extends by linearity to arbitrary initial states $\varrho \in \mathcal{S}$. We conclude:

> **Time Evolution in Open Quantum Systems**
>
> In an open quantum system (that we see as a closed bi-partite quantum system composed by our system S and its environment E) the time evolution of states is given by a *quantum channel*, that is, by a linear map of the form
>
> $$\varrho \mapsto \mathcal{E}(\varrho) \equiv \sum_a M_a \varrho\, M_a^\dagger, \qquad (7.171)$$
>
> where the operators $M_a \in \mathcal{B}(\mathcal{H}_S)$, called the *Kraus operators* of the channel $\mathcal{E}$, satisfy the completeness equation
>
> $$\sum_a M_a^\dagger M_a = 1 \qquad (7.172)$$

Quantum channels are also called *trace-preserving completely positive maps*, (TPCP maps) because of the properties which characterize them.

Fact 7.6 *As a map $\mathcal{E}\colon \mathcal{B}(\mathcal{H}) \to \mathcal{B}(\mathcal{H})$ a quantum channel satisfies:*

QC1   linearity: $\mathcal{E}(\alpha\varrho_1 + \beta\varrho_2) = \alpha\,\mathcal{E}(\varrho_1) + \beta\,\mathcal{E}(\varrho_2)\, for\ \alpha,\ \beta \in \mathbb{C};$
QC2 $\mathcal{E}$ *commutes with* $\dagger$: $\mathcal{E}(\varrho^\dagger) = \mathcal{E}(\varrho)^\dagger$, *hence preserve Hermiticity;*
QC3 $\mathcal{E}$ *preserves positivity:* $\varrho \geq 0 \Rightarrow \mathcal{E}(\varrho) \geq 0;$
QC4 $\mathcal{E}$ *preserves traces:* $\mathrm{Tr}\,\mathcal{E}(\varrho) = \mathrm{Tr}\varrho.$

In particular a quantum channel $\mathcal{E}$ preserves the subspace $\mathcal{S} \subset \mathcal{B}(\mathcal{H})$ of quantum states and hence maps states into states: $\mathcal{E}\colon \mathcal{S} \to \mathcal{S}.$

However to get an "if and only if" characterization of the quantum channel maps we need to replace the *positivity* property **QC3** by the stronger condition of *complete positivity*. This will be done in the next subsection. Before going to that, we make some general considerations.

Quantum Decoherence
It follows from Eqs. (7.171) and (7.172) that the time-evolution $\mathcal{E}_t$ is unitary if and only if there is *only one* non-zero Kraus operator. When the evolution is non-unitary we may start with a pure state and it will evolve into a *mixed one* as a result of its interactions with the environment. A process in which a pure state $|\psi_0\rangle \in \mathcal{H}_S$ becomes *mixed* is said to produce *(quantum) decoherence* because it blurs the coherent quantum interference arising from complex linear combinations of state *vectors*: in Chap. 1 we used interference of photons to conclude that the quantum states must be represented by vectors in a complex space—conversely when states *are no longer represented by vectors,* the interference fringes fade away and the photons behave *incoherently*. Quantum decoherence happens all the time in Nature, because all real systems have an environment. Quantum decoherence

destroys quantum interference. Very roughly speaking, its "typical" upshot is to effectively "average" over the phases of certain "stationary" states $\{|n\rangle\}$

$$
|\psi\rangle \equiv \sum_n c_n |n\rangle \rightsquigarrow \sum_{m,n} c_m\, c_n^* \oint \frac{\mathrm{d}\phi_m\, \mathrm{d}\phi_n}{(2\pi)^2} e^{\mathrm{i}\phi_m} |m\rangle \langle n| e^{-\mathrm{i}\phi_n} =
$$
$$
= \sum_n |c_n|^2\, |n\rangle\langle n|,
$$
(7.173)

a time-evolution which is produced, say, by the quantum channel with Kraus operators (here $N \equiv \dim \mathcal{H}$; for simplicity we assume $\{|n\rangle\}$ to be a complete orthonormal system)

$$
M_k = \frac{1}{2^{N/2}} \sum_{n=1}^{N} (-1)^{k_n} |n\rangle\langle n|, \qquad \sum_{k \in (\mathbb{Z}/2\mathbb{Z})^N} M_k^\dagger M_k = 1,
$$
(7.174)

$$
\text{where} \qquad k \equiv (k_1, \ldots, k_N) \in (\mathbb{Z}/2\mathbb{Z})^N .
$$

The "phase average" destroys the coherent interference pattern encoded in the relative phases of the complex coefficients $\{c_n\}$. From the viewpoint of the observables, the phase average is equivalent to

$$
\langle m|\mathcal{O}|n\rangle \equiv \mathcal{O}_{m,n} \rightsquigarrow \oint \frac{\mathrm{d}\phi_m\, \mathrm{d}\phi_n}{(2\pi)^2}\, e^{-\mathrm{i}\phi_m} \langle m|\mathcal{O}|n\rangle\, e^{\mathrm{i}\phi_n} = \delta_{mn}\, \mathcal{O}_{n,n}.
$$
(7.175)

which (crudely speaking) makes the algebra of observables effectively "commutative". Quantum decoherence explains why so many physical systems look to "behave classically" (in certain respects).

Non-Uniqueness of the Kraus Operators

The Kraus representation (7.171) of the channel $\mathcal{E}$ is *not unique*. Indeed to produce it we used an arbitrary orthonormal basis $\{|a\rangle\}$ of the environment Hilbert space $\mathcal{H}_E$. We could have used a different basis, $\{|\tilde{a}\rangle\}$ related to $\{|a\rangle\}$ by a unitary matrix $\langle \tilde{b}|V|a\rangle$ with $V \in \mathcal{U}(\mathcal{H}_E)$

$$
|a\rangle = \sum_{\tilde{b}} |\tilde{b}\rangle\langle \tilde{b}|V|a\rangle,
$$
(7.176)

and then

$$
U(t)(|\psi_0\rangle \otimes |a_0\rangle) = \sum_a M_a |\psi_0\rangle \otimes |a\rangle =
$$
$$
= \sum_a \sum_{\tilde{b}} \langle \tilde{b}|V|a\rangle\, M_a |\psi_0\rangle \otimes |\tilde{b}\rangle = \sum_{\tilde{b}} \tilde{M}_{\tilde{b}} |\psi_0\rangle \otimes |\tilde{b}\rangle,
$$
(7.177)

where the new Kraus operators $\tilde{M}_{\tilde{b}}$ are related to the old ones M_a by

$$\tilde{M}_{\tilde{b}} = \sum_a \langle \tilde{b}|V|a\rangle M_a. \tag{7.178}$$

Semigroup Law
Clearly two channels may be composed to give a third channel

$$\mathcal{E}_1 \circ \mathcal{E}_2(\varrho) = \sum_{a_1,a_2} M_{1,a_1} M_{2,a_2}\, \varrho\, M_{2,a_2}^\dagger M_{1,a_1}^\dagger \tag{7.179}$$

so that they form a *semigroup*. Indeed, the time evolution from today to tomorrow can be composed with the evolution from tomorrow to the day after tomorrow, to produce a 48-hour time evolution.

Irreversibility
While the quantum channels form a semigroup, they *do not* form a group in general, since typically they have *no inverse.*

Fact 7.7 *A quantum channel $\mathcal{E}$ is invertible if and only if it is unitary:*

$$\mathcal{E}: \varrho \mapsto U\varrho U^\dagger, \qquad U \in \mathcal{U}(\mathcal{H}). \tag{7.180}$$

Proof Assume that our channel $\mathcal{E}$, with Kraus operators $\{M_a\}$, has an inverse $\mathcal{E}^{-1}$ which is also a channel with Kraus operators $\{\tilde{M}_{\tilde{a}}\}$. Then, for all pure state $|\psi\rangle \in \mathcal{H}_S$,

$$|\psi\rangle\langle\psi| = \mathcal{E}^{-1} \circ \mathcal{E}(|\psi\rangle\langle\psi|) = \sum_{a,\tilde{a}} \tilde{M}_{\tilde{a}} M_a |\psi\rangle\langle\psi| M_a^\dagger \tilde{M}_{\tilde{a}}^\dagger. \tag{7.181}$$

The RHS is a sum of positive terms; adapting the proof of Lemma 7.1, we infer that the equality (7.181) requires each term in the RHS to be proportional to the LHS, i.e.

$$\tilde{M}_{\tilde{a}} M_a = \lambda_{\tilde{a}a} \cdot \mathbf{1} \qquad \lambda_{\tilde{a}a} \in \mathbb{C}, \ \ \forall\, \tilde{a}, a, \tag{7.182}$$

and then, using the completeness relation,

$$M_b^\dagger M_a = M_b^\dagger \Big(\sum_{\tilde{a}} \tilde{M}_{\tilde{a}}^\dagger \tilde{M}_{\tilde{a}} \Big) M_a = \sum_{\tilde{a}} \lambda_{\tilde{a}b}^* \lambda_{\tilde{a}a} \cdot \mathbf{1} \equiv \mu_{ba} \cdot \mathbf{1}, \tag{7.183}$$

$\mu_{ba} \in \mathbb{C}$ while the diagonal numbers $\mu_{aa} \geq 0$ with equality iff $M_a = 0$. The polar decomposition of M_a is

$$M_a = U_a \sqrt{M_a^\dagger M_a} = \mu_{aa}^{1/2} U_a, \tag{7.184}$$

and now (7.183) yields

$$M_b^\dagger M_a = (\mu_{aa}\,\mu_{bb})^{1/2}\, U_b^\dagger U_a = \mu_{ba} \cdot \mathbf{1}, \tag{7.185}$$

that is, for all a, b

$$U_a = \frac{\mu_{ba}}{(\mu_{aa}\mu_{bb})^{1/2}}\, U_b, \tag{7.186}$$

and hence all Kraus operators M_a are proportional to the *same* unitary operator U, i.e. $\mathcal{E}$ is an unitary map as in Eq. (7.180). $\qquad\qquad\square$

Discussion Fact 7.7 shows that the time evolution of an open quantum system is *non-reversible*—due to the interactions with its environment—*even if* the basic laws of Quantum Mechanics are time-reversal symmetric. In other words, the interactions with the environment produce a "friction" which generates dissipative effects that lead to quantum decoherence. Decoherence is, of course, irreversible. The dissipation is associated to both quantum and statistical fluctuations. The precise relation between fluctuations and dissipation is described by the quantum *dissipation-fluctuation theorem* which is stated and proven in chap. 6 of [20].

Stinespring Dilation
Purification means that a state $\varrho \in \mathcal{B}(\mathcal{H})$ of an open quantum system can always be made into a pure state by introducing an auxiliary "environment" Hilbert space $\mathcal{H}_E$ and going to the bigger Hilbert space $\mathcal{H} \otimes \mathcal{H}_E$. There is a corresponding statement for the quantum channels $\mathcal{E}_t$ which yield the time-evolution of the open system.

Fact 7.8 (Stinespring Dilation) *All linear map $\mathcal{E}\colon \mathcal{B}(\mathcal{H}) \to \mathcal{B}(\mathcal{H})$ of the form (7.171), where the Kraus operators M_a satisfy the completeness relation (7.172), has a dilation ($\equiv$ extension) to a unitary transformation $U \in \mathcal{U}(\mathcal{H} \otimes \mathcal{H}_E)$ acting on a suitable bigger Hilbert space $\mathcal{H} \otimes \mathcal{H}_E$.*

This is essentially the converse of the construction in Eqs. (7.163)–(7.170) which produced the Kraus representation of the quantum channel. The total system which includes all the environment is a closed system, and its time evolution is unitary. This total unitary evolution is the *Stinespring dilation* of the quantum channel $\mathcal{E}$.

7.4.1 Complete Positivity

We return to the issue of *complete positivity* of a quantum channel. By definition this property means that the channel remains positive when we consider our system to be a subsystem of *any* larger system. To formalize the notion, we start with a channel

$$\mathcal{E}\colon \mathcal{B}(\mathcal{H}_A) \to \mathcal{B}(\mathcal{H}_A), \tag{7.187}$$

enlarge the Hilbert space to $\mathcal{H}_A \otimes \mathcal{H}_B$, and identify $\mathcal{E}$ with the linear map

$$(\mathcal{E} \otimes \mathbf{1}_B): \mathcal{B}(\mathcal{H}_A \otimes \mathcal{H}_B) \to \mathcal{B}(\mathcal{H}_A \otimes \mathcal{H}_B). \tag{7.188}$$

If the map $\mathcal{E}$ remains positive for all such extensions of the Hilbert space, we say that $\mathcal{E}$ is *completely positive*. Quantum channels of the form (7.171) are completely positive since, if M_a are Kraus operators for $\mathcal{E}$, $M_a \otimes \mathbf{1}_B$ are Kraus operators for $\mathcal{E} \otimes \mathbf{1}_B$ for all Hilbert spaces $\mathcal{H}_B$.

Example 7.2 (A Positive But Not Completely Positive Linear Map) Consider the linear map $T: \mathcal{B}(\mathcal{H}) \to \mathcal{B}(\mathcal{H})$ given by transposition

$$T: |i\rangle\langle j| \mapsto |j\rangle\langle i|. \tag{7.189}$$

It is clearly positive. But now consider two copies of the system with Hilbert space $\mathcal{H} \otimes \mathcal{H}$, and let $T \otimes \mathbf{1}$ act on the pure state

$$|\Phi\rangle = \sum_i |i\rangle \otimes |i\rangle, \tag{7.190}$$

where $\{|i\rangle\}$ is an orthonormal basis of $\mathcal{H}$. One has

$$T \otimes \mathbf{1}: |\Phi\rangle\langle\Phi| \equiv \sum_{i,j} (|i\rangle\langle j|) \otimes (|i\rangle\langle j|) \mapsto \sum_{i,j} (|j\rangle\langle i|) \otimes (|i\rangle\langle j|). \tag{7.191}$$

The RHS is the operator which interchanges the two copies of the system. $(T \otimes 1)^2$ is the identity, so $T \otimes 1$ has eigenvalue $+1$ (resp. -1) for states symmetric (resp. anti-symmetric) under the exchange. Thus $T \otimes \mathbf{1}$ has negative eigenvalues and T is *not* totally positive.

The conclusions of our physical discussion may be translated in a rigorous mathematical theorem:

Theorem 7.9 *A linear map $\mathcal{E}: \mathcal{B}(\mathcal{H}) \to \mathcal{B}(\mathcal{H})$ is completely positive and trace preserving if and only if it has the Kraus form (7.171):*

$$\varrho \to \mathcal{E}(\varrho) \equiv \sum_a M_a \varrho M_a^\dagger, \tag{7.192}$$

and the Kraus operators M_a satisfy the completeness equation

$$\sum_a M_a^\dagger M_a = 1. \tag{7.193}$$

That is: the TPCP maps are precisely the Kraus quantum channels.

For a proof see e.g. §. 10.3 of [2] or chap. 3 of the Lecture Notes [10].

7.4.2 Measures and POVM

Finally we discuss the measurement of physical quantities in open quantum systems. We perform the measure of an observable $\mathcal{O}$ which has a set R of possible values (we think of R as *discrete* for simplicity). In the case of closed systems, the measurement of $\mathcal{O}$ is described by its spectral family of orthogonal projectors $\{P_a\}_{a \in R}$:

$$P_a\, P_b = \delta_{ab}\, P_a, \qquad P_a^\dagger = P_a, \qquad \sum_{a \in R} P_a = 1. \tag{7.194}$$

When our system is in the pure state $|\psi\rangle$, the measurement outcome a has probability

$$\mathsf{Prob}(a) = \langle\psi|P_a|\psi\rangle, \tag{7.195}$$

and after a measure with result a the system is in the state

$$\frac{P_a|\psi\rangle}{\|P_a|\psi\rangle\|}. \tag{7.196}$$

Hence, if we *do not record* its result, the measurement of $\mathcal{O}$ produces an ensemble of states with density matrix

$$\varrho_\mathrm{m} = \sum_{a \in R} \mathsf{Prob}(a)\, \frac{P_a|\psi\rangle\langle\psi|P_a}{\|P_a|\psi\rangle\|^2} = \sum_{a \in R} P_a|\psi\rangle\langle\psi|P_a. \tag{7.197}$$

By linearity this result extends to mixed states. The measurement of $\mathcal{O}$ modifies the state according to the rule

$$\varrho \mapsto \sum_a P_a \varrho\, P_a. \tag{7.198}$$

Next consider the interaction between our system and the measurement apparatus. (The "apparatus" contains the experimentalist, the laboratory, and everything else). To fix the ideas, we visualize the apparatus as a *pointer* which moves on a scale with marks labelled by the possible outcomes $a \in R$. The pointer is also a quantum system in its own right, so, when we see it in the position a along the scale, it really means that the pointer is in a specific quantum state that we call $|a\rangle \in \mathcal{H}_\mathrm{app}$. When the pointer is on mark a—that is, when the experimental apparatus is in its quantum state $|a\rangle \in \mathcal{H}_\mathrm{app}$—we bluntly assert "to have measured" the observable $\mathcal{O}$ for our system S getting as result the value a. Therefore the measurement is just some quantum process taking place in the combined Hilbert space $\mathcal{H} \otimes \mathcal{H}_{\mathrm{app}}$ which is described by a quantum channel

$$|\psi\rangle \otimes |0\rangle \to \sum_{a \in R} M_a |\psi\rangle \otimes |a\rangle, \qquad \sum_a M_a^\dagger M_a = 1, \qquad (7.199)$$

where $|\psi\rangle \in \mathcal{H}$ is the original state of our system and $|0\rangle$ is the initial state of the apparatus before the measure. The probability of the outcome a is then the probability of the apparatus ending in the state $|a\rangle \in \mathcal{H}_{\mathrm{app}}$, that is,

$$\mathsf{Prob}(a) = \sum_{b,c \in R} \langle b| \otimes \langle \psi | M_b^\dagger (1 \otimes P_a) M_c |\psi\rangle \otimes |c\rangle = \langle \psi | M_a^\dagger M_a |\psi\rangle =$$

$$= \| M_a |\psi\rangle \|^2 = \mathrm{Tr}(M_a \varrho\, M_a^\dagger) \equiv \mathrm{Tr}(\varrho\, E_a),$$

$$(7.200)$$

where

$$E_a \overset{\text{def}}{=} M_a^\dagger M_a \in \mathcal{B}(\mathcal{H}). \qquad (7.201)$$

By linearity the last expression in the RHS of (7.200) applies to all states $\varrho \in \mathcal{B}(\mathcal{H})$ not just to pure states.

The *measurement operators* $\{E_a\}_{a \in R}$ form a complete set of Hermitian non-negative operators. They satisfy the properties:

MO1 *Hermiticity:* $E_a = E_a^\dagger$;
MO2 *Positivity:* $E_a \geq 0$.
MO3 *Completeness:* $\sum_{a \in R} E_a = 1$.

Definition 7.10 A partition of the identity operator by non-negative operators

$$\sum_a E_a = 1 \qquad (7.202)$$

is called a *positive operator-valued measure,* or POVM for short.

The term "measure" reflects the fact that the spectrum of our observable $\mathcal{O}$ may have continuous components, so that in general we have a measure $E(\lambda)$ on the spectrum $\sigma(\mathcal{O}) \subset \mathbb{R}$ which is valued in non-negative operators. This generalizes the closed system ideal measurements, where the E_a's reduce the projectors P_a's in Eq. (7.194), which for a general observable $\mathcal{O}$ is replaced by its *spectral family* $\{P_\mathcal{O}(\lambda)\}$ (in the sense of Definition 2.3) described by the spectral theorem (Theorem 2.2). Thus the only difference between *ideal* measurements in closed systems, and *actual* measurements in real quantum systems, is that the measure on $\sigma(\mathcal{O})$ valued in projectors, $\mathrm{d}P_\mathcal{O}(\lambda)$, is replaced by a measure $\mathrm{d}E_\mathcal{O}(\lambda)$, still on $\sigma(\mathcal{O})$, but valued in the more general class of non-negative Hermitian operators which satisfy the same completeness condition.

However now the first Eq. (7.194) may not hold, i.e. *the measurement operators need not to be mutually orthogonal.* Therefore, when we repeat the measure a second time immediately after a first measure, we may get a different answer. The conditional probability of getting the outcome b in the second measure, when we got a in the first one, is

$$\text{Prob}(b|a) = \frac{\|M_b M_a |\psi\rangle\|^2}{\|M_a |\psi\rangle\|^2} \tag{7.203}$$

which is not zero for $b \neq a$ and general $|\psi\rangle$ unless $M_b M_a = \delta_{ab} M_a$, a condition which, in view of the completeness condition (7.192), says that $\{M_a\}_a$ is a family of orthogonal projectors as in (7.194). Indeed,

$$M_a = \left(\sum_b M_b^\dagger M_b \right) M_a = M_a^\dagger M_a \tag{7.204}$$

which shows that M_a is Hermitian, and then $M_a^2 = M_a^\dagger M_a = M_a$ is a projector.

TPCP maps and POVM are strictly related: any TPCP defines a PVOM by the correspondence

$$M_a \mapsto E_a \equiv M_a^\dagger M_a, \tag{7.205}$$

while a purification ψ_a of the non-negative operators E_a defines a TPCP map with Kraus operators $M_a = \psi_a^\dagger$.

7.5　Quantum Entanglement and Its Wonders

We know that a pure state $|\Psi\rangle$ of a bi-partite system with Hilbert space $\mathcal{H}_A \otimes \mathcal{H}_B$ is *entangled* iff the Schmidt number of the partial trace state

$$\varrho_A \equiv \text{Tr}_{\mathcal{H}_B}(|\Psi\rangle\langle\Psi|) \in \mathcal{B}(\mathcal{H}_A) \tag{7.206}$$

(hence of ϱ_B) is greater than 1. We assume (with no essential loss) that

$$\dim \mathcal{H}_B = \dim \mathcal{H}_A = n. \tag{7.207}$$

We say that the state $|\Psi\rangle$ is *maximally entangled* iff

$$\varrho_A = \frac{1}{n}\mathbf{1}_A, \qquad n \equiv \dim \mathcal{H}_A \tag{7.208}$$

which also implies $\varrho_B = \frac{1}{n}\mathbf{1}_B$ since the general purification $|\Psi\rangle$ of $\varrho_A \equiv \frac{1}{n}\mathbf{1}_A$ has the form

$$|\Psi\rangle = \frac{1}{n}\sum_{i=j=1}^{n}|i\rangle \otimes |j\rangle, \tag{7.209}$$

where $\{|i\rangle\}$ (resp. $\{|j\rangle\}$) is an orthonormal basis in $\mathcal{H}_A$ (resp. $\mathcal{H}_B$). The maximally entangled state $\frac{1}{n}\mathbf{1}_A$ is also called the *ignorance state* since the probability of *any* outcome $\{a_1, \cdots, a_n\}$ of the measurement of a generic[14] observable $\mathcal{O}_A$ of our subsystem (whose spectral projectors we write as $\{P_{a_i}\}$) has equal probability

$$\mathsf{Prob}(a_i) = \mathrm{Tr}_{\mathcal{H}_A}(P_{a_i}\varrho_A) = \frac{1}{n}\mathrm{Tr}(P_{a_i}) = \frac{1}{n}, \tag{7.210}$$

as we may expect a priori when we have no information at all on the state.

We now discuss a pair of prototypical examples of the subtle quantum effects produced by entanglement. These examples are the most frequently used in the applications (and discussions) of entanglement.

7.5.1   Two Qubits Entanglement: EPR States

We consider a quantum system consisting of two qubits: their Hilbert space is the tensor product of two copies of the spin-$\frac{1}{2}$ representation $V_{1/2} \simeq \mathbb{C}^2$ (cf. Sect. 7.1.1). The orthonormal basis of $V_{1/2}$ consisting of eigenvectors of $S_3 = \frac{1}{2}\sigma_3$ will be written $\{|0\rangle, |1\rangle\}$ as for the Fermi oscillator, cf. Sect. 4.12.2. This is also the standard notation in Quantum Information Theory (cf. Eq. (7.41)).

In the two qubit Hilbert space $V_{1/2} \otimes V_{1/2}$ we have four mutually orthogonal states all of which are *maximally entangled*

$$|\phi^{\pm}\rangle = \frac{1}{\sqrt{2}}\Big(|0\rangle \otimes |0\rangle \pm |1\rangle \otimes |1\rangle\Big)$$
$$|\psi^{\pm}\rangle = \frac{1}{\sqrt{2}}\Big(|0\rangle \otimes |1\rangle \pm |1\rangle \otimes |0\rangle\Big). \tag{7.211}$$

Note that the state $|\psi^-\rangle$ is antisymmetric under the exchange of the two qubits, so in the decomposition of $\mathfrak{su}(2)$-modules

$$V_{1/2} \otimes V_{1/2} = \big(V_{1/2} \odot V_{1/2}\big) \oplus \big(V_{1/2} \wedge V_{1/2}\big) \equiv V_1 \oplus V_0, \tag{7.212}$$

---

[14] "Generic" here means that all eigenvalues of the Hermitian operator $\mathcal{O}_A$ are distinct.

it corresponds to the state of *zero total spin,* while the other three states form a basis of the spin-1 representation space V_1. The four entangled vectors (7.211) are a basis of $V_{1/2}^{\otimes 2}$ formed by simultaneous eigenvectors of the *complete system* of commuting observables formed by the two operators

$$\sigma_3 \otimes \sigma_3, \quad \text{and} \quad \sigma_1 \otimes \sigma_1. \tag{7.213}$$

Both observables have eigenvalues ± 1 and hence square to 1. The eigenvalue of the first operator is called the *parity bit,* while the eigenvalue of the second operator is the *phase bit.* These two classical bits are encoded in the above quantum states as:

$$
\begin{aligned}
(\sigma_3 \otimes \sigma_3)|\phi^\pm\rangle &= |\phi^\pm\rangle, & (\sigma_3 \otimes \sigma_3)|\psi^\pm\rangle &= -|\psi^\pm\rangle \\
(\sigma_1 \otimes \sigma_1)|\phi^\pm\rangle &= \pm|\phi^\pm\rangle, & (\sigma_1 \otimes \sigma_1)|\psi^\pm\rangle &= \pm|\psi^\pm\rangle.
\end{aligned}
\tag{7.214}
$$

The four states (7.211) are then automatically eigenvectors also of

$$\sigma_2 \otimes \sigma_2 = -(\sigma_1 \otimes \sigma_1)(\sigma_3 \otimes \sigma_3). \tag{7.215}$$

The maximally entangled states of two qubits are called *EPR states*, after Einstein, Podolsky and Rosen [21] who first discussed the *gedanken* experiment which follows.

The EPR *Gedanken* Experiment
In this experiment somebody prepares the two qubit system in a maximal entangled state, and then sends its first qubit to Alice and the second one to Bob who live in two different Countries far away from each other. Alice and Bob can perform measurements only on their respective qubit, and have no access to the full system. An example of a real physical system which realizes this set-up is a spin zero particle which decays into two spin-$\frac{1}{2}$ particles (or two polarized photons). In the rest frame of the original particle, the total angular momentum is zero for both the initial and final state, so the resulting two qubit system must be in the maximally entangled state $|\psi^-\rangle$. In the rest frame of the decaying spin-0 particle, the two spin-$\frac{1}{2}$ particles move fast with opposite momenta, and after a while they will be largely separated in space. Neglecting the effects of the environment, their spins will remain entangled in the $|\psi^-\rangle$ state since angular momentum is conserved.

Acting locally on their respective spin ($\equiv$ qubit), neither Alice nor Bob can get any information about the identity of the full state. Indeed to identify the state they should be able to measure the two commuting operators $\sigma_3 \otimes \sigma_3$ and $\sigma_1 \otimes \sigma_1$, while Alice can measure only observables of the form

$$\sigma_a^{(A)} \equiv \sigma_a \otimes \mathbf{1} \qquad (a = 1, 2, 3) \tag{7.216}$$

and Bob only those of the form $\sigma_a^{(B)} \equiv \mathbf{1} \otimes \sigma_a$. What they *can do* locally in their respective Country is *to manipulate* the full state. For instance Alice may apply σ_3 to her qubit making

$$|\phi^\pm\rangle \leftrightarrow |\phi^\mp\rangle, \qquad |\psi^\pm\rangle \leftrightarrow |\psi^\mp\rangle. \tag{7.217}$$

In facts either Alice or Bob can perform a local unitary transformation U to her/his qubit that changes one maximally entangled state to any other maximal entangled state. However these transformations will not modify the partial trace density matrices

$$\varrho_A = \varrho_B = \frac{1}{2}\mathbf{1}, \tag{7.218}$$

hence they cannot know how the state was modified by the other person.

Next we assume that Alice and Bob can phone to each other, exchanging *classical* information. They can agree to measure, say, spins along the 3-rd direction, so that they measure, respectively, $\sigma_3^{(A)}$ and $\sigma_3^{(B)}$. Since these two operators commute with $\sigma_3 \otimes \sigma_3$, they can recover the first half of the information about the total state i.e. the $\sigma_3 \otimes \sigma_3$ eigenvalue (the parity bit). But, since $\sigma_3^{(A)}$ and $\sigma_3^{(B)}$ do not commute with $\sigma_1 \otimes \sigma_1$, they cannot recover the second half of the information (the phase bit).

Suppose the two qubit state is $|\psi^-\rangle$. If Alice measures σ_3 on her spin and finds $+1$ (resp. -1), and then Bob also measures σ_3, he gets -1 (resp. $+1$) *with certainty.* Thus the results of two experiments—performed at space-like distance one from the other—may be *correlated at the quantum level,* even if no interaction between the two systems exists. However Alice cannot use this quantum correlation to send faster-than-light messages to Bob, so the quantum correlations arising from entanglement yield no paradoxes with causality.

Hidden-Variable Theory   Despite the absence of logical contradictions, Einstein found disturbing the idea that an operation (experiment) performed here may affect the description of the state far away at a space-like distance, and proposed an alternative paradigm known as the *hidden variable theory.*

In the hidden variable proposal the "probabilistic" nature of the quantum measurements is not fundamental but *emergent:* it arises from the existence of hidden degrees of freedom $\{\lambda_i\}$ that we are not able to detect experimentally, whose values and dynamics are out of our control. The result $a(\lambda_i)$ of the measurement of an observable A will depend on the actual values λ_i of these hidden variables. We do not know anything about them, let alone their value, except that we know that they behave classically, so that their values λ_i are *well defined,* even if we cannot determine them. We parametrize our ignorance of their dynamics in a *classical* probability distribution $p(\lambda_i)$ for their values λ_i. The observed value a of the standard quantum observable A depends on the values of the hidden variables λ_i, hence a effectively becomes a random variable, since we are forced by our ignorance to average over the distribution $p(\lambda_i)$ of the degrees of freedom we do not control. We stress that hidden variable theory leads to observables whose values are random in the sense of *classical* probability theory, which is radically different from the quantum probabilistic description, as we are going to see. In particular, in Quantum Physics the probabilistic description is *fundamental* and not emergent, and it cannot be explained in any classical way. Indeed, quantum mechanically the value of an observable *is not defined* if we don't observe it, while classically the value is intrinsically well defined even though we are able to determine it only statistically. One easily sees that, as long as we work with states with *zero entanglement*, we can always construct an *ad hoc* hidden variable theory which reproduces the results of experiments as they are predicted by the quantum theory. However *no* such classical construction works in presence of entanglement, as we are going to see. This is what Schrödinger meant in the opening quotation of this chapter: the new phenomena associated with entanglement represent the most dramatic deviation of Quantum Mechanics from the classical description of reality, and they cannot be explained by *any* classical model however subtle and intricate.

What is more, there are technologically doable experiments which do concretely realize in the laboratory the slightly idealized experiments we are going to describe: hence quantum entanglement (and the fundamental axioms and paradigms of Quantum Physics) may be put to an experimental verification against the prediction of alternative proposals (such as hidden variables). The bottom line is that the "counterintuitive" predictions of Quantum Physics get confirmed with high accuracy by all experiments, while the "intuitive" classical ideas get ruled out.

The Experiment   To understand the difference between Quantum Mechanics and classical probability, we discuss the EPR experiment in some detail.

As mentioned before, a laboratory prepares zillions of copies of the entangled state $|\psi^-\rangle$ (say as products of lots of spin-0 particle decays), and sends the first qubit of each copy to Alice and the second qubit to Bob in their different Countries. In their respective laboratories Alice and Bob have an experimental device which allows them to measure three different observables $\mathcal{O}_1$, $\mathcal{O}_2$, and $\mathcal{O}_3$, each one taking the two values ± 1. Alice and Bob are allowed to perform *only one* of the three measurements on each copy of their qubit (since the $\mathcal{O}_a$'s do not commute), but they may freely choose which one to measure on each copy.

Independently of each other, Alice and Bob measure one observable of their choice in each copy of their qubit and keep a record of the results:

<table>
<tr><td colspan="3" align="center">Logbook of Alice</td><td colspan="3" align="center">Logbook of Bob</td></tr>
<tr><td>copy</td><td align="center">observable</td><td>result</td><td>copy</td><td align="center">observable</td><td>result</td></tr>
<tr><td align="center">1</td><td align="center">$\mathcal{O}_2$</td><td align="center">$+1$</td><td align="center">1</td><td align="center">$\mathcal{O}_3$</td><td align="center">-1</td></tr>
<tr><td align="center">2</td><td align="center">$\mathcal{O}_1$</td><td align="center">$+1$</td><td align="center">2</td><td align="center">$\mathcal{O}_1$</td><td align="center">$+1$</td></tr>
<tr><td align="center">3</td><td align="center">$\mathcal{O}_3$</td><td align="center">-1</td><td align="center">3</td><td align="center">$\mathcal{O}_3$</td><td align="center">-1</td></tr>
<tr><td align="center">4</td><td align="center">$\mathcal{O}_3$</td><td align="center">$+1$</td><td align="center">4</td><td align="center">$\mathcal{O}_2$</td><td align="center">$+1$</td></tr>
<tr><td align="center">⋮</td><td align="center">⋮</td><td align="center">⋮</td><td align="center">⋮</td><td align="center">⋮</td><td align="center">⋮</td></tr>
</table>

$$\tag{7.219}$$

Comparing their logbooks they find the following peculiar regularity: if in the n-th copy they both measure the *same* observable—whichever one of the three—they find the *same* result, either both $+1$ or both -1. See their findings for copies 2 and 3 in (7.219). They repeat the experiment zillions of times to be sure, and the rule is confirmed 100% of the time. Thinking classically, Alice and Bob conclude that they can learn the value of *two* observables for each copy, even if they don't commute. Their strategy is elementary: Alice measures $\mathcal{O}_1$ and Bob measures (say) $\mathcal{O}_2$ and then they phone each other their results. At the end Alice will know the value of $\mathcal{O}_1$ that she measured herself, and also know what *she would have got, had she measured* $\mathcal{O}_2$, from Bob's phone call.

Are they *right?* Is this protocol a valid strategy to get the value of two non-commuting observables at once?

In a classical world it would be a correct methodology. Let us review the *classical* story. Classically the value of the observables is defined intrinsically (even if unknown to us). When it is known that their value is the same if either Alice or Bob measures them, we are allowed to simplify the notation and write the observables simply as $\mathcal{O}_a$ instead of distinguishing them in $\mathcal{O}_a^{(A)}$ vs. $\mathcal{O}_a^{(B)}$. Then one would have a probability distribution $P(\mathcal{O}_1, \mathcal{O}_2, \mathcal{O}_3)$ for the results of the measurements. The probabilities $P(\mathcal{O}_1, \mathcal{O}_2, \mathcal{O}_3)$ are non-negative, and the sum of the probabilities over all possible outcomes is 1

$$\sum_{\mathcal{O}_1, \mathcal{O}_2, \mathcal{O}_3 \in \{\pm 1\}} P(\mathcal{O}_1, \mathcal{O}_2, \mathcal{O}_3) = 1. \tag{7.220}$$

Let us consider the classical probability that the measures of two different observables yield the same value:

$$\mathbf{P}(\mathcal{O}_1 = \mathcal{O}_2) \equiv P(+, +, +) + P(+, +, -) + P(-, -, +) + P(-, -, -) \tag{7.221}$$

$$\mathbf{P}(\mathcal{O}_2 = \mathcal{O}_3) \equiv P(+, +, +) + P(-, +, +) + P(+, -, -) + P(-, -, -) \tag{7.222}$$

$$\mathbf{P}(\mathcal{O}_3 = \mathcal{O}_1) \equiv P(+, +, +) + P(+, -, +) + P(-, +, -) + P(-, -, -), \tag{7.223}$$

so that

$$\mathbf{P}(\mathcal{O}_1 = \mathcal{O}_2) + \mathbf{P}(\mathcal{O}_2 = \mathcal{O}_3) + \mathbf{P}(\mathcal{O}_3 = \mathcal{O}_1) = $$
$$= 1 + 2\,P(+, +, +) + 2\,P(-, -, -) \geq 1, \tag{7.224}$$

indeed, thinking classically, given *three* random variables which may take *only two* values, at least *two* variables should have the same value: this obvious fact is known as the *pigeon-hole lemma*.

Alice and Bob perform the experiment to a very high accuracy: Alice measures the observable $\mathcal{O}_a$ and Bob the observable $\mathcal{O}_{a+1 \bmod 3}$ on each copy of the state, for more than a zillion times, and evaluate the probabilities that they get the same result (either $+1$ or -1) for the diverse measures. To their surprise, they find

$$\mathbf{P}(\mathcal{O}_1 = \mathcal{O}_2) = \mathbf{P}(\mathcal{O}_2 = \mathcal{O}_3) = \mathbf{P}(\mathcal{O}_3 = \mathcal{O}_1) \approx \frac{1}{4}, \tag{7.225}$$

so that

$$\mathbf{P}(\mathcal{O}_1 = \mathcal{O}_2) + \mathbf{P}(\mathcal{O}_2 = \mathcal{O}_3) + \mathbf{P}(\mathcal{O}_3 = \mathcal{O}_1) \approx \frac{3}{4} < 1. \tag{7.226}$$

Clearly the result of their experiment cannot be explained by *any* classical probabilistic theory, no matter how elaborated: *Alas,* classically there is no way to evade the pigeon-hole lemma!

In particular adding a classical hidden sector will not help. Since the very counterintuitive result (7.226) is actually obtained in experiments [22, 23], we conclude that the real world behaves in a *very non-classical* way.

Let us see how the result (7.226) is produced in Quantum Mechanics as a consequence of the fact that Alice and Bob qubits were prepared in the maximally entangled state $|\psi^-\rangle$. Recall that $|\psi^-\rangle$ is the state with zero total spin. Hence

$$(\sigma_i^{(A)} + \sigma_i^{(B)})|\psi^-\rangle = 0 \qquad \text{for } i = 1, 2, 3. \tag{7.227}$$

The experimental device used by Alice and Bob allows them to measure the three operators

$$\begin{aligned}
\mathcal{O}_a^{(A)} &= \ \mathbf{n}_a \cdot \boldsymbol{\sigma}^{(A)} \ \overset{\text{def}}{=} \ e^{ia\pi\sigma_2/3}\sigma_1 e^{-ia\pi\sigma_2/3} \otimes \mathbf{1}, \qquad a = 1, 2, 3, \\
\mathcal{O}_a^{(B)} &= -\mathbf{n}_a \cdot \boldsymbol{\sigma}^{(B)} \ \overset{\text{def}}{=} \ -\mathbf{1} \otimes e^{ia\pi\sigma_2/3}\sigma_1 e^{-ia\pi\sigma_2/3},
\end{aligned} \tag{7.228}$$

where $\boldsymbol{\sigma} \equiv (\sigma_1, \sigma_2, \sigma_3)$ are the Pauli matrices and $\mathbf{n}_a$ are three unit vectors in the 1-3 plane forming $2\pi/3$ angles between them[15] so that

$$\mathbf{n}_a \cdot \mathbf{n}_b = \begin{cases} 1 & a = b \\ -\frac{1}{2} & a \neq b. \end{cases} \tag{7.229}$$

Therefore, when Alice and Bob both measure the same observable $\mathcal{O}_a$ they get the same result because Eq. (7.227) gives

$$\mathcal{O}_a^{(A)}|\psi^-\rangle = \mathcal{O}_a^{(B)}|\psi^-\rangle \quad \text{for } a = 1, 2, 3. \tag{7.230}$$

---

[15] That is: if we identify the plane with $\mathbb{C}$, the $\mathbf{n}_a \in \mathbb{C}$ are the three cubic roots of 1.

Notice that Alice cannot measure two different observables, $\mathcal{O}_a^{(A)} \neq \mathcal{O}_b^{(A)}$, on the same copy of the qubit since they do not commute. The same applies to Bob.

The projector associated to the outcome $(-1)^\alpha$ when Alice (resp. $(-1)^\beta$ when Bob) measures the observable $\mathcal{O}_a^{(A)}$ (resp. $\mathcal{O}_a^{(B)}$) is

$$
\begin{aligned}
\mathcal{P}(\alpha)_a^{(A)} &= \frac{1}{2}\left(\mathbf{1} + (-1)^\alpha\, \mathbf{n}_a \cdot \boldsymbol{\sigma}^{(A)}\right), \qquad \alpha = 0, 1 \\[2mm]
\mathcal{P}(\beta)_a^{(B)} &= \frac{1}{2}\left(\mathbf{1} - (-1)^\beta\, \mathbf{n}_a \cdot \boldsymbol{\sigma}^{(B)}\right), \qquad \beta = 0, 1,
\end{aligned}
\tag{7.231}
$$

hence the probability that Alice finds $(-1)^\alpha$ and Bob $(-1)^\beta$ when measuring, respectively, the observables $\mathcal{O}_a^{(A)}$ and $\mathcal{O}_b^{(B)}$ is

$$
\begin{aligned}
\mathbf{P}(\alpha, \beta)_{a,b} &= \langle \psi^- | \mathcal{P}(\alpha)_a^{(A)} \mathcal{P}(\beta)_b^{(B)} | \psi^- \rangle = \\[2mm]
&= \frac{1}{4} \langle \psi^- | \left(1 + (-1)^\alpha\, \mathbf{n}_a \cdot \boldsymbol{\sigma}^{(A)}\right)\left(1 + (-1)^\beta\, \mathbf{n}_b \cdot \boldsymbol{\sigma}^{(A)}\right) | \psi^- \rangle \\[2mm]
&= \frac{1}{8} \operatorname{Tr}_A\left[\left(1 + (-1)^\alpha\, \mathbf{n}_a \cdot \boldsymbol{\sigma}^{(A)}\right)\left(1 + (-1)^\beta\, \mathbf{n}_b \cdot \boldsymbol{\sigma}^{(A)}\right)\right] \\[2mm]
&= \frac{1}{4}\left(1 + (-1)^{\alpha+\beta}\, \mathbf{n}_a \cdot \mathbf{n}_b\right) = \begin{cases} \frac{1}{4}\left(1 + (-1)^{\alpha+\beta}\right) & a = b \\[2mm] \frac{1}{4}\left(1 - \frac{(-1)^{\alpha+\beta}}{2}\right) & a \neq b \end{cases}
\end{aligned}
\tag{7.232}
$$

where in the second line we used (7.227) and in the third line Eq. (7.7) together with $\varrho_A = \frac{1}{2}\mathbf{1}$ for the maximally entangled state $|\psi^-\rangle$. Therefore the probability of getting the same result when measuring two different observables is

$$
\sum_{\alpha=\beta=0}^{1} \frac{1}{4}\left(1 - \frac{(-1)^{\alpha+\beta}}{2}\right) = \frac{1}{4}
\tag{7.233}
$$

which is precisely the result (7.225) found by Alice and Bob in their experiment.

We conclude:

E1   due to the phenomenon of quantum entanglement between the different subsystems of a closed system, the quantum measures behave quite differently from any classical theory of probabilistic outcomes;

E2   the difference between quantum probability and classical probability—hence the quantum phenomenon of entanglement—may be checked in real experiments, thus demonstrating the correctness of the quantum paradigm: its peculiar properties are *physical realities* (hence non contradictory).

Indeed, actual experiments were performed [22, 23], and the quantum description of reality fully confirmed. Moreover

E3   quantum entanglement opens new opportunities: something which is impossible to do—or extremely hard to do—according to the classical rules, becomes doable thanks to the new effects produced by entanglement. Hence entanglement should be seen as a *resource* which allows us to perform tasks otherwise too hard. In addition to the usual resources: time, energy, etc. we have the new, deeply quantum, resource *entanglement* which has an actual economic value.

We shall not go through the ways we may use the new resource to perform hard tasks. This is the job of Quantum Computation, Quantum Computers, Quantum Algorithms, Quantum Communication, and all that stuff, topics and futuristic technologies which are far beyond the scope of this introductory textbook. We shall content ourselves with another remarkable example, and then will say a few words about Quantum Information Theory.

7.5.2 Three Qubits Entanglement: GHZM Analysis

As a further illustration we consider the GHZM (*gedanken*) experiment [24–27]. For a nice discussion of it, see [28].

In this experiment one subtle guy produces a zillion copies of an entangled state of three qubits, $|\Upsilon\rangle$, and then sends the first qubit of each copy to Alice, the second one to Bob, and the third one to Charlie, three experimentalists who work in different Countries. Each one of them may choose to measure on her/his part of the n-th copy either one of two different observables, X or Y, which both take the two values ± 1. We write X_1, X_2, X_3 (resp. Y_1, Y_2, Y_3) for the result of the measure of X (resp. Y) by, respectively, Alice, Bob, and Charlie on their part of a given copy of the entangled state $|\Upsilon\rangle$.

To make the procedure well defined, the three experimentalists agree on the following protocol: a device chooses randomly between A, B, and C for each copy of the state. If A is chosen, Alice measures the quantity X and the other two experimentalists measure Y. If B is chosen, Bob will measure X while Alice and Charlie go for Y, and so on. After performing several zillions measurements following this protocol, they discover that—while the single results are pretty random—there is a *remarkable regularity* that they state as the

Rule *If we measure two Y's and one X in each copy, the product of our results is always $+1$ independently of which copy and of who happens to measure X. That is,*

$$X_1 Y_2 Y_3 = Y_1 X_2 Y_3 = Y_1 Y_2 X_3 = +1. \tag{7.234}$$

The validity of this **Rule** was confirmed in everyone of the several zillion repetitions of the experiment in the agreed format.

Now Alice, Bob, and Charlie want to interpret their result theoretically in order to make *predictions* for future experiments. They are experimentalists, with little understanding of the subtleties of quantum entanglement, and they argue as follows:

Naive Prediction *Using the **Rule** we can predict the result of the modified experiment where everyone of us measures the observable X. The product of our findings for each copy of the state must be:*

$$X_1 X_2 X_3 = (X_1 Y_2 Y_3)(Y_1 X_2 Y_3)(Y_1 Y_2 X_3) = +1 \qquad (7.235)$$

where the second equality follows from (7.234) and the first one follows from the fact that the square of each variable Y_a is $+1$

$$Y_a^2 = +1, \quad a = 1, 2, 3. \qquad (7.236)$$

They perform the modified experiment and, to their bewilderment, they get

$$X_1 X_2 X_3 = -1 \qquad (7.237)$$

each one of the zillion times they repeat the measurement!!

As in the EPR experiment, the peculiar result (7.237) arises from the non-classical correlations between the space-like separated measures of Alice, Bob, and Charlie produced by quantum entanglement. The measure of her qubit by Alice is not independent of the measures performed by Bob and Charlie on theirs, despite the fact that the several measures are performed at large space-like distances. Their **Naive Prediction** was based on the implicit assumption that the correlations between their measures are classical, whereas the smart guy, who originally prepared the copies of the entangled state $|\Upsilon\rangle$, made them with subtle quantum correlations for the malicious purpose of puzzling the three poor experimentalists.

Let us explain what he did. The smart guy prepared the three qubits in the entangled state

$$|\Upsilon\rangle = \frac{1}{\sqrt{2}}\Big(|0\rangle \otimes |0\rangle \otimes |0\rangle - |1\rangle \otimes |1\rangle \otimes |1\rangle\Big) \qquad (7.238)$$

which is a simultaneous eigenstate of the four commuting observables

$$\sigma_1^{(1)} \otimes \sigma_2^{(2)} \otimes \sigma_2^{(3)} \qquad \sigma_2^{(1)} \otimes \sigma_1^{(2)} \otimes \sigma_2^{(3)}$$
$$\sigma_2^{(1)} \otimes \sigma_2^{(2)} \otimes \sigma_1^{(3)} \qquad \sigma_1^{(1)} \otimes \sigma_1^{(2)} \otimes \sigma_1^{(3)} \qquad (7.239)$$

with, respectively, eigenvalues $+1$, $+1$, $+1$ and -1. These four observables are not independent

$$
\begin{aligned}
(\sigma_1^{(1)} \otimes \sigma_2^{(2)} \otimes \sigma_2^{(3)})(\sigma_2^{(1)} \otimes \sigma_1^{(2)} \otimes \sigma_2^{(3)})(\sigma_2^{(1)} \otimes \sigma_2^{(2)} \otimes \sigma_1^{(3)}) = \\
= -\sigma_1^{(1)} \otimes \sigma_1^{(2)} \otimes \sigma_1^{(3)}.
\end{aligned}
\tag{7.240}
$$

The first three operators corresponds, respectively, to the measures

$$
X_1 Y_2 Y_3, \quad Y_1 X_2 Y_3, \quad Y_1 Y_2 X_3,
\tag{7.241}
$$

while the last one to $X_1 X_2 X_3$. Since $|\Upsilon\rangle$ is an eigenvector of all four observables, measuring them we get the corresponding eigenvalue with absolute certainty. However, since the *individual* measures $\sigma_i^{(a)}$ do not commute, the functional relation between the eigenvalues of the various observables is not the one expected on classical grounds. As an effect of entanglement, the actual result is the *opposite* one of what would be predicted by a classical reasoning, independently of the complexity of the underlying classical system. This experiment rules out all "hidden variables" explanation of quantum phenomena.

7.6   Information: Classical Theory

To set the stage, before going to quantum information theory and the pivotal role of von Neumann entropy in it, we briefly review its classical counterpart (largely due to Shannon [29]). The two central problems of Shannon theory are:

Pr1   How much can a message be *compressed* or, equivalently, how redundant is the information in a typical message?

Pr2   At what *rate* can we communicate reliably over a *noisy* channel, that is, how much redundancy must be incorporated into a message to protect it against errors?

Shannon answers question **Pr1** in his *source coding theorem* (a.k.a. the *noiseless coding theorem*) and **Pr2** in the *noisy channel coding theorem*. Both questions concern redundancy of information, and Shannon's key insight is that *entropy* is a way to quantify this redundancy.

Message Compression

A message is a *string of letters,* where each letter ℓ is taken from an alphabet L consisting in a set of k letters

$$
\ell \in L \equiv \{1, \ldots, k\}.
\tag{7.242}
$$

In our idealized set-up the message is produced by a *source* which picks each letter ℓ according to a probability distribution $p(\ell)$ with

$$p(\ell) \geq 0, \qquad \sum_{\ell=1}^{k} p(\ell) = 1. \tag{7.243}$$

The letters of a message are independent and identically distributed (i.i.d.) random variables. Then, if the source emits an n-letter message, the particular string

$$\boldsymbol{\ell} = \ell_1 \ell_2 \cdots \ell_n \in L^n \tag{7.244}$$

occurs with probability

$$p(\boldsymbol{\ell}) = \prod_{i=1}^{n} p(\ell_i). \tag{7.245}$$

We see the alphabet L equipped with the probability distribution $\{p(\ell)\}$ as a statistical *ensemble*. In this set-up Shannon's first problem becomes:

Problem Pr1 *Given an n-letter message $\boldsymbol{\ell}$ with $n \ggg 1$, is it possible to compress it to a shorter message $\boldsymbol{\ell}_{\mathrm{sh}}$ that conveys essentially the same information?*

It turns out that the compression is possible provided the ensemble L is not uniformly random (i.e. provided the letters are not all equiprobable). To make a precise statement we need to specify what we mean by "essentially the same information": the probability of an error in the transmission of a sufficiently long message should be negligible, that is, less than ϵ or

$$(\text{probability of success}) > 1 - \epsilon, \tag{7.246}$$

where ϵ may be taken as small as we please for sufficiently long messages.

Let us formalize this notion. In a typical string of n letters, with $n \ggg 1$, the letter ℓ will appear approximately $n\,p(\ell)$ times. For large n the number of such typical strings behaves as

$$\frac{n!}{\prod_{\ell}(np(\ell))!} \approx 2^{n\,S(L)}, \tag{7.247}$$

where

$$S(L) \stackrel{\mathrm{def}}{=} -\sum_{\ell \in L} p(\ell) \log_2 p(\ell) \tag{7.248}$$

($\log_2(\cdot)$ is the logarithm in base[16] 2) and we used the Stirling asymptotic formula [30, 31] for the factorial

$$\log m! = m \log m - m + O(\log m) \qquad m \gg 1. \tag{7.249}$$

Definition 7.11 The quantity $S(L)$ in Eq. (7.248) is called the *Shannon entropy* of the ensemble (alphabet) L.

Lemma 7.4 *The entropy* $S(L)$

$$S(L) = - \sum_{\ell=1}^{k} p(\ell) \log_2 p(\ell) \tag{7.250}$$

satisfies the bounds

$$0 \le S(L) \le \log_2 k \tag{7.251}$$

The lower bound is saturated iff $p(\ell) = \delta_{\ell \ell_0}$ *for some* ℓ_0 *(that is, only the letter* ℓ_0 *appears in all messages[17]) and the upper bound is saturated iff* $p(\ell) = 1/k$ *for all* $\ell \in L$ *(that is, the letters are* equidistributed).

Proof We maximize the entropy over the probabilities $p(\ell)$'s which satisfy the constraint (7.243) using the Lagrange multiplier method:

$$\frac{\partial}{\partial p(\ell)} \left[- \sum_j p(j) \log_2 p(j) + \lambda \left(\sum_j p(j) - 1 \right) \right] =$$
$$= - \log_2 p(\ell) - \frac{1}{\log 2} + \lambda = 0 \tag{7.252}$$

and $S(L)$ is maximal when the probability $p(\ell)$ is independent of ℓ. $\square$

From Eq. (7.248) we see that the quantity $S(L)$ has the interpretation of the expectation value of the random variable $- \log_2 p(\ell)$ in the ensemble L

$$S(L) \overset{\text{def}}{=} \left\langle - \log_2 p \right\rangle_L . \tag{7.253}$$

---

[16] In information theory it is customary to use logarithms in base 2 instead of the natural ones so that the information gets measured in bits. In physics we use natural logarithms: hence the two notions of entropy differ by a normalization factor

$$S_{\text{phy}} = \log 2 \cdot S_{\text{inf}}$$

[17] When $S(L) = 0$ the messages are rather boring and convey no essential information.

A n-letter message $\ell_1 \ell_2 \cdots \ell_n$ is an n-element sample of the ensemble L. The central limit theorem [32] implies that for all $\epsilon, \delta > 0$ and *any* i.i.d. random variable x the average over an n-element sample $\{x_1, \ldots, x_n\}$ in the ensemble L will differ by less than δ from its expectation value $\langle x \rangle_L$

$$\left| \frac{1}{n} \sum_{i=1}^{n} x_i - \langle x \rangle_L \right| < \delta \tag{7.254}$$

with probability $1 - \epsilon$ for n large enough. We apply the theorem to the random variable $- \log_2 p(\ell)$. A n-letter message ℓ is called δ-*typical* iff

$$\left| -\frac{1}{n} \log_2 p(\ell) - S(L) \right| \equiv \left| -\frac{1}{n} \sum_{i=1}^{n} \log_2 p(\ell_i) - S(L) \right| < \delta. \tag{7.255}$$

The central limit theorem then says that, for n large enough, an n-letter message will be δ-typical with probability $(1 - \epsilon)$ i.e. *almost certainly.*

How many δ-typical message are there? We improve the rough estimate (7.247), using the rigorous bound (7.255). An easy computation yields the following bounds for the number N_{typ} of (sufficiently long) δ-typical strings

$$(1 - \epsilon)\, 2^{n(S(L)-\delta)} \leq N_{\text{typ}} \leq 2^{n(S(L)+\delta)}. \tag{7.256}$$

Now we compress the information carried by the n-letter string ℓ. We first introduce a *block code*, that is, we number the δ-typical strings by the integers $1, 2, \ldots, N_{\text{typ}}$, and instead of transmitting the n-letters of the string ℓ we transmit its number $n(\ell) \in \{1, \ldots, N_{\text{typ}}\}$. The original message consisted of $n \log_2 k$ bites, whereas the compressed message requires slightly more than $n\, S(L)$ bites. As $n \to \infty$ the *compression rate* is

$$\begin{array}{c} \text{compression} \\ \text{rate} \end{array} \equiv R = \frac{S(L)}{\log_2 k} + o(1) \tag{7.257}$$

From Lemma 7.4 we see that $R < 1$ unless the letters are equidistributed. In all other cases we may compress our message into a fraction of its original number of bits with a success probability $> (1 - \epsilon)$ where ϵ can be taken very small for large n. On the other hand, the lower bound in Eq. (7.256) says that we cannot do any better: R is the *optimal compression ratio* and if we use a smaller number of bits the probability of success will go to zero exponentially for large n. These statements are the *Shannon source coding theorem.*

Mutual Information and Conditional Entropy

Now we consider *two* sources of information with their respective alphabets L, M. The two sources may be correlated, so we see them as a *combined ensemble LM* in which a pair of letters $(\ell, m) \in L \times M$ appears with probability $p(\ell, m)_{LM}$. The two sources are independent iff

$$p(\ell, m)_{LM} = p_1(\ell)\, p_2(m) \tag{7.258}$$

and correlated otherwise. We denote L (resp. M) the marginal ensemble with probabilities

$$p(\ell)_L = \sum_{m \in M} p(\ell, m)_{LM} \qquad \left(\text{resp. } p(m)_M = \sum_{\ell \in L} p(\ell, m)_{LM} \right). \tag{7.259}$$

When the sources are correlated, reading a message $m \in M^n$ from the second source reduces our ignorance about the message $\ell \in L^n$ from the first source, and therefore we must be able to compress the message ℓ even further than the compression rate R in Eq. (7.257).

Now we make this idea precise. Consider the length-n combined message

$$\boldsymbol{\ell m} = \ell_1 \ell_2 \cdots \ell_n m_1 m_2 \cdots m_n \in (LM)^n, \tag{7.260}$$

and its two constituent messages

$$\boldsymbol{\ell} = \ell_1 \ell_2 \cdots \ell_n \in L^n, \quad \text{and} \quad \boldsymbol{m} = m_1 m_2 \cdots m_n \in M^n, \tag{7.261}$$

with their respective probabilities

$$p(\boldsymbol{\ell m})_{LM} = \prod_{i=1}^{n} p(\ell_i, m_i), \tag{7.262}$$

$$p(\boldsymbol{\ell})_L = \prod_{i=1}^{n} p(\ell_i)_L, \qquad p(\boldsymbol{m})_M = \prod_{i=1}^{n} p(m_i)_M. \tag{7.263}$$

Definition 7.12 The sequence $\boldsymbol{\ell m}$ is *jointly δ-typical* iff it is δ-typical with respect to all three ensembles LM, L and M, that is,

$$\left| -\frac{1}{n} \log_2 p(\boldsymbol{\ell})_L - S(L) \right| < \delta, \quad \left| -\frac{1}{n} \log_2 p(\boldsymbol{m})_M - S(M) \right| < \delta, \tag{7.264}$$

$$\left| -\frac{1}{n} \log_2 p(\boldsymbol{\ell m})_{LM} - S(LM) \right| < \delta. \tag{7.265}$$

The central limit theorem guarantees that a joint sequence will be δ-typical with probability $(1 - \epsilon)$ for n large enough. We conclude that, when the combined message $\boldsymbol{\ell m}$ is jointly δ-typical, the *conditional probability* of the constituent message $\boldsymbol{\ell}$ if we know the message $\boldsymbol{m}$

$$p(\boldsymbol{\ell}|\boldsymbol{m}) \stackrel{\text{def}}{=} \frac{p(\boldsymbol{\ell},\boldsymbol{m})_{LM}}{p(\boldsymbol{m})_M} \tag{7.266}$$

satisfies the bounds

$$-n(S(L|M)+2\delta) \leq \log_2 p(\boldsymbol{\ell}|\boldsymbol{m}) \leq -n(S(L|M)-2\delta), \tag{7.267}$$

where $S(L|M)$ is the *conditional entropy*

$$S(L|M) \stackrel{\text{def}}{=} S(LM) - S(M) =$$
$$= \left\langle -\log_2 p(\ell,m)_{LM} + \log_2 p(m)_M \right\rangle_{LM} \equiv \left\langle -\log_2 p(\ell|m) \right\rangle_{LM}. \tag{7.268}$$

The conditional entropy measures our residual ignorance about $\boldsymbol{\ell}$ when we know $\boldsymbol{m}$. If the combined message is jointly typical (which is the case with high probability for large n), the number of possible values of $\boldsymbol{\ell}$ consistent with the known value of $\boldsymbol{m}$ is at most

$$2^{n(S(L|M)+2\delta)}. \tag{7.269}$$

Therefore we can transmit $\boldsymbol{\ell}$, with the high success probability $(1 - \epsilon)$, using just $S(L|M) + o(1)$ bits per letter. We cannot do better: with a lower number of bits the success probability will go to zero exponentially as $n \to \infty$. In conclusion: the conditional entropy $S(L|M)$ is the number of additional bits per letter required to specify both messages $\boldsymbol{\ell}$ and $\boldsymbol{m}$ when $\boldsymbol{m}$ is known.

The *gain* in information on L from the knowledge of M is given by the reduction in the number of bits per letter needed to specify $\boldsymbol{\ell} \in L^n$

$$I(L; M) \stackrel{\text{def}}{=} S(L) - S(L|M) =$$
$$= S(L) + S(M) - S(LM) = S(M) - S(M|L) = I(M; L) \tag{7.270}$$

called the *mutual information*. The mutual information is symmetric in L and M: the information about L we learn from the knowledge of M is equal to the information on M we get from knowing L. In Eq. (7.253) we wrote Shannon's entropy as the expectation value in the ensemble L of the random variable $- \log_2 p(\ell)_L$. The corresponding expression for the mutual information reads

$$I(L; M) = \left\langle \log_2 \frac{p(\ell,m)_{LM}}{p(\ell)_L \, p(m)_M} \right\rangle_{LM}. \tag{7.271}$$

Learning M cannot reduce my knowledge of L, so $I(L; M)$ is *non-negative*. Equivalently: the Shannon entropy is *sub-additive*

$$S(LM) \leq S(L) + S(M), \tag{7.272}$$

see property **SE7** in Sect. 7.6.1 below. In facts we have the stronger inequality

$$I(L; MN) \geq I(L; M) \tag{7.273}$$

which states the obvious point that the correlations of the ensemble L with the combined ensemble MN are at least as strong as the correlations with M alone, see property **SE10** (*strong sub-additivity*) in next subsection.

7.6.1 *Properties of Shannon Entropy*

Let P be an ensemble of n "letters" with probability distribution $\boldsymbol{p} = (p_1, \ldots, p_n)$. The Shannon entropy of P

$$S(P) \equiv S(\boldsymbol{p}) \equiv S(p_1, \ldots, p_n) \overset{\text{def}}{=} -\sum_{i=1}^{n} p_i \log_2 p_i \tag{7.274}$$

has the following basic properties **SE1-SE10**:

SE1   *Continuity:* $S(\boldsymbol{p})$ is a continuous function of the probabilities p_i;
SE2   *Expansibility:* $S(p_1, \ldots, p_n, 0) = S(p_1, \ldots, p_n)$;
SE3   *Positivity:* $S(\boldsymbol{p}) \geq 0$ with equality iff $p_i = \delta_{ii_0}$ for some i_0;
SE4   *Upper bound:* $S(p_1, \ldots, p_n) \leq \log_2 n$, with equality iff P is equidistributed, i.e. $p_i = 1/n$ for all i;
SE5   *Concavity:* For all $0 \leq \lambda \leq 1$,

$$S(\lambda\, \boldsymbol{p}_1 + (1 - \lambda)\boldsymbol{p}_2) \geq \lambda\, S(\boldsymbol{p}_1) + (1 - \lambda)\, S(\boldsymbol{p}_2); \tag{7.275}$$

SE6   *Monotonicity:* We have a combined ensemble LM with probability $p(\ell, m)_{LM}$ and the associated marginal distributions L and M with probabilities

$$p(\ell)_L = \sum_m p(\ell, m)_{LM} \quad \text{and} \quad p(m)_M = \sum_\ell p(\ell, m)_{LM}. \tag{7.276}$$

For the combined ensemble LM

$$S(LM) \geq S(L), \tag{7.277}$$

that is, our ignorance on the full system cannot be smaller than our ignorance about just a part of it. Equivalently: our ignorance about the full system cannot *increase* if we get information about one of its parts;

SE7   *Sub-additivity:* The Shannon entropy of the combined ensemble LM is not larger than the sum of the entropies of the two sub-ensembles

$$S(LM) \leq S(L) + S(M), \tag{7.278}$$

with equality if and only if the two distributions are not correlated i.e. iff

$$p(\ell, m)_{LM} = p(\ell)_L \, p(m)_M; \tag{7.279}$$

SE8   *Schur concavity:* Recall the partial order $\prec$ between probability distributions $\boldsymbol{p} = (p_1, p_2, \ldots, p_n)$ we defined in Sect. 7.3.3:

$$\boldsymbol{p} \prec \boldsymbol{q} \;\Leftrightarrow\; \sum_{i=1}^{k} p_i^{\downarrow} \leq \sum_{i=1}^{k} q_i^{\downarrow} \;\text{ for }\; k = 1, 2, \ldots, n. \tag{7.280}$$

A function f is said to be *Schur concave* iff

$$\boldsymbol{p} \prec \boldsymbol{q} \;\Rightarrow\; f(\boldsymbol{p}) \geq f(\boldsymbol{q}). \tag{7.281}$$

The entropy function $S(\boldsymbol{p})$ is Schur concave. Note that $\boldsymbol{p} \prec \boldsymbol{q}$ ($\boldsymbol{q}$ majorizes $\boldsymbol{p}$) means that "$\boldsymbol{p}$ is at least as random as $\boldsymbol{q}$": indeed Lemma 7.3 states that $\boldsymbol{p} \prec \boldsymbol{q}$ if and only if $\boldsymbol{p} = D\boldsymbol{q}$ for some *doubly stochastic* matrix D. Schur concavity of $S(\boldsymbol{p})$ then says that an ensemble with more randomness has higher entropy;

SE9   *Coarse graining recursion:* We coarse grain the probability distribution L by grouping together several outcomes. Concretely, we construct a new "block" ensemble $B = \{b_1, \ldots, b_r\}$ whose s-th element b_s is the subset

$$b_s = \{\ell_{k_{s-1}}, \ell_{k_{s-1}+1}, \ldots, \ell_{k_s}\} \subset \{\ell_1, \ldots, \ell_n\} \tag{7.282}$$

for some integers

$$1 \equiv k_0 < k_1 < k_2 < \cdots < k_{r-1} < k_r \equiv n. \tag{7.283}$$

The probabilities $\{q_1, \ldots, q_r\}$ of $\{b_1, \ldots, b_r\}$ are given by

$$q_1 = \sum_{i=1}^{k_1} p_i, \quad q_2 = \sum_{i=k_1+1}^{k_2} p_i, \quad \cdots, \quad q_r = \sum_{k_{r-1}}^{n} p_i. \tag{7.284}$$

The entropy of the original ensemble L is the sum of the entropy of the coarse grained ensemble B plus the expectation value on this ensemble of the random variable $S(b_s) \equiv S(\pi_s)$ given by the internal entropy of the s-th element $b_s \in B$ seen as an ensemble on its own right with probability distribution

$$\pi_s = \left\{ \frac{p_{k_{s-1}+1}}{q_s}, \cdots, \frac{p_{k_s}}{q_s} \right\} \quad s = 1, 2, \ldots, r. \tag{7.285}$$

The coarse graining recursion then takes the form

$$S(L) = S(B) + \sum_{s=1}^{r} q_s \, S(\pi_s). \tag{7.286}$$

Property **SE7**, together with **SE1** and **SE3**, uniquely determines Shannon entropy up to overall normalization.

SE10   *Strong sub-additivity:* We consider an ensemble LMN with three parts L, M, N. Then the Shannon entropy satisfies the inequality

$$S(LMN) + S(M) \leq S(LM) + S(MN) \tag{7.287}$$

which reduces to sub-additivity **SE7** when $M = \varnothing$. Written in terms of *mutual information*, strong sub-additivity reads

$$I(L; MN) \geq I(L; M) \tag{7.288}$$

which reflects the intuitive fact that L has at least as much correlation with the combined sector MN than with M alone. The quantity

$$I(L; N|M) \stackrel{\text{def}}{=} I(L; MN) - I(L; M) \tag{7.289}$$

is called *conditional mutual information,* and the strong sub-additivity property of Shannon entropy states that it is non-negative.

Proofs

SE1 and **SE2** are obvious. For **SE3** and **SE4** see Lemma 7.4. **SE5** holds because the function $f(z) = -z \log_2 z$ is concave. **SE6** is a special case of **SE9** where we start with the ensemble $P = LM$ and block together the pairs $(\ell_i, m_j) \in LM$ with the same ℓ_i so that $Q \equiv L$. Then

$$S(LM) = S(L) + \sum_{\ell} p(\ell) \, S(E_\ell) \geq S(L) \tag{7.290}$$

since the probabilities $p(\ell)$ and the internal entropy of the blocks $S(E_\ell)$ are non-negative. To show **SE7** we recall the Jensen's inequality:

Lemma 7.5 (Jensen Inequality) *Let $F(z)$ be a convex function of the real variable z. Then for all sequences of real numbers $\{x_m\}_{m=1}^n$ and $\{w_m\}_{m=1}^n$, with $w_m \geq 0$ and not all zero, we have*

$$F\left(\frac{\sum_m w_m x_m}{\sum_m w_m}\right) \leq \frac{\sum_m w_m F(x_m)}{\sum_m w_m}. \tag{7.291}$$

We apply this inequality to the convex function $F(z) = z \log_2 z$ with

$$w_m = p(m)_M \equiv \sum_\ell p(\ell, m)_{LM} \quad \text{and} \quad x_m = \frac{p(\ell, m)_{LM}}{p(m)_M} \tag{7.292}$$

getting (from now on we omit the subscripts to simplify the notation)

$$\left(\sum_m p(m)\frac{p(\ell, m)}{p(m)}\right) \log_2\left(\sum_m p(m)\frac{p(\ell, m)}{p(m)}\right) \leq$$
$$\leq \sum_m p(m)\frac{p(\ell, m)}{p(m)} \log_2\left(\frac{p(\ell, m)}{p(m)}\right). \tag{7.293}$$

Simplifying and summing over ℓ,

$$-S(L) \leq -S(LM) - \sum_{\ell,m} p(\ell, m) \log_2 p(m) = -S(LM) + S(M), \tag{7.294}$$

which is the sub-additivity inequality. To show **SE8** we quote

Lemma 7.6 (Schur [2]) *A function $F(p_1, \ldots, p_n)$ is Schur concave iff it is invariant under permutations of the p_i's and*

$$(p_1 - p_2)\left(\frac{\partial F}{\partial p_1} - \frac{\partial F}{\partial p_2}\right) \leq 0. \tag{7.295}$$

It is elementary to check that Eq. (7.295) holds for $F(p_i) = -\sum_i p_i \log_2 p_i$. Next we prove that **SE1**, **SE3**, and **SE9** imply the formula (7.274) hence they fully characterize Shannon entropy (up to overall normalization). By continuity it is enough to show the formula for and ensemble P with rational probabilities, $p_i = k_i/N$ with k_i positive integers such that $\sum_i k_i = N$. We can obtain the ensemble P by coarse graining an equidistributed ensemble P_0 with probabilities $\{1/N, \ldots, 1/N\}$. We write $A(m)$ for the entropy of an equidistributed ensemble where the probabilities are all equal $1/m$. E.g. the entropy of P_0 is $A(N)$. The subensemble E_i is also equidistributed with probability $1/k_i$, so $S(E_i) = A(k_i)$ and the recursion (7.286) reads

$$A(N) = S(P) + \sum_i p_i\, A(k_i) \tag{7.296}$$

In the special case that $k_i = k$ for all i, so $N = km$, and the original ensemble was also equidistributed with probabilities $1/m$. In this case the recursion specializes to

$$A(km) = A(m) + A(k) \tag{7.297}$$

whose solution is $A(k) = \text{const.}\log_2 k$. We choose the overall normalization constant to be 1. Then Eq. (7.296) becomes

$$S(P) = -\sum_i p_i\, \log_2(p_i N) + \log_2 N = -\sum_i p_i \log_2 p_i, \tag{7.298}$$

which is Shannon's formula. Finally, we show **SE10**. We write the conditional mutual information $I(L; N|M)$ using Eq. (7.271) (omitting subscripts)

$$
\begin{aligned}
I(L; MN) - I(L; M) &= -\left\langle \log_2 \frac{p(\ell)\, p(m, n)}{p(\ell, m, n)} + \log_2 \frac{p(\ell, m)}{p(\ell)\, p(m)} \right\rangle \\[2mm]
&= -\left\langle \log_2 \left[\frac{p(\ell, m)}{p(m)} \frac{p(m, n)}{p(m)} \frac{p(m)}{p(\ell, m, n)} \right] \right\rangle \\[2mm]
&= \left\langle \log_2 \frac{p(\ell, n|m)}{p(\ell|m)\, p(n|m)} \right\rangle \\[2mm]
&= \sum_m p(m)\, I(L; N|m) \geq 0
\end{aligned}
\tag{7.299}
$$

where in the last line we used the fact that, for each fixed m,

$$p(\ell, n|m) \equiv \frac{p(\ell, m, n)}{p(m)}, \tag{7.300}$$

is a normalized joint probability distribution with a non-negative mutual information.

7.6.2   Noisy Channel Coding Theorem

Alice sends messages to Bob through a *noisy* channel: Bob receives the letter m with probability $p(m|\ell)$ when Alice sends the letter ℓ. Alice plans to transmit *reliably* a message to Bob by using the channel $n \gg 1$ times. To achieve the goal of an accurate transmission they must use a code. For example: Alice might send the same bit k times, and Bob uses a majority vote of the k noisy bits he receives to decode the message. A good code must be *asymptotically reliable*, that is, the probability of an error should go to zero in the limit that the number n of bits sent by Alice goes to infinity. The second problem solved by Shannon is

Problem Pr2 *Are there asymptotically reliable codes for a given noisy channel? If yes, what is the transmission rate R of an optimal code which is asymptotically reliable? (R is defined as the number of bits to be sent per bit of the original message as the number of bits $n \to \infty$.)*

Shannon's noisy channel coding theorem answers the question **Pr2**. All noisy channels admit asymptotically reliable codes as long as there is some correlation between the input ℓ and the output m. The optimal transmission rate R is given by the *channel capacity*

$$C = \max_L I(M; L) \tag{7.301}$$

where $I(M; L)$ is the mutual information computed with the conditional probability $p(m|\ell)$ of the given channel and the maximization is over all probability distributions $p(\ell)_L$. For more details and a proof of the theorem see e.g. the Lecture Notes by Preskill [10].

7.7   von Neumann Entropy

Next we consider *Quantum Information*. The source prepares messages of n "letters", where now the "letters" are chosen from an ensemble of quantum states. In other words, the quantum alphabet is a set of quantum states

$$\left\{ \rho(\ell) \in \mathcal{S} \subset \mathcal{B}(\mathcal{H}) \right\}_{\ell \in L}, \tag{7.302}$$

the state $\rho(\ell)$ occurring with a given probability $p(\ell)$

$$p(\ell) \geq 0, \qquad \sum_{\ell \in L} p(\ell) = 1. \tag{7.303}$$

The probability of a particular outcome of a measurement on a "letter" chosen from this ensemble L—when the observer has no knowledge about which letter was prepared—is determined by the density operator of the ensemble

$$\varrho = \sum_{\ell} p(\ell) \, \rho(\ell). \tag{7.304}$$

Let $\mathcal{O}$ be an observable and $\boldsymbol{E} = \{E_a\}$ the associated POVM. The probability that a measurement of $\mathcal{O}$ yields the outcome a is (cf. Sect. 7.4.2)

$$\mathsf{Prob}(a) = \mathrm{Tr}(\varrho\, E_a). \tag{7.305}$$

The *von Neumann entropy* $S(\varrho)$ of the density operator ($\equiv$ state) ϱ is

$$S(\varrho) \overset{\text{def}}{=} -\mathrm{Tr}(\varrho\, \log \varrho) \tag{7.306}$$

where in Physics one takes "log" to stand for the natural logarithm, while in Information Theory the convention is to use logarithms[18] in base 2.

Since ϱ is Hermitian non-negative with $\mathrm{Tr}\,\varrho = 1$, we can find an orthonormal basis $\{|e_m\rangle\}$ of eigenvectors such that[19]

$$\varrho = \sum_{m=1}^{n} \lambda_m |e_m\rangle\langle e_m|, \qquad \langle m|m'\rangle = \delta_{m,m'}, \tag{7.307}$$

with eigenvalues satisfying $0 \le \lambda_m \le 1$ and $\sum_m \lambda_m = 1$. The von Neumann entropy is then

$$S(\varrho) = -\sum_{m=1}^{n} \lambda_m \log \lambda_m, \tag{7.308}$$

that is, *the von Neumann entropy of ϱ is the Shannon entropy of its spectrum $\{\lambda_m\}$.* Let us list some of the properties of the von Neumann entropy $S(\varrho)$ (some of which are analogue to properties of Shannon's entropy):

vNE1   *Unitary invariance:* the von Neumann entropy is invariant under unitary transformations

$$S(U\varrho\, U^{-1}) = S(\varrho) \qquad \forall\, U \in \mathcal{U}(\mathcal{H}); \tag{7.309}$$

vNE2   The von Neumann entropy $S(\varrho)$ is a continuous function of the eigenvalues $\{\lambda_i\}$ of ϱ;

vNE3   Suppose that ϱ has n non-zero eigenvalues. The von Neumann entropy satisfies the bound

$$0 \le S(\varrho) \le \log n, \tag{7.310}$$

---

[18] Cf. Footnote 16.

[19] To simplify the formulae, we write them assuming the spectrum of all operators to be discrete. In Quantum Information theory the Hilbert spaces are finite-dimensional, so this condition is automatically fulfilled. The extension to more general spectra is straightforward.

where the lower bound is saturated by a pure state, $\varrho = |\psi\rangle\langle\psi|$, while the upper bound is saturated by the *maximally mixed state* $\mathbf{1}/n$ (i.e. by a state whose purification is a maximally entangled pure state);

vNE4 *Concavity:* if $\sigma_i \geq 0$ and $\sum_i \sigma_i = 1$

$$S\left(\sum_i \sigma_i \varrho_i\right) \geq \sum_i \sigma_i \, S(\varrho_i). \tag{7.311}$$

vNE5 *Recursion relation:* Suppose that the density matrices ϱ_i $(i = 1, \ldots, n)$ have support in orthogonal subspaces $\mathcal{H}_i$ of the Hilbert space $\mathcal{H} = \oplus_i \mathcal{H}_i$, that is, they satisfy

$$\varrho_i \, \varrho_j = 0 \ \text{ for } i \neq j, \tag{7.312}$$

and let P be a classical ensemble with probabilities $\boldsymbol{p} = \{p_1, \ldots, p_n\}$. Then $\varrho \equiv \sum_i p_i \, \varrho_i$ is a density matrix and

$$S(\varrho) = S(\boldsymbol{p}) + \sum_i p_i \, S(\varrho_i), \tag{7.313}$$

where $S(\boldsymbol{p})$ is the classical Shannon entropy of P. Again, the recursion relation plus **vNE1**, **vNE2**, and **vNE3** uniquely characterize the von Neumann entropy;

vNE6 *Sub-additivity:* we have a bi-partite Hilbert space $\mathcal{H}_1 \otimes \mathcal{H}_2$ with a density matrix $\varrho_{12} \in \mathcal{B}(\mathcal{H}_1 \otimes \mathcal{H}_2)$ and partial trace density matrices

$$\varrho_1 = \mathrm{Tr}_{\mathcal{H}_2}\varrho_{12} \in \mathcal{B}(\mathcal{H}_1), \quad \varrho_2 = \mathrm{Tr}_{\mathcal{H}_1}\varrho_{12} \in \mathcal{B}(\mathcal{H}_2). \tag{7.314}$$

Then

$$S(\varrho_{12}) \leq S(\varrho_1) + S(\varrho_2). \tag{7.315}$$

As in the classical case, the *quantum mutual information* of the two parts A and B of a system with state $\varrho_{AB} \in \mathcal{H}_A \otimes \mathcal{H}_B$ is defined as

$$I(A; B) \stackrel{\text{def}}{=} S(\varrho_A) + S(\varrho_B) - S(\varrho_{AB}) \tag{7.316}$$

(here $\varrho_A \equiv \mathrm{Tr}_B\varrho_{AB}$ and $\varrho_B \equiv \mathrm{Tr}_A\varrho_{AB}$). $I(A; B)$ is *non-negative* since the von Neumann entropy is sub-additive. The quantum mutual information $I(A; B)$ is symmetric in A, B and zero only for a product state

$$\varrho_{AB} = \varrho_A \otimes \varrho_B; \tag{7.317}$$

vNE7   *Araki-Lieb triangle inequality:* In the bi-partite set-up of **vNE6** we have

$$|S(\varrho_1) - S(\varrho_2)| \leq S(\varrho_{12}); \tag{7.318}$$

vNE8   *Bipartite pure states:* In particular, when the state ϱ_{12} of the bipartite system is pure $S(\varrho_{12}) = 0$ and hence

$$S(\varrho_1) = S(\varrho_2). \tag{7.319}$$

Indeed, as we saw in Corollary 7.2, the two partial traces have the same non-zero eigenvalues, hence the same entropy;

vNE9   *Strong sub-additivity:* in a three-partite quantum system, with Hilbert space $\mathcal{H} = \mathcal{H}_1 \otimes \mathcal{H}_2 \otimes \mathcal{H}_3$, we have

$$S(\varrho_{123}) + S(\varrho_2) \leq S(\varrho_{12}) + S(\varrho_{23}). \tag{7.320}$$

We may rewrite this inequality in the form

$$S(\varrho_1) + S(\varrho_{23}) - S(\varrho_{123}) \geq S(\varrho_1) + S(\varrho_2) - S(\varrho_{12}). \tag{7.321}$$

which is an inequality for the quantum mutual informations between the parts of a tripartite system with subsystems A, B, C

$$I(A; BC) \geq I(A; B) \tag{7.322}$$

which states that we learn not less about A knowing BC than knowing only B;

vNE10   *Equivalent form of strong sub-additivity:* in a three-partite system

$$S(\varrho_{12}) \leq S(\varrho_1) + S(\varrho_2) \leq S(\varrho_{13}) + S(\varrho_{23}). \tag{7.323}$$

vNE11   *Mixing inequality:* Suppose we have a quantum ensemble of states $\{|\psi_j\rangle\}$ *not necessarily orthogonal* (nor linear independent) where the j-th state has probability $p_j \geq 0$ and $\sum_j p_j = 1$. This quantum ensemble is represented by the density matrix

$$\varrho = \sum_j p_j \, |\psi_j\rangle\langle\psi_j|. \tag{7.324}$$

Then (here $\boldsymbol{p} \equiv (p_1, \ldots, p_n)$)

$$S(\varrho) \leq S(\boldsymbol{p}), \tag{7.325}$$

where the RHS is the classical Shannon entropy of the probabilities $\boldsymbol{p}$. The inequality is saturated iff the states $|\psi_j\rangle$ are orthogonal, so that the p_j's in Eq. (7.324) coincide with the eigenvalues of ϱ.

Before going to the proofs, let us make some general remarks about the *crucial differences* between the von Neumann quantum entropy $S(\varrho)$ and the Shannon classical entropy $S(\boldsymbol{p})$.

The Shannon entropy has the property **SE6**, *monotonicity,*

$$S(LM) \geq \max\{S(L), S(M)\}. \tag{7.326}$$

This inequality *fails* for the von Neumann entropy. For instance consider the case of a bipartite *pure* quantum state

$$\varrho_{AB} = |\psi\rangle\langle\psi| \in \mathcal{B}(\mathcal{H}_A \otimes \mathcal{H}_B). \tag{7.327}$$

The entropy of the full state vanishes, $S(\varrho_{AB}) = 0$, while

$$S(\varrho_A) = S(\varrho_B) \geq 0. \tag{7.328}$$

The entropy of the global system vanishes because we know the quantum state $|\psi\rangle$ fully. But we have incomplete informations about the subsystems, and our ignorance is measured by $S(\varrho_A) = S(\varrho_B)$. The crucial point is that for a *quantum* system—but not for a classical one—information can be encoded in the correlations among the parts of our system, while these correlations cannot be seen looking at the single subsystems. As a result, the von Neumann *conditional entropy,*

$$S(A|B) = S(\varrho_{AB}) - S(\varrho_B), \tag{7.329}$$

may be *negative.* E.g. when ϱ_{AB} is an entangled pure state

$$S(A|B) = -S(\varrho_A) \equiv -S(\varrho_B) < 0. \tag{7.330}$$

In a sense this equation says that in the quantum case "knowing" the subsystem B of an entangled pure state makes us "more than certain" about the subsystem A. This clearly shows that quantum information behaves quite differently from the classical one, and hence it opens new possibilities that may have terrific applications to real-world technology. Again: *quantum entanglement is a resource!*

In terms of the conditional entropy the strong sub-additivity reads

$$S(A|B) \geq S(A|BC) \tag{7.331}$$

which says, roughly speaking, that our ignorance about A when we know only B is no smaller than our ignorance about A when we know both B and C. This property is subtler than the corresponding classical statement since the quantum "ignorance" may sometimes be negative, as we saw above.

7.7.1   *Math Lemmas and* **Proofs**

Before sketching the proofs of the properties **vNE1–vNE11** of von Neumann entropy $S(\varrho)$, we introduce some general math tools.

Some Matrix Inequalities
Recall that a smooth function $f : \mathbb{R} \to \mathbb{R}$ is *convex* iff

$$f(x) - f(y) \geq (x - y) f'(y). \tag{7.332}$$

Lemma 7.7 (Klein Inequality)

(1) If $f : \mathbb{R} \to \mathbb{R}$ is a convex function and A, B are Hermitian operators

$$\mathrm{Tr}\big[f(A) - f(B) \big] \geq \mathrm{Tr}\big[(A - B)\, f'(B) \big] \tag{7.333}$$

(2) Specialize to the function $t \mapsto t \log t$

$$\mathrm{Tr}\big[A \log A - A \log B \big] \geq \mathrm{Tr}[A - B] \tag{7.334}$$

with equality if and only if $A = B$.

Proof We use orthonormal eigenbases for the two operators

$$A|a_i\rangle = a_i|a_i\rangle, \qquad B|b_j\rangle = b_j|b_j\rangle. \tag{7.335}$$

Then

$$
\begin{aligned}
\langle a_i \big[f(A) - f(B) - (A - B)f'(B)|a_i\rangle &= \\
= f(a_i) - \sum_j |\langle a_i|b_j\rangle|^2 [f(b_j) + (a_i - b_j)f'(b_j)] &= \\
= \sum_j |\langle a_i|b_j\rangle|^2 \big[f(a_i) - f(b_j) - (a_i - b_j)f'(b_j) \big], &
\end{aligned}
\tag{7.336}
$$

where we used that the matrix $\langle a_i|b_j\rangle$ is unitary. The RHS is non-negative since f is convex.   □

Lemma 7.8 (Golden-Thompson Inequality) *For self-adjoint operators A and B*

$$\mathrm{Tr}\, e^{A+B} \leq \mathrm{Tr}(e^A e^B), \tag{7.337}$$

with equality if and only if A and B commute. (Cf. [19] eq.(IX.19)).

Proof For all matrix M and positive integer m

$$\left|\mathrm{Tr}(M^{2m})\right| \leq \mathrm{Tr}(MM^{\dagger})^{m} \tag{7.338}$$

which implies for all pair of matrices M, N and positive integer k

$$\left|\mathrm{Tr}(MN)^{2^{k}}\right| \leq \mathrm{Tr}\left[(M^{\dagger}MNN^{\dagger})^{2^{k-1}}\right]. \tag{7.339}$$

Setting $M = \exp(A/2^{k})$, $N = \exp(B/2^{k})$, with A, B Hermitian, we get

$$\mathrm{Tr}(e^{A/2^{k}}e^{B/2^{k}})^{2^{k}} \leq \mathrm{Tr}(e^{A/2^{k-1}}e^{B/2^{k-1}})^{2^{k-1}} \tag{7.340}$$

and, by inverse induction on k,

$$\mathrm{Tr}(e^{A/2^{k}}e^{B/2^{k}})^{2^{k}} \leq \mathrm{Tr}(e^{A}e^{B}) \quad \text{for all } k \in \mathbb{N}. \tag{7.341}$$

Hence, using the Trotter formula (Theorem 2.5),

$$\mathrm{Tr}\, e^{A+B} = \lim_{k\to\infty} \mathrm{Tr}(e^{A/2^{k}}e^{B/2^{k}})^{2^{k}} \leq \mathrm{Tr}(e^{A}e^{B}). \tag{7.342}$$

$\square$

Theorem 7.10 (Lieb) *Let R, S, T, be* positive *operators. Then*

$$\mathrm{Tr}\, e^{\log R - \log S + \log T} \leq \int_{0}^{\infty} du\, \mathrm{Tr}\left[R\frac{1}{S+u}T\frac{1}{S+u}\right]. \tag{7.343}$$

This is a deep theorem. For a proof see [33] or [34].

Properties of von Neumann Entropy: *Proofs*
The two properties **vNE1**,**vNE2** are obvious, and **vNE3** follows from properties **SE3**, **SE4** of Shannon's entropy applied to the eigenvalues of ϱ. **vNE4**: we write

$$\varrho = t\varrho_1 + (1-t)\varrho_2. \tag{7.344}$$

Klein inequality (7.334) with $B = \varrho$ and A equal, respectively, to ϱ_1, ϱ_2, yields

$$\mathrm{Tr}(\varrho_1 \log \varrho_1) \geq \mathrm{Tr}(\varrho_1 \log \varrho), \quad \mathrm{Tr}(\varrho_2 \log \varrho_2) \geq \mathrm{Tr}(\varrho_2 \log \varrho). \tag{7.345}$$

Then

$$\mathrm{Tr}(\varrho \log \varrho) = t\,\mathrm{Tr}(\varrho_1 \log \varrho) + (1-t)\mathrm{Tr}(\varrho_2 \log \varrho) \leq$$
$$\leq t\,\mathrm{Tr}(\varrho_1 \log \varrho_1) + (1-t)\mathrm{Tr}(\varrho_2 \log \varrho_2). \tag{7.346}$$

vNE5: the density matrices ϱ_i commute so they can be diagonalized simultaneously. In the diagonal basis Eq. (7.313) reduces to **SE9**.

vNE6: in Klein's inequality (7.334) we take $A = \varrho_{12}$ and $B = \varrho_1 \otimes \varrho_2$:

$$\mathrm{Tr}(\varrho_{12} \log \varrho_{12}) \geq \mathrm{Tr}[\varrho_{12}(\log \varrho_1 \otimes \mathbf{1} + \mathbf{1} \otimes \log \varrho_2)] =$$
$$= \mathrm{Tr}(\varrho_1 \log \varrho_1) + \mathrm{Tr}(\varrho_2 \log \varrho_2). \tag{7.347}$$

(A second proof is obtained by the specialization of **vNE9** to $\mathcal{H}_2 \simeq \mathbb{C}$).

vNE8 is equivalent to Corollary 7.2. To prove **vNE7** we apply the *purification trick* which allows to produce new entropy inequalities from old ones. By **vNE8** we can always purify the density matrix $\varrho_A \in \mathcal{H}(\mathcal{H}_A)$ to a pure state

$$\varrho_{AB} \in \mathcal{B}(\mathcal{H}_A \otimes \mathcal{H}_B). \tag{7.348}$$

Then $S(\varrho_A) = S(\varrho_B)$. As ϱ_A we take ϱ_{12} in **vN6** whose purification we write as $\varrho_{123} \in \mathcal{B}(\mathcal{H}_1 \otimes \mathcal{H}_2 \otimes \mathcal{H}_3)$. Then

$$S(\varrho_3) = S(\varrho_{12}) \leq S(\varrho_1) + S(\varrho_2) = S(\varrho_{23}) + S(\varrho_2), \tag{7.349}$$

i.e. renumbering the factor spaces

$$S(\varrho_{12}) \geq S(\varrho_2) - S(\varrho_1). \tag{7.350}$$

Interchanging $\varrho_1 \leftrightarrow \varrho_2$ we get the triangle inequality (7.318). Next we prove strong subadditivity **vNE9**. We start from Klein's inequality (7.334) with

$$A = \varrho_{123} \quad \text{and} \quad \log B = \log \varrho_{12} - \log \varrho_2 + \log \varrho_{23}. \tag{7.351}$$

Then

$$-S(\varrho_{123}) + S(\varrho_{12}) - S(\varrho_2) + S(\varrho_{23}) =$$
$$= \mathrm{Tr}_{123}\Big(\varrho_{123} \log \varrho_{123} - \varrho_{123}\big(\log \varrho_{12} - \log \varrho_2 + \log \varrho_{23}\big)\Big) \geq$$
$$\geq \mathrm{Tr}_{123}\Big(\varrho_{123} - e^{\log \varrho_{12} - \log \varrho_2 + \log \varrho_{23}}\Big) \geq$$
$$\geq \mathrm{Tr}_{123}\left[\varrho_{123} - \int_0^\infty du\left(\varrho_{12}\frac{1}{\varrho_2 + u}\varrho_{23}\frac{1}{\varrho_2 + u}\right)\right] = \tag{7.352}$$
$$= \mathrm{Tr}_{123}\varrho_{123} - \int_0^\infty du\,\mathrm{Tr}_2\left(\varrho_2\frac{1}{\varrho_2 + u}\varrho_2\frac{1}{\varrho_2 + u}\right) =$$
$$= \mathrm{Tr}_{123}\,\varrho_{123} - \mathrm{Tr}_2\,\varrho_2 = 0.$$

where in the fourth line we used Lieb's theorem (7.343). We now show that **vNE10** is equivalent to **vN9**: to see this we use again the purification trick. We purify the state ϱ_{123} to $\varrho_{1234} = |\psi\rangle\langle\psi|$. Now Eq. (7.320) can be rewritten as

$$S(\varrho_4) + S(\varrho_2) \leq S(\varrho_{12}) + S(\varrho_{14}) \tag{7.353}$$

By relabeling the factor spaces we get (7.323).

Finally we consider **vNE11**. The von Neumann entropy of ϱ is the Shannon entropy of its eigenvalues $\lambda = (\lambda_1, \ldots, \lambda_n)$. In view of the Schur concavity of Shannon entropy, the statement follows from Schrödinger's mixing theorem which yields the majorization

$$\boldsymbol{p} \prec \boldsymbol{\lambda}. \tag{7.354}$$

7.7.2   Quantum Rényi Entropy

The quantum Rényi entropies are a family of entropy functions parametrized by a non-negative parameter q

$$S_q(\varrho) \stackrel{\text{def}}{=} \frac{1}{1-q} \log \mathrm{Tr}\varrho^q. \tag{7.355}$$

These functions also measure our "ignorance" about the quantum state ϱ. We list some of their properties without proofs (see e.g. [2]).

The quantum Rényi entropies non-negative functions and are additive for product states

$$S_q(\varrho_1 \otimes \varrho_2) = S_q(\varrho_1) + S_q(\varrho_2). \tag{7.356}$$

When $0 < q \leq 1$ the Rényi entropies are also concave

$$S_q(t\varrho_1 + (1-t)\varrho_2) \geq t S_q(\varrho_1) + (1-t)S_q(\varrho_2), \quad 0 \leq t \leq 1. \tag{7.357}$$

For all q, S_q is zero for a pure state and maximal for the ignorance state $\varrho = \boldsymbol{1}/n$

$$S_1(\boldsymbol{1}/n) = \log n. \tag{7.358}$$

In the limit $q \to 1$ we get back the von Neumann entropy: indeed

$$\log \mathrm{Tr}\varrho^q = (q-1)\mathrm{Tr}(\varrho \log \varrho) + O((q-1)^2). \tag{7.359}$$

The Rényi entropy with $q = 2$ is related to the *purity* of the state which measures how far the state is from the maximal mixed state

$$\text{purity}(\varrho) \stackrel{\text{def}}{=} D\left(\varrho, \frac{\mathbf{1}}{n}\right)^2_{\text{HS}} + \frac{1}{n} \equiv \text{Tr}\,\varrho^2 \qquad (7.360)$$

where $D(\cdot, \mathbf{1}/n)_{\text{HS}}$ is the Hilbert-Schmidt distance from the maximally mixed state. One has

$$\text{purity}(\varrho) = \exp\left(-S_2(\varrho)\right). \qquad (7.361)$$

$S_0(\varrho)$ is the *Hartley entropy* equal to the logarithm $\log r$ of the Schmidt rank r of the state ϱ.

When $q_1 \le q_2$ we have the inequalities

$$S_{q_2}(\varrho) \le S_{q_1}(\varrho), \qquad \frac{q_2 - 1}{q_2} S_{q_2}(\varrho) \ge \frac{q_1 - 1}{q_1} S_{q_1}(\varrho) \qquad (7.362)$$

in particular the Hartley entropy is the largest Rényi entropy. The quantum Rényi entropies satisfy a *weak* version of subadditivity

$$S_q(\varrho_1) - S_0(\varrho_2) \le S_q(\varrho_{12}) \le S_q(\varrho_1) + S_0(\varrho_2) \qquad (7.363)$$

where $S_0(\varrho)$ is the Hartley entropy.

The Rényi entropies are often more easy to compute than other entropy functions and hence are widely used in the literature.

7.8 Quantum Relative Entropy

The *quantum relative entropy* of two states ϱ_1 and ϱ_2 is

$$S(\varrho_1 \| \varrho_2) \stackrel{\text{def}}{=} \text{Tr}\left[\varrho_1(\log \varrho_1 - \log \varrho_2)\right] \qquad (7.364)$$

which is continuous and finite unless ϱ_2 has some zero eigenvalue. For diagonal matrices $S(\varrho_1 \| \varrho_2)$ reduces to the classical relative entropy which measures how different two probability distributions p, q are.

We can write the relative entropy in the integral form

$$S(\varrho_1 \| \varrho_2) = \int_0^\infty du\, \text{Tr}\left(\varrho_1 \frac{1}{\varrho_2 + u}(\varrho_1 - \varrho_2)\frac{1}{\varrho_2 + u}\right). \qquad (7.365)$$

We list its main properties:

RE1   *Unitary invariance:* $S(U\varrho_1 U^\dagger || U\varrho_2 U^\dagger)$ for all $U \in \mathcal{U}(\mathcal{H})$;
RE2   *Positivity:* $S(\varrho_1||\varrho_2) \geq 0$ with equality iff $\varrho_1 = \varrho_2$. This is simply the Klein inequality (7.334) with $A = \varrho_1$ and $B = \varrho_2$;
RE3   *Strong positivity:*

$$S(\varrho_1||\varrho_2) \geq \frac{1}{2}\mathrm{Tr}(\varrho_1 - \varrho_2)^2; \tag{7.366}$$

RE4   *Joint convexity:* for all $0 \leq t \leq 1$

$$S(t\varrho_1 + (1-t)\varrho_2 || t\tilde\varrho_1 + (1-t)\tilde\varrho_2) \leq t\,S(\varrho_1||\tilde\varrho_1) + (1-t)S(\varrho_2||\tilde\varrho_2); \tag{7.367}$$

RE5   *Monotonicity under partial trace:*

$$S(\varrho_1||\tilde\varrho_2) \equiv S(\mathrm{Tr}_2\varrho_{12}||\mathrm{Tr}_2\tilde\varrho_{12}) \leq S(\varrho_{12}||\tilde\varrho_{12}) \tag{7.368}$$

RE6   *Monotonicity under completely positive maps:* If $\mathcal{E}$ is a completely positive map

$$S(\mathcal{E}(\varrho_1)||\mathcal{E}(\varrho_2)) \leq S(\varrho_1||\varrho_2). \tag{7.369}$$

Proofs   **RE3** follows from Klein's inequality (7.334) applied to the function

$$f : [0, 1] \to \mathbb{R}, \qquad t \mapsto t \log t - \frac{t^2}{2} \tag{7.370}$$

which is convex in the interval $[0, 1]$. **RE4**, **RE5**, **RE6** are equivalent to each other as well as with strong sub-additivity for von Neumann entropy **vNE9**. For instance, assume **RE5**:

$$-S(\varrho_{123}) + S(\varrho_{12}) = \mathrm{Tr}\Big(\varrho_{123}\big(\log\varrho_{123} - \log(\varrho_{12} \otimes \mathbf{1})\big)\Big) \equiv$$

$$\equiv S(\varrho_{123}||\varrho_{12} \otimes \mathbf{1}) \overset{\mathbf{RE5}}{\geq} S(\varrho_{23}||\varrho_2 \otimes \mathbf{1}) = \tag{7.371}$$

$$= \mathrm{Tr}\Big(\varrho_{23}\big(\log\varrho_{23} - \log(\varrho_2 \otimes \mathbf{1})\big)\Big) = -S(\varrho_{23}) + S(\varrho_2)$$

which is equivalent to (7.320). For a direct proof of **RE4** see **Theorem IX.6.5** in [19].

7.9   Quantum Information Theory

We limit ourselves to a brief discussion of the very basic results of Quantum Information theory to give a flavor of the subject which is beyond an introductory course in Quantum Mechanics. We refer the reader to the textbooks on the subject [2–9] and the nice Lecture Notes by Preskill [10].

Monotonicity of Mutual Information
Strong sub-additivity of von Neumann entropy, written in terms of quantum mutual information as in Eq. (7.322), implies an important property of quantum mutual information:

QMI   Quantum mutual information is monotonic under the action of quantum channels $\mathcal{E} \colon \mathcal{B}(\mathcal{H}_B) \to \mathcal{B}(\mathcal{H}_B)$

$$I(A; B) \geq I(A; \mathcal{E}(B)), \tag{7.372}$$

that is, explicitly,

$$\begin{aligned} S(\varrho_A) + S(\varrho_B) - S(\varrho_{AB}) &\geq \\ &\geq S(\varrho_A) + S(\mathcal{E}(\varrho_B)) - S(\mathcal{E}(\varrho_{AB})). \end{aligned} \tag{7.373}$$

Said in plain English: *a quantum channel $\mathcal{E}$ acting on the system B cannot increase the mutual information of A with B.*

Proof of **QMI** We know from Sect. 7.4 that $\mathcal{E}$ has a unitary extension, its Stinespring dilation $\mathcal{U}$:

$$\mathcal{U} \colon \mathcal{B}(\mathcal{H}_B \otimes \mathcal{H}_E) \to \mathcal{B}(\mathcal{H}_B \otimes \mathcal{H}_E) \tag{7.374}$$

$$\mathcal{E}(\varrho_B) = \mathrm{Tr}_E\, \mathcal{U}\big(\varrho_B \otimes |*\rangle\langle *|\big) \tag{7.375}$$

for a suitable environment Hilbert space $\mathcal{H}_E$ and reference pure state $|*\rangle$ in $\mathcal{H}_E$. For all operators $A \in \mathcal{B}(\mathcal{H}_B \otimes \mathcal{H}_E)$ we write $\tilde{A}$ for its unitary evolution $\mathcal{U}(A)$.

Since $\mathcal{U}$ is unitary, it does not change the non-zero eigenvalues of the density operators, so it preserves the von Neumann entropy; this implies

$$S(\varrho_B) = S(\mathcal{U}(\varrho_B)) \equiv S(\tilde{\varrho}_{BE}), \tag{7.376}$$

$$S(\varrho_{AB}) = S\big((\mathbf{1} \otimes \mathcal{U})(\varrho_{AB})\big) \equiv S(\tilde{\varrho}_{ABE}), \tag{7.377}$$

$$\text{where} \quad \tilde{\varrho}_{BE} \equiv \mathcal{U}(\varrho_{BE}), \quad \varrho_{BE} \equiv \varrho_B \otimes |*\rangle\langle *|, \tag{7.378}$$

and then

$$I(A; B) \equiv S(\varrho_A) + S(\varrho_B) - S(\varrho_{AB}) =$$
$$= S(\varrho_A) + S(\tilde{\varrho}_{BE}) - S(\tilde{\varrho}_{ABE}) \equiv \tag{7.379}$$
$$\equiv I(A; \mathcal{U}(\varrho_{BE})) \geq I(A, \mathcal{E}(\varrho_B)),$$

where the last inequality follows from Eq. (7.322). $\qquad\qquad \square$

7.9.1 Quantum Source Coding

Finally we consider the quantum counterpart to the classical Shannon theorem on noiseless source coding. The quantum source coding theorem is due to Schumacher [35].

The quantum "alphabet" is an ensemble $\mathcal{L}$ of (normalized) quantum pure states $\{|\ell\rangle\}$ where the ℓ-th state ("qu-letter") has probability $p(\ell)$. The ensemble $\{|\ell\rangle, p(\ell)\}_{\ell \in \mathcal{L}}$ realizes the single-qu-letter density operator

$$\varrho \equiv \varrho(\mathcal{L}) = \sum_{\ell} p(\ell) |\ell\rangle\langle\ell|. \tag{7.380}$$

If the states $\{|\ell\rangle\}$ are pairwise orthogonal, hence distinguishable with certainty, the situation is similar to the classical one. The new quantum behavior arises when the states $|\ell\rangle$, while normalized, are *not* mutually orthogonal hence not fully distinguishable.

The n-qu-letter messages have a density operator

$$\varrho^{\otimes n} \equiv \overbrace{\varrho \otimes \cdots \otimes \varrho}^{n \text{ factors}} \in \mathcal{B}(\mathcal{H}^{\otimes n}). \tag{7.381}$$

We want a quantum code which allows to compress the quantum message in a Hilbert space smaller than $\mathcal{H}^{\otimes n}$, paying a small cost ϵ in terms of the fidelity of the message. Schumacher theorem describes the *optimal compression* of a quantum message: for a message of length n we need a Hilbert space $\mathcal{H}$ of dimension just[20]

$$\dim \mathcal{H} = 2^{n\, S(\varrho) + o(1)} \leq \dim \mathcal{H}^{\otimes n} \equiv (\dim \mathcal{H})^n. \tag{7.382}$$

---

[20] Here we measure the entropy measure in (qu)bits i.e. $S(\varrho) = -\mathrm{Tr}(\varrho \, \log_2 \varrho)$.

We conclude that the von Neumann entropy is the number of qubits of quantum information carried per qu-letter of the quantum message. Compression is always possible unless ϱ is the maximally mixed state

$$\varrho_{\max} \equiv \frac{1}{\dim \mathcal{H}} \tag{7.383}$$

in which case the inequality (7.382) is saturated.

To prove the theorem we have to show that, for n large, the density matrix $\varrho^{\otimes n}$ has *nearly all* of its support in a sub-space

$$\mathcal{H} \subset \mathcal{H}^{\otimes n} \tag{7.384}$$

of the Hilbert space of messages, where the dimension of $\mathcal{H}$ approaches $2^{nS(\varrho)}$ asymptotically for large n. This result follows directly from the corresponding classical statement. We may realize ϱ using an "orthonormal qu-alphabet $\mathcal{L}_{\mathrm{eig}}$", that is, in terms of an ensemble of orthonormal pure states, namely its eigenstates $|e_i\rangle$, with probabilities given by the corresponding eigenvalues λ_i (cf. Sect. 7.3.3). When transmitting messages in this orthonormal qu-alphabet (i.e. orthonormal basis $\{|e_i\rangle\}$) our source of quantum information behaves effectively as a classical source: it produces messages which are tensor products of ϱ eigenstates,

$$|e_{i_1}\rangle \otimes |e_{i_2}\rangle \otimes \cdots \otimes |e_{i_n}\rangle, \tag{7.385}$$

each one with a probability equal to the product of the corresponding eigenvalues

$$\mathsf{Prob}\left(\bigotimes_{k=1}^{n} |e_{i_k}\rangle\right) = \prod_{k=1}^{n} \lambda_{i_k}. \tag{7.386}$$

Mimicking was with did in Sect. 7.6 for the classical case, we define the δ-typical subspace $\mathcal{H}_{\delta,n}$ for a given n and δ,

$$\mathcal{H}_{\delta,n} = \mathbb{C}\text{-span}\left\{|e_{i_1}\rangle \otimes |e_{i_2}\rangle \otimes \cdots \otimes |e_{i_n}\rangle \; : \; \left| -\frac{1}{n}\log_2 \prod_{k=1}^{n} \lambda_{i_k} - S(\varrho) \right| \leq \delta\right\} \tag{7.387}$$

i.e. $\mathcal{H}_{\delta,n}$ is the sub-space of $\mathcal{H}^{\otimes n}$ spanned by the eigenvectors of $\varrho^{\otimes n}$ with eigenvalues λ in the range

$$2^{-n(S(\varrho)-\delta)} \geq \lambda \geq 2^{-n(S(\varrho)+\delta)}. \tag{7.388}$$

Let

$$\mathbf{P} \colon \mathcal{H}^{\otimes n} \to \mathcal{H}_{\delta,n} \hookrightarrow \mathcal{H}^{\otimes n} \tag{7.389}$$

be the orthogonal projection on the δ-typical subspace $\mathcal{H}_{\delta,n}$. The trace

$$\mathrm{Tr}(\mathbf{P}\varrho^{\otimes n}) \tag{7.390}$$

is the total probability that an n-qu-letter message is δ-typical. Applying the central limit theorem to the classical probability distribution given by the eigenvalues of ϱ, Eq. (7.386), we conclude that

$$\mathrm{Tr}(\mathbf{P}\varrho^{\otimes n}) > 1 - \epsilon \tag{7.391}$$

for any $\epsilon > 0$, provided n is large enough. Comparing (7.386) with (7.388) we get the estimate of the dimension of the δ-typical subspace for large n

$$(1 - \epsilon)2^{n(S(\varrho)-\delta)} \leq \dim \mathcal{H}_{\delta,n} \leq 2^{n(S(\varrho)+\delta)}. \tag{7.392}$$

A *quantum code* c is an isomorphism of Hilbert spaces

$$\mathsf{c}\colon \mathcal{H} \to (\mathbb{C}^2)^{n\,S(\varrho)+o(1)} \tag{7.393}$$

which maps each state $|\mu\rangle$ in the typical subspace $\mathcal{H} \subset \mathcal{H}^{\otimes n}$ into a quantum message $\mathsf{c}|\mu\rangle$ consisting of just $n\,S(\varrho) + o(1)$ qubits.

Alice shares her quantum code c with Bob. To send a quantum message to Bob in the most economic way, Alice first makes the measurement $\mathbf{P}$ that projects the input message $|\mu_{\mathrm{in}}\rangle \in \mathcal{H}^{\otimes n}$ onto either $\mathcal{H}$ or $\mathcal{H}^{\perp}$. The outcome is $\mathcal{H}$ with probability $1 - \epsilon$: In this case Alice takes the projected state

$$|\mu\rangle \equiv \mathbf{P}|\mu_{\mathrm{in}}\rangle \in \mathcal{H} \tag{7.394}$$

and codes it into $n\,S(\varrho) + O(1)$ qubits using c

$$|\mu\rangle \rightsquigarrow \mathsf{c}|\mu\rangle \in (\mathbb{C}^2)^{n\,S(\varrho)+o(1)}, \tag{7.395}$$

and then she sends the resulting compressed message $\mathsf{c}|\mu\rangle$ to Bob. The outcome $|\mu_{\mathrm{in}}\rangle \in \mathcal{H}^{\perp}$ has probability $O(\epsilon)$, and Alice doesn't care much what to do in this rare event, and just sends some random message.

Bob receives the message. With probability $1 - \epsilon$ what he gets is a genuine compressed message of the form $\mathsf{c}|\mu\rangle$. He decompresses it by applying the inverse isomorphism c^{-1}

$$\mathsf{c}^{-1}(\mathsf{c}|\mu\rangle) = |\mu\rangle \in \mathcal{H} \subset \mathcal{H}^{\otimes n}. \tag{7.396}$$

We wish to estimate the fidelity of the quantum comunication protocol used by Alice and Bob. Let Alice's n qu-letter input pure state message be

$$|\mu_{\mathrm{in}}\rangle \equiv |\ell\rangle = |\ell_{i_1}\rangle \otimes |\ell_{i_2}\rangle \otimes \cdots \otimes |\ell_{i_n}\rangle \in \mathcal{H}^{\otimes n}. \tag{7.397}$$

After coding, transmission, and decoding, Bob reconstructs the state

$$\varrho_B(\ell) = \mathbf{P}|\ell\rangle\langle\ell|\mathbf{P} + \langle\ell|(1-\mathbf{P})|\ell\rangle\,\rho_{\text{Junk}}, \tag{7.398}$$

where ρ_{Junk} is the decoding of the arbitrary message sent by Alice when the first measure produced a state in $\mathcal{H}^\perp$. The fidelity of the decoded message $\varrho_B(\ell)$ with respect to the original input $|\ell\rangle$ is

$$F\big(|\ell\rangle\langle\ell|, \varrho_B(\ell)\big) = \langle\ell|\varrho_B(\ell)|\ell\rangle, \tag{7.399}$$

cf. Eq. (7.136). The actual fidelity of the transmission depends on the specific message $|\ell\rangle$, and to estimate the reliability of the protocol it is appropriate to compute the *average* $\overline{F}$ of the fidelity F over the ensemble $\mathcal{L}^n$ of all possible n qu-letter messages

$$
\begin{aligned}
\overline{F} &= \sum_\ell p(\ell) F\big(|\ell\rangle\langle\ell|, \varrho_B(\ell)\big) = \sum_\ell p(\ell)\langle\ell|\varrho_B(\ell)|\ell\rangle = \\
&= \sum_\ell p(\ell)\langle\ell|\mathbf{P}|\ell\rangle\langle\ell|\mathbf{P}|\ell\rangle + \sum_\ell p(\ell)\langle\ell|(1-\mathbf{P})|\ell\rangle\langle\ell|\rho_{\text{Junk}}|\ell\rangle \geq \\
&\geq \sum_\ell p(\ell)\big(\langle\ell|\mathbf{P}|\ell\rangle\big)^2 \geq \sum_\ell p(\ell)\Big(2\langle\ell|\mathbf{P}|\ell\rangle - 1\Big) = \\
&= 2\,\mathrm{Tr}(\varrho^{\otimes n}\mathbf{P}) - 1 \geq 2(1-\epsilon) - 1 = 1 - 2\epsilon,
\end{aligned}
\tag{7.400}
$$

where in the third line we used that $\varrho_{\text{Junk}} \geq 0$, $(1-\mathbf{P}) \geq 0$ and then the inequality $z^2 \geq 2z - 1$ which holds for all $z \in \mathbb{R}$. Since ϵ and δ may be chosen as small as we please, Eq. (7.400) shows that we may compress the message down to

$$n(S(\varrho) + o(1)) \tag{7.401}$$

qubits with a fidelity that becomes arbitrarily good as $n \to \infty$. One can show that this is the optimal compression rate. Indeed, if the dimension of the code subspace is $\approx 2^{n(S(\varrho)-\delta')}$ the average fidelity would go to zero exponentially as $n \to \infty$. That the optimal compression rate for a quantum message is given by

$$R_{\text{optimal}} = S(\varrho) + o(1) \tag{7.402}$$

is the content of Schumacher's *quantum source coding theorem*.

References

1. E. Schrödinger, Discussion of probability relations between separated systems. Proc. Camb. Phi. Soc. **31**, 555 (1935)
2. I. Bengtsson, K. Zyczkowski, *Geometry of Quantum States. An Introduction to Quantum Entanglement*, 2nd edn. (Cambridge University Press, Cambridge, 2017)
3. D. Petz, *Quantum Information Theory and Quantum Statistics*. Theoretical and Mathematical Physics (Springer, Berlin, 2008)
4. M. Hayashi, S. Ishizaka, A. Kawachi, G. Kimura, T. Ogawa, *Introduction to Quantum Information Science*. Graduate Texts in Physics (Springer, Berlin, 2015)
5. M. Hayashi, *Quantum Information. An Introduction* (Springer, Berlin, 2006)
6. L. Diósi, *A Short Course in Ouantum Information Theory. An Approach From Theoretical Physics*. Lecture Notes in Physics, vol. 827, 2nd edn. (Springer, Berlin, 2011)
7. G. Jaeger, *Entanglement, Information, and the Interpretation of Quantum Mechanics* (Springer, Berlin, 2009)
8. M. Hayashi, *A Group Theoretic Approach to Quantum Information* (Springer, Berlin, 2017)
9. M.M. Wilde, *Quantum Information Theory* (Cambridge University Press, Cambridge, 2013)
10. J. Preskill, *Quantum Information*. Lecture Notes at Caltech. Available on-line at http://www.theory.caltech.edu/~preskill/ph229
11. G. Auletta, M. Fortunato, G. Parisi, *Quantum Mechanics* (Cambridge University Press, Cambridge, 2009)
12. H. Năstase, *Quantum Mechanics. A Graduate Course* (Cambridge University Press, Cambridge, 2023)
13. W. Hunziker, A note on symmetry operations in quantum mechanics. Helv. Phys. Acta **45**, 233 (1972)
14. S. Helgason, *Differential Geometry and Symmetric Spaces* (Academic Press New-York/London, 1962), 2nd edn. (1978)
15. M.M. Postnikov, *Geometry VI. Riemannian Geometry* (Springer, Berlin, 2001)
16. S. Cecotti, *Analytic Mechanics. A Concise Textbook* (Springer, Berlin, 2024)
17. S. Cecotti, *Supersymmetric Field Theories. Geometric Structures and Dualities* (Cambridge University Press, Cambridge, 2015)
18. L.P. Hughston, R. Jozsa, W.K. Wooters, A complete classification of quantum ensembles having a given density matrix. Phys. Lett. **A283**, 14 (1993)
19. R. Bhatia, *Matrix Analysis*. Graduate Texts in Mathematics, vol. 169 (Springer, Berlin, 1997)
20. S. Cecotti, *Statistical Mechanics. A Concise Advanced Textbook* (Springer, Berlin, 2024)
21. A. Einstein, B. Podolsky, N. Rosen, Can quantum-mechanical description of physical reality be considered complete? Phys. Rev. **47**, 777 (1935)
22. S.J. Freedman, J.F. Clauser, Experimental text of local hidden-variable theories. Phys. Rev. Lett. **28**, 938–941 (1972)
23. M.D. Reid, D.F. Walls, Violation of classical inequalities in quantum optics. Phys. Rev. **A34**, 1260–1276 (1986)
24. N.D. Mermin, Is the moon there when nobody looks? Phys. Today **38**, 38 (1985)
25. D.M. Greenberger, M.A. Horne, A. Shimony, A. Zeilinger, Bell's theorem without inequalities. Am. J. Phys. **58**, 1131–1143 (1990)
26. N.D. Mermin, Quantum mysteries revisited. Am. J. Phys. **58**, 731–734 (1990)
27. N.D. Mermin, What's wrong with these elements of reality? Phys. Today **43**, 9–11 (1990)
28. S. Coleman, Sidney Coleman's Dirac Lecture "Quantum Mechanics in Your Face". arXiv:2011.12671
29. C.E. Shannon, A mathematical theory of communication. Bell Syst. Tech. J. **27**, 379–423, 623–656 (1948)
30. G.E. Andrews, R. Askey, R. Roy, *Special Functions*. Encyclopedia of Mathematics and Its Applications, vol. 71 (Cambridge University Press, Cambridge, 2009)

31. F.W. Olver, D.M. Lozier, R.F. Boisvert, C.W. Clark, W. Charles (eds.), *NIST Handbook of Mathematical Functions* (Cambridge University Press, Cambridge, 2010). Available on-line at https://dlmf.nist.gov
32. R.B. Schinazi, *Probability with Statistical Applications*, 3rd edn. (Birkhäuser, Basel, 2022)
33. E.H. Lieb, Convex trace functions and the Wigner-Yanase-Dyson conjecture. Adv. Math. **11**, 267–288 (1973)
34. M.B. Ruskai, Inequalities for quantum entropy: a review with conditions for equality. arXiv:quant-ph/0205064
35. B. Schumacher, Quantum coding. Phys. Rev. **A54**, 2614 (1996)

Chapter 8
Methods, Techniques, Approximation Schemes

In the previous chapters we described the general mathematical framework of Quantum Mechanics and introduced several *exact* methods to solve quantum dynamical systems. While these methods are elegant and powerful, the typical realistic quantum system cannot be solved in exact closed form, and one should revert to approximate methods. While the emphasis in this textbook is on exact methods, in this chapter we present a survey of the main *analytic*[1] approximation schemes. We stress that exact and approximate methods are not mutually exclusive: a typical computation consists of several steps, and some of them may be performed in an exact fashion and other ones using a suitable approximation. One must optimize the procedure in the sense of carrying out exactly a maximal number of steps. Knowing a large assortment of exact techniques is then crucial even to deal with quantum problems which do not admit exact full integration. This is most evident in the large N limit of the $O(N)$ model (Sect. 8.6) where one uses several exact manipulations to set the path integral in a form amenable to asymptotic analysis as $N \to \infty$.

The techniques presented in this chapter are so-called *controlled* approximations, meaning that they are *exact methods* for asymptotic expansions (in the sense of Poincaré) of the relevant quantum amplitudes.

8.1 Semiclassical Approximation I: Generalities

We start with the semiclassical approximation[2] which we discuss from a variety of viewpoints using diverse techniques. Mathematically speaking, the method produces *asymptotic expansions* of the relevant physical quantities as $\hbar \to 0$ or

---

[1] *Analytic* as contrasted to *numeric*.

[2] Also called as the WKB approximation after Wentzel, Kramers and Brillouin.

S. Cecotti, *Quantum Mechanics*, UNITEXT for Physics,
https://doi.org/10.1007/978-3-031-98824-0_8

rather, as $1/\hbar \to \infty$. There is a more sophisticate approach, known as *exact WKB* [1], which (under appropriate conditions) resums the semiclassical expansion (using methods from *resurgence* [2] and *Picard-Lefshetz theory* [3]) and computes the *exact* quantum amplitudes. These sophisticate procedures, however, are beyond the level of an introductory textbook, and we content ourselves of introducing the semiclassical analysis as a mere *approximation scheme* instead of an exact procedure.

Physically $\hbar$ is an universal constant, with the dimension

$$[\hbar] = [\text{energy}] \times [\text{time}] \equiv [\text{action}], \tag{8.1}$$

whose value we cannot tune at our will. There is no real process that may lead to a vanishing $\hbar$, so the reader may wonder what is the physical meaning of studying the asymptotic limit $1/\hbar \to \infty$. The point is that there are many concrete situations where the physical quantities which have the dimension $[\text{energy}] \times [\text{time}]$ have *huge values* when measured in the natural units where $\hbar = 1$. For instance, we know from Chap. 4 that the third component L_3 of the orbital angular momentum $\mathbf{L}$, and its square $L^2 \equiv \mathbf{L} \cdot \mathbf{L}$, take the values

$$L_3 = m\,\hbar, \qquad L^2 = \ell(\ell+1)\hbar^2, \qquad \hbar \approx 10^{-34}\,\text{J} \cdot s \tag{8.2}$$

with m, ℓ integers. Human scale rotating bodies have angular momenta which are typically of order 1 when measured in $\text{J} \cdot s$, so the integers m, ℓ are of order 10^{34}, and the relative quantum indeterminacy of the three components of $\mathbf{L}$, due to their non-commutativity, is

$$\frac{\Delta L_1}{|\mathbf{L}|} \cdot \frac{\Delta L_2}{|\mathbf{L}|} \geq \frac{1}{2} \frac{|L_3|\,\hbar}{L^2} = \frac{|m|}{2\ell(\ell+1)} = O(10^{-34}), \tag{8.3}$$

where we used the general indetermination formula (2.420). We see from (8.3) that the angular momenta $\mathbf{L}$ of the objects of everyday life have, for all practical purposes, deterministic sharp values, and may be treated as classical commuting quantities to a fantastic accuracy. The same story holds for the harmonic oscillator: $2E/\omega$ is $\hbar$ times the quantum number $2n+1$ which is of order 10^{34} for a macroscopic pendulum. When the quantum numbers m, ℓ, n, etc. are large, but not *that* huge, the classical behavior is only approximate, and we are interested in understanding the small deviations of the observables from their classical pattern. This is the purpose of the semiclassical *approximation* (as contrasted to the *exact* WKB analysis).

Mathematically we describe the situations where the quantum numbers, such as m, are *large* as the double scaling limit where we take m to have asymptotically large values while rescaling the units we use from the fundamental ones, where $\hbar = 1$, to (say) the international system based on the meter, kilogram, and second, so that we effectively keep the *numerical value* of $L_3 \equiv m\,\hbar$ fixed when taking the asymptotic limit $m \to \infty$. This is equivalent to keeping L_3 fixed why sending the *numerical value* of $\hbar$ to zero since we measure it in terms of larger and larger units of action, while the relevant physical quantities, such as L_3, have values $O(1)$ when measured in the hugely rescaled units. This conceptual issue being clarified, for the rest of the chapter we shall loosely speak of the asymptotic analysis in the limit $\hbar \to 0$. The large $1/\hbar$ asymptotics of a physical quantity is called its *semiclassical value* (or expression).

In Sects. 8.2–8.5 we consider the asymptotic expansion for $1/\hbar \to \infty$ from several viewpoints with diverse applications in mind. We start from the path integral approach which is, perhaps, the most intuitive one.

8.2 Semiclassical Approximation II: Path Integrals

We consider a (imaginary time) path integral of the general form

$$\int_{\mathfrak{P}} [\mathrm{d}x] \, \exp\left[-\frac{1}{\hbar} \int_{t_i}^{t_f} (L_E + \text{sources}) \, \mathrm{d}t \right], \tag{8.4}$$

where $\mathfrak{P}$ is a space of paths $x : [t_i, t_f] \to \mathcal{M}$ satisfying the appropriate boundary conditions (b.c.) at the initial and final times t_i and t_f. For instance:

(1) Dirichlet b.c. for paths representing the heat kernel;
(2) periodic b.c. for the partition function at temperature β^{-1} (with $\beta \equiv t_f - t_i$);
(3) free boundary conditions at $(t_i, t_f) \equiv (-\infty, +\infty)$ for the generating functional of time-ordered correlation functions on the ground state.[3] The action now contains the relevant source terms.

Our task is to produce an asymptotic expansion of (8.4) as $1/\hbar$ becomes large. For the sake of comparison we first consider finite-dimensional ordinary integrals.

Laplace Asymptotic Method for Integrals
We want the compute the large $1/\hbar$ asymptotic behavior of a finite-dimensional integral of the form

$$\int_{\mathcal{M}} \mathrm{d}^n x \, e^{-\frac{1}{\hbar} S(x)}, \tag{8.5}$$

where $S(x)$ is a smooth function in $\mathbb{R}^n$ (the "action"). The following procedure, usually referred to as the *saddle point method*, was introduced by Laplace [4]. Suppose, for simplicity, that the real analytic function $S(x)$ has a single minimum

---

[3] When there are more than one ground state, one needs extra care in defining the generating functional. We leave the details to the reader.

at the point x_0 which lays in the interior of $\mathcal{M}$. x_0 is then a *critical point* for the function $S(x)$, i.e.

$$\left.\frac{\partial S}{\partial x^i}\right|_{x=x_0} = 0. \tag{8.6}$$

We also assume that this minimum is non-degenerate in the sense that the Hessian matrix at x_0 is non-singular

$$\det S_{ij} \neq 0, \quad \text{where} \quad S_{ij} \equiv \left.\frac{\partial^2 S}{\partial x^i \, \partial x^j}\right|_{x=x_0}. \tag{8.7}$$

The critical points of the function $S(x)$ are also called *saddle points* and Eq. (8.6) the *saddle point equation*.

Away from x_0 the integrand of the rescaled integral

$$e^{-\frac{1}{\hbar}S(x_0)} \int_{\mathcal{M}} \mathrm{d}^n x \, e^{-\frac{1}{\hbar}(S(x)-S(x_0))} \tag{8.8}$$

is exponentially small for large $1/\hbar$. Then the integral is given by the contribution of a small neighborhood of x_0 up to an exponentially small error. Expanding in Taylor series around the minimum

$$e^{-\frac{1}{\hbar}S(x_0)} \int_{\mathcal{M}} \mathrm{d}^n x \, \exp\left[-\frac{S_{ij}}{2\hbar}(x^i - x_0^i)(x^i - x_0^i) - \right.$$
$$\left. -\frac{S_{ijk}}{3!\,\hbar}(x^i - x_0^i)(x^j - x_0^j)(x^k - x_0^k) + \cdots\right] =$$
$$= \hbar^{n/2} e^{-\frac{1}{\hbar}S(x_0)} \int_{\mathcal{M}'} \mathrm{d}^n y \, e^{-\frac{1}{2}S_{ij}y_i y_j} \times$$
$$\times \sum_{m=0}^{\infty} \frac{(-1)^m}{m!} \left(\frac{\sqrt{\hbar}}{3!}S_{ijk}y^i y^j y^k + \frac{\hbar}{4!}S_{ijkl}y^i y^j y^k y^l + \cdots\right)^n \tag{8.9}$$

where

$$S_{i_1\ldots i_k} = \partial_{i_1}\cdots\partial_{i_k}S|_{x=x_0} \quad \text{and} \quad y^i = (x - x_0^i)/\sqrt{\hbar}. \tag{8.10}$$

As $\hbar \to 0$ we may replace $\mathcal{M}'$ with a small neighborhood of x_0 (diffeomorphic to an open set $U \subset \mathbb{R}^n$), and then U with the full $\mathbb{R}^n$, at the price of an exponentially small error. Our integral (8.5) then reduces to the form

$$\hbar^{n/2} e^{-\frac{1}{\hbar}S(x_0)} \int_{\mathbb{R}^n} \mathrm{d}^n y \, e^{-\frac{1}{2}S_{ij}y^i y^j}\left(1 + \sum_{m\geq 1} \hbar^m A_{i_1,\ldots i_{2m}} y^{i_1}\cdots y^{i_{2m}}\right) \tag{8.11}$$

since the terms with an odd number of y's integrate to zero by symmetry. The numerical coefficients $A_{i_1 \dots i_{2m}}$ may be found by expanding out the sum in the RHS of (8.9). The leading term in the asymptotic expansion is given by the Gaussian integral

$$\int_M d^n x \, e^{-\frac{1}{\hbar} S(x)} = \hbar^{n/2} e^{-\frac{1}{\hbar} S(x_0)} \int_{\mathbb{R}^n} d^n y \, e^{-\frac{1}{2} S_{ij} y^i y^j} \left(1 + O(\hbar) \right) =$$
$$= (2\pi \hbar)^{n/2} e^{-\frac{1}{\hbar} S(x_0)} (\det S_{ij})^{-1/2} \left(1 + O(\hbar) \right). \tag{8.12}$$

The full asymptotic expansion of (8.5) is then

$$(2\pi \hbar)^{n/2} e^{-\frac{1}{\hbar} S(x_0)} (\det S_{ij})^{-1/2} \left(1 + \sum_{m \geq 1} \hbar^m A_{i_1 \dots i_{2m}} \langle y^{i_1} \cdots y^{i_{2m}} \rangle \right), \tag{8.13}$$

where $\langle \cdots \rangle$ is the expectation value computed with the Gaussian probability measure

$$\sqrt{\det(S_{ij})} \, e^{-\frac{1}{2} S_{ij} y^i y^j} \frac{d^n y}{(2\pi)^{n/2}}. \tag{8.14}$$

The Wick theorem[4] (Theorem 6.7) states that

$$\langle y_{i_1} \cdots y_{i_{2m}} \rangle = \sum_{\text{pairings } p} \prod_{\{i_j, i_k\} \in p} (S^{-1})_{i_j i_k} \tag{8.15}$$

where $(S^{-1})_{ij}$ is the inverse matrix of the Hessian S_{ij} at the saddle point, and the sum is over all *pairings* p of the indices $i_1, \dots, i_{2m}$, that is, over all ways of decomposing the $2m$-element set $\{i_1, \dots, i_{2m}\}$ in m sets $\{i_j, i_k\}$ with two elements ($\equiv$ pairs); the product is then over all pairs in each pairing p. The Wick theorem yields a combinatorial algorithm to compute the order $\hbar^k$ correction to the leading contribution (8.12) for all k's.

Example 8.1 (Stirling Formula) Euler's Gamma-function is defined as

$$\Gamma(s) = e^{s \log s} \int_{-\infty}^{+\infty} e^{-s(e^x - x)} \, dx, \tag{8.16}$$

---

[4] Known as the *Isserlis theorem* in probability theory.

and its large s asymptotic may be obtained by the Laplace technique. The function $e^x - x$ has a unique minimum at $x = 0$ with critical value and Hessian both equal 1. Using (8.12) we get the main terms in the Stirling asymptotic formula

$$\Gamma(s) = e^{s \log s - s} \sqrt{\frac{2\pi}{s}} \left(1 + O(1/s)\right) \quad \text{as } s \to \infty. \tag{8.17}$$

When the action $S(x)$ has more than one (absolute) minimum we have to sum over the contributions from the small neighborhoods of each of them. If the minimum is attained on a submanifold $\mathcal{N} \subset \mathcal{M}$ of positive dimension, the sum is replaced by an integral over $\mathcal{N}$, and the Hessian S_{ij} gets restricted to the normal bundle of $\mathcal{N}$ in $\mathcal{M}$.

Example 8.2 ("Ultralocal" Anharmonic Oscillator) Consider the integral

$$\int\limits_{-\infty}^{+\infty} dx \, e^{-s(-\frac{1}{2}x^2 + \frac{1}{4}x^4)} = \frac{\pi}{2} e^{s/8} \left(I_{\frac{1}{4}}(s/8) + I_{-\frac{1}{4}}(s/8) \right), \tag{8.18}$$

where $I_\nu(z)$ is the modified Bessel function of the first kind with index ν [5, 6]. Now the action has two absolute minima at $x = \pm 1$. The asymptotic estimate of the RHS for large s yields

$$\int_{-\infty}^{+\infty} dx \, e^{-s(-\frac{1}{2}x^2 + \frac{1}{4}x^4)} \approx 2 \sqrt{\frac{\pi}{s}} \, e^{s/4} \tag{8.19}$$

in agreement with the known asymptotics of the Bessel functions for large argument [6, 7].

Semiclassical Heat Kernels

Next we consider the path integral representation of the heat kernel as given by the Feynman-Kac formula

$$\langle x_2 | e^{-HT/\hbar} | x_1 \rangle = \int\limits_{(x_1,0)}^{(x_2,T)} [dx] \, \exp\left[-\frac{1}{\hbar} \int_0^T dt \, L_E(x, \dot{x}) \right], \tag{8.20}$$

where we integrate over all paths in $x : [0, T] \to \mathcal{M}$ with b.c.

$$x(0) = x_1, \qquad x(T) = x_2. \tag{8.21}$$

We wish to get the asymptotic expansion of the RHS of (8.20) as $1/\hbar \to \infty$ using Laplace's strategy. Again, up to exponentially small corrections, the functional measure is concentrated in a functional neighborhood of the paths $x(t)$ which minimize the action functional

$$S_E[x] = \int_0^T dt\, L_E(x, \dot{x}). \tag{8.22}$$

In view of the variational characterization of the Lagrange equations of motion in classical mechanics ([8] chap. 4), the paths which realize the local minima of $S_E[x]$ are the classical solutions for the *imaginary time* Lagrangian L_E, i.e. for the classical Lagrangian with the sign of the potential *inverted* $V(x) \rightsquigarrow -V(x)$

$$L_E(x, \dot{x}) = \frac{1}{2} m(x)_{ij}\, \dot{x}^i \dot{x}^j + V(x). \tag{8.23}$$

There may be more than one solution to the "wrong-sign" classical equations of motion which starts at x_1 at time zero and ends in x_2 at time T. In particular, when $\mathcal{M}$ is not simply-connected there are typically solutions in each homotopy component of the space of paths with fixed endpoints at x_1, x_2. For *non-generic* values of x_1, x_2 there may also be a continuous family of solutions: e.g. if our system is a free particle moving on S^2 and x_1, x_2 are, respectively, the North and the South pole, the one-parameter family of geodesics

$$(\theta(t), \phi(t)) = (\pi t/T, \phi_0) \qquad 0 \le \phi_0 < 2\pi, \tag{8.24}$$

are all solutions satisfying the boundary conditions. For *non-generic* T there may be no classical solution. Think of the harmonic oscillator with T equal the period $2\pi/\omega$: if $x_1 \ne x_2$ there is no solution satisfying the boundary conditions. For these reasons, we adopt the following

Computing Strategy When computing matrix elements $\langle x_2 | e^{-HT/\hbar} | x_1 \rangle$ we always take the endpoints x_1, x_2 and the time interval T to be *sufficiently generic*, in order to avoid inessential complications. The value of the heat kernel at special points is then recovered by taking the appropriate limit.

We consider the (generic) case where the classical solutions are *isolated*, so that the semiclassical heat kernel is given by a sum of contributions from the functional neighborhoods of each classical solution satisfying the appropriate boundary conditions. We focus on one such solution $y(t)$ and write

$$x(t) = y(t) + \xi(t), \tag{8.25}$$

where the *quantum fluctuation* $\xi(t)$ is a section[5] of the normal bundle to the curve $y \subset \mathcal{M}$ with $\xi(0) = \xi(T) = 0$. Then we expand the action functional in powers of the quantum fluctuation $\xi(t)$

$$\frac{1}{\hbar} S_E[y + \xi] = \frac{1}{\hbar} S_E[y] + \frac{1}{\hbar} \int_0^T dt \, \frac{\delta S_E[y]}{\delta y(t)} \xi(t) +$$

$$+ \frac{1}{2\hbar} \int_0^T dt \, ds \, \xi(t) \frac{\delta^2 S_E[y]}{\delta y(t) \delta y(s)} \xi(s) + O(\xi^3) \qquad (8.26)$$

$$= \frac{1}{\hbar} S_E[y] + \frac{1}{2\hbar} \int_0^T dt \, \xi \mathbf{D} \xi + \frac{1}{\hbar} O(\xi^3),$$

where $S_E[y]$ is the Hamilton principal function (for the "wrong sign" potential), and in the last line we used that the term linear in $\xi(t)$ vanishes because $y(t)$ is a solution to the equation of motion

$$\frac{\delta S_E[y]}{\delta y(t)} = 0, \qquad (8.27)$$

i.e. a *saddle point* of the action functional. The 2nd order differential operator $\mathbf{D}$ in the last line of (8.26) is the *second variation operator* (see [8, 9]). Rescaling the quantum fluctuation as $\phi = \xi / \sqrt{\hbar}$ we get

$$\frac{1}{\hbar} S_E[y + \xi] = \frac{1}{\hbar} S_E[y] + \frac{1}{2} \int_0^T dt \, \phi \mathbf{D} \phi + O(\sqrt{\hbar} \, \phi^3) \qquad (8.28)$$

and we see that the contribution from a neghborhood of the isolated wrong-sign solution $y(t)$ has the form

$$\mathcal{N} e^{-\frac{1}{\hbar} S_E[y]} (\text{Det } \mathbf{D})^{-1/2} \left(1 + \sum_{k=1}^{\infty} A_k \hbar^k \right) \qquad (8.29)$$

where the power-law sub-leading corrections $A_k \hbar^k$ may be computed using the Wick combinatoric algorithm in terms of the Green's function

$$\langle x | \mathbf{D}^{-1} | y \rangle \qquad (8.30)$$

---

[5] More pedantically: a section of the pull-back to $[0, T]$ of the normal bundle N of the trajectory in $\mathcal{M}$ via the map $y : t \mapsto y(t) \in \mathcal{M}$.

of the differential operator D. The Green's function can be computed by anyone of the several methods described in Chap. 6. As always, N stands for an overall constant where we absorb all irrelevant factors. The factor

$$\exp\left(-\frac{1}{\hbar} S_E[y]\right) \tag{8.31}$$

is called the "classical" heat kernel, while the first quantum correction to its logarithm, given by

$$-\frac{1}{2}\log\operatorname{Det} D \equiv -\frac{1}{2}\operatorname{Tr}\log D, \tag{8.32}$$

is called the *one-loop contribution* to the quantum action. More generally, the term of order $\hbar^k$ in the expansion

$$-\log\int [dx]\, e^{-\frac{1}{\hbar} S_E[x]} \approx \frac{1}{\hbar}\, S_E[y] + \frac{1}{2}\log\operatorname{Det} D + \sum_{k\geq 1} B_k\, \hbar^k \tag{8.33}$$

is called the $(k+1)$-*loop correction*. The "loop" terminology arises from Feynman's graphs [10], a technique we shall study in Sect. 8.9.

The heat kernel is a sum of contributions of the form (8.29) from each classical solution. When $\mathcal{M}$ is not simply-connected the sum involves topological weights as discussed in Sect. 6.4. In the exact WKB method also *complex valued* classical solutions should be kept into proper account (recall that the path integral should be thought of as a contour integral in the complex domain).

Remark 8.1 The path integral computation of the Mehler kernel in Sect. 6.6 shows that the one-loop semiclassical approximation is exact for the harmonic oscillator. The same is true whenever the path integral is Gaussian (i.e. quadratic in the paths $x(t)^i$).

Integrable Systems: Functional Determinant
Most often one focuses on the leading asymptotic behavior of the heat kernel neglecting the higher loop terms $A_k\hbar^k$ in Eq. (8.29) whose computation is straightforward but rather tedious. Then the only non-classical computation we need to perform is the functional determinant of the second variation operator D, which for the moment we assume *free of zero modes* (as it is generically the case). When the classical version of the system is *integrable*, we may write closed analytic expressions for the classical solution $y(t)$, the Hamilton principal function $S_E[y]$, and the operator D. In this situation the first quantum correction, i.e. the determinant of D, may also be computed exactly.

Separation of variables reduces solving a classical integrable model to a sequence of one-dimensional systems. We may therefore limit ourselves to the study of (time-independent) one-dimensional systems. The real-time Lagrangian has the form

$$L = \frac{1}{2}m\dot{x}^2 - V(x). \tag{8.34}$$

Henceforth we set $m = 1$ by choice of units. The imaginary-time Lagrangian is then

$$L_E = \frac{1}{2}\dot{x}^2 + V(x). \tag{8.35}$$

If $y(t)$ is a solution of the imaginary-time equations of motion

$$\left.\frac{\delta S_E[x]}{\delta x(t)}\right|_{x(t)=y(t)} \equiv -\ddot{y}(t) + V'(y(t)) = 0, \tag{8.36}$$

the second variation operator $\boldsymbol{D}$ is

$$\delta(t-s)\,\boldsymbol{D} \overset{\text{def}}{=} \left.\frac{\delta^2 S_E[x]}{\delta x(t)\,\delta x(s)}\right|_{x(t)=y(t)} = \delta(t-s)\left(-\frac{\mathrm{d}^2}{\mathrm{d}t^2} + V''(y(t))\right), \tag{8.37}$$

acting on functions $\psi\colon [0, T] \to \mathbb{R}$ satisfying the Dirichlet conditions at the endpoints. Such determinants of second-order differential operators were computed in Chap. 6. There we proved the formula

$$\mathrm{Det}\,\boldsymbol{D}_{[0,T]} = \psi_1(0)\,\psi_2(T) - \psi_2(0)\,\psi_1(T), \tag{8.38}$$

where $\{\psi_1(t),\,\psi_2(t)\}$ is a basis of solutions to the homogeneous equation $\boldsymbol{D}\psi = 0$ with unit Wronskian, $W[\psi_1, \psi_2] = 1$. Using formula (6.405) we can write

$$\mathrm{Det}\,\boldsymbol{D}_{[0,T]} = \psi_1(0)\,\psi_1(T)\int_0^T \frac{\mathrm{d}s}{\psi_1(s)^2}, \tag{8.39}$$

where $\psi_1(t)$ is *any* particular solution of the homogeneous equation with $\psi_1(0) \neq 0$ and we assumed the integral to converge, that is, that the function $\psi_1(t)$ has no zeros in the interval $[0, T]$.

The classical system is time-independent. This means that if $y(t)$ is a solution, so is $y(t+c)$. Then

$$0 = \frac{\partial}{\partial c}\left(-\ddot{y}(t+c) + V'(y+c)\right)\Big|_{c=0} = -\frac{\mathrm{d}^2}{\mathrm{d}t^2}\dot{y} + V''(y(t))\dot{y}, \tag{8.40}$$

i.e. $\dot{y}(t)$ is a particular solution of the homogeneous equation, and Eq. (8.41) yields a quadrature formula for the determinant in terms of the classical solution $y(t)$

$$\text{Det } \boldsymbol{D}_{[0,T]} = \dot{y}(0)\,\dot{y}(T)\int_0^T \frac{ds}{\dot{y}(s)^2} \tag{8.41}$$

provided $\dot{y}(0)$ is *not zero*, i.e. if $y(0)$ is not a turning point of the classical motion ([8] chap. 5), and the integral converges, i.e. $\dot{y}(t)$ has no zeros in $[0, T]$. When $y(0)$ is a turning point, i.e. $\dot{y}(0) = 0$ while $\ddot{y}(0) \neq 0$, we may still use the formula (6.392) in the form

$$\text{Det } \boldsymbol{D}_{[0,T]} = \frac{\dot{y}(T)}{\ddot{y}(0)} \equiv \frac{\dot{y}(T)}{V'(y(0))} \tag{8.42}$$

which can also be recovered from Eq. (8.41) by taking the limit

$$\lim_{\epsilon \to 0} \text{Det } \boldsymbol{D}_{[\epsilon,T]} \tag{8.43}$$

using the L'Hôpital rule. The case $\dot{y}(0) = \ddot{y}(0) = 0$ cannot happen for *finite* time intervals $[0, T]$ but it may arise in the asymptotic limit $T \to \infty$.

Example 8.3 (Quantum Pendulum) We compute the semiclassical heat kernel for the one-dimensional system with configuration space the circle $\theta \sim \theta + 2\pi$ and Euclidean action[6]

$$S_E[\theta] = \int_0^T \left(\frac{1}{2}\dot{\theta}^2 - \omega^2 \cos\theta\right)dt. \tag{8.44}$$

that is, the quantum pendulum. The homotopy class of paths from θ_1 to θ_2 are classified by an integer v (the *winding number*) defined by the equality

$$\int_0^T \dot{\theta}\,dt = \theta_2 - \theta_1 + 2\pi v, \quad \text{where} \quad 0 \leq \theta_1, \theta_2 < 2\pi. \tag{8.45}$$

The classical (imaginary-time) equations of motion are

$$\ddot{\theta} - \omega^2 \sin\theta = 0, \tag{8.46}$$

whose general solution is

$$\theta(t) = \pi + 2\,\text{am}\left(\frac{\omega}{\kappa}(t + c), \kappa\right) \tag{8.47}$$

$$\dot{\theta}(t) = \frac{2\omega}{\kappa}\text{dn}\left(\frac{\omega}{\kappa}(t + c), \kappa\right), \tag{8.48}$$

---

[6] Recall that the Euclidean action differs from the real-time one by the sign of the potential.

where $\mathsf{am}(z, \kappa)$ (resp. $\mathsf{dn}(z, \kappa)$) is the *Jacobi amplitude function of modulus κ* (resp. *delta-amplitude*) [6, 11]. The two integration constants κ and c parametrize the solutions of the ODE (8.46). The shift by π in Eq. (8.47) reflects the inversion of the sign of the potential in imaginary time. One has

$$\kappa = -\frac{1}{\cos(\theta_*/2)} \tag{8.49}$$

where θ_* is the turning point. Equation (8.47) may also be written as

$$\cos\frac{\theta(t)}{2} = -\mathsf{sn}\left(\frac{\omega}{\kappa}(t+c), \kappa\right) = -\frac{1}{\kappa}\mathsf{sn}\left(\omega(t+c), \frac{1}{\kappa}\right) \tag{8.50}$$

where $\mathsf{sn}(z, \kappa)$ is the *Jacobi sin-amplitude function* of modulus κ [6, 11]. The quadratic operator describing the quantum fluctuations around the solution $\theta(t)$ is

$$-\frac{\mathrm{d}^2}{\mathrm{d}t^2} + \omega^2\cos\theta(t) = -\frac{\mathrm{d}^2}{\mathrm{d}t^2} - \omega^2 + 2\omega^2\cos^2\frac{\theta(t)}{2} =$$
$$= -\frac{\mathrm{d}^2}{\mathrm{d}t^2} - \omega^2 + 2\omega^2\,\mathsf{sn}^2\left(\frac{\omega}{\kappa}(t+c), \kappa\right). \tag{8.51}$$

Changing variable to $x = \omega(t+c)/\kappa$, the operator reduces to the *Lamé* one [6]

$$\boldsymbol{D} \equiv -\frac{\mathrm{d}^2}{\mathrm{d}x^2} - \kappa^2 + 2\kappa^2\,\mathsf{sn}^2(x, \kappa). \tag{8.52}$$

From Eq. (8.40) we know that the Lamé equation $\boldsymbol{D}\psi(x) = 0$ has the two linearly independent solutions[7]

$$\psi_1(x) = \mathsf{dn}(x, \kappa) \tag{8.53}$$

$$\psi_2(x) = \frac{1}{\kappa^2 - 1}\left(\kappa^2\,\mathsf{sn}(x, \kappa)\,\mathsf{cn}(x, \kappa) - \mathscr{E}(x, \kappa)\,\mathsf{dn}(x, \kappa)\right) \tag{8.54}$$

where $\mathsf{cn}(x, \kappa)$, $\mathsf{dn}(x, \kappa)$, and

$$\mathscr{E}(x, \kappa) \equiv \int_0^x \mathsf{dn}^2(t, \kappa)\,\mathrm{d}t \tag{8.55}$$

are, respectively, the Jacobi cosine-amplitude, the delta-amplitude, and the epsilon functions [6]. To get the solution (8.54) we used the formula (6.404) and the integral 22.16.20 in [6]. The functions (8.53), (8.54) satisfy

$$\psi_1(0) = 1, \quad \psi_1'(0) = 0, \quad \psi_2(0) = 0, \quad \psi_2'(0) = 1, \tag{8.56}$$

---

[7] $\psi_1(x)$ is the Lamé polynomial $dE_1^0(x, \kappa)$ [6].

and their Wroskian is 1 (by construction)

$$W[\psi_1, \psi_2] \equiv \psi_1(t)\psi_2'(t) - \psi_2(t)\psi_1'(t) = 1. \tag{8.57}$$

In conclusion

$$\mathrm{Det}(\boldsymbol{D})_{[0,T]} = \psi_1\left(\frac{\omega c}{\kappa}\right)\psi_2\left(\frac{\omega(T+c)}{\kappa}\right) - \psi_2\left(\frac{\omega c}{\kappa}\right)\psi_1\left(\frac{\omega(T+c)}{\kappa}\right). \tag{8.58}$$

8.2.1 The Van Vleck Determinant

More generally, we consider a system with n degrees of freedom having a classical analogue, say with the Euclidean Lagrangian

$$L_E(x, \dot{x}) = \frac{1}{2}m_{ij}\,\dot{x}^i\dot{x}^j + V(x). \tag{8.59}$$

Let $y^i(t)$ be a (imaginary time) solution for $t \in [0, T]$ with boundary conditions

$$y^i(0) = x_1^i, \qquad y^i(T) = x_2^i. \tag{8.60}$$

We wish to give a simple explicit formula for the functional determinant of the 2nd variation operator

$$\boldsymbol{D}_{ij} = -\frac{\mathrm{d}}{\mathrm{d}t}m_{ij}\frac{\mathrm{d}}{\mathrm{d}t} + \frac{\partial^2 V}{\partial x^i\,\partial x^j}\bigg|_{x=y(t)} \tag{8.61}$$

known as the *Van Vleck(-Pauli-Morette) formula.* Our strategy is to apply Corollary 6.3. To this end, we consider the family $\{z^i(t, x^j, p_k)\}$ of all classical solutions which at $t = 0$ start from the point $(x^i, p_j) \in \mathcal{W}$ in phase space

$$z^i(t) \equiv z^i(t, x^j, p_k), \qquad -\frac{\mathrm{d}}{\mathrm{d}t}(m_{ij}\dot{z}^j) + \frac{\partial V(z)}{\partial z^j} = 0 \tag{8.62}$$

$$\text{with}\quad z^i(0) = x^i \qquad \text{and}\quad \dot{z}^i(0) = m^{ij}p_j,$$

where m^{ij} is the inverse of the Jacobi metric m_{ij}. The final point of the trajectory originating at $(x^i, p_j) \in \mathcal{W}$ will be denoted as

$$x_2^i \equiv z^i(T, x^j, p_k). \tag{8.63}$$

By construction, the matrix

$$\mathbf{\Phi}(t)^i{}_j \stackrel{\text{def}}{=} \left. \frac{\partial z^i(t)}{\partial p_k} m_{kj} \right|_{\substack{x^i = y^i(0) \\ p_i = m_{ij}\dot{y}^j(0)}} \tag{8.64}$$

satisfies

$$\mathbf{D}_{ij}\mathbf{\Phi}(t)^j{}_k = 0, \qquad \mathbf{\Phi}(0)^i{}_j = 0, \qquad \dot{\mathbf{\Phi}}(t)^i{}_j = \delta^i{}_j. \tag{8.65}$$

The determinant defined by the rule and overall normalization in Corollary 6.3 is

$$\text{Det } \mathbf{D} = \det \mathbf{\Phi}(T) = \det m \det \left(\frac{\partial x_2^i}{\partial p(0)_j} \right) = \tag{8.66}$$

$$= \det m \left[\det \left(\frac{\partial p(0)_j}{\partial x_2^i} \right) \right]^{-1}$$

Now let

$$S(x_1, x_2; T) \equiv \int_0^T L_E(y, \dot{y}) \, dt \tag{8.67}$$

be the Hamilton principal function (i.e. the action evaluated on the solution $y(t)$ as a function of its endpoints). One has ([8], §. 4.2)

$$p_i(0) = -\frac{\partial S}{\partial x_1^i}, \tag{8.68}$$

and we get the Van Vleck determinant

$$\left(\frac{\text{Det } \mathbf{D}}{\det m} \right)^{-1} = (-1)^n \det \left[\frac{\partial^2 S}{\partial x_1^i \, \partial x_2^j} \right], \tag{8.69}$$

(usually the constant factor $\det m$ is absorbed in the overall factor $\mathcal{N}$).

8.2.2   *Maslov Index and All That*

The *Maslov index* $\mu(\mathbf{D})$ of the second variation operator $\mathbf{D}$ around the classical solution $y(t)$ is, by definition, the Morse index of the action functional $S_E[x]$ at its critical point $y(t)$, namely the number of negative eigenvalues of the functional Hessian ($\equiv$ second variation operator)

$$\langle t|\boldsymbol{D}|s\rangle = \left.\frac{\delta^2 S_E[x]}{\delta x(t)\,\delta x(s)}\right|_{x(t)=y(t)} \tag{8.70}$$

that is, the (finite) dimension of the subspace in function space along which the action *decreases* from the critical value $S_E[y]$. In [8] chap. 5 it was stated that $\mu(\boldsymbol{D})$ is the number of points in $\{y(t)\}_{[0,T]}$ conjugated to the starting point $y(0)$. Now

Fact 8.1 *Let* $y\colon [0, T] \to \mathcal{M}$ *be a classical solution (in imaginary time) of the one-dimensional system (8.35) and let* $\dot{y}(t)$ *be the corresponding velocity. Let* $v(T)$ *be the number of* turning points *crossed by the particle along its trajectory* $y(t)$ *for* $t \in (0, T)$. *The Maslov index* $\mu(\boldsymbol{D})$ *of the second variation operator (i.e. the Morse index of* y *as a critical function of* $S_E[x]$) *satisfies*

$$\mu(\boldsymbol{D}) = v(T) \quad or \quad v(T) - 1. \tag{8.71}$$

Proof We know that $\dot{y}$ satisfies the homogeneous equation $\boldsymbol{D}\dot{y} = 0$. Let $z(t)$ be a second solution such that $W[\dot{y}, z] = 1$. The solution of the homogeneous equation

$$J(t) = \dot{y}(0)z(t) - z(0)\,\dot{y}(t) \tag{8.72}$$

satisfies $J(0) = 0$, $\dot{J}(0) = 1$. The solution $J(t)$ is then a *Jacobi field* along the solution $y(t)$, and the points along the trajectory $y(t_k)$ such that $J(t_k) = 0$ are—by definition—the conjugates of the point $y(0)$ along this trajectory.[8] Now

$$\mathrm{Det}(\boldsymbol{D})_{[0,T]} = J(T). \tag{8.73}$$

By Fact 6.8 we have that $\mu(\boldsymbol{D})$ is the number of zeros of the Jacobi field $J(t)$ for $t \in (0, T)$ not counting the zeros at the endpoints. By definition, this is number of conjugate points to $y(0)$ along the trajectory $y(t)$. This proves Theorem 4.1 of [8] in the special case of one-dimensional systems. Let

$$0 \equiv t_0 < t_1 < \cdots < t_k < T \tag{8.74}$$

be the list of zeros of the solution $J(t)$ of the 2nd order homogeneous ODE

$$\boldsymbol{D}J(t) = 0. \tag{8.75}$$

---

[8] See [9] for the particular case where the motion is free, i.e. the classical trajectories are the geodesics of the configuration manifold.

By the Sturm-Liouville Theorem 3.5(2) the other solution $\dot{y}(t)$ has a single zero in the interval $(0, t_1)$, another one in the interval (t_1, t_2) etcetera. It may or may not have a zero in the interval (t_k, T). Hence

$$\#(\text{turning points along the trajectory}) = \mu(D) \text{ or } \mu(D) + 1. \tag{8.76}$$

$$\square$$

8.3   Semiclassical Approximation III: Instanton Calculus

The main ingredient of the semiclassical approximation are the imaginary-time classical solutions $y(t)$, that is, the classical motions in the *inverted* potential $- V(x)$. A particular class of such solution—called *instantons*—play a special role in Quantum Mechanics and even a more fundamental one in Quantum Field Theory [12]. We start our discussion by outlining the situations where they arise to explain their physical relevance.

To be specific, we focus on a quantum system in a (connected) configuration space $\mathcal{M}$ defined by the imaginary-time Lagrangian

$$L_E(x^i) = \frac{1}{2} g_{ij} \dot{x}^i \dot{x}^j + V(x^i), \qquad V(x) \geq 0, \tag{8.77}$$

where the additive constant in $V(x)$ is chosen so that its minimum value is exactly zero. We say that the point $x_0 \in \mathcal{M}$ is a *classical ground state* iff x_0 is an *absolute minimum* of $V(x)$, i.e. iff $V(x_0) = 0$. For simplicity we assume that the classical ground states are isolated and *non-degenerate,* meaning that the Hessian

$$H_{ij}(x_0) \overset{\text{def}}{=} \left. \frac{\partial^2 V}{\partial x^i \, \partial x^j} \right|_{x=x_0} \tag{8.78}$$

is non-degenerate,[9] hence positive-definite since x_0 is a minimum of $V(x)$.

There may be more than one[10] classical ground states $x_0^{(a)}$ $(a = 1, \ldots, N)$. The simplest example of a system with several classical ground states is the one-dimensional *double-well* potential

$$V(x) = \frac{\omega^2}{8\sigma^2}(x^2 - \sigma^2)^2, \qquad x \in \mathbb{R} \tag{8.79}$$

---

[9] x_0 is a non-degenerate critical point of the function $V(x)$ in the sense of Morse theory [13, 14].
[10] Even continuous families of them, if we relax the assumption of isolated non-degenerate critical points.

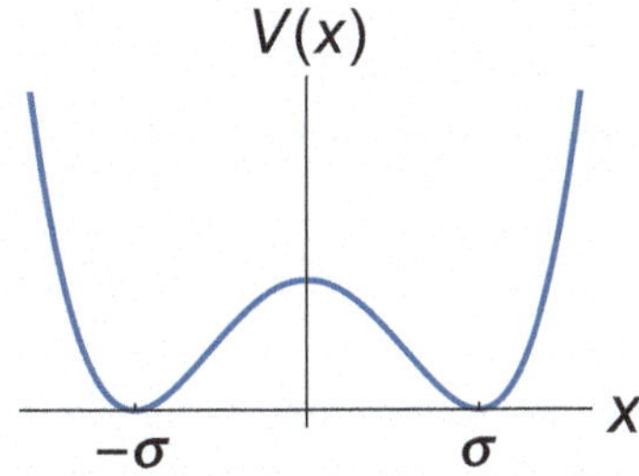

Fig. 8.1 The double-well potential. The classical ground states are the *real-time* stationary solutions where the particle remains forever in one of the two minima of the potential at $x = \pm\sigma$. These solutions have zero classical energy

which has two classical ground states at $x = \pm\sigma$ both with Hessian ω^2: see Fig. 8.1.

In the very crudest semiclassical approximation, when $\hbar$ is very small, each classical ground state at $x = x_0^{(a)}$ behaves as the ground state of the harmonic oscillator which describes the small fluctuations around $x_0^{(a)}$ with the crude quadratic potential

$$V(x)_{\text{crude}} \approx \frac{1}{2} H_{ij}(x_0^{(a)})\,(x - x_0)^i (x - x_0^{(a)})^j, \qquad \text{for } x \approx x_0^{(a)}, \tag{8.80}$$

whose energy, in this most rudimentary approximation, is given by the zero-point energy of the oscillators

$$E_{\text{crude}}(x_0^{(a)}) \approx \frac{1}{2}\,\hbar\,\text{tr}\,\sqrt{H(x_0^{(a)})}, \tag{8.81}$$

where $\sqrt{H(x_0^{(a)})}$ is the square-root of the positive matrix (cf. Sect. 7.1.3)

$$H_{rs}(x_0^{(a)}) = e_r^i(x_0^{(a)}) H_{ij}(x_0^{(a)}) e_s^j(x_0^{(a)}) \quad \text{where} \quad e_r^i(x) e^j(x)_r = g^{ij}(x). \tag{8.82}$$

$e_r^i(x)$ is the inverse vielbein of the metric $g_{ij}(x)$. When, as in the double-well example (8.79), the Hessian $H_{rs}(x_0^{(a)})$ is the same in all classical ground states $x_0^{(a)}$, the corresponding "crude" quantum states have the same energy to the first order in $\hbar$. In facts in the double-well we have the stronger condition that the two classical ground states are related by an exact symmetry, namely *parity*

$$\mathscr{P}: x \mapsto -x. \tag{8.83}$$

In this case the "crude" ground state energies are degenerate to *all* orders in an expansion of their energies as an asymptotic power series in $\hbar$.

However this cannot be the full story, not even at a qualitative level. We know from Sect. 6.13 that a system with a Lagrangian of the form (8.77) has a *unique* quantum ground state, and therefore the energies of the various classical ground states $x_0^{(a)}$ cannot remain degenerate when we switch on a non-zero $\hbar$ however small: the quantum effects *split* their energy levels. We are aware of what happens in the double-well model: it has the parity symmetry $\mathscr{P}$, and the wave function of the ground state must be *even* (i.e. symmetric under $x \leftrightarrow -x$) with *no zero*, while the first excited state is *odd* with precisely *one zero* at $x = 0$. Then, if

$$\psi_{\pm}(x) \approx C \exp\left[-\frac{1}{2\hbar}\omega(x \mp \sigma)^2 \right], \qquad \psi_{\pm}(-x) = \psi_{\mp}(x), \tag{8.84}$$

are the crudest-approximation wave-functions of the two lowest energy states with respective support around $x = \pm\sigma$, we expect (roughly)

$$\psi(x)_{\text{ground}} \approx \frac{1}{\sqrt{2}}(\psi(x)_+ + \psi(x)_-), \tag{8.85}$$

$$\psi(x)_{1^{\text{st}}\text{exc.}} \approx \frac{1}{\sqrt{2}}(\psi(x)_+ - \psi(x)_-). \tag{8.86}$$

To confirm our expectations, we need to compute the energy splitting

$$\Delta E \equiv E_{1^{\text{st}}\text{exc.}} - E_{\text{ground}} \tag{8.87}$$

in the asymptotic regime $\hbar \approx 0$ and check that it is positive and smaller than *any* power of $\hbar$, that is, *exponentially small* as $\hbar \to 0$.

The proof of Theorem 6.4 clarifies the physical mechanism which splits the energy levels of the two classically degenerate states $\psi_{\pm}(x)$. Let $x = \pm\sigma$ be the two isolated classical ground states interchanged by the symmetry $x \leftrightarrow -x$. The matrix element

$$\langle \sigma | e^{-\beta H/\hbar} | -\sigma \rangle, \tag{8.88}$$

of the imaginary-time evolution operator, is *non zero* since it receives contributions from paths going from $-\sigma$ to $+\sigma$ whose functional measure is small but non zero. These off-diagonal matrix elements split the eigenvalues of $\exp(-\beta H/\hbar)$, hence of H. After analytic continuation to real time, the matrix element (8.88) becomes the amplitude for the tunneling under the potential barrier which separates the two classical ground states (cf. Fig. 8.1): due to the tunnelling effect, a particle—whose wave-function at time $t = 0$ has support in a small neighborhood of the first classical ground state at $-\sigma$—will have a small but non zero probability of being found near the other classical ground state $+\sigma$ at any time $t > 0$. This could not happen if the energies of the two states remain degenerate. If the particle is localized at time $t = 0$ around $x = -\sigma$ with initial wave-function $\psi(x)_-$, its state at time $t > 0$ will be

$$\frac{1}{2}e^{-E_0 t/\hbar}\left(1 + e^{-\Delta E\, t/\hbar}\right)\psi(x)_- + \frac{1}{2}e^{-E_0 t/\hbar}\left(1 - e^{-\Delta E\, t/\hbar}\right)\psi(x)_+, \tag{8.89}$$

and the probability of finding the particle around $x = +\sigma$ at later times would vanish if $\Delta E = 0$. Since the tunneling amplitude is exponentially suppressed as $\hbar \to 0$, the energy splitting ΔE must also be exponentially small.

For definiteness we focus on the double-well example even if the story is pretty general as the reader doubtless realizes. To select the state of lowest energy in the sector of *even* (resp. *odd*) parity states, we take the imaginary time interval $\beta \to \infty$, and exploit the fact that the limit

$$\lim_{\beta \to \infty} \frac{1}{2}(1 \pm \mathscr{P})e^{-\beta H/\hbar}, \tag{8.90}$$

with the upper (resp. lower) sign, is the projector on the lowest lying state in the even (resp. odd) sector (up to overall normalization). Hence, to understand the splitting ΔE, we need the behavior of the matrix element (8.88) for β *large*.

Instanton Solutions

In the asymptotic limit $\hbar \to 0$ the functional measure of the paths connecting $-\sigma$ to $+\sigma$ is concentrated in a tiny neighborhood of the classical solutions of the Euclidean equations of motion. When taking the large time limit $\beta \to \infty$, with $\hbar$ small but non-zero, we must focus on the solutions $y(t)$ which start at $t = -\infty$ in the classical ground state $-\sigma$ and end for $t = +\infty$ at $+\sigma$. For β large, but finite, the contribution to the path integral from such a solution is of the order

$$\exp\left(-\frac{1}{\hbar}S_E[y(t)]\right) \equiv \exp\left(-\frac{1}{\hbar}\int_{-\beta/2}^{+\beta/2}\left(\frac{1}{2}\dot{y}^2 + V(y)\right)dt\right). \tag{8.91}$$

Hence only the solutions $y: (-\beta/2, \beta/2) \to \mathbb{R}$ whose action $S_E[y]$ remains *finite* as $\beta \to \infty$ yield a non-zero contribution to the relevant amplitude. Since $V(y) \geq 0$, the action may be finite for $\beta = \infty$ only if both the velocity $\dot{y}$ and the potential $V(y)$ are (essentially[11]) zero for almost all times $t \in \mathbb{R}$. However $\dot{y}$ cannot be identically zero because

$$\int_{-\infty}^{+\infty} \dot{y}\,dt = 2\sigma \neq 0. \tag{8.92}$$

The velocity $\dot{y}$ and the potential $V(y)$ are essentially non-zero only for a small time interval centered at some $t = t_0$. Outside this interval both $\dot{y}$ and $V(y)$ vanish up to exponentially small corrections, i.e.

$$|\dot{y}|, \; V(y) = O\left(e^{-c|t-t_0|}\right) \qquad c > 0. \tag{8.93}$$

---

[11] "Essentially zero for almost all times" means that the domain $\{t \in \mathbb{R}: L_E(x(t)) > \epsilon\} \subset \mathbb{R}$ is compact for all $\epsilon > 0$.

In other words, the action

$$S_E[y(t)] = \int_{-\infty}^{+\infty} L_E(\dot{y}(t), y(t))\, dt \tag{8.94}$$

of the relevant solution $y(t)$ is "localized" at one particular instant t_0 along the time axis. This property motivates the name of this important class of classical solutions. We give their formal definition.

Definition 8.1 An *instanton* is a solution $y \colon (-\infty, +\infty) \to \mathcal{M}$ of the imaginary-time classical equations of motion, with *finite Euclidean action* $S_E[y]$, which interpolates between two classical ground states, x_0 and $\tilde{x}_0$, as $t \to \pm\infty$, that is,

$$\lim_{t \to -\infty} y(t) = x_0, \qquad \lim_{t \to +\infty} y(t) = \tilde{x}_0 \quad \text{with} \quad V'(x_0) = V'(\tilde{x}_0) = 0. \tag{8.95}$$

We choose an orientation of the imaginary-time paths. For a one-dimensional system the orientation is given by the sign of the "topological charge"

$$\int_{-\infty}^{+\infty} \dot{y}\, dt = \tilde{x}_0 - x_0. \tag{8.96}$$

Finite-action imaginary-time solutions with positive (resp. negative) orientation are called *instantons* (resp. *anti-instantons*). Informally we think of them as particles/anti-particles and see t_0 as their "position" along the line $\mathbb{R}$ parametrized by the imaginary time t.

For a general one-dimensional system with Hamiltonian of the form

$$H = \frac{p^2}{2} + V(x), \tag{8.97}$$

the instantons are the same functions of t as the *limitation motions*[12] for the inverted potential $-V(x)$, see Fig. 8.2. In other words, the instanton solutions are the motions of a particle in the inverted potential $-V(x)$ with *zero energy,*[13] that is, they are given explicitly by the curve

$$t - c = \pm \int_{*}^{y(t)} \frac{dx}{\sqrt{2(E + V(x))}} \bigg|_{E=0} \equiv \pm \int_{*}^{y(t)} \frac{dx}{\sqrt{2\,V(x)}}, \tag{8.98}$$

---

[12] Cf. [8], p. 111.

[13] Recall that in our normalization the maximum value of $-V$ is zero.

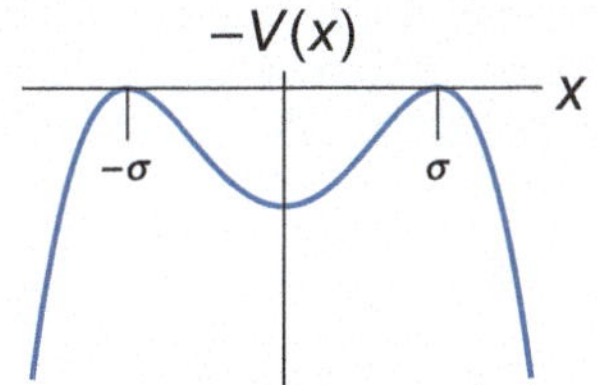

Fig. 8.2 The effective potential entering in the imaginary time equations of motion of the double-well system

where c is an integration constant and $*$ a chosen convenient reference point with $* = y(c)$. The overall sign $\pm$ distinguishes instantons from anti-instantons. Near the classical ground state $\tilde{x}_0$ we have

$$V(x) \approx \frac{1}{2}V''(\tilde{x}_0)\,(x - \tilde{x}_0)^2 + \cdots , \qquad (8.99)$$

so that, for large t where $x(t) \approx \tilde{x}_0$, we have

$$\tilde{\omega}_0(t - c) \approx -\int_*^{y(t)} \frac{\mathrm{d}x}{x - \tilde{x}_0} \quad \Rightarrow \quad y(t) \approx \tilde{x}_0 + \mathrm{e}^{-\tilde{\omega}_0(t - t_0)}, \qquad (8.100)$$

where

$$\tilde{\omega}_0 \equiv \sqrt{V''(\tilde{x}_0)}, \qquad (8.101)$$

and t_0 is an integration constant. We see from Eq. (8.100) that for late times

$$t \gtrsim t_0 + \frac{1}{\tilde{\omega}_0} \qquad (8.102)$$

the particle lays in the final classical ground state $\tilde{x}_0$ modulo exponentially small corrections. Likewise, at the early times $t \lesssim t_0 - 1/\omega_0$ the particle is essentially in the initial classical ground state x_0. The action $S_E[x]$ is concentrated in time around the instant $t = t_0$ inside the finite interval

$$t_0 - \frac{1}{\omega_0} \lesssim t \lesssim t_0 + \frac{1}{\tilde{\omega}_0}. \qquad (8.103)$$

Remark 8.2 We stress that Eq. (8.98) yields not just *one* instanton solution, but an one-parameter family of them parametrized by the central "instant" $t_0 \in \mathbb{R}$. The freedom in the choice of t_0 reflects the invariance of the classical equations of motion under time translations, a symmetry which is broken by the boundary conditions for β finite, but gets restored in the limit $\beta \to \infty$.

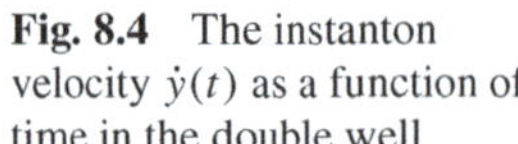

Fig. 8.3   The instanton solution $y(t)$ for the double-well (8.79)

Fig. 8.4   The instanton velocity $\dot{y}(t)$ as a function of time in the double well

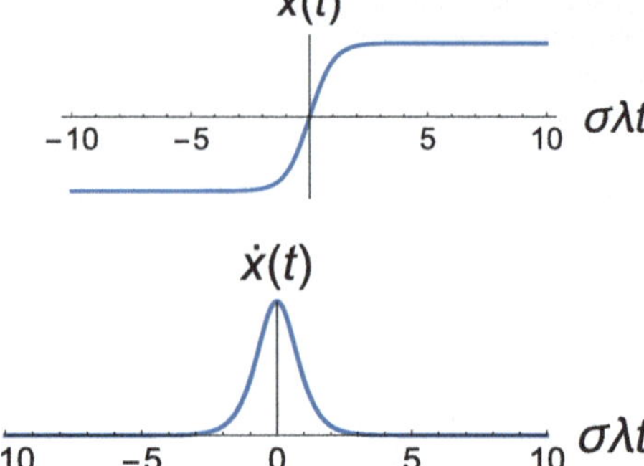

Example 8.4 (Instantons in the Double-Well)   For the potential (8.79) we get the one-parameter family of finite-action solutions

$$y(t) = \pm \sigma \tanh\big(\omega(t - t_0)/2\big) \tag{8.104}$$

($+$ for the instanton, $-$ for the anti-instanton). The solution $y(t)$ and its velocity $\dot{y}(t)$ are plotted in Figs. 8.3 and 8.4 for $t_0 = 0$. We see from the figures that the particle is in a classical ground state to very high precision except for a small interval of time around the center t_0 of the instanton. The action is finite

$$S_1 \equiv \frac{1}{2} \int\limits_{-\infty}^{+\infty} \Big(\dot{y}^2 + \frac{\omega^2}{4\sigma^2}(y^2 - \sigma^2)^2\Big)\mathrm{d}t = \int\limits_{-\infty}^{+\infty} \Big(\dot{y}^2 - E\Big)\mathrm{d}t =$$

$$= \int\limits_{-\infty}^{+\infty} \dot{y}^2 \,\mathrm{d}t = \frac{2}{3}\omega\sigma^2. \tag{8.105}$$

Example 8.5 (Instantons in the Pendulum)   The potential of the pendulum is the periodic function

$$V(\theta) = 1 - \cos\theta, \tag{8.106}$$

and the instantons (8.98) take the form

$$\theta(t) = \pm 2 \arctan\big(\exp(t - t_0)\big). \tag{8.107}$$

with finite action $S_1 = 2$.

The Zero Mode of the 2nd Variation Operator

We claim that the second variation operator around the instanton solution $y(t)$

$$\boldsymbol{D} = -\frac{\mathrm{d}^2}{\mathrm{d}t^2} + V''(y(t)) \equiv -\frac{\mathrm{d}^2}{\mathrm{d}t^2} + W(t) \tag{8.108}$$

is positive semi-definite with precisely *one* zero mode, namely the time derivative $\dot{y}(t)$. Indeed in Eq. (8.40) is was shown that $\dot{y}(t)$ is a solution to the homogenous equation, $\boldsymbol{D}\dot{y} = 0$, while we have seen in Eq. (8.100) that $\dot{y}(t)$ vanishes exponentially as $t \to \pm\infty$, so $\dot{y}(t)$ is a *normalizable* eigenfunction of $\boldsymbol{D}$ associated to the zero eigenvalue, hence a genuine zero mode of $\boldsymbol{D}$. Moreover from the equation

$$\dot{y}(t) = \pm\sqrt{2\,V(y(t))} \tag{8.109}$$

we see that $\dot{y}(t)$ vanishes only at infinity, i.e. $\dot{y}$ is a normalizable eigenfunction of $\boldsymbol{D}$ without zeros, and hence its eigenvalue, zero, is the *lowest one* by the Sturm oscillation theorem. This shows that $\boldsymbol{D}$ is non-negative. We conclude that the instantons are marginally stable solutions.

As it is clear from Eq. (8.40), the existence of the zero mode $\dot{y}$ reflects the invariance of the system (8.97) under translations in time. If $y(t)$ is an instanton solution, $y(t - t_0)$ is another such solution (with the same action) for all $t_0 \in \mathbb{R}$. In conclusion: we have a one parameter family of classical solutions interpolating between the two classical ground states as $t \to \pm\infty$, all with the same action. To get the semiclassical approximation to the tunneling amplitude between the two classical ground states we must integrate over this family of saddle points. However the are many other saddles that we have to keep into account, as we are going to see.

8.3.1 *Dilute Gas of Instantons*

We wish to understand the quantum effects associated with the instantons. To this end, we specialize our discussion to the double-well system (8.79) and compute the tunnelling matrix element

$$\langle \sigma \,|\mathrm{e}^{-\beta H/\hbar}|-\sigma\rangle \tag{8.110}$$

for large β in the semiclassical regime $\hbar \approx 0$. The classical ground states $x = \pm\sigma$ are highly non-generic configurations: to avoid trouble, following our **Computing Strategy** for semiclassical heat kernels (cf. Sect. 8.2), we start from a slightly different *generic* matrix element

$$\langle \sigma(1 - \epsilon) \,|\mathrm{e}^{-\beta H/\hbar}|-\sigma(1 - \epsilon)\rangle \tag{8.111}$$

Fig. 8.5 The solutions contributing to the matrix element (8.111) in the semiclassical limit

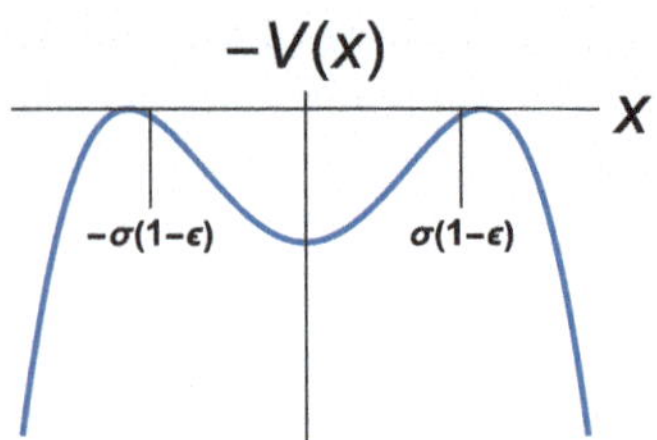

and send $\epsilon \to 0$ only at the end of the computation. For instance, if we approach the double limit $\epsilon \to 0,\ \beta \to \infty$ along the precise curve

$$\beta = \frac{4}{\omega}\operatorname{arctanh}(1 - \epsilon) \approx -\frac{2}{\omega}\Big(\log(\epsilon/2) + \frac{\epsilon}{2} + \cdots\Big), \tag{8.112}$$

the solution which satisfies the Dirichlet boundary condition

$$x(0) = -\sigma(1 - \epsilon), \quad x(\beta) = \sigma(1 - \epsilon) \tag{8.113}$$

is the instanton (8.104) with $t_0 = \beta/2$. This solution describes an imaginary-time motion where our particle starts at $-\sigma(1 - \epsilon)$, rolls down the valley between the two maxima of the inverted potential (Fig. 8.5), crosses its minimum in 0 at maximal speed, and then goes uphill to reach the final point $\sigma(1 - \epsilon)$.

But the instanton is not the only solution satisfying (8.113). There are solutions $x(t)$ where the particle starts from $-\sigma(1 - \epsilon)$, goes down the valley and then upwards, overpasses the ending point $\sigma(1 - \epsilon)$, reaches a turning point x_* in the region

$$\sigma(1 - \epsilon) < x_* < \sigma, \tag{8.114}$$

where its velocity $\dot{x}$ becomes zero, and then the particle returns back, reaching $\sigma(1 - \epsilon)$ from above. But, it needs not to stop there: it may go down the valley, then up to the symmetric turning point at $-x_*$,

$$-\sigma < -x_* < -\sigma(1 - \epsilon), \tag{8.115}$$

then go down the valley a second time, and then up toward the final point $\sigma(1 - \epsilon)$. Again, it needs not to stop there: clearly there may be many different solutions in which the particle goes back and forth several times through the valley, inverting the direction of its motion at the turning points $\pm x_*$ located at distance $O(\epsilon)$ from the classical ground states $\pm \sigma$. The (imaginary-time) energy of such a solution is

$$E_* = -\frac{\omega^2}{8\sigma^2}(x_*^2 - \sigma^2)^2. \tag{8.116}$$

Explicitly, the imaginary-time solutions of the (inverted) double well with energy E_* are[14]

$$x(t) = x_* \operatorname{sn}\!\left(\frac{\omega}{2\sigma}\sqrt{2\sigma^2 - x_*^2}\,(t - c); k\right), \qquad k^2 \equiv \frac{x_*^2}{2\sigma^2 - x_*^2}. \tag{8.117}$$

$x(t)$ reduces to the instanton (8.104) in the limit $x_* \to \sigma$ (cf. §. 22.5(ii) of [6]). We conclude that there are several solutions which start at $-\sigma(1 - \epsilon)$ and end at $+\sigma(1 - \epsilon)$ in the normalized time

$$\tau \equiv \frac{\omega}{2}\sqrt{2 - \frac{x_*^2}{\sigma^2}}\,\Delta t \equiv \frac{1}{2}\sqrt{\frac{2}{1 + k^2}}\,\omega\,\Delta t, \tag{8.118}$$

equal to one of the following (depending on the class of the path)

$$2L + 4nK, \quad 2K + 4nK, \quad 2K + 4nK, \quad 4K - 2L + 4nK, \tag{8.119}$$

where n is a non-positive integer (the "period number" of the solution), K is the complete elliptic integral of modulus k,

$$K(k) \equiv \int_0^1 \frac{dt}{\sqrt{1 - t^2}\sqrt{1 - k^2 t^2}} = -\frac{1}{2}\log\!\left(\frac{1 - k}{8}\right)(1 + O(1 - k^2)), \tag{8.120}$$

and

$$L = \int_0^{\sigma(1-\epsilon)/x_*} \frac{dt}{\sqrt{1 - t^2}\sqrt{1 - k^2 t^2}}. \tag{8.121}$$

The four possible classes of solutions correspond to the particle which exits (enters) its initial (final) point from below (above). As $\epsilon \to 0$, $L \to K$ and the distinction between the 4 classes of solutions disappears. Only the period number n distinguishes the inequivalent solutions in the limit. For a fixed $\epsilon > 0$ all these solutions may satisfy the boundary conditions

$$x\!\left(\tfrac{1}{2}(1 \pm 1)\beta\right) = \pm\sigma(1 - \epsilon) \tag{8.122}$$

of the Feynman-Kac formula, by choosing a suitable energy E_*, i.e. for a specific inversion point x_* in the range (8.114) which corresponds to an elliptic modulus $k^2 \equiv k^2(x_*)$ in the range

$$\frac{1 - 2\epsilon + \epsilon^2}{1 + 2\epsilon - \epsilon^2} \leq k^2 \leq 1. \tag{8.123}$$

---

[14] $\operatorname{sn}(z; k)$ is Jacobi's *sin-amplitude* function of modulus k [6].

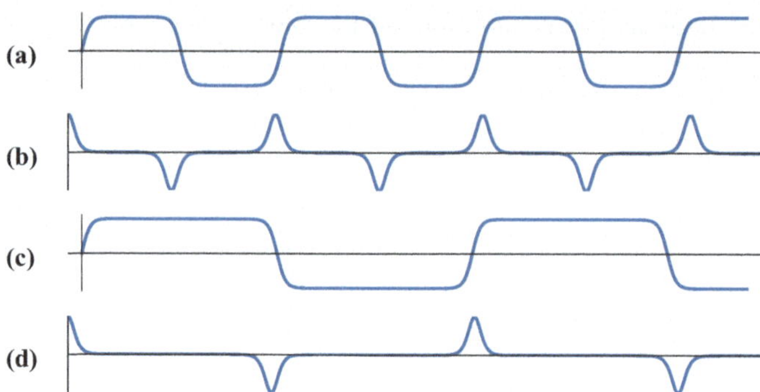

Fig. 8.6 (**a**) The *exact* solution $x(t)$ of the double well with energy $E_* = -0.125 \cdot 10^{-4} \omega^2 \sigma^2$; (**b**) the velocity $\dot{x}(t)$ of the solution in (**a**); (**c**) the *exact* solution of the double well with energy $E_* = -0.125 \cdot 10^{-9} \omega^2 \sigma^2$; (**d**) the velocity of the solution in (**c**)

For instance, the first solution with $n = 0$ is the instanton (8.104) which corresponds to $\epsilon = 0$. Increasing ϵ we get other solutions with larger n. In facts, for large β, a solution with period number n is obtained for an ϵ of order

$$\epsilon \approx 2\,\mathrm{e}^{-\omega\beta/(4n)}. \tag{8.124}$$

The bottom line is that taking the limits in the correct order, that is, first the limit $\beta \to \infty$, with $\epsilon \lll 1$ fixed, and then $\epsilon \to 0$, we get classical solutions of the form (8.117) of all period numbers n. Examples of exact solutions of this kind are drawn in Fig. 8.6.

The interpretation of these solutions for $\beta = \infty$ and $\epsilon \lll 1$ becomes evident by comparing Fig. 8.6 with Figs. 8.3 and 8.4: up to exponentially small corrections, the exact solution is a sequence of widely separated instantons and anti-instantons. More precisely:

DG1   for the matrix element $\langle \sigma | e^{-\beta H/\hbar} | - \sigma \rangle |_{\beta \to \infty}$ we have $n + 1$ instantons interlaced with n anti-instantons;

DG2   for $\langle \sigma | e^{-\beta H/\hbar} | \sigma \rangle |_{\beta \to \infty}$ we have n instantons interlaced with n anti-instantons.

For a fixed n, the distances between one instanton and the following anti-instanton are of the order

$$\Delta t \approx \frac{4}{\omega} K(k) \approx -\frac{2}{\omega} \log\!\left(\frac{\epsilon}{4}\right) \tag{8.125}$$

which diverge as $\epsilon \to 0$. In other words: in the limit that computes the large β behavior of the matrix element (8.110), the set of classical solutions which contribute takes the form of *an extremely diluted gas of instantons and anti-instantons*. Moreover: configurations with an arbitrary large number of instantons/anti-instantons contribute subjected only to the topological constraint

that the number of instantons minus the number of anti-instantons must be equal to
the net topological charge

$$
\int_{-\infty}^{+\infty} \frac{\dot{x}\,dt}{2\sigma} =
\begin{cases}
\pm 1 & \text{for } \langle \pm\sigma | e^{-\beta H} \mp \sigma \rangle \\
0 & \text{for } \langle \pm\sigma | e^{-\beta H} | \pm \sigma \rangle.
\end{cases}
\tag{8.126}
$$

In conclusion: If we are interested only in the amplitudes between the classical vacua

$$
\langle \pm\sigma | e^{-\beta H} | \mp \sigma \rangle, \qquad \langle \pm\sigma | e^{-\beta H} | \pm \sigma \rangle,
\tag{8.127}
$$

as $\beta \to \infty$ (and asymptotically small $\hbar$) we may use the *dilute gas approximation*,
where instantons/anti-instantons are seen as a gas of non-interacting particles freely
moving on the line. This *dilute gas* approximation gives the correct asymptotic
behavior.

The instanton gas is *non-interacting* because in the relevant limit it gets infinitely
diluted, see Eq. (8.125). Of course, for *finite* ϵ and β there are interactions between
instantons: this reflects the fact that a sequence of instantons/anti-instantons is not
an exact classical solution for finite ϵ, β, that is, their action is not a minimum
of the functional $S_E[x]$; therefore $\delta S/\delta x$ is small but non-zero, and the gradient
of the action measures the effective force between instantons and anti-instantons.
However, as illustrated by Fig. 8.6, as $\epsilon \to 0$ this force vanishes, and the free gas
approximation becomes fully justified, indeed exact as an asymptotic statement for
$\hbar \to 0$. It only remains the topological restriction (8.126): instantons and anti-
instantons should alternate along the time line: this is akin to an overall "charge
neutrality" of the instanton gas. While we demonstrated the validity of the dilute
gas approximation in the example of the double-well, it is pretty clear that the
conclusions hold in full generality.

Sometimes one expresses the result of the above analysis by saying that the gas
of instantons are *solutions at infinity in path space*. What this means is that they are
the limit as $\epsilon \to 0$ of genuine solutions, as it is clear from Fig. 8.6. Technically they
belong to the compactification of the space of classical solutions, and in computing
path integrals we should always first compactify our integration spaces if we don't
want to forget relevant contributions. In plain English: whenever there are "solutions
at infinity", we have to keep them into account in the semiclassical analysis.

Conclusion: Diluted Gas of Instantons
To the leading semiclassical approximation, the asymptotic behavior for
$\beta \to \infty$ of the matrix elements between classical ground states

$$
\langle \sigma | e^{-\beta H/\hbar} | -\sigma \rangle \quad \text{resp.} \quad \langle -\sigma | e^{-\beta H/\hbar} | -\sigma \rangle
\tag{8.128}
$$

(continued)

have the form

$$\sum_{n=0}^{\infty} I_{2n+1} \quad \text{resp.} \quad \sum_{n=0}^{\infty} I_{2n}, \tag{8.129}$$

where I_m is the contribution from a dilute one-dimensional gas of m instantons/anti-instantons "moving" in the segment $[0, \beta]$. Instantons and anti-instantons alternate along the line. The action of m dilute instantons/anti-instantons is k times the action S_1 of a single instanton, so that

$$I_m = O(e^{-mS_1/\hbar}) \quad \text{as } \hbar \to 0. \tag{8.130}$$

The computation of the semiclassical amplitudes using dilute gases of instantons is an art called *instanton calculus*. The classic reference is [12]; for a more detailed discussion see e.g. [15]; for mathematically more sophisticate treatments and rigorous justifications see [16, 17].

Dilute Gas vs. Zero-Modes

We need to understand the zero-modes of the second variation operator D around an *exact* classical solution $y(t)_m$—which looks to extreme accuracy as a sequence of m instantons/anti-instatons[15]—in the limit of β large and ϵ small. For definiteness we consider the exact "m-instanton" solution $y(t)_m$ in the time interval $[0, \beta]$ whose turning point $x_* \approx \sigma$ solves the transcendental equation

$$2m\, K(k(x_*)) = \frac{1}{2}\sqrt{\frac{2}{1 + k(x_*)^2}}\, \omega\beta, \tag{8.131}$$

where $k(x_*)$ is the function in the second equation (8.117), and such that

$$y(0) = -x_*, \qquad y(\beta) = (-1)^{m+1} x_*. \tag{8.132}$$

The velocity $\dot{y}_m$ is a solution to the equation $D\dot{y}_m = 0$ which vanishes whenever $y(t)_m$ is a turning point $\pm x_*$. By construction it vanishes at $t = 0$ and $t = \beta$, hence it is a genuine zero mode of the operator D in the interval $[0, \beta]$ with the Dirichlet boundary conditions. $\dot{y}(t)_m$ has, in addition, precisely $(m - 1)$ zeros in the interior of the interval $(0, \beta)$. Indeed from Fig. 8.6b, d it is evident that $\dot{y}(t)_k$ has one zero in each interval separating an instanton from the nearby anti-instanton: in fact the zeros of the velocity $\dot{y}(t)_k$ are just the turning points which separate time intervals where the velocity is positive ("instantons") from intervals where $\dot{y}(t)_k$ is negative

---

[15] Cf. the exact solutions in Fig. 8.6.

("anti-instantons"). Then the Sturm oscillation theorem says that there are $(m-1)$ eigenfunctions of D with negative eigenvalues which satisfy the Dirichlet boundary condition.

The $m-1$ negative modes have a clear interpretation. In the limit $\beta \to \infty$ (with m fixed!) the exact zero mode $\dot{y}(t)_m$ becomes the shift of the overall center-of-mass of the instanton gas, and its presence reflects the invariance under overall time translations at $\beta = \infty$. The $m-1$ negative modes are then associated with the *relative* positions of the instantons, that is, to the lengths of the $m-1$ intervals separating them. They are negative since there is a deformation of the solution which decreases the action by reducing the distance between the components of the instanton gas: in the limit that the distance becomes zero the instanton/anti-instanton annihilate and the action lowers from $\approx m S_1$ to $\approx (m-2) S_1$. However, the action becomes independent of these separations asymptotically as $\beta \to \infty$ (equivalently $\epsilon \to 0$) and we end up with m zero modes in the double limit. Explicitly, the m component asymptotic solution has the form

$$y(t)_m \approx \sum_{s=1}^{m} (-1)^s y(t - t_s), \tag{8.133}$$

where $y(t)$ is the instanton solution, and t_s is the center of the s-th instanton/anti-instanton which have wide separations

$$t_{s+1} - t_s \gg (\omega)^{-1}. \tag{8.134}$$

The s-th negative mode is

$$\psi_s(t) \approx \dot{y}(t - t_s). \tag{8.135}$$

$\psi_s(t)$ has support around t_s and vanishes exponentially away from the center t_s of the s-th instanton. Computing $D\psi_s$ we see that the negative eigenvalues of D also go to zero exponentially in the relevant limit. The bottom line is that we have one zero-mode per instanton/antinstanton in the dilute gas which is localized at its center.

Functional Determinants in the Dilute Gas Limit

Consider a differential operator D of the form (8.108) which acts on the space of functions in the interval $[-T, T]$, with a smooth potential of the form

$$W(t) = \omega^2 - F(t),$$

$$\text{where } F(t) \geq 0 \text{ and } F(t) = 0 \text{ for } |t| > a, \tag{8.136}$$

which corresponds to a potential well of width $2a$ and shape $F(t)$ localized at $t = 0$ along the real line. We wish to compute its determinant in the asymptotic limit $T \to \infty$. In particular we may assume $T \gg a$. The potential $W(t)$ of the operator

Fig. 8.7 The shifted potential $W(t) - \omega^2$ in the second variation operator around an instanton of the double well model

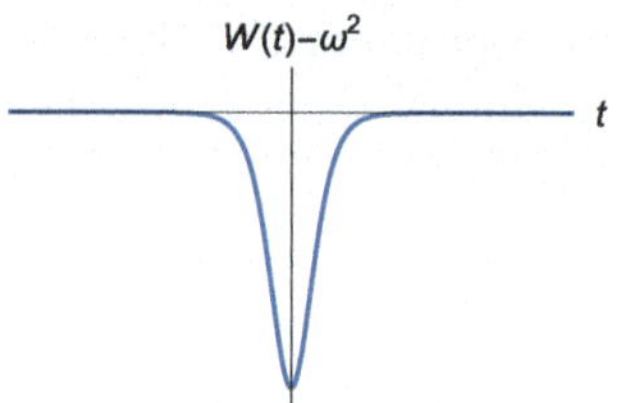

$\boldsymbol{D}$ around the instanton has the form (8.136) where $F(t)$ is exponentially small away from a deep well centered at t_0. See Fig. 8.7.

Let $\psi(t)$ be the solution to the linear ODE $\boldsymbol{D}\psi = 0$ such that

$$\psi(-T) = 0, \quad \dot{\psi}(-T) = 1. \tag{8.137}$$

For $t < -a$, $\psi(t)$ is

$$\psi(t) = \frac{1}{\omega} \sinh[\omega(t + T)], \tag{8.138}$$

while for $t > a$ it takes the form

$$\psi(t) = \frac{A}{2\omega} \exp[\omega(t + T)] - \frac{B}{2\omega} \exp[-\omega(t + T)], \tag{8.139}$$

for some connection constants A, B which do not depend on t and T. The functional determinant is then

$$\mathrm{Det}\, \boldsymbol{D}_{[-T,T]} = \frac{A}{2\omega} \exp[2\omega T] - \frac{B}{2\omega} \exp[-2\omega T] \approx \frac{A}{2\omega} e^{2\omega T}, \tag{8.140}$$

where the RHS holds for large T. Now consider a sequence of m widely separated, identical potential wells, that is, a potential of the form

$$W(t)_{(m)} = \omega^2 - \sum_{k=0}^{m-1} F(t - kT) \tag{8.141}$$

in the time interval $[-T, mT]$ with separations $T \gg a$. The 2nd variation operator for an *exact* solution with period number n which starts at $-\sigma(1 - \epsilon)$ and ends at $\sigma(1 - \epsilon)$ (resp. $-\sigma(1 - \epsilon)$) has the form

$$\boldsymbol{D}_{(m)} = -\frac{\mathrm{d}^2}{\mathrm{d}t^2} + W(t)_{(m)} \tag{8.142}$$

with $m = 2n + 1$ (resp. $m = 2n$) up to corrections which are exponentially small *uniformly* in the whole time interval.

Away from the wells (i.e. almost everywhere as $T \approx \infty$), the exact solution to the homogeneous equation $\boldsymbol{D}_{(m)}\psi = 0$ reads

$$\psi(t) = \begin{cases} \dfrac{A^k}{2\omega}\exp[\omega(t+T)] - \dfrac{B^k}{2\omega}\exp[-\omega(t+T)] \\[2mm] \text{for } (k-1)T + a < t < kT - a, \quad k = 1, 2, \ldots, m, \end{cases} \tag{8.143}$$

with the same connection coefficients A, B as in the case of a single well. We stress that formula (8.143) is valid even if the separations between the successive wells are randomly distributed, as long as they are all larger than $2a$.

We conclude that for T very large the determinant of the operator (8.142) in a dilute sequence of m potential wells is

$$\text{Det } \boldsymbol{D}_{(m)} \approx \frac{1}{2\omega}\, A^m\, \mathrm{e}^{(m+1)\omega T} \equiv \frac{1}{2\omega}\, \mathrm{e}^{\omega\beta}\, A^m, \tag{8.144}$$

where $\beta = (m+1)T$ is the total length of the time interval, and the constant A is computed by the single well amplitude

$$A = 2\omega\, \mathrm{e}^{-\omega\beta}\, \text{Det } \boldsymbol{D}_{(1)}. \tag{8.145}$$

In summary: in the dilute limit of wells, the determinant factorizes in the product of single wells determinants, except for the kinematical prefactor $1/2\omega$. This reflects the obvious physical fact that at *infinite dilution* all gases become free. The determinant of $\boldsymbol{D}$ in the dilute gas of m instantons/anti-instantons, has the form in Eqs. (8.141), (8.142) up to corrections which are exponentially small as $\beta \to \infty$, $\epsilon \to 0$. However the presence of zero modes requires some extra care.

When (8.136) is the potential in the 2nd variation around a single instanton

$$W(t)_{(1)} \equiv V''(y(t)), \qquad t = \int_0^{y(t)} \frac{\mathrm{d}x}{\sqrt{2\,V(x)}}, \tag{8.146}$$

the operator $\boldsymbol{D}_{(1)}$ has one zero-mode, and in Eq. (8.145) the determinant must be replaced by the product over the non-zero modes $\text{Det}^*\boldsymbol{D}_{(1)}$.

The 2nd variation, around an exact solution which goes to a gas of m instantons in the limit $\beta \to \infty$, $\epsilon \to 0$, has one zero mode and $(m-1)$ negative eigenvalues which become zero in the limit, and the determinant (8.144) should be replaced by the product of the eigenvalues which remain bounded away from zero in the double limit, i.e. over the positive eigenvalues only. Writing

$$\text{Det}(\boldsymbol{D}_{(m)} + z)\Big|_{\substack{\text{dilute}\\ \text{limit}}} = z^m\, \text{Det}^*(\boldsymbol{D}_{(m)})\Big(1 + O(z)\Big), \tag{8.147}$$

and applying the previous argument to $\boldsymbol{D}_{(m)} + z$, we conclude that also the product of the positive eigenvalues factorizes in the dilute limit (up to the prefactor $1/2\omega$),

$$\mathrm{Det}^* \boldsymbol{D}_{(m)} = \frac{1}{2\omega} A_*^m \, \mathrm{e}^{\omega\beta}, \qquad A_* = \left.\frac{\partial A}{\partial z}\right|_{z=0}, \tag{8.148}$$

while the zero modes get traded for the integration over the position t_s of the centers of the m instantons. Indeed, from Eq. (8.133) we see that the variation of the m instanton solution $y(t)_m$ when we make $t_s \rightsquigarrow t_s + \delta t_s$ is

$$\delta y(t)_m \approx (-1)^{s+1} \psi_s(t) \, \delta t_s. \tag{8.149}$$

Looking at their explicit form (8.135), we see that the emergent zero modes $\psi_s(t)$ are bound states trapped in a single potential well of $W(t)_{(m)}$, hence insensitive to the shape of $W(t)_{(m)}$ far away from their own well.

The β dependent factor in (8.144) has a clear meaning. The semiclassical amplitude is proportional to

$$(\mathrm{Det}\, \boldsymbol{D}_{(m)})^{-1/2} = \sqrt{2\omega}\, A^{-m/2}\, \mathrm{e}^{-\omega\beta} \equiv \sqrt{2\omega}\, A^{-m/2}\, \mathrm{e}^{-E_0 \beta/\hbar} \tag{8.150}$$

with

$$E_0 = \frac{1}{2}\omega\hbar, \tag{8.151}$$

the zero point energy of the small fluctuation around the classical ground configurations which is the leading (and crudest) approximation to the quantum energies of the ground states.

8.3.2   One-Instanton Contribution

The dilute-limit instanton calculus reduces the computation of the leading semiclassical large-time amplitude to the evaluation of the one instanton contribution. We now focus on the computation of the one instanton term I_1 in the sum (8.129) in the leading asymptotic approximation for $\hbar$ small. We consider an arbitrary one-dimensional system of the form (8.97) which admits instantons with the shape (8.98) that interpolate between two classical ground states x_1 and x_2 whose energy levels are not split to order $O(\hbar)$ i.e. with

$$V(x_1) = V(x_2) = 0, \qquad V''(x_1) = V''(x_2) = \omega^2. \tag{8.152}$$

As before, we write the integration path in the form

$$x(t) = y(t - t_0) + \xi(t) \tag{8.153}$$

where $y(t)$ is the instanton solution and $\xi(t)$ is the "quantum fluctuation" around it. Since there is a family of solutions parametrized by t_0, we have to integrate over t_0, i.e. over the position of the instanton in the time line. However we must be careful not to count twice the same path: under a variation δt_0 of the position of the instanton center, the path changes

$$\delta x(t) = -\dot{y}(t - t_0)\,\delta t_0 \tag{8.154}$$

and this variation may be compensated by a change in the fluctuation $\xi(t)$ proportional to the zero mode $\dot{y}$. Thus, to avoid double counting, we have to integrate only over fluctuations which are orthogonal to the zero mode. The leading order Gaussian integral over the orthogonal fluctuations around a given instanton gives

$$(\mathrm{Det}^* \boldsymbol{D}_{(1)})^{-1/2} e^{-S_1/\hbar}\left(1 + O(\hbar)\right), \tag{8.155}$$

where

$$\boldsymbol{D}_{(1)} = -\frac{\mathrm{d}^2}{\mathrm{d}t^2} + V''(y(t - t_0)). \tag{8.156}$$

From Eq. (6.425) we get the formula

$$\mathrm{Det}^* \boldsymbol{D}_{(1)} = \frac{-1}{\ddot{y}(-\beta/2)\,\ddot{y}(\beta/2)} \int_{-\beta/2}^{\beta/2} \dot{y}_0(s)^2\,\mathrm{d}s \approx \frac{-S_1}{\ddot{y}(-\beta/2)\,\ddot{y}(\beta/2)} \tag{8.157}$$

where the second equality holds asymptotically as $\beta \to \infty$. From the asymptotic behavior (8.100) we get

$$\left(\ddot{y}(-\beta/2)\,\ddot{y}(\beta/2)\right)^{-1} \approx -C^{-2} e^{\omega\beta}, \quad \omega\beta \gg 1 \tag{8.158}$$

for some constant C which depends on the particular model. For instance, for the solution (8.104) we have

$$C = 2\sigma\omega^2. \tag{8.159}$$

Then

$$\mathrm{Det}^* \boldsymbol{D}_{(1)} = C^2\, S_1\, e^{\omega\beta}. \tag{8.160}$$

In conclusion

$$(\mathrm{Det}^* \boldsymbol{D}_{(1)})^{-1/2} = \frac{C\, e^{-\omega\beta/2}}{\sqrt{S_1}}. \tag{8.161}$$

Finally we have to integrate over the family of instantons parametrized by t_0. In doing this we need to use the correct measure on the family. We write

$$\dot{y}(t - t_0)\,\delta t_0 = \psi_0(t - t_0)\,\mathrm{d}a_0 \equiv \frac{\dot{y}(t - t_0)}{(\int \dot{y}^2\,\mathrm{d}s)^{1/2}}\,\mathrm{d}a_0 \tag{8.162}$$

where ψ_0 is the properly normalized zero-mode and a_0 its coefficient in the expansion of the fluctuation $\xi(t)$ in eigenfunctions of $\boldsymbol{D}_{(1)}$

$$\xi(t) \equiv x(t) - y(t - t_0) = \sum_k a_k\,\psi_k(t). \tag{8.163}$$

Since the functional measure has the form

$$\prod_k \frac{\mathrm{d}a_k}{\sqrt{2\pi\hbar}}, \tag{8.164}$$

when we replace the integral over a_0 by the integral over t_0 we get the Jacobian factor

$$\frac{\mathrm{d}a_0}{\sqrt{2\pi\hbar}} = \sqrt{\frac{S_1}{2\pi\hbar}}\,\mathrm{d}t_0 \quad \text{because} \quad \int_{-\infty}^{+\infty} \dot{y}^2\mathrm{d}s = S_1. \tag{8.165}$$

The integrand does not depend on t_0, and the integration from 0 to β just produces a factor β. We get (asymptotically for large β and $1/\hbar$)

$$I_1 = \frac{\beta}{\sqrt{2\pi\hbar}}C\,\mathrm{e}^{-S_1/\hbar}\,\mathrm{e}^{-\frac{1}{2}\omega\beta} = \sqrt{\frac{\omega}{\pi\hbar}}\,\beta K \mathrm{e}^{-S_1/\hbar}\,\mathrm{e}^{-\frac{1}{2}\omega\beta} \tag{8.166}$$

where $K = C/\sqrt{2\omega}$ is a model-dependent $O(1)$ constant.

8.3.3   *Energy Splitting in the Double-Well Potential*

To get the contribution I_m from m instantons/anti-instantons we have to integrate over their positions t_s whose order

$$-\beta/2 < t_1 < t_2 < \cdots < t_m < \beta/2, \quad \beta \to \infty \tag{8.167}$$

is fixed by the topological constraint. Replacing the integral over the m zero modes with the integral over the positions, introduces a Jacobian factor which is just the

m-th power of the one-instanton Jacobian computed above. Then the integrand is independent of the t_s's, and the integration over them produces a factor

$$\int_{-\beta/2}^{\beta/2} dt_1 \int_{t_1}^{\beta/2} dt_2 \int_{t_2}^{\beta/2} \cdots \int_{t_{m-1}}^{\beta/2} dt_m = \frac{\beta^m}{m!}. \tag{8.168}$$

Putting everything together we get

$$\langle \sigma | e^{-\beta H/\hbar} | \sigma \rangle = \sqrt{\frac{\omega}{\pi \hbar}}\, e^{-\frac{1}{2}\omega\beta} \sum_{k=0}^{\infty} \frac{1}{(2k)!} \beta^{2k} K^{2k} e^{-2k S_1/\hbar} \tag{8.169}$$

$$\langle \sigma | e^{-\beta H/\hbar} | -\sigma \rangle = \sqrt{\frac{\omega}{\pi \hbar}}\, e^{-\frac{1}{2}\omega\beta} \sum_{k=0}^{\infty} \frac{1}{(2k+1)!} \beta^{2k+1} K^{2k+1} e^{-(2k+1) S_1/\hbar}, \tag{8.170}$$

and then

$$\frac{1}{2}\langle \sigma | e^{-\beta H/\hbar} (1 \pm \mathscr{P}) | \sigma \rangle = \frac{1}{2}\sqrt{\frac{\omega}{\pi \hbar}} \exp\left[-\left(\frac{\omega}{2} \mp K e^{-S_1/\hbar} \right) \beta \right], \tag{8.171}$$

where the dependence on β exponentiates as expected. The energy of the ground state (upper sign) and the first excited state (lower sign) then are (to the present approximation)

$$E_{\text{ground}} = \frac{1}{2}\omega\hbar - K\hbar\, e^{-S_1/\hbar} \tag{8.172}$$

$$E_{\text{1st exc.}} = \frac{1}{2}\omega\hbar + K\hbar\, e^{-S_1/\hbar}, \tag{8.173}$$

and the energy splitting is

$$\Delta E = 2\hbar\, K\, e^{-S_1/\hbar}. \tag{8.174}$$

Thinking of the ground state amplitude

$$\frac{1}{2}\langle \sigma | e^{-\beta H/\hbar} (1 + \mathscr{P}) | \sigma \rangle \equiv \mathscr{Z} \tag{8.175}$$

as the partition function of the Grand canonical ensemble $\mathscr{Z}$ for the ideal instanton gas with fugacity K, we see that the average number of instantons is

$$\langle N \rangle = K \frac{\partial}{\partial K} \log \mathscr{Z} = \beta K e^{-S_1/\hbar}, \tag{8.176}$$

hence the distance between them is of order $O(e^{S_1/\hbar})$ which is exponentially larger than the instanton size $O(1/\omega)$ for small $\hbar$. Again, this shows that the dilute gas approximation is exact in the asymptotic sense.

8.4 Semiclassical Approximation IV: Schrödinger Equation

Up to now we discussed the semiclassical approximation in the framework of path integrals where the methods are highly intuitive. However the WKB method was originally introduced as an approximation scheme for the Schrödinger equation and it is time to discuss the small $\hbar$ limit in that language.

We focus on a Schrödinger equation of the form

$$i\hbar \frac{\partial \psi}{\partial t} = -\frac{\hbar^2}{2m\sqrt{g}} \frac{\partial}{\partial x^i}\left(\sqrt{g}\, g^{ij} \frac{\partial \psi}{\partial x^j}\right) + V(x^i)\psi. \tag{8.177}$$

In the elementary[16] WKB method, one looks for an asymptotic solution of the form

$$\psi(x) = \exp(iS(x)/\hbar)\left(A_0 + \sum_{k\geq 1} \hbar^k A_k(x)\right) \tag{8.178}$$

where the series is merely asymptotic, that is, not necessarily convergent to the actual solution. Replacing (8.178) by a convergent *transeries* is the job of exact WKB. Plugging (8.178) in (8.177) and equating the two sides order by order in powers of $\hbar$, we get a sequence of equations for the function $S(x)$ and the coefficients $A_k(x)$. We already saw in Chap. 3 that the order $\hbar^0$ and $\hbar^1$ equations are, respectively, the classical Hamilton-Jacobi equation

$$\frac{\partial S}{\partial t} + \frac{1}{2m} g^{ij} \frac{\partial S}{\partial x^i} \frac{\partial S}{\partial x^j} + V(x,t) = 0, \tag{8.179}$$

and the leading order continuity equation

$$\frac{\partial A_0^2}{\partial t} + \nabla_i(A_0^2\, v^i) = 0 \quad \text{where} \quad v^i \equiv \frac{1}{m} g^{ij} \frac{\partial S}{\partial x^j}. \tag{8.180}$$

When the system is time-independent we may separate the time variable and consider the stationary Schrödinger equation

$$\frac{\hbar^2}{2m} \Delta\phi + V\phi = E\phi. \tag{8.181}$$

---

[16] As contrasted to the *exact* one.

We look for a solution of the stationary equation of the form

$$\phi(x) = \exp(iS(x)/\hbar)\Big(\sum_{k=0} B_k \hbar^k\Big) \tag{8.182}$$

where now $S(x)$ is Hamilton's characteristic function.

Regime of Validity of the WKB Approximation

One important issue is to understand the limits of validity of the WKB approximation for the Schrödinger wave function $\phi(x)$. Rewriting Eq. (8.181) in the form

$$\begin{aligned}
2m(E - V(x)) &= \hbar^2 \frac{\Delta\phi}{\phi} = \\[2mm]
&= g^{ij}\frac{\partial S}{\partial x^i}\frac{\partial S}{\partial x^j} - \frac{i\hbar}{B_0} g^{ij}\nabla_i(B_0^2 \partial_i S) + O(\hbar^2),
\end{aligned} \tag{8.183}$$

we see that the semiclassical approximation is justified when in the RHS the classical $O(\hbar^0)$ term is larger than the other ones. Since $\partial S/\partial x^i = p_i$ is the classical momentum, this condition reads

$$|p_i| \gg \frac{\hbar}{L}, \tag{8.184}$$

where L is a typical geometric length of the problem. The momentum equals $2\pi\hbar/\lambda$, where λ is the wave-length of the quantum wave-function, and we conclude that the semiclassical approximation is valid whenever the quantum wave-length λ is small with respect to the typical length scale L in the problem. Notice that this is exactly the condition found in [8] for the validity of geometric optics as a description of the electromagnetic waves. This fact is not a coincidence: the math technique for the high-frequency asymptotic solutions to PDEs may be applied to the Maxwell equations, producing the *eikonal equation* of geometric optics, as well as to the Schrödinger PDE leading to the WKB approximation. In point of fact the eikonal equation in vacuum is just the relativistic Hamilton-Jacobi equation which describes the classical motion of massless particles.

Consider, say, a one-dimensional particle moving in the interval $[0, L]$. The wave-function length λ is then the typical distance between two nodes of $\psi(x)$, and the condition $\lambda \ll L$ is equivalent to the number n of zeros being huge $n \gg 1$. By the oscillation theorem, the number of zeros n is the quantum number which enumerates the successive energy levels E_n, so the condition $\lambda \ll L$ of validity of the WKB approximation is equivalent to asking that the discrete quantum numbers, such as n, ℓ, etc., take huge values, in full agreement with the discussion in Sect. 8.1.

However $\partial S/\partial x^i$ is a function of the position $x \in \mathcal{M}$, hence, even when the quantum numbers n, ℓ are large, there are regions in the configuration space $\mathcal{M}$ where the exact wave-function is poorly approximated by the WKB formulae. This has crucial consequences, as we shall see in Sect. 8.4.1.

One-Dimensional Systems

We consider a one-dimensional Schrödinger equation of the form

$$\hbar^2 \frac{d^2\psi}{dx^2} + 2m(E - V(x))\psi = 0. \tag{8.185}$$

As we already stressed in Chap. 3, it is convenient to replace the second order linear differential equation (8.185) by a first order non-linear one. For the present purposes it is convenient to use the *Riccati version* of the trick. Writing

$$W(x) = \hbar \frac{d}{dx} \log \psi(x) \tag{8.186}$$

reduces (8.185) to the 1st order quadratic ODE[17]

$$W^2 + \hbar W' = 2m(V(x) - E). \tag{8.187}$$

We look for a solution of this Riccati equation in the form of a power series in $\hbar$

$$W(x) = W_0 + \hbar W_1 + \hbar^2 W_2 + \cdots + \hbar^k W_k + \cdots \tag{8.188}$$

Plugging in this series in (8.187), and equating the coefficients of each power of $\hbar$, we get the equations

$$
\begin{array}{lll}
\text{order} & \text{equation} & \\
\hbar^0 & W_0^2 = 2m(V(x) - E) & \\
\hbar^1 & 2W_0 W_1 + W_0' = 0 & \\
\hbar^2 & 2W_0 W_2 + W_1^2 + W_1' = 0 & \\
\vdots & \vdots \qquad \vdots & \\
\hbar^k & 2W_0 W_k + \sum_{\ell=1}^{k-1} W_\ell W_{k-\ell} + W_{k-1}' = 0,
\end{array}
\tag{8.189}
$$

which may be solved recursively in k. At order zero we have

$$W_0(x) = \pm\sqrt{2m(V(x) - E)} = \pm i\, p(x), \tag{8.190}$$

where $p(x)$ is the momentum of a classical particle of energy E in the position x. The function $W_0(x)$ is real (resp. imaginary) when x is in the classically forbidden (resp. allowed) region. Then, recursively,

$$\text{1st step} \qquad W_1(x) = -\frac{1}{2}\frac{W_0'}{W_0} = -\frac{1}{2}\frac{d}{dx}\log W_0(x) \tag{8.191}$$

---

[17] Here and below we write W' for the derivative of $W(x)$.

$$\text{2nd step}\qquad W_2(x) = -\frac{1}{2}\frac{W_1^2 + W_1'}{W_0} = \frac{1}{4}\frac{W_0''}{W_0^2} - \frac{3}{8}\frac{(W_0')^2}{W_0^3} \tag{8.192}$$

$$\text{3rd step}\qquad W_3(x) = -\frac{W_0^{(3)}}{8W_0^2} + \frac{3}{4}\frac{W_0''W_0'}{W_0^4} - \frac{3}{4}\frac{(W_0')^3}{W_0^5} \tag{8.193}$$

and so on. At the k-th recursive step we have

$$W_k(x) = -\frac{1}{2W_0}\left(\sum_{\ell=1}^{k-1} W_\ell W_{k-\ell} + W_{k-1}'\right), \tag{8.194}$$

where all quantities in the RHS have been computed in previous recursive steps. The two-line Mathematica instruction

```
W[0, x_] := Subscript[W, 0][x]
W[k_, x_] := -Together[(Sum[W[l, x] W[k - 1, x], {1, 1, k - 1}]
  + D[W[k - 1, x], x])/(2 W[0, x])]
```

computes the explicit expression of $W_k(x)$ to any desired order. In particular, to the first order

$$\hbar\frac{d}{dx}\log\psi(x) \equiv W(x) \approx \pm\sqrt{2m(V(x) - E)} - \frac{\hbar}{4}\frac{d}{dx}\log(V(x) - E) \tag{8.195}$$

that is,

$$\psi(x) \approx C\big(V(x) - E\big)^{-1/4}\exp\left(\pm\frac{1}{\hbar}\int_*^x \sqrt{2m(V(y) - E)}\,dy\right). \tag{8.196}$$

Classically Integrable Systems

The solution to the stationary Hamilton-Jacobi equation

$$\frac{1}{2m}g^{ij}\partial_i S\,\partial_j S + V(x) = E \tag{8.197}$$

for a classical system with n degrees of freedom may be written as

$$S(x) = \sum_{i=1}^n \int_{x_0}^x p_i\,dx^i, \tag{8.198}$$

where the integral is along a path which solves the classical equations of motion, hence which lays on the energy hypersurface

$$\left\{\frac{1}{2m}g^{ij}p_i p_j + V(x) = E\right\} \subset \mathcal{W} \tag{8.199}$$

in the phase space $\mathcal{W}$. When the system is classically integrable things simplify dramatically. The phase space $\mathcal{W}$ of an integrable system has a Liouville fibration

$$\mathcal{W} \to B \tag{8.200}$$

whose fibers $\mathcal{W}_b$ are Lagrangian submanifolds while the coordinates of the base B are conserved quantities. Hence the motion takes place in the n-dimensional fiber $\mathcal{W}_b$. The fibers are Lagrangian, that is,

$$d(p_i\, dx^i\big|_{\mathcal{W}_b}) = 0, \tag{8.201}$$

so that the integral in (8.198) does not depend on the actual path $\gamma \subset \mathcal{W}_b$ but only on its homology class in the fiber. Hence the leading term in the WKB expansion

$$\log \psi(x) = \frac{i}{\hbar} \int^x p_i\, dq^i + \cdots = \frac{i}{\hbar} \int^\theta I_i\, d\theta^i + \cdots =$$
$$= \frac{i}{\hbar}\left(I_i\, \theta^i + \text{const}\right) + \cdots \tag{8.202}$$

where in the second equality we change the canonical variables from (p_i, q^j) to the action-angle ones (I_i, θ^j) and the last equality follows since the I_i's are conserved quantities.

8.4.1 Semiclassical Connection Formulae

The connection formulae are a fundamental tool in the study of linear differential equations especially emphasized by Riemann. For a typical linear ODE of the Sturm-Liouville form, Eq. (3.157), defined in some connected one-dimensional manifold $\mathcal{M}$, we typically have several disconnected domains $R_i \subset \mathcal{M}$ where we can find a basis of *reliable* approximate solutions $\psi^a(x)_i$

$$\psi^a(x)_i \approx \psi(x)^a_{i,\text{exact}} \quad \text{in } R_i \quad a = 1, 2, \tag{8.203}$$

where $\{\psi^1_{i,\text{exact}}, \psi^2_{i,\text{exact}}\}$ is a local basis of solutions in R_i normalized so that

$$W[\psi^1_{i,\text{exact}}, \psi^2_{i,\text{exact}}] = 1. \tag{8.204}$$

Typically there is one global exact solution $\psi(x)_{\text{exact}}$ which satisfies the physical conditions to be a valid wave-function: in particular, $\psi(x)_{\text{exact}}$ should vanish exponentially in all classically forbidden regions. Then, in each trustworthy region R_i, we have

$$\psi(x)_{\text{exact}} \approx C_{i,1}\, \psi^1(x)_i + C_{i,2}\, \psi^2(x)_i \quad \text{in } R_i, \tag{8.205}$$

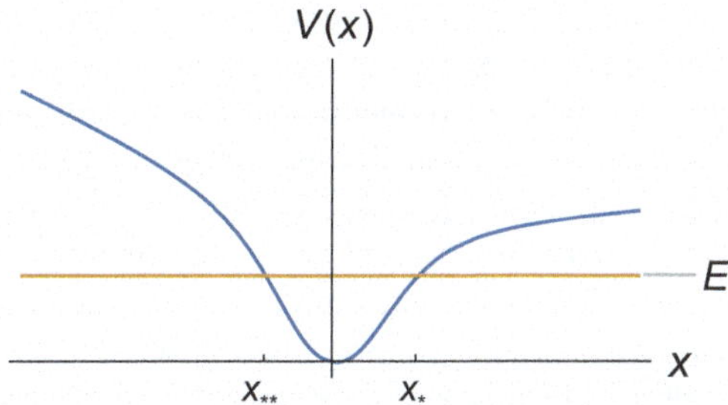

Fig. 8.8 A 1-dimensional system with turning points x_* and x_{**}

for suitably coefficients $C_{i,a}$. In the classical forbidden regions we know that the appropriate linear combination is the unique one (up to overall normalization) which vanishes exponentially.

The *connection coefficient problem* asks to determine the coefficients $C_{i,a}$ in all regions R_i so that we have reliable approximations to the proper wave-function $\psi(x)_{\text{exact}}$ in all regions R_i. The problem boils down to finding the matrices which yield the changes of basis between the locally defined bases of solutions $\{\psi^1_{i,\text{exact}}, \psi^2_{i,\text{exact}}\}$. With our normalization (8.204) these matrices belong to $SL(2, \mathbb{R})$.

In our present physical set-up the connection problem takes the following form. Suppose we want to compute, in the semiclassical approximation, the wave-function for the one-dimensional system with the potential in Fig. 8.8. To the right of the turning point x_* there is the classical forbidden region. The wave function should go to zero exponentially as we go deeper and deeper in that region. Therefore of the two approximate WKB solutions in the region $V(x) > E$

$$\frac{C}{(V(x) - E)^{1/4}} \exp\left(-\frac{1}{\hbar} \int^x \sqrt{2m(V(y) - E)}\, dy\right), \tag{8.206}$$

$$\frac{C'}{(V(x) - E)^{1/4}} \exp\left(\frac{1}{\hbar} \int^x \sqrt{2m(V(y) - E)}\, dy\right) \tag{8.207}$$

only the first one is acceptable. In the classically allowed region $E > V(x)$ the physical solution must be a linear combination of the two oscillatory solutions, and, since the wave function is real, it must have the form

$$\frac{B}{p(x)^{1/4}} \exp\left(\frac{i}{\hbar} \int^x p(y)\, dy\right) + \frac{B^*}{p(x)^{1/4}} \exp\left(-\frac{i}{\hbar} \int^x p(y)\, dy\right) \tag{8.208}$$

where

$$p(y) \equiv \sqrt{2m(E - V(y))}. \tag{8.209}$$

The connection problem requires to write the complex coefficient B that yields the solution which extends to the region $x < x_*$ the good solution in the region $x > x_*$ which decreases exponentially. The reason why the problem is not totally trivial, is that the semiclassical approximation is *not reliable* near a turning point, where $p(x_*) = 0$, hence it makes no sense to require that the two approximate solutions from the left and the right of x_* glue smoothly there.

There are many ways of computing the connection coefficients. One method is to use an alternative approximation scheme which is valid in the region $x \approx x_*$ and match this locally good approximate solution on the right with (8.206) and on the left with (8.208). E.g. one may use the exact solution for the approximate potential

$$V(x)_{\mathrm{appr}} = V(x_*) + V'(x_*)(x - x_*) \quad x \approx x_* \tag{8.210}$$

A better approach is to realize that under mild conditions the Schrödinger equation may be extended to an ODE in the complex domain, $x \in \mathbb{C}$, in such a way that the solutions behave well away from certain "bad" loci as the turning points. Therefore, we may analytically continue the solution from, say, the classically allowed region to the classically forbidden domain along a contour $\gamma \subset \mathbb{C}$ which stays away from the turning point, so that the WKB approximation is valid everywhere along γ. The analytic continuation along γ then computes the connection coefficients (in the WKB approximation).

However the story is more subtle than it may look at first sight for two reasons. First, in the WKB *approximation* (as contrasted to the *exact* WKB approach) we work *asymptotically* i.e. only keep the dominant contribution, the one associated with the (complex) solution x of the Hamilton-Jacobi equation with *minimal* $\mathrm{Im}(S[x])$. Now the analytic continuation of the asymptotic solution does not give the asymptotics of the analytically continued solution. This phenomenon[18] implies that we have to take into account the analytic continuations over *all* possible contours because a relevant contribution may be subleading along a particular path and hence we may easily miss it. Second reason: the complex analytic continuation of a solution is typically *multivalued* in the complex plane, and we need to take linear combinations of the several solutions in order to produce univalued wave-functions for real x.

Let us apply the analytic continuation method to the connection problem between the two sides of the turning point x_* of a general one dimensional system with a potential as in Fig. 8.8. In Fig. 8.9 we have three contours that go around the turning point x_*.

---

[18] Called the *Stokes phenomenon*.

Fig. 8.9 Contours in the complex x-plane which go around the turning point x_*. The two contours on the left (blue and resp. purple) connect points in the classically allowed region to points in the forbidden one, while the trajectory on the right goes back to the classically allowed domain

In a small disk D_* centered at a generic[19] turning point we have

$$2m(V(x) - E) \approx c^2(x - x_*), \qquad c > 0. \tag{8.211}$$

For $x \in D_* \setminus \{x_*\}$ we have the two linear independent WKB solutions

$$\psi(x)_\pm \approx \frac{1}{(x - x_*)^{1/4}} \exp\left(\pm \frac{c}{\hbar} \int^x (y - x_*)^{1/2}\,dy\right)$$

$$\approx \frac{1}{(x - x_*)^{1/4}} \exp\left(\pm \frac{2c}{3\hbar}(x - x_*)^{3/2}\right). \tag{8.212}$$

The argument of the exponential is an analytic function away from a branch cut originating at x_*, and the same holds for the pre-factor $(x - x_*)^{-1/4}$. We may take the cut along $(-\infty, x_*)$ in the real axis. The analytic continuation of the good solution in the classically forbidden region

$$\psi(x)_0 \approx \frac{C}{(x - x_*)^{1/4}} \exp\left(-\frac{c}{\hbar} \int^x (y - x_*)^{1/2}\,dy\right) \tag{8.213}$$

(C real) to the classically allowed region in the upper half plane (blue contour in Fig. 8.9) is then

$$C\,e^{-i\pi/4} \frac{1}{\sqrt{|p(x)|}^{-1/4}} \exp\left(\frac{i}{\hbar}\left|\int^x p(y)\,dy\right|\right) \tag{8.214}$$

while the analytic continuation in the lower half plane (purple contour) yields

$$C\,e^{i\pi/4} \frac{1}{\sqrt{|p(x)|}^{-1/4}} \exp\left(-\frac{i}{\hbar}\left|\int^x p(y)\,dy\right|\right). \tag{8.215}$$

---

[19] That is, not the endpoint of a limitation motion.

The wave-function must be univalued along the real axis for $x < x_*$; this requires that the analytic continuation along the red path in Fig. 8.9 gives back the same function. To have that property, the solution must be the sum of (8.214) and (8.215) which then is automatically real, as expected on general grounds. In the classically allowed region the WKB wave-function then is

$$\frac{2C}{\sqrt{|p|}} \cos\left(\frac{1}{\hbar}\left|\int^x p(y)\,dy\right| - \frac{\pi}{4}\right). \tag{8.216}$$

In particular when going around the turning point along the red contour of Fig. 8.9 the phase of the solution changes by an additional $-\pi/2$.

Relation to the Maslov Index

From the above discussion we see that the phase of the wave function in the classically allowed region has the form

$$\frac{1}{\hbar}\int^x p\,dx - \text{(constant shift)} \tag{8.217}$$

and not just the action $\int^x p\,dx$ of the classical solution divided by $\hbar$. The shift is a local contribution at the turning points, i.e. at the zeros of the velocity $\dot{x}$. Therefore the shift is proportional to the number of zeros of $\dot{x}$ encountered along the classical trajectory $x(t)$. This number is the Maslov index μ of the trajectory (ending at x), cf. Sect. 8.2.2. Hence in the classically allowed region

$$\text{(phase of the WKB wave-function)} = \frac{1}{\hbar}\int^x p_i\,dq^i - \frac{\pi}{2}\mu. \tag{8.218}$$

8.5 Semiclassical Approximation V: Bohr-Sommerfeld

The above results (8.218) on the phase of the WKB wave-function leads us to a formula for the semiclassical discrete energy levels E_n known as the *Bohr-Sommerfeld quantization formula*. This is the version of quantization which was used historically before a complete formulation of Quantum Mechanics was set up by Schrödinger and Heisenberg.

Suppose that we have a discrete energy level which may be described semiclassically. We know from Chap. 3 that a discrete energy corresponds to a bound state, so at the leading order in the WKB approximation its wave-function is given by a solution of the Hamilton-Jacobi equation which describes a classical *bounded* motion in a compact region of the phase space $\mathcal{W}$. If, in addition, the classical system is integrable in the sense of Liouville [8], the motion must be quasi-periodic. In this situation, the separation of variables essentially reduces the problem to a sequence of one-dimensional systems. Then, for simplicity, we focus on the one-dimensional case, leaving the straightforward extension to a general solvable model to the reader.

In one dimension there are two kinds of periodic motions: librations and rotations [8]. In a *libration* the particle moves to the right until it reaches a turning point, then gets reflected back to the left until it reaches the opposite turning point, and then returns to the original position after one period time. The typical examples of libration are the oscillations of a harmonic oscillator. In a *rotation,* instead, the velocity of the particle never vanishes, and there are no turning points. The typical example is a particle moving freely in a circle S^1. The crucial difference between the two cases is that in a libration the Maslov index μ of a closed orbit is two times the number of its periods, while for a rotation $\mu = 0$. In both situations the WKB wave-function in the classically allowed region is

$$\frac{C}{\sqrt{p(x)}} \exp\left(\frac{i}{\hbar} \int_*^x p\, dq - \frac{i\pi}{2}\mu\right) + \text{(complex conjugate)}. \tag{8.219}$$

The condition that the wave function is *univalued* requires that the total phase accumulated along one period of the classical motion is a multiple of 2π, that is,

$$\frac{1}{\hbar} \oint p\, dq - \frac{\pi\mu}{2} = 2n\pi \quad n \in \mathbb{Z}, \tag{8.220}$$

where the integral is over one period of the classical motion and μ is its Maslov index, i.e. the number of turning points in one period. By construction, the integral in (8.220) is 2π times the classical adiabatic invariant [8]

$$I \equiv \oint \frac{p\, dq}{2\pi}, \tag{8.221}$$

also known as the *action* variable. Equation (8.220) then gives the Bohr-Sommerfeld quantization condition for the adiabatic invariants/action variables

$$I = \begin{cases} (n + \frac{1}{2})\hbar, & n \in \mathbb{N} \quad \text{libration} \\ n\,\hbar, & n \in \mathbb{Z} \quad \text{rotation.} \end{cases} \tag{8.222}$$

The difference between $\mathbb{N}$ and $\mathbb{Z}$ arises since in the case of rotation we have two possible signs for the never-vanishing velocity, while in the libration I may be defined to be non-negative.

Example 8.6 (Harmonic Oscillator) We know [8] that the adiabatic invariant I for an harmonic oscillator of frequency ω is

$$I = \frac{E}{\omega}. \tag{8.223}$$

Then the Bohr-Sommerfeld formula yields the WKB energy levels

$$E = (n + \tfrac{1}{2})\hbar\omega \quad n = 0, 1, 2, \ldots, \tag{8.224}$$

which, in this case, coincides with the exact result. We already know that the semiclassical approximation is exact for the harmonic oscillator from the path integral approach.

Example 8.7 (Free Motion in S^1) In this case $I = J$, the angular momentum, and the Bohr-Sommerfeld formula yields

$$J = n\hbar \quad n \in \mathbb{Z}, \tag{8.225}$$

which is also exact.

8.6   Large N Methods

Suppose we have the quantum system with Euclidean action

$$S_E[x(t); m^2, \lambda]_N =$$

$$= \frac{1}{2} \int dt \sum_{i=1}^{N} \left(\dot{x}_i \dot{x}_i + m^2 x_i x_i \right) + \frac{\lambda}{2} \int dt \left(\sum_{i=1}^{N} x_i x_i \right)^2, \tag{8.226}$$

called the $O(N)$-invariant anharmonic oscillator or the $O(N)$-*model*, for short. We are specifically interested in the $O(N)$-invariant operator

$$X(t) \overset{\text{def}}{=} x_i(t)\, x_i(t), \tag{8.227}$$

and we wish to compute the generating functional for its (time-ordered) correlations

$$Z[J(t)]_N = \sum_{n \geq 0} \frac{1}{n!} \int dt_1 \cdots dt_n \, \mathbf{T}\Big(X(t_1) \cdots X(t_n) \Big) J(t_1) \cdots J(t_n) \tag{8.228}$$

which is given by

$$Z[J(t)]_N = \mathcal{N} \int [dx] \, e^{-S_E[x(t), M(t)^2, \lambda]_N} \tag{8.229}$$

where

$$M(t)^2 = m^2 + 2 J(t). \tag{8.230}$$

The path integral (8.229) cannot be computed exactly: to make progress we must rely on a controlled approximation scheme, that is, we must evaluate (8.229) in a suitable asymptotic regime. We may use semiclassical methods, but for some physical application the semiclassical approximation is not good enough. An alternative approach is the *large-N approximation*. The functional $Z[J(t)]_N$ depends on the number N of operators (fields) x_i. The idea of the large-N method is to write the RHS of (8.229) as an asymptotic expansion in the small parameter $1/N$ to describe the large N behavior of $Z[J(t)]_N$. To get a smooth limit as $N \to \infty$, we need to rescale the coupling λ and consider the double scaling limit

$$N \to \infty \quad \text{with} \quad N\lambda = g \quad \text{kept fixed.} \tag{8.231}$$

Then, writing

$$x_i = \sqrt{N}\, y_i, \tag{8.232}$$

the action S_E becomes

$$\frac{N}{2} \int dt \sum_{i=1}^{N} \left(\dot{y}_i \dot{y}_i + M^2 y_i y_i \right) + \frac{Ng}{2} \int dt \left(\sum_{i=1}^{N} y_i y_i \right)^2. \tag{8.233}$$

We have the identity

$$\exp\left[-\frac{Ng}{2} \int dt \left(\sum_i y_i^2 \right)^2 \right] = \int [dz]\, e^{\frac{Ng}{2} \int dt [z^2 - 2z \sum_i y_i y_i]}, \tag{8.234}$$

where the RHS is defined by analytic continuation to imaginary z. Then we have

$$Z_N = \int [dy_i\, dz] \exp\left[-\frac{N}{2} \int dt \left(\sum_{i=1}^{N} (\dot{y}_i^2 + (M^2 + 2gz)y_i^2) - gz^2 \right) \right]. \tag{8.235}$$

Now the action is quadratic in the y_i's, and we may perform explicitly the Gaussian integral over these N fields getting

$$\int [dy_i] \exp\left[-\frac{N}{2} \int dt \left(\sum_{i=1}^{N} (\dot{y}_i^2 + (M^2 + 2gz)y_i^2) \right) \right] = \tag{8.236}$$

$$= \exp\left(-\frac{N}{2} \log \Delta[M^2 + 2gz] \right),$$

where the functional $\Delta[\phi(t)]$ is defined as

$$\Delta[\phi(t)] \overset{\text{def}}{=} \text{Det}\left(-\frac{\mathrm{d}^2}{\mathrm{d}t^2} + \phi(t)\right) \tag{8.237}$$

with the functional determinant computed in the space of functions $\mathbb{R} \to \mathbb{R}$. We remain with the path integral in the field z

$$Z_N = \int [\mathrm{d}z]\, e^{-\frac{N}{2}\left(\log \Delta[M^2+2gz]-g\int \mathrm{d}t\, z^2\right)} \equiv \int [\mathrm{d}\phi]\, e^{-N S[\phi]_{\text{eff}}} \tag{8.238}$$

$$\phi(t) \equiv M(t)^2 + 2\,g\,z(t) \equiv m^2 + 2\,J(t) + 2\,g\,z(t). \tag{8.239}$$

Now the full dependence of (8.238) on the number N of x_i's is given by the overall coefficient in front of the effective action $S[\phi]_{\text{eff}}$

$$S[\phi]_{\text{eff}} = \frac{1}{2}\log \Delta[\phi] - \frac{1}{8g}\int \mathrm{d}t\,(\phi(t) - M^2)^2 \equiv \int \mathrm{d}t\, L_E[\phi]. \tag{8.240}$$

The explicit expression of $S[\phi]_{\text{eff}}$ for all temperatures is computed in the next subsection.

The path integral (8.238) is now well defined for all real positive N, not just for the integers. We managed to extend the path integral from N integer to real, and now we are entitled to treat the number of degrees of freedom N as a real parameter.

Up to now all our manipulation were *exact*. We reduced the path integral over the large number N of degrees of freedom x^i to a path integral over the single field ϕ at the price of replacing the original polynomial action functional (8.226) by the more complicated one (8.240). Luckily, the new action—while intricate and non-local in time—has a special form which suggests a natural strategy for an approximate evaluation.

Let us change notation and write the real parameter N as $\hbar^{-1}$. Equation (8.238) becomes

$$\int [\mathrm{d}\phi]\, e^{-\frac{1}{\hbar} S[\phi]_{\text{eff}}} \tag{8.241}$$

which is the general form of the path integral in Quantum Mechanics before setting $\hbar = 1$. As $\hbar \to 0$, that is, as $N \to \infty$, the path integral gives back the classical theory, as we saw in Sect. 8.2. Therefore, to the leading order in $1/N$ the path integral is given by the integrand evaluated at its saddle point ϕ_0

$$-\log \int [\mathrm{d}\phi]\, e^{-N S[\phi]_{\text{eff}}} = N S[\phi_0]_{\text{eff}} + \frac{1}{2}\log \text{Det}\left[\frac{\delta^2 S[\phi]_{\text{eff}}}{\delta\phi^2}\bigg|_{\phi=\phi_0}\right] + O\left(\frac{1}{N}\right) \tag{8.242}$$

where ϕ_0 is the solution to the "classical" equation of motion

$$\frac{\delta S_{\text{eff}}}{\delta \phi(t)} = 0. \tag{8.243}$$

In conclusion: if we are interested only in observables which are $O(N)$-invariant, in the asymptotic limit $N \to \infty$ we may replace the original quantum model with a classical theory.

The "classical" equations (8.243) look still rather formidable. However consider the $O(N)$-invariant observables normalized so that their limit as $N \to \infty$ is smooth

$$Y(t) \equiv \frac{1}{N} X(t). \tag{8.244}$$

Their *connected* correlation n-functions are

$$\langle Y(t_1) \cdots Y(t_n) \rangle_{\text{conn}} = \frac{1}{N^{n-1}} \frac{\delta^n (S[\phi]_{\text{eff}})_{\text{saddle}}}{\delta J(t_1) \cdots \delta J(t)_n} + O\left(\frac{1}{N^n}\right), \tag{8.245}$$

so that in the strict limit $N = \infty$ only the one-point connected correlations are non-zero, or, in other words the ordinary correlations on the ground state factorize

$$\langle Y(t_1) \cdots Y(t_n) \rangle = \langle Y \rangle^n. \tag{8.246}$$

We conclude that to get all correlations functions of (rescaled) $O(N)$-invariant observables we need only to compute the one-point function of Y, which is invariant under time-translation. Hence we may take the source $J(t)$, hence $M^2(t)$, to be time-independent *without loss*. The original path integral (8.229) now has a time-independent Lagrangian. The solution $\phi_0(t)$ corresponding to the ground state should respect all symmetries of the problem (since there is no spontaneous breaking in this class of quantum systems), and hence must be invariant under time translation, that is, a constant. Now, when ϕ is constant

$$\log \Delta[\phi] = \text{Tr} \log\left(-\frac{d^2}{dt^2} + \phi\right) = \int dt \int_{-\infty}^{+\infty} \frac{d\omega}{2\pi} \log(\omega^2 + \phi), \tag{8.247}$$

the inner integral is divergent, but we know how to regularize it: taking the derivative

$$\frac{d}{d\phi} \log \Delta[\phi] = \int dt \int_{-\infty}^{+\infty} \frac{d\omega}{2\pi} \frac{1}{\omega^2 + \phi} = \int dt \frac{1}{2\sqrt{\phi}} \tag{8.248}$$

and integrating back, we get

$$\log \Delta[\phi] = \int dt \left(\sqrt{\phi} + \text{const}\right). \tag{8.249}$$

The integration constant is absorbed in the overall normalization $\mathcal{N}$, and no physical quantity depends on it. Hence the effective Lagrangian $L_E[\phi]$, when evaluated on a constant field, reduces to

$$L_E[\phi]\Big|_{\phi \text{ constant}} = \frac{1}{2}\sqrt{\phi} - \frac{1}{8g}(\phi - M^2)^2 \tag{8.250}$$

The saddle point equation, restricted to the space of constant ϕ's, is then

$$\frac{\mathrm{d}}{\mathrm{d}\phi}L_E[\phi]\Big|_{\phi \text{ constant}} = 0, \tag{8.251}$$

and $\sqrt{\phi_0}$ is the *unique*[20] real positive solution to the cubic equation

$$\sqrt{\phi_0}\left((\sqrt{\phi_0})^2 - M^2\right) = g. \tag{8.252}$$

Plugging in the solution of this equation in (8.242) we get the leading order approximation to the log of the generating function for $O(N)$-invariant correlations. We get

$$\langle Y \rangle \equiv \frac{1}{N}\langle X \rangle = 2\frac{\partial L_E|_{\text{saddle}}}{\partial M^2}\Big|_{M^2=m^2} = \frac{1}{4g}(\phi - m^2)\Big|_{\text{saddle}} = \frac{1}{4\sqrt{\phi_0}}\Big|_{M^2=m^2}. \tag{8.253}$$

8.6.1 Exact Effective Action at Finite Temperature

Although it is not necessary to solve the model at $N = \infty$, we give the explicit formula for the effective action (8.240) at finite temperature, i.e. taking $\Delta[\phi]$ to be the determinant over functions periodic of period β. Here N is any (finite) integer.

The Stochastic Form of a Positive Hill Determinant
By a positive Hill operator we mean a 2nd order differential operator

$$D[\phi] = -\frac{\mathrm{d}^2}{\mathrm{d}t^2} + \phi(t) \tag{8.254}$$

where the real function $\phi(t)$ is periodic of period β with positive mean

$$\int_0^\beta \phi(t)\,\mathrm{d}t > 0. \tag{8.255}$$

---

[20] That the algebraic equation (8.252) has precisely *one* real positive solution follows from Descartes rule of sign together with $M^2 > 0$ and $g > 0$.

Let Det $D[\phi]$ be the determinant of $D[\phi]$ computed on the periodic functions of period β. We claim that the following formula holds for all s

$$\frac{1}{2^{2s+1}}(\mathrm{Det}\, D[\phi])^s =$$

$$= \int [\mathrm{d}\sigma]\, \delta[\dot{\sigma} + \sigma^2 - \phi]\, \Theta\left(\int_0^\beta \sigma(t)\, \mathrm{d}t\right)\left[\sinh\left(\frac{1}{2}\int_0^\beta \sigma(t)\, \mathrm{d}t\right)\right]^{2s+1} \tag{8.256}$$

Indeed, using the monodromy method for periodic determinants, we get the identity

$$\delta[\dot{\sigma} + \sigma^2 - \phi](\mathrm{Det}\, D[\phi])^s = \delta[\dot{\sigma} + \sigma^2 - \phi](\mathrm{Det}\, D[\dot{\sigma} + \sigma^2])^s =$$

$$= \delta[\dot{\sigma} + \sigma^2 - \phi]\, \mathrm{Det}\left[-\frac{\mathrm{d}^2}{\mathrm{d}t^2} + \sigma^2 + \frac{\mathrm{d}\sigma}{\mathrm{d}t}\right]^s =$$

$$= \delta[\dot{\sigma} + \sigma^2 - \phi]\, \mathrm{Det}\left[\left(\frac{\mathrm{d}}{\mathrm{d}t} + \sigma\right)\left(-\frac{\mathrm{d}}{\mathrm{d}t} + \sigma\right)\right]^s = \tag{8.257}$$

$$= \delta[\dot{\sigma} + \sigma^2 - \phi]\left|\mathrm{Det}\left(-\frac{\mathrm{d}}{\mathrm{d}t} + \sigma\right)\right|^{2s} =$$

$$= \delta[\dot{\sigma} + \sigma^2 - \phi]\left|2\sinh\left(\frac{1}{2}\int_0^\beta \sigma(t)\, \mathrm{d}t\right)\right|^{2s}$$

which however is non-trivial only if the support of the delta-functional is not empty, that is, if there are real periodic solutions to the Riccati ODE

$$\dot{\sigma} + \sigma^2 = \phi. \tag{8.258}$$

This requires the positivity condition (8.255). If σ is a periodic solution of the Riccati equation, there is a second one of the form

$$\tilde{\sigma} = \sigma + \frac{1}{u} \tag{8.259}$$

where u satisfies the linear ODE

$$\dot{u} - 1 - 2\sigma u = 0. \tag{8.260}$$

Then

$$\int_0^\beta \tilde{\sigma}(t)\, \mathrm{d}t = \int_0^\beta \left(\sigma + \frac{1}{u}\right)\mathrm{d}t = -\int_0^\beta \left(\sigma + \frac{\dot{u}}{u}\right)\mathrm{d}t = -\int_0^\beta \sigma\, \mathrm{d}t, \tag{8.261}$$

We select the periodic solution with positive integral that we write as $\mathring{\sigma}(t)$. Then

$$
\delta[\dot{\sigma} + \sigma^2 - \phi]\,\Theta\!\left(\int_0^\beta \sigma(t)\,dt\right) =
$$

$$
= \delta[\sigma - \mathring{\sigma}]\,\mathrm{Det}\left[\frac{\delta\sigma}{\delta\phi}\right]\,\Theta\!\left(\int_0^\beta \mathring{\sigma}(t)\,dt\right) = \tag{8.262}
$$

$$
= \frac{1}{2}\delta[\sigma - \mathring{\sigma}]\,\Theta\!\left(\int_0^\beta \mathring{\sigma}(t)\,dt\right)\left[\sinh\!\left(\frac{1}{2}\int_0^\beta \mathring{\sigma}(t)\,dt\right)\right]^{-1}
$$

Plugging (8.257), (8.262) in (8.256) and performing the trivial integral in σ we get the claim.

The Exact Effective Action of the $O(N)$-Model at Finite T
From Eq. (8.238) we have

$$
Z_N \equiv \int [d\phi]\,e^{\frac{N}{8g}\int_0^\beta dt\,(\phi - M^2)^2}\,(\mathrm{Det}\,D[\phi])^{-N/2} =
$$

$$
= \int [d\phi\, d\sigma]\,\delta[\dot{\sigma} + \sigma^2 - \phi]\,\Theta\!\left(\int_0^\beta \sigma\,dt\right)\left[2\sinh\!\left(\frac{1}{2}\int_0^\beta \sigma\,dt\right)\right]^{1-N}
$$

$$
= \int [d\sigma]\,e^{\frac{N}{8g}\int_0^\beta [\dot{\sigma}^2 + (\sigma^2 - M^2)^2]dt}\,\Theta\!\left(\int_0^\beta \sigma\,dt\right)\left[2\sinh\!\left(\frac{1}{2}\int_0^\beta \sigma\,dt\right)\right]^{1-N} \tag{8.263}
$$

where we used the fact that the integrals of total derivatives vanish for periodic boundary conditions. We conclude that the effective action in terms of the new field σ reads

$$
S[\sigma]_{\mathrm{eff}} = -\frac{N}{8g}\int_0^\beta dt\,(\dot{\sigma}^2 + (\sigma^2 - M^2)^2) + (N-1)\log\left|\sinh\!\left(\frac{1}{2}\int_0^\beta \sigma\,dt\right)\right|. \tag{8.264}
$$

It is easy to check that in the zero temperature limit $\beta \to \infty$ we get the same result as before.

8.7 Perturbation Theory I: Time-Independent Methods

Time-independent perturbation theory may be applied to "quasi-integrable" quantum systems, that is, systems whose Hamiltonian H is time-independent of the form

$$
H = H_0 + \epsilon\, H_I \tag{8.265}
$$

where H_0, called the *unperturbed* Hamiltonian, is explicitly solvable, while ϵH_I called the *perturbation* (or the *interaction Hamiltonian*) is small, being formally of order ϵ, where ϵ is a small parameter. The basic idea of the time-independent perturbation theory is to solve the stationary Schrödinger equation

$$(H_0 + \epsilon H_I)|E\rangle = E|E\rangle \tag{8.266}$$

order by order in a (formal) power series in ϵ. The notion of a "small" perturbation works better when H_I is a *bounded* operator acting on the Hilbert space $\mathcal{H}$, which is often the case in the applications. When H_I is a *compact* operator the situation is even better, and the series typically converges.

We look for a solution to (8.266) in the form of two power series

$$|E\rangle = |E_0\rangle + \epsilon|\psi_1\rangle + \epsilon^2|\psi_2\rangle + \cdots \tag{8.267}$$

$$E = E_0 + \epsilon a_1 + \epsilon^2 a_2 + \cdots \tag{8.268}$$

We are particularly interested in the perturbed energy eigenvalues $E(\epsilon)$. A continuous energy spectrum will not change much (except near its boundary) and the most interesting application is when the unperturbed energy eigenvalue E_0 is a discrete one. Therefore we assume

Assumption E_0 is an isolated point in the spectrum of the unperturbed Hamiltonian H_0 while its E_0-eigenspace $\mathcal{H}_{E_0} \subset \mathcal{H}$ has finite dimension d.

Plugging in (8.267), (8.268) into (8.266), and equating the coefficients of ϵ^n order by order, we get the sequence of equations

$$(H_0 - E_0)|E_0\rangle = 0 \tag{8.269}$$

$$(H_0 - E_0)|\psi_1\rangle = (a_1 - H_I)|E_0\rangle \tag{8.270}$$

$$(H_0 - E_0)|\psi_2\rangle = (a_1 - H_I)|\psi_1\rangle + a_2|E_0\rangle \tag{8.271}$$

$$\vdots \qquad \vdots$$

$$(H_0 - E_0)|\psi_n\rangle = (a_1 - H_I)|\psi_{n-1}\rangle + a_n|E_0\rangle + \sum_{m=1}^{n-2} a_{n-m}|\psi_m\rangle \tag{8.272}$$

that we solve recursively in the order $n = 0, 1, 2, \ldots$.

The first equation (8.269) just says that $E_0 \in \sigma_p(H_0)$ while the zero-order eigenstate $|E_0\rangle$ is an element of the unperturbed eigenspace $\mathcal{H}_{E_0}$. However the zero-order eigenstate $|E_0\rangle$ must be a *specific* linear combination of the elements of $\mathcal{H}_{E_0}$ as we are going to see.

We fix an orthonormal basis

$$\{|E_0\rangle_1, |E_0\rangle_1, \ldots, |E_0\rangle_d\} \qquad (d \equiv \dim \mathcal{H}_{E_0}) \tag{8.273}$$

of the unperturbed energy eigenspace $\mathcal{H}_{E_0}$. The zero-order state $|E_0\rangle$ has the form

$$|E_0\rangle = |E_0\rangle_k\, c_k \tag{8.274}$$

for some complex coefficients c_k (sum over repeated indices implied). Multiplying the second equation (8.270)

$$(H_0 - E_0)|\psi_1\rangle = (a_1 - H_I)|E_0\rangle_k\, c_k \tag{8.275}$$

by the eigen-bra $_h\langle E_0|$ we get

$$a_1\, c_h = {}_h\langle E_0|H_I|E_0\rangle_k\, c_k, \tag{8.276}$$

that is, the first order corrections a_1 to the eigenvalue E_0 (whose multiplicity is d) are the eigenvalues of the $d \times d$ matrix

$$(H_I)_{hk} \stackrel{\text{def}}{=} {}_h\langle E_0|H_I|E_0\rangle_k, \tag{8.277}$$

while the unperturbed eigenstates are the linear combinations (8.274) which diagonalize the matrix (8.277). Indeed, when the energy level E_0 of the unperturbed Hamiltonian H_0 is degenerate, $d > 1$, the perturbation H_I will typically split the energy levels, lifting the degeneration, so that at the first order we have d *distinct* eigenvalues $E_k(\epsilon)$ with one-dimensional eigenspaces. As $\epsilon \to 0$ the perturbed eigenstates go into d particular linear combinations of the $|E_0\rangle_k$ with coefficients given by the eigenvectors of the matrix (8.277).

The case of a non-degenerate eigenvalue of the unperturbed Hamiltonian, $d = 1$, is particularly simple since in this case

$$E(\epsilon) = E_0 + \epsilon\,\langle E_0|H_I|E_0\rangle + O(\epsilon^2) \tag{8.278}$$

i.e. the first order term in the expansion (8.268) of the eigenvalue is just the expectation value of the perturbation $\epsilon\, H_I$ in the non-degenerate eigenstate $|E_0\rangle$.

When c_k is an eigenvector of the $d \times d$ matrix (8.277), and a_1 is the corresponding eigenvalue, the RHS of (8.275) is orthogonal to $\mathcal{H}_{E_0}$. After restriction to the orthogonal complement $\mathcal{H}_{E_0}^\perp$, the operator $(H_0 - E_0)$ becomes invertible. Let $\{|i, s\rangle\}$ be an orthonormal basis of $\mathcal{H}_{E_0}^\perp$ satisfying

$$H_0|i, s\rangle = E_i|i, s\rangle, \qquad E_i \neq E_0. \tag{8.279}$$

We have

$$(H_0 - E_0)^{-1}\Big|_{\mathcal{H}_{E_0}^\perp} = \sum_{i,s} \frac{|i, s\rangle\langle i, s|}{E_i - E_0}. \tag{8.280}$$

The general solution of the linear equation (8.275) is the sum of a particular solution plus the general solution of the homogeneous equation i.e.

$$|\psi_1\rangle = -\sum_{i,s} \frac{|i,s\rangle\langle i,s|H_I|E_0\rangle}{E_i - E_0} + |\Phi_1\rangle, \qquad |\Phi_1\rangle \in \mathcal{H}_{E_0} \tag{8.281}$$

where the second term reflects our freedom to redefine the basis of $\mathcal{H}_{E_0}$ by a ϵ-dependent linear transformation of the form

$$|E_0\rangle_k \to |E_0\rangle_h\big(\delta_{hk} + \epsilon\, M_{hk} + \cdots\big). \tag{8.282}$$

Locally in parameter space,[21] for ϵ near zero, we may choose $|\Phi_1\rangle = 0$, so that

$$|\psi_1\rangle \in \mathcal{H}_{E_0}^{\perp}. \tag{8.283}$$

However the ambiguity $|\Phi_1\rangle$ cannot be eliminated at the global level in the parameter space; this term leads to important subtle phenomena (the *Berry phase*) that will be discussed in Sect. 8.11 below. These global effects are of no consequence for perturbation theory where the parameter ϵ takes value in a parametrically small interval centered at zero.[22]

Next we go to the third equation (8.271) which yields the $O(\epsilon^2)$ corrections. Repeating the same steps as before, we get[23]

$$\langle E_0|(a_1 - H_I)|\psi_1\rangle + a_2 = 0 \quad \Rightarrow \quad a_2 = \langle E_0|H_I|\psi_1\rangle - a_1\langle E_0|\Phi_1\rangle. \tag{8.284}$$

Plugging in Eq. (8.281) we get the final answer

$$a_2 = -\sum_{i,s} \frac{\langle E_0|H_I|i,s\rangle\langle i,s|H_I|E_0\rangle}{E_i - E_0}. \tag{8.285}$$

Note that the RHS is independent of the ambiguity $|\Phi_1\rangle$. Plugging (8.285) in (8.271), the RHS becomes an element of $\mathcal{H}_{E_0}^{\perp}$. Using the formula (8.280) for the inverse in the subspace $\mathcal{H}_{E_0}^{\perp}$ of the operator in the LHS of (8.271), we get

$$|\psi_2\rangle = \sum_{i,s} \frac{|i,s\rangle\langle i,s|(a_1 - H_I)|\psi_1\rangle}{E_i - E_0} + |\Phi_2\rangle,, \qquad |\Phi_2\rangle \in \mathcal{H}_{E_0}, \tag{8.286}$$

where, again, the ambiguity $|\Phi_2\rangle$ may be set to zero for ϵ small.

---

[21] By "parameter space" we mean the space of all physical parameters entering in the Hamiltonian.

[22] That is, ϵ is a mere formal parameter in the sense of formal power series.

[23] Assuming that the degeneracy had being fully lifted by the first order perturbation. Otherwise the RHS of (8.285) should be seen as a matrix in the space of degenerate eigenvectors, and one has to diagonalize it as before.

More generally, at order n we have to solve the Eq. (8.272). All quantities in the RHS are know from the recursive solution at previous orders, except for the n-th order correction a_n to the eigenvalue. We also assume that we have fixed the ambiguities in the previous orders by setting $|\Phi_k\rangle = 0$ for $k = 1, 2, \ldots, n - 1$.

Multiplying Eq. (8.272) by $\langle E_0|$ we get the equations

$$a_n = \langle E_0|H_I|\psi_{n-1}\rangle \tag{8.287}$$

and

$$|\psi_n\rangle = -\sum_{i,s} \frac{|i,s\rangle\langle i,s|H_I|\psi_{n-1}\rangle}{E_i - E_0} + \sum_{i,s} |i,s\rangle \sum_{k=1}^{n-1} a_{n-k} \frac{\langle i,s|\psi_k\rangle}{E_i - E_0}, \tag{8.288}$$

modulo, as always, the ambiguity $|\Phi_n\rangle \in \mathcal{H}_{E_0}$. In practice the perturbative procedure may be quite tedious, but in principle it may be carried on explicitly to arbitrarily high orders. Luckily for most practical purposes the first few orders suffice.

Example: The Zeeman Effect
Consider a hydrogen atom in a magnetic field of intensity B in the direction $\hat{z}$. The Hamiltonian is

$$
\begin{aligned}
H &= \frac{1}{2m}\left(p_x + \frac{By}{2}\right)^2 + \frac{1}{2m}\left(p_y - \frac{Bx}{2}\right)^2 + \frac{p_z^2}{2m} - \frac{\alpha}{r} - \frac{\hbar B}{2m}\sigma_3 = \\
&= \left[\frac{\mathbf{p}^2}{2m} - \frac{\alpha}{r}\right] - \frac{B}{2m}\left(L_z + \hbar\sigma_3\right) + O(B^2),
\end{aligned}
\tag{8.289}
$$

where the term proportional to σ_3 arises from the magnetic moment of the electron which is proportional to its spin $\mathbf{s} = \hbar\boldsymbol{\sigma}/2$. The Schrödinger equation with the Hamiltonian (8.289) cannot be solved in closed form. However, when the magnetic field B is very small, we may use perturbation theory to compute the energy levels with high precision. To the first order in perturbation theory, we may neglect the term quadratic in B in the Hamiltonian, and take as perturbation H_I the term linear in B. This example is a bit peculiar, since when we omit the quadratic term, the resulting Hamiltonian

$$H_0 + BH_I \equiv \frac{\mathbf{p}^2}{2m} - \frac{\alpha}{r} - \frac{B}{2m}\left(L_z + \hbar\sigma_3\right) \tag{8.290}$$

becomes exactly solvable[24] since we have three conserved quantities in involution H_0, L^2 and L_z, while σ_3 commutes with H. We write

---

[24] To get a more precise evaluation of the energy levels, one may take $H_0 + BH_I$ as the unperturbed Hamiltonian and the term quadratic in B as the perturbation.

$$|n, \ell, m, \pm\rangle \tag{8.291}$$

for the simultaneous eigenstates of these four operators (n being the principal quantum number). We know that the n-th unperturbed eigenvalue E_n has degeneracy $d_n = n^2$ given by the possible values of ℓ and m i.e.

$$0 \le \ell \le n - 1 \quad \text{and} \quad |m| \le \ell \quad \Rightarrow \quad d_n \equiv \sum_{\ell=0}^{n-1} (2\ell + 1) = n^2. \tag{8.292}$$

In the basis (8.291) the first order perturbation H_I is already diagonal, so the shifts in the energy level are simply

$$\Delta E_{n,\ell,m,\pm} = -\frac{B}{2m}(m \pm \hbar) + O(B^2). \tag{8.293}$$

Example: Stark Effect

Consider a hydrogen atom in a constant electric field of intensity $\mathcal{E}$ directed along $\hat{z}$ with Hamiltonian

$$H = H_0 + \mathcal{E}H_I \equiv \left[\frac{\mathbf{p}^2}{2m_e} - \frac{\alpha}{r} \right] + \mathcal{E}z \tag{8.294}$$

In Sect. 5.5.4 we saw that this system is separable having three commuting conserved quantities. However, when $\mathcal{E}$ is small, it may be more convenient to study it using perturbation theory: we take the hydrogen atom Hamiltonian H_0 as the unperturbed one and $H_I = z$. The full Hamiltonian commutes with L_z, and hence we may look for energy eigenvalues with fixed eigenvalue m of L_z. However now L^2 is *not* conserved.

To the first order in $\mathcal{E}$, the change in the energy levels is obtained by diagonalizing the perturbing operator z in the subspace of common eigenvectors of H_0 and L_z

$$\langle n, \ell', m|z|n, \ell, m\rangle \equiv Z(n, m)_{\ell',\ell}, \qquad |m| \le \ell, \ell' \le n - 1 \tag{8.295}$$

with n, m fixed. Since $|n, \ell, m\rangle$ belongs to the irreducible representation V_ℓ of $SO(3)$, and z to the vector representation V_1, we have

$$z|n, \ell, m\rangle \in V_1 \otimes V_\ell = V_{\ell+1} \oplus V_{\ell-1}, \tag{8.296}$$

and the matrix element (8.295) is non-zero only for $\ell' = \ell \pm 1$. Thus, to the first order in $\mathcal{E}$, there is no shift in the ground state energy ($n = 1$), while for the first excited state ($n = 2$) only the $Z(2, 0)$ matrix is non-zero

$$M(2, 0) = \begin{pmatrix} 0 & A \\ A^* & 0 \end{pmatrix}, \qquad A = \langle 2, 0, 0|z|2, 1, 0\rangle, \tag{8.297}$$

leading to a shift in energy

$$\Delta E_{n=2,m=0,\pm} = \pm \mathcal{E}|A|. \tag{8.298}$$

8.8 Perturbation Theory II: Time-Dependent Methods

The time independent perturbative methods may be used only for time independent systems with discrete energy levels. The time-dependent perturbation theory may be applied to all quantum system—time-dependent as well as time-independent—with discrete or continuous energy spectrum.

The purpose of the technique is to compute the time-evolution operator $U(t)$

$$i\hbar \frac{dU(t)}{dt} = H(t)U(t), \tag{8.299}$$

$$U(0) = \mathbf{1}, \tag{8.300}$$

$$U(t)U(t)^{\dagger} = U(t)^{\dagger}U(t) = \mathbf{1}, \tag{8.301}$$

for a Hamiltonian of the form

$$H(t) = H_0 + \epsilon\, H_I(t), \tag{8.302}$$

where we suppose to know already the *unperturbed* time-evolution

$$U_0(t) = e^{-iH_0 t/\hbar}, \tag{8.303}$$

and assume ϵ to be a small parameter. In facts, this problem was already discussed and solved in Chap. 2 under the names of *interaction picture* and **T**-*ordered exponentials*. Let us summarize the story. We write

$$U(t) = e^{-iH_0 t/\hbar}\, V(t). \tag{8.304}$$

The Schrödinger equation satisfied by the unitary operator $V(t)$ is

$$i\hbar\, \dot{V} = \epsilon\, \hat{H}_I(t)V, \tag{8.305}$$

where $\hat{H}_I$ is the *interaction picture operator*

$$\hat{H}_I(t) \stackrel{\text{def}}{=} e^{iH_0 t/\hbar}\, H_I(t)\, e^{-iH_0 t/\hbar}. \tag{8.306}$$

The solution of (8.305) is given by a **T**-ordered exponential

$$
V(t) = \mathbf{T}\exp\left(-\frac{i\epsilon}{\hbar}\int_0^t \hat{H}_I(s)\,ds\right) = 1 - \frac{i\epsilon}{\hbar}\int_0^t \hat{H}_I(s)\,ds -
$$
$$
- \frac{\epsilon^2}{2\hbar^2}\int_0^t \hat{H}_I(s_1)\,ds_1 \int_0^{s_1}\hat{H}_I(s_2)\,ds_2 + \cdots
$$
(8.307)

The expansion (8.307) of the **T**-order exponential may be seen as a series in powers of the small parameter ϵ: therefore it coincides with the perturbative expansion of the evolution operator. To order ϵ^k in the perturbative expansion, we keep only the first $k+1$ terms in the RHS of (8.307).

For many concrete problems the first perturbative correction suffices; to the leading order the time evolution operator takes the form

$$
U(t) = e^{-iH_0 t/\hbar}\left(1 - \frac{i\epsilon}{\hbar}\int_0^t \hat{H}_I(s)\,ds + O(\epsilon^2)\right).
$$
(8.308)

Transition Rates: *The Golden Rule*

Suppose that our system starts at $t = 0$ in an initial eigenstate $|E_i\rangle$ of the unperturbed Hamiltonian H_0 and we want to compute the *transition rate per unit time*, $R_{i\to f}$, to some final eigenstate $|E_f\rangle \neq |E_i\rangle$:

$$
R_{i\to f} \stackrel{\text{def}}{=} \lim_{T\to\infty}\frac{1}{T}\left|\langle E_f|U(T)|E_i\rangle\right|^2.
$$
(8.309)

We focus on the most important situation for the real-world applications: the interaction $H_I(t)$ oscillates in time with frequency ω i.e.

$$
\epsilon\, H_I(t) = 2\sin(\omega t)\, K
$$
(8.310)

for some constant Hermitian operator K. A typical instance is an electromagnetic wave of frequency ω which hits a hydrogen atom: we have a Hamiltonian of the form (8.294) where now the electric field $\mathcal{E}\propto\sin(\omega t)$. In this example the transition rate $R_{i\to f}$ gives the probability that the atom "jumps" from the energy level E_i to the energy level E_f (per unit time) by effect of the electromagnetic radiation hitting it.

To the first order in perturbation theory (for $E_f \neq E_i$)

$$
\langle E_f|U(T)|E_i\rangle = -\frac{i}{\hbar}\int_0^T e^{iE_f t/\hbar}\sin(\omega t)\langle E_f|K|E_i\rangle\,e^{-iE_i t/\hbar}\,dt =
$$
$$
= -\frac{1}{\hbar}\langle E_f|K|E_i\rangle\int_0^T\left(e^{i(E_f-E_i+\hbar\omega)t/\hbar} - e^{i(E_f-E_i-\hbar\omega)t/\hbar}\right)dt,
$$
(8.311)

and the transition rate (to the first order in perturbation theory) is

$$R_{i \to f} \equiv \frac{1}{\hbar^2} |\langle E_f | K | E_i \rangle|^2 \times$$

$$\times \lim_{T \to \infty} \frac{1}{T} \left(\int_0^T \left(e^{i(E_f - E_i + \hbar\omega)t/\hbar} - e^{i(E_f - E_i - \hbar\omega)t/\hbar} \right) dt \right)^2 . \tag{8.312}$$

The limit in the second line should be computed with some care. The integral inside the big parenthesis becomes as $T \to \infty$

$$2\pi\hbar \Big(\delta(E_f - E_i + \hbar\omega) - \delta(E_f - E_i - \hbar\omega) \Big) \tag{8.313}$$

so, naively, the RHS of (8.312) looks proportional to the square of the Dirac δ-function which is **not** defined. Let's evaluate carefully the square of the integral in (8.312) in the asymptotic limit $T \to \infty$:

$$\left(\int_0^T e^{i(E_f - E_i + \hbar\omega)t/\hbar} - e^{i(E_f - E_i - \hbar\omega)t/\hbar} \right) dt \right)^2 \approx$$

$$\approx 2\pi\hbar \Big(\delta(E_f - E_i + \hbar\omega) - \delta(E_f - E_i - \hbar\omega) \Big) \times$$

$$\times \int_0^T e^{i(E_f - E_i + \hbar\omega)t/\hbar} - e^{i(E_f - E_i - \hbar\omega)t/\hbar} \right) dt = \tag{8.314}$$

$$= 2\pi\hbar \, \delta(E_f - E_i + \hbar\omega) \int_0^T (1 - e^{-2i\omega t}) dt + (\omega \leftrightarrow -\omega) \approx$$

$$\approx 2\pi\hbar \, T \Big(\delta(E_f - E_i + \hbar\omega) + \delta(E_f - E_i - \hbar\omega) \Big).$$

The overall factor T cancels with the factor $1/T$ in Eq. (8.312), giving a finite limit as $T \to \infty$. Finally we get

$$R_{i \to f} = \frac{2\pi}{\hbar} |\langle E_f | K | E_i \rangle|^2 \Big(\delta(E_f - E_i + \hbar\omega) + \delta(E_f - E_i - \hbar\omega) \Big) \tag{8.315}$$

a fundamental result known as the *Fermi golden rule.*

Physical Interpretation   Let us pause a little while to explain the physical meaning of the golden rule. We return to the previous example where we have a hydrogen atom in the initial energy eigenstate $|E_i\rangle$. We hit it with an electromagnetic wave of frequency ω and amplitude $|\mathcal{E}|$. If the amplitude is sufficiently small, we can study the process in perturbation theory. The atom may absorb a photon from the incident wave and jump to a higher energy level. Since the photon has energy $\omega\hbar$, by conservation of energy the atom jumps to the energy level

$$E_f = E_i + \omega\hbar. \tag{8.316}$$

This process corresponds to the second delta-function in Eq. (8.315) which, as we see, just enforces the conservation of energy. The first delta-function then represents the time-reversed process: the atom is initially in an exited energy level E_i and decays to a lower energy level E_f by emitting a photon of energy $\omega = (E_i - E_f)/\hbar$. Energy is conserved in both processes. The amplitude for the emission/absorption of the photon is given by the square of the matrix element of the perturbing operator K, see Eq. (8.315). The two processes have the same probability by time-reversal symmetry.

Of course, there are also processes involving the absorption/emission of two or more photons. It is obvious from the above derivation that these multiple-photon processes are described by higher order terms in the perturbative expansion, an n-photon process first appears at the n-th order.

Selection Rules

To introduce the notion of *selection rule,* and to clarify its physical meaning and relevance, we focus on the previous example where we have a hydrogen atom in presence of an electric field which oscillates at the frequency ω with Hamiltonian

$$H(t) = \left(\frac{\mathbf{p}^2}{2m_e} - \frac{\alpha}{r} \right) + \mathcal{E}\sin(\omega t)\, z \tag{8.317}$$

The operator in the large parenthesis is the unperturbed Hamiltonian H_0, and we assume $\mathcal{E}$ to be parametrically small, so a perturbative treatment is justified. We suppose that the initial state is a simultaneous eigenstate of the three commuting observables H_0, L^2, and L_z, with respective quantum numbers n, ℓ, and m (we neglect the spin, since it is going to play no role). In units where $\hbar = m_e = 1$ the unperturbed energy levels are (cf. Chap. 5)

$$E_n = -\frac{\alpha^2}{2}\frac{1}{n^2}, \qquad n = k + \ell + 1, \quad k, \ell = 0, 1, 2, \dots. \tag{8.318}$$

The initial state is $|n, \ell, m\rangle$ and the final one is $|n', \ell', m'\rangle$. A *selection rule* is a criterion to decide whether a transition

$$|n, \ell, m\rangle \longrightarrow |n', \ell', m'\rangle, \tag{8.319}$$

which is consistent with the conservation of energy

$$-\frac{\alpha^2}{2}\frac{1}{(n')^2} + \frac{\alpha^2}{2}\frac{1}{n^2} = \omega, \tag{8.320}$$

is *allowed* or *forbidden,* i.e. whether its probability is *non-zero* (at first order in perturbation theory). From the Fermi golden rule we see that a transition (8.319) is forbidden (at the leading order) iff

$$\langle n', \ell', m'|z\, |n, \ell, m\rangle = 0 \tag{8.321}$$

We already studied this equation for the (static) Stark effect in Eqs. (8.295), (8.296). Borrowing from that discussion we conclude

Fact 8.2 (Selection Rules for the Hydrogen Atom) *The transition (8.319) in the hydrogen atom is forbidden (to first order) unless*

$$m' = m, \qquad \ell' = \ell \pm 1. \tag{8.322}$$

Similar rules may be established for other physical processes. They reflect how the perturbation H_I transforms under the symmetries of the unperturbed Hamiltonian.

Diabatic Limit

An important special class of time-dependent perturbations are the *adiabatic processes:*[25] in these systems the perturbation $H_I(t)$ depends on time, but its variation is very very slow,

$$\frac{\mathrm{d}}{\mathrm{d}t} H_I(t) \lll \omega H_I, \tag{8.323}$$

where ω is a typical frequency of the unperturbed system. Adiabatic processes are so important that we dedicate a full section to them, see Sect. 8.10. For the sake of comparison, here we describe the opposite situation: the *diabatic* limit. While in the adiabatic limit the change of the Hamiltonian $H(t)$ with time is extremely slow, in the diabatic limit the variation is quite rapid: in the strict diabatic limit the variation happens suddenly at the sharp time $t = 0$, that is, the time-dependent Hamiltonian has the form

$$H(t) = \begin{cases} H_0 & t < 0 \\ H_1 & t > 0. \end{cases} \tag{8.324}$$

The time evolution kernel is (for $t_1 < t_2$)

$$U(t_2, t_1) = U_1(t_2, t_1)\, U_0(t_2, t_1), \tag{8.325}$$

where

$$U_1(t_2, t_1) = \exp\left(\frac{-\mathrm{i} H_1}{\hbar} \left(\max\{0, t_2\} - \max\{0, t_1\} \right) \right) \tag{8.326}$$

$$U_0(t_2, t_1) = \exp\left(\frac{-\mathrm{i} H_0}{\hbar} \left(\min\{0, t_2\} - \min\{0, t_1\} \right) \right). \tag{8.327}$$

---

[25] For adiabatic processes in classical mechanics see [8]; for adiabatic processes in thermal physics, see [18].

8.9 Perturbation Theory III: Path Integral Methods

Modern perturbation theory starts from the path integral formulation of Quantum Mechanics. For brevity we focus on the simplest situation: a one-dimensional system with Euclidean Lagrangian of the form

$$L_E(x, \dot{x}) = \frac{1}{2}(\dot{x}^2 + \omega^2 x^2) + \lambda\, W(x) \equiv L_0(x, \dot{x}) + \lambda\, W(x) \tag{8.328}$$

where $\lambda > 0$ is a small parameter and $W(x)$ is some function such that the potential

$$V(x) = \frac{\omega^2}{2}x^2 + \lambda\, W(x) \tag{8.329}$$

is bounded below for λ small enough. The quadratic expression $L_0(x, \dot{x})$ is the *unperturbed* (or *free*) Lagrangian. A prototypical important example is

$$W(x) = \frac{x^4}{4!}, \tag{8.330}$$

BOX: Why the Perturbative Series Is (Usually) Divergent: The Dyson Argument

In most situations the perturbative power series is only asymptotic, and this should be seen as good news because *were it convergent* its sum will describe inconsistent physics. Consider the basic example (8.328) where $W(x)$ is a polynomial of degree larger than 2. Suppose that the power series has a positive convergence radius $\rho > 0$. Then the sum of the series defines the correlation functions—hence the theory—also for λ negative as long as $|\lambda| < \rho$. But the Hamiltonian is not bounded below for λ negative: when λ is small and negative, the ground state cannot exist, and the ground-state correlation functions make no sense. Thus assuming $\rho > 0$ we get a contradiction. The same holds for the finite-dimensional analogue of path integrals. Consider the ordinary integral

$$\int_{-\infty}^{+\infty} \frac{dx}{\sqrt{2\pi}} e^{-x^2/2 - \lambda x^4} \overset{!}{=} \sum_{n \geq 0} \frac{(-\lambda)^n}{n!} \int\!\!\int_{-\infty}^{+\infty} \frac{dx}{\sqrt{2\pi}} e^{-x^2/2} x^{4n} = \sum_{n \geq 0} \frac{(-\lambda)^n}{n!}(4n - 1)!!$$

While the integral in the LHS is convergent for $\lambda > 0$, the power series in the RHS is badly divergent as it should be, since the integral diverges for all negative λ

where the factor 1/4! (which one may absorb in the definition of λ, if so wishes) is chosen for later convenience. In this example the potential $V(x)$ is bounded below iff $\lambda \geq 0$: when $\lambda < 0$ there is *no* ground state. This means that the ground-state correlation functions cannot be analytic in λ at $\lambda = 0$: see the discussion in the off-text **BOX** on page 611.

From the path integral perspective, perturbation theory aims to provide a controlled approximation scheme to compute the imaginary-time s-point correlation functions

$$\langle x(t_1) \cdots x(t_s) \rangle = \frac{1}{Z} \int [\mathrm{d}x]\, \mathrm{e}^{-\int \mathrm{d}t\, L_E(x,\dot{x})}\, x(t_1) \cdots x(t_s) \tag{8.331}$$

or, equivalently, their generating functional

$$\begin{aligned} Z[J] &= \mathcal{N} \int [\mathrm{d}x]\, \mathrm{e}^{-\int \mathrm{d}t \left(L_E(x,\dot{x}) + J(t)x(t) \right)} \equiv \\[2mm] &\equiv \mathcal{N} \int [\mathrm{d}x]\, \mathrm{e}^{-\int \mathrm{d}t \left(L_0(x,\dot{x}) + \lambda\, W(x) + J(t)x(t) \right)}. \end{aligned} \tag{8.332}$$

When $\lambda = 0$, the path integral is Gaussian, and can be computed explicitly by the techniques of Chap. 6. When $W(x)$ and λ are generic, computing $Z[J]$ exactly is hard or even impossible. However when $\lambda \ll 1$ is small, it looks reasonable to treat the term $\lambda\, W(x)$ as a small "perturbation" of the unperturbed (Euclidean) Lagrangian $L_0(x, \dot{x})$, and write the correlation functions (8.331) in the form of an asymptotic power series in λ. The expansion of the correlation functions in a *naive* power series in λ is formally obvious

$$\langle x(t_1) \cdots x(t_s) \rangle =$$

$$= \frac{1}{Z} \sum_{n=0}^{\infty} \frac{(-\lambda)^n}{n!} \int [\mathrm{d}x]\, \mathrm{e}^{-\int \mathrm{d}t\, L_E} \left(\int \mathrm{d}s\, W(x(s)) \right)^n x(t_1) \cdots x(t_s) =$$

$$= \frac{1}{Z} \sum_{n=0}^{\infty} \frac{(-\lambda)^n}{n!} \int \mathrm{d}s_1 \cdots \int \mathrm{d}s_n \Big\langle W(s_1) \cdots W(s_n)\, x(t_1) \cdots x(t_s) \Big\rangle_0 \tag{8.333}$$

where $\langle \cdots \rangle_0$ stands for the correlation functions computed in the unperturbed theory in absence of sources (also called the *free* correlations), $W(s)$ is a short-hand for $W(x(s))$, and

$$Z = \sum_{n=0}^{\infty} \frac{(-\lambda)^n}{n!} \int \mathrm{d}s_1 \cdots \int \mathrm{d}s_n \Big\langle W(s_1) \cdots W(s_n) \Big\rangle_0 \tag{8.334}$$

is the perturbative path integral without insertions, normalized to be 1 in the unperturbed theory, i.e.

$$Z_0 \equiv Z(\lambda)\big|_{\lambda=0} = 1. \tag{8.335}$$

Graphical Form of the Wick Theorem

The correlators in the RHS of (8.333) and (8.334) are the ones of the Gaussian model with Euclidean Lagrangian L_0: by the Wick theorem (Theorem 6.7) they are written in terms of the free propagator ($\equiv$ the Green's function of the differential operator in the quadratic Lagrangian L_0)

$$G(t', t) = \langle t' | \left(-\tfrac{d^2}{dt^2} + \omega^2 \right)^{-1} | t \rangle = \int \frac{dk}{2\pi} \frac{e^{ik(t'-t)}}{k^2 + \omega^2} = \frac{e^{-\omega|t-t'|}}{2\omega}. \tag{8.336}$$

It is usually convenient to work in the "energy representation" i.e. with the Fourier transform $G(k)$ of the Green function

$$G(k) = \frac{1}{k^2 + \omega^2}, \tag{8.337}$$

called the *propagator*. Explicitly: the unperturbed correlation function

$$\langle x(t_1) \cdots x(t_s) \rangle_0 \tag{8.338}$$

is zero when s is odd, while when $s = 2m$ is even, it is given by

$$\sum_{p \in \mathfrak{P}} \prod_{\{i,j\} \in p} G(t_i, t_j) \tag{8.339}$$

where $\mathfrak{P}$ is the set of the $(2m - 1)!!$ pairings of $2m$ elements, i.e. the list of ways of writing the set $\{1, 2, \ldots, 2m\}$ as the disjoint union of m sets $\{i, j\}$ of two elements (cf. Theorem 6.7).

The Wick formula for the free correlation function (8.339) has a convenient graphical representation. Each term p in the sum (8.339) is represented by a graph constructed as follows: one draws $2m$ nodes labelled, respectively, by $1, 2, \ldots, 2m$ and connects by a link the nodes whose labels belong to the same pair in p. Figure 8.10 shows the $3 \cdot 5 = 15$ graphs which represent the free 6-time function. The 3 graphs representing the 4-time function

$$\langle x(t_1) \, x(t_2) \, x(t_3) \, x(t_4) \rangle_0 \tag{8.340}$$

Fig. 8.10 The 15 diagrams which yield the Wick expansion of the free 6-times correlation $\langle x(t_1)x(t_2)\cdots x(t_6)\rangle_0$. The nodes represent the 6 times $t_1,\ldots,t_6$, and an edge connecting two points t_i, t_j represents a Green's function factor $G(t_i,t_j)$ in (8.339)

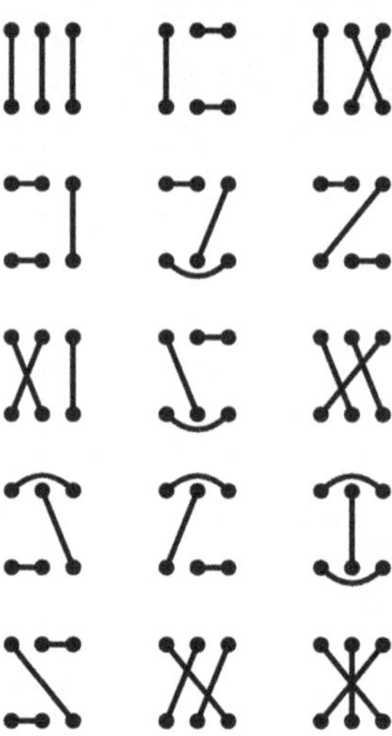

are obtained from the ones in the first row of the **Figure** by deleting the leftmost vertical link. For mere notational convenience we fix an orientation for the links; if l is a link, we write $h(l)$ and $t(l)$ for, respectively, the head and the tail node of l. All the expressions are independent of the chosen orientation which is a mere book-keeping device. Then

$$\left\langle x(t_1)\cdots x(t_s)\right\rangle_0 = \sum_{\Gamma\in\mathfrak{G}}\prod_{l\in\Gamma}\int\frac{\mathrm{d}k_l}{2\pi}\frac{e^{ik_l(t_{h(l)}-t_{t(l)})}}{k_l^2+\omega^2} \tag{8.341}$$

where $\mathfrak{G}$ is the set of graphs[26] which are in one-to-one correspondence with the Wick pairings $\mathfrak{P}$. The product is over the links l in each graph $\Gamma\subset\mathfrak{G}$. Each link l is the graphical representation of a propagator factor

$$(k_l^2+\omega^2)^{-1}, \tag{8.342}$$

and we say that k_l is the *energy flowing along* l (in the direction from tail to head).

---

[26] We stress that $\mathfrak{G}$ is the set of *unoriented* graphs; we then choose (arbitrarily) a conventional orientation for each of them, but the sum is over the underlying unoriented graphs.

8.9.1 Feynman Graphs

The graphical version of the Wick theorem leads to a graphical representation of the several terms in the perturbative expansion of the correlation functions known as *Feynman graphs* (or *diagrams*).

As a basic example we consider the anharmonic oscillator with perturbation

$$W(x(t)) = \frac{x(t)^4}{4!}. \tag{8.343}$$

To compute the several terms in its perturbative expansion (8.333), we have preliminarily to determine the free correlation functions with extra insertions of the operator $x(t)^4$. The insertion of this operator in a free correlation function is equivalent to inserting four x's at the same time t. In the Wick graphs Γ we identify the four nodes labelled by the same t, which amounts to gluing them together in a single node of valency 4, called an interaction *vertex*. A graph Γ which represents a term in the Wick expansion of the free correlator

$$\left\langle W(s_1) \cdots W(s_n)\, x(t_1) \cdots x(t_s) \right\rangle_0 \tag{8.344}$$

then contains n such 4-vertices. A node of Γ which is not a vertex is called *external*. A graph for (8.344) contains s external nodes in correspondence with the original insertions $x(t_1), \ldots, x(t_s)$. An external node is connected to the rest of the graph by a single link that we call an *external leg*. A *link* which is not external is called an *internal line*. For instance, identifying the 4 nodes of the 3 graphs representing the free 4-point function, we get 3 copies of the connected graph

$$\equiv \Gamma_{1,\mathrm{vac}} \tag{8.345}$$

with one vertex (the purple node), 2 internal lines, no external legs, and 2 *loops*, that is, bounded disks in the plane cut out by the links.

As a further example, if we identify the 4 leftmost nodes in the 15 graphs representing the free 6-time function, Fig. 8.10, we get the two graphs

$$\tag{8.346}$$

the first one with multiplicity 3 and the second one with multiplicity 12. The second graph is connected, while the first one has two disconnected components.

The number L of loops in a *connected* graph is related to the numbers V of its vertices, P of its links, and E of external nodes, by the Euler characteristic formula

$$L = 1 + P - V - E, \tag{8.347}$$

as the reader may check in the above examples.

Via this graphical construction, all terms in the RHS of Eqs. (8.333), (8.334) can be written as sums of contributions from graphs with $V = n$ vertices of valency 4 and $E = s$ external nodes. These graphs are the *Feynman diagrams*.

Consider the contribution $A(\Gamma)$ of a particular graph Γ to the function

$$\frac{1}{4!} \int dt \, \langle x(t)^4 \cdots \rangle_0, \tag{8.348}$$

where $\cdots$ stands for arbitrary operator insertions. We write $\star \in \Gamma$ for the vertex associated with the $x(t)^4$ insertion that we singled out. $A(\Gamma)$ has the schematic form

$$A(\Gamma) = \int dt \, \exp\left[it\left(\sum_{h(j)=\star} k_j - \sum_{t(j)=\star} k_j \right) \right] (\cdots) \prod_{l \in \Gamma} \frac{dk_l}{2\pi}, \tag{8.349}$$

where $(\cdots)$ is a t-independent expression. Integrating in t we get

$$\int \prod_{l \in \Gamma} \frac{dk_l}{2\pi} \, 2\pi \, \delta\left(\sum_{h(j)=\star} k_j - \sum_{t(j)=\star} k_j \right)(\cdots). \tag{8.350}$$

The δ-function ensures that the total energy flowing in the vertex $\star$, $\sum_{h(j)=\star} k_j$ is equal to the total energy flowing out of it, $\sum_{t(j)=\star} k_j$, i.e. *energy is conserved at each interaction vertex*. Next consider a graph Γ which contributes to the Fourier transformed correlation

$$\int dt_s \, e^{it_s k_s} \langle \cdots x(t_s) \rangle_0. \tag{8.351}$$

The s-th node is connected to the graph by a single link (an *external leg*) l_s. The Fourier transform (8.351) fixes the energy flowing in this external leg to be k_s. We conclude that not all energies $\{k_l\}_{l \in \Gamma}$ which propagate in the P links l of a graph Γ are independent. Let Γ be a connected graph with V vertices and E external nodes which contributes to

$$\int d^V s \int d^E t \, \exp\left[-i \sum_{a=1}^{E} t_a k_a \right] \langle x(s_1)^4 \ldots x(s_V)^4 x(t_1) \ldots x(t_E) \rangle_0 \tag{8.352}$$

The k_l's are constrained by the conservation law of total energy at each one of the V vertices, while the energies k_a of the E external legs are fixed. The full graph conserves energy, hence the total energy flowing into it must be zero: therefore $A(\Gamma)$ must be proportional to

$$2\pi \, \delta\left(\sum_a k_a\right) \tag{8.353}$$

and in a non-zero connected amplitude only $E-1$ external energies are independent. In conclusion

$$\#\left(\begin{array}{c}\textbf{independent energies}\\ \textbf{circulating in } \Gamma\end{array}\right) = P - V - (E-1) = L \tag{8.354}$$

i.e. *we have one independent energy k_l per loop $l \in \Gamma$*. To get the contribution from Γ to the quantum amplitude we have to integrate over these L undetermined energies.

Feynman Rules

All perturbative amplitudes are a sum of the contributions of a number of Feynman graphs. There are two issues to address:

(1) lay down the rules to compute the contribution $A(\Gamma)$ of each graph Γ;

(2) understand the combinatorics of how the several graphs contribute to the physical quantities at each order in λ, i.e. the multiplicity of the various graphs.

We first present the rules for the value of the single graph (the *Feynman rules*) and then address the combinatorial issue (see rule **FR6**).

We consider a slightly more general model: we assume the interaction $\lambda \, W(x)$ to be a sum of couplings of the form

$$\lambda \, g_m \frac{x^m}{m!}, \quad \text{that is,} \quad W(x) = \lambda \sum_m g_m \frac{x^m}{m!} \tag{8.355}$$

for various $m \in \mathbb{N}$. The g_m's are real parameters (the "couplings"). E.g. in the example (8.330) we had a single non-zero coupling with $m = 4$. Now the graphs contain different kinds of vertices of valency m one per each non-zero term in $W(x)$. We focus on a graph Γ with V vertices and E external nodes, and work in the energy space, i.e. compute directly the Fourier transforms of the correlation functions. The contribution from Γ to the amplitude is then a function $A(\Gamma; k_1, \ldots, k_E)$ of the energies entering[27] at the external nodes. The *Feynman rules* are:

---

[27] An energy k_a *exiting* from the a-th node is the same as a energy $-k_a$ entering into it. The notion of entering/exiting is interchanged by a flip of the conventional orientation of the external leg. E.g. in the rightmost edge of the graph (8.357) k is the energy *exiting* from the rightmost external node.

FR1 $A(\Gamma; k_1, \ldots, k_E)$ has an overall factor

$$2\pi \, \delta\left(\sum_{a=1}^{E} k_a\right),\tag{8.356}$$

where k_a is the energy entering from the a-th external node. This factor expresses overall energy conservation;

FR2   the energy k_p flowing along the p-th link of Γ is a linear combination of the E external energies k_a and of the L energies k_l flowing in the loops; the linear combinations k_p are determined by the condition that energy is conserved at all vertices. For instance, in the following 3-valent graph, with all edges oriented from left to right (including the external ones!), the valid energy assignments are as indicated

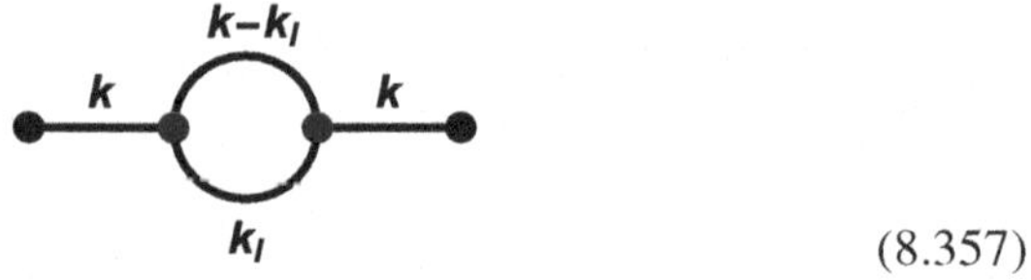

$$\tag{8.357}$$

where k is the external node energy and k_l is the energy flowing in the loop;

FR3   we have to integrate over the loop energies k_l with the measure

$$\int \left(\text{integrand}\right) \prod_{l \in L} \frac{\mathrm{d}k_l}{2\pi}\tag{8.358}$$

FR4   each vertex of Γ with valency m contributes a factor λg_m to the integrand. In particular we have an overall factor λ^V, and a graph with V vertices contributes to order V in perturbation theory;

FR5   each link $p \in \Gamma$ contributes a propagator factor to the integrand of (8.358). The integrand then reads

$$\left(\text{integrand}\right) = \lambda^V \prod_m (g_m)^{V_m} \prod_{p \in P} \frac{1}{k_p^2 + \omega^2},\tag{8.359}$$

where k_p are the linear combinations of the k_l's and k_a's described in rule **FR2**, and V_m is the number of vertices of valency m;

FR6   the contribution of each graph Γ is multiplied by a combinatorial factor which arises from the number of ways the particular graph Γ is produced by the Wick algorithm together with the explicit factors $(1/m!)^{V_m}$ from the vertices (8.355), and the overall $1/n! \equiv 1/V!$ in the order n term from the expansion of the exponential in Eq. (8.333). Keeping all these factors into account, one can show[28] that the resulting combinatoric factor is

$$\text{combinatoric factor of } \Gamma = \frac{1}{|S(\Gamma)|} \tag{8.360}$$

where $S(\Gamma)$ is the automorphism group of the graph Γ, that is, $|S(\Gamma)|$ is the number of permutations of the edges and vertices of Γ which fix the external nodes and send the graph to itself.

$\hbar$ as the Loop Counting Parameter If we do not set $\hbar = 1$, we have the factor $1/\hbar$ in front of the full action. The derivative with respect to the sources now insert in the correlation the dual operator with the normalization $\mathcal{O}/\hbar$. In this normalization each vertex (including the valency-1 vertices associated to the sources) produces a factor $1/\hbar$, while each propagator introduces a factor $\hbar$. Hence the amplitude of a connected graph with $V + E$ nodes and P edges scales as

$$\hbar^{P-V-E} = \hbar^{L-1}. \tag{8.361}$$

Fact 8.3 *The L-loop contribution to the* connected *amplitudes scales as* $\hbar^{L-1}$.

This explains why in Sect. 8.2 we called the $O(\hbar^0)$ term in the semiclassical expansion of the logarithm of the path integral, namely $-\frac{1}{2} \log \boldsymbol{D}$, the "one-loop" term. Indeed, from the graphical viewpoint this contribution is the sum of one-loop connected diagrams with no external legs.

8.9.2 The Vacuum Amplitude Z

To keep the formulae readable and short, we return to the basic example of the anharmonic oscillator with $W(x) = x^4/4!$. Our conclusions below are valid for a general perturbation $W(x)$ with the obvious modifications.

The normalization factor Z, defined in Eq. (8.334)—called the *vacuum amplitude*—is the sum of the contributions from Feynman graphs without external legs (the *vacuum diagrams*):

$$Z = 1 - \frac{\lambda}{4!} \int_0^\beta dt \, \langle x(t)^4 \rangle_0 + O(\lambda^2) =$$

$$= 1 - \frac{3\lambda}{4!} \int_0^\beta dt \, \langle x(t)x(t_1) \rangle_0 \langle x(t_2)x(t_3) \rangle_0 \prod_{i=1}^{3} \delta(t - t_i) dt_i + O(\lambda^2) = \tag{8.362}$$

$$= 1 - \frac{\lambda}{8}\beta \left(\int \frac{dk}{2\pi} \frac{1}{k^2 + \omega^2} \right)^2 + O(\lambda^2) = 1 - \frac{\lambda\beta}{32\omega^2} + O(\lambda^2),$$

---

[28] Easy proof left to the reader.

where we took the time interval to have the finite length β in order to regularize the energy conservation factor $2\pi\,\delta(0)$ which is ill-defined in absence of external legs

$$2\pi\,\delta(0) \xrightarrow{\text{reg.}} \int_0^\beta dt\, e^{it(0)} = \beta. \tag{8.363}$$

This regularization factor will drop out of all observable quantities.

The term of order λ in (8.362) is the contribution from the graph (8.345) (counted with the appropriate multiplicity). For easy of presentation, from now on we use the graph Γ as a symbol for the corresponding amplitude $A(\Gamma)$, i.e. we do not distinguish a graph Γ and the expression $A(\Gamma)$ obtained by applying the Feynman rules to it.

At the next order, $n = 2$, the graphs have two insertions of the interaction $x^4/4!$: there are one disconnected graph and two connected ones. The connected ones are

$$\tag{8.364}$$

while the disconnected $n = 2$ graph consists of two copies of (8.345). The contribution to the vacuum amplitude Z, Eq. (8.334), from this disconnected graph is simply

$$\frac{1}{2!}\left(\;\bigcirc\!\!\bigcirc\;\right)^2 \tag{8.365}$$

where the factor $1/2!$ is the factor $1/n!$ from the expansion of the exponential for $n = 2$ (cf. Eq. (8.334)). More generally, at order n we have *inter alia* a graph which consists of n disconnected copies of (8.345) weighted by the combinatorial factor $1/n!$, hence the sum of all graphs made of disconnected copies of (8.345) is simply

$$\sum_{n=0}^{\infty} \frac{1}{n!}\left(\;\bigcirc\!\!\bigcirc\;\right)^n \equiv \exp\left(\;\bigcirc\!\!\bigcirc\;\right) \tag{8.366}$$

We can go on adding the contributions from higher order graphs. Repeating the argument, is easy to see that

Fact 8.4 *The vacuum amplitude Z—which is the sum of all the vacuum diagrams— is just equal to the* exponential of the sum of all **connected** *vacuum diagrams*

$$Z = \exp\left(\;\bigcirc\!\!\bigcirc\; + \;\bigcirc\!\!\bigcirc\!\!\bigcirc\; + \;\ominus\; + \cdots\right) \tag{8.367}$$

namely, we have the combinatorial identity

$$\exp\left[\begin{array}{c}\textbf{sum of all connected}\\\textbf{vacuum graphs}\end{array}\right] = \left[\begin{array}{c}\textbf{sum of all vacuum graphs with}\\\textbf{proper combinatorial factors}\end{array}\right] = Z \qquad (8.368)$$

For a different viewpoint about this **Fact**, more directly related to the path integral arguments in Chap. 6, see Sect. 8.9.5 below.

8.9.3   Correlation Functions

We return to the perturbative expansion of the correlation functions of the anharmonic oscillator, Eq. (8.333) with $W(x) = x^4/4!$. The expansion contains the overall normalization factor Z^{-1} and has the schematic structure:

$$2m\text{-time correlation function} = Z^{-1}\sum_n A_n \qquad (8.369)$$

$$\text{where } A_n = \left(\begin{array}{c}\textbf{sum of graphs with } n \textbf{ vertices and } 2m \textbf{ external legs}\\\textbf{with their proper combinatorial factors}\end{array}\right) \qquad (8.370)$$

As an example, we consider the 2-time function, $m = 1$. At order zero in λ, Z is just 1, while only one graph contributes to Γ_0, the one with two external nodes connected by a link: this is the 2-time function of the unperturbed theory at $\lambda = 0$. At first order in λ, the amplitude A_1 is the sum of two graphs, $\Gamma_{1,\text{dis}}$ disconnected, and $\Gamma_{1,\text{con}}$ connected:

$$(8.371)$$

The disconnected amplitude $\Gamma_{1,\text{dis}}$ is equal to the zero order amplitude Γ_0 times the first order contribution $\Gamma_{1,\text{vac}}$ to Z, Eq. (8.345):

$$\Gamma_{1,\text{dis}} = \Gamma_{1,\text{vac}} \times \Gamma_0. \qquad (8.372)$$

Thus

$$2\text{-time function} = Z^{-1}\left(\Gamma_0 + \Gamma_{1,\text{vac}} \times \Gamma_0 + \Gamma_{1,\text{con}} + \cdots\right)$$

$$= \exp\left(-\Gamma_{1,\text{vac}} + \cdots\right)\left(\Gamma_0 + \Gamma_{1,\text{vac}} \times \Gamma_0 + \Gamma_{1,\text{con}} + \cdots\right) \qquad (8.373)$$

$$= \Gamma_0 + \Gamma_{1,\text{con}} + \cdots$$

where the disconnected graph in (8.371) cancels against the normalization factor Z^{-1}. Repeating the argument order by order in the graphical expansion, we see that the only effect of the factor Z^{-1} is to cancel all disconnected graphs which contain vacuum graph components. We conclude:

Fact 8.5 *The correlation functions are given by the sum over the diagrams neglecting the ones which contain vacuum connected components since they are cancelled by the Z^{-1} factor:*

$$\frac{1}{Z}\left[\begin{array}{c}\text{sum of all graphs}\\\text{with } E \text{ external legs}\end{array}\right] = \left[\begin{array}{c}\text{sum of all graphs with } E \text{ external legs}\\\text{with no vacuum connected component}\end{array}\right] \tag{8.374}$$

8.9.4   Connected Correlation Functions

As emphasized in Chap. 6 (see also [18]), instead of the ordinary correlation functions it is often more convenient and illuminating to work with the *connected* correlations whose combinatoric definition is given in [18]. The path integral arguments of Chap. 6 yield the following generalization of Eq. (8.368):

Fact 8.6 *The connected correlation is given by the sum over the connected graphs.*

$$\left[\begin{array}{c}E-\text{time connected}\\\text{correlation functions}\end{array}\right] = \left[\begin{array}{c}\text{sum of connected graphs}\\\text{with } E \text{ external legs}\end{array}\right] \tag{8.375}$$

This graphic characterization explains why these correlations are called "connected".

8.9.5   Generating Functionals

The generating functional of the correlation functions $Z[J]$ (resp. the generating functional $W[J] = \log Z[J]$ of the *connected* correlations) may be seen as the vacuum amplitude (resp. the vacuum free energy) in presence of a valency 1 interaction vertex

$$\underset{}{\overset{k}{\rule{3cm}{0.4pt}\!\bullet}} \equiv \hat{J}(k) \tag{8.376}$$

where $\hat{J}(k)$ is the Fourier transform of the source $J(t)$. In particular this relates the rule (8.368) to the discussion of the source insertions in Chap. 6.

8.10 Adiabatic Theorem and Approximation

Historical Background As reviewed in §. 10.3 of [8], in 1911 the founding Fathers of Modern Physics (Einstein, Planck, Poincaré, Lorentz, …) met at the first Solvay conference to discuss the newly formulated "quantum hypothesis". They proposed that, in bound motions of classically integrable systems, the action variables $\{I_i\}$ must be quantized in integer multiples of $\hbar$. Their idea later became the Bohr-Sommerfeld semiclassical quantization rule, which follows from the $\hbar \to 0$ limit of the Schrödinger equation, as we proved in Sect. 8.5. Their argument in 1911 was quite different: indeed the Schrödinger equation was formulated only 15 years later. The Fathers used the deep physical insight that a (conserved) quantized quantity cannot jump in an adiabatic process, so the classical quantities which get discretized at the quantum level must be *adiabatic invariants*. The action variables $\{I_i\}$ are the basic adiabatic invariants of classical physics. The $\{I_i\}$'s generate the ring $\mathfrak{Ab}$ of classical adiabatic invariants, and they are singled out between all systems of generators of this ring by a natural integral structure: their differentials $\{dI_i\}$ are unique only up to linear redefinitions with *integral* coefficients [8]. The proposed quantization is then consistent with both the physical and arithmetical properties of the $\{I_i\}$'s.

We ask: *were Einstein, Planck, Poincaré, Lorentz, … right in claiming that the eigenvalues of the discrete spectrum of (say) the Hamiltonian cannot jump in an adiabatic process?*

That their physical insight was perfectly correct is the content of the *adiabatic theorem* of Quantum Mechanics.

Math Formalization

Let $\{H(s)\}_{s\in\mathbb{R}}$ be a smooth family of Hamiltonians acting on the Hilbert space $\mathcal{H}$. The family smoothly interpolates between two constant Hamiltonians $H_0,\ H_1$

$$H(s) = \begin{cases} H_0 & s < 0 \\ H_1 & s > 1, \end{cases} \tag{8.377}$$

called, respectively, the *initial* and *final* Hamiltonian. We consider the time-dependent Hamiltonian $H(t/T)$ which at early times $t < 0$ (resp. late times $t > T$) reduces to the time-independent Hamiltonian H_0 (resp. H_1). The transition between the early and late time dynamics takes a time T. The *adiabatic limit* is $T \to \infty$. In this limit the change of the Hamiltonian with time is so slow that it takes an infinite time to go from H_0 to H_1.

We require the Hamiltonians $\{H(s)\}$ to have a discrete spectrum $\{E_n(s)\}$ with eigenvectors

$$H(s)|n, a; s\rangle = E_n(s)|n, a; s\rangle, \quad \langle m, a; s|n, b; s\rangle = \delta_{m,n}\,\delta_{a,b}, \quad s \in \mathbb{R} \tag{8.378}$$

separated by a gap from the continuous spectrum (if present at all). The dimensions of the energy eigenspaces are finite, but otherwise arbitrary. In most situations the discrete energy levels are non-degenerate for generic s, but there are important special examples where generic degeneration is unavoidable.[29] To avoid cluttering, in the rest of this section we omit writing the labels a, b which distinguish the states with the same energy level; in all formulae we leave implicit the sum over these omitted labels. In addition we require that the energy eigenvalues $\{E_n(s)\}$ *do not cross* each other, i.e.

$$E_n(s) - E_m(s) \neq 0 \quad \text{for } m \neq n \text{ and all } s \in \mathbb{R}, \tag{8.379}$$

and also that the gap with the continuous spectrum does not close. The dimensions of the discrete level eigenspaces are then independent of s. All these conditions are automatically satisfied by the bound states of one dimensional systems (see Chap. 3), and hence also by the discrete-spectrum Hamiltonians $H(s)$ whose Schrödinger equations are separable in a fixed system of coordinates for all s. In technical terms we subsume all the required conditions in the assumption that

$$\frac{\langle m; s| \frac{\partial H}{\partial s} |n; s\rangle}{E_m(s) - E_n(s)}, \qquad n \neq m \tag{8.380}$$

is bounded for all $s \in \mathbb{R}$ and $m \neq n$. We stress that this is a mild regularity condition equivalent to the statement that $\{|n; s\rangle\}$ is a smooth family of vectors for all n and s. Indeed if $\frac{\partial}{\partial s}|n; s\rangle \in \mathcal{H}$ for all n and s, then Eq. (8.380) holds: for the argument see Eq. (8.390) below. It is convenient to choose the phases of the eigenvectors $|n; s\rangle$ so that

$$\langle n; s | \frac{\partial}{\partial s} |n; s\rangle = 0. \tag{8.381}$$

We shall comment on the meaning of this condition in Sect. 8.11 where we also discuss the limitations of its validity.

Theorem 8.7 (Adiabatic Theorem) *Let $\{H(s)\}$ be a family of Hamiltonians as above, and let $\{|n, s\rangle\}$ be the corresponding family of energy eigenstates (8.378) (label a implicit) which satisfy condition (8.380). Consider the time-dependent Schrödinger equation*

$$i\hbar \frac{d}{dt} |\psi, t\rangle = H(t/T)|\psi, t\rangle \tag{8.382}$$

---

[29] The typical example is a smooth family $\{H(s)\}$ of supersymmetric Hamiltonians with Witten index $|\Delta| > 1$ [20]. The dimension of the zero-energy eigenstate is at least $|\Delta|$ for all s by invariance of the index under continuous deformations of the Hamiltonian.

with initial condition

$$|\psi, 0\rangle = |n, 0\rangle \tag{8.383}$$

an energy eigenstate of the early-time Hamiltonian H_0

$$H_0|n, 0\rangle = E_n(0)|n, 0\rangle. \tag{8.384}$$

The asymptotic solution of (8.382) in the adiabatic limit $T \to \infty$ is

$$|\psi, t\rangle = \exp\left(-\frac{i}{\hbar} \int_0^t E_n(t'/T)\, dt'\right)|n; t/T\rangle, \tag{8.385}$$

that is: in the adiabatic limit the system is at **each instant** t in the n-th *instantaneous eigenstate $|n; t/T\rangle$ of the instantaneous Hamiltonian $H(t/T)$, and the asymptotic dynamics is fully encoded in the phase*

$$\exp\left(-\frac{i}{\hbar} \int_0^t E_n(t'/T)\, dt'\right). \tag{8.386}$$

Remark 8.3 When the derivatives $\partial H(s)/\partial s$ and $\partial^2 H(s)/\partial s^2$ are bounded operators, while the gaps $E_n(s) - E_m(s)$ are bounded away from zero, one may state a more precise result: there is a constant K such that, for all $\epsilon > 0$, for $T > K/\epsilon$ one has

$$\left\| |\psi, t\rangle - \exp\left(-\frac{i}{\hbar} \int_0^t E_n(t'/T)\, dt'\right)|n; t/T\rangle \right\| < \epsilon. \tag{8.387}$$

Proof Taking the derivative with respect to s of (8.378) (here $\dot{H}(s) \equiv \partial H(s)/\partial s$)

$$\dot{H}(s)|n; s\rangle + H(s)\frac{\partial}{\partial s}|n; s\rangle = \frac{\partial E_n(s)}{\partial s}|n; s\rangle + E_n(s)\frac{\partial}{\partial s}|n; s\rangle, \tag{8.388}$$

and then multiplying by $\langle m; s|$, we get

$$\langle m; s|\dot{H}(s)|n; s\rangle = \Big(E_n(s) - E_m(s)\Big)\langle m; s|\frac{\partial}{\partial s}|n; s\rangle + \frac{\partial E_n(s)}{\partial s}\,\delta_{mn}, \tag{8.389}$$

that is,

$$\langle m; s|\frac{\partial}{\partial s}|n; s\rangle = \begin{cases} \dfrac{\langle m; s|\dot{H}(s)|n; s\rangle}{E_n(s) - E_m(s)} & \text{for } m \neq n \\[2mm] 0 & \text{for } m = n. \end{cases} \tag{8.390}$$

We rewrite the Schrödinger equation (8.382) in the form

$$i\hbar \frac{d}{ds}|\psi; s\rangle = T H(s)|\psi; s\rangle, \tag{8.391}$$

and expand the solution $|\psi, s\rangle$ in the complete set $\{|m; s\rangle\}$ in the form

$$|\psi, s\rangle = \sum_m \exp\left(-\frac{iT}{\hbar}\int_0^s E_m(s')\,ds'\right)|m; s\rangle\, c_m(s) \tag{8.392}$$

where $c_m(s)$ are suitably coefficients which satisfy the early time condition

$$c_m(s) = \delta_{mn} \quad \text{for } s < 0. \tag{8.393}$$

Now

$$0 = e^{\frac{iT}{\hbar}\int_0^s E_k ds'}\langle k; s|\left(\frac{d}{ds} + \frac{i}{\hbar}T H(s)\right)|\psi, t\rangle =$$
$$= \dot{c}_k + \sum_m e^{-\frac{iT}{\hbar}\int_0^s (E_m - E_k)ds'}\langle k; s|\frac{\partial}{\partial s}|m; s\rangle\, c_m, \tag{8.394}$$

that is, using (8.390),

$$\dot{c}_k(s) = \sum_{m \neq k} \exp\left(-\frac{iT}{\hbar}\int_0^s \left(E_m(s') - E_k(s')\right)ds'\right)\frac{\langle k; s|\dot{H}(s)|m; s\rangle}{E_k(s) - E_m(s)}\, c_m(s). \tag{8.395}$$

This is an equation of the general form

$$\dot{c}_k(s) = A(s)_{km}\, c_m(s), \qquad c_k(0) = \delta_{kn}, \tag{8.396}$$

whose solution is the **T**-ordered exponential

$$c_k(s) = \delta_{kn} + \int_0^s ds_1\, A(s_1)_{kn} +$$
$$+ \int_0^s ds_1\, A(s_1)_{kj}\int_0^{s_1} ds_2\, A_{jn}(s_2) + \cdots \tag{8.397}$$

(sum over repeated indices implied). We know that the series converges (by the assumption (8.380)). Its sum yields the *exact solution* of the time-dependent system (8.382) which is valid for all T. Now we change gears and look for the asymptotic limit as $T \to \infty$ of the exact solution we got. Consider, say, the first non-trivial term in the expansion (8.397)

$$
\int_0^s ds_1\, A(s_1)_{kn} = (1 - \delta_{kn}) \times
$$
$$
\times \int_0^s ds_1 \exp\left(-\frac{iT}{\hbar}\int_0^{s_1}\left(E_n(s') - E_k(s')\right)ds'\right)\frac{\langle k; s_1|\dot{H}(s_1)|n; s_1\rangle}{E_k(s_1) - E_n(s_1)}
\tag{8.398}
$$

Since the integrand is bounded, the large T asymptotics of this integral may be evaluated by the stationary phase method. The condition that there is no level crossings, Eq. (8.379), then says that there are *no* saddle-points; in this case the dominant contribution to the integral comes from the region near the lower boundary of the integration region $s_1 \simeq 0$. For large T this contribution is at most of order $O(1/T)$.[30] The same argument may be repeated recursively for the higher order terms in the convergent expansion (8.397) of the **T**-ordered exponential. We conclude that, asymptotically for large T, the solution is just

$$
c_k(s) = \delta_{kn} \quad \text{for all } s
\tag{8.399}
$$

up to corrections $O(1/T)$. This is the statement of the adiabatic theorem. $\qquad\square$

8.10.1 *The Adiabatic Approximation*

When the Hamiltonian $H(t)$ varies very slowly with time—but not *infinitely* slowly—i.e. the time-scale T of its variation is large but finite, if we replace the exact time-evolution $U(t)$ by its adiabatic version

$$
U(t)_{\text{adia}} = \sum_n |n; t\rangle\, e^{-\frac{i}{\hbar}\int_0^t E_n(s)ds}\langle n; 0|
\tag{8.400}
$$

we make an error of at most order $O(1/T)$. In many concrete situations this may be a tiny error, or at least an acceptable one. Replacing the actual evolution by its asymptotic form (8.400) is the *adiabatic approximation*.

8.11 The Berry Phase

In the previous section we set

$$
\langle n; t|\frac{\partial}{\partial t}|n; t\rangle = 0,
\tag{8.401}
$$

---

[30] A simple way to see this is to compute the limit $T \to \infty$ *á la* Cesaro as the mean.

deferring a discussion of the meaning and limitation of this condition to the present section. The statement that in the adiabatic limit the system remains in an instantaneous eigenstate of $H(t)$ is true without imposing (8.401), but the phase[31] will have a more complicated expression than in Eq. (8.386). Now we discuss the geometric meaning of condition (8.401), understand under which conditions it can be enforced, and study what happens in the general case when it cannot be imposed *globally*. To put things in the proper perspective, we start by recalling some basic facts from Differential Geometry.

Sub-bundle Connection

Let $\mathcal{V} \to M$ be a smooth vector bundle over a manifold M. Recall that a *connection on* $\mathcal{V}$ is a linear map [19]

$$\nabla: \mathcal{V} \to \Omega^1 \otimes \mathcal{V}, \qquad \psi \mapsto \nabla\psi, \tag{8.402}$$

which maps local sections ψ of $\mathcal{V}$ into 1-forms on M with coefficients in $\mathcal{V}$, and such that, for all local functions f on M,

$$\nabla(f\psi) = f\,\nabla\psi + \mathrm{d}f \wedge \psi. \tag{8.403}$$

The connection ∇ may be extended to a linear map of degree $+1$ acting on k-forms on M with coefficients in $\mathcal{V}$

$$\nabla: \Omega^\bullet \otimes \mathcal{V} \to \Omega^{\bullet+1} \otimes \mathcal{V}. \tag{8.404}$$

Then Eq. (8.403) yields

$$\nabla^2(f\psi) = f\nabla^2\psi, \qquad \nabla^2: \Omega^\bullet \otimes \mathcal{V} \to \Omega^{\bullet+2} \otimes \mathcal{V} \tag{8.405}$$

i.e. ∇^2 commutes with multiplication by functions, and hence is not a differential operator but a section

$$\nabla^2 \equiv \mathcal{F} \in \Omega^2 \otimes \mathrm{End}(\mathcal{V}) \tag{8.406}$$

called the *curvature* of ∇. When $\mathcal{V}$ has a Hermitian fiber metric $h(\cdot, \cdot)$, we say that the connection ∇ is *compatible with the metric* iff

$$\mathrm{d}h(\psi_1, \psi_2) = h(\nabla\psi_1, \psi_2) + h(\psi_1, \nabla\psi_2), \tag{8.407}$$

in which case $\mathcal{F}$ is skew-Hermitian, $\mathcal{F}^\dagger = -\mathcal{F}$.

---

[31] Actually it is just a phase only when the level is non-degenerate; otherwise the coefficient is an unitary matrix acting on the implicit indices distinguishing the states of the same energy. For examples and discussions see [20] and references therein.

Let $S \subset V$ be a vector sub-bundle. The connection ∇ on V induces a connection ∇^S on V, called the *sub-bundle connection* [19], by the following construction. Let ϕ a local section of S. ϕ is, in particular, a section of V, hence $\nabla \phi$ makes sense: it is a local section of $\Omega^1 \otimes V$. Let $P: V \to S$ be the projector on the sub-bundle and define

$$\nabla^S \phi = P \nabla \phi. \tag{8.408}$$

$\nabla^S : \Omega^\bullet \otimes S \to \Omega^{\bullet+1} \otimes S$ satisfies (8.403), so it is a connection on S. The connection ∇^S is compatible with the metric h (restricted to S) whenever the projection P is orthogonal.

Example 8.8   An important example of this construction is the Levi-Civita connection of a sub-manifold of the flat space. Consider a manifold $M \subset \mathbb{R}^n$ embedded in $\mathbb{R}^n$, and equip M with the Riemannian metric induced by the Euclidean metric of the ambient space $\mathbb{R}^n$. The tangent bundle of $\mathbb{R}^n$ is the trivial bundle

$$T\mathbb{R}^n = \mathbb{R}^n \times \mathbb{R}^n \xrightarrow{\;p_2\;} \mathbb{R}^n \tag{8.409}$$

equipped with the constant Euclidean metric and the compatible trivial connection d (the exterior derivative). The pull-back of the trivial bundle $T\mathbb{R}^n$ to M has the orthogonal decomposition

$$T\mathbb{R}^n|_M = TM \oplus N_M \tag{8.410}$$

where TM is the tangent bundle to M and N_M its normal bundle in $\mathbb{R}^n$. $T\mathbb{R}^n|_M$ is a trivial bundle with fiber $\mathbb{R}^n$, constant metric, and trivial connection d. Hence we have a sub-bundle connection ∇ on TM induced by the ambient trivial connection

$$\nabla \phi = \mathrm{d}\phi\big|_{TM} \tag{8.411}$$

which is metric for the induced Riemannian metric g_{ij} on M and torsionless by construction; hence ∇ is the Levi-Civita metric[32] of g_{ij}. While the connection on the full bundle $T\mathbb{R}^n\big|_M$ is trivial, and in particular flat since its curvature $\mathrm{d}^2 \equiv 0$, the sub-bundle connection ∇ is non-trivial in general and its curvature $\nabla^2 \neq 0$. As an example, think of the unit sphere in $\mathbb{R}^3$ whose curvature is positive.

Generalizing this example, we consider a general sub-bundle $S \hookrightarrow V$, where the ambient vector bundle $V \to M$ is equipped with the trivial connection d and a compatible Hermitian metric h. We have the decomposition

$$\mathrm{d} = \nabla + \nabla^\perp + \rho + \rho^\perp : \Omega^k \otimes V \to \Omega^{k+1} \otimes V \tag{8.412}$$

---

[32] Recall that the Levi-Civita connection is the *unique* connection on TM which is both torsionless and compatible with the Riemannian metric on TM.

where

$$\nabla: \Omega^k \otimes \mathcal{S} \to \Omega^{k+1} \otimes \mathcal{S}, \qquad \nabla^\perp: \Omega^k \otimes \mathcal{S}^\perp \to \Omega^{k+1} \otimes \mathcal{S}^\perp$$
$$\rho: \Omega^k \otimes \mathcal{S} \to \Omega^{k+1} \otimes \mathcal{S}^\perp \qquad \rho^\perp: \Omega^k \otimes \mathcal{S}^\perp \to \Omega^{k+1} \otimes \mathcal{S}. \tag{8.413}$$

Here ∇ (resp. $\nabla^\perp$) is a connection on $\mathcal{S}$ (resp. $\mathcal{S}^\perp$) while ρ (resp. $\rho^\perp$) is a section of $\Omega^1 \otimes \mathrm{Hom}(\mathcal{S}, \mathcal{S}^\perp)$ (resp. $\Omega^1 \otimes \mathrm{Hom}(\mathcal{S}^\perp, \mathcal{S})$). From the identity

$$0 = \mathrm{d}^2 = (\nabla + \nabla^\perp + \rho + \rho^\perp)^2 \tag{8.414}$$

we get

$$0 = P(\nabla + \nabla^\perp + \rho + \rho^\perp)^2 P = \nabla^2 + \rho^\perp \rho, \tag{8.415}$$

so that the curvature of the sub-bundle connection ∇ is

$$\nabla^2 = -\rho^\perp \rho, \tag{8.416}$$

which is typically non-zero.

Let $\{\phi_a(s)\}$ be an orthonormal local trivialization of $\mathcal{S}$, i.e. a set of orthonormal local sections of $\mathcal{S}$ in the coordinate patch $\mathcal{U} \subset M$ which yields a basis in each fiber $\mathcal{S}_s$ ($s \in \mathcal{U}$) of the sub-bundle satisfying

$$h\big(\phi_a(s), \phi_b(s)\big) = \delta_{ab}. \tag{8.417}$$

We identify the local section $\phi(s) = c^a(s)\phi_a(s)$ of $\mathcal{S}$ with the vector of functions $(c^1(s), \ldots, c^m(s))$ on $\mathcal{U}$. Here $m = \mathrm{rank}\,\mathcal{S}$. Then the connection ∇ takes the explicit form

$$\nabla c^a = \mathrm{d}c^a + A^a{}_b c^b \tag{8.418}$$

where the connection coefficients $A^a{}_b$ are the components of a local 1-form with coefficients in $\mathrm{End}(\mathcal{S})$ written in the basis $\{\phi_a(s)\}$. From the definition of the sub-bundle connection, Eq. (8.408), we get the explicit formula

$$A^a{}_b = h(\phi_a, \mathrm{d}\phi_b). \tag{8.419}$$

Changing trivialization of $\mathcal{S}$ will modify the explicit form of the connection $A^a{}_b$: this is a gauge transformation in the usual sense of Yang-Mills theory. However the change of gauge is a mere modification of the way we represent the connection: the intrinsic geometry (and correspondingly the physical observables) are independent of the choice of gauge/trivialization.

Sub-bundle Constructions in Quantum Mechanics

Suppose we have a smooth family of quantum systems, with a fixed separable Hilbert space $\mathcal{H}$, defined by a Hamiltonian

$$H(u_1, \ldots, u_\ell) \tag{8.420}$$

which depends smoothly on a number of parameters $(u_1, \ldots, u_\ell) \equiv u$ that we see as local coordinates in some parameter space $\mathscr{C}$. We assume the spectrum of $H(u)$ to be discrete (this is automatic if, say, the Hamiltonians $H(u)$ have a fixed *compact* configuration space $\mathcal{M}$). We allow the energy eigenvalues to be degenerate, but assume the energy eigenspaces to be finite-dimensional (as it happens for $\mathcal{M}$ compact). We focus on the n-th energy level $E_n(u)$. The gaps between this level and the neighbouring ones may close at some loci $\mathscr{B} \subset \mathscr{C}$ in the parameter space. If this happens, we redefine our parameter space by replacing $\mathscr{C}$ with a connected component of the complement $\mathscr{C} \setminus \mathscr{B}$, that is, we "cut away" the "bad" locus $\mathscr{B}$ where a relevant spectral gap closes.[33] We assume that the redefined $\mathscr{C}$ is non-trivial, i.e. has positive dimension.

This being understood, over the "good locus" $\mathscr{C}$ we have the trivial Hilbert bundle

$$\mathcal{H} \equiv \mathcal{H} \times \mathscr{C} \to \mathscr{C} \tag{8.421}$$

which is a vector bundle with typical fiber $\mathcal{H}$ equipped with the *constant* Hermitian metric along the fibers given by the Hilbert space product

$$\langle -, - \rangle \colon \mathcal{H} \to \Omega^0(\mathscr{C}) \tag{8.422}$$

and with the trivial flat connection, compatible with the Hermitian metric $\langle -, - \rangle$,

$$\mathrm{d} \colon \mathcal{H} \otimes \Omega^k(\mathscr{C}) \to \mathcal{H} \otimes \Omega^{k+1}(\mathscr{C}), \tag{8.423}$$

$$\mathrm{d} \colon |\psi\rangle \mapsto d|\psi\rangle, \qquad \mathrm{d}^2 = 0. \tag{8.424}$$

The spectral theorem yields a decomposition of the fibers of $\mathcal{H}$

$$\mathcal{H}_u = \bigoplus_{m \geq 0} \ker\left(H(u) - E_m(u) \right) \equiv \bigoplus_{m \geq 0} \mathcal{H}_m(u). \tag{8.425}$$

By our construction of the "good" $\mathscr{C}$

$$\dim \mathcal{H}_n(u) = k_n \quad \text{constant for } u \in \mathscr{C}. \tag{8.426}$$

---

[33] In the case that the theory is supersymmetric with at least 4 supercharges, and we focus on the ground energy level, the bad locus is empty $\mathscr{B} = \varnothing$ since the number of ground states is protected by a refined version of the Witten index [20].

Hence we have a well-defined rank-k_n sub-bundle of the trivial Hilbert bundle $\mathcal{H}$ for each[34] n

$$\mathcal{H}_n \hookrightarrow \mathcal{H}, \qquad (\mathcal{H}_n)_{\boldsymbol{u}} \equiv \mathcal{H}_n(\boldsymbol{u}). \tag{8.427}$$

Locally in $\mathscr{C}$ we may choose a trivialization of $\mathcal{H}_n$, that is, a smooth family of orthonormal[35] frames in the fibers of $\mathcal{H}_n$

$$\big\{ |a, \boldsymbol{u}\rangle \big\}_{a=1,\dots,k_n} \tag{8.428}$$

$$H(\boldsymbol{u})|a, \boldsymbol{u}\rangle = E_n(\boldsymbol{u})|a, \boldsymbol{u}\rangle, \tag{8.429}$$

$$\langle a, \boldsymbol{u}|b, \boldsymbol{u}\rangle = \delta_{ab}. \tag{8.430}$$

We write

$$P = \sum_b |b, \boldsymbol{u}\rangle\langle b, \boldsymbol{u}| \tag{8.431}$$

for the bundle orthogonal projection $P : \mathcal{H} \to \mathcal{H}_n$.

Just as for the tangent bundle of curved manifolds isometrically embedded in flat $\mathbb{R}^n$ (cf. Example 8.8), the sub-bundle $\mathcal{H}_n \to \mathscr{C}$ needs not to be trivial. The trivial connection d on $\mathcal{H}$ induces a sub-bundle connection ∇ on $\mathcal{H}_n$. Let

$$|\psi, \boldsymbol{u}\rangle = \sum_a |a, \boldsymbol{u}\rangle \, f_a(\boldsymbol{u}) \tag{8.432}$$

be a general local section of $\mathcal{H}_n$. In the local trivialization (8.428), by definition,

$$\sum_a |a, \boldsymbol{u}\rangle \, \nabla f_a = \nabla|\psi, \boldsymbol{u}\rangle \equiv P\mathrm{d}|\psi, \boldsymbol{u}\rangle =$$
$$= \sum_a |a, \boldsymbol{u}\rangle \, \mathrm{d}f_a + \sum_{a,b} |a, \boldsymbol{u}\rangle\langle a, \boldsymbol{u}|\mathrm{d}|b, \boldsymbol{u}\rangle \, f_b(\boldsymbol{u}) \tag{8.433}$$

that is,

$$\nabla f_a = \mathrm{d}f_a + A(\boldsymbol{u})_a{}^b \, f_b, \tag{8.434}$$

with connection coefficient (in the chosen orthonormal trivialization i.e. gauge)

$$A(\boldsymbol{u})_a{}^b \stackrel{\text{def}}{=} \langle a, \boldsymbol{u}|\mathrm{d}|b, \boldsymbol{u}\rangle. \tag{8.435}$$

---

[34] The "good locus" $\mathscr{C}$ depends on n.

[35] Often it is convenient to choose a more general trivialization which is not orthonormal, see [20].

Since the connection ∇ is compatible with the metric given by the Hilbert norm, A is a $\mathfrak{u}(k_n)$-*valued* connection over the "good" parameter manifold $\mathscr{C}$. In the "generic" case the energy levels are non-degenerate, and we get an Abelian $U(1)$ connection. In the supersymmetric case the connection is typically non-Abelian [20].

Definition 8.2 The connection ∇ induced on the bundle $\mathcal{H}_n$ of n-th level states by the trivial connection on the Hilbert bundle

$$\mathcal{H} \equiv \mathcal{H} \times \mathscr{C} \xrightarrow{\pi} \mathscr{C} \tag{8.436}$$

is called the *Berry connection*.

Consequences of the Berry Connection
Let

$$\gamma \equiv \left\{ u(s) \right\}_{s=0}^{s=1} \subset C, \qquad u(0) = u(1) = u_0, \tag{8.437}$$

be a closed path in the parameter space based at $u_0 \in \mathscr{C}$. We start from the quantum system defined by the Hamiltonian $H(u_0)$ in an initial eigenstate $|a, u_0\rangle$ with energy $E_n(u_0)$. Then we adiabatically modify the couplings u along the curve γ, producing the slowly-varying time-independent Hamiltonian

$$H\big(u(t/T)\big) \quad T \text{ very large}, \tag{8.438}$$

which eventually returns back to the original system with Hamiltonian $H(u_0)$. In view of Eq. (8.435), to implement the condition (8.401) along the path γ we must redefine our trivialization ($\equiv$ change the Berry gauge) as

$$|a; s\rangle = \mathbf{T} \exp\left(-\int_0^s A(u(s')) \cdot \dot{u}(s')\, ds' \right)_{ab} |b, u(s)\rangle, \tag{8.439}$$

in other words: *we need to parallel transport the state along the curve γ using the Berry connection.* When, after the very large time T, we get back to our original system with Hamiltonian $H(u_0)$, we end up in the final state

$$\exp\left(-iT \int_0^1 E_n(u(s))\, ds \right) |a; 1\rangle =$$

$$= \exp\left(-iT \int_0^1 E_n(u(s))\, ds \right) W(\gamma)_{ab} |b, u_0\rangle \tag{8.440}$$

with the extra action of the $k_n \times k_n$ matrix

$$W(\gamma) = P \exp\left(-\int_\gamma A\right) \stackrel{\text{def}}{=}$$

$$\stackrel{\text{def}}{=} \mathbf{T} \exp\left(-\int_0^1 A(\boldsymbol{u}(s')) \cdot \dot{\boldsymbol{u}}(s')\, \mathrm{d}s'\right) \in \mathrm{Aut}(\mathcal{H}_n(\boldsymbol{u}_0)), \tag{8.441}$$

namely our state got an additional rotation by the *holonomy element* of the connection $\nabla = \mathrm{d} + A$ along the closed path $\gamma \subset \mathscr{C}$.

$W(\gamma) \not\equiv 1$ unless the connection is trivial, i.e. unless A may be set to zero by a change of trivialization (in physics language: unless A is *pure gauge*). This is automatically true if the parameter space $\mathscr{C}$ is one dimensional and contractible, as we (implicitly) assumed in the proof of the adiabatic theorem. When $W(\gamma) \neq 1$ the non-trivial holonomy produces new *observable* quantum phenomena akin to the Bohm-Aharonov effect in presence of an electromagnetic connection with non-trivial holonomy (cf. Sect. 3.11).

Definition 8.3 The holonomy element $W(\gamma)$ of the Hilbert sub-bundle connection ∇ is the *Berry phase* (also called *geometric phase*) of the n-th energy level.

The name "phase" reflects the typical case, where the n-th level is non-degenerate ($k_n = 1$), then A is a $U(1)$ connection, and the holonomy elements are phases. In the general case $W(\gamma) \in U(k_n)$.

When the parameter space $\mathscr{C}$ has dimension larger than 1, the Berry connection ∇ may have, and typically has, a non-zero curvature

$$\mathcal{F} \equiv \nabla^2 \neq 0 \tag{8.442}$$

in the parameter space $\mathscr{C}$. In presence of a *Berry curvature*, $\mathcal{F} \neq 0$, the final state vector $|n, \boldsymbol{u}_2\rangle$—obtained by an adiabatic process where we vary very slowly the couplings in the Hamiltonian from $\boldsymbol{u}_1$ to $\boldsymbol{u}_2$—depends on the particular path γ we follow in the coupling space $\mathscr{C}$ even if the energy levels are the same along the two paths $\boldsymbol{u}(s)$ and $\boldsymbol{u}'(s)$ i.e. even iff

$$E_n(\boldsymbol{u}(s)) = E_n(\boldsymbol{u}'(s)) \quad \text{for all } s. \tag{8.443}$$

To illustrate the ubiquitousness of the Berry cruvature in Quantum Physics, we show that it is already present in the very simplest non-trivial quantum system.

Example 8.9 (The General 2-State System) We saw in Chap. 4 that a general 2-state system with Hilbert space $\mathbb{C}^2$ has Hamiltonian

$$H(\mathbf{u}) = \mathbf{u} \cdot \sigma \qquad \mathbf{u} \in \mathbb{R}^3 \tag{8.444}$$

where $\boldsymbol{\sigma} = (\sigma_1, \sigma_2, \sigma_3)$ are the Pauli matrices. The energy eigenvalues are non-degenerate except at the "bad" point $\mathscr{B} \equiv \{0\}$, so our "good" parameter space is

$$\mathscr{C} \equiv \mathbb{R}^3 \setminus \{0\} \simeq \mathbb{R}_{>0} \times S^2. \tag{8.445}$$

The overall scale of energy may always be set to 1 by a choice of units, so in Eq. (8.444) we may restrict to unit vectors $|\mathbf{u}| = 1$ with no loss. The parameter space $\mathscr{C}$ then reduces to S^2 with spherical coordinates (θ, ϕ)

$$\mathbf{u} = \big(\cos\phi \sin\theta,\ \sin\phi \sin\theta,\ \cos\theta\big), \tag{8.446}$$

which parametrizes the 2-dimensional family of Hamiltonians

$$H(\phi, \theta) = \begin{pmatrix} \cos\theta & e^{-i\phi}\sin\theta \\ e^{i\phi}\sin\theta & -\cos\theta \end{pmatrix} \tag{8.447}$$

whose eigenvalues are ± 1 independently of (θ, ϕ). The corresponding orthonormal eigenstates are

$$|+, \phi, \theta\rangle = \begin{pmatrix} \cos\frac{\theta}{2}\, e^{-i\phi/2} \\ \sin\frac{\theta}{2}\, e^{i\phi/2} \end{pmatrix}, \qquad |-, \phi, \theta\rangle = \begin{pmatrix} -\sin\frac{\theta}{2}\, e^{i\phi/2} \\ \cos\frac{\theta}{2}\, e^{-i\phi/2} \end{pmatrix} \tag{8.448}$$

The Berry $U(1)$ connection on the sub-bundle spanned by $|+, \phi, \theta\rangle$ is

$$\langle +, \phi, \theta | d | +, \phi, \theta \rangle = -\frac{i}{2}\cos\theta\, d\phi, \tag{8.449}$$

which is identical to the gauge field produced on the unit sphere $S^2 \subset \mathbb{R}^3$ by a monopole of unit charge at the origin of $\mathbb{R}^3$. We conclude

Fact 8.8 *The Berry bundle (with connection) of energy eigenstates of the 2-state system is a copy of the* Hopf *bundle* $S^3 \to S^2$ *hence topologically non-trivial.*

8.12 Born-Oppenheimer Technique

The Born-Oppenheimer approximation is yet another technique based on the adiabatic theorem. We shall be sketchy, outlining the logic of the method, without entering in the details. To make the idea more intuitive, we shall work in the path integral formulation, even if the original formulation was in the context of the Schrödinger equation.

Suppose that we have a quantum system described by the Euclidean Lagrangian

$$L_E = \frac{m}{2}\dot{x}^i\dot{x}^i + \frac{M}{2}\dot{y}^a\dot{y}^a + V(x^i, y^a), \tag{8.450}$$

where we divided the degrees of freedom ("fields") in two sets denoted x^i and y^a, respectively. The path integral with sources takes the form (we leave the indices implicit)

$$Z[K, J] = \int [\mathrm{d}y]\, e^{-\int\left(\frac{M}{2}\dot{y}\dot{y}+Ky\right)\mathrm{d}t} \int [\mathrm{d}x]\, e^{-\int\left(\frac{m}{2}\dot{x}\dot{x}+V(x,y)+Jx\right)\mathrm{d}t}. \tag{8.451}$$

Let us consider the asymptotic limit $M \to \infty$ in which the degrees of freedom y^a become extremely heavy. In this limit the support of the functional measure localizes in a tiny neighborhood of the locus $\{\dot{y} = 0\}$. Therefore, when performing the inner path integral in $[\mathrm{d}x]$, the y's appearing in the potential $V(x, y)$ may be treated as slowly varying *parameters* in a Hamiltonian

$$H(y) = \sum_i \frac{p_i^2}{2m} + V(x^i, y^a) \tag{8.452}$$

which governs the dynamics of the "fast" degrees of freedom x^i in a constant background of the "slow" ones y^a. In other words, in the limit $M \to \infty$ we may replace the inner path integral in (8.451) by its adiabatic limit. The inner path integral becomes

$$\exp\left(-\int_0^T F[y; J]\,\mathrm{d}t\right) = \int [\mathrm{d}x]\, e^{-\int_0^T\left(\frac{m}{2}\dot{x}\dot{x}+V(x,y)+Jx\right)\mathrm{d}t}\Bigg|_{\substack{y\ \text{constant}\\ T\approx\infty}} \tag{8.453}$$

where $F[y; J]$ is the free energy (in presence of the sources J) for the quantum system with degrees of freedom x^i and Hamiltonian $H(y)$ for *constant* $y = (y^a)$. Then, asymptotically for large M,

$$Z[J(t), K(t)] \approx \int [\mathrm{d}y]\, e^{-\int\left(\frac{M}{2\hbar}\dot{y}\dot{y}+Ky+F[y(t); J]\right)\mathrm{d}t} =$$

$$= \left\langle e^{-\int\left(K(t)y(t)+F[y(t); J]\right)\mathrm{d}t} \right\rangle_{\!\text{free}} \tag{8.454}$$

where, for all functional $\mathfrak{F}[y]$, we set

$$\left\langle \mathfrak{F}[y] \right\rangle_{\!\text{free}} \equiv \int [\mathrm{d}y]\, e^{-\int \frac{M}{2\hbar}\dot{y}\dot{y}\,\mathrm{d}t}\, \mathfrak{F}[y]. \tag{8.455}$$

We reinserted $\hbar$ explicitly in the expression (8.455) to emphasize that the limit $M \to \infty$ may be also seen as the semiclassical limit $\hbar \to 0$ for the degrees of freedom y^a. Therefore the final step in the computation may be performed with the techniques explained in the first sections of this chapter.

Physically, the Born-Oppenheimer method is based on the division of the degrees of freedom into the *fast* ones x^i and the *slow* ones y^a, whose velocities satisfy

$$|\dot{y}| \lll |\dot{x}|. \tag{8.456}$$

We may equivalently think in terms of a separation into *light* degrees of freedom x^i and *heavy* ones y^a. The idea of the method is to reduce the problem to a computation in the system of the fast degrees of freedom, with the slow degrees of freedom treated as slowly varying *parameters* to which the adiabatic approximation is applicable. The method becomes exact in the limit $M \to \infty$ i.e. $|\dot{y}| \to 0$. In practice one has to evaluate only the path integral in the RHS of (8.453) which is much simpler than the original one (8.451).

The observation after Eq. (8.455) may be stated in a more suggestive way: *informally, we treat the x^i's as quantum degrees of freedom coupled to the classical ones y^a*. Thus the Born-Oppenheimer formalism leads to a kind of "hybrid" between classical and quantum mechanics.

An important class of physical systems with this fast/slow decomposition are the molecules. A molecule consists of n very heavy nuclei of mass M_Z and k light electrons of mass m_e with

$$\frac{M_Z}{m_e} = O(10^4). \tag{8.457}$$

The coordinates of the nuclei then play the role of the y's and those of the electrons the role of the x's. To understand the structure of molecules is the original problem for which the technique was introduced by Born and Oppenheimer [21].

References

1. K. Iwaki, T. Nakanishi, Exact WKB analysis and cluster algebras. J. Phys. A: Math. Theor. **47**, 474009 (2014). arXiv:1401.7094
2. J. Ecalle, *Les Fonctions Resurgentes*, vols. I–III (Publ. Math. Orsay, 1981)
3. A. Behtash, G.V. Dunne, T. Schäfer, T. Sulejmanpasic, M. Ünsal, Toward Picard-Lefschetz theory of path integrals, complex saddles and resurgence. Ann. Math. Sci. Appl. **02**, 95–212 (2017). arXiv:1510.03435 [hep-th]
4. J.D. Murray, *Asymptotic Analysis* (Springer, Berlin, 1984)
5. G.E. Andrews, R. Askey, R. Roy, *Special Functions*. Encyclopedia of Mathematics and Its Applications, vol. 71 (CUP, Cambridge, 2009)
6. F.W. Olver, D.M. Lozier, R.F. Boisvert, C.W. Clark, W. Charles (eds.), *NIST Handbook of Mathematical Functions* (CUP, Cambridge, 2010). Available online at https://dlmf.nist.gov

7. I.S. Gradshteyn, I.M. Ryzhik, *Table of Integrals, Series, and Products*, 7th edn. (Elsevier, Amsterdam, 2007)
8. S. Cecotti, *Analytic Mechanics. A Concise Textbook* (Springer, Berlin, 2024)
9. J. Jost, *Riemannian Geometry and Geometric Analysis*. UniversiText, 7th edn. (Springer, Berlin, 2017)
10. M. Veltman, *Diagrammatica. The Path to Feynman Diagrams* (CUP, Cambridge, 1994)
11. T. Takebe, *Elliptic Integrals and Elliptic Functions* (Springer, Berlin, 2023)
12. S. Coleman, The use of instantons, in *The Whys of Subnuclear Physics*, ed. by A. Zichicchi. Erice Lectures 1977 (Plenum Press, New York, 1979)
13. J. Milnor, *Morse Theory* (Princeton University Press, Princeton, 1963)
14. M.W. Hirsch, *Differential Topology*. Graduate Texts in Mathematics, vol. 33 (Springer, Berlin, 1976)
15. H. Kleinert, *Path Integrals in Quantum Mechanics, Statistics, Polymer Physics, and Financial Markets*, 5th edn. (World Scientific, Singapore, 2009)
16. J. Zinn-Justin, U.D. Jentschura, Multi-instantons and exact results I: conjectures, WKB expansions, and instanton interactions. Ann. Phys. **313**, 197–267 (2004). arXiv:quant-ph/0501136; J. Zinn-Justin, U.D. Jentschura, Multi-instantons and exact results II: specific cases, higher-order effects, and numerical calculations. Ann. Phys. **313**, 269–325 (2004). arXiv:quant-ph/0501137; U.D. Jentschura, A. Surzhykov, J. Zinn-Justin, Multi-instantons and exact results. III: unification of even and odd anharmonic oscillators. Ann. Phys. **325**, 1135–1172 (2010). arXiv:1001.3910; U.D. Jentschura, J. Zinn-Justin, Multi-instantons and exact results. IV: path integral formalism. Ann. Phys. **326**, 2186–2242 (2011)
17. M. Marino, R. Schiappa, M. Weiss, Multi-instantons and multi-cuts. J. Math. Phys. **50**, 052301 (2009). arXiv:0809.2619
18. S. Cecotti, *Statistical Mechanics. A Concise Advanced Textbook* (Springer, Berlin, 2024)
19. J.-P. Demailly, *Complex Analytic and Differential Geometry*. Book online https://www-fourier.ujf-grenoble.fr/demailly/manuscripts/agbook.pdf
20. S. Cecotti, C. Vafa, Topological anti-topological fusion. Nucl. Phys. B **367**, 359–461 (1991)
21. M. Born, J.R. Oppenheimer, Zur Quantentheorie der Molekeln. Ann. Phys. (Leipzig) **84**, 457–484 (1927)

Index

Symbols